Heat Conduction Using Green's Functions

SECOND EDITION

Series in Computational and Physical Processes in Mechanics and Thermal Sciences

A Series of Reference Books and Textbooks

Series Editors

W. J. Minkowycz

Mechanical and Industrial Engineering
University of Illinois at Chicago
Chicago, Illinois

E. M. Sparrow

Mechanical Engineering
University of Minnesota, Twin Cities
Minneapolis, Minnesota

Numerical Heat Transfer and Fluid Flow, *Suhas V. Patankar*

Heat Conduction Using Green's Functions, Second Edition, *Kevin D. Cole, James V. Beck, A. Haji-Sheikh, and Bahman Litkouhi*

Numerical Heat Transfer, T.M. Shih

Finite Element Analysis in Heat Transfer, *Gianni Comini, Stefano Del Guidice, and Carlo Nonino*

Computer Methods for Engineers, *Yogesh Jaluria*

Computational Fluid Mechanics and Heat Transfer, Second Edition, *John C. Tannehill, Dale A. Anderson, and Richard H. Pletcher*

Computational Grids, *Graham F. Casey*

Modern Computational Methods, *Herbert A. Konig*

The Intermediate Finite Element Method: Fluid Flow and Heat Transfer Applications, *Juan C. Henrich and Darrell W. Pepper*

Modeling and Dynamics of Regenerative Heat Transfer, *A. John Willmott*

Computational Heat Transfer, Second Edition, *Yogesh Jaluria and Kenneth Torrance*

The Finite Element Method: Basic Concepts and Applications, Second Edition, *Darrell W. Pepper and Juan C. Heinrich*

Computational Methods in Heat and Mass Transfer, *Pradip Majumdar*

Series in Computational and Physical Processes in Mechanics and Thermal Sciences

Heat Conduction Using Green's Functions

SECOND EDITION

Kevin D. Cole
A. Haji-Sheikh
James V. Beck
Bahman Litkouhi

CRC Press
Taylor & Francis Group
Boca Raton London New York

CRC Press is an imprint of the
Taylor & Francis Group, an **informa** business

Taylor & Francis
6000 Broken Sound Parkway NW, Suite 300
Boca Raton, FL 33487-2742

International Standard Book Number: 978-1-4398-1354-6 (Hardback)

Library of Congress Cataloging-in-Publication Data

Heat conduction using Green's functions / Kevin D. Cole ... [et al.]. -- 2nd ed.
p. cm. -- (Series in computational and physical processes in mechanics and thermal sciences)
Includes bibliographical references and index.
ISBN 978-1-4398-1354-6
1. Heat--Conduction--Mathematics. 2. Green's functions. 3. Heat transfer. I. Cole, K. D. II. Title.

TJ260.H3854 2010
536'.23--dc22 2010000316

Visit the Taylor & Francis Web site at
http://www.taylorandfrancis.com

and the CRC Press Web site at
http://www.crcpress.com

Contents

Preface to the First Edition

The purpose of this book is to simplify and organize the solution of heat conduction and diffusion problems and to make them more accessible. This is accomplished using the method of Green's functions, together with extensive tables of Green's functions and related integrals. The tables of Green's functions were first compiled as a supplement to a first-year graduate course in heat conduction taught at Michigan State University. The book was originally envisioned as a reference volume, but it has grown into a heat conduction treatise from a Green's function perspective.

There is enough material for a one-semester course in analytical heat conduction and diffusion. There are worked examples and student problems to aid in teaching. Because of the emphasis on Green's functions, some traditional topics such as Fourier series and Laplace transform methods are treated somewhat briefly; this material could be supplemented according to the interest of the instructor. The book can also be used as a supplementary text in courses on heat conduction, boundary value problems, or partial differential equations of the diffusion type.

We hope the book will be used as a reference for practicing engineers, applied mathematicians, physicists, geologists, and others. In many cases, a heat conduction or diffusion solution may be assembled from tabulated Green's functions rather than derived. The book contains the most extensive set of Green's functions and related integrals that is currently available for heat conduction and diffusion.

The book is organized on a geometric basis because each Green's function is associated with a unique geometry. For each of the three coordinate systems—Cartesian, cylindrical, and spherical—there is a separate appendix of Green's functions named Appendix X, Appendix R, and Appendix RS, respectively. Each of the Green's functions listed is identified by a unique alphanumeric character that begins with either X, R, or RS to denote the x, r, or the spherical r coordinate, respectively. It is important for the reader to know something about this numbering system to use the tables of Green's functions. A more detailed numbering system, which covers both Green's functions and temperature solutions, is discussed in Chapter 2. We find the numbering system very helpful in identifying exactly which solution is under discussion, and all of the solutions discussed in the text are listed in Appendix N indexed according to the numbering system.

The level of treatment is intended for senior and first-year graduate students in engineering and mathematics. We have emphasized solution of problems rather than theorems and proofs, which are generally omitted. A prerequisite is an undergraduate course in ordinary differential equations. A previous introduction to the method of separation of variables for partial differential equations is also important.

The first nine chapters of the book are written with senior engineering students in mind. The Introduction contains background information on heat conduction and brief derivations of the heat conduction equations. Chapters 1 through 5 introduce Green's functions for transient heat conduction in one-dimensional bodies. The Cartesian coordinate system is emphasized in this section as an aid to learning. Steady-state problems are treated as a special case of the transient solution in Section 3.5 and 3.6. Chapters 6 through 9 are devoted to the solution of problems in the rectangular, cylindrical, and spherical coordinate systems. Transient problems are emphasized and steady problems are treated briefly in separate sections for each coordinate system (Sections 6.9, 8.7, and 9.8). Chapters 10 and 11 introduce the Galerkin-based Green's function method, which combines the efficient analysis of the Green's function method with the flexibility of geometry afforded by numerical methods. Chapter 12 introduces the unsteady surface element method, a numerical method that involves the matching of analytical solutions at the boundaries of bodies in contact.

No other book on Green's functions combines introductory material, worked examples, and extensive tables of Green's functions. Important books that contain some of this material include *Heat Conduction* by M. N. Ozisik (Wiley, New York, 1980), *Conduction of Heat in Solids* by H. S. Carslaw and J. C. Jaeger (Oxford, London, 1959), *Methods of Theoretical Physics* by P. M. Morse and H. Feshbach (McGraw-Hill, New York, 1953), *Elements of Green's Function and Propagation* by G. Barton (Oxford, London, 1989), *Green's Functions and Transfer Function Handbook* by A. G. Butkovskiy (Halsted Press, New York, 1982), *Application of Green's Functions in Science and Engineering* by M. D. Greenberg (Prentice-Hall, Englewood Cliffs, New Jersey, 1971), and *Green's Functions: Introductory Theory with Applications* by G. F. Roach (Van Nostrand Reinhold, NewYork, 1970).

James Beck would like to express his appreciation to the National Science Foundation for support over the years that has aided in the development of this work. Particularly important is the support related to the unsteady surface element method in which Dr. Ned Keltner of Sandia National Laboratories has also had a very influential part.

Kevin Cole would like to acknowledge support from the Engineering Foundation that has contributed to this project. Thanks also go to many students in heat conduction classes who have read the manuscript and have made many suggestions over the years.

A. Haji-Shiekh would like to acknowledge support from the National Science Foundation, under the directorship of Win Aung and Richard O. Buckius, who were instrumental in the development of the Galerkin-based integral method. Special thanks also to Win Aung who recognized the potential of the Galerkin-based integral method even before the work began. Thanks also to my wife who spent many hours typing and proofreading the manuscript, and to David Lou, former chairman of the Mechanical Engineering Department at UTA, for his encouragement.

Special thanks to the staff at Hemisphere for their competent handling of an equation-filled book. The authors take full responsibility for any errors that may remain in the book, but because this contains many new solutions we invite readers to send us any errors that they may find. Concerning errors please contact Kevin Cole, Department of Mechanical Engineering, P. O. Box 880656, University of Nebraska—Lincoln, Lincoln, NE 68588-0656 (402-472-5857). We will compile a list of errata and make it available to interested readers.

J. V. Beck, K. D. Cole, A. Haji-Shiekh, B. Litkouhi

Preface to the Second Edition

Since the first edition was published, there is considerable evidence of continued interest in Green's functions (GFs). There have been several new books on GFs for heat conduction and diffusion, including *Green's Functions in Applied Mechanics* by Y. A. Melnikov (WIT Press, Southhampton, 1995), *Green's Functions with Applications* by D. G. Duffy (Chapman and Hall/CRC Press, 2001), *Diffusion-Wave Fields: Mathematical Methods and Green Functions*, by A. Mandelis (Springer, New York, 2001), *Handbook of Green's Functions and Matrices* by V. D. Seremet and V. D. Sheremet (WIT Press, Southhampton, 2002), as well as several books on GFs applied to quantum physics. The number of research papers on GFs published in 2009 has more than doubled compared to the year the first edition was published. The four of us have continued to find new GFs and to apply them in our research. The second edition reflects our conviction that although Green discovered them in the nineteenth century, the functions bearing his name remain relevant to twenty-first century engineers and scientists.

For the second edition all chapters have been reviewed and updated. Based on our research and our classroom experience with this material, several chapters have been extensively revised. Chapter 1 has been expanded to provide a better introduction to Green's functions, both steady and unsteady, and a section on the Dirac delta function has been added. Chapter 4 now includes a discussion of the eigenfunction expansion method. Chapter 5 has been rewritten to include sections on the convergence speed of series solutions, the importance of alternate GF, and intrinsic verification, which is an important new tool for obtaining correct numerical values from analytical solutions. The chapters on cylindrical geometries from the first edition have been combined into one (Chapter 7), and the chapter on spherical geometries has been renumbered (Chapter 8). Several new examples and new figures have been added to Chapters 6, 7, and 8 on rectangular, cylindrical, and spherical geometries, respectively. A new chapter has been added on the subject of steady-periodic heat conduction (Chapter 9). The extensive appendices of GF and related functions, a central feature of the first edition, have been expanded to include three new appendices: the Dirac Delta Function (Appendix D); the Laplace Transform (Appendix L); and Properties of Common Materials (Appendix P). Two appendices have been renamed: Appendix F for Functions and Series; and Appendix I for Integrals.

One of the goals of the first edition was to make GF more accessible, and towards this end one of us (Cole) created an Internet site called the Green's Function Library (www.greensfunction.unl.edu) . The GF Library is the online companion site for the second edition. This web-searchable collection of GFs, based on the appendices in this book, is organized by differential equation, by geometry, and by boundary condition. Each GF is also identified and cataloged according to our GF numbering system. The GF Library also contains explanatory material, references, and links to related sites. Since it was created in 1999, the GF Library has received many thousands of visitors from all over the world.

Many students have made suggestions for this book, as have several readers who contacted us through the GF Library, and we thank all of them for this assistance. A special thanks to the students who prepared improved figures for the second edition, including Andrei Vaipan, Stuart Douglas, and Monchai Duangpanya.

This book contains an unusually large number of functions and solutions, each of which carries a risk of typographical error. We have diligently worked to remove such errors and we take full responsibility for any that remain. If you find an error, please check if it appears in the error list posted at the GF Library. If it is not listed there, please contact us through the GF Library or contact Kevin Cole (402-472-5857 or kcole1@unl.edu).

Kevin D. Cole
James V. Beck
A. Haji-Sheikh
Bahman Litkouhi

Authors

Kevin D. Cole received his MS in aerospace engineering and mechanics from the University of Minnesota in 1979 and his PhD in mechanical engineering from Michigan State University in 1986. Dr. Cole has held several positions in academia and industry and is currently associate professor of mechanical engineering at the University of Nebraska–Lincoln. Dr. Cole is active in writing and reviewing in the areas of heat conduction and thermal measurements. He is the creator of the Green's Function Library Internet site.

James V. Beck received his SM in mechanical engineering from MIT in 1957 and his PhD from Michigan State University in 1964. Dr. Beck is currently professor emeritus of mechanical engineering at Michigan State University. He has been honored with the MSU Distinguished Faculty Award and the ASME Heat Transfer Memorial Award. He is the originator of the Inverse Problems Symposium and is the inventor, with Professor Litkouhi, of the numbering system for heat conduction solutions. Dr. Beck has contributed to the field of heat transfer with numerous referred journal articles and books.

A. Haji-Sheikh received his MS in ME, MA in Mathematics from the University of Michigan and a PhD in 1965 from the University of Minnesota. In 1966, he joined the Department of Mechanical Engineering at the University of Texas at Arlington, and is currently a professor and member of the Distinguished Scholars Academy. His contributions to heat conduction include the floating random walk in Monte Carlo method, Green's function in two-step models, inverse problems, and Galerkin-based integral methods. He is a registered PE in the state of Texas, a fellow of ASME, a recipient of the ASME Memorial Award in Science.

Bahman Litkouhi received his MS and PhD from Michigan State University and is presently professor and graduate program director of the Mechanical Engineering Department at Manhattan College. Dr. Litkouhi is a registered professional engineer in the state of New York and a member of the American Society of Mechanical Engineers. He has authored several technical publications in heat transfer and has served as an industrial consultant.

Nomenclature

a	geometrical dimension (m)
A	area (m^2)
b	geometrical dimension (m)
B	Biot number, usually hL/k
c	specific heat (J/kg/K)
ds'_j	differential element of body surface S_j
dv'	differential element of volume
f	boundary term
f_i	basis function (Chapter 10)
$F(\mathbf{r})$	initial temperature distribution
g	energy generation per unit time per unit volume
G	Green's function
h	heat transfer coefficient ($W{\cdot}m^{-2}{\cdot}K^{-1}$)
H	Heaviside unit step function
i, j	indices
J	joules, unit of energy
k	thermal conductivity (W/m/K)
K	fundamental heat conduction solution
K	kelvin, unit of temperature
L	slab thickness (m)
m	fin effect parameter; also meter, unit of length
n_j	outward normal unit vector
N_m	norm (Chapter 4)
$\mathbf{q}$	heat flux (W/m^2)
Q	heat energy (J)
r	radial coordinate (m)
$\mathbf{r}$	position vector
$\mathbf{r}'$	position vector; also dummy variable
s	Laplace transform parameter (Chapter 4)
S_j	surface
t	time (s)
T	temperature (K)
T_0	initial temperature distribution
u	cotime (s)
$\mathbf{U}$	velocity (m/s)
$\mathbf{V}, w$	velocity (m/s)
W	watt, one joule per second
x, y, z	Cartesian coordinates
X_m	eigenfunction

Greek symbols

α	thermal diffusivity (m^2/s)
β_m	eigenvalue
γ_n	eigenvalue
δ	Dirac delta function
ρ	density (kg/m^3)
σ	propagation speed for heat transfer (m/s)
τ	time; also dummy variable (s)
ϕ, Φ	angular coordinate (azimuth) for cylindrical and spherical coordinates
θ	polar angle coordinate, spherical coordinates
ψ_n	eigenvalue (Chapter 10)

1 Introduction to Green's Functions

1.1 INTRODUCTION

Green's functions (GFs), named after English physicist George Green (1793–1841), are powerful tools for obtaining solutions of linear heat conduction problems. They also apply to the solution of many other phenomena described by linear differential equations. A GF is a basic solution of a specific differential equation with homogeneous boundary conditions; it is a building block from which many useful solutions may be constructed. For transient heat conduction, a GF describes the temperature distribution caused by an instantaneous, local heat pulse.

This book contains an extensive set of exact GFs for the heat conduction equation in Cartesian, cylindrical, and spherical coordinates. By utilizing these tabulated GFs, solutions of many heat conduction problems can be obtained in a straightforward and efficient manner. In many cases, the formal solutions can be written directly in terms of integrals which can be evaluated either exactly using integrals provided herein or approximately using numerical integration. Compared to the usual analytical methods, the GF method with tabulated GFs requires a lower level of mathematical ability for the solution of partial differential equations.

The GF method is related to other methods for solving heat conduction problems. The classic methods of heat conduction, such as the method of separation of variables and the Laplace transform method, may be used to derive GFs (as in Chapter 4). Approximate methods of finding GFs developed by Haji-Sheikh and Lakshminarayanan (1987) and Haji-Sheikh (1988) are also discussed (see Chapter 10). In addition to solution procedures, the GF method also provides greater understanding of the nature of diffusion processes, including heat conduction in porous media.

GFs have been used in the solution of heat conduction for many decades, for example in the classic books by Morse and Feshbach (1953) and Carslaw and Jaeger (1959). The purpose of this book is to provide a single text containing the following components: a careful derivation of the GF solution equation; a systematic and practical approach to the solution of diffusive-type problems; and, an extensive compilation of GFs. Other books contain some of these components: Ozisik (1993) has a fine derivation of the GF solution equation; Butkovskiy (1982) provides a catalog of many GFs; and Carslaw and Jaeger (1959) also list some GFs. Other important books on GFs are Roach (1970), Greenberg (1971), Stakgold (1979), Barton (1989), and Duffy (2001).

1.1.1 Advantage of the Green's Function Method

There is ample motivation for the use of GFs in linear heat conduction. One advantage of GFs is that they are flexible and powerful. The *same* GF for a given geometry and a given set of homogeneous boundary conditions is a building block for the temperature distribution resulting from (a) space-variable initial temperature distribution, (b) time- and space-variable boundary conditions, and (c) time- and space-variable volume energy generation.

A second advantage of the GF method is the systematic solution procedure. Many GFs have been derived and are tabulated in this book, so the derivation of the GF may be omitted in many cases. Eigenfunctions and eigenconditions need not be developed. In these cases the solution can be written immediately in terms of the GFs. The saving of effort and reduced possibility of errors are particularly important for two- and three-dimensional geometries. The systematic solution procedure also allows for construction of families of closely related solutions for checking purposes. This can greatly improve one's confidence in computed numerical values.

A third advantage is that two- and three-dimensional GFs can be found, for transient cases, by simple multiplication of one-dimensional GFs for the rectangular coordinate system for most of the boundary conditions considered in this book. The limitations of the multiplicative property are that the differential equation must be linear, the body must be spatially uniform (homogeneous), and the geometry must be "orthogonal." An orthogonal geometry is one for which any boundary is located where only one coordinate is a constant, such as $x = 0$ or $y = W$, and no boundary is defined by a relationship such as $x + y = C$. A further discussion of nonorthogonal bodies is given in Chapter 11. The multiplicative construction of two- and three-dimensional GF can result in great simplification in solving temperature problems, and provides a very compact means for cataloging GFs for these cases. For certain two-dimensional cases involving cylindrical coordinates, multiplication of the GFs can also be used.

A fourth advantage is that the GF solution equation has an alternative form which can improve the convergence of series solutions which arise from heating at a boundary (nonhomogeneous boundary conditions). Slow convergence of series solutions, which require that a very large number of terms be evaluated, can cause lengthy computer-evaluation times, and can reduce numerical accuracy by excessive round-off error. When it applies, the alternative formulation of the GF solution equation can greatly reduce the number of series terms needed for an accurate numerical evaluation.

A fifth advantage of the GF method is intrinsic verification. That is, solutions constructed from GF contain within them the means to check that computed numerical values are correct. As an example of intrinsic verification, when a time-varying solution contains a steady term and a transient term, at early time there is a region in which these terms must sum to zero. In this region these terms may be checked, one against the other. Several types of intrinsic verification are given in Chapter 5.

A sixth advantage of the GF method is time partitioning, which can reduce the number of series terms needed to obtain an accurate solution. Time partitioning is a general method that arises naturally from the GF method, and can provide accurate values for temperature using only a few terms of the infinite series. Time partitioning is introduced in Chapter 5.

1.1.2 Scope of This Chapter

The purposes of this chapter are to introduce GFs in one-dimensional heat conduction and to provide insight and motivation. Some basic information on heat conduction is also provided. More rigorous aspects, such as derivation of the GF solution equation, are deferred to later chapters. In Sections 1.1 through 1.5 some basic information on heat conduction is given, including the heat conduction equation applied to a point (differential equation) and to a control volume (integral equation). In Section 1.6 the Dirac delta function, the foundation of every GF, is introduced. In Section 1.7 a steady GF is derived for the one-dimensional wall. Sections 1.8 through 1.12 give an introduction to the transient one-dimensional GF, first in the infinite body, then the semi-infinite body and the flat plate. In Section 1.13 the properties common to transient GFs are given. Sections 1.14 through 1.17 provide additional topics that briefly indicate how the GF method can be applied to a broader scope of engineering problems, including heterogeneous bodies, anisotropic bodies, moving bodies, bodies with fins, and non-Fourier heat conduction.

1.2 HEAT FLUX AND TEMPERATURE

In a solid body that contains variations of temperature, heat flow proceeds from a region of high temperature to a region of low temperature. The term heat flow is the rate of energy transfer (in Joules per second, or J/s) associated with the vibrational energy of atoms and molecules in the body. Heat flux is the heat flow per unit area at any point in the body. Heat conduction theory is the relationship between heat flux and temperature in a solid body; it also applies to liquids and gases when there is no bulk motion of the fluid.

Heat flux cannot be measured directly, but its effects can be indirectly observed. At the surface of a solid body the heat flux can sometimes be observed as an effect on the surroundings, such as the melting of ice, the warming of a well stirred water bath, or the vaporization of water at a certain rate. Inside a solid body, the heat flux can be deduced from the temperature distribution, and then only if the relationship between temperature and heat flux is thoroughly understood.

In a solid body with a steady temperature gradient, heat flux has a magnitude and a direction and it is denoted by vector $\vec{q}$. The component of heat flux, in a direction of coordinate x, for example, is

$$q_x = -k\frac{\partial T}{\partial x} \tag{1.1}$$

where parameter k is the thermal conductivity with units W/(m K). In general the thermal conductivity may be a function of temperature. The negative sign implies that heat always flows in the direction of reducing temperature. Similarly in the in y- and z-directions,

$$q_y = -k\frac{\partial T}{\partial y}; \quad q_z = -k\frac{\partial T}{\partial z} \tag{1.2}$$

This is Fourier's law of heat conduction. Fourier's law applies to any body that is homogeneous (the same substance all the way through), isotropic (heat flows equally well in any direction), and of macroscale size (not too small). Non-Fourier heat conduction, appropriate for very small bodies and for very short-duration heat conduction events, is discussed in Section 1.17.

1.3 DIFFERENTIAL ENERGY EQUATION

The differential energy equation is derived in this section for homogeneous isotropic bodies. The rectangular (x, y, z) coordinate system is used for simplicity.

The energy equation, also called the heat conduction equation, is based on the conservation of energy. Consider a small parallelepiped shaped control volume in a stationary, homogeneous, and isotropic body. The control volume is located at point (x, y, z) in the body and has volume $dV = dx\,dy\,dz$. See Figure 1.1. A form of the first law of thermodynamics gives the energy balance on the control volume:

$$\begin{pmatrix}\text{Net rate of}\\ \text{heat flow in}\end{pmatrix} + \begin{pmatrix}\text{rate of}\\ \text{energy generation}\end{pmatrix} = \begin{pmatrix}\text{rate of}\\ \text{energy storage}\end{pmatrix} \tag{1.3}$$

Each term in this rate equation has units of energy/time (J/s or watts). The three terms in this equation will be examined one at a time.

Net rate of heat flow in. There are six faces on the control volume through which heat can enter or leave. Heat flux is positive in the positive coordinate directions, and each heat flux multiplied by the area of the face gives the correct units of watts. Figure 1.2 shows the flow of heat in the x-direction, where $q_x A_x$ has the units of

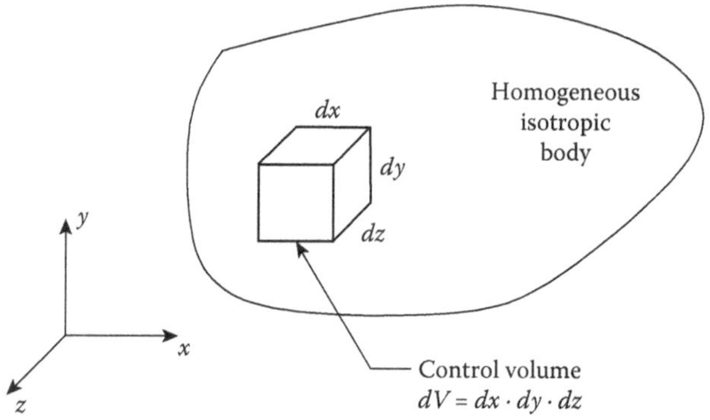

FIGURE 1.1 Control volume.

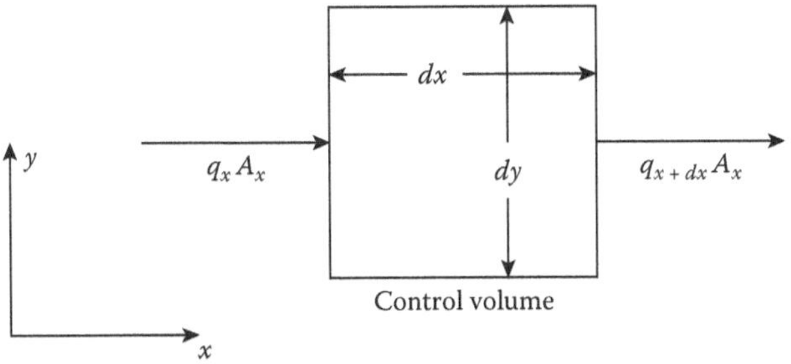

FIGURE 1.2 Flow of heat in the x direction.

watts. The net flow of heat is the difference between the inflow and the outflow $(q_x A_x - q_{x+dx} A_x)$. For all three directions and all six faces of the control volume,

$$\begin{pmatrix} \text{Net rate of} \\ \text{heat flow in} \end{pmatrix} = (q_x - q_{x+dx})A_x + (q_y - q_{y+dy})A_y + (q_z - q_{z+dz})A_z \quad (1.4)$$

Rate of energy generation. Energy generation is energy that affects the temperature throughout the volume of the body. It is distinguished from energy that enters the body through the boundaries. Energy generation can come from electrical resistance heating inside the body, from chemical reaction (for example, concrete generates heat when curing), or from absorption of radiation (nuclear, microwave, or other electromagnetic energy). The energy generation may vary from place to place in the body and it may vary with time. The energy generation may also be simply equal to zero. It is given the symbol $g(x, y, z, t)$ with units W/m^3 (rate of energy generation per unit volume). For the control volume, then,

$$(\text{Rate of energy generation}) = g(x, y, z, t)\, dx\, dy\, dz \quad (1.5)$$

Rate of energy storage. A change in the storage of energy is defined by a change in the specific internal energy (a thermodynamic quantity) which is given by $c\delta T$ for solid bodies. Here c is the specific heat [J/(kg K)] and δT is the change in temperature. The rate of specific energy storage (per unit mass) is given by the time derivative $c\partial T / \partial t$. The partial derivative on time is used because T also depends on position (x, y, z). Multiply the time rate of change of specific internal energy by the density and the volume to obtain watts:

$$(\text{Rate of energy storage}) = \rho c \frac{\partial T}{\partial t}\, dx\, dy\, dz \quad (1.6)$$

To place the energy equation in differential form, the control volume will be made arbitrarily small. Then, the heat flux at the faces located at $x + dx$, $y + dy$, and $z + dz$ can be related to the heat flux at x, y, and z by the first term of a Taylor series,

TABLE 1.1
One Term of Taylor Series for q

Direction	Flux	Area
x	$q_{x+dx} = q_x + \dfrac{\partial q_x}{\partial x} dx$	$A_x = dy\,dz$
y	$q_{y+dy} = q_y + \dfrac{\partial q_y}{\partial y} dy$	$A_y = dx\,dz$
z	$q_{z+dz} = q_z + \dfrac{\partial q_z}{\partial z} dz$	$A_z = dx\,dy$

according to Table 1.1. When the table values are substituted into Equation 1.4, the energy equation can be assembled from Equations 1.4 through 1.6 in the form

$$-\frac{\partial q_x}{\partial x} - \frac{\partial q_y}{\partial y} - \frac{\partial q_z}{\partial z} + g(x, y, z, t) = \rho c \frac{\partial T}{\partial t} \tag{1.7}$$

Now, applying Fourier's law yields

$$\frac{\partial}{\partial x}\left(k\frac{\partial T}{\partial x}\right) + \frac{\partial}{\partial y}\left(k\frac{\partial T}{\partial y}\right) + \frac{\partial}{\partial z}\left(k\frac{\partial T}{\partial z}\right) + g(x, y, z, t) = \rho c \frac{\partial T}{\partial t} \tag{1.8}$$

This is the energy equation for a homogeneous isotropic body. Properties c and k may depend upon the temperature and therefore may vary with position in the body.

In the special case when the thermal conductivity does not depend on position (for example, when the temperature gradients are not too large), the energy equation can be written as

$$\frac{\partial^2 T}{\partial x^2} + \frac{\partial^2 T}{\partial y^2} + \frac{\partial^2 T}{\partial z^2} + \frac{1}{k} g(x, y, z, t) = \frac{1}{\alpha}\frac{\partial T}{\partial t} \tag{1.9}$$

where $\alpha = k/(\rho c)$ is the thermal diffusivity (m^2/s). This form of the energy equation is extensively studied in this book.

The energy equation, developed here in the rectangular coordinate system, can be cast in other orthogonal coordinate systems, as follows. A general vector form of Fourier's law is given by

$$\mathbf{q} = -k\nabla T \tag{1.10}$$

where ∇T is the gradient of the temperature and $\mathbf{q}$ is the heat flux vector. A vector form of the energy equation that is independent of coordinate system is given by (see Ozisik, 1993, pp. 3–6 for a derivation)

$$-\nabla \cdot \mathbf{q} + g(\mathbf{r}, t) = \rho c \frac{\partial T}{\partial t} \tag{1.11}$$

where $\nabla \cdot \mathbf{q}$ is the divergence of the heat flux. The energy equation in any coordinate system can be found by substituting the correct form of the divergence and gradient

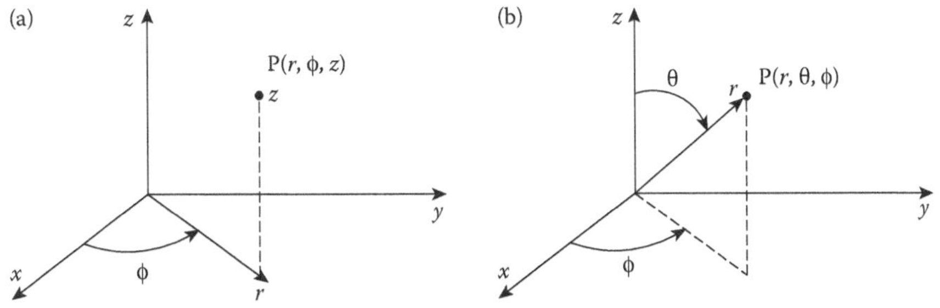

FIGURE 1.3 (a) Cylindrical coordinate system. (b) Spherical coordinate system.

operators for that particular coordinate system; the cylindrical and spherical forms are given next.

Energy equation in cylindrical coordinates. In the cylindrical coordinate system shown in Figure 1.3a the energy equation is

$$\frac{1}{r}\frac{\partial}{\partial r}\left(kr\frac{\partial T}{\partial r}\right)+\frac{1}{r^2}\frac{\partial}{\partial \phi}\left(k\frac{\partial T}{\partial \phi}\right)+\frac{\partial}{\partial z}\left(k\frac{\partial T}{\partial z}\right)+g=\rho c\frac{\partial T}{\partial t} \tag{1.12}$$

or for k = constant

$$\frac{\partial^2 T}{\partial r^2}+\frac{1}{r}\frac{\partial T}{\partial r}+\frac{1}{r^2}\frac{\partial^2 T}{\partial \phi^2}+\frac{\partial^2 T}{\partial z^2}+\frac{g}{k}=\frac{1}{\alpha}\frac{\partial T}{\partial t} \tag{1.13}$$

Energy equation in spherical coordinates. In the spherical coordinate system shown in Figure 1.3b, the energy equation is

$$\frac{1}{r^2}\frac{\partial}{\partial r}\left(kr^2\frac{\partial T}{\partial r}\right)+\frac{1}{r^2\sin\theta}\frac{\partial}{\partial \theta}\left(k\sin\theta\frac{\partial T}{\partial \theta}\right)+\frac{1}{r^2\sin^2\theta}\frac{\partial}{\partial \phi}\left(k\frac{\partial T}{\partial \phi}\right)+g=\rho c\frac{\partial T}{\partial t} \tag{1.14}$$

or for k = constant,

$$\frac{1}{r}\frac{\partial^2 (rT)}{\partial r^2}+\frac{1}{r^2\sin\theta}\frac{\partial}{\partial \theta}\left(\sin\theta\frac{\partial T}{\partial \theta}\right)+\frac{1}{r^2\sin^2\theta}\frac{\partial^2 T}{\partial \phi^2}+\frac{g}{k}=\frac{1}{\alpha}\frac{\partial T}{\partial t}$$

1.4 BOUNDARY AND INITIAL CONDITIONS

This book is concerned with solutions to the energy equation as they apply to problems in engineering and physics. The mathematical form of the solutions (such as GFs) are determined by the boundary conditions, that is, the value of the temperature (or its derivative) at the boundaries of the heat conducting body. The combination of the energy equation, the specific boundary conditions, and the initial condition is called a boundary value problem. Most of this book is concerned with orthogonal bodies, whose boundaries are located where one coordinate is a constant, such as

$x = 0$ or $x = L$. Where possible, the coordinate system is chosen so that the body of interest may be treated as an orthogonal body. (Nonorthogonal bodies are discussed in Chapter 10.)

The number of boundary conditions for a boundary value problem depends on the form of the energy equation and the geometry of the system under consideration. For example, the two-dimensional energy equation in the rectangular coordinate system,

$$\frac{\partial^2 T}{\partial x^2} + \frac{\partial^2 T}{\partial y^2} + \frac{1}{k} g(x, y, z, t) = \frac{1}{\alpha} \frac{\partial T}{\partial t} \tag{1.15}$$

requires five conditions: two each for the boundaries on x and y and one initial condition. Boundary conditions typically have the form

$$k_i \frac{\partial T}{\partial n_i} + h_i T = f_i(r_i, t) \tag{1.16}$$

where all quantities are evaluated at the ith boundary. Here r_i is the location of the ith boundary in a specific coordinate system and n_i is the outward unit-normal vector at the boundary. Initial conditions have the form

$$T(r_i, t = 0) = F(r_i) \tag{1.17}$$

Boundary conditions and initial conditions are discussed in detail in Chapter 2.

1.5 INTEGRAL ENERGY EQUATION

In this section, the integral energy equation is derived for heat transfer in a solid. The solid may be moving but it may not change shape. There are no changes in the shape of the body during heating due to thermal expansion; the subject of thermal stresses is beyond our scope.

The derivation starts with a system, which is a body or portion of a body that is identified for study. The system may move and exchange energy with its surroundings. The first law of thermodynamics for a system can be written as

$$\frac{\delta Q}{dt} = +\frac{\delta W}{dt} + \frac{dE}{dt} \tag{1.18}$$

where δW and δQ denote path dependent quantities. Each term in Equation 1.18 can be described in words by

$$\frac{\delta Q}{dt} = \text{Rate of heat addition to the system at the boundaries}$$
$$\frac{\delta W}{dt} = \text{Rate of work done by the system on its surroundings}$$
$$\frac{dE}{dt} = \text{Rate of energy accumulation inside the system}$$

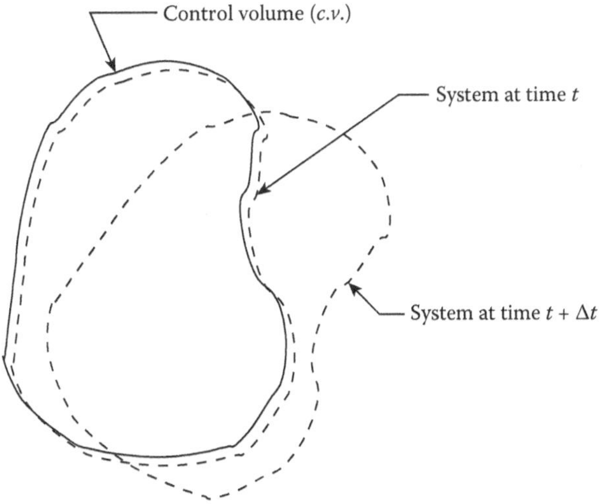

FIGURE 1.4 Relation between moving system and fixed control volume.

The thermal energy of the system E is given by the internal energy of the system,

$$E = mu \tag{1.19}$$

where m is mass of the system in kilograms (kg), and u is the internal energy per unit mass, J/kg. Kinetic and potential energy are neglected compared to thermal energy.

The next step is to relate the system to a control volume with the Reynolds transport theorem (see Currie, 2002 or White, 2006). The control volume is fixed in space and has fixed shape and fixed boundaries. At the moment of interest, time t, the system and the control volume occupy the same region. At a later time, $t + \Delta t$, the system has moved away from the fixed control volume. Refer to Figure 1.4. A statement of the Reynolds transport theorem for the change of energy in the system is

$$\frac{dE}{dt} = \frac{\partial}{\partial t}\int_{c.v.} u\rho\, dv + \int_{c.s.} \rho u(\mathbf{V}\cdot\hat{\mathbf{n}})\, dA \tag{1.20}$$

where c.v. denotes the control volume, c.s. denotes the surface of the control volume (control surface), dv is an element of volume, ρ is density, $\mathbf{V}$ is the velocity vector, and $\hat{\mathbf{n}}$ is an outward drawn unit normal vector. Equation 1.20 relates the energy in the system at time t to that in the control volume.

Next, replace dE/dt with the first law of thermodynamics, Equation 1.18,

$$\left.\frac{\delta Q}{dt}\right|_{sys} - \left.\frac{\delta W}{dt}\right|_{sys} = \frac{\partial}{\partial t}\int_{c.v.} u\rho\, dv + \int_{c.s.} \rho u(\mathbf{V}\cdot\hat{\mathbf{n}})\, dA \tag{1.21}$$

The terms of Equation 1.21 will next be examined separately. The first term of Equation 1.21 relates to energy traveling across the control surface and can be given by

$$\frac{\delta Q}{dt} = \int_{c.s.} (-\mathbf{q}\cdot\hat{\mathbf{n}})\, dA \tag{1.22}$$

where $\mathbf{q}$ is the heat flux crossing the control surface in W/m^2. It can include conduction and radiation,

$$\mathbf{q} = \mathbf{q}_{cond} + \mathbf{q}_{radiation} \tag{1.23}$$

but not any term caused by fluid flow for any element inside the body. When a body is "lumped" in some way so that a solid-fluid boundary is included in the control volume, then a convection-related term may enter. Otherwise, for any element in a solid body or porous body, the only two modes of heat transfer are conduction and radiation.

The $\delta W/dt$ term in Equation 1.21 relates to the rate of work done by the system on the surroundings and could be composed of a number of parts:

$$\frac{\delta W}{dt} = \begin{cases} \text{Shaft work + flow work + viscous work} \\ \text{+ electrical work + nuclear work + chemical work,} \\ \text{all acting on the surroundings.} \end{cases}$$

For a solid body that does not change shape, there is no shaft work, flow work or viscous work. The electrical, nuclear, and chemical work are all combined together as volume energy generation, denoted with symbol g:

$$\frac{\delta W}{dt} = -\int_{c.v.} g\, dv \tag{1.24}$$

The volume energy generation term has units of W/m^3 ; g is positive for heat produced in the body; g may vary with position in the body, and it may vary with time.

Next consider the third term of Equation 1.21 for a fixed control volume in a solid $[\rho \neq \rho(t)]$

$$\frac{\partial}{\partial t}\int_{c.s.} u\rho\, dv = \int_{c.s.} \rho \frac{\partial u}{\partial t}\, dv \tag{1.25}$$

That is, the time derivative bypasses the volume integral because the density and the volume are constant with respect to time.

Next the internal energy will be related to the temperature. Let $\nu = \rho^{-1}$ where ν is the specific volume. From thermodynamics, internal energy can be a function of two independent thermodynamic quantities. Let u be a function of temperature T and specific volume ν, both of which are functions of position vector $\mathbf{r}$ and time t or

$$u = u(T(\mathbf{r}, t), \nu(\mathbf{r}, t)) \tag{1.26}$$

Then using the chain rule for differentiation gives

$$\frac{\partial u}{\partial t} = \left.\frac{\partial u}{\partial T}\right|_{\nu} \frac{\partial T}{\partial t} + \left.\frac{\partial u}{\partial \nu}\right|_{T} \frac{\partial \nu}{\partial t} \tag{1.27}$$

In a solid, density is not a function of time, so that $\partial\nu/\partial t$ is equal to zero. Also, from the definition of the specific heat at constant volume,

$$c_{\nu} = \left.\frac{\partial u}{\partial T}\right|_{\nu} \tag{1.28}$$

In an incompressible solid, the specific heat at constant volume is the same as at constant pressure or

$$c_v = c_p = c \tag{1.29}$$

Substitute Equations 1.28 and 1.29 into Equation 1.27

$$\frac{\partial u}{\partial t} = c\frac{\partial T}{\partial t} \tag{1.30}$$

so the third term of Equation 1.21 is given by

$$\int_{c.v.} \rho c \frac{\partial T}{\partial t}\, dv \tag{1.31}$$

Notice that the specific heat can be a function of position and temperature, $c = c(\mathbf{r}, T)$. In particular, note that c is *not* inside the derivative with respect to time.

Then Equations 1.22, 1.24, and 1.31 can be substituted into Equation 1.21 to give the general form of the **integral energy equation for an incompressible solid,**

$$\int_{c.s.} (-\mathbf{q}\cdot\hat{\mathbf{n}})\, dA + \int_{c.v.} g\, dv = +\int_{c.v.} \rho c \frac{\partial T}{\partial t}\, dv + \int_{c.s.} \rho u(\mathbf{V}\cdot\hat{\mathbf{n}})\, dA \tag{1.32}$$

This equation is valid for $\rho = \rho(\mathbf{r})$ and $c = c(\mathbf{r}, t)$.

Many forms of the heat conduction equation can be derived from this equation, including general partial differential equations and also lumped capacitance equations. If the control volume is taken to represent a thin region at a boundary, then Equation 1.32 can be used to obtain boundary conditions.

1.6 DIRAC DELTA FUNCTION

The Dirac delta function (sometimes called the *unit impulse function*) plays a central role in the method of GFs. In this section we define the Dirac delta function in terms of those properties important to the GF method. Strictly speaking, the Dirac delta function is a *generalized function*; see Duffy (2001, pp. 5–14) for a discussion of this viewpoint.

The Dirac delta function $\delta(x)$ is defined to be zero when $x \neq 0$, and infinite at $x = 0$ in such a way that the area under the function is unity. A concise definition is the following: given nonzero numbers $\eta_1 > 0$ and $\eta_2 > 0$,

$$\delta(x) = 0 \text{ if } x \neq 0; \text{ and, } \int_{-\eta_1}^{\eta_2} \delta(x)\, dx = 1. \tag{1.33}$$

Some of the properties of the Dirac delta function are given next.

Sifting property. Given function $f(x)$ continuous at $x = x'$,

$$\int_a^b f(x')\,\delta(x - x')\, dx' = \begin{cases} f(x) & \text{if } a < x < b \\ 0 & \text{if } (a, b) \text{ does not contain } x \end{cases} \tag{1.34}$$

When integrated, the product of any (well-behaved) function and the Dirac delta yields the function evaluated where the Dirac delta is singular. The sifting property also applies if the arguments x and x' are exchanged.

Relation to unit step function. The integral of the Dirac delta function may be related to the unit step function, as follows:

$$\int_{-\infty}^{t} \delta(\tau)\,d\tau = H(t) \tag{1.35}$$

where $H(t)$ is the Heaviside unit step function defined as

$$H(t) = \begin{cases} 0 & \text{if } t < 0 \\ 1 & \text{if } t > 0 \end{cases}$$

The derivative of the unit-step function, then, gives the Dirac delta function:

$$\frac{dH(t-\tau)}{dt} = \delta(t-\tau)$$

Note that this derivative is singular at $t - \tau = 0$.

Units. Since the definition of the Dirac delta requires that the product $\delta(x)dx$ is dimensionless, the units of the Dirac delta are the inverse of those of its argument. That is, $\delta(x)$ has units meters^{-1}, and $\delta(t)$ has units sec^{-1}. Later, when two- and three-dimensional cases are discussed, the Dirac delta function will be used in the form $\delta(\mathbf{r} - \mathbf{r}')dv'$ where dv' is differential volume; therefore the units of $\delta(\mathbf{r})$ are inverse volume. This is particularly important in cylindrical and spherical coordinates.

These properties are also listed in Table 1.2. More information on the Dirac delta function, including a proof of the sifting property, is given in Appendix D.

TABLE 1.2
Basic Properties of the Dirac Delta Function

1.	$\delta(x - x') = \begin{cases} \infty, & \text{at } x = x' \\ 0, & \text{otherwise} \end{cases}$
2.	$\int_{-\infty}^{\infty} \delta(x - x')\,dx' = 1$
3.	$\int_{-\infty}^{\infty} F(x')\delta(x - x')\,dx' = F(x)$, the sifting property
4.	$\frac{dH(t-\tau)}{dt} = \delta(t - \tau)$, where H is the unit step
5.	$\delta(t - \tau)$ has units of s^{-1} $\delta(x - x')$ has units of m^{-1} $\delta(\mathbf{r} - \mathbf{r}')$ has units such that $\delta(\mathbf{r} - \mathbf{r}')\,dv'$ has no units

1.7 STEADY HEAT CONDUCTION IN ONE DIMENSION

In this section one-dimensional steady heat conduction will be discussed to introduce the concept of GFs. Steady heat conduction is described by an ordinary differential equation, and the GF has a simple form.

Steady heat conduction in the one-dimensional slab body is described by the following energy equation:

$$\frac{d^2T}{dx^2} + \frac{g(x)}{k} = 0; \quad 0 < x < L \tag{1.36}$$

In the slab body ($0 < x < L$) there are two boundaries, therefore two boundary conditions are also needed. For the present discussion a specific geometry will be studied; other combinations of boundary conditions will be given later. Suppose the boundary conditions are given by

$$T|_{x=0} = T_1 \tag{1.37}$$

$$\left.\frac{dT}{dx}\right|_{x=L} = 0 \tag{1.38}$$

The temperature at $x = 0$ is a specified value (first kind), and the slope of the temperature is specified at $x = L$ (second kind). The solution of this steady heat conduction problem will be sought in two different ways.

1.7.1 Solution by Integration

For this steady case, the temperature may be found by integrating the energy equation two times. This is best demonstrated by a specific example.

Suppose the energy generation is spatially uniform, that is, $g(x) = g_0$. (By the numbering system discussed in Chapter 2, this is case $X12B10G1$.) Integrate the energy equation once

$$\frac{dT}{dx} = -\frac{g_0}{k}x + C_1$$

and again to find the general solution:

$$T(x) = -\frac{g_0}{k}\frac{x^2}{2} + C_1 x + C_2. \tag{1.39}$$

Constants of integration C_1 and C_2 are found by applying the boundary conditions, first at $x = 0$,

$$T|_{x=0} = 0 + 0 + C_2 \Longrightarrow C_2 = T_1$$

and then at $x = L$

$$\left.\frac{dT}{dx}\right|_{x=L} = -\frac{g_0}{k}L + C_1 \Longrightarrow C_1 = \frac{g_0 L}{k}$$

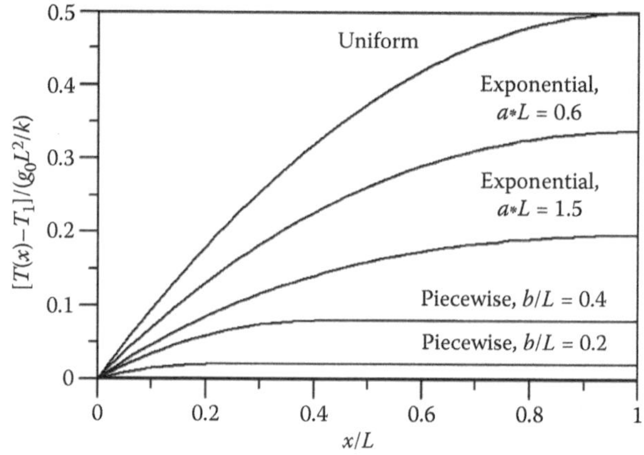

FIGURE 1.5 Steady temperature in the slab caused by different distributions of internal heat generation. The left side of the slab has a specified temperature and the right side is insulated.

Using these constants, the solution for this specific example is

$$T(x) = -\frac{g_0 x^2}{2k} + \frac{g_0 L}{k} x + T_1$$

or in normalized form,

$$\frac{T(x) - T_1}{g_0 L^2 / k} = \frac{x}{L} - \frac{1}{2}\left(\frac{x}{L}\right)^2 \qquad (1.40)$$

This solution is plotted in Figure 1.5 (uniform generation case). Note that the boundary conditions are clearly satisfied: at $x = 0$ where $T - T_1 = 0$; and, at $x = L$ where the slope of temperature is zero (insulated condition).

This solution was found by direct integration, which is appropriate for finding a single solution. Suppose, however, that another solution is needed for a nonuniform generation term, such as $g(x) = g_0 e^{-ax}$. Then the entire solution procedure would have to be repeated. The method of GFs, introduced in the next section, can be used to find the temperature caused by various $g(x)$ functions without re-solving the entire problem.

1.7.2 Solution by Green's Function

In this section the steady one-dimensional problem discussed above will be solved by the method of GF. The first step is to find the GF appropriate for the temperature problem. The GF, G, associated with this specific temperature problem satisfies the following *auxiliary problem*:

$$\frac{d^2 G}{dx^2} + \delta(x - x') = 0; \quad 0 < x < L \qquad (1.41)$$

$$G|_{x=0} = 0 \tag{1.42}$$

$$\left.\frac{dG}{dx}\right|_{x=L} = 0 \tag{1.43}$$

The differential equation for G is similar to the temperature equation except that the generation term has been replaced by a unit impulse function (the Dirac delta function). That is, the GF is the unit-impulse response. Note that the boundary conditions are of the *same kind* as the temperature problem, that is, the first kind at $x = 0$ and the second kind at $x = L$. (This is case $X12$ in the numbering system discussed in Chapter 2.) However, the boundary conditions for G are homogeneous (equal to zero); this is important so that any number of GF may be superposed, but the boundary conditions remain unchanged. The GF depends on two variables, the observation location x and the heat-source location x'.

The GF G for this case will be derived presently. However, it is instructive at this point to postulate the temperature solution. If the GF is known, the temperature $T(x)$ is given by

$$T(x) - T_1 = \frac{1}{k}\int_{x'=0}^{L} g(x')G(x,x')\,dx' \tag{1.44}$$

(A full discussion of this temperature expression is given in Chapter 3.) This integral is a summing up of a large number of unit-impulse responses, each of a size determined by $g(x')$, in order to produce the desired temperature response. The temperature caused by several different functions $g(x')$ can be studied merely by repeating the integration, without repeating the entire solution.

Now the GF will be derived. Break the domain $(0 < x < L)$ into two regions at $x = x'$, then the differential equation for G takes on the following form:

$$\text{(a)}\qquad 0 < x < x'; \qquad \frac{d^2G_a}{dx^2} = 0$$

$$\text{(b)}\qquad x' < x < L; \qquad \frac{d^2G_b}{dx^2} = 0 \tag{1.45}$$

Because the Dirac delta function is zero everywhere except at $x = x'$, this approach has removed the singularity from the differential equation. Then the solutions for G_a and G_b may be found by integrating the above equations twice:

$$\text{(a)}\qquad G_a = C_1x + C_2$$

$$\text{(b)}\qquad G_b = C_3x + C_4 \tag{1.46}$$

The four constants introduced by integration can be found from four conditions. The first two are the boundary conditions from the original domain:

$$\text{(i)}\qquad G_a|_{x=0} = 0 \tag{1.47}$$

$$\text{(ii)}\qquad \left.\frac{dG_b}{dx}\right|_{x=L} = 0 \tag{1.48}$$

The third condition comes from requiring that solutions G_a and G_b match at $x = x'$:

$$\text{(iii)} \qquad G_a|_{x=x'} = G_b|_{x=x'} \tag{1.49}$$

The fourth condition comes from integrating Equation 1.41, the original differential equation for G, from $(x' - \epsilon)$ to $(x' + \epsilon)$ for some small $\epsilon > 0$. That is,

$$\int_{x'-\epsilon}^{x'+\epsilon} \frac{d^2G}{dx^2} dx = -\int_{x'-\epsilon}^{x'+\epsilon} \delta(x - x')\, dx$$

$$\left.\frac{dG}{dx}\right|_{x'-\epsilon}^{x'+\epsilon} = -1$$

Note that the singularity in the Dirac delta function has been removed by integration. Now in the limit as $\epsilon \to 0$ we have the *jump condition*

$$\text{(iv)} \qquad \left.\frac{dG_b}{dx}\right|_{x'} - \left.\frac{dG_a}{dx}\right|_{x'} = -1 \tag{1.50}$$

The jump condition describes the slope of the GF at $x = x'$. With these four conditions, it is now possible to seek the four constants. Applying conditions (i) through (iv) to Equation 1.46 gives

$$\begin{aligned} &\text{(i)} && C_1 \cdot 0 + C_2 = 0 \\ &\text{(ii)} && C_3 = 0 \\ &\text{(iii)} && C_1 \cdot x' + C_2 = C_3 \cdot x' + C_4 \\ &\text{(iv)} && C_3 - C_1 = -1 \end{aligned} \tag{1.51}$$

An algebraic solution gives $C_1 = 1$, $C_2 = 0$, $C_3 = 0$, and $C_4 = x'$. Substitute these values back into the general solution, Equation 1.46, to give

$$G(x, x') = \begin{cases} x; & x < x' \\ x'; & x' < x \end{cases} \tag{1.52}$$

This is the steady one-dimensional GF for this case. A plot of this GF is given in Figure 1.6 which displays the four conditions discussed above, specifically: the value of G is zero at $x = 0$; the slope of G is zero at $x = L$; function G is piecewise continuous; and, the slope of G contains a jump at $x = x'$.

The GF is specific to the shape of the body (slab) and the kind of boundary conditions present (the first kind at $x = 0$ and the second kind at $x = L$). Although this GF was derived for instructional purposes, in many cases the GF is given elsewhere in this book so that the derivation is not needed. (This GF is given as case $X12$ in Table X.3, Appendix X.)

Next the steady GF given above is used in the integral equation to find the temperature.

Uniform generation. For spatially uniform energy generation, the temperature is given by the integral expression, Equation 1.44, with $g(x) = g_0$, and with the GF given by Equation 1.52:

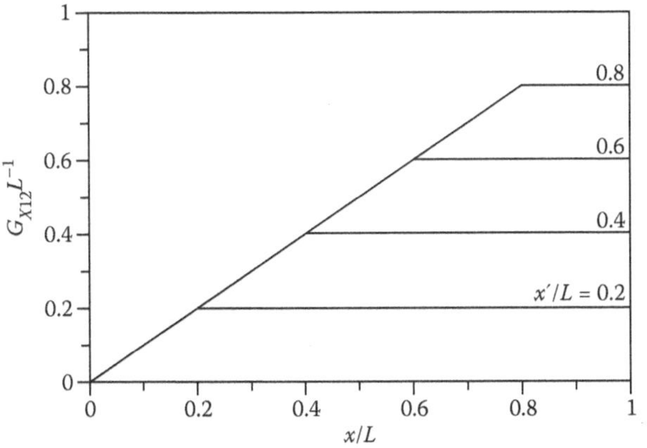

FIGURE 1.6 Steady Green's function for $G = 0$ at $x = 0$ and $dG/dx = 0$ at $x = L$ (case $X12$).

$$\begin{aligned}
T(x) - T_1 &= \frac{1}{k}\int_{x'=0}^{L} g_0\, G(x, x')\, dx' \\
&= \frac{1}{k}\int_{x'=0}^{x} g_0\, x'\, dx' + \frac{1}{k}\int_{x'=x}^{L} g_0\, x\, dx' \\
&= \frac{g_0}{k}\left.\frac{(x')^2}{2}\right|_{x'=0}^{x} + \frac{g_0}{k}\, x\, x'\Big|_{x'=x}^{L} \\
&= \frac{g_0}{k}\left[\frac{x^2}{2} + x(L - x)\right] = \frac{g_0 L^2}{k}\left[\frac{x}{L} - \frac{1}{2}\left(\frac{x}{L}\right)^2\right]
\end{aligned} \tag{1.53}$$

Because the GF is piecewise continuous the integral has been split at $x' = x$, and the correct form of the GF must be used in each interval. This result is identical to the direct-integration solution presented earlier in Equation 1.40.

Exponentially varying generation. For heating that decays exponentially, $g(x) = g_0 e^{-ax}$, and using the same GF as before, the temperature is given by

$$\begin{aligned}
T(x) - T_1 &= \frac{1}{k}\int_{x'=0}^{L} g_0 e^{-ax'}\, G(x, x')\, dx' \\
&= \frac{1}{k}\int_{x'=0}^{x} g_0 e^{-ax'}\, x'\, dx' + \frac{1}{k}\int_{x'=x}^{L} g_0 e^{-ax'}\, x\, dx' \\
&= \frac{g_0}{k}\left[\frac{e^{-ax'}}{a^2}(-1 - ax')\right]\Bigg|_{x'=0}^{x} + \frac{g_0}{k}\frac{x}{a} - e^{-ax'}\Big|_{x'=x}^{L} \\
&= \frac{g_0}{ka^2}\left[1 - e^{-ax} - axe^{-aL}\right]
\end{aligned} \tag{1.54}$$

This heating condition is a reasonable description of microwave absorption in a solid. In the limit as $a \to 0$, the temperature curve approaches the uniform-generation case (see Figure 1.5).

Piecewise constant generation. For piecewise constant energy generation given by

$$g(x) = \begin{cases} g_0; & 0 < x < b \\ 0; & b < x < L \end{cases}$$

the temperature integral is given by

$$T(x) - T_1 = \frac{1}{k}\int_{x'=0}^{b} g_0\, G(x, x')\, dx' + \frac{1}{k}\int_{x'=b}^{L} 0 \cdot dx'$$

Using the same GF as before, the temperature expression must be evaluated in two pieces.

(i) For $0 < x < b$ the temperature is given by

$$\begin{aligned} T(x) - T_1|_{x<b} &= \frac{1}{k}\int_{x'=0}^{x} g_0\, x'\, dx' + \frac{1}{k}\int_{x'=x}^{b} g_0\, x\, dx' \\ &= \frac{g_0 x^2}{2k} + \frac{g_0 x}{k}(b - x) = \frac{g_0 b^2}{k}\left[\frac{x}{b} - \frac{1}{2}\left(\frac{x}{b}\right)^2\right] \end{aligned}$$

(ii) and for $x > b$, only the $x > x'$ part of the GF is needed:

$$T(x) - T_1|_{x>b} = \frac{1}{k}\int_{x'=0}^{b} g_0\, x'\, dx' = \frac{g_0 b^2}{2k}$$

The full temperature expression is given by

$$T(x) - T_1 = \begin{cases} \dfrac{g_0 b^2}{k}\left[\dfrac{x}{b} - \dfrac{1}{2}\left(\dfrac{x}{b}\right)^2\right]; & 0 < x < b \\ \dfrac{g_0 b^2}{2k}; & b < x < L \end{cases} \qquad (1.55)$$

A plot of this temperature is given in Figure 1.5 for $b/L = 0.4$ and $b/L = 0.2$.

In this section the GF method was introduced in a discussion of steady, one-dimensional heat conduction. The GF method involves three components: the boundary value problem for the temperature; the auxiliary problem for G; and, the integral expression for the temperature. For elementary problems such as this, the GF method offers some flexibility over direct integration. Greater advantages arise for more challenging problems, such as transient heat conduction, discussed in the next section, and for two- and three-dimensional heat conduction, discussed in later chapters.

1.8 GF IN THE INFINITE ONE-DIMENSIONAL BODY

In their 1959 book on heat conduction, Carslaw and Jaeger simply state the GF for the one-dimensional infinite body, without derivation, and then show that is satisfies the heat equation. We choose to derive this GF which is also called the fundamental heat conduction solution.

1.8.1 Auxiliary Problem for G

The transient GF for the one-dimensional infinite body satisfies the following set of equations:

$$\frac{\partial^2 G}{\partial x^2} - \frac{1}{\alpha}\frac{\partial G}{\partial t} = -\frac{1}{\alpha}\,\delta(x-x')\,\delta(t-\tau) \tag{1.56}$$

$$G(x,t\,|\,x',\tau) = 0 \text{ for } t-\tau < 0 \tag{1.57}$$

$$G(x\to\pm\infty,t\,|\,x',\tau) \quad \text{is bounded} \tag{1.58}$$

The above equations, Equations 1.56 through 1.58, define the auxiliary problem for the one-dimensional infinite body. The Green's function G is the response to an impulsive, planar heat source of infinitesimal thickness, described by the product of two Dirac delta functions, one for space and one for time. Factor $1/\alpha$ which premultiplies the δ-functions is used to set the units of G to m^{-1} for the one-dimension case. Initially the GF is zero until $t > \tau$, and far away from the heat source the value of G is bounded.

The GF for the infinite body will be derived with the Laplace transform method, and a brief discussion of the Laplace transform method is given here. Later in Chapter 4 additional GF are also found with this method.

1.8.2 Laplace Transform, Brief Facts

The Laplace transform of function $f(t)$ is defined by

$$\mathscr{L}[f(t)] = \int_0^\infty e^{-st} f(t)\,dt \tag{1.59}$$

The properties of the Laplace transform needed for this discussion are given next; see Appendix L for further information on Laplace transforms.

Notation. The overbar is used to denote the transformed function,

$$\mathscr{L}[f(t)] = \overline{f}(s)$$

and the inverse Laplace transform is denoted

$$f(t) = \mathscr{L}^{-1}[\overline{f}(s)]$$

Linear. The Laplace transform is a linear operator. If a and b are constants then

$$\mathscr{L}[af(t)+bg(t)] = a\overline{f}(s) + b\,\overline{g}(s) \tag{1.60}$$

Transform of derivative. Using the definition of the Laplace transform and integration by parts, the transform of a derivative is

$$\begin{aligned}\mathscr{L}\left[\frac{d}{dt}f(t)\right] &= \int_0^\infty e^{-st}\left[\frac{d}{dt}f(t)\right]dt\\ &= f(t)\,e^{-st}\Big|_0^\infty + s\int_0^\infty e^{-st} f(t)\,dt\end{aligned}$$

Therefore

$$\mathscr{L}\left[\frac{d}{dt}f(t)\right] = s\overline{f}(s) - f(0) \tag{1.61}$$

1.8.3 Derivation of the GF

Now the auxiliary equation for G will be solved. Applying the Laplace transform to the auxiliary problem (Section 1.8.1), with $\tau = 0$, gives the following relations:

$$\frac{d^2\overline{G}}{dx^2} - \left[\frac{s}{\alpha}\overline{G} - 0\right] = -\frac{1}{\alpha}\,\delta(x - x')\,e^0 \tag{1.62}$$

$$\overline{G}(x \to \pm\infty, t\,|\,x') \quad \text{is bounded} \tag{1.63}$$

Note that the sifting property of the Dirac delta function has been applied to the impulsive heating term. Since the GF is zero until after the impulsive heating occurs, there is no loss of generality in setting $\tau = 0$ so that the impulsive heating occurs at $t = 0$. The resulting ordinary differential equation for $\overline{G}$ will be solved by splitting the infinite body into two regions at $x = x'$ in order to remove $\delta(x - x')$ from the differential equation. That is, seek solutions $\overline{G}_a$ and $\overline{G}_b$ that satisfy:

$$\begin{aligned}
&\text{(a)} \qquad -\infty < x < x'; \qquad \frac{d^2\overline{G}_a}{dx^2} - \sigma^2\overline{G}_a = 0 \\
&\text{(b)} \qquad x' < x < +\infty; \qquad \frac{d^2\overline{G}_b}{dx^2} - \sigma^2\overline{G}_b = 0
\end{aligned} \tag{1.64}$$

where $\sigma^2 = s\,/\,\alpha$. Then the general solution in each region may be stated in the form of exponentials:

$$\begin{aligned}
&\text{(a)} \qquad \overline{G}_a = C_1e^{\sigma x} + C_2e^{-\sigma x} \\
&\text{(b)} \qquad \overline{G}_b = C_3e^{\sigma x} + C_4e^{-\sigma x}
\end{aligned} \tag{1.65}$$

There are four constants, requiring four conditions. The first two conditions are the boundary conditions from the auxiliary problem:

$$\text{(i)} \qquad \overline{G}_a\big|_{x=-\infty} \quad \text{is bounded} \tag{1.66}$$

$$\text{(ii)} \qquad \overline{G}_b\big|_{x=+\infty} \quad \text{is bounded} \tag{1.67}$$

The third condition is the requirement that the two solutions match at $x = x'$:

$$\text{(iii)} \qquad \overline{G}_a\big|_{x=x'} = \overline{G}_b\big|_{x=x'} \tag{1.68}$$

The fourth condition comes from integrating the original differential equation for $\overline{G}$, from $(x' - \epsilon)$ to $(x' + \epsilon)$ for some small $\epsilon > 0$. That is,

$$\int_{x'-\epsilon}^{x'+\epsilon} \frac{d^2\overline{G}}{dx^2}\,dx - \frac{s}{\alpha}\int_{x'-\epsilon}^{x'+\epsilon} \overline{G}\,dx = -\int_{x'-\epsilon}^{x'+\epsilon} \frac{\delta(x - x')}{\alpha}\,dx$$

Evaluate integrals to obtain

$$\left.\frac{d\overline{G}_b}{dx}\right|_{x'+\epsilon} - \left.\frac{d\overline{G}_a}{dx}\right|_{x'-\epsilon} - \frac{s}{\alpha}\int_{x'-\epsilon}^{x'+\epsilon} \overline{G}\,dx = -\frac{1}{\alpha}$$

Finally, take the limit as $\epsilon \to 0$ to eliminate the remaining integral.

$$\text{(iv)} \qquad \left.\frac{d\overline{G}_b}{dx}\right|_{x'} - \left.\frac{d\overline{G}_a}{dx}\right|_{x'} = -\frac{1}{\alpha} \tag{1.69}$$

This is the *jump condition* which provides information on the slope of the GF at $x = x'$. Apply the above four conditions to the general solution for $\overline{G}$:

$$\begin{aligned}
&\text{(i)} && C_1 e^{-\infty} + C_2 e^{+\infty} \ \text{ is bounded} \implies C_2 = 0\\
&\text{(ii)} && C_3 e^{\infty} + C_4 e^{-\infty} \ \text{ is bounded} \implies C_3 = 0\\
&\text{(iii)} && C_1 e^{\sigma x'} = C_4 e^{-\sigma x'}\\
&\text{(iv)} && -C_4 \sigma e^{-\sigma x'} - C_1 \sigma e^{\sigma x'} = -1/\alpha
\end{aligned} \tag{1.70}$$

An algebraic solution of the last two equations gives $C_1 = e^{-\sigma x'}/(2\sigma\alpha)$ and $C_4 = e^{\sigma x'}/(2\sigma\alpha)$ so that the specific solution for $\overline{G}$ may be written

$$\overline{G} = \begin{Bmatrix} \dfrac{1}{2\sigma\alpha} e^{-\sigma(x'-x)}; & x < x' \\ \dfrac{1}{2\sigma\alpha} e^{-\sigma(x-x')}; & x > x' \end{Bmatrix} = \frac{e^{-\sigma|x-x'|}}{2\sigma\alpha} \tag{1.71}$$

This is the GF in Laplace transform space. The next step is to invert this expression into the time domain, with the use of appropriate tables of Laplace transform pairs. To put $\overline{G}$ in a form listed in tables, let $k = |x - x'|/\sqrt{\alpha}$. Then with $\sigma = \sqrt{s/\alpha}$, $\overline{G}$ takes on the form

$$\overline{G} = \frac{1}{2\sqrt{\alpha}} \frac{e^{-k\sqrt{s}}}{\sqrt{s}}$$

whose inverse Laplace transform is given by (see Appendix L, Table L.1, number 43):

$$\mathscr{L}^{-1}\left(\frac{1}{2\sqrt{\alpha}} \frac{e^{-k\sqrt{s}}}{\sqrt{s}}\right) = \begin{cases} \dfrac{1}{2\sqrt{\alpha}} \dfrac{1}{\sqrt{\pi t}} \exp\left(-\dfrac{(x-x')^2}{4\alpha t}\right); & t > 0 \\ 0; & t < 0 \end{cases}$$

For this development, the impulsive heating occurs at $t = 0$. The impulsive heating time may be shifted to occur at time τ, without loss of generality, by replacing t by $t - \tau$. That is, the GF for the one-dimensional infinite body is given by:

$$G(x, t \,|\, x', \tau) = \begin{cases} \dfrac{1}{\sqrt{4\pi\alpha(t-\tau)}} \exp\left(-\dfrac{(x-x')^2}{4\alpha(t-\tau)}\right); & t > \tau \\ 0; & t < \tau \end{cases} \tag{1.72}$$

There are alternate ways to derive this GF, including the spatial Fourier transform (Barton, 1989, p. 181) and combined Laplace and Fourier transforms (Duffy, 2001, p. 181). In the next section this GF will be used to find temperature.

1.9 TEMPERATURE IN AN INFINITE ONE-DIMENSIONAL BODY

The GF for an infinite body will now be used to find the temperature. We seek the temperature in an infinite, one-dimensional, constant-property body with initial temperature $F(x)$ and volumetric energy generation $g(x,t)$ (with units W/m^3). This temperature satisfies the following equations:

$$\frac{\partial^2 T}{\partial x^2} + \frac{1}{k} g(x,t) = \frac{1}{\alpha}\frac{\partial T}{\partial t} \tag{1.73a}$$

$$T(x,0) = F(x) \tag{1.73b}$$

$$T(x \to \pm\infty, t) \quad \text{is bounded} \tag{1.73c}$$

Here T is temperature (K), x is position (m), t is time (s), k is thermal conductivity (W/m/K), and α is the thermal diffusivity (m^2/s).

1.9.1 Green's Function Solution Equation

The temperature $T(x,t)$ that is a solution to the above equations may be formally stated with the GF Solution Equation (a full discussion is given in Chapter 3).

$$T(x,t) = \int_{x'=-\infty}^{\infty} G(x,t|x',0)\, F(x')\, dx' + \frac{\alpha}{k}\int_{\tau=0}^{t}\int_{x'=-\infty}^{\infty} G(x,t|x',\tau)\, g(x',\tau)\, dx'\, d\tau \tag{1.74}$$

There are two integral terms in this temperature expression, one containing the initial condition $F(x)$ and the other containing the volumetric energy source $g(x,t)$. Each integral term can be considered to be the solution of a separate problem, one caused by $F(x)$ and one by $g(x,t)$, which are superimposed (i.e., added together) to form the complete solution. It is important to note that when F and g are substituted into the above integrals, the coordinate dependence takes the form $F(x')$ and $g(x',\tau)$, associated with the variables of integration.

1.9.2 Fundamental Heat Conduction Solution

Depending on the geometry and boundary conditions, there are many expressions for the GF $G(x,t|x',\tau)$. The particular form of GF for an infinite one-dimensional body, derived in the previous section, is the fundamental heat conduction solution (Cannon, 1984, p. 33), which we give the special symbol $K(x-x',t-\tau)$:

$$K(x-x',t-\tau) = \begin{cases} \dfrac{1}{\sqrt{4\pi\alpha(t-\tau)}} \exp\left[-\dfrac{(x-x')^2}{4\alpha(t-\tau)}\right]; & t-\tau \geq 0 \\ 0; & t-\tau < 0 \end{cases} \tag{1.75}$$

(In the numbering system introduced in Chapter 2, this is case $X00$.) The fundamental heat conduction solution, $K(x-x',t-\tau)$, has several important properties:

First, $K(x-x',t-\tau)$ satisfies the heat conduction equation given by Equation 1.73a for $g(x,t)=0$ for t greater than zero. See Problem 1.16 at the end of the chapter.

Second, $K(x-x',t-\tau)$ is always equal to or greater than zero for $(t-\tau)$ greater than zero,

$$K(x-x',t-\tau) \geq 0, \quad \text{for } (t-\tau) > 0 \tag{1.76a}$$

Third, the integral of $K(x-x',t-\tau)$ over $-\infty < x' < \infty$ is unity for all x values and for all times $(t-\tau) > 0$,

$$\int_{x'=-\infty}^{\infty} K(x-x',t-\tau)\,dx' = 1, \quad \text{for } (t-\tau) > 0 \tag{1.76b}$$

and is equal to zero for times $(t-\tau) < 0$,

$$\int_{x'=-\infty}^{\infty} K(x-x',t-\tau)\,dx' = 0, \text{ for } (t-\tau) < 0 \tag{1.76c}$$

Fourth, the value of $K(x-x',t-\tau)$ is unchanged if $x-x'$ is replaced by $x'-x$,

$$K(x-x',t-\tau) = K(x'-x,t-\tau) \tag{1.76d}$$

Fifth, the limit of the integral as x approaches x' from below is 1/2

$$\lim_{x\uparrow x'} \int_0^t \frac{\partial K(x-x',t-\tau)}{\partial x}\,d\tau = \frac{1}{2} \tag{1.76e}$$

and approaching x' from above is $-1/2$

$$\lim_{x\downarrow x'} \int_0^t \frac{\partial K(x-x',t-\tau)}{\partial x}\,d\tau = -\frac{1}{2} \tag{1.76f}$$

Depending on the geometry and the boundary conditions, there are many expressions for the GF, $G(x,t|x',\tau)$, but there is only one GF for the case of an infinite body, and a convenient form of it is given by Equation 1.75.

It is instructive to examine a plot of $K(x-x',t-\tau)$. Figure 1.7 shows $K(x-x',t-\tau)$ as a function of $x-x'$ for various values of $\alpha(t-\tau)$. As $(t-\tau)$ goes to zero, the $K(\cdot)$

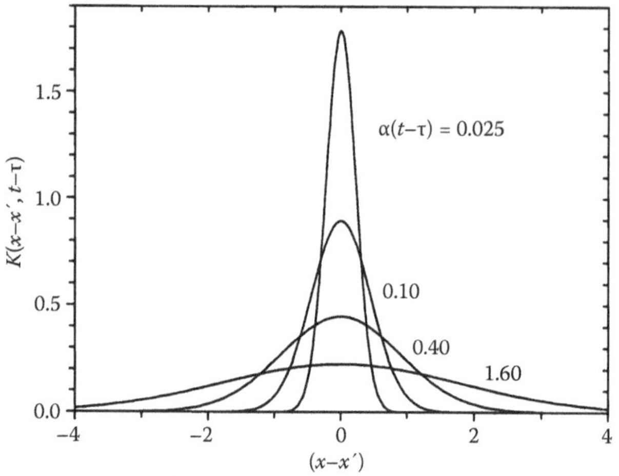

FIGURE 1.7 Fundamental heat conduction solution, $K(x - x', t - \tau)$.

function approaches the Dirac delta function. Each curve in Figure 1.7 has the bell shape of the Gaussian distribution. At all times $t > \tau$, the area underneath a curve in Figure 1.7 is unity as given by Equation 1.76b. As time $t - \tau$ increases, the $K(\cdot)$ function spreads out and the maximum decreases.

The temperature distribution in an infinite body ($-\infty < x < \infty$) for the initial temperature distribution $F(x)$ and the volumetric energy generation of $g(x, t)$ is found using Equation 1.75 in Equation 1.74. The result is

$$T(x,t) = \int_{-\infty}^{\infty} [4\pi\alpha t]^{-1/2} \exp\left[-\frac{(x-x')^2}{4\alpha t}\right] F(x')\,dx' + \frac{\alpha}{k}\int_{\tau=0}^{t}\int_{x'=-\infty}^{\infty} [4\pi\alpha(t-\tau)]^{-1/2} \times \exp\left[-\frac{(x-x')^2}{4\alpha(t-\tau)}\right] g(x',\tau)\,dx'\,d\tau \quad (1.77)$$

Some examples of the use of Equation 1.77 are given next.

Example 1.1:

Find the temperature distribution for the case of

$$F(x) = \begin{cases} T_1, & \text{for } c < x < d \\ 0, & \text{otherwise} \end{cases}$$

$$g(x,t) = 0 \quad \text{for all } x$$

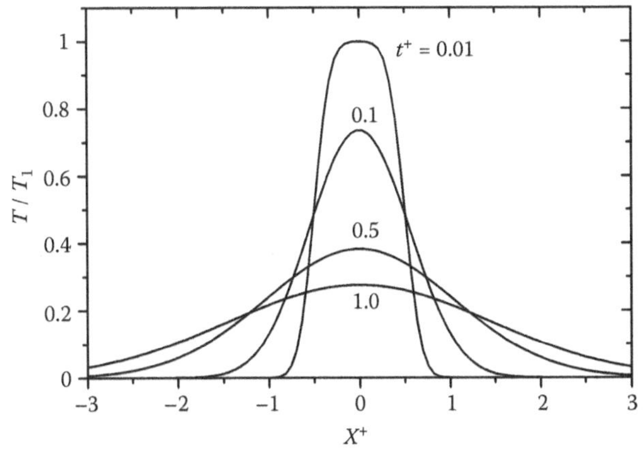

FIGURE 1.8 Temperature distribution for nonuniform initial temperature in an infinite body.

Solution

The solution for T is obtained by using Equation 1.77 with $F(x') = T_1$, for $c < x' < d$ and $F(x') = 0$ otherwise. The result is

$$\begin{aligned} T(x,t) &= \int_c^d K(x-x',t)\,T_1\,dx' \\ &= T_1 \int_c^d [4\pi\alpha t]^{-1/2} \exp\left[-\frac{(x-x')^2}{4\alpha t}\right] dx' \end{aligned} \tag{1.78}$$

Using the substitution $u = (x - x')/(4\alpha t)^{1/2}$, this integral can be written as

$$T(x,t) = \frac{T_1}{\pi^{1/2}} \int_{(x-d)/(4\alpha t)^{1/2}}^{(x-c)/(4\alpha t)^{1/2}} e^{-u^2}\,du \tag{1.79a}$$

$$T(x,t) = \frac{T_1}{2}\left\{\text{erf}\left[\frac{x-c}{(4\alpha t)^{1/2}}\right] - \text{erf}\left[\frac{x-d}{(4\alpha t)^{1/2}}\right]\right\} \tag{1.79b}$$

$$T(x,t) = \frac{T_1}{2}\left\{\text{erfc}\left[\frac{x-d}{(4\alpha t)^{1/2}}\right] - \text{erfc}\left[\frac{x-c}{(4\alpha t)^{1/2}}\right]\right\} \tag{1.79c}$$

where the error function, erf(·), and the complementary error function, erfc(·) = 1 − erf(·), are defined by

$$\text{erf}(z) = \frac{2}{\pi^{1/2}} \int_0^z e^{-u^2}\,du \tag{1.80a}$$

$$\text{erfc}(z) = \frac{2}{\pi^{1/2}} \int_z^\infty e^{-u^2}\,du \tag{1.80b}$$

These functions commonly occur in transient heat conduction. Some relations involving these functions are given in Appendix E.

Equation 1.79 is plotted in Figure 1.8 for $t^+ = \alpha t/(d-c)^2 = 0.01, 0.1, 0.5,$ and 1 as a function of $x^+ = (x - x_m)/(d-c)$ where x_m is the mean x value which is $(c+d)/2$. In this case, the temperature distribution can be written as

$$\frac{T}{T_1} = \frac{1}{2}\left\{\text{erfc}\left[\frac{x^+ - 0.5}{(4t^+)^{1/2}}\right] - \text{erfc}\left[\frac{x^+ + 0.5}{(4t^+)^{1/2}}\right]\right\} \tag{1.81}$$

The temperature distribution is affected first near the edges of the step change of the initial temperature distribution and, as the dimensionless time increases, the effect penetrates further.

Example 1.2:

Find the temperature distribution for the case of $a < b < c < d$ and

$$F(x) = \begin{cases} T_0 & \text{for } a < x < b \\ T_1 & \text{for } c < x < d \\ 0 & \text{otherwise} \end{cases}$$
$$g(x,t) = 0 \quad \text{for all } x$$

Solution

The solution can be found as in Example 1.1 by integrating over the two nonzero regions of $F(x)$ or by using Equation 1.79c as a building block (i.e., let $T_1 \to T_0$, $d \to b$, and $c \to a$). Using either procedure results in

$$\begin{aligned} T(x,t) = {} & \frac{T_0}{2}\left\{\text{erfc}\left[\frac{x-b}{(4\alpha t)^{1/2}}\right] - \text{erfc}\left[\frac{x-a}{(4\alpha t)^{1/2}}\right]\right\} \\ & + \frac{T_1}{2}\left\{\text{erfc}\left[\frac{x-d}{(4\alpha t)^{1/2}}\right] - \text{erfc}\left[\frac{x-c}{(4\alpha t)^{1/2}}\right]\right\} \end{aligned} \tag{1.82}$$

Two interesting special cases can be obtained from Equation 1.82. One of these is for $b \to -c$, $a \to -d$, and $T_0 \to T_1$. The resulting solution is

$$\begin{aligned} T(x,t) = {} & \frac{T_1}{2}\left\{\text{erfc}\left[\frac{x+c}{(4\alpha t)^{1/2}}\right] - \text{erfc}\left[\frac{x+d}{(4\alpha t)^{1/2}}\right]\right. \\ & \left. + \text{erfc}\left[\frac{x-d}{(4\alpha t)^{1/2}}\right] - \text{erfc}\left[\frac{x-c}{(4\alpha t)^{1/2}}\right]\right\} \end{aligned} \tag{1.83}$$

This solution is symmetric about $x = 0$. See Figure 1.9a for the initial temperature distribution.

Substitution of $-x$ for x in Equation 1.82 and use of the Appendix E identity of $\text{erfc}(-z) = 2 - \text{erfc}(z)$ reveals the symmetry, which can also be noted in Figure 1.9a. This condition of symmetry can also be expressed mathematically by $\partial T/\partial x = 0$ at $x = 0$; $\partial T/\partial x = 0$ is sometimes called the insulation condition. In other words, the solution for a semi-infinite body ($x > 0$) which is insulated at $x = 0$ can be found from the infinite solution if the temperature distribution is made symmetric about $x = 0$.

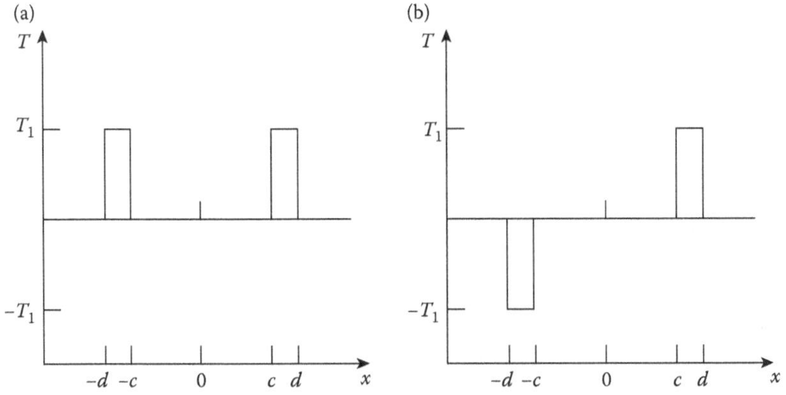

FIGURE 1.9 Initial temperature distribution for Example 1.2.

The other special case is for $b \to -c$, $a \to -d$, and $T_0 \to -T_1$, which has the solution

$$T(x,t) = -\frac{T_1}{2}\left\{\operatorname{erfc}\left[\frac{x+c}{(4\alpha t)^{1/2}}\right] - \operatorname{erfc}\left[\frac{x+d}{(4\alpha t)^{1/2}}\right] - \operatorname{erfc}\left[\frac{x-d}{(4\alpha t)^{1/2}}\right] + \operatorname{erfc}\left[\frac{x-c}{(4\alpha t)^{1/2}}\right]\right\} \quad (1.84)$$

This expression has the value of zero at $x = 0$ and is antisymmetric about the $x = 0$ axis. The zero temperature boundary condition is called the homogeneous isothermal condition. See Figure 1.9b for the initial temperature distribution for this case.

Example 1.3:

Find the temperature distribution in the infinite body for the case of

$$F(x) = 0 \quad \text{for all } x$$
$$g(x,t) = q_{x0}\delta(x - x_0)$$

where q_{x0} has units of W/m^2, the same as those for heat flux. Refer to Table 1.2 for properties of the Dirac delta function.

Solution

The solution for the temperature is obtained by using Equation 1.74 with $F(x') = 0$,

$$T(x,t) = \frac{\alpha}{k}\int_{\tau=0}^{t}\int_{x'=-\infty}^{\infty}[4\pi\alpha(t-\tau)]^{-1/2} \times \exp\left[-\frac{(x-x')^2}{4\alpha(t-\tau)}\right] q_{x0}\delta(x'-x_0)\,dx'\,d\tau \quad (1.85)$$

$$T(x,t) = \frac{\alpha q_{x0}}{k}\int_{\tau=0}^{t}[4\pi\alpha(t-\tau)]^{-1/2}\exp\left[-\frac{(x-x_0)^2}{4\alpha(t-\tau)}\right] d\tau \quad (1.86)$$

because the only contribution to the integral is at $x' = x_0$. Using integral 9 in Table I.6, Appendix I gives,

$$T(x,t) = \frac{q_{x0}}{k}(\alpha t)^{1/2}\text{ierfc}\left[\frac{|x - x_0|}{(4\alpha t)^{1/2}}\right] \tag{1.87}$$

where ierfc(z) is given by (see Appendix E)

$$\text{ierfc}(z) = \int_z^\infty \text{erfc}(u)\,du = \pi^{-1/2}\exp(-z^2) - z\,\text{erfc}(z) \tag{1.88}$$

The identity ierfc(∞) = 0 is needed to evaluate the above integral. Notice that Equation 1.87 is symmetric about $x = x_0$. The maximum temperature is finite, occurs at $x = x_0$, and can be evaluated using ierfc(0) $= \pi^{1/2}$ to find

$$T_{max}(x_0, t) = q_{x0}\left(\frac{t}{\pi k \rho c}\right)^{1/2} \tag{1.89}$$

1.10 TWO INTERPRETATIONS OF GREEN'S FUNCTIONS

Two different physical interpretations of $G(\cdot)$ can be found from the GF solution equation, Equation 1.74, and are described below. The first physical interpretation of $G(\cdot)$ is the temperature distribution caused by a particular initial condition and the second interpretation is the temperature distribution for an instantaneous heat source.

The first physical interpretation is associated with the first term in Equation 1.74 and is the solution $T(x,t)$ for the problem

$$\frac{\partial^2 T}{\partial x^2} = \frac{1}{\alpha}\frac{\partial T}{\partial t}; \quad -\infty < x < \infty; \quad t > 0 \tag{1.90a}$$

$$T(x,0) = F(x) \tag{1.90b}$$

If the initial temperature distribution is zero everywhere except at x_0 where it is equal to F'_0 times the Dirac delta function (see Table 1.2),

$$F(x) = F'_0\,\delta(x - x_0) \tag{1.91}$$

then the solution of Equations 1.90a and 1.91 is

$$T(x,t) = F'_0\,G(x,t|x_0,0) \tag{1.92}$$

Hence, the GF $G(x,t|x',0)$ can be interpreted as being the temperature distribution in the body that is the result of the initial temperature being zero everywhere except at point x_0 where there is a Dirac delta in the temperature distribution of magnitude $F'_0 = 1$ K-m (kelvin-meter). The units of $G(\cdot)$ and $K(\cdot)$ are both reciprocal length m^{-1}; the unit for $\delta(x - x_0)$ is also m^{-1}.

The second physical interpretation of a GF is the temperature caused by an instantaneous heat source at time t_0 and position x_0 and of strength $H\rho c$. For this case, the volumetric energy generation term in the heat equation, Equation 1.73a, becomes

$$g(x,t) = H\rho c\,\delta(x - x_0)\,\delta(t - t_0) \tag{1.93}$$

where H has the units of K-m; $\delta(x - x_0)$ has the unit m^{-1}; $\delta(t - t_0)$ has the units s^{-1}; and ρc has the units of $\text{J/m}^3\text{/K}$. These units are consistent with those of the volume energy generation g, which are W/m^3. The symbol g given by Equation 1.93 represents the amount of energy that is released at $x = x_0$ and at $t = t_0$. It can be visualized as the energy associated with an instantaneous plane source in the direction normal to the x-axis. It is also like an instantaneous (pulsed) laser sheet being released at $x = x_0$ and at time t_0. For this case, the describing differential equation is

$$\frac{\partial^2 T}{\partial x^2} + \frac{1}{k} H\rho c\, \delta(x - x_0)\, \delta(t - t_0) = \frac{1}{\alpha}\frac{\partial T}{\partial t}; \quad -\infty < x < \infty; \quad t > t_0 \tag{1.94}$$

and the initial temperature distribution is zero,

$$T(x, t_0) = 0; \quad -\infty < x < \infty \tag{1.95}$$

The solution for the temperature is zero until time $t = t_0$. After time t_0, the solution for $T(x, t)$ given by Equation 1.74 is

$$T(x,t) = \frac{\alpha}{k} \int_{\tau=0}^{t} \int_{x'=-\infty}^{\infty} G(x, t|x', \tau) H\rho c$$
$$\times\, \delta(x' - x_0)\delta(\tau - t_0)\, dx'\, d\tau \tag{1.96a}$$

that yields

$$T(x,t) = H\, G(x, t|x_0, t_0) \tag{1.96b}$$

Notice that in using Equation 1.74 for $g(x, t)$, it is necessary to replace x by x' and t by τ. The major point, however, is that the GF is equal to the temperature rise for the instantaneous plane heat source given by Equation 1.93 with $H = 1$ K-m.

These two alternate ways of thinking about transient GFs are important. In the first interpretation, the GF is equal to the temperature resulting from an initial temperature distribution that is zero everywhere except at the location of the Dirac delta function with strength of 1 K-m. In the second interpretation, the GF is equal to the temperature rise due to an instantaneous plane source with a strength of one K-m times ρc.

1.11 TEMPERATURE IN SEMI-INFINITE BODIES

A semi-infinite body is described by a body occupying the region $x \geq 0$. Although it represents an idealized body extending to positive infinity, it is a good model for many problems. A finite body of thickness L can be represented by a semi-infinite body, $0 < x < \infty$, when the boundary condition at $x = L$ does not influence the temperature distributions near $x = 0$. This happens for the small dimensionless times of $\alpha t / L^2 < 0.05$. Isothermal and insulation boundary conditions at $x = 0$ can be constructed from the infinite region solutions. The examples of Section 1.9 illustrate these points; also see Figure 1.9.

Temperature solutions for a semi-infinite body with an isothermal surface and an insulated surface can be obtained using the fundamental heat conduction solution, given by Equation 1.75. The homogeneous isothermal case is for the surface temperature (at $x = 0$) held at 0 degrees. A prescribed temperature at a boundary is called a boundary condition of the first kind. If the prescribed temperature is zero, the boundary condition is termed homogeneous. A prescribed heat flux at a surface is called a boundary condition of the second kind; if this heat flux is zero, the surface is said to be insulated and the boundary condition is also homogeneous. Both boundary conditions, the first and second kinds, are now considered by utilizing the concept of superposition which is valid because the defining equations are linear.

1.11.1 Boundary Condition of the First Kind

Consider a homogeneous boundary condition of the first kind (specified temperature) for a semi-infinite body,

$$\frac{\partial^2 T}{\partial x^2} = \frac{1}{\alpha}\frac{\partial T}{\partial t}; \quad 0 < x < \infty; \quad t > 0 \tag{1.97}$$

$$T(x,0) = F(x) \tag{1.98}$$

$$T(0,t) = 0 \tag{1.99}$$

See Figure 1.10a for the geometry. The solution to this problem is the same as for an infinite body with the initial temperature $T(x,0)$ equal to $F(x)$ for $x > 0$ and equal to $-F(-x)$ for $x < 0$; see Figure 1.9b. Then the first term of Equation 1.74 with $G(x,t|x',0) = K(x-x',t)$ gives

$$T(x,t) = \int_{x'=0}^{\infty} K(x-x',t)\,F(x')\,dx' - \int_{x'=-\infty}^{0} K(x-x',t)\,F(-x')\,dx' \tag{1.100a}$$

In the second integral, replace $-x'$ by x'' to get

$$\begin{aligned} T(x,t) &= \int_{x'=0}^{\infty} K(x-x',t)\,F(x')\,dx' - \int_{x''=0}^{\infty} K(x+x'',t)\,F(x'')\,dx'' \\ &= \int_{x'=0}^{\infty} [K(x-x',t) - K(x+x',t)]\,F(x')\,dx' \end{aligned} \tag{1.100b}$$

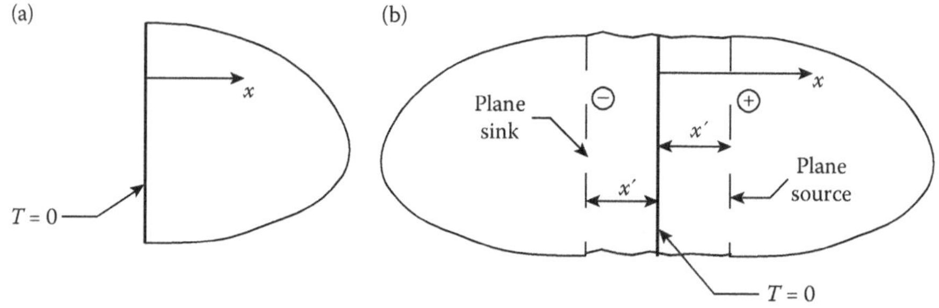

FIGURE 1.10 (a) Semi-infinite body with an isothermal boundary. (b) Semi-infinite body with $T = 0$ at $x = 0$ simulated by an infinite body with source at x' and sink at $-x'$.

since x' and x'' are dummy variables. Notice that the domain of $0 \leq x' \leq \infty$ is included in the integral of Equation 1.100b. This equation can be written in terms of a new Green's function,

$$T(x,t) = \int_{x'=0}^{\infty} G(x,t|x',0)\,F(x')\,dx' \tag{1.101a}$$

where the new GF is equal to

$$G(x,t|x',\tau) = K(x-x',t-\tau) - K(x+x',t-\tau) \tag{1.101b}$$

$$= [4\pi\alpha(t-\tau)]^{-1/2}\left\{\exp\left[-\frac{(x-x')^2}{4\alpha(t-\tau)}\right] - \exp\left[-\frac{(x+x')^2}{4\alpha(t-\tau)}\right]\right\};\quad t-\tau \geq 0 \tag{1.101c}$$

This GF represents the physical problem of an instantaneous plane source of strength $H = 1$ K-m times ρc and at location x' and at time τ in a semi-infinite body with zero boundary conditions and zero initial conditions. The GF satisfies the following equations:

$$\frac{\partial^2 G}{\partial x^2} + \frac{1}{\alpha}\delta(x-x_0)\delta(t-t_0) = \frac{1}{\alpha}\frac{\partial G}{\partial t};\quad 0 < x < \infty;\quad t > 0 \tag{1.102}$$

$$G(0,t|x',\tau) = 0;\qquad G(\infty,t|x',\tau) = 0 \tag{1.103}$$

$$G(x,0|x',\tau) = 0; \tag{1.104}$$

Equation 1.102 is obtained from Equation 1.94 by replacing H by 1 and T by G. The presence of a sink at $x = -x'$, shown in Figure 1.10b, ensures that G is equal to zero at $x = 0$. The GF given by Equation 1.101c is plotted in Figure 1.11. The curves are given for constant values of $\alpha(t-\tau)/x^2$ equal to 0.025, 0.05, 0.25, 1.0, and 4.0 versus

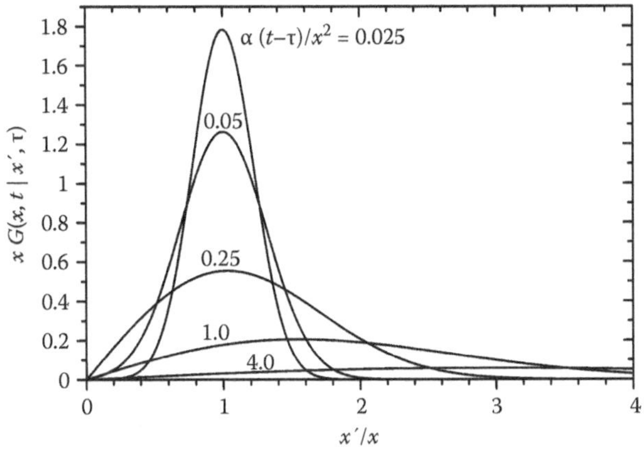

FIGURE 1.11 GF for semi-infinite body with isothermal condition of $G = 0$ at $x = 0$.

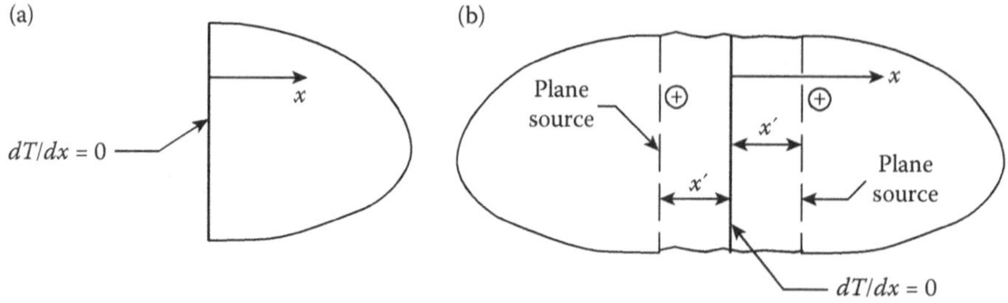

FIGURE 1.12 (a) Semi-infinite body with an insulated boundary. (b) Semi-infinite body with $\partial T / \partial x = 0$ at $x = 0$ simulated by an infinite body with source at x' and at $-x'$.

x'/x; the same curves are obtained for fixed values of $\alpha(t-\tau)/x'^2$ versus x/x'. The GF is little affected by the isothermal boundary condition for $\alpha(t-\tau)/x^2 < 0.05$. For larger dimensionless times, the maximum G moves to larger x'/x values and its magnitude decreases.

1.11.2 Boundary Condition of the Second Kind

Next consider the case of the insulated surface (the boundary condition of the second kind). See Figure 1.12a. This case can be treated in a similar manner as the homogeneous isothermal case. The differential equation, Equation 1.97, and the initial condition, Equation 1.98, are the same, but the boundary condition is

$$\left.\frac{\partial T}{dx}\right|_{x=0} = 0$$

which is a condition associated with symmetry about $x = 0$. (Other coordinate systems, such as radial, may not have symmetry for $\partial T / \partial r = 0$.)

The solution for the temperature can be obtained by using Equation 1.74 (which is for $-\infty < x < \infty$) by making the initial temperature distribution symmetric, that is, equal to $F(x)$ for $x > 0$ and equal to $F(-x)$ for $x < 0$. Then using Equation 1.74 with $G(x,t|x',0) = K(x-x',t)$ gives

$$T(x,t) = \int_{x'=0}^{\infty} K(x-x',t)\,F(x')\,dx' + \int_{x'=-\infty}^{0} K(x-x',t)\,F(-x')\,dx' \qquad (1.105)$$

Replacing $-x'$ in the second integral by x'' and then combining into a single integral gives

$$T(x,t) = \int_{x'=0}^{\infty} K(x-x',t)\,F(x')\,dx' + \int_{x''=\infty}^{0} K(x+x'',t)\,F(x'')(-dx'') \qquad (1.106a)$$

$$T(x,t) = \int_{x'=0}^{\infty} [K(x - x', t) + K(x + x', t)]\, F(x')\, dx' \qquad (1.106b)$$

$$\text{or,} \quad T(x,t) = \int_{x'=0}^{\infty} G(x, t|x', 0)\, F(x')\, dx' \qquad (1.106c)$$

where $G(\cdot)$ is given by

$$G(x, t|x', \tau) = K(x - x', t - \tau) + K(x + x', t - \tau) \qquad (1.107a)$$

$$= [4\pi\alpha(t-\tau)]^{-1/2} \left\{ \exp\left[-\frac{(x - x')^2}{4\alpha(t-\tau)} \right] + \exp\left[-\frac{(x + x')^2}{4\alpha(t-\tau)} \right] \right\}; \quad t - \tau \geq 0 \qquad (1.107b)$$

This expression is the GF for a semi-infinite body insulated at $x = 0$. This solution can be also visualized as the result of superimposing two sources, one at $x = x'$ and the other at $x = -x'$. See Figure 1.12b. The GF given by Equation 1.107b is shown in Figure 1.13 which shows $xG(\cdot)$ versus x'/x for $x \neq 0$; if $x = 0$, the $G(\cdot)$ function given by Equation 1.107b is twice as large as the infinite-body GF shown in Figure 1.7. As for the boundary condition of the first kind, the GF in Figure 1.13 is unaffected by the $\partial T/\partial x = 0$ boundary condition at $x' = 0$ for $\alpha(t-\tau)/x^2 < 0.05$. Unlike that case, however, the maximum G moves to $x'/x = 0$ as the dimensionless time increases. Moreover, this case has G values (for the same x and t's) that are always as large or larger than the $G = 0$ at $x' = 0$ case, Figure 1.11; the effect is most noticeable for $\alpha(t-\tau)/x^2 = 0.25$ to 4.0.

The method of deriving the GF given by Equation 1.107b is related to the method of images for deriving the GFs, which is discussed in greater depth in Chapter 4.

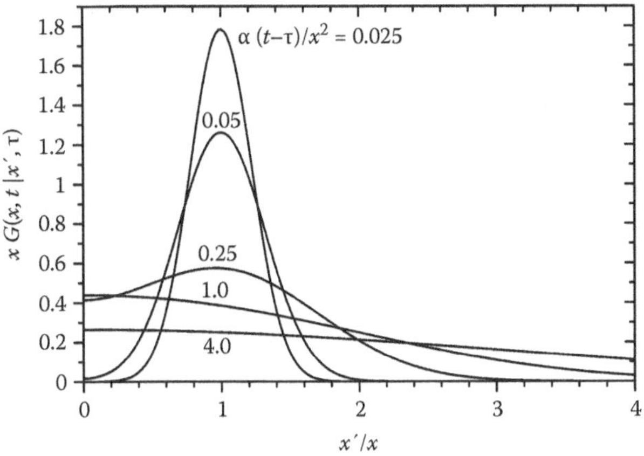

FIGURE 1.13 GF for semi-infinite body with insulation condition of $\partial G/\partial x = 0$ at $x = 0$.

Example 1.4:

Find the temperature distribution in a semi-infinite body with initial temperature T_0 that has $T = 0$ suddenly applied at the boundary. The temperature satisfies

$$\frac{\partial^2 T}{\partial x^2} = \frac{1}{\alpha}\frac{\partial T}{\partial t}; \quad x > 0; \quad t > 0 \tag{1.108a}$$

$$T(0,t) = 0; \quad T(\infty, t) \to T_0 = \text{constant} \tag{1.108b}$$

$$T(x,0) = T_0; \quad x > 0 \tag{1.108c}$$

Solution

This problem has the boundary condition of the first kind and the solution is given by Equation 1.101a with $G(\cdot)$ given by Equation 1.101c.

$$\begin{aligned} T(x,t) &= \int_{x'=0}^{\infty} [K(x-x',t) - K(x+x',t)]\, T_0\, dx' \\ &= T_0 \left[\frac{1}{2}\operatorname{erfc}\left(\frac{x-x'}{(4\alpha t)^{1/2}}\right) + \frac{1}{2}\operatorname{erfc}\left(\frac{x+x'}{(4\alpha t)^{1/2}}\right)\right]_{x'=0}^{\infty} \\ &= T_0 \left[\left(1 - \frac{1}{2}\operatorname{erfc}\frac{x}{(4\alpha t)^{1/2}}\right) + \left(0 - \frac{1}{2}\operatorname{erfc}\frac{x}{(4\alpha t)^{1/2}}\right)\right] \\ &= T_0\left[1 - \operatorname{erfc}\frac{x}{(4\alpha t)^{1/2}}\right] = T_0 \operatorname{erf}\frac{x}{(4\alpha t)^{1/2}} \end{aligned} \tag{1.109}$$

This solution is plotted in Figure 1.14 versus $z = x/(4\alpha t)^{1/2}$; also shown is erfc(z). The variation of temperature is most pronounced for $x/(4\alpha t)^{1/2}$ less than 1.0.

Example 1.5:

Find the temperature distribution in the semi-infinite body initially at zero temperature, and temperature T_0 is suddenly applied at boundary $x = 0$. The temperature satisfies

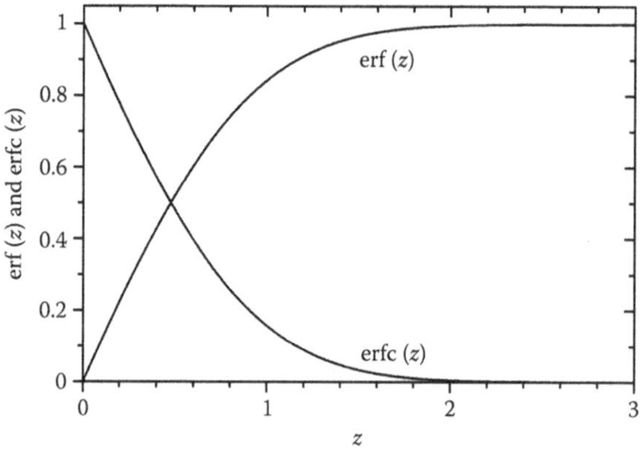

FIGURE 1.14 Error function (erf) and complementary error function (erfc).

$$\frac{\partial^2 T}{\partial x^2} = \frac{1}{\alpha}\frac{\partial T}{\partial t}; \quad x > 0; \quad t > 0 \tag{1.110a}$$

$$T(0,t) = T_0; \qquad T(\infty, t) \to 0 \tag{1.110b}$$

$$T(x,0) = 0; \qquad x > 0 \tag{1.110c}$$

Solution

This problem does not have a homogeneous isothermal boundary condition but a related problem does. Define the new variable,

$$\theta = T_0 - T$$

so that

$$\frac{\partial^2 \theta}{\partial x^2} = \frac{1}{\alpha}\frac{\partial \theta}{\partial t}; \quad x > 0; \quad t > 0 \tag{1.111a}$$

$$\theta(0,t) = 0; \qquad \theta(\infty, t) \to T_0 \tag{1.111b}$$

$$\theta(x,0) = T_0; \qquad x > 0 \tag{1.111c}$$

the solution for θ is given by Equation 1.109 so that the solution for T is

$$T(x,t) = T_0 \operatorname{erfc} \frac{x}{(4\alpha t)^{1/2}} \tag{1.112}$$

which also is shown in Figure 1.14 as the erfc(z) curve.

1.12 FLAT PLATES

The construction of the GF by superposition of the plane sources and sinks in an infinite body, as discussed in the previous section for the geometry of semi-infinite body, can also be extended to the finite geometry of the flat plate. This approach is an application of the method of images (Carslaw and Jaeger, 1959, p. 273) which is discussed in more detail in Chapter 4. Even though the method of images can be employed to construct the GFs (from the fundamental heat conduction solution) for the geometry of the flat plate, there are many cases for which the GFs cannot be readily obtained by this method; in particular, cases that involve boundary conditions other than the first and second kinds. A more general approach for construction of the GFs is through the use of an auxiliary problem.

1.12.1 TEMPERATURE FOR FLAT PLATES

The temperature problem that motivates the study of the one-dimensional GF for the geometry of the flat plate is

$$\frac{\partial^2 T}{\partial x^2} + \frac{1}{k} g(x,t) = \frac{1}{\alpha}\frac{\partial T}{\partial t} \tag{1.113}$$

with boundary conditions

$$k_i \left.\frac{\partial T}{\partial n_i}\right|_{x_i} + h_i T|_{x_i} = f_i(t) \tag{1.114}$$

where n_i is a unit normal coordinate directed outward from the body at the boundary. The subscript i is either 1 or 2 to represent the two boundaries. Thus, x_1 and x_2 are the locations of the left and right boundaries, respectively. The initial condition is

$$T(x,0) = F(x) \tag{1.115}$$

The boundary condition, Equation 1.114, represents three different kinds of boundary conditions by the choice of k_i, h_i, and f_i. These three boundary conditions are commonly studied and are called the first, second, and third kinds.

The first kind of boundary condition (also called the Dirichlet condition) is obtained from Equation 1.114 by setting $k_i = 0$ and $h_i = 1$ to get the prescribed surface temperature

$$T(x_i,t) = f_i(t) \tag{1.116}$$

where f_i can also be simply zero. The second kind of boundary condition (also called the Neumann condition) is prescribed surface heat flux

$$k_i \left.\frac{\partial T}{\partial n_i}\right|_{x_i} = f_i(t) \tag{1.117}$$

which becomes an insulated boundary if $f_i(t) = 0$.

The third kind of boundary condition is a convective boundary condition (also called the Robin condition) given by Equation 1.114, where $f_i(t)$ is usually $h_i T_\infty$. The most familiar form of this boundary condition is then

$$-k_i \left.\frac{\partial T}{\partial n_i}\right|_{x_i} = h_i(T|_{x_i} - T_\infty) \tag{1.118}$$

where T_∞ is the constant or time-varying ambient temperature.

1.12.2 Auxiliary Problem for Flat Plates

The GF associated with the temperature given by Equations 1.113 through 1.115 is the solution to the auxiliary equation,

$$\frac{\partial^2 G}{\partial x^2} + \frac{1}{\alpha}\delta(x - x')\delta(t - \tau) = \frac{1}{\alpha}\frac{\partial G}{\partial t} \tag{1.119a}$$

subject to the homogeneous boundary conditions

$$k_i \left.\frac{\partial G}{\partial n_i}\right|_{x_i} + h_i G|_{x_i} = 0; \quad i = 1, 2 \tag{1.119b}$$

and zero initial condition

$$G(x, t|x', \tau) = 0; \quad \text{when } t < \tau \tag{1.119c}$$

(Equation 1.119a is similar to Equation 1.94 with $T \rightarrow G$ and $H \rightarrow 1$.) The auxiliary equation for any GF is identical to the original heat conduction equation except for the energy generation term, which is a Dirac delta function at location x' and at time τ. The one-dimensional GF G, defined by Equation 1.119a, has units of m^{-1}. This is apparent from the units of the energy generation term in Equation 1.119a [$\delta(x - x')\delta(t - \tau)/\alpha$, which has units of m^{-3}]. The homogeneous boundary conditions for the auxiliary equation are the *same kinds* as for the original problem.

1.13 PROPERTIES COMMON TO TRANSIENT GREEN'S FUNCTIONS

The properties common to GF for transient heat conduction are summarized below.

1. The GF obeys the auxiliary equation.
2. The GF is a solution of the heat conduction problem having the same geometry but having homogeneous boundary conditions of the *same kind* as the original heat conduction problem.
3. The GF obeys the causality relation: $G \geq 0$ in the domain R for $t - \tau \geq 0$; and, $G = 0$ in the domain R for $t - \tau < 0$.
4. The GF obeys the reciprocity relation: $G(x, t|x', \tau) = G(x', -\tau|x, -t)$.
5. The time dependence of G is always $t - \tau$, so a one-dimensional GF could be written $G(x, x', t - \tau)$.
6. In rectangular coordinates, the transient GF has units of: m^{-1} for one-dimensional problems; m^{-2} for two-dimensional problems; and m^{-3} for three-dimensional problems.

Every GF is a solution to an auxiliary equation with homogeneous boundary conditions. The GF is always positive or zero, because it is the temperature caused by a positive heat pulse. The causality relation relates to the idea that the GF is the response at time t and location x to a pulse of heat occurring at time τ and at location x'. In a real (or causal) system, there can be no response before the pulse of heat occurs.

The reciprocity relation can be understood from the auxiliary equation, Equation 1.119a. Exchanging x and x' in the auxiliary equation leaves the sign of the solution unchanged because of the second derivative with respect to x. However, exchanging t and τ changes the sign of the solution, because of the first derivative with respect to t. Spatial orientation has no preferred direction in heat conduction, but time does have a preferred direction.

1.14 HETEROGENEOUS BODIES

A body composed of two or more parts with different thermal conductivities is called a heterogeneous body (also called a nonhomogeneous body). Fourier's law may apply to each homogeneous subregion of such a body, but the interface where the conductivity changes must be treated with special techniques, two of which are discussed in this book. In Chapter 11 the Galerkin-based GF method is applied to a body with an inclusion. In Chapter 12 the surface element method is applied to two homogeneous bodies in thermal contact.

1.15 ANISOTROPIC BODIES

Many bodies of engineering interest do not conduct heat equally well in all directions and are called anisotropic bodies. Laminates, crystals, fiber/matrix composites, and wood are among the materials that have preferred directions of heat flow. For example, wood conducts heat along the grain more readily than across the grain.

Conductivity matrix. For anisotropic bodies, a generalized form of Fourier's law is used that includes a thermal conductivity matrix. For example, in rectangular coordinates, the conductivity matrix is given by

$$\begin{bmatrix} k_{11} & k_{12} & k_{13} \\ k_{21} & k_{22} & k_{23} \\ k_{31} & k_{32} & k_{33} \end{bmatrix} \tag{1.120}$$

and the components of heat flux vector are given by

$$q_i = \sum_{j=1}^{3} k_{ij} \frac{\partial T}{\partial x_j} \tag{1.121}$$

The energy equation for anisotropic bodies contains cross derivatives and its solution is not covered here; refer to Carslaw and Jaeger (1959, p. 38) and Ozisik (1993, Chapter 15).

Orthotropic bodies. The conductivity matrix depends on the orientation of the coordinate system in the body. If the coordinate system is parallel to three mutually perpendicular preferred directions of heat conduction, then the geometry is said to be *orthotropic* and the coordinate system lies along the principal axes of heat conduction. In an orthotropic body the conductivity matrix has a diagonal form,

$$\begin{bmatrix} k_{11} & 0 & 0 \\ 0 & k_{22} & 0 \\ 0 & 0 & k_{33} \end{bmatrix} \tag{1.122}$$

Wood is an example of an orthotropic body in the particular cylindrical coordinate system (r, ϕ, z) corresponding to the direction of the rays, rings, and axis of the tree (Carslaw and Jaeger, 1959, p. 41).

The energy equation for orthotropic bodies does not contain any cross derivatives and it can be transformed into the standard isotropic energy equation by a suitable choice of new spatial coordinates. This transformation is given in the next section.

1.16 TRANSFORMATIONS

There are several heat transfer equations that may be converted, through a transformation, into the familiar heat conduction equation. These transformations extend the heat conduction solutions discussed in this book to a broader range of heat transfer problems.

Three transformations are presented for heat transfer in orthotropic bodies, in moving solids, and in fins.

1.16.1 Orthotropic Bodies

An orthotropic body, introduced in the previous section, has direction-dependent thermal properties whose principal values are aligned with the coordinate axes. In this section a transformation is given to convert the orthotropic heat conduction equation to the usual heat conduction equation.

The heat conduction equation in Cartesian coordinates for an orthotropic body is given by

$$k_{11}\frac{\partial^2 T}{\partial x^2} + k_{22}\frac{\partial^2 T}{\partial y^2} + k_{33}\frac{\partial^2 T}{\partial z^2} + g(x, y, z, t) = \rho c\frac{\partial T}{\partial t} \tag{1.123}$$

Define stretched coordinate axes of the form

$$x_1 = x\left(\frac{k}{k_{11}}\right)^{1/2}; \quad y_1 = y\left(\frac{k}{k_{22}}\right)^{1/2}; \quad z_1 = z\left(\frac{k}{k_{33}}\right)^{1/2} \tag{1.124}$$

where k is a reference conductivity. Replace these scaled coordinates into Equation 1.123 to show that the orthotropic heat conduction equation can be written

$$k\left(\frac{\partial^2 T}{\partial x^2} + \frac{\partial^2 T}{\partial y^2} + \frac{\partial^2 T}{\partial z^2}\right) + g(x, y, z, t) = \rho c\frac{\partial T}{\partial t} \tag{1.125}$$

That is, the heat conduction equation in an orthogonal body has been transformed into the isotropic heat conduction equation. The boundary conditions must also be adjusted (see Problem 3.17 at the end of Chapter 3).

The reference conductivity is not arbitrary, it must be chosen so that the original differential volume is equal to the scaled differential volume. For the 3D Cartesian case, the differential volume scales according to

$$dx\,dy\,dz = \frac{(k_{11}k_{22}k_{33})^{1/2}}{k^{3/2}}dx_1\,dy_1\,dz_1$$

and the requirement that $dv = dv_1$ causes

$$k = (k_{11}k_{22}k_{33})^{1/3}$$

This requirement may be extended to other orthogonal coordinate systems.

1.16.2 Moving Solids

Heat conduction in moving solids can arise because the solid is moving, as during an extrusion process, or when a fixed solid contains a moving heat source. If the problem is formulated with the coordinate system attached to the moving heat source, then a velocity term appears in the partial differential equation for heat transfer. In this section a transformation is given for converting the moving-solid heat transfer equation into the (usual) heat conduction equation.

Consider a solid moving with bulk velocity U_1, U_2, and U_3 in the x-, y-, and z-directions, respectively. Velocities U_1, U_2, and U_3 are assumed to be constant, known quantities. The temperature in the moving body is described, for constant thermal properties, by

$$k\left(\frac{\partial^2 T}{\partial x^2}+\frac{\partial^2 T}{\partial y^2}+\frac{\partial^2 T}{\partial z^2}\right)+g(x,y,z,t)$$
$$=\rho c\left(\frac{\partial T}{\partial t}+U_1\frac{\partial T}{\partial x}+U_2\frac{\partial T}{\partial y}+U_3\frac{\partial T}{\partial z}\right) \tag{1.126}$$

The transformation

$$T(x,y,z,t)=W(x,y,z,t)\exp\left(\frac{U_1 x}{2\alpha}-\frac{U_1^2 t}{4\alpha}\right)$$
$$\times\exp\left(\frac{U_2 y}{2\alpha}-\frac{U_2^2 t}{4\alpha}\right)\exp\left(\frac{U_3 z}{2\alpha}-\frac{U_3^2 t}{4\alpha}\right) \tag{1.127}$$

allows the moving-body heat transfer equation to be written

$$\frac{\partial^2 W}{\partial x^2}+\frac{\partial^2 W}{\partial y^2}+\frac{\partial^2 W}{\partial z^2}+\frac{1}{k}g^*(x,y,z,t)=\frac{1}{\alpha}\frac{\partial W}{\partial t} \tag{1.128}$$

where g^* is given by

$$g^*(x,y,z,t)=g(x,y,z,t)\exp\left(-\frac{U_1 x}{2\alpha}+\frac{U_1^2 t}{4\alpha}\right)$$
$$\times\exp\left(-\frac{U_2 y}{2\alpha}+\frac{U_2^2 t}{4\alpha}\right)\exp\left(-\frac{U_3 z}{2\alpha}+\frac{U_3^2 t}{4\alpha}\right) \tag{1.129}$$

This transformation relocates the effect of the convective heat transfer terms to the internal-heating term. This transformation only applies to transient heat transfer, as the time-derivative term has an active part in the transformation. The boundary conditions must also be transformed for a complete solution to the problem.

1.16.3 Fin Term

The fin term appears in the heat conduction equation to describe heat loss (or gain) that is proportional to temperature. This term, named for the convective heat loss from a fin, can also be used to represent heat loss by radiation (linearized), or heat generation by chemical reaction.

The heat conduction equation with the fin term is given by

$$\nabla^2 T - m^2(T - T_\infty) + \frac{1}{k} g(\mathbf{r}, t) = \frac{1}{\alpha} \frac{\partial T}{\partial t} \tag{1.130}$$

where m^2 is a constant and T_∞ may be an external fluid temperature. Let $\theta = T - T_\infty$ and define the transformation by

$$\theta(\mathbf{r}, t) = W(\mathbf{r}, t) \exp\left(-m^2 \alpha t\right) \tag{1.131}$$

Upon replacing the transform into the heat equation, the result is

$$\nabla^2 W + \frac{1}{k} g^*(\mathbf{r}, t) = \frac{1}{\alpha} \frac{\partial W}{\partial t} \tag{1.132}$$

where

$$g^*(\mathbf{r}, t) = g(\mathbf{r}, t) \exp\left(+m^2 \alpha t\right) \tag{1.133}$$

With this transformation the effect of the fin term is shifted into the energy generation term. The boundary conditions are similarly affected. The fin transformation can be used simultaneously with the moving-solid transformation (see homework problem 3.28). As with the moving-solid transformation, the fin transformation only works with transient heat transfer.

1.17 NON-FOURIER HEAT CONDUCTION

Fourier's law of heat conduction describes heat transfer very accurately in most applications. However, it predicts that heat introduced at one point in a body is instantaneously transmitted throughout the body. Of course, the size of the predicted temperature response is vanishingly small far from the heat pulse, but with Fourier's law the speed of propagation is infinite. In our post-Einstein world, an infinite speed of propagation is not physically reasonable. This means that Fourier's law may not be accurate in a brief time period after the heat pulse. For very short times, for very short distances, or for temperatures very near zero kelvin, a relation other than Fourier's law is needed to describe energy transport. The application to microscale or nanoscale heat transfer is presently an active area of research. Two relations for non-Fourier heat transfer are briefly introduced here.

One relation between temperature and heat flux that allows for a finite speed of heat propagation is given by Ozisik and Vick (1984)

$$\frac{\alpha}{\sigma^2} \frac{\partial \mathbf{q}}{\partial t} + \mathbf{q} = -k \nabla T \tag{1.134}$$

where σ is the propagation speed for heat transfer and σ^2/α is the relaxation time for the heat flux to begin after a temperature gradient is imposed on the body. Conversely, the heat flow does not cease immediately after the temperature gradient is removed but dies away over a short period of time.

An energy equation that embodies the above finite propagation of heat may be found by taking the divergence of Equation 1.134,

$$\frac{\alpha}{\sigma^2}\frac{\partial}{\partial t}[\nabla \cdot \mathbf{q}] + \nabla \cdot \mathbf{q} = -\nabla \cdot (k\nabla T) \tag{1.135}$$

Now solve the vector energy equation, Equation 1.11, for $\nabla \cdot \mathbf{q}$,

$$\nabla \cdot \mathbf{q} = g(\mathbf{r}, t) - \rho c\frac{\partial T}{\partial t} \tag{1.136}$$

and substitute $\nabla \cdot \mathbf{q}$ into Equation 1.135. After some rearranging, the result is

$$\nabla \cdot (k\nabla T) + \left[g(\mathbf{r}, t) + \frac{\alpha}{\sigma^2}\frac{\partial g}{\partial t}\right] = \rho c\frac{\partial T}{\partial t} + \frac{k}{\sigma^2}\frac{\partial^2 T}{\partial t^2} \tag{1.137}$$

This heat conduction equation includes a finite speed of heat propagation. There are two additional terms that do not appear when Fourier's law is used. One is a time derivative of the energy generation $g(\mathbf{r}, t)$. The other term is a second derivative of temperature with respect to time. This is a wave term and the wave speed is σ. This wave term is said to be hyperbolic in time, and equations of this type are sometimes described as a hyperbolic heat conduction equation. In the limiting case of infinite propagation speed, Equation 1.137 reduces to the classic diffusive energy equation.

Strictly speaking, the above energy transport equation applies to materials that are crystalline and nonelectrically conducting, in which heat is transferred as vibrational lattice energy. In the language of quantum mechanics, lattice energy is transferred in discrete quanta called phonons. In metals, heat conduction is carried both by phonons and by free electrons. For transport of energy in metals, Qui and Tien (1992) proposed that the electron temperature T_e and the lattice temperature T_l be different and are related by the following relation

$$T_e(t) = T_l(t) + \frac{C_l}{\Gamma}\frac{\partial T_l(t)}{\partial t} \tag{1.138}$$

where C_l is the capacitance of the lattice and Γ is the electron–phonon coupling factor. Experimental and theoretical values of Γ are collected from different sources and given by Qui and Tien (1992, Table 1). For this application, the non-Fourier energy equation is

$$-\nabla \cdot \mathbf{q}(\mathbf{r}, t) = C\frac{\partial T_l(\mathbf{r}, t)}{\partial t} + \frac{C_e C_l}{\Gamma}\frac{\partial^2 T_l(\mathbf{r}, t)}{\partial t^2} \tag{1.139}$$

where $C = C_e + C_l$ and C_e is the electron capacitance. The combined effect of electron and phonon energy transport is discussed by Tzou (1997). A GF solution for the energy transport through the combined effects of electron transport and phonon transport is given by Hays-Stang and Haji-Sheikh (1999).

PROBLEMS

1.1 Calculate the gradient, ∇T, in the coordinate system given:

(a) $T = 3x + 4y^3 + e^{-2z}$, rectangular (x, y, z).
(b) $T = 3r^2 + 4\cos\phi + 2z$, cylindrical (r, ϕ, z).
(c) $T = 2r\phi + 2\phi\cos\theta$, spherical (r, ϕ, θ).
(d)

$$T = 2x + \sum_{m=1}^{\infty} \frac{\cos(m\pi y)}{m^3} e^{-m^2\pi^2 t}, \text{ rectangular } (x, y, z).$$

1.2 Show by direct computation that $(1/r)$ is a solution of Laplace's equation in two ways. That is, show that

$$\nabla^2 \left(\frac{1}{r}\right) = 0$$

(a) in Cartesian coordinates where $r = \sqrt{x^2 + y^2 + z^2}$, and,
(b) directly in spherical coordinates.

1.3 Write out the energy equation

$$\nabla^2 T = \frac{1}{\alpha} \frac{\partial T}{\partial t}$$

for the following special cases. Be sure to drop terms that are zero.

(a) $T = T(r, t)$ for a long cylinder.
(b) $T = T(r, z, t)$ for a thin film on a large surface with axisymmetry.
(c) $T = T(x, y, t)$ for a rectangular body.
(d) $T = T(r, \theta)$ for a rocket nose cone with axisymmetry, where θ is the polar angle with $\theta = 0$ along the axis of the rocket.

1.4 Show that each of the following functions satisfies the heat equation,

$$\alpha \nabla^2 T = \frac{\partial T}{\partial t}$$

(a) $T = e^{-14\alpha\pi^2 t} \sin(\pi x)\cos(3\pi y)\sin(2\pi z)$.
(b) $T = \exp\left(29\alpha\pi^2 t + \pi(3x + 2y + 4z)\right)$.
(c) $T = x^2 + y^2 - 2z^2 - 3x - 5y + 6z + 1$.

1.5 Repeat the derivation of the differential energy equation in Section 1.3 but for the cylindrical coordinate system. The control volume has the form $dV = r d\phi\, dr\, dz$, and your result should agree with Equation 1.12.

1.6 Show that under the assumption of a very small control volume $dv = dxdydz$, the integral energy equation, Equation 1.32, can be used to derive the differential energy equation, Equation 1.8. Use the divergence theorem:
Given vector field $\vec{C}$ in a control volume,

$$\int_{c.s.} \hat{n} \cdot \vec{C}\, dA = \int_{c.v.} \nabla \cdot \vec{C}\, dv$$

where $\nabla \cdot \vec{C}$ is the divergence of $\vec{C}$ and $\hat{n}$ is the outward normal on the control surface.

1.7 Apply the integral energy equation to find the equation for a lumped body, by applying the following assumptions: uniform thermal properties; spatially uniform temperature ($T = T(t)$ only); insulated body surface; and, spatially uniform internal energy generation ($g = g(t)$ only).

1.8 Convection at a solid surface is described by Newton's law of cooling, $q = h(T_{surface} - T_{fluid})$. Using this expression for surface convection in the integral energy equation, Equation 1.32, derive the convection boundary condition at surface $x = 0$ in a semi-infinite body. Use a very thin control volume (take the limit as thickness $\rightarrow 0$) that encloses the body surface. The result should agree with Equation 1.16. What is f_i in this case?

1.9 Show that the steady GF solution equation, Equation 1.44, satisfies the steady heat equation, Equation 1.36, by direct substitution.

1.10 Derive the steady GF for the slab with the following boundary conditions:

$$G(x = 0) = 0$$
$$G(x = L) = 0$$

Check your answer with case $X11$ in Table X.3 in Appendix X.

1.11 Derive the steady GF for the slab with the following boundary conditions:

$$\text{at } x = 0, \quad \partial G \,/\, \partial x = 0$$
$$\text{at } x = L, \quad k\partial G \,/\, \partial x + hG = 0$$

Check your answer with case $X23$ in Table X.3 in Appendix X.

1.12 Using the GF from Problem 1.11, find the steady temperature in the slab caused by uniform energy generation. That is, find the temperature that satisfies the following equations:

$$\frac{\partial^2 T}{\partial x^2} + \frac{g_0}{k} = 0$$
$$\partial T \,/\, \partial x = 0 \quad \text{at } x = 0$$
$$k\partial T \,/\, \partial x + hT = 0 \quad \text{at } x = L$$

1.13 Show that the Dirac delta function has the following properties, where a is a nonzero constant and function $f(t)$ is continuous at the origin. Note: The delta function is defined by its integral behavior, so that by an equation such as $\delta(-t) = \delta(t)$ we mean that

$$\int_{-\infty}^{\infty} f(t)\,\delta(-t)\,dt = \int_{-\infty}^{\infty} f(t)\,\delta(t)\,dt$$

and you have to verify that both sides of the equation reduce to $f(0)$.

(a) $\delta(-t) = \delta(t)$

(b) $\delta(at) = \delta(t) / |a|$

(c) $\int_{-\infty}^{t} \delta(\tau)\,d\tau = H(t)$

1.14 The Dirac delta function can be used to define derivatives of discontinuous functions. Find the derivative of $|x|$, $\sin|x|$, and $\cos|x|$. (Hint: let $|x| = x\ \mathrm{sign}(x)$.)

1.15 Find the Laplace transform of the following functions by direct integration of the definition of the Laplace transform. Here a and b are constants. If you use an integral table, give a detailed reference.

(a) a
(b) $a + bt$
(c) e^{at}
(d) $\sin(at)$
(e) $\delta(t - a)$

1.16 By substituting $K(x-x', t-\tau)$ into Equation 1.90a for $T(x,t)$, verify that $K(x - x', t - \tau)$ is a solution. What is the initial condition?

1.17 Given the following heat conduction problem,

$$\frac{\partial^2 T}{\partial x^2} = \frac{1}{\alpha}\frac{\partial T}{\partial t} = 0, \quad 0 < x < L$$
$$T(x = 0, t) = T_0$$
$$T(x = L, t) = T_0$$
$$T(x, t = 0) = T_1$$

use normalized variables given by

$$x^+ = x\,/\,L; \quad t^+ = \alpha t\,/\,L^2; \quad \theta = \frac{T(x,t) - T_0}{T_1 - T_0}$$

to restate the problem with x^+, t^+, and θ in place of x, t, and T.

1.18 Verify the identity for ierfc, Equation 1.88, using integration by parts.

1.19 Investigate the behavior of the approximation of erfc(x) given by

$$\pi^{-1/2}\exp(-x^2)\left[\frac{1}{x} - \frac{1}{2x^3} + \frac{1\cdot 3}{2^2x^5} - \frac{1\cdot 3\cdot 5}{2^3x^7}\cdots\right]$$

for a given $x > 1$ as the number of terms is increased. Verify that the error is less in absolute value than the last term retained.

1.20 Find the temperature distribution in a semi-infinite body with the initial temperature given by

$$T = x \ \text{ for } 0 < x \le 1 \text{ and } T = 0 \text{ for } x > 1$$

The surface temperature at $x = 0$ is maintained at zero temperature. (Appendix I may be helpful.)

1.21 Find the temperature in a semi-infinite body with the initial temperature given by

$$T = T_1\frac{x^2}{L^2} + T_0 \text{ for } 0 < x \le L \text{ and } T = T_0 \text{ for } x > L$$

The surface at $x = 0$ is insulated. (Appendix I may be helpful.)

1.22 The temperature due to a specified heat flux boundary condition (nonhomogeneous boundary condition of the second kind) in a semi-infinite body may be found by using a planar heat source located at the surface. Find the temperature resulting from a volumetric heat source given by

$$g(x,t) = q_0\delta(x-0)$$

Also, find the heat flux through the point $x = a$ inside the body.

1.23 Derive the below expression for the heat flux at x starting with $T(x,t)$ given by Equation 1.87,

$$q(x,t) = \frac{q_{x0}}{2}\text{sign}(x-x_0)\text{erfc}\left[\frac{|x-x_0|}{(4\alpha t)^{1/2}}\right]$$

REFERENCES

Barton, G., 1989, *Elements of Green's Functions and Propagation*, Oxford University Press, London.

Butkovskiy, A. G., 1982, *Green's Functions and Transfer Functions Handbook*, Halsted Press, New York.

Cannon, J. R., 1984, *The One-Dimensional Heat Equation*, Addison-Wesley, Menlo Park, CA.

Carslaw, H. S. and Jaeger, J. C., 1959, *Conduction of Heat in Solids*, 2nd edn, Oxford University Press, New York.

Currie, I. G., 2002, *Fundamental Mechanics of Fluids*, 3rd edn, Marcel Dekker, New York, Chapter 1.

Duffy, D. G., 2001, *Green's Functions with Applications*, Chapman and Hall, Boca Raton, FL.

Greenberg, M. D., 1971, *Application of Green's Functions in Science and Engineering*, Prentice-Hall, Englewood Cliffs, N.J.

Haji-Sheikh, A., 1988, Heat diffusion in heterogeneous media using heat-flux-conserving basis functions, *J. Heat Transfer*, vol. 110, pp. 276–282.

Haji-Sheikh, A. and Lakshminarayanan, R., 1987, Integral solution of diffusion equation: Part 2—Boundary conditions of second and third kinds, *J. Heat Transfer*, vol. 109, pp. 557–562.

Hays-Stang, K. J. and Haji-Sheikh, A., 1999, A unified solution for heat conduction in thin films, *Int. J. Heat Mass Transfer*, vol. 42, pp. 455–465.

Morse, P. M. and Feshbach, H., 1953, *Methods of Theoretical Physics*, McGraw-Hill, New York.

Ozisik, M. N., 1993, *Heat Conduction*, John Wiley, New York.

Ozisik, M. N. and Vick, B., 1984, Propagation and reflection of thermal waves in a finite medium, *Int. J. Heat Mass Transfer*, vol. 27, pp. 1845–1854.

Qui, T. Q. and Tien, C. L. 1992, Short-pulse laser heating on metals, *Int. J. Heat Mass Transfer*, vol. 35, pp. 719–726.

Roach, G. F., 1970, *Green's Functions, Introductory Theory with Applications*, Van Nostrand Reinhold, New York.

Stakgold, I., 1979, *Green's Functions and Boundary Value Problems*, Wiley-Interscience, New York.

Tzou, D. K., 1997, *Macro- to Microscale Heat Transfer*, Taylor and Francis, New York, Chapter 1.

White, F. M., 2006, *Viscous Fluid Flow*, 3rd edn, McGraw Hill, New York, Chapter 3.

2 Numbering System in Heat Conduction

2.1 INTRODUCTION

The number of exact solutions in transient heat conduction and diffusion is extremely large and is growing. These solutions are needed for thermal modeling of various devices, as test cases for finite difference/element programs, and as influence functions for the unsteady surface element method (see Chapter 12). Solutions are given in many different papers, government reports, and industry reports. Because of the lack of organization of the solutions, it was frequently easier to rederive a solution than to search for it. With the advent of the internet and inexpensive computer storage, the development of specialized data bases has become practical, and they exist in medicine, law, and many other fields.

There is considerable variation among existing numbering systems. Most identification numbers have meaning only when a look-up table, or key, is consulted. For example, the Chemical Abstracts Service (CAS) assigns numbers to chemical compounds in the order they are discovered, so the CAS number itself contains no technical information on the compound. Many numbering systems contain some useful information in the numbers themselves. For example in a postal zip code, the first one or two digits can indicate a general location. A few numbering systems embody a great deal of information in the number itself. The number system of Butkovskiy (1982) and Butkovskiy and Pustylnikov (1993) identifies differential equations in the form (p, q, r). Integer p denotes the number of spatial dimensions in the domain, integer q denotes the highest derivative with respect to time, and integer r denotes the highest derivative with respect to space.

The purpose of this chapter is to present a number system for heat conduction and diffusion for which the number itself contains a great deal of information. Such a system not only simplifies the construction of a computer data base such as the Green's Function Library (Cole, 2009) but it makes locating existing solutions less tedious and lowers the effort needed to derive new solutions. The number system was first proposed by Beck and Litkouhi (1985) and other discussions are given in Beck (1984, 1986).

The numbering system covers basic geometries such as plates, cylinders, and spheres. Irregular geometries such as plates with several randomly spaced holes are not covered in the numbering system. This book deals mainly with solutions for temperature-independent thermal properties, but the numbering system can be employed for nonlinearities caused by temperature-variable properties.

The numbering system is specifically developed for transient diffusion and heat conduction. The same concepts, however, are applicable to other fields, such as convective heat transfer, fluid mechanics, and wave phenomena. Steady state is covered because it is included by the more general transient notation.

The plan of this chapter is first to give the numbering system for geometry and boundary conditions in Section 2.2. Section 2.3 provides boundary condition modifiers to describe the time and/or space variations of the nonhomogeneous term at a boundary. Section 2.4 gives an initial temperature distribution numbering system, and Section 2.5 provides a numbering system to treat interfaces between bodies. Section 2.6 gives a numbering system for the volumetric energy generation term $g(x,t)$, and then Section 2.7 gives some examples of the numbering system. The chapter concludes with Section 2.8, further discussion of advantages of the numbering system.

We recognize that not all readers will share our enthusiasm for the heat conduction numbering system. However, it is important that readers have some knowledge of the numbering system in order to use the extensive appendices of Green's functions (GFs) in this book. Most of the book will be accessible to the reader with a working knowledge of Section 2.2 on the numbering system for geometry and boundary conditions. Some readers may prefer to read Section 2.2 and then jump ahead to Chapter 3 on the Green's function solution equation (GFSE). Later these readers can return to Chapter 2 to learn more about the numbering system as the need arises.

2.2 GEOMETRY AND BOUNDARY CONDITION NUMBERING SYSTEM

For the rectangular coordinate system, the symbol X is used to denote the x-coordinate; Y is used to denote the y-direction; and Z is used to denote the z-direction. For a two-dimensional problem involving x and y-coordinates, X and Y are used; for a three-dimensional problem, X, Y, and Z are used. The three-dimensional equation for transient conduction with constant, isotropic thermal conductivity k is

$$k\left(\frac{\partial^2 T}{\partial x^2}+\frac{\partial^2 T}{\partial y^2}+\frac{\partial^2 T}{\partial z^2}\right)=\rho c\frac{\partial T}{\partial t} \tag{2.1}$$

For the cylindrical coordinates, r, ϕ, x, the symbol R is for r, Φ is for the angle ϕ, and X is for the axial coordinate. For constant k, the three-dimensional equation is

$$k\left[\frac{1}{r}\frac{\partial}{\partial r}\left(r\frac{\partial T}{\partial r}\right)+\frac{1}{r^2}\frac{\partial^2 T}{\partial \phi^2}+\frac{\partial^2 T}{\partial x^2}\right]=\rho c\frac{\partial T}{\partial t} \tag{2.2}$$

For spherical coordinates, r, ϕ, θ, the symbols are RS, Φ, Θ, respectively. The symbol RS is used to denote the radial-spherical coordinate direction. The angle ϕ for both the cylindrical and spherical coordinates goes from 0 to 2π.

Six different boundary conditions are given and are numbered 0, 1, 2, 3, 4, and 5. See Table 2.1.

The *first* kind of boundary condition is the prescribed temperature at boundary i,

$$T(\mathbf{r}_i,t)=f_i(\mathbf{r}_i,t) \tag{2.3}$$

TABLE 2.1
Types of Boundary Conditions

Notation	Name of Boundary Condition	Description of Boundary Condition
0	Zeroth kind (natural)	No physical boundary
1	Dirichlet	Prescribed temperature, Equation 2.3
2	Neumann	Prescribed heat flux, Equation 2.4
3	Robin	Convective condition, Equation 2.6
4	Fourth kind (Carslaw)	Thin film, no convection, Equation 2.7
5	Fifth kind (Jaeger)	Thin film, convection, Equation 2.8

where $f_i(\mathbf{r}_i, t)$ is the space- and time-dependent surface temperature. For a one-dimensional case at $x = 0$, $f_i(\cdot)$ can be a function of time only, such as $T(0, t) = f_1(t)$. For a two-dimensional case with coordinates x, y, at $x = x_1$, $T(x_1, y, t) = f_1(y, t)$.

The *second* kind of boundary condition is prescribed heat flux,

$$k \left.\frac{\partial T}{\partial n_i}\right|_{\mathbf{r}_i} = f_i(\mathbf{r}_i, t) \tag{2.4}$$

where n_i is an outward pointing normal. For a one-dimensional case of boundaries at $x_1 = 0$ and $x_2 = L$, $n_1 = -x$ and $n_2 = x$; the boundary conditions are

$$-k \left.\frac{\partial T}{\partial x}\right|_{x=0} = f_1(t) \qquad k \left.\frac{\partial T}{\partial x}\right|_{x=L} = f_2(t) \tag{2.5a, b}$$

and $f_1(t)$ and $f_2(t)$ are heat fluxes directed toward the surfaces.

The *third* kind is a convective boundary condition,

$$k \left.\frac{\partial T}{\partial n_i}\right|_{\mathbf{r}_i} + h_i T\big|_{\mathbf{r}_i} = f_i(\mathbf{r}_i, t) \tag{2.6}$$

where h_i is the heat transfer coefficient and $f_i(\mathbf{r}_i, t)$ is usually equal to $h_i T_\infty$ with T_∞ being the ambient temperature, but $f_i(\mathbf{r}_i, t)$ can also include a prescribed heat flux.

The *fourth* kind is for a thin film at a surface with a prescribed heat flux $f_i(\cdot)$,

$$k \left.\frac{\partial T}{\partial n_i}\right|_{\mathbf{r}_i} = f_i(\mathbf{r}_i, t) - (\rho c b)_i \left.\frac{\partial T}{\partial t}\right|_{\mathbf{r}_i} \tag{2.7}$$

The product $(\rho c b)_i$ is for the film at the ith surface, and b_i is its thickness. A physical example of this type of boundary condition is heat transfer into a large ceramic object with a thin metal coating on the surface. The temperature distribution in the metal coating may be neglected across the small thickness b_i because the thermal conductivity of the metal is large compared to the ceramic, but storage of thermal energy in the metal coating may not be neglected. This boundary condition can also describe a surface film composed of a well-stirred fluid with heat capacity of $(\rho c_p b)_i$.

The *fifth* kind of boundary condition is for a thin film permitting heat losses from the film by convection,

$$k \left. \frac{\partial T}{\partial n_i} \right|_{\mathbf{r}_i} + h_i T \big|_{\mathbf{r}_i} = f_i(\mathbf{r}_i, t) - (\rho c b)_i \left. \frac{\partial T}{\partial t} \right|_{\mathbf{r}_i} \tag{2.8}$$

The boundary condition of the fifth kind is physically identical to the fourth kind except that instead of a specified heat flux on the thin film at the surface there is a specified heat transfer coefficient h.

Another important case is the *zeroth* kind. It is for conditions for which there is no physical boundary; it is sometimes called a natural boundary condition. It includes several cases, one of which is in the rectangular coordinates when a boundary extends to infinity. For example, a semi-infinite body that is convectively heated at $x = 0$ is denoted $X30$. Another case is for the center of radial cylindrical and spherical bodies that are solid. A solid cylinder with a prescribed surface heat flux is denoted $R02$. The case associated with a convective boundary condition at $r = a$ and a spherical domain outside $r = a$ is denoted $RS30$. Another case is for a thin annular ring which is denoted $\Phi 00$.

Cases included by this numbering system are organized in Figures 2.1 through 2.3; notice that the structural arrangement of each of these cases is different, with the radial coordinate having the largest number of distinct cases and the angular, the least. Figure 2.1 is for the Cartesian coordinate x and includes 21 distinct cases; others such as $X12$ can be listed but these can be found by a simple change of coordinates (i.e., $x \to L - x$, where L is the plate thickness). Notice that the cylindrical radial chart shown in Figure 2.2 includes 26 cases because the $RI0$ ($I = 1, \ldots, 5$) geometries are quite different from the $R0I$ geometries, the former being the infinite region bounded

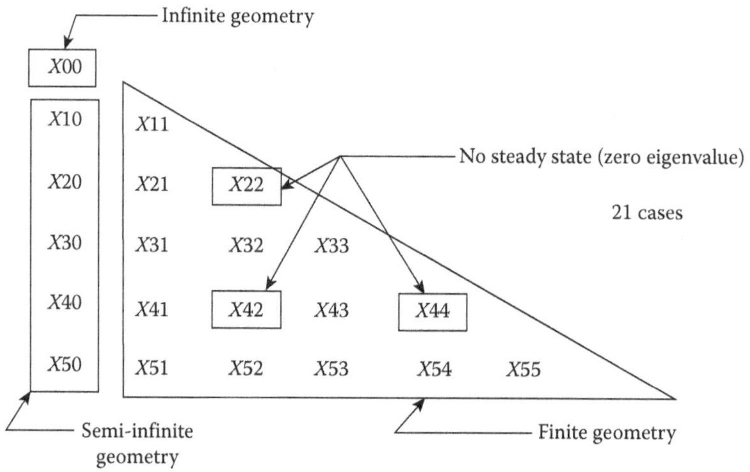

FIGURE 2.1 Distinct cases for one-dimensional Cartesian geometries.

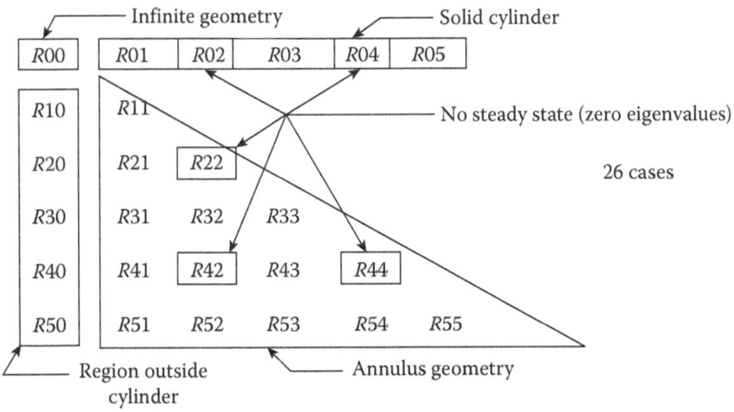

FIGURE 2.2 Distinct cases for one-dimensional cylindrical radial geometries.

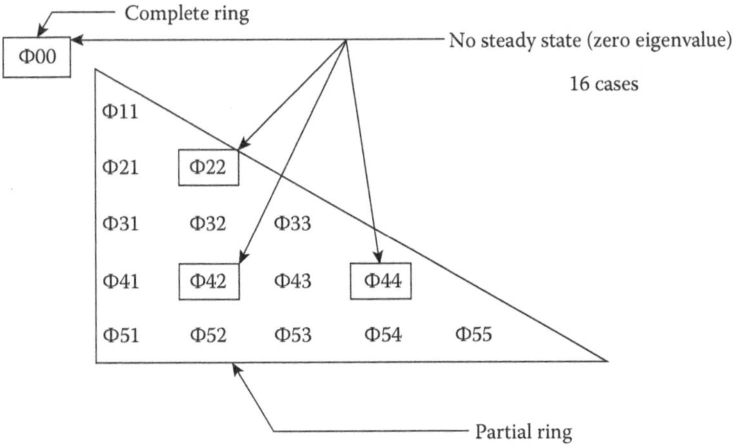

FIGURE 2.3 Distinct cases for ring geometries.

internally by the radius $r = a$ and the latter for solid cylinders of radius a. For annular geometries with boundary radii of a and b, neither I nor J in RIJ are equal to zero. The spherical radial cases $RSIJ$ is similar to Figure 2.2 with R replaced by RS. For the cylindrical coordinate ϕ and small changes in r, a ring is obtained; cases are displayed in Figure 2.3. The special case in Figure 2.3 is for a complete ring. There are neither $\Phi 0I$ nor $\Phi I0$ cases with $I \neq 0$. Except for the $\Phi 00$ case, the ΦIJ cases in Figure 2.3 have similar mathematical solutions as the corresponding XIJ cases of Figure 2.1.

There are three special finite-body cases in Figure 2.1 which (usually) have no steady state, namely $X22$, $X42$, and $X44$. There are five such special cases in Figure 2.2 and four in Figure 2.3. Mathematically, these cases are associated with zero eigenvalues. From a physical perspective, these cases do not have a steady state for time-independent values of $f_i(\cdot)$ in Equation 2.4 or 2.7 (unless there is the special case of zero net heat added). The $\Phi00$ case is unique since there are no physical boundaries; however, in this case (and the special finite bodies cases) there is no steady state for a constant volume source in the respective bodies.

For the infinite and semi-infinite geometries of Figures 2.1 and 2.2, i.e., the first column in both figures, steady state is not usually attained in finite times.

2.3 BOUNDARY CONDITION MODIFIERS

The boundary conditions of the first through fifth kinds are denoted as indicated in Section 2.2 but the time and/or space variation must also be specified. This means that the function $f_i(\mathbf{r}_i, t)$ in Equations 2.3, 2.4, 2.6 through 2.8 must be described. For one-dimensional cases, f_i can be only a function of time. The one-dimensional case is first considered and then the two- and three-dimensional cases are discussed.

For one-dimensional cases, the function $f_i(t)$ includes zero (denoted $B0$), constant with time ($B1$) (actually a step increase at $t = 0$), linear with time ($B2$), some power other than 1 of t ($B3$), exponentials ($B4$), two or more step changes ($B5$), and sinusoids ($B6$). See Table 2.2. Only the basic cases are given specific notation. Solutions permitting an arbitrary time variation are indicated by a dash (–).

For one-, two-, or three-dimensional bodies, the geometry and boundary condition descriptors are followed by the boundary condition modifier BIJ. An example is $X12B14$ where the $B14$ indicates that the boundary condition of the first kind (prescribed T) at $x = 0$ is nonzero constant and the boundary condition of the second kind (prescribed q) at $x = L$ has an exponential dependence on time. In general, two indices follow B but there are exceptions. Only one index is needed when there is a boundary condition of the zeroth kind such as $X20B1$ or $R03B1$, where the $B1$'s

TABLE 2.2
Types of Time- and Space-Variable Function at Boundary Conditions

Notation	Time-Variable Boundary Function	Notation	Space-Variable Boundary Function (Two-Dimensional)
B-	Arbitrary $f(t)$	Bx-	Arbitrary $f(x)$
$B0$	$f(t) = 0$		
$B1$	$f(t) = C$		
$B2$	$f(t) = Ct$	$Bx2$	$f(x) = Cx$
$B3$	$f(t) = Ct^p, \quad p > 1$	$Bx3$	$f(x) = Cx^p, p > 1$
$B4$	$f(t) = \exp(-at)$	$Bx4$	$f(x) = \exp(-ax)$
$B5$	Step changes in $f(t)$	$Bx5$	Step changes in $f(x)$
$B6$	$\sin(\omega t + E)$, $\cos(\omega t + E)$	$Bx6$	$\sin(\omega x + E)$, $\cos(\omega x + E)$

describe the nonzero boundary conditions. If both boundaries are of the zeroth kind (e.g., $X00$, $R00$, and $\Phi00$), then the B modifier is not used.

For two-dimensional cases the variation of $f(\cdot)$ at a boundary can be a function of space as well as time. For a two-dimensional problem involving x- and y-coordinates and at a y surface, $f(\cdot)$ could be a function of x alone, a function of t alone, or a function of x and t. If $f = f(x)$, then the boundary condition is denoted BxI, $I = 2, \ldots, 6$ (since $I = 0$ and 1 are not needed here). If $f = f(x,t)$, then the notation $B(xItJ)$ (where I is for x and J for t) can be used. Generalization to three-dimensional cases is direct; for example, $f = f(x,z,t)$ has the modifier $B(xIzJtK)$ with appropriate values of I, J, and K corresponding to x, z, and t. The parentheses are used to enclose notation for a single boundary.

2.4 INITIAL TEMPERATURE DISTRIBUTION

The initial temperature distribution is given in general coordinates by

$$T(\mathbf{r},0) = F(\mathbf{r}) \tag{2.9}$$

and for a one-dimensional case with x being the coordinate,

$$T(x,0) = F(x) \tag{2.10}$$

A numbering system for $F(\cdot)$ is given that is analogous to that for the boundary conditions. The letter T is followed by digits $0, 1, \ldots, 7$, as shown in Table 2.3. The coordinate r in Table 2.3 represents any single space coordinate such as r, x, or ϕ. Figure 2.4 displays some one-dimensional cases and gives the numbers including the notation for the initial temperature distribution. For two- and three-dimensional cases, see Figures 2.5 and 2.6 which are discussed in Section 2.6. For steady state problems, the initial condition index T and the associated digit are not used.

TABLE 2.3
Types of Space-Variable Initial Conditions

Notation	Single Space-Variable Initial Condition
T-	Arbitrary $F(r)$
$T0$	$F(r) = 0$
$T1$	$F(r) = C$
$T2$	$F(r) = Cr$
$T3$	$F(r) = Cr^p$, p not 0 or 1
$T4$	$F(r) = \exp(-ar)$
$T5$	Step changes in $F(r)$
$T6$	$\sin(\omega r + E)$, $\cos(\omega r + E)$
$T7$	Dirac delta function, $\delta(r - r_0)$

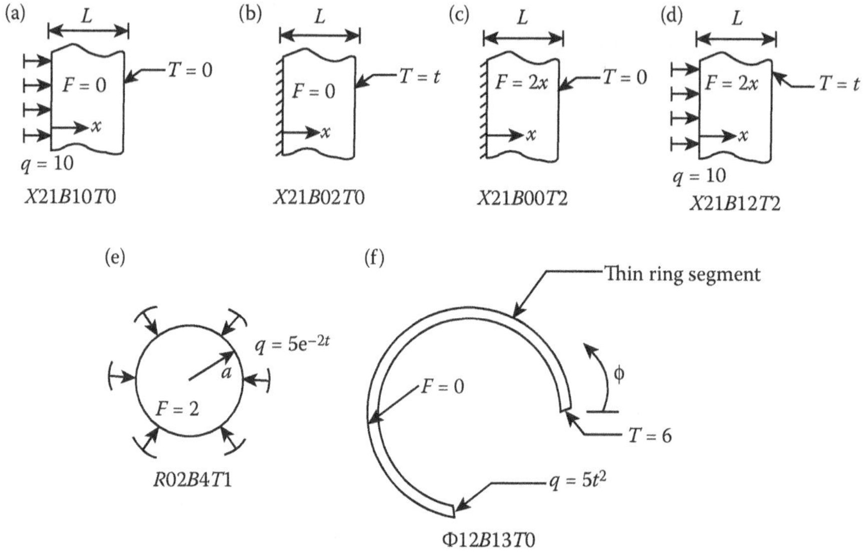

FIGURE 2.4 Some one-dimensional examples of numbering system.

2.5 INTERFACE DESCRIPTORS

The numbering system also applies to composite bodies. The interface conditions are denoted in a manner similar to the boundary conditions. For perfect contact, a capital C is used for the interface. For example, a plate perfectly bonded to another one, with prescribed temperatures on either side is denoted, $X1B$-$CX1B$-T-, for arbitrary time-variation of the surface temperatures and arbitrary initial temperature distribution.

For other conditions the letter C is followed by a single digit; see Table 2.4. The notation $C2$ is used to denote a perfect contact with a heat source at the interface; since heat flux is involved, it is analogous to the boundary condition of the second kind, hence the use of 2. The notation $C3$ is used to denote an imperfect contact at location r_i with a contact conductance of h_c at the interface (analogous to the boundary condition of the third kind)

$$-k \left.\frac{\partial T}{\partial n_i}\right|_{r_i^-} = h_c(T_{r_i^-} - T_{r_i^+}) = -k \left.\frac{\partial T}{\partial n_i}\right|_{r_i^+} \tag{2.11}$$

The $C4$ case is for a thin film (or well-stirred fluid) in perfect contact at the interface,

$$-k \left.\frac{\partial T}{\partial n_i}\right|_{r_i^-} = (\rho cb)_i \left.\frac{\partial T}{\partial t}\right|_{r_i^+} - k \left.\frac{\partial T}{\partial n_i}\right|_{r_i^+} \tag{2.12}$$

where $(\rho cb)_i$ is for the thin film or well-stirred fluid.

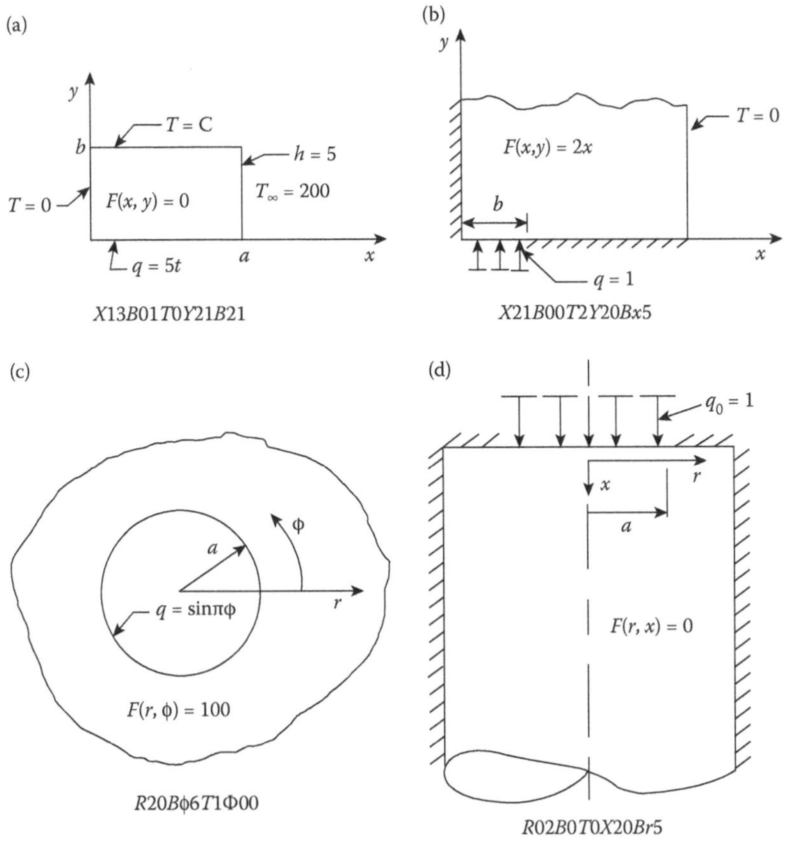

FIGURE 2.5 Two-dimensional examples of numbering system.

2.6 NUMBERING SYSTEM FOR *g(x, t)*

A notation for the geometry and for the boundary conditions is given in previous sections. In this section, extensions to the numbering system are given to classify the volumetric source term $g(x, t)$.

The notation for the volumetric source term $g(x, t)$ is indicated by a capital G followed by up to four modifiers to denote the x and t dependence. The notation is $GxItJ$, where xI represents the x dependence, and tJ represents the time dependence of the volume source term. The values I and J can assume the values $0, 1, 2, \ldots, 7$, or the dash (–) to represent different functions. See Table 2.5 for a listing of notation for the source term.

Several examples of the notation for the source term are presented below. For a source term of the form

$$g(x, t) = 10xt$$

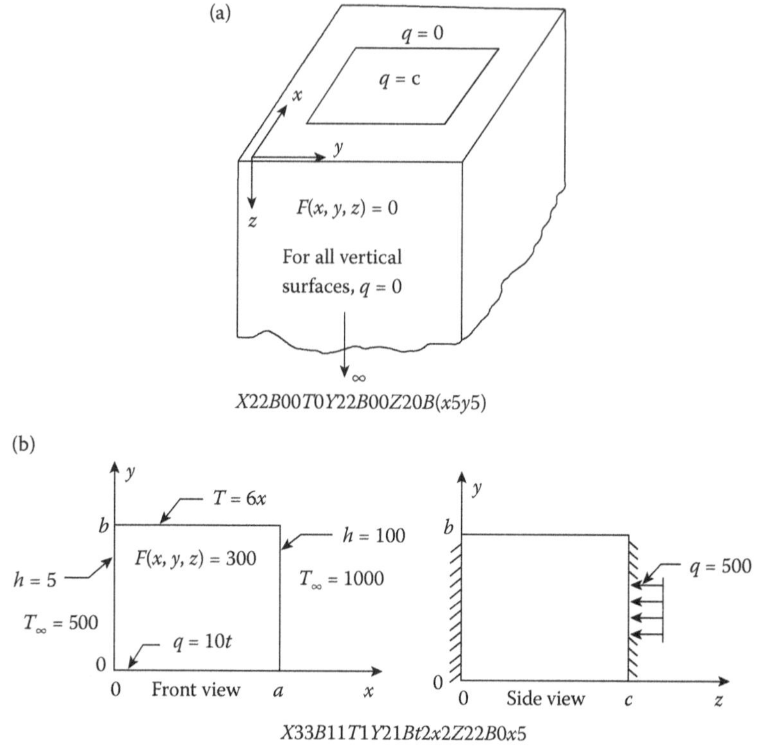

FIGURE 2.6 Three-dimensional examples of numbering system.

TABLE 2.4
Types of Interface Conditions

Notation	Description of Interface Condition
C	Perfect contact
$C2$	Perfect contact with source at interface
$C3$	Finite contact conductance
$C4$	Thin film at interface, perfect contact

the notation is $Gx2t2$. A source term of the form

$$g(x,t) = 10x$$

is denoted $Gx2t1$, or simply $Gx2$, since the modifier $t1$ is not needed to state that $g(x,t)$ does not depend on time. An even simpler case is

$$g(x,t) = 2$$

TABLE 2.5
Notation for Time-Variable Source Terms

Notation	Time Variation
Gt-	Arbitrary, $g(t)$
$Gt0$	$g(t) = 0$
$Gt1$	$g(t) = C$
$Gt2$	$g(t) = Ct$
$Gt3$	$g(t) = Ct^p$, $p \neq 0$ or 1
$Gt4$	$g(t) = \exp(-at)$
$Gt5$	Step changes in $g(t)$
$Gt6$	$\sin(\omega t + E)$, $\cos(\omega t + E)$
$Gt7$	Dirac delta function, $\delta(t - t_0)$

which is denoted $Gx1t1$ or more simply, $G1$. For the case where $g(x,t)$ is composed of a sum of several terms, such as

$$g(t) = a_0 + a_1 t + a_2 t^2$$

the notation is $Gx1t(1,2,3)$ or $Gt(1,2,3)$. Due to linearity of the heat conduction problem, the solution to the problem with source term $Gt(1,2,3)$ can be found as the sum of three problems,

$$Gt(1,2,3) = Gt1 + Gt2 + Gt3$$

2.7 EXAMPLES OF NUMBERING SYSTEM

The proposed numbering system can be used to describe a very large number of cases. Some one-dimensional cases are shown in Figure 2.4. The first four cases of Figure 2.4 are for the same basic case of $X21$. Figure 2.4a depicts a plate with a constant heat flux at $x = 0$ (boundary condition of the second kind) and $T = 0$ at $x = L$ (condition of the first kind). The initial temperature is zero. The number for this case is $X21B10T0$ where the 1 following B is for $q = C$ at $x = 0$ and the 0 following $B1$ is for the $T = 0$ condition at $x = L$. See Table 2.2. The problem of Figure 2.4b has an insulated surface at $x = 0$, a linear time variation of temperature at $x = L$ and a zero initial temperature; its number is $X21B02T0$. The two in $B02$ is for the linear time variation at $x = L$. Figure 2.4c has $f = 0$ at both boundaries but the initial temperature is a linear function of x and thus is denoted $X21B00T2$. The case shown by Figure 2.4d includes all the nonzero f_i and F values of Figure 2.4a, b, and c.

A cylindrical radial case is shown in Figure 2.4e. Depicted is a solid cylinder with a heat flux of exponential form at $r = a$ and the initial temperature is a constant. Figure 2.4f is for a segment of a thin ring.

Some two-dimensional cases are illustrated in Figure 2.5. A rectangular plate is shown in Figure 2.5a. The number description in the x-direction is similar to that for a one-dimensional case and it is then followed by the one in the y-direction. Since the initial temperature is known to be zero, it is redundant to repeat this information with the y-direction notation. Another two-dimensional case is shown in Figure 2.5b; it is for a plate that is finite in the x-direction and semi-infinite in the y-direction. For the x-direction, the boundary conditions are of the second and first kinds and are homogeneous, but the initial temperature distribution is linear with x; thus this part of the notation is $X21B00T2$. For the y-direction, there is a step increase in q at $x = 0$ and a step decrease at $x = b$, and there is no physical boundary for large y. Hence, the notation in the y-direction is $Y20Bx5$ where the $Bx5$ notation is for the steps in q in the x-direction at the $y = 0$ boundary. There is no y-direction dependence of the initial temperature so it is omitted in the notation.

A case of a body outside the cylindrical radius of $r = a$ is shown by Figure 2.5c. There is a sinusoidal variation with ϕ of the surface heat flux and the initial temperature distribution is constant. The notation is $R20B\phi6T1\Phi00$. The $B\phi6$ describes the boundary condition at $r = a$ and no index is needed for $r \to \infty$ where there is no physical boundary.

Figure 2.5d displays a semi-infinite cylinder that is insulated at all surfaces except at the center at the top where a circular heat flux is applied. The initial temperature is zero. The number for this case is $R02B0T0X20Br5$ where the $Br5$ notation is used because the heat flux is not constant with r but can be considered to have a step increase at $r = 0$ and a step decrease $r = a$. If the heat flux were over the circular region shown and also varied as ct in time, $Br5$ would be replaced by $B(r5t2)$ where the parentheses are used to denote that both conditions apply at the same boundary.

The numbering system readily extends to three-dimensional cases such as given in Figure 2.6. The first case is for a semi-infinite rod that is insulated on all surfaces except there is a constant heat flux over a rectangular region at $z = 0$. The case of a rectangular block is shown in Figure 2.6b, where front and side views are shown.

2.8 ADVANTAGES OF NUMBERING SYSTEM

There are several types of advantages of the numbering system. The first relates to a data base of conduction solutions. The second relates to an algebra that can be given for linear problems. The last major advantage relates to use of the method in conjunction with GFs to obtain solutions for linear problems; full explanation is deferred until after Chapter 3.

2.8.1 Data Base in Transient Heat Conduction

One of the obvious advantages of a numbering system is that it facilitates the organizing of a data base. A structure is provided that makes the storage of solutions easier. Also important is that it greatly reduces the effort in locating solutions. Instead of relying on imprecise verbal titles of papers (or abstracts) to describe a particular problem, a search based on the notation given herein can be much more direct and less prone to miss related solutions.

TABLE 2.6
Some One-Dimensional Cases in Carslaw and Jaeger (1959)

Number	Page	Equation	Comments
$X00T5$	54	3	$T(x,0) = T_0, -a < x < a; T(x,0) = 0, \|x\| > a$
$X10B1T0$	60	10	
$X10B3T0$	305	6	$T(0,t) = T_0 t^{n/2}, n = 1, 2, \ldots$
$X11B00T1$	96	6	
$R01B0T1$	199	5	
$R01B1T0$	331	3	Small time solution

The numbering system has been utilized to catalog most of the solutions of Carslaw and Jaeger. An example of a portion of data base for some solutions is given in Table 2.6. A more complete tabulation is available on the Green's Function Library internet site (Cole, 2009). Table 2.6 gives numbers of some one-dimensional cases from Carslaw and Jaeger (1959). The first column contains the number; the second and third columns give the page and equation numbers of the reference; and the last column contains some comments.

2.8.2 Algebra for Linear Cases

For linear cases, several kinds of algebraic manipulations are possible. This brief discussion can include only a few possibilities.

One case involves boundary conditions of the zeroth, first, and third kinds and the uniform initial temperature distribution. An example is

$$[X10B1T0|_{T(0,t)=T_0}] = T_0[1 - (X10B0T1|_{T(x,0)=1})] \tag{2.13}$$

where T_0 is a constant.

In addition to relating boundary conditions and the initial temperature, the notation suggests a method of superimposing solutions. The number of nonzero values of the indices following B and T give the number of superposition problems that can be formed; this is the number of "forcing" terms. An example is provided by the first four cases of Figure 2.4. The Figure 2.4d case is the sum of the first three cases,

$$X21B12T2 = X21B10T0 + X21B02T0 + X21B00T2 \tag{2.14}$$

Notice that $B12$ contains two nonzero digits and $T2$ contains one; hence, the case of Figure 2.4d can be given as the sum of three problems. The same superposition principles can be used for the two-dimensional problem of Figure 2.5a.

Another type of superposition is possible for more than one forcing term at a boundary. An example is for the Figure 2.4a case with

$$q = 10 + 5t \tag{2.15}$$

The temperature solution can be written as

$$T|_{q=10+5t} = 10\,[X21B10T0|_{q=1}] + 5\,[X21B20T0|_{q=t}] \tag{2.16}$$

Another aspect of the algebra for the numbering system is that it can aid in identifying the number of explicit dimensions of a problem. A plate is a three-dimensional object but the temperature distribution can be an explicit function of only one or two coordinates. Boundary conditions of the zeroth, second, and fourth kinds have the potential of reduction in the number of dimensions while the first, third, and fifth kinds do not. However, for reduction in the number of the dimensions, both boundaries in a given direction must be homogeneous and there cannot be any explicit dependence of the initial temperature or g in that direction.

As an example, consider the case of a cube which is at zero initial temperature and there is no volumetric energy source. At time zero, each surface is heated with a constant heat flux (which may or may not be the same for each face). The number for this case is $X22B11Y22B11Z22B11T0$ and the solution is equal to the sum of six one-dimensional problems,

$$\begin{aligned} X22B11Y22B11Z22B11T0 &= X22B10T0 + X22B01T0 + Y22B10T0 \\ &\quad + Y22B01T0 + Z22B10T0 + Z22B01T0 \end{aligned} \tag{2.17}$$

This reduction of dimensions on the right side of Equation 2.17 is because the typical three-dimensional problem of $X22B10Y22B00Z22B00T0$ reduces to

$$X22B10Y22B00Z22B00T0 = X22B10Y22B00T0 = X22B10T0 \tag{2.18}$$

Note that the $Y22B00$ and the $Z22B00$ conditions have boundary conditions of the second kind and are homogeneous.

An example that does not reduce in the same manner is for a cube initially at $T = 0$ and subjected to a step increase in temperature on each surface (i.e., a constant temperature with time and over the surface). The number and algebra for this case are

$$\begin{aligned} &X11B11Y11B11Z11B11T0 \\ &\quad = X11B10Y11B00Z11B00T0 + X11B01Y11B00Z11B00T0 \\ &\qquad + X11B00Y11B10Z11B00T0 + X11B00Y11B01Z11B00T0 \\ &\qquad + X11B00Y11B00Z11B10T0 + X11B00Y11B00Z11B01T0 \end{aligned} \tag{2.19}$$

Each of these problems is three-dimensional although simplifications in the solutions result because the problems are similar. If each surface of the cube is subjected to the same temperature condition (or even convective boundary condition), the GF solution leads to further simplifications. For example, if the cube is initially at temperature T_0, and suddenly immersed in a fluid at $T_\infty = 0$ with the same h on each surface, the temperature distribution is given by

$$\begin{aligned} &X33B00Y33B00Z33B00T1 \\ &\quad = T_0[X33B00T1|_{F=1}][Y33B00T1|_{F=1}][Z33B00T1|_{F=1}] \end{aligned} \tag{2.20}$$

This is related to the multiplication of solutions associated with one-dimensional solutions which are discussed in undergraduate heat transfer textbooks.

The possible uses of this numbering system for transient heat conduction and diffusion are numerous and these can be considerably expanded beyond what is outlined in this book.

PROBLEMS

2.1 Give the numbering system designation for Example 1.1 of Chapter 1.

2.2 Give two numbers for Equation 1.83 that are valid for $x > 0$.
(Answer: $X00T5$ and $X20B0T5$)

2.3 Give two numbers for Equation 1.84 that are valid for $x > 0$.

2.4 Give the numbering system designation for Example 1.3 of Chapter 1.

2.5 Give the number for the temperature given by Equation 1.92 and by Equation 1.96b; these are two interpretations of the relation between temperature and the Green's function.

2.6 Give the numbering system designation for Example 1.4 of Chapter 1.

2.7 Give the numbering system designation for Problem 1.22 of Chapter 1.

2.8 Give the number for the problem with the same geometry and boundary condition shown in Figure 1.12a with the initial temperature being a constant and with a constant volumetric energy source.

2.9 Give the number for Figure 2.4d with $F = 6$, $q = 2$. At $x = L$, $T = 5 + 2\sin 4t$.

2.10 Using the numbering system for conduction, give the numbers for the following one-dimensional cases, each of which satisfies the partial differential equation,

$$\frac{\partial C}{\partial t} = D\frac{\partial^2 C}{\partial x^2} \qquad 0 < x < L \qquad t > 0$$

with conditions:

(a) $C(0,t) = C_0$, $C(L,t) = 0$, $C(x,0) = 6\sin 2\pi x / L$.

(b) $+\partial C / \partial x = 0$ at $x = 0$, $C(L,t) = C_0$, $C(x,0) = C_1$.

(c) $C(0,t) = 3 + 4t^2$, $C(L,t) = \cos 2t$, $C(x,0) = \cos 2x$.

2.11 Write the describing differential equation, boundary conditions, and initial condition for the problem denoted $X24B21G1T0$.

2.12 For the partial differential equation for cylindrical heat flow with volume energy generation, give the numbers for the following cases.

(a) A solid cylinder is initially at a uniform temperature and is suddenly plunged into a fluid at a temperature of T_∞; where $g = 0$.

(b) A hollow cylinder is initially at a uniform temperature is insulated at the inner surface and is heated by a constant heat flux at the outer surface; $g = 5$.

(c) The region is that outside the radius of $r = a$ and a constant heat flux exists at $r = a$. The initial temperature is T_0 and $g = 0$.

(d) The geometry is the same as shown in Figure 2.5d but q at $x = 0$ is $\sin \pi r / a$ for $r < a$ and zero for larger values of r. The initial temperature is a function of r and ϕ.

REFERENCES

Beck, J. V., 1984, Green's function solution for transient heat conduction problems, *Int. J. Heat Mass Transfer*, vol. 27, pp. 1235–1244.

Beck, J. V., 1986, Green's functions and numbering system for transient heat conduction, *AIAA J.*, vol. 24, pp. 327–333.

Beck, J. V. and Litkouhi, B., 1985, Heat conduction numbering system, ASME paper 85-WA/HT-63.

Butkovskiy, A. G., 1982, *Green's Functions and Transfer Function Handbook*, Ellis Horwood Limited, Chichester, U.K.

Butkovskiy, A. G. and Pustylnikov, L. M., 1993, *Characteristics of Distributed-Parameter Systems*, Kluwer, Dordrecht, Netherlands.

Carslaw, H. S. and Jaeger, J. C., 1959, *Conduction of Heat in Solids*, 2nd edn, Oxford University Press, New York.

Cole, K. D., Green's Function Library, http:\\www.greensfunction.unl.edu, accessed August 2009.

3 Derivation of the Green's Function Solution Equation

3.1 INTRODUCTION

The Green's function solution equation (GFSE) for transient heat conduction is derived in this chapter in several forms. First, the one-dimensional form for rectangular coordinates is derived for boundary conditions of the first, second, and third kinds. This form is easy to understand and examples are included to demonstrate how the equation is applied. Second, the GFSE is derived in a general three-dimensional form that applies to rectangular, cylindrical, and spherical coordinates. An even more general form of the GFSE is derived in Chapter 10; it covers the case of nonhomogeneous materials. Third, an alternative form particularly appropriate for nonhomogeneous boundary conditions is given. Fourth, a steady-state form is given and, finally, the GFSE is given for moving solids.

This chapter contains background material that, although important, is not essential to the application of the Green's functions (GF) method. One can begin with the GFSE, choose the correct GF, evaluate the integrals, and find the solution for temperature. However, an understanding of the GFSE will lead to a greater understanding of the GFs themselves.

This chapter covers the derivation of the one-dimensional GFSE in Section 3.2 and a general vector-based form in Section 3.3. Section 3.4 contains an alternative form of the GFSE (AGFSE) which may be helpful for nonhomogeneous boundary conditions when slow convergence is a concern. Section 3.5 covers the m^2T term which is associated with fins. Section 3.6 covers the steady-state GSFE as a limit of the transient case. Finally, Section 3.7 contains a derivation of the GFSE for moving solids.

3.2 DERIVATION OF THE ONE-DIMENSIONAL GREEN'S FUNCTION SOLUTION EQUATION

The one-dimensional GFSE for rectangular coordinates is derived in this section. The one-dimensional form of the GFSE is free of vector calculus, so one can gain intuition about the GF method with a minimum of notation. The derivation makes use of the properties of GFs, and the result is an expression for the temperature that fully exploits the linear property of the heat conduction equation.

The boundary value problem for the temperature in a one-dimensional rectangular geometry is given in Section 1.12, by Equations 1.113 through 1.115 as

$$\frac{\partial^2 T}{\partial x^2} + \frac{1}{k} g(x,t) = \frac{1}{\alpha}\frac{\partial T}{\partial t} \qquad t > 0 \tag{3.1}$$

$$k_i \left.\frac{\partial T}{\partial n_i}\right|_{x_i} + h_i T|_{x_i} = f_i(t) \quad t > 0 \quad \text{and } i = 1,2 \tag{3.2}$$

$$T(x,0) = F(x) \tag{3.3}$$

This is the problem that we are trying to solve with the GF method. In general, Equation 3.2 describes convection boundary conditions (boundary conditions of the third kind), but temperature or heat flux boundary conditions may be obtained by taking $k_i = 0$ or $h_i = 0$, respectively, on surfaces $i = 1$ or $i = 2$.

The derivation of the GFSE begins with the auxiliary boundary value problem for the GF that corresponds to the above temperature problem. The auxiliary boundary value problem is very similar to the boundary value problem for the temperature with two important differences: first, the energy generation term in the differential equation for the GF is a Dirac delta function; and second, the boundary conditions and the initial conditions for the GF are homogeneous. The auxiliary boundary value problem was previously discussed in Section 1.12 and is given by

$$\frac{\partial^2 G}{\partial x^2} + \frac{1}{\alpha}\,\delta(x - x')\,\delta(t - \tau) = \frac{1}{\alpha}\frac{\partial G}{\partial t} \qquad t > \tau \tag{3.4a}$$

$$k_i \left.\frac{\partial G}{\partial n_i}\right|_{x_i} + h_i G|_{x_i} = 0 \qquad i = 1,2 \tag{3.4b}$$

$$G(x,t|x',\tau) = 0 \qquad t < \tau \tag{3.4c}$$

Next, the reciprocity relation (Section 1.13)

$$G(x,t|x',\tau) = G(x',-\tau|x,-t)$$

is applied to the auxiliary equation (3.4a) to give

$$\frac{\partial^2 G}{\partial x'^2} + \frac{1}{\alpha}\,\delta(x' - x)\,\delta(t - \tau) = -\frac{1}{\alpha}\frac{\partial G}{\partial \tau} \tag{3.5}$$

Notice the minus sign on the time derivative. The next step is to write the original heat conduction equation for T in terms of x' and τ. That is, write Equation 3.1 with a simple change of variables: replace x by x' and replace t by τ to give

$$\frac{\partial^2 T}{\partial x'^2} + \frac{1}{k} g(x',\tau) = \frac{1}{\alpha}\frac{\partial T}{\partial \tau} \tag{3.6}$$

Multiply Equation 3.6 by $G(x,t|x',\tau)$, multiply Equation 3.5 by $T(x',\tau)$, and then subtract Equation 3.5 from Equation 3.6 to get

$$G\frac{\partial^2 T}{\partial x'^2} - T\frac{\partial^2 G}{\partial x'^2} + \frac{G}{k} g(x',\tau) - \frac{T}{\alpha}\delta(x' - x)\,\delta(t - \tau) = \frac{1}{\alpha}\frac{\partial (TG)}{\partial \tau} \tag{3.7}$$

Integrate Equation 3.7) with respect to x' over the domain $0 \le x' \le L$, and integrate with respect to τ from 0 to $t + \epsilon$, where ϵ is a small positive number. The result is

$$\begin{aligned}
&\int_{\tau=0}^{t+\epsilon} d\tau \int_{x'=0}^{L} \left(G \frac{\partial^2 T}{\partial x'^2} - T \frac{\partial^2 G}{\partial x'^2} \right) dx' \\
&\quad + \frac{1}{k} \int_{\tau=0}^{t+\epsilon} d\tau \int_{x'=0}^{L} g(x', \tau)\, G(x, t | x', \tau)\, dx' - \frac{1}{\alpha} T(x, t) \\
&= \frac{1}{\alpha} \int_{x'=0}^{L} [TG]_{\tau=0}^{\tau=t+\epsilon} dx'
\end{aligned} \tag{3.8}$$

Note that the properties of the Dirac delta function give the term $T(x' = x, \tau = t)$ on the left-hand side of this equation. This equation can be solved for $T(x,t)$ to give

$$\begin{aligned}
T(x,t) = &- \int_{x'=0}^{L} [TG]_{\tau=0}^{\tau=t+\epsilon} dx' \\
&+ \frac{\alpha}{k} \int_{\tau=0}^{t+\epsilon} d\tau \int_{x'=0}^{L} g(x', \tau)\, G(x, t | x', \tau)\, dx' \\
&+ \alpha \int_{\tau=0}^{t+\epsilon} d\tau \int_{x'=0}^{L} \left(G \frac{\partial^2 T}{\partial x'^2} - T \frac{\partial^2 G}{\partial x'^2} \right) dx'
\end{aligned} \tag{3.9}$$

This is the GFSE for one-dimensional rectangular coordinates. The three terms on the right-hand side of Equation 3.9 will next be examined and simplified one at a time.

The first term of Equation 3.9 can be simplified because $G(x, t | x', t + \epsilon) = 0$ from the causality relation. That is, G is zero because $t - \tau = t - (t + \epsilon) = -\epsilon < 0$; there is zero response before the impulse occurs. Also, $T(x', 0)$ can be replaced by the initial condition, given by Equation 3.3. Thus, the first term of the GF equation represents the effect of the initial condition, and it is written

$$\int_{x'=0}^{L} F(x')\, G(x, t | x', 0)\, dx' \tag{3.10}$$

The second term in Equation 3.9 is the effect of the volume energy generation. This term will not be simplified any further at this point.

The third term of Equation 3.9 can be simplified with integration by parts. (The analogous step in the three-dimensional derivation involves Green's theorem.) Consider just the integral on x' from this third term, and integrate by parts to get

$$\begin{aligned}
\int_{x'=0}^{L} \left(G \frac{\partial^2 T}{\partial x'^2} - T \frac{\partial^2 G}{\partial x'^2} \right) dx' &= G \frac{\partial T}{\partial x'} \bigg|_{x'=0}^{x'=L} - \int_{x'=0}^{L} \frac{\partial G}{\partial x'} \frac{\partial T}{\partial x'} dx' \\
&\quad - T \frac{\partial G}{\partial x'} \bigg|_{x'=0}^{x'=L} + \int_{x'=0}^{L} \frac{\partial T}{\partial x'} \frac{\partial G}{\partial x'} dx' \\
&= G \frac{\partial T}{\partial x'} \bigg|_{x'=0}^{x'=L} - T \frac{\partial G}{\partial x'} \bigg|_{x'=0}^{x'=L}
\end{aligned} \tag{3.11}$$

Note that the two integrals in Equation 3.11 cancel.

If the boundary conditions are of the second or third kinds, then the boundary conditions for T and G can be used to evaluate $\partial T / \partial x'$ and $\partial G / \partial x'$ at the boundaries. Equations 3.2 and 3.4b can be written as

$$\left.\frac{\partial G}{\partial n_i'}\right|_{x'=x_i} = -\frac{h_i}{k_i} G|_{x'=x_i} \tag{3.12}$$

$$\left.\frac{\partial T}{\partial n_i'}\right|_{x'=x_i} = \frac{f_i(\tau)}{k_i} - \frac{h_i}{k_i} T|_{x'=x_i} \tag{3.13}$$

The notation n_i is for the *outward normal* from the body. Substitute these boundary conditions into Equation 3.11 to get

$$\begin{aligned} G\left.\frac{\partial T}{\partial x'}\right|_{x'=0}^{x'=L} - T\left.\frac{\partial G}{\partial x'}\right|_{x'=0}^{x'=L} &= \left[\frac{f_i(\tau)}{k_i} G - \frac{h_i}{k_i} T\, G\right]_{x'=L} - \left[-\frac{f_i(\tau)}{k_i} G + \frac{h_i}{k_i} T\, G\right]_{x'=0} \\ &\quad - \left(-\frac{h_i}{k_i} T\, G\right)_{x'=L} + \left(\frac{h_i}{k_i} T\, G\right)_{x'=0} \\ &= \sum_{i=1}^{2} \frac{f_i(\tau)}{k_i} G|_{x'=x_i} \end{aligned} \tag{3.14}$$

Note that the terms that involve T cancel. The summation over $i = 1, 2$ is meant to cover all the possibilities for the boundary conditions of one-dimensional bodies. The total number of boundary terms is two for a finite body ($0 \leq x \leq L$). (The derivation also applies for semi-infinite and finite bodies. The semi-infinite one-dimensional body requires only one boundary term, and the infinite body does not require any boundary terms.)

If the boundary conditions are of the first kind, Equation 3.11 takes a different form. At the boundaries, $G = 0$ and $T = f_i(\tau)$, so that

$$G\left.\frac{\partial T}{\partial x'}\right|_{x'=0}^{x'=L} - T\left.\frac{\partial G}{\partial x'}\right|_{x'=0}^{x'=L} = -\sum_{j=1}^{2} f_j(\tau) \left.\frac{\partial G}{\partial n_j'}\right|_{x'=x_j} \tag{3.15}$$

Again, the summation over $j = 1, 2$ is used to represent the contribution from both boundaries.

The last step in the derivation of the GF equation is to take the limit of Equation 3.9 as $\epsilon \to 0$. Then, $t + \epsilon$ can be replaced by t in the equation, without altering the conclusions that are drawn from $\epsilon > 0$. Finally, Equation 3.9 is combined with the simplified terms given by Equations 3.10, 3.11, 3.14, and 3.15 to give the desired result

$$\begin{aligned} T(x,t) &= \int_{x'=0}^{L} G(x,t|x',0)\, F(x')\, dx' \qquad \text{(for the initial condition)} \\ &\quad + \frac{\alpha}{k} \int_{\tau=0}^{t} d\tau \int_{x'=0}^{L} g(x',\tau)\, G(x,t|x',\tau)\, dx' \qquad \text{(for energy generation)} \end{aligned}$$

$$+\alpha \int_{\tau=0}^{t} d\tau \sum_{i=1}^{2} \left[\frac{f_i(\tau)}{k_i} G(x, t|x_i', \tau) \right] \quad \text{(for boundary conditions of the second and third kinds)}$$

$$-\alpha \int_{\tau=0}^{t} d\tau \sum_{i=1}^{2} \left[f_i(\tau) \frac{\partial G}{\partial n_i'}\bigg|_{x'=x_i} \right] \quad \text{(for boundary conditions of the first kind only)} \qquad (3.16)$$

This is the desired GFSE which applies to one-dimensional transient heat conduction in the rectangular coordinate system. The one-dimensional body is assumed to be homogeneous and to have constant properties (independent of temperature and position).

Each term in the GFSE must have the units of temperature. In the first term, $F(x')$ has units of temperature, so the product $G dx'$ must be dimensionless for the units to be correct, therefore the one-dimensional GF has units of m^{-1}. In the second term, $g(x', \tau)$ has units of W/m^3, so the product $(\alpha / k) g(x', \tau)\, d\tau$ has units of temperature, as it should. In the third term, $f_i(\tau)$ has units of W/m^2 (heat flux), so the product $(\alpha / k_i)\, f_i(\tau)\, G d\tau$ has the units of temperature. Finally, in the fourth term, $f_i(\tau)$ has units of temperature, so the product $\alpha(\partial G / \partial n_i)\, d\tau$ is dimensionless.

In the usual cases discussed in this book, the boundary terms $f_i(t)$ are known. There are special cases when $T(x, t)$ is known from measurements, and $f_i(t)$ is the unknown. This is called the inverse heat conduction problem (Beck et al., 1985). In this case, Equation 3.16 is considered to be an integral equation because the unknown, $f_i(t)$, is inside the integral.

Each $G(\cdot)$ term in Equation 3.16 represents the same GF, which is mathematically unique for each set of boundary conditions. For example, in a geometry with $X12$ boundary conditions, the correct GF to use in Equation 3.16 is the $X12$ GF, as in the following example.

Example 3.1:

For the geometry shown in Figure 3.1, the boundary conditions for $T(x, t)$ are

$$T(0, t) = T_0 \qquad (3.17)$$

$$+k \left.\frac{\partial T}{\partial x}\right|_{x=L} = q(t) \qquad (3.18)$$

This is an example of the $X12$ geometry. The initial condition is

$$T(x, 0) = F(x) \qquad (3.19)$$

and there is no energy generation in this case. If the $X12$ GF is assumed to be a known function named $G_{X12}(x, t|x', \tau)$, what is the appropriate form of the GFSE?

Solution

The GFSE is a sum of the various effects that contribute to the temperature $T(x, t)$. The contribution of the initial condition is given by the first term from Equation 3.16,

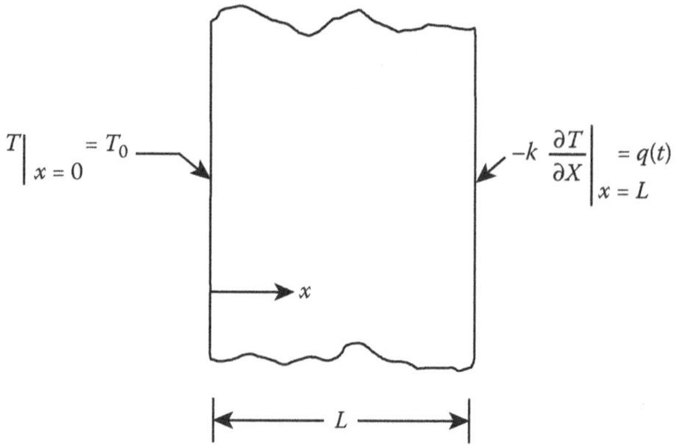

FIGURE 3.1 Slab body geometry for Example 3.1: $X12$ case.

$$\int_{x'=0}^{L} G_{X12}(x,t|x',0)\,F(x')\,dx' \tag{3.20}$$

The boundary condition at $x = 0$ is of the first kind. This boundary condition contributes to the temperature according to the last term of Equation 3.16 where $f_i(\tau) = T_0$. This term is

$$-\alpha \int_{\tau=0}^{t} \left(-T_0 \frac{\partial G_{X12}}{\partial x'}\bigg|_{x'=0} \right) d\tau \tag{3.21}$$

Notice the minus sign that appears because $\partial/\partial n_i' = -\partial/\partial x'$ at $x = 0$; the outward pointing normal n_i is in the minus x-direction for the $x = 0$ surface.

The boundary condition at $x = L$ is of the second kind. This boundary condition contributes to the temperature according to the third term of Equation 3.16, where $f_i(\tau) = q(\tau)$,

$$\alpha \int_{\tau=0}^{t} \frac{q(\tau)}{k} G_{X12}(x,t|L,\tau)\,d\tau \tag{3.22}$$

The temperature $T(x, t)$ is the sum of these three effects, or

$$\begin{aligned} T(x,t) = &\int_{x'=0}^{L} G_{X12}(x,t|x',0)\,F(x')\,dx' \\ &+ \alpha \int_{\tau=0}^{t} T_0 \frac{\partial G_{X12}}{\partial x'}\bigg|_{x'=0} d\tau \\ &+ \alpha \int_{\tau=0}^{t} \frac{q(\tau)}{k} G_{X12}(x,t|L,\tau)\,d\tau \end{aligned} \tag{3.23}$$

which is the GFSE for this example.

Notice that $G_{X12}(x,t|x',\tau)$ in each term is evaluated at the time or location appropriate to that term in the GF equation. For example, in the initial condition term, G_{X12} is evaluated at $\tau = 0$. In the term for the left-side boundary condition, $\partial G_{X12}/\partial x'$ is evaluated at $x' = 0$.

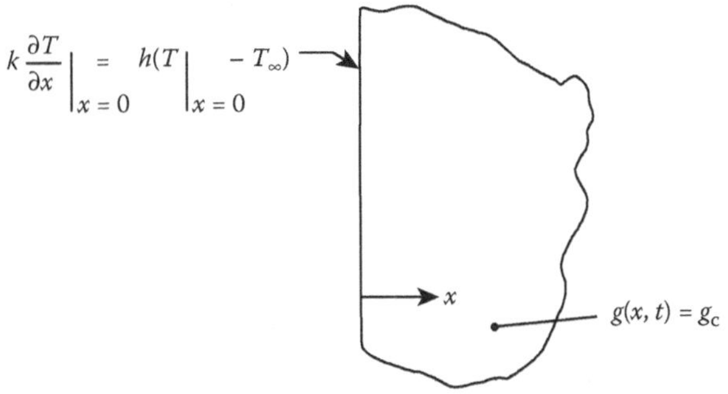

FIGURE 3.2 Semi-infinite body with convection at the boundary and internal energy generation. Geometry for Example 3.2.

Example 3.2:

The one-dimensional semi-infinite body shown in Figure 3.2 has a convection boundary condition given by

$$-k\frac{\partial T}{\partial x}\bigg|_{x=0} = h(T_\infty - T|_{x=0}) \tag{3.24}$$

where T_∞ is the ambient temperature. This is the $X30$ geometry. The volume heat generation is given by $g(x, t) = g_c$, where g_c is a constant. The heat conduction equation is thus given by

$$\frac{\partial^2 T}{\partial x^2} + \frac{1}{k}g_c = \frac{1}{\alpha}\frac{\partial T}{\partial t} \tag{3.25}$$

The initial condition is

$$T(x, 0) = F(x) \tag{3.26}$$

If the $X30$ GF is assumed to be a known function denoted $G_{X30}(x, t|x', \tau)$, what is the appropriate form of the GFSE?

Solution

There are three terms that contribute to the temperature $T(x, t)$: the initial condition, the volume heat generation, and the convection boundary condition. The effect of the boundary at infinity does not require an explicit term, because it is already included in the correct GF, denoted $G_{X30}(\cdot)$. The temperature for this case is given by the GFSE

$$\begin{aligned} T(x, t) &= \int_{x'=0}^{\infty} G_{X30}(x, t|x', 0)\, F(x')\, dx' \\ &+ \frac{\alpha}{k}\int_{\tau=0}^{t}\int_{x'=0}^{\infty} g_c\, G_{X30}(x, t|x', \tau)\, dx'd\tau \\ &+ \alpha\int_{\tau=0}^{t} \frac{hT_\infty}{k} G_{X30}(x, t|0, \tau)\, d\tau \end{aligned} \tag{3.27}$$

Note that the integrals on x' in the first two terms are evaluated over the entire body, $0 \le x' \le \infty$. This is an extension of Equation 3.16 to the semi-infinite case. (A similar extension to the $X00$, or infinite body case, is to evaluate the x' integral over $-\infty \le x' \le \infty$.)

3.3 GENERAL FORM OF THE GREEN'S FUNCTION SOLUTION EQUATION

In this section the GFSE will be derived in a general form for an additional term in the heat conduction equation (the m^2T term) and for two additional boundary conditions. This general form of the GFSE can be applied to three-dimensional geometries in any orthogonal coordinate system. The rectangular, cylindrical, or spherical coordinate systems are treated in this book.

3.3.1 Temperature Problem

The partial differential equation that describes transient, multidimensional, linear heat conduction in a homogeneous isotropic body is,

$$\nabla^2 T + \frac{1}{k} g(\mathbf{r}, t) - m^2 T = \frac{1}{\alpha}\frac{\partial T}{\partial t} \quad \text{in region } R \text{ and } t > 0 \qquad (3.28)$$

The thermal conductivity k and thermal diffusivity α are both constant with position, time, and temperature. Any orthogonal coordinate $\mathbf{r}$ can be used in Equation 3.28. The $g(\mathbf{r}, t)$ term represents space- and time-variable volume energy generation.

The m^2T term could represent side heat losses for a fin; m^2 can be a function of $\mathbf{r}$ but not t. (The m^2T term is not needed for the three-dimensional treatment of a fin.) If there is a component of volume energy generation g that is linearly proportional to temperature, it should be included in the m^2T term which could then encompass the effects of electric heating and dilute chemical reactions; in such cases m^2 could be either positive or negative. An example of transient conduction involving the m^2T term is given in Section 3.5.

The initial temperature distribution is expressed by

$$T(\mathbf{r}, 0) = F(\mathbf{r}) \qquad (3.29)$$

The boundary conditions for Equation 3.28 have the general form

$$k_i \frac{\partial T}{\partial n_i} + h_i T = f_i(\mathbf{r}_i, t) - (\rho c b)_i \frac{\partial T}{\partial t} \qquad t > 0 \qquad (3.30)$$

where the temperature T and its derivatives are evaluated at the boundary surface S_i, and $\mathbf{r}_i$, denotes the boundary. The spatial derivative $\partial/\partial n_i$ denotes differentiation along an *outward* drawn normal to the boundary surface S_i, $i = 1, 2, \ldots, s$. The heat transfer coefficient, h_i, and $(\rho c b)_i$ can vary with position on S_i but are independent of temperature and time. The boundary condition given by Equation 3.30 includes the possibility of a high conductivity surface film of thickness b_i. There is a negligible

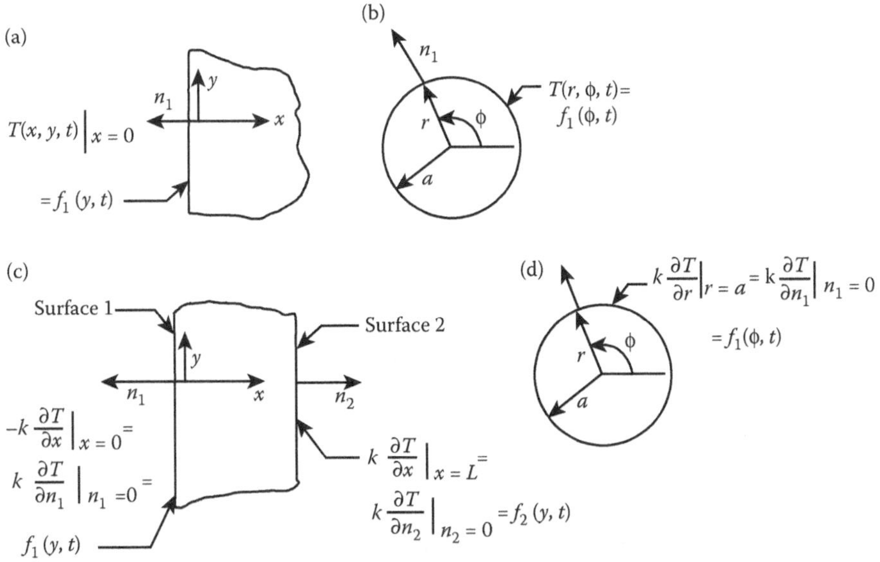

FIGURE 3.3 Examples of boundary conditions of the first and second kinds: (a) first kind of boundary condition at $x = 0$; (b) first kind of boundary condition at $r = a$; (c) second kind of boundary condition at $x = 0$ and L, rectangular coordinates; (d) second kind of boundary condition at $r = a$, cylindrical coordinates.

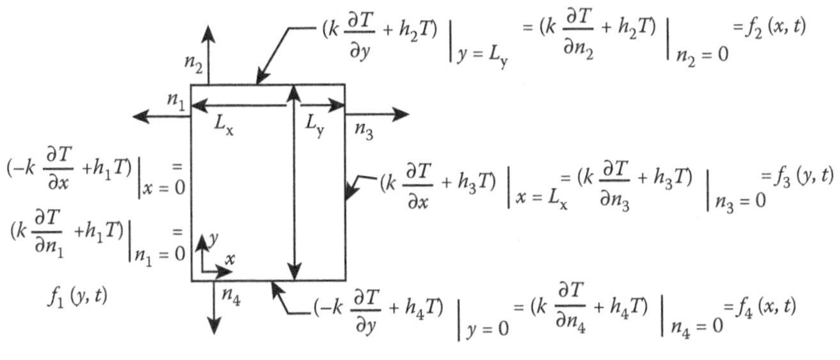

FIGURE 3.4 Examples of convection boundary conditions (third kind) on rectangular body.

temperature gradient through the film and there is no heat flux parallel to the surface inside the film. Five different boundary conditions can be obtained from Equation 3.30 by setting $k_i = 0$ or k, $h_i = 0$ or h, and also $b = 0$ or nonzero.

Figure 3.3 shows some examples of boundary conditions of the first and second kinds. Figures 3.4 and 3.5 show some examples of boundary conditions of the third kind and fifth kind, respectively. The five different boundary conditions are discussed in Chapter 2.

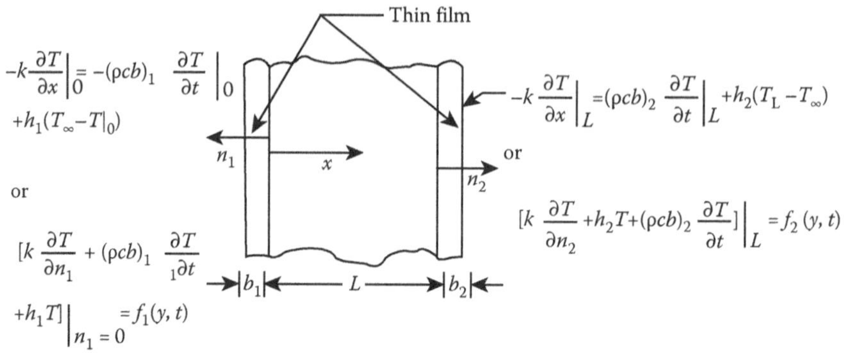

FIGURE 3.5 Examples of the film boundary condition (fifth kind).

3.3.2 Derivation of the Green's Function Solution Equation

The GFSE is derived using Equations 3.28 through 3.30 and also an auxiliary problem for an instantaneous heat source inside the body. The solution to the auxiliary problem is $G(\mathbf{r}, t|\mathbf{r}', \tau)$, where the instantaneous source is located at position $\mathbf{r}'$ and at time τ; $\mathbf{r}$ is the location at which the temperature is observed at time t. There can be a nonzero response at $\mathbf{r}$ only if $t - \tau > 0$. The auxiliary problem has homogeneous boundary conditions and a zero initial temperature.

The derivation of the general GFSE begins with the reciprocity relation of GF,

$$G(\mathbf{r}, t|\mathbf{r}', \tau) = G(\mathbf{r}', -\tau|\mathbf{r}, -t) \tag{3.31}$$

substituted into the auxiliary equation, resulting in

$$\nabla_0^2 G + \frac{1}{\alpha}\delta(\mathbf{r} - \mathbf{r}')\,\delta(t - \tau) - m^2 G = -\frac{1}{\alpha}\frac{\partial G}{\partial \tau} \qquad t > \tau \tag{3.32}$$

$$G(\mathbf{r}', -\tau|\mathbf{r}, -t) = 0 \qquad t < \tau \tag{3.33}$$

$$k_i \frac{\partial G}{\partial n_i'} + h_i G = (\rho cb)_i \frac{\partial G}{\partial \tau} \qquad t > \tau \tag{3.34}$$

where ∇_0^2 is the Laplacian operator for the $\mathbf{r}'$ coordinates and the minus sign on the right side in Equation 3.32 is a result of Equation 3.31, with t being replaced by $-\tau$. Next, the temperature equation 3.28, can be written in terms of $\mathbf{r}'$ and τ as

$$\nabla_0^2 T + \frac{1}{k} g(\mathbf{r}', \tau) - m^2 T = \frac{1}{\alpha}\frac{\partial T}{\partial \tau} \tag{3.35}$$

Multiply Equation 3.35 by G, multiply Equation 3.32 by T, and subtract Equation 3.32 from Equation 3.35 to get

$$(G\nabla_0^2 T - T\nabla_0^2 G) + \frac{g(\mathbf{r}', \tau)}{k} G - \frac{1}{\alpha}\delta(\mathbf{r} - \mathbf{r}')\,\delta\,(t - \tau) T = \frac{1}{\alpha}\frac{\partial(GT)}{\partial \tau} \tag{3.36}$$

Integrate this equation with respect to $\mathbf{r}'$ over the total region R, and integrate with respect to τ from 0 to $t+\epsilon$, where ϵ is an arbitrarily small positive number. This yields

$$\int_{\tau=0}^{t+\epsilon}\int_R \alpha\,(G\nabla_0^2 T - T\nabla_0^2 G)\,dv'd\tau + \int_{\tau=0}^{t+\epsilon}\int_R \frac{\alpha}{k} G\, g(\mathbf{r}',\tau)\,dv'd\tau - T(\mathbf{r},t)$$
$$= \int_R [GT]_{\tau=0}^{t+\epsilon} dv' \tag{3.37}$$

where dv' is a volume element in the region R. By rearranging the above equation, the temperature distribution in the body is

$$T(\mathbf{r},t) = -\int_R [GT]_{\tau=0}^{t+\epsilon} dv' + \int_{\tau=0}^{t+\epsilon}\int_R \frac{\alpha}{k} G\, g(\mathbf{r}',\tau)\,dv'd\tau$$
$$+ \int_{\tau=0}^{t+\epsilon}\int_R \alpha(G\nabla_0^2 T - T\nabla_0^2 G)\,dv'\,d\tau \tag{3.38}$$

The left side of this equation is the temperature distribution in the body at location $\mathbf{r}$ and at time t. The right side of this equation is now examined term by term.

The first term on the right side of Equation 3.38 can be simplified because $G(\mathbf{r},t|\mathbf{r}',t+\epsilon) = 0$ by the causality relation; the effect cannot begin before the instantaneous source. Also, at $\tau = 0$, the temperature distribution $T(\mathbf{r}',0)$ is the initial temperature distribution $F(\mathbf{r})$. Hence, the first right side term of Equation 3.38 becomes

$$\int_R G(\mathbf{r},t|\mathbf{r}',0)\, F(\mathbf{r}')\,dv' \tag{3.39}$$

For transient heat conduction in a body, this is the effect of the initial temperature distribution on the transient temperature distribution.

The second term on the right side of Equation 3.38 arises from the volume energy generation $g(\mathbf{r},t)$. This term will not be simplified further.

The third term on the right side of Equation 3.38 represents the contribution of all the boundary conditions. This term can be simplified with Green's theorem to change the volume integral to a surface integral (see homework problem 3.1). The result is

$$\int_{\tau=0}^{t+\epsilon}\int_R \alpha(G\nabla_0^2 T - T\nabla_0^2 G)\,dv'\,d\tau$$
$$= \int_{\tau=0}^{t+\epsilon}\sum_{i=1}^{s}\int_{S_i} \alpha\left(G\left.\frac{\partial T}{\partial n_i'}\right|_{\mathbf{r}'=\mathbf{r}_i'} - T\left.\frac{\partial G}{\partial n_i'}\right|_{\mathbf{r}'=\mathbf{r}_i'}\right) ds_i'\,d\tau \tag{3.40}$$

where $\partial/\partial n_i'$ denotes differentiation along an outward drawn normal to the boundary surface S_i and ds_i is an area element of S_i.

The integrand from Equation 3.40 can be expressed in terms of the boundary conditions of the heat conduction equation and the auxiliary GF equation. If the boundary conditions are of the second, third, fourth, or fifth kind, then the boundary conditions for T and G can be used to evaluate $\partial T/\partial n_i'$ and $\partial G/\partial n_i'$ at the boundaries. Equations 3.30 and 3.34 can be written as

$$\left.\frac{\partial G}{\partial n_i'}\right|_{\mathbf{r}'=\mathbf{r}_i'} = -\frac{h_i}{k}G|_{\mathbf{r}'=\mathbf{r}_i'} + \frac{(\rho c b)_i}{k}\frac{\partial G}{\partial \tau} \tag{3.41}$$

$$\left.\frac{\partial T}{\partial n_i'}\right|_{\mathbf{r}'=\mathbf{r}_i'} = \frac{f_i(\mathbf{r}_i',\tau)}{k} - \frac{h_i}{k}T|_{\mathbf{r}'=\mathbf{r}_i'} - \frac{(\rho c b)_i}{k}\frac{\partial T}{\partial \tau} \tag{3.42}$$

Multiplying the boundary condition Equation 3.42 by the GF, multiplying Equation 3.41 by the temperature and subtracting yields,

$$\left[G\left.\frac{\partial T}{\partial n_i'}\right|_{\mathbf{r}'=\mathbf{r}_i'} - T\left.\frac{\partial G}{\partial n_i'}\right|_{\mathbf{r}'=\mathbf{r}_i'}\right] = \frac{f_i(\mathbf{r}_i',\tau)}{k}G - \frac{(\rho c b)_i}{k}\left(T\frac{\partial G}{\partial \tau} + G\frac{\partial T}{\partial \tau}\right)$$

$$= \frac{f_i(\mathbf{r}_i',\tau)}{k}G - \frac{(\rho c b)_i}{k}\frac{\partial (GT)}{\partial \tau} \tag{3.43}$$

Replace Equation 3.43 into Equation 3.40 to obtain for boundary conditions of the second through fifth kinds:

$$\int_{\tau=0}^{t+\epsilon}\sum_{i=1}^{s}\int_{S_i}\alpha\left(G\left.\frac{\partial T}{\partial n_i'}\right|_{\mathbf{r}'=\mathbf{r}_i'} - T\left.\frac{\partial G}{\partial n_i'}\right|_{\mathbf{r}'=\mathbf{r}_i'}\right)ds_i'\,d\tau$$

$$= \alpha\int_{\tau=0}^{t+\epsilon}\sum_{i=1}^{s}\int_{S_i}\frac{f_i(\mathbf{r}_i',\tau)}{k}G(\mathbf{r},t|\mathbf{r}_i',\tau)\,ds_i'\,d\tau$$

$$+\alpha\sum_{i=1}^{s}\int_{S_i}\frac{(\rho c b)_i}{k}G(\mathbf{r},t|\mathbf{r}_i',0)\,F(\mathbf{r}')\,ds_i' \tag{3.44}$$

Note that the integral over τ has been evaluated for the term $\partial(GT)/\partial\tau$.

For a boundary condition of the first kind the right side of Equation 3.40 takes a different form. At the boundary, G is zero and T is $f_i(\mathbf{r}_i, t)$ for boundary conditions of the first kind. Then, the right-hand side of Equation 3.40 becomes

$$-\alpha\int_{\tau=0}^{t+\epsilon}\sum_{j=1}^{s}\int_{S_j} f_j(\mathbf{r}_j',\tau)\left.\frac{\partial G}{\partial n_j'}\right|_{\mathbf{r}'=\mathbf{r}_j'} ds_j'\,d\tau \tag{3.45}$$

for boundary conditions of the first kind.

The final step in the derivation of the GFSE is to take the limit of Equation 3.38 as $\epsilon \to 0$. Then, $t+\epsilon$ can be replaced by t in the equation without altering the conclusions drawn from $\epsilon > 0$. The derivation is completed by combining Equation 3.38 with the simplified terms given by Equations 3.39, 3.44, and 3.45 to give the important *general GFSE for heat conduction* for homogeneous bodies:

$$T(\mathbf{r},t) = T_{\text{in}}(\mathbf{r},t) + T_g(\mathbf{r},t) + T_{\text{b.c.}}(\mathbf{r},t) \tag{3.46a}$$

which contains three terms, one for the initial conditions, one for the volumetric energy source, and one for the nonhomogeneous boundary conditions. The initial temperature contribution term is

$$T_{\text{in}}(\mathbf{r},t) = \int_R G(\mathbf{r},t|\mathbf{r}',0)F(\mathbf{r}')\,dv' \qquad \text{(for all boundary conditions)}$$

$$+\,\alpha \sum_{i=1}^{s} \int_{S_i} \frac{(\rho c b)_i}{k} G(\mathbf{r},t|\mathbf{r}_i',0)F(\mathbf{r}_i')ds_i' \qquad \begin{array}{l}\text{(for boundary conditions}\\ \text{of the fourth and fifth kinds)}\end{array} \tag{3.46b}$$

The term for the volumetric energy generation inside the body is

$$T_g(\mathbf{r},t) = \int_{\tau=0}^{t} \int_R \frac{\alpha}{k} G(\mathbf{r},t|\mathbf{r}',\tau)\, g(\mathbf{r}',\tau)\, dv'd\tau \tag{3.46c}$$

The term for the boundary conditions contains two types of expressions, one for boundary conditions of the second through fifth kinds and the other is for boundary conditions of the first kind. The term for the boundary conditions is

$$T_{\text{b.c.}}(\mathbf{r},t) = \alpha \int_{\tau=0}^{t} \sum_{i=1}^{s} \int_{S_i} \frac{f_i(\mathbf{r}_i',\tau)}{k} G(\mathbf{r},t|\mathbf{r}_i',\tau)\, ds_i'\, d\tau$$

(for boundary conditions of the second through fifth kinds)

$$-\alpha \int_{\tau=0}^{t} \sum_{j=1}^{s} \int_{S_j} f_j(\mathbf{r}_j',\tau) \left.\frac{\partial G}{\partial n_j'}\right|_{\mathbf{r}'=\mathbf{r}_j'} ds_j'\, d\tau$$

(for boundary conditions of the first kind only) (3.46d)

This equation has two parts because the boundary condition of the first kind must be treated in a different manner than the others.

Equation 3.46 applies to any orthogonal coordinate system if the correct form for ds and dv are used. See Table 3.1 for the differential elements ds_i, and dv for rectangular, cylindrical, and spherical coordinates systems.

The total number of terms considered between the i and j summations is exactly s, that is, the heat flux boundary conditions (second, third, fourth, and fifth kinds) and temperature boundary conditions (first kind) are mutually exclusive on a given boundary. For a one-dimensional boundary, $0 \le s \le 2$; for a two-dimensional geometry, $0 \le s \le 4$; and, for a three-dimensional geometry, $0 \le s \le 6$. The number of boundary conditions s includes only conditions at "real" boundaries; it does not include a boundary condition at $x \to \infty$ for a semi-infinite body, for example.

Equation 3.46 is the main result of this chapter and is a general form of the GFSE. See Chapter 10 for a general form that applies to nonhomogeneous bodies.

Example 3.3:

Consider a two-dimensional rectangular region starting at $x = a_1$ and extending to $x = a_2$ in the x-direction and starting at $y = b_1$ and going to $y = b_2$ in the y-direction. See Figure 3.6.

a. Formulate and discuss the problem for boundary conditions of zeroth, first, second, and third kinds.
b. Give the appropriate form of the GFSE for this problem.

TABLE 3.1
Quantities ds_i' and dv' for the Transient GFSE for Three Coordinate Systems

Coordinate System	Example of Geometry	Coordinates	ds_i'	dv'	Units of G[a]
Rectangular					
	Slab	x	1^b	dx'	m^{-1}
	Rectangle	x, y	dx' or dy'	$dx' dy'$	m^{-2}
	Parallelepiped	x, y, z	$dx' dy', dx'dz'$ or $dy'dz'$	$dx'dy'dz'$	m^{-3}
Cylindrical					
	Infinite cylinder	r	$^b2\pi r_i$	$2\pi r'dr'$	m^{-2}
	Thin shell	ϕ	$^b\delta$ (thin-shell thickness)	$\delta a\, d\phi'$ (a = shell radius)	m^{-2}
	Finite cylinder	r, z	$2\pi r_i dz'$ or $2\pi r'dr'$	$2\pi r'dr'dz'$	m^{-3}
	Wedge	r, ϕ	dr' or $r_i d\phi'$	$r'dr'd\phi'$	m^{-2}
Spherical					
	Sphere	r	$^b4\pi r_i^2$	$4\pi(r')^2dr'$	m^{-3}
	Conical section of sphere	r, θ	$2\pi(r')2dr' \sin\theta_i$ or $2\pi r_i^2 \sin\theta' d\theta'$	$2\pi(r')^2dr' \sin\theta' d\theta'$	m^{-3}

[a] Units of G are such that $G\,dv'$ is dimensionless for heat conduction.
[b] No integral on S_i.

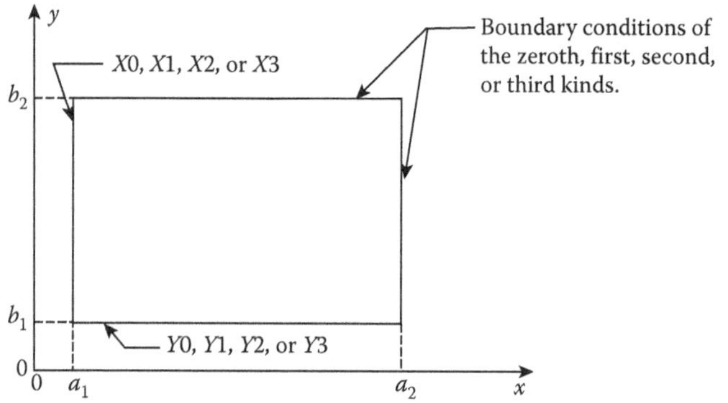

FIGURE 3.6 Two-dimensional rectangular body, geometry for Example 3.3.

Solution

(a) *Formulation of the problem.* The describing partial differential equation is

$$k\left(\frac{\partial^2 T}{\partial x^2} + \frac{\partial^2 T}{\partial y^2}\right) + g(x, y, t) = \rho c \frac{\partial T}{\partial t} \tag{3.47}$$

and the boundary conditions are either of the zeroth, first, second, or third kinds. For the boundary condition of the zeroth kind, there is actually no boundary. For the $X20Y10$ case, for example, it is convenient to set $a_1 = 0$ and $b_1 = 0$ and to note that $a_2 \to \infty$ and $b_2 \to \infty$, and hence no boundary source term, $f(\cdot)$, enters for $x = a_2 \to \infty$ and $y = b_2 \to \infty$.

The boundary condition at $x = a_1$ can be written as

$$-k_{a1}\frac{\partial T(a_1, y, t)}{\partial x} + h_{a1}T(a_1, y, t) = f_{a1}(y, t) \tag{3.48}$$

for boundary conditions of the first, second, and third kinds. For boundary conditions of the first kind, the k_{a1}, h_{a1}, and $f_{a1}(\cdot)$ terms are

$$k_{a1} = 0 \qquad h_{a1} = 1 \qquad f_{a1}(y, t) = T_{a1}(y, t) \tag{3.49}$$

where $T_{a1}(y, t)$ is the prescribed temperature history at $x = a_1$. For the boundary condition of the second kind, the values are

$$k_{a1} = k \qquad h_{a1} = 0 \qquad f_{a1}(y, t) = q_{a1}(y, t) \tag{3.50}$$

where k is the thermal conductivity of the solid, and $q_{a1}(\cdot)$, is the prescribed heat flux at $x = a_1$. For the boundary condition of the third kind, these terms in Equation 3.48 are

$$k_{a1} = k \qquad h_{a1} = h_{a1}(y) \qquad f_{a1}(y, t) = h_{a1}(y)T_{\infty a1}(y, t) \tag{3.51}$$

where $h_{a1}(y)$ is the heat transfer coefficient at $x = a_1$ and $T_{\infty a1}(y, t)$ is the ambient temperature at $x = a_1$. In general, the usual GF approach permits h to be a function of position but not time.

In Equation 3.49, the $f(\cdot)$ function represents prescribed temperatures at a boundary. Functions $f(\cdot)$ can depend on time and position. For boundary conditions of the second kind, $f(\cdot)$ in Equations 3.50 is a prescribed heat flux q. For a boundary condition of the third kind, $f(\cdot)$ in Equations 3.51 is a prescribed variation of hT_∞ where h can be a function of position (not time) and T_∞ can be a function of time and position. The boundary conditions at the $x = a_2$ surface are similar to the one at $x = a_1$

$$k_{a2}\frac{\partial T(a_2, y, t)}{\partial x} + h_{a2}T(a_2, y, t) = f_{a2}(y, t) \tag{3.52}$$

This equation also applies for the first, second, and third kind of boundary conditions by suitable choice of k_{a2}, h_{a2}, and f_{a2} in a manner similar to that in Equations 3.49 through 3.51.

The boundary condition at $y = b_1$ is

$$-k_{b1}\frac{\partial T(x, b_1, t)}{\partial y} + h_{b1}T(x, b_1, t) = f_{b1}(x, t) \tag{3.53}$$

and the boundary condition at $y = b_2$ is

$$k_{b2}\frac{\partial T(x, b_2, t)}{\partial y} + h_{b2}T(x, b_2, t) = f_{b2}(x, t) \tag{3.54}$$

In order to complete the statement of the problem, the initial temperature distribution is needed,

$$T(x, y, 0) = F(x, y) \tag{3.55}$$

where $F(x, y)$ is the temperature distribution at $t = 0$.

(b) *Two-dimensional GFSE.* The two-dimensional GFSE can be written as

$$\begin{aligned} T(x, y, t) = &\int_{x'=a_1}^{a_2} \int_{y'=b_1}^{b_2} G_{XIJ}(x, t|x', 0)\, G_{YMN}(y, t|y', 0)\, F(x', y')\, dx'dy' \\ &+ \frac{\alpha}{k} \int_{\tau=0}^{t} \int_{x'=a_1}^{a_2} \int_{y'=b_1}^{b_2} G_{XIJ}(x, t|x', \tau)\, G_{YMN}(y, t|y', \tau) \\ &\times g(x', y', \tau)\, dy'dx'd\tau + I_{x'=a_1} + I_{x'=a_2} + I_{y'=b_1} + I_{y'=b_2} \end{aligned} \tag{3.56}$$

The notation G_{XIJ} refers to the GF specific to the rectangular coordinate type of boundary condition on the boundaries $x = a_1$ and $x = a_2$. Similarly, YMN refers to the GF for the type of boundary conditions at $y = b_1$ and b_2. The last four terms denoted I in Equation 3.56 depend on the type of boundary condition, of the zeroth, first, second, or third kinds. There are four I terms, one for each boundary. For a boundary condition of the zeroth kind, the associated I term is equal to zero. For boundary condition of type 1, specified temperature, at $x' = a_1$, the $I_{x'=a_1}$ term for the boundary at $x' = a_1$ is

$$\begin{aligned} I_{x'=a_1} = \alpha &\int_{\tau=0}^{t} \int_{y'=b_1}^{b_2} \left(-\frac{\partial G_{XIJ}(x, t|a_1, \tau)}{\partial n'} \right) \\ &\times G_{YMN}(y, t|y', \tau)\, f_{x1}(y', \tau)\, dy'\, d\tau \end{aligned} \tag{3.57a}$$

and for boundary conditions of second or third kinds term $I_{x'=a_1}$ is

$$\begin{aligned} I_{x'=a_1} = \frac{\alpha}{k} &\int_{\tau=0}^{t} \int_{y'=b_1}^{b_2} G_{XIJ}(x, t|a_1, \tau) \\ &\times G_{YMN}(y, t|y', \tau)\, f_{x1}(y', \tau)\, dy'd\tau \end{aligned} \tag{3.57b}$$

For the $I_{x'=a_2}$ term, the same expressions as given in Equation 3.57a are used with a_1 in $G_{XIJ}(\cdot)$ replaced by a_2, $f_{a1}(\cdot)$ by $f_{a2}(\cdot)$. For $x' = a_1$ in Equation 3.57b, $\partial n'$ is $-\partial x'$, while for $x' = a_2$, $\partial n'$ is $\partial x'$.

The terms $I_{y'=b_1}$ and $I_{y'=b_2}$ may be found in a manner similar to Equation 3.57.

Example 3.4:

Give the appropriate form of the GFSE for a three-dimensional rectangular parallelepiped region of $a_1 \le x \le a_2, b_1 \le y \le b_2, c_1 \le z \le c_2$, with the initial temperature of $F(x, y, z)$.

Solution

The three-dimensional GFSE can be written as

$$\begin{aligned} T(x, y, z, t) = &\int_{x'=a_1}^{a_2} \int_{y'=b_1}^{b_2} \int_{z'=c_1}^{c_2} G_{XIJ}(x, t|x', 0)\, G_{YKL}(y, t|y', 0) \\ &\times G_{ZMN}(z, t|z', 0)\, F(x', y', z')\, dx'dy'dz' \end{aligned}$$

$$+\frac{\alpha}{k}\int_{\tau=0}^{t}\int_{x'=a_1}^{a_2}\int_{y'=b_1}^{b_2}\int_{z'=c_1}^{c_2}$$
$$\times\, G_{XIJ}(x,t|x',\tau)\, G_{YKL}(y,t|y',\tau)$$
$$\times\, G_{ZMN}(z,t|z',\tau)\, g(x',y',z',\tau)\, dx'dy'dz'd\tau$$
$$+ I_{x'=a_1} + I_{x'=a_2} + I_{y'=b_1} + I_{y'=b_2} + I_{z'=c_1} + I_{z'=c_2} \tag{3.58}$$

where the I's can be found as in Example 3.3.

3.4 ALTERNATIVE GREEN'S FUNCTION SOLUTION EQUATION

In some cases, the use of GFs for nonhomogeneous boundary conditions can yield slowly converging solutions. Some of these cases can be modified to produce better-behaved solutions by using an alternative GFSE (AGFSE). A brief derivation is given in this section and a more complete derivation is given in Section 10.3.

The derivation begins with a known solution, $T^*(\mathbf{r},t)$, to the problem

$$\nabla^2 T^* - m^2 T^* = -\frac{g^*(\mathbf{r},t)}{k} \qquad \text{in region } R \tag{3.59}$$

with the general boundary condition of

$$k_i \frac{\partial T^*}{\partial n_i}\bigg|_{\mathbf{r}_i} + h_i T^*\bigg|_{\mathbf{r}_i} = f_i(\mathbf{r}_i,t) - (\rho c b)_i \frac{\partial T^*}{\partial t}\bigg|_{\mathbf{r}_i} \tag{3.60}$$

Notice that the boundary conditions are nonhomogeneous and contain the same prescribed source term $f_i(\mathbf{r},t)$ that is in the $T(\mathbf{r},t)$ problem. In addition, Equation 3.59 contains the arbitrary source term of $g^*(\mathbf{r},t)$, which in some cases is set equal to zero and in others a particular choice, such as $g^* = g$, simplifies the problem; it does not have to correspond to $g(\mathbf{r},t)$.

Let the solution to the usual transient heat conduction problem be made equal to

$$T(\mathbf{r},t) = T^*(\mathbf{r},t) + T'(\mathbf{r},t) \tag{3.61}$$

Then a solution is desired for $T'(\mathbf{r},t)$,

$$T'(\mathbf{r},t) = T(\mathbf{r},t) - T^*(\mathbf{r},t) \tag{3.62}$$

which must satisfy

$$\nabla^2 T' + \frac{1}{k}[g(\mathbf{r},t) - g^*(\mathbf{r},t)] - m^2 T' - \frac{1}{\alpha}\frac{\partial T^*}{\partial t} = \frac{1}{\alpha}\frac{\partial T'}{\partial t} \qquad \text{in } R \tag{3.63}$$

with the initial condition

$$T'(\mathbf{r},0) = F(\mathbf{r}) - T^*(\mathbf{r},0) \tag{3.64}$$

and the general boundary condition (at $r = r_i$)

$$k_i \frac{\partial T'}{\partial n_i} + h_i T' = -(\rho c b)_i \frac{\partial T'}{\partial t} \tag{3.65}$$

which is now homogeneous. From the above, it can be seen that the solution to the $T'(\mathbf{r},t)$ problem can be obtained by using the GFSE given by Equation 3.46 but using a modified initial condition, a modified volume energy generation term, and homogeneous boundary conditions. Using Equation 3.46 for $T'(\mathbf{r},t)$ and then using Equation 3.61 yields the AGFSE for $T(\mathbf{r},t)$:

$$T(\mathbf{r},t) = T^*(\mathbf{r},t) + \int_R G(\mathbf{r},t|\mathbf{r}',0)\,[F(\mathbf{r}') - T^*(\mathbf{r}',0)]dv'$$

$$+\,\alpha\sum_{i=1}^{s}\int_{S_i}\frac{(\rho cb)_i}{k_i}G(\mathbf{r},t|\mathbf{r}',0)\,[F(\mathbf{r}_i') - T^*(\mathbf{r}_i',0)]ds_i'$$

(for boundary conditions of the fourth and fifth kinds only)

$$+\,\frac{\alpha}{k}\int_{\tau=0}^{t}\int_R G(\mathbf{r},t|\mathbf{r}',\tau)\left[g(\mathbf{r}',\tau) - g^*(\mathbf{r}',\tau) - \rho c\frac{\partial T^*(\mathbf{r}',\tau)}{\partial\tau}\right]dv'd\tau \tag{3.66}$$

Example 3.5:

Consider the problem of a plate with the boundary and initial conditions

$$T(0,t) = T_0$$
$$T(L,t) = T_0 + (T_L - T_0)\sin\omega t$$
$$T(x,0) = T_0$$

where T_0 and T_L are constants and ω is the frequency of oscillation of the temperature at $x = L$. Solve this problem using the standard GFSE and AGFSE.

Solution

The standard form of the GFSE is used first. In this solution (and the alternative form) it is convenient to solve for $\theta = [T(x,t) - T_0]$ with conditions

$$\theta(0,t) = 0$$
$$\theta(L,t) = (T_L - T_0)\sin\omega t$$
$$\theta(x,0) = 0$$

By solving this problem rather than the $T(x,t)$ problem, the nonhomogeneous boundary condition and nonzero initial conditions are replaced by the easier zero conditions. For this problem, the solution using Equation 3.46 for $\theta(x,t)$ has a nonzero term only for the boundary condition at $x = L$,

$$\theta(x,t) = -\alpha\int_0^t \frac{\partial G_{X11}(x,t|L,\tau)}{\partial n'}f(\tau)\,d\tau$$

$$= \alpha\int_0^t \frac{2\pi}{L^2}\sum_{m=1}^{\infty} e^{-m^2\pi^2\alpha(t-\tau)/L^2}m(-1)^{m+1}\sin\left(m\pi\frac{x}{L}\right)(T_L - T_0)\sin\omega\tau\,d\tau$$

$$= (T_L - T_0)\frac{2\pi\alpha}{\omega L^2}\sum_{m=1}^{\infty}\frac{m(-1)^{m+1}\sin(m\pi x/L)}{D_m^2 + 1}$$

$$\times\,(e^{-m^2\pi^2\alpha t/L^2} + D_m\sin\omega t - \cos\omega t) \tag{3.67a}$$

where

$$D_m = \frac{m^2\pi^2\alpha}{\omega L^2} \tag{3.67b}$$

Here the derivative of the GF, $\partial G_{X11}/\partial n'$, has been taken from Appendix X, Equation X11.12. The integral on τ is given by

$$\int_0^t e^{-m^2\pi^2\alpha(t-\tau)/L^2} \sin\omega\tau \, d\tau = \frac{1}{\omega(D_m^2+1)}\left(e^{-m^2\pi^2\alpha t/L^2} + D_m \sin\omega t - \cos\omega t\right) \tag{3.68}$$

The expression given by Equation 3.67a contains two parts, a steady-periodic part and a transient part. The steady-periodic part persists in time and is periodic. The expression is not a rapidly convergent one, however. Notice that there is a term in the numerator proportional to m^3 and in the denominator to m^4; this results in terms that are proportional to m^{-1}. Series with terms that are proportional to m^{-1} typically converge very slowly, if at all. An indication of difficulty is observed for the location of $x = L$, because $\sin m\pi = 0$ but this value gives $\theta(L, t) = 0$ which is not equal to the given boundary condition. This seeming contradiction is related to the convergence problem.

Consider now the use of AGFSE. The $T^*(x, t)$ solution is obtained by solving Equation 3.59 in the form

$$\frac{\partial^2 T^*}{\partial x^2} = 0 \tag{3.69}$$

and the boundary conditions

$$T^*(0, t) = T_0 \qquad T^*(L, t) = T_0 + (T_L - T_0)\sin\omega t \tag{3.70}$$

The solution for $T^*(x, t)$ is

$$T^*(x, t) = T_0 + (T_L - T_0)\frac{x}{L}\sin\omega t \tag{3.71}$$

Now Equation 3.66 is used. The first integral has no contribution because

$$F(x') - T^*(x', 0) = T_0 - (T_0 + 0) = 0$$

The second integral is not present because the boundary conditions are not the fourth or fifth kinds. Then, Equation 3.66 gives

$$\begin{aligned} T(x,t) &= \left[T_0 + (T_L - T_0)\frac{x}{L}\sin\omega t\right] \\ &\quad - \frac{\alpha}{k}\int_0^t \int_{x'=0}^{L} G_{X11}(x, t|x', \tau)\rho c \frac{\partial T^*(x', \tau)}{\partial \tau} d\tau \, dx' \\ &= T_0 + (T_L - T_0)\frac{x}{L}\sin\omega t \\ &\quad - \frac{2}{L}\int_{\tau=0}^{t}\int_{x'=0}^{L}\sum_{m=1}^{\infty} e^{-m^2\pi^2\alpha(t-\tau)/L^2}\sin\left(m\pi\frac{x}{L}\right) \\ &\quad \times \sin\left(m\pi\frac{x'}{L}\right)(T_L - T_0)\frac{x'}{L}\omega\cos\omega\tau \, dx' \, d\tau \end{aligned}$$

$$
\begin{aligned}
&= T_0 + (T_L - T_0)\frac{x}{L}\sin\omega t \\
&+ \frac{2}{\pi}(T_L - T_0)\sum_{m=1}^{\infty}\frac{\sin(m\pi x/L)(-1)^m}{m(D_m^2+1)} \\
&\times\left(D_m\cos\omega t + \sin\omega t - D_m e^{-\frac{m^2\pi^2\alpha t}{L^2}}\right)
\end{aligned}
\tag{3.72}
$$

where $D_m = m^2\pi^2\alpha/(\omega L^2)$. In contrast with Equation 3.67, which has terms proportional to m^{-1}, Equation 3.72 has terms proportional to m^{-3} for large m. Equation 3.72 converges rapidly and has no convergence problems; it also gives the correct result at $x = L$. The issue of convergence speed is discussed further in Chapter 5.

In general, the alternative GFSE is preferred over the standard form for nonhomogeneous boundary conditions when the large time form of the GF is used. This is particularly true for boundary conditions of the first kind and when results near the boundaries are needed. Notice, for this example, however, that two integrations were required for the alternative form but only one for the standard form. The large-time GFs have the time- and space-dependent components in separate terms, such as

$$
\exp\left(-\frac{\beta_m^2\alpha(t-\tau)}{L^2}\right) \quad \text{and} \quad \sin\frac{\beta_m x}{L}
$$

while the short-time GFs have the t and x together, such as

$$
\exp\left[-\frac{(2mL + x - x')^2}{4\alpha(t-\tau)}\right]
$$

When the short-time GFs for nonhomogeneous boundary conditions are used, the standard form may be better than the alternative form of the GFSE because fewer integrations are needed.

Another way to approach this solution would be to treat the steady-periodic portion directly with the steady-periodic techniques discussed in Chapter 9. For many steady-periodic problems in one spatial dimension, the solution has a nonseries form, completely avoiding the issue of series convergence.

3.5 FIN TERM m^2T

The fin approximation may be applied in geometries with one dimension that is thin and if the temperature distribution in the thin-axis direction is approximately uniform (lumped). In this case, the energy equation may be simplified by replacing the diffusion term corresponding to the thin-axis direction by the term m^2T, called the fin term. In general, the fin parameter m can be a function of position $\mathbf{r}$, but not a function of time. The fin term can also be used to represent volume heat generation that is proportional to temperature, such as electric heating or dilute chemical reactions.

The GF method applies to fin problems even though the GFSE for the transient temperature, Equation 3.46, does not explicitly involve the m^2 term. In the GFSE there are terms for the boundary conditions, the energy generation, and the initial condition, but there is no term for fins. The dependence of the solution on m^2 is hidden in the GF so that a different GF must be found when the fin term is present in the differential equation. The dependence of the GF on the fin term m^2 may be seen explicitly in the auxiliary equation, Equation 3.32.

In this section, transient and steady GFs are discussed for the special case of a *spatially constant* fin term. In the case when m^2 is not spatially constant, the GF may be quite complicated if it can be found at all; in this event, the Galerkin-based GF method discussed in Chapter 10 is recommended.

3.5.1 Transient Fin Problems

All the transient GFs listed in this book are for the $m^2 = 0$ case. However, these same transient GFs can also be used for the case of a *spatially constant* m^2 when the following transformation is applied to the temperature.

Let $W(\mathbf{r},t)$ be a new dependent variable, related to $T(\mathbf{r},t)$ by

$$T(\mathbf{r},t) = W(\mathbf{r},t)\exp(-m^2\alpha t) \tag{3.73}$$

where m^2 is constant. Substitute this relation into the heat conduction equation, Equation 3.28, and multiply the equation by $e^{+m^2\alpha t}$. The result is

$$\nabla^2 W + \frac{1}{k} g(\mathbf{r},t) e^{m^2\alpha t} = \frac{1}{\alpha}\frac{\partial W}{\partial t} \tag{3.74}$$

The m^2 term has canceled out so the transformed variable $W(\mathbf{r},t)$ may be found using GFs that do not involve the $m^2 T$ term. Then the transformation can be inverted to find the original temperature $T(\mathbf{r},t)$. The transformation does not work on steady-state problems at all because the time derivative is involved in canceling the $m^2 T$ term. For steady-state problems with the fin term, a separate set of GFs must be used; see Section 3.5.2.

Transient problems that involve the fin term can be quite complex, and although the transformation allows a familiar set of transient GFs to be applied to those problems, the complexity of the solution has not been removed but has been shifted to the energy generation term and the boundary conditions. The boundary conditions for the transformed variable W involve the term $e^{+m^2\alpha t}$.

The initial condition and boundary conditions for $W(\mathbf{r},t)$ can be found by carefully applying the transformation. The initial condition for W is given by

$$W(\mathbf{r},0) = F(\mathbf{r})e^0 = F(\mathbf{r}) \tag{3.75}$$

which is unchanged. The boundary conditions will be examined according to kind. The boundary condition of the first kind is

$$T(\mathbf{r}_i,t) = f_i(\mathbf{r}_i,t) \tag{3.76}$$

Using the relationship that defines the new variable W, the boundary condition of the first kind becomes

$$W(\mathbf{r}_i,t) = f_i(\mathbf{r}_i,t)\exp(m^2\alpha t) \tag{3.77}$$

The boundary condition of the second kind becomes

$$k\left.\frac{\partial W}{\partial n_i}\right|_{\mathbf{r}_i} = f_i(\mathbf{r}_i,t)\exp(m^2\alpha t) \tag{3.78}$$

The boundary condition of the third kind becomes

$$k\left.\frac{\partial W}{\partial n_i}\right|_{\mathbf{r}_i} = h_i[f_i(\mathbf{r}_i,t)\exp(m^2\alpha t) - W(\mathbf{r}_i,t)] \tag{3.79}$$

The boundary conditions of the fourth and fifth kinds are more affected. Using Equation 3.73 in Equation 3.30 gives

$$k_i\left.\frac{\partial W}{\partial n_i}\right|_{\mathbf{r}_i} + \left.[h_i - (\rho cb)_i m^2\alpha]W(\mathbf{r},t)\right|_{\mathbf{r}_i} = f_i(\mathbf{r}_i,t)\exp(m^2\alpha t) - (\rho cb)_i\left.\frac{\partial W}{\partial t}\right|_{\mathbf{r}_i} \tag{3.80}$$

Notice the extra coefficient $(\rho cb)_i m^2\alpha$ that appears with h_i in this equation.

In summary, for the case of the transient heat conduction equation with the m^2T term for m^2 *constant*, the GFs for the transformed variable $W(\mathbf{r},t)$ are exactly the same as for the $m^2 = 0$ case, but the energy generation term is now multiplied by $e^{+m^2\alpha t}$, and the boundary conditions are different. For boundary conditions of the fourth and fifth kinds, h_i is replaced by $h_i - (\rho cb)_i m^2\alpha$ at the ith boundary. In addition, for each of the five types of boundary conditions, $f_i(\mathbf{r}_i,t)$ in Equation 3.46d is replaced by $f_i(\mathbf{r}_i,t)\exp(m^2\alpha t)$. After the GF solution for $W(\mathbf{r},t)$ is obtained, $T(\mathbf{r},t)$ is simply obtained by multiplying $W(\mathbf{r},t)$ by $\exp(-m^2\alpha t)$ as given in Equation 3.73.

Example 3.6: *X*11 Case with Fin Term

A thin fin of uniform cross-section is initially at temperature T_∞ and the $x = 0$ end of the fin is suddenly set to temperature T_0. Derive the one-dimensional fin equation and find the GF solution for the temperature in the fin if the heat transfer coefficient for side heat losses is constant and the $x = L$ end of the fin is maintained at T_∞.

Solution

(a) *Differential equation.* The fin geometry is shown in Figure 3.7. The fin has thickness $\delta \ll L$ so that the temperature varies only in the x-direction. The differential equation for the fin may be found by considering the control volume of length dx at location x. The energy balance for the control volume given by the integral energy equation (Equation 1.32) could be used.

$$w\delta[q(x) - q(x+dx)] - (2w\,hdx)(T - T_\infty) = \rho c(w\delta\,dx)\frac{\partial T}{\partial t} \tag{3.81}$$

where w is the width of the fin, $q(\cdot)$ is heat flux (W/m^2), and h is the constant heat transfer coefficient. Divide the energy balance by the volume of the control volume ($w\delta\,dx$) to get

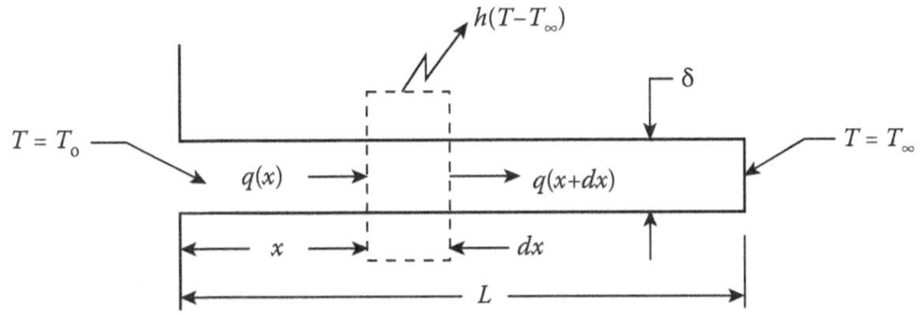

FIGURE 3.7 One-dimensional fin with constant cross section for Examples 3.6 and 3.7.

$$-\frac{q(x+dx)-q(x)}{dx} - \frac{2h}{\delta}(T - T_\infty) = \rho c \frac{\partial T}{\partial t} \tag{3.82}$$

The heat flux terms may be replaced by a derivative in the limit as $dx \to 0$. Replace the heat flux terms by Fourier's law $q(x) = -k\,\partial T/dx$ to give

$$k\frac{\partial^2 T}{\partial x^2} - \frac{2h}{\delta}(T - T_\infty) = \rho c \frac{\partial T}{\partial t} \tag{3.83}$$

Finally, divide by k and introduce a new variable $\Theta(x,t) = (T - T_\infty)$ to make the equation homogeneous:

$$\frac{\partial^2 \Theta}{\partial x^2} - m^2\Theta = \frac{1}{\alpha}\frac{\partial \Theta}{\partial t} \tag{3.84}$$

where now $m^2 = 2h/(\delta k)$ with units (meters)$^{-2}$. This is the differential equation for a fin of uniform cross-section. The initial and boundary conditions are

$$\Theta(x,0) = 0$$
$$\Theta(0,t) = T_0 - T_\infty$$
$$\Theta(L,t) = 0$$

(b) *Green's function solution.* The boundary value problem for $\Theta(x,t)$ may be transformed according to Equation 3.73 for the GF solution. The transformed boundary value problem for $W(x,t)$ is given by

$$\frac{\partial^2 W}{dx^2} = \frac{1}{\alpha}\frac{\partial W}{\partial t}$$

$$W(x,0) = 0$$
$$W(0,t) = (T_0 - T_\infty)e^{m^2\alpha t} \tag{3.85}$$
$$W(L,t) = 0$$

The transient temperature is driven by the boundary condition at $x = 0$ and the solution is given by the GF method as

$$W(x,t) = \alpha \int_{\tau=0}^{t} (T_0 - T_\infty)e^{m^2\alpha\tau} \left.\frac{\partial G_{X11}}{\partial x'}\right|_{x'=0} d\tau \tag{3.86}$$

Note that the boundary condition is introduced into the integral as a function of dummy variable τ. The function G_{X11} and its derivative is given in Appendix X so the solution is

$$W(x,t) = \alpha \int_{\tau=0}^{t} (T_0 - T_\infty) e^{m^2\alpha\tau} \frac{2\pi}{L^2} \times \sum_{n=1}^{\infty} e^{-n^2\pi^2\alpha(t-\tau)/L^2} n \sin\frac{n\pi x}{L} d\tau \tag{3.87}$$

Be careful to distinguish fin parameter m from the summation index n, and to distinguish integration variable τ from time t. The integral on τ may be carried out to give the transformed solution:

$$W(x,t) = (T_0 - T_\infty) 2\pi \sum_{n=1}^{\infty} (e^{m^2\alpha t} - e^{-n^2\pi^2\alpha t/L^2}) n \times \sin\left(\frac{n\pi x}{L}\right)(m^2L^2 + n^2\pi^2)^{-1}$$

Finally the temperature in the fin may be found by the inverse transform $\Theta = W\exp(-m^2\alpha t)$, or,

$$\Theta(x,t) = T(x,t) - T_\infty = (T_0 - T_\infty)2\pi \times \sum_{n=1}^{\infty}\left(1 - e^{-m^2\alpha t} e^{-n^2\pi^2\alpha t/L^2}\right) n \times \sin\left(\frac{n\pi x}{L}\right)(m^2L^2 + n^2\pi^2)^{-1} \tag{3.88}$$

In the limit as $t \to \infty$, the series converges to the steady-state solution, but the series converges slowly (like $1/n$). A better form of the steady solution can be found by using a steady GF directly as shown in Section 3.6.

3.5.2 Steady Fin Problems in One Dimension

The W transformation discussed in Section 3.5.1 does not apply to steady fin problems because the W transformation relies on the time derivative $\partial T/\partial t$ to cancel the fin term from the differential equation. Many steady fin solutions exist in the literature and methods other than GFs may be appropriate.

Steady fin problems may be solved with the steady GF method if the steady-fin GF can be found. An example of a steady fin problem is given in the next section. A list of steady-fin GFs in rectangular coordinates is given in Appendix X, Tables X.2 and X.4, for the special case m^2 = constant. Steady-fin GF for *radial-cylindrical coordinates* are given in Chapter 9, Equation 9.21; these were developed for steady-periodic conditions, but apply to annular fins of uniform thickness.

3.6 STEADY HEAT CONDUCTION

In this section, steady GFs are presented through their relationship with the transient GFs. The steady-state GFSE is stated in a general form.

TABLE 3.2
Units of Steady and Transient GFs in Cartesian Coordinates

Geometry	Units of Transient GF	Units of Steady GF
One dimension	m^{-1}	m
Two dimensions	m^{-2}	1 (dimensionless)
Three dimensions	m^{-3}	m^{-1}

3.6.1 Relationship between Steady and Transient Green's Functions

The steady GF is the limit as $t \to \infty$ of the time integral of the transient GF:

$$G(\mathbf{r}|\mathbf{r}') = \lim_{t\to\infty} \int_{\tau=0}^{t} \alpha\, G(\mathbf{r}, t|\mathbf{r}', \tau)\, d\tau \tag{3.89}$$

This relationship may be regarded as the definition of the steady GF and it is one way to find the steady GF if the transient GF is known. For two- and three-dimensional geometries, this relationship is useful; refer to Section 4.7.3 on the limit method. For one-dimensional geometries, it is better to find the steady GF directly from the auxiliary equation for G as discussed earlier in Section 1.7.2.

The limit in Equation 3.89 does not exist for all geometries. Specifically, for geometries with all boundaries insulated the usual GF does not exist. However, in these cases a pseudo-GF can be used instead, as discussed later in Section 4.7.2.

In Equation 3.89, the transient GF is multiplied by the term $\alpha d\tau$ with units (m^2), so the steady GF has different units than the transient GFs which depend on the dimensionality of the geometry under discussion. The relationship between units of steady and transient GF in Cartesian coordinates are given in Table 3.2.

3.6.2 Steady Green's Function Solution Equation

In this section, the steady GFSE is stated in a general form. The steady GFSE may be derived as the limit of the transient GFSE as $t \to \infty$ because the steady temperature is simply the transient temperature in the limit as $t \to \infty$. The steady GFSE may also be derived directly from the boundary value problem for the temperature and from the auxiliary equation for the GF in a manner parallel to that for the transient GFSE presented in Section 3.3; this derivation is given as Problem 3.18 at the end of the chapter.

The partial differential equation that describes steady, multidimensional, linear heat conduction is

$$\nabla^2 T + \frac{1}{k} g(\mathbf{r}) - m^2 T = 0 \qquad \text{in region } R \tag{3.90}$$

where ∇^2 is the Laplacian operator in the appropriate coordinate system. The thermal conductivity k is constant with position and temperature. The $m^2 T$ term could represent side heat losses for a fin; in general m^2 can be a function of $\mathbf{r}$. (The m^2T term is not needed for the three-dimensional treatment of fins.)

The steady boundary conditions for Equation 3.90 have the general form

$$k_i \frac{\partial T}{\partial n_i} + h_i T = f_i(\mathbf{r}_i) \tag{3.91}$$

where the temperature T and its derivatives are evaluated at the boundary surface S_i, and $\mathbf{r}_i$ denotes the location of the boundary. The spatial derivative $\partial / \partial n_i$ denotes differentiation along an *outward* drawn normal to the boundary surface S_i, $i = 1, 2, \ldots, S$. The heat transfer coefficient h_i can vary with position on S_i but is independent of temperature. Three different boundary conditions can be obtained from Equation 3.91 by setting $k_i = 0$ or k, and by setting $h_i = 0$ or h. Boundary conditions of type 4 or 5 involve energy storage $\partial T / \partial t$ and therefore do not appear in steady problems.

The steady GF satisfies the auxiliary equation

$$\nabla^2 G + \delta(\mathbf{r} - \mathbf{r}') - m^2 G = 0 \tag{3.92}$$

$$k_i \frac{\partial G}{\partial n_i'} + h_i G = 0 \tag{3.93}$$

If the GF is known for a geometry, the steady temperature may be found from the steady-state GFSE:

$$\begin{aligned} T(\mathbf{r}) = & \int_R \frac{1}{k} G(\mathbf{r}|\mathbf{r}') g(\mathbf{r}') dv' && \text{(for internal energy generation)} \\ & + \sum_{i=1}^{s} \int_{S_i} \frac{f_i(\mathbf{r}_i')}{k_i} G(\mathbf{r}|\mathbf{r}_i') ds_i' && \text{(for boundary conditions of the second and third kind)} \\ & - \sum_{j=1}^{s} \int_{S_j} f_j(\mathbf{r}_j') \left. \frac{\partial G}{\partial n_j'} \right|_{\mathbf{r}'=\mathbf{r}_j'} ds_j' && \text{(for boundary condition of the first kind only)} \end{aligned} \tag{3.94}$$

Next, a steady example is given that includes the fin term. Other examples of steady heat transfer are given in Sections 6.9, 7.13, and 8.8.

Example 3.7:

Steady fin of constant cross-section with specified temperatures on the ends. Find the steady temperature in a fin with equation and boundary conditions given by

$$\begin{aligned} \frac{d^2 T}{dx^2} - m^2 (T - T_\infty) = 0 \qquad 0 < x < L \\ T(0) = T_0 \\ T(L) = T_\infty \end{aligned} \tag{3.95}$$

where T_0 and T_∞ are constant temperatures, and $m^2 = 2h/(k\delta)$, as shown in Figure 3.7. This is a fin of constant cross-section and the number of this case is $X11$.

Solution

Define variable $\Theta(x) = T(x) - T_\infty$ to simplify the temperature relations:

$$\frac{d^2\Theta}{dx^2} - m^2\Theta = 0 \qquad \Theta(0) = T_0 - T_\infty \qquad \Theta(L) = 0 \tag{3.96}$$

The GF for the X11 case is given in Appendix X, Table X.3 as

$$G(x|x') = \frac{e^{-m(2L-|x-x'|)} - e^{-m(2L-x-x')} + e^{-m|x-x'|} - e^{-m(x+x')}}{2m\left(1 - e^{-2mL}\right)} \tag{3.97}$$

The boundary-condition term of the steady GFSE, Equation 3.94, gives

$$T(x) - T_\infty = -(T_0 - T_\infty)\left.\frac{dG}{dn'}\right|_{x'=0} = (T_0 - T_\infty)\left.\frac{dG}{dx'}\right|_{x'=0} \tag{3.98}$$

Because the boundary term is evaluated at $x' = 0$, the $x > x'$ form of the above GF must be used to give

$$\frac{T(x) - T_\infty}{T_0 - T_\infty} = \frac{e^{-mx} - e^{-m(2L-x)}}{(1 - e^{-2mL})} \tag{3.99}$$

The usual fin solutions, as given in heat transfer texts, are found by direct solution of Equation 3.96 with independent solutions cosh and sinh (for example Nellis and Klein, 2009). To show that the above temperature expression may be restated with hyperbolic trig functions, rearrange as follows:

$$\frac{T(x) - T_\infty}{T_0 - T_\infty} = \frac{e^{-mx} - e^{-m(2L-x)}}{(1 - e^{-2mL})}\,\frac{e^{mL}/2}{e^{mL}/2} = \frac{\sinh m(L-x)}{\sinh mL}$$

However it is expressed, the shape of the temperature distribution is a decreasing exponential.

3.7 MOVING SOLIDS

3.7.1 Introduction

Moving solid problems occur in many cases in heat conduction. These problems can be the result of a solid moving past a heating condition, such an extruded wire moving out of a die and being cooled by convection and radiation. Another case is a physically fixed solid with a moving heat source, such as a moving laser source on the surface of a plate. A third case can result from a moving surface, such as the ablating surface of a reentry heat shield. In each of these cases, it frequently is convenient to formulate the problem so that the coordinate system is attached to the heat source which causes a velocity term to appear in the partial differential equation of heat conduction. The equations usually must be derived using a control volume approach as discussed in Chapter 1.

These problems can be one-, two-, or three-dimensional. An example of a one-dimensional problem is a moving circular die that is convectively cooled and lumped in the radial direction. That is, the temperature is only a function of the axial coordinate (not radial also) and possibly time. The describing equation can be given as

$$k\frac{\partial^2 T}{\partial x^2} - \frac{2h}{a}(T - T_\infty) = \rho c\left(\frac{\partial T}{\partial t} + U\frac{\partial T}{\partial x}\right) \tag{3.100}$$

where the thermal conductivity is assumed to be independent of temperature, the coordinate system is fixed at the die with the wire moving at a velocity of U in the positive x-direction, and a is the wire radius. It is possible to have a steady state (actually, called a quasisteady state) in this problem with respect to the die. In that case, the time derivative disappears in Equation 3.100. The expression "quasisteady state" is used because the temperature at any location fixed in the body varies with time, even though the temperature at a location fixed with respect to the die does not depend on time.

Another problem is for a small laser beam heating the surface of a plate. One way to visualize the problem is for the beam to be stationary and the plate to be moving in the x-, y-, and z-directions with velocities of U_1, U_2, and U_3, respectively. Another way is to visualize that the beam is moving in the $-U_1$, $-U_2$, and $-U_3$ directions, in other words, just opposite to the previous way. In both cases, the coordinate system is fixed on the beam. The describing equation can be given as

$$k\left(\frac{\partial^2 T}{\partial x^2} + \frac{\partial^2 T}{\partial y^2} + \frac{\partial^2 T}{\partial z^2}\right) = \rho c\left(\frac{\partial T}{\partial t} + U_1\frac{\partial T}{\partial x} + U_2\frac{\partial T}{\partial y} + U_3\frac{\partial T}{\partial z}\right) \tag{3.101}$$

The velocities U_1, U_2, and U_3 are assumed to be known. Again, a quasisteady state exists for the coordinates fixed on the beam and the velocities being steady, although the temperature varies with time for a fixed point in the plate. To simplify the problem, assume that the beam is moving in the negative x-direction while the plate is fixed (or equivalently, the beam is fixed and the plate is moving in the positive direction), then the equation becomes

$$k\left(\frac{\partial^2 T}{\partial x^2} + \frac{\partial^2 T}{\partial y^2} + \frac{\partial^2 T}{\partial z^2}\right) = \rho c\left(\frac{\partial T}{\partial t} + U_1\frac{\partial T}{\partial x}\right) \tag{3.102}$$

A further simplification occurs when the velocity U_1 is sufficiently large that the U_1 term in Equation 3.102 is much larger than the second derivative with respect to the x term, resulting in the second derivative in the x term being negligible. If, further, there is a quasisteady state, then Equation 3.102 simplifies to

$$k\left(\frac{\partial^2 T}{\partial y^2} + \frac{\partial^2 T}{\partial z^2}\right) = \rho c U_1\frac{\partial T}{\partial x} \tag{3.103}$$

This equation is interesting because it is the same parabolic type as the heat conduction equation, but now the time is replaced by x/U_1. This is an important point, but it is not the main thrust of this section.

3.7.2 Three-Dimensional Formulation

The emphasis in this section is to develop a method to treat moving solid problems in a manner that the same GF and GFSE can be used, with appropriate modifications. We consider the describing equation

$$k\left(\frac{\partial^2 T}{\partial x^2}+\frac{\partial^2 T}{\partial y^2}+\frac{\partial^2 T}{\partial z^2}\right)=\rho c\left(\frac{\partial T}{\partial t}+V\frac{\partial T}{\partial x}\right) \tag{3.104}$$

where V is the velocity in the positive direction of the solid through a fixed control volume. (A more general equation is considered in the problems at the end of this chapter.) The boundary conditions can be of the first kind such as

$$T(0,y,z,t)=T_{x1}(y,z,t) \tag{3.105}$$

$$T(L,y,z,t)=T_{x2}(y,z,t) \tag{3.106}$$

and the second and third kinds,

$$-k\left.\frac{\partial T}{\partial x}\right|_{x=0}=h_{x1}[T_{x\infty1}(y,z,t)-T(0,y,z,t)]+q_{x1}(y,z,t) \tag{3.107a}$$

$$-k\left.\frac{\partial T}{\partial x}\right|_{x=L}=h_{x2}[T(L,y,z,t)-T_{x\infty2}(y,z,t)]-q_{x2}(y,z,t) \tag{3.107b}$$

or equivalently,

$$-k\left.\frac{\partial T}{\partial x}\right|_{x=0}+h_{x1}T(0,y,z,t)=h_{x1}T_{x\infty1}(y,z,t)+q_{x1}(y,z,t)$$

$$=f_{x1}(y,z,t) \tag{3.108a}$$

$$k\left.\frac{\partial T}{\partial x}\right|_{x=L}+h_{x2}T(L,y,z,t)=h_{x2}T_{x\infty2}(y,z,t)+q_{x2}(y,z,t)$$

$$=f_{x2}(y,z,t) \tag{3.108b}$$

Notice the definition of f_{x1} and f_{x2} implied by these equations.

The initial condition is

$$T(x,y,z,0)=F(x,y,z) \tag{3.109}$$

These equations and boundary conditions are transformed using

$$T(x,y,z,t)=W(x,y,z,t)\exp\left(\frac{Vx}{2\alpha}-\frac{V^2}{4\alpha}t\right) \tag{3.110}$$

where $W(x,y,z,t)$ is the velocity transformation and is described by

$$k\left(\frac{\partial^2 W}{\partial x^2}+\frac{\partial^2 W}{\partial y^2}+\frac{\partial^2 W}{\partial z^2}\right)=\rho c\frac{\partial W}{\partial t} \tag{3.111}$$

with boundary conditions of the first kind

$$W(0,y,z,t) = T_{x1}(y,z,t)\, e^{V^2 t/(4\alpha)} \tag{3.112}$$

$$W(L,y,z,t) = T_{x2}(y,z,t)\, e^{-VL/(2\alpha)+(V^2 t)/(4\alpha)} \tag{3.113}$$

or boundary conditions of the second or third kinds,

$$-k\left.\frac{\partial W}{\partial x}\right|_{x=0} + h_{xe1} W\big|_{x=0} = f_{x1}(y,z,t) e^{V^2 t/(4\alpha)} \tag{3.114}$$

$$h_{xe1} = h_{x1} - \frac{kV}{2\alpha} \tag{3.115}$$

$$k\left.\frac{\partial W}{\partial x}\right|_{x=L} + h_{xe2} W\big|_{x=L} = f_{x2}(y,z,t) e^{-VL/(2\alpha)+V^2 t(4\alpha)} \tag{3.116}$$

$$h_{xe2} = h_{x2} + \frac{kV}{2\alpha} \tag{3.117}$$

Notice the effective heat transfer coefficient definitions in Equations 3.115 and 3.117. Also notice that the boundary condition of the second kind turns into one of the third kind; this means that the G_{X2-} and G_{X-2} GFs are transformed to the G_{X3-} and G_{X-3} GFs with the effective h values being $-kV/2\alpha$ and $kV/2\alpha$, respectively.

The initial condition for W is obtained from Equations 3.109 and 3.110, given by

$$W(x,y,z,0) = F(x,y,z)\exp\left(-\frac{Vx}{2\alpha}\right) \tag{3.118}$$

This concludes the formulation of the W problem. It now remains to obtain the solution to the W problem and then to use Equation 3.110 to get the T solution.

The GFSE can be written as

$$T(x,y,z,t) = \exp\left(\frac{Vx}{2\alpha} - \frac{V^2 t}{4\alpha}\right)[W_{\text{in}}(x,y,z,t) + W_{\text{bc1}}(x,y,z,t) + W_{\text{bc2,3}}(x,y,z,t)] \tag{3.119}$$

where $W_{\text{in}}(\cdot)$ is for the initial condition, $W_{bc1}(\cdot)$ is for boundary conditions only of the first kind, and $W_{bc2,3}(\cdot)$ is for boundary conditions of the second and third kinds. It is important to note that there can only be one boundary condition at a given boundary, but it can be of the first or second or third kinds. The second and third kinds are treated in a similar manner. The boundary condition of the zeroth kind (no physical boundary) does not have an explicit term in Equation 3.119.

Each of the W terms in Equation 3.119 is now considered separately. The expression for $W_{\text{in}}(\cdot)$ is

$$W_{\text{in}}(x,y,z,t) = \int_{x'=0}^{L}\int_{y'}\int_{z'} G_{X--}(x,t|x',0)G_{Y--}(y,t|y',0) \times G_{Z--}(z,t|z',0)\, e^{-Vx'/(2\alpha)} F(x',y',z')\, dx'\, dy'\, dz' \tag{3.120}$$

The dashes in X-- can be 1, 2, or 3; the second dash could also be 0, but then the upper limit L must be changed to infinity. The first dash in Y-- and Z-- can be 1, 2, or 3, while the second dash can be 0, 1, 2, or 3. If the problem is two-dimensional with x- and y-coordinates, then the dependence on z and z' disappears. If the problem is one-dimensional with x being the only coordinate, Equation 3.120 becomes

$$W_{\text{in}}(x,t) = \int_{x'=0'}^{L} G_{X\text{--}}(x,t|x',0)\, e^{-Vx'/(2\alpha)} F(x')\, dx' \tag{3.121}$$

Consider next the boundary conditions of the first kind. There could be all boundary conditions of this kind in a given problem or none might be present. Also the problem might be only one- or two-dimensional. For reasons of brevity and clarity, only the $x = 0$ and $x = L$ boundaries are explicitly considered to be of the first kind for the three-dimensional case, resulting in the $W_{\text{bc1}}(x,y,z,t)$ expression of

$$\begin{aligned} W_{\text{bc1}}(x,y,z,t) = {} & \alpha \int_{\tau=0}^{t} \int_{y'} \int_{z'} \frac{\partial G_{X1\text{-}}(x,t|0,\tau)}{\partial x'} G_Y(y,t|y',\tau)\, G_Z(z,t|z',\tau) \\ & \times T_{x1}(y',z',\tau) e^{V^2\tau/(4\alpha)} d\tau\, dy'\, dz' \\ & - \alpha \int_{\tau=0}^{t} \int_{y'} \int_{z'} \frac{\partial G_{X-1}(x,t|L,\tau)}{\partial x'} G_Y(y,t|y',\tau)\, G_Z(z,t|z',\tau) \\ & \times T_{x2}(y',z',\tau) e^{V^2\tau/(4\alpha)} e^{-VL/(2\alpha)}\, d\tau\, dy'\, dz' \end{aligned} \tag{3.122}$$

where the Y and Z notation subscripts have omitted the -- symbols. Recall that boundary conditions of the second kind have been transformed to those of the third kind. If there are boundary conditions of the first kind at the y boundaries as well as at the x boundaries, then in addition to the two terms in Equation 3.122, two more terms are added with the integration now on x', z' and τ, the x' derivative replaced with one with respect to y', and the appropriate boundary temperature used. For a one-dimensional problem in the x-direction, Equation 3.122 reduces to

$$\begin{aligned} W_{\text{bc1}}(x,t) = {} & \alpha \int_{\tau=0}^{t} \frac{\partial G_{X1-}(x,t|0,\tau)}{\partial x'} T_{x1}(\tau) e^{V^2\tau/(4\alpha)}\, d\tau \\ & - \alpha \int_{\tau=0}^{t} \frac{\partial G_{X-1}(x,t|L,\tau)}{\partial x'} T_{x2}(\tau)\, e^{-VL/(2\alpha)+V^2\tau/(4\alpha)}\, d\tau \end{aligned} \tag{3.123}$$

Consider next boundary conditions of the second and/or third kinds. Again for brevity, only the x-direction boundary conditions are treated. The result for $W_{bc2,3}(x,y,z,t)$ is

$$\begin{aligned} W_{\text{bc2,3}}(x,y,z,t) = {} & \frac{\alpha}{k} \int_{\tau=0}^{t} \int_{y'} \int_{z'} G_{X3\text{-}}(x,t|0,\tau) G_Y(y,t|y',\tau) \\ & \times G_Z(z,t|z',\tau)\, f_{x1}(y',z',\tau) e^{V^2\tau/(4\alpha)}\, d\tau\, dy'\, dz' \\ & + \frac{\alpha}{k} \int_{\tau=0}^{t} \int_{y'} \int_{z'} G_{X-3}(x,t|L,\tau)\, G_Y(y,t|y',\tau) \\ & \times G_Z(z,t|z',\tau)\, f_{x2}(y',z',\tau) e^{-VL/(2\alpha)+V^2\tau/(4\alpha)}\, d\tau\, dy'\, dz' \end{aligned} \tag{3.124}$$

The GFs used above, $[G_{X\text{--}}(x,t|x',\tau), G_{Y\text{--}}(y,t|y',\tau), G_{Z\text{--}}(z,t|z',\tau)]$, are tabulated in the appendices and can be used, along with the eigenconditions. There are some changes, however. The boundary condition of the second kind is transformed to the third kind, while the first and third kinds remain the same. For both the second and third kinds, however, the h_1 and h_2 values are replaced by other values. At $x = 0$, for boundary conditions of the second kind (having $G_{X2\text{-}}$), G becomes $G_{X3\text{-}}$, and the h_1 values become

$$h_1 \rightarrow h_1 - \frac{kV}{2\alpha} = -\frac{kV}{2\alpha} \tag{3.125}$$

where h_1 on the right is zero for boundary conditions of the second kind. For $x = L$ with G_{X-2}, G goes to G_{X-3} and h_2 goes to

$$h_2 \rightarrow h_2 + \frac{kV}{2\alpha} = \frac{kV}{2\alpha} \tag{3.126}$$

Hence, at $x = 0$ for $V > 0$, the effective h is decreased while it is increased at $x = L$. If the velocity is in the negative direction, these relations are changed.

Example 3.8:

A large body is initially at the temperature T_i, and then its surface at $x = 0$ is suddenly decreased to zero. The body is porous and a fluid is flowing through so that the describing partial differential equation is

$$k\frac{\partial^2 T}{\partial x^2} = \rho c\left(\frac{\partial T}{\partial t} + V\frac{\partial T}{\partial x}\right) \tag{3.127}$$

The body can be considered to be semi-infinite ($0 < x < \infty$) since it is said to be large. The boundary and initial conditions are

$$T(0, t) = 0 \tag{3.128}$$

$$T(x \rightarrow \infty, t) \rightarrow T_i \tag{3.129}$$

$$T(x, 0) = T_i \tag{3.130}$$

Solution

Only the initial condition gives a contribution so that Equations 3.119 and 3.120 are needed. The number of this case is $XV10B0T1$. The equations become

$$T(x, t) = \exp\left(\frac{Vx}{2\alpha} - \frac{V^2 t}{4\alpha}\right) \times \int_{x'=0}^{\infty} G_{X10}(x, t|x', 0)e^{-Vx'/(2\alpha)} T_i dx' \tag{3.131}$$

The $G_{X10}(x, t|x', 0)$ GF can be found in Appendix X and is equal to

$$G_{X10}(x, t|x', 0) = (4\pi\alpha t)^{-1/2}\left\{\exp\left[-\frac{(x - x')^2}{4\alpha t}\right] - \exp\left[-\frac{(x + x')^2}{4\alpha t}\right]\right\} \tag{3.132}$$

Integrals of the type

$$I_1 = \int_{x'=0}^{\infty} \exp\left[-\frac{(x-x')^2}{4\alpha t}\right] \exp\left(-\frac{Vx'}{2\alpha}\right) dx'$$
$$= (\pi\alpha t)^{1/2} \exp\left(\frac{V^2 t}{4\alpha} - \frac{Vx}{2\alpha}\right)$$
$$\times \operatorname{erfc}\left[\frac{(\alpha t)^{1/2} V}{2\alpha} - \frac{x}{(4\alpha t)^{1/2}}\right] \quad (3.133)$$

and another integral of the same type is needed with x replaced by $-x$. This integral can be evaluated by completing the square or by using Table I.6, integral 1, Appendix I. Then using Equation 3.133 in Equation 3.131 gives

$$T(x,t) = \frac{T_i}{2}\left\{\operatorname{erfc}\left[\frac{(\alpha t)^{1/2} V}{2\alpha} - \frac{x}{(4\alpha t)^{1/2}}\right] - e^{Vx/\alpha} \operatorname{erfc}\left[\frac{(\alpha t)^{1/2} V}{2\alpha} + \frac{x}{(4\alpha t)^{1/2}}\right]\right\} \quad (3.134)$$

For the case of positive V and $t \to \infty$, the steady-state temperature $T(x,\infty)$ goes to zero, while for a negative $V(=-U)$ and $t \to \infty$, $T(x,\infty)$ goes to

$$T(x,\infty) = T_i\left[1 - \exp\left(-\frac{Ux}{\alpha}\right)\right] \quad (3.135)$$

where erfc $(-\infty) = 2$ is used. Equation 3.135 is also valid for steady-state ablation in which a solid is being decomposed at its heated surface by intense heating and is moving at a constant velocity; x would be measured from the ablating surface and T and T_i would be interpreted as the temperature differences from the ablation temperature.

PROBLEMS

Note: Unless otherwise requested, the explicit forms of the GFs are not needed; simply using the notation $G_{X12}(\cdot)$, for example, is sufficient.

3.1 For a vector **A**, Green's theorem is usually stated

$$\iiint \nabla \cdot \mathbf{A}\, dv = \iint \mathbf{A} \cdot \mathbf{n}\, ds$$

where **n** is the outward normal. Use this form of Green's theorem to establish the following identities:

(a) $\iiint \{\Phi\nabla^2\Phi + |\nabla\Phi|^2\} dv = \iint \Phi(\nabla\Phi)\cdot\mathbf{n}\, ds$

(b) $\iiint \{\Psi\nabla^2\Phi - \Phi\nabla^2\Psi\} dv = \iint [\Psi(\nabla\Phi)\cdot\mathbf{n} - \Phi(\nabla\Psi)\cdot\mathbf{n}] ds$

3.2 Demonstrate for $XIJ(I, J = 1, 2, 3,$ and $4)$ that

$$T(x,t) = \int_{x'=0}^{L} G_{XIJ}(x,t|x',0) F(x')\, dx'$$

is the solution to the equation

$$\alpha\frac{\partial^2 T}{\partial x^2} = \frac{\partial T}{\partial t}$$

with the initial condition of $T(x,t) = F(x)$ and appropriate homogeneous boundary conditions. Use Equation 3.4a in your solution.

3.3 A plate has the boundary conditions given by

$$T(0,t) = T_0(t) \qquad \text{and} \qquad T(L,t) = T_i$$

and the initial condition $T(x,0) = T_i$. Give the solution for the temperature in terms of the appropriate GF. Only one integral should be in the solution.

3.4 A semi-infinite region, $0 \le x \le \infty$, is initially at temperature $F(x)$. For times $t > 0$, boundary surface at $x = 0$ is kept at zero temperature and heat is generated within the solid at the rate of $g(x,t)$. Give the expression for the temperature distribution in terms of GFs.

3.5 A semi-infinite region, $0 \le x \le \infty$, is initially at zero temperature. For times $t > 0$, boundary surface at $x = 0$ is heated by a constant heat flux q_0. Heat is generated within the solid at the rate of $g_0 =$ constant from $x = a$ to b. Give the GF solution equation for the temperature distribution.

3.6 Give the GF solution to the problems in Problem 2.10.

3.7 Give the Green's function solution for determining the temperature in a concrete driveway (modeled as a one-dimensional semi-infinite solid) that is exposed to a convective surface heating condition with heat transfer coefficient h_s, plus a net radiative heat input of $q(t)$. The ambient temperature is assumed to be varying with time and is given by $T_\infty(t)$. At time zero, there is a nonuniform initial temperature-distribution given by $F(x)$.

3.8 Give the GF solution to the problem denoted $X23B10Y13B00T$-G- and also give the describing differential equation, boundary, and initial conditions.

3.9 A plane wall is suddenly subjected to a step change in temperature at $x = 0$ to temperature of 100°C and the initial temperature is 50°C. The $x = L$ boundary is exposed to a convection condition with an h of 10 W / m^2 °C and a fluid temperature of $50+50\sin(5t)$ °C. Obtain three different expressions for the temperature distribution in terms of the appropriate G_x (which should not be given explicitly). The three different expressions are found by different treatments of the initial condition and the boundary conditions.

3.10 A cube is initially at the temperature $F(x,y,z)$ and the surfaces are exposed to a fluid at temperature T_∞, which is a constant, and a heat transfer coefficient h. Give an expression using GF for $T(x,y,z,t)$.

3.11 A solid cylinder of radius a in a nuclear reactor is initially at the temperature $F(r)$. It is cooled by a fluid at $T_\infty(t)$ and has a heat transfer coefficient of h. Give a mathematical statement of the problem and also the number using the number system of Chapter 2. Find the solution in terms of GFs.

3.12 Solve Problem 3.11 also with a volumetric heat source due the nuclear reactions of $g(r) = g_0 \exp[-(a-r)/R]$ where R is a constant.

3.13 The alternative GF solution equation involves the quantity T^*, defined by

$$\nabla^2 T^* - m^2 T^* = \frac{1}{k} g^*(\mathbf{r}, t)$$

Give a physical interpretation of T^*, then in one-dimensional rectangular coordinates find a general solution for T^* for the following cases for $m^2 = 0$:

(a) $g^*(\mathbf{r}, t) = g_1$, a constant
(b) $g^* = x$
(c) $g^* = e^{-ax}$

3.14 Using the notation $G(r, \theta, \phi, t|r', \theta', \phi', \tau)$ for the GF, write the GF solution equation for the temperature in an infinite body in spherical polar coordinates. The initial condition is $F(r, \theta, \phi)$ and the volume energy generation is $g(r, \theta, \phi, t)$.

3.15 Using this name $G(r, \phi, z, t|r', \phi', z', \tau)$ for the GF, write the GF solution equation for the temperature in a half cylinder, $0 \le r \le a, 0 \le \phi \le \pi, 0 \le z \le L$. The boundary conditions are homogeneous, the initial condition is $F(r, \phi, z)$ and the volume energy generation is $g(r, \phi, z, t)$.

3.16 Repeat the derivation of Section 3.3 for the same problem but the right-hand side replaced by

$$\frac{1}{\alpha} u(\mathbf{r}) \frac{\partial T}{\partial t}$$

The function $u(\mathbf{r})$ could represent a velocity term for a flow problem if the second derivative in the flow direction were dropped and t were replaced by the coordinate in the flow direction. Show that the GF solution equation is the same as Equation 3.46 except $u(\mathbf{r}')$ is also inside the first integral of Equation 3.46b.

3.17 An orthotropic plate is a model for aligned-fiber composite materials. For a two-dimensional orthotropic body, the thermal conductivity has two components (and only two), such as k_x and k_y for the x- and y-directions, respectively. Consider the problem of

$$k_x \frac{\partial^2 T}{\partial x^2} + k_y \frac{\partial^2 T}{\partial y^2} = \rho c \frac{\partial T}{\partial t}$$

$$-k_x \left.\frac{\partial T}{\partial x}\right|_{x=0} = q_{x0}(y, t) \quad T(a, y, t) = T_a(y, t) \quad T(x, 0, t) = 0$$

$$-k_y \left.\frac{\partial T}{\partial y}\right|_{y=b} = h_{yb}[T(x, b, t) - T_\infty(x, t)]$$

The objective is to obtain a GF solution equation for this case by using the transformation given below.

(*a*) By using the transformation $y' = y(k_x|k_y)^{1/2}$, show that the problem can be transformed to

$$k_x \frac{\partial^2 T}{\partial x^2} + k_x \frac{\partial^2 T}{\partial y'^2} = \rho c \frac{\partial T}{\partial t}$$

$$-k_x \frac{\partial T}{\partial x}\bigg|_{x=0} = q_{x0}(y', t) \quad T(a, y', t) = T_a(y', t) \quad T(x, 0, t) = 0$$

$$-k_x \frac{\partial T}{\partial y'}\bigg|_{y'=b'} = h'_{yb}[T(x, b', t) - T_\infty(x, t)]$$

where $b' = b(k_x / k_y)^{1/2}$ and $h'_{yb} = h_{yb}(k_x / k_y)^{1/2}$.

(*b*) By comparing the above problem with those previously given, obtain a GF solution equation. (It is not necessary to completely rederive the GF solution equation.) Leave in a form that does not contain the GFs in explicit form.

(*c*) Give the GF(s) for this problem.

3.18 Derive the steady-state GF solution equation, Equation 3.94, from first principles.

3.19 Derive Equation 3.100 using the control volume equation from Chapter 1.

3.20 Using the relationship between steady and unsteady GF, (Equation 3.89), show how the unsteady GF solution equation reduces to the steady GF solution equation in the limit as $t \to \infty$.

3.21 Repeat Example 3.7 with added constant energy generation in the body: $g(x, t) \to g_0$.

3.22 Repeat Example 3.7 with the boundary condition $x \to L$ given by

$$k \frac{\partial T(x = L)}{\partial x} + h[T(x = L) - T_\infty] = 0$$

3.23 Show that if $m = ax$ in the equation

$$\frac{\partial^2 T}{\partial x^2} - m^2 T = \frac{1}{\alpha} \frac{\partial T}{\partial t}$$

that the W transformation (Equation 3.73) does not eliminate the $m^2 T$ term.

3.24 Give the solution in terms of GFs for the moving long circular die described by Equation 3.100 for T_∞ equal to a constant and the boundary condition at $x = 0$ of $T = T_0$. The initial temperature is $F(x)$.

3.25 Give the solution using GFs for the problem denoted $XV23B11T-$.

3.26 Use the alternative GF solution equation to obtain $T(x, t)$ for

$$\alpha \frac{\partial^2 T}{\partial x^2} = \frac{\partial T}{\partial t} + V \frac{\partial T}{\partial x}$$

$$-k \frac{\partial T}{\partial x} = h(T(0, t) - T_\infty) \qquad \text{at } x = 0$$

$$T = 0 \qquad \text{at } x = L \qquad T(x, 0) = 0$$

3.27 The GF for the hyperbolic energy equation is defined by

$$\nabla^2 G - \frac{1}{\alpha}\frac{\partial G}{\partial t} - \frac{1}{\sigma^2}\frac{\partial^2 G}{\partial t^2} = -\frac{\delta(\mathbf{r}-\mathbf{r}')\,\delta(t-\tau)}{\alpha}$$

Derive the GF solution equation for the hyperbolic energy equation in the infinite body.

3.28 Show that the equation

$$\frac{\partial^2 T}{\partial x^2} + \frac{\partial^2 T}{\partial y^2} + \frac{\partial^2 T}{\partial z^2} + \frac{g(x,y,z,t)}{k} - m^2 T$$
$$= \frac{1}{\alpha}\left[\frac{\partial T}{\partial t} + u\frac{\partial T}{\partial x} + \mathrm{v}\frac{\partial T}{\partial y} + w\frac{\partial T}{\partial z}\right]$$

by using the transformation

$$T(x,y,z,t) = W(x,y,z,t)\exp\left[\frac{ux}{2\alpha} - \left(\frac{u^2}{4\alpha} + m^2\alpha\right)t\right]$$
$$\times \exp\left[\frac{\mathrm{v}y}{2\alpha} - \frac{\mathrm{v}^2 t}{4\alpha}\right]\exp\left[\frac{wz}{2\alpha} - \frac{w^2 t}{4\alpha}\right]$$

can be written as

$$\frac{\partial^2 W}{\partial x^2} + \frac{\partial^2 W}{\partial y^2} + \frac{\partial^2 W}{\partial z^2} + \frac{H(x,y,z,t)}{k} = \frac{1}{\alpha}\frac{\partial W}{\partial t}$$

where H is defined to be

$$H = g(x,y,z,t)\exp\left[-\frac{ux}{2\alpha} + \left(\frac{u^2}{4\alpha} + m^2\alpha\right)t\right]$$
$$\times \exp\left[-\frac{\mathrm{v}y}{2\alpha} + \frac{\mathrm{v}^2 t}{4\alpha}\right]\exp\left[-\frac{wz}{2\alpha} + \frac{w^2 t}{4\alpha}\right]$$

REFERENCES

Beck, J. V., Blackwell, E., and St. Clair, C. R., Jr., 1985, *Inverse Heat Conduction, Ill-Posed Problems*, John Wiley, New York.

Nellis, G. and Klein, S., 2009, *Heat Transfer*, Cambridge University Press, Cambridge, U.K., p. 104.

4 Methods for Obtaining Green's Functions

4.1 INTRODUCTION

Although the Green's function (GF) approach represents a powerful and flexible method for solving heat conduction and diffusion problems, it is necessary to have mathematical expressions for the GFs. Many GFs are known; Appendixes X, R, and RS provide listings of GFs in a systematic form for rectangular, cylindrical, and spherical coordinates, respectively. The purpose of this chapter is to demonstrate several methods of obtaining *exact* expressions for the GFs. Galerkin-based GFs for composite bodies and other difficult cases are discussed in Chapters 10 and 11. Once the GF is known for a given problem, the general solution of the problem can be written down immediately using the GF solution equations given in Chapter 3; integrations may still be needed, but the integrations can be performed numerically, if not analytically.

For many problems involving finite bodies, the GF expressions have two different forms: the small-cotime GF and the large-cotime GF. Various solution techniques are used to determine the different forms. The small-cotime and large-cotime forms of the GF are mathematically equivalent and both apply for $t \geq 0$; however, depending on the practical applications, one may be preferred to the other. Applications of the small-cotime and large-cotime GF are discussed in more detail in Chapter 5.

In Chapter 1, we saw that the appropriate GF for a given problem is the solution to the corresponding homogeneous auxiliary problem. Consequently, the GFs themselves can be found by classic mathematical methods. In this chapter, several different approaches for obtaining the GFs are discussed and illustrated through various examples. The first method uses sources and sinks in an infinite body for construction of the GF in a finite planar body. This method, which is known as the method of images, is illustrated in Section 4.2. The next method utilizes the Laplace transform. Many small-cotime GFs are derived from the Laplace transform solutions of the heat conduction equation. This approach is discussed in Section 4.3. The third method uses the separation of variables technique. Many large-cotime GFs are obtained through this procedure. The method of separation of variables (and its relation to the GF) is discussed in Section 4.4. Section 4.5 shows that certain two- and three-dimensional GFs can be found by simple multiplication of the corresponding one-dimensional GFs. The method of eigenvalue expansion is discussed in Section 4.6. Finally, Section 4.7 covers steady-state GFs and their relationship with transient GFs.

4.2 METHOD OF IMAGES

The method of images for rectangular coordinates is based on the construction of a transient GF for a finite body from the transient GF for an infinite body (the fundamental heat conduction solution). This method readily applies to transient problems with boundary conditions of the zeroth, first, and second kinds. A few steady GF may be found with the method of images for boundaries of the first kind only (see Barton, 1989, pp. 127).

Earlier in Section 1.11 the method of images was used to find transient GFs for a semi-infinite body. Here the method of images is used to find transient GFs for four different flat plate cases. They are denoted $X11$, $X12$, $X21$, and $X22$ in the heat conduction numbering system. Transient planar sources and sinks are used for these one-dimensional cases, but transient line and point sources can also be used in two- or three-dimensional geometries.

The temperature solutions for each of the cases mentioned above with an initial temperature of $F(x)$ and homogeneous boundary conditions is given by

$$T(x,t) = \int_{x'=0}^{L} G\left(x,\, t|x',0\right)\, F(x')\, dx' \tag{4.1}$$

The integration is over the domain 0 to L. Four $G(\cdot)$ functions can be constructed by superimposing the plane source solution for an *infinite* body (the fundamental heat conduction solution). See Figure 4.1 for the location of these plane sources (which are denoted by the plus signs) or sinks (which are denoted by the minus signs). The physical locations of the sources or sinks are at positions included by the equations

$$z^- = 2nL + x - x' \qquad n = \ldots, -2, -1, 0, 1, 2, \ldots \tag{4.2a}$$

$$z^+ = 2nL + x + x' \qquad n = \ldots, -2, -1, 0, 1, 2, \ldots \tag{4.2b}$$

One of the simplest cases to visualize is the $X22$ case which has two insulated boundary conditions; these boundary conditions can be modeled by symmetric images or reflections. The result is a series of sources (not sinks) at the z^- and z^+ locations given by Equation 4.2a and b. As a consequence, the $X22$ GF has only positive components as given in Table 4.1.

Another case is denoted $X11$ and is shown at the top of Figure 4.1. Notice that each image (at $x = 0, \pm L, \pm 2L, \ldots$) must have the opposite sign to the adjacent one in order to have a zero contribution at the common boundary. This leads to the distribution of signs shown in the $X11$ case in Figure 4.1 and the $X11$ GF given in Table 4.1. A similar procedure is followed in the $X12$ and $X21$ cases shown in Figure 4.1. The boundaries at $x = 0, \pm 2L, \pm 4L, \ldots$ are repeated as are those at $x = \pm L, \pm 3L, \ldots$; as a consequence, the symmetric condition (boundary condition of the second kind) has the same sign on both sides of a boundary and the antisymmetric condition (boundary condition of the first kind) is modeled by a source on one side and a sink on the other.

The cases shown in Figure 4.1 have the GFs that are tabulated in Table 4.1 as the last five cases, with the last case being a general form containing all of the previous four cases. There are summations that extend from $n = -\infty$ to $n = +\infty$, but only a few terms are needed for small dimensionless times; this is discussed further in the next paragraph. A more extensive table of GFs for Cartesian coordinates is given in Appendix X.

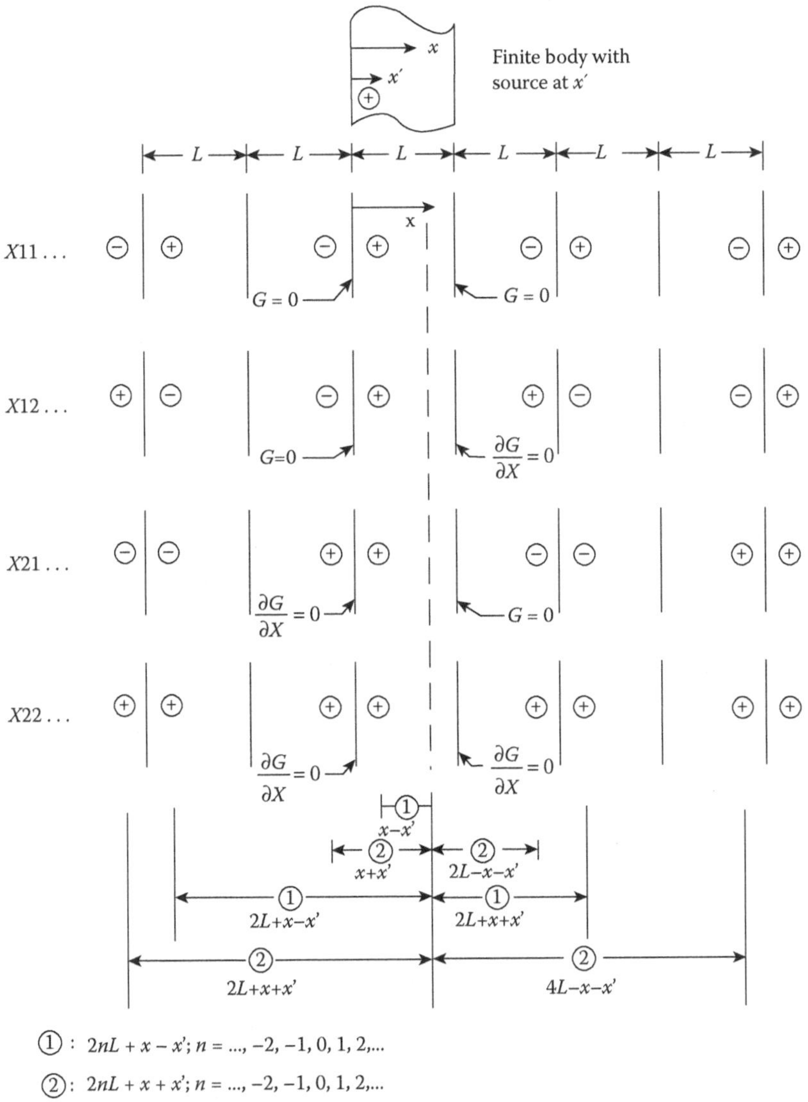

FIGURE 4.1 Location of sources (+) and sinks (−) for finite-body GFs created from infinite-body GFs by the method of images.

It is instructive to see how many terms in the $X11$, $X12$, $X21$, and $X22$ cases are needed for small dimensionless times, $\alpha(t-\tau)/L^2$. Consider the typical term, $K(2nL+x\pm x', t-\tau)\cdot L$, which is plotted in Figures 4.2 and 4.3. Results for the dimensionless time of 0.025 are plotted in the first figure and for the dimensionless time of 0.1 in the second figure. The function $K\cdot L$ is plotted versus $(x-x')/L$ or $(x+x')/L$, where $(x-x')/L$ can vary from -1 to $+1$, and $(x+x')/L$ can vary

TABLE 4.1
Green's Functions Formed from Fundamental Heat Conduction Solution

Case	Green's Function
$X00$	$K(x-x',t-\tau)=[4\pi\alpha(t-\tau)]^{-1/2}\exp[-(x-x')^2/4\alpha(t-\tau)]$
$X10$	$K(x-x',t-\tau)-K(x+x',t-\tau)$
$X20$	$K(x-x',t-\tau)+K(x+x',t-\tau)$
$X11$	$\sum_{n=-\infty}^{\infty}[K(2nL+x-x',t-\tau)-K(2nL+x+x',t-\tau)]$
$X12$	$\sum_{n=-\infty}^{\infty}(-1)^n[K(2nL+x-x',t-\tau)-K(2nL+x+x',t-\tau)]$
$X21$	$\sum_{n=-\infty}^{\infty}(-1)^n[K(2nL+x-x',t-\tau)+K(2nL+x+x',t-\tau)]$
$X22$	$\sum_{n=-\infty}^{\infty}[K(2nL+x-x',t-\tau)+K(2nL+x+x',t-\tau)]$
XIJ	$\sum_{n=-\infty}^{\infty}(-1)^{(I+J)n}[K(2nL+x-x',t-\tau)+(-1)^I K(2nL+x+x',t-\tau)]$, $I,J=1,2$

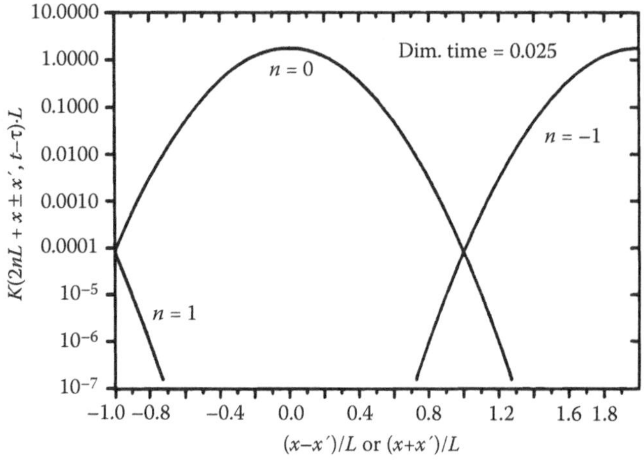

FIGURE 4.2 Function $K(2nL+x\pm x',t-\tau)\cdot L$, a component of the small-cotime GF, at dimensionless time $\alpha(t-\tau)/L^2=0.025$.

from 0 to 2. For $\alpha(t-\tau)/L^2=0.025$, the maximum $K\cdot L$ value is almost 2. See Figure 4.2. For terms with values at least 0.0001 (0.005% of the maximum), the $n=0$ term is needed for $(x-x')/L$ between -1 and 1, and for the $(x+x')/L$ term for 0 to 1. The $n=-1$ term is needed only for $(x+x')/L$ between 1 and 2. For the larger time of $\alpha(t-\tau)/L^2=0.1$, Figure 4.3 shows that for terms being less than 0.005% of the maximum, the $K\cdot L$ terms for $(x-x')/L$ are needed for $n=0$ (region of -1 to 1), $n=1$ (region of -1 to 0), and $n=-1$ (region of 0 to 1). The $K\cdot L$

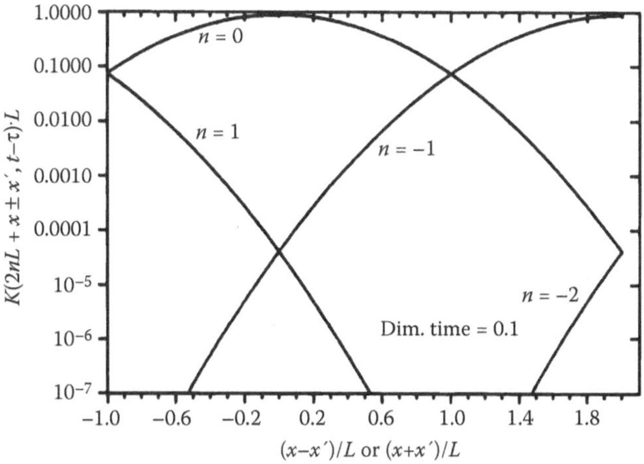

FIGURE 4.3 Function $K(2nL + x \pm x', t - \tau) \cdot L$, a component of the small-cotime GF, at dimensionless time $\alpha(t - \tau)/L^2 = 0.10$.

terms for $(x + x')/L$ are needed for $n = 0$ (region of 0 to 2) and $n = -1$ (region of 0 to 2). For other criteria regarding the magnitude of terms that are neglected, the number of required terms could be greater or smaller. The major point is that for small dimensionless times such as $\alpha(t - \tau)/L^2 < 0.025$, only two terms are needed for $K(\cdot)$, one for $n = 0$ and the other for $n = -1$.

4.3 LAPLACE TRANSFORM METHOD

The Laplace transformation is a powerful tool in the solution of linear ordinary and partial differential equations, and has accordingly been applied to many heat conduction problems (Carslaw and Jaeger, 1959; Arpaci, 1966; Luikov, 1968; Ozisik, 1993). The method is particularly well suited for the solution of one-dimensional time-dependent problems. The process of solution consists of three main steps. First, the time variable is removed from the problem by means of Laplace transformation, resulting in a simpler equation than the original equation. Next, the new equation is solved in the transformed space; and finally, the solution of the new equation is transformed back to obtain the solution to the original problem. Since a brief introduction to the Laplace transform method was given earlier in Section 1.8, the present discussion is intended mainly to illustrate various approaches for obtaining the GFs. For a more comprehensive presentation of the Laplace transform method applied to heat conduction problems, see Carslaw and Jaeger (1959, Chapters 12, 13, and 15).

In this section, we first present a brief definition of the Laplace transformation. An example problem is given next, to demonstrate the application of the method to a typical heat conduction problem by employing a table of transform pairs. Then, the method is utilized for the determination of the GFs through the use of three examples.

4.3.1 Definition

Consider a function $f(t)$ for $t \geq 0$. This function can be multiplied by e^{-st} and integrated with respect to t from zero to infinity. Then, if the resulting integral exists, it is a function of the parameter s; that is,

$$\overline{f}(s) = \int_0^{\infty} e^{-st} f(t)\,dt \tag{4.3}$$

The function $\overline{f}(s)$ is called the Laplace transform of the function $f(t)$, and is denoted by $\mathscr{L}[f(t)]$. The original function $f(t)$ is called the inverse transform of $\overline{f}(s)$ and is denoted by

$$f(t) = \mathscr{L}^{-1}[\overline{f}(s)] \tag{4.4}$$

Both functions $f(t)$ and $\overline{f}(s)$ are called a Laplace transform pair, and knowledge of either one enables the other to be recovered. A list of properties of the Laplace transform is given in Appendix L.

An important step in the process of solving a problem by Laplace transforms is that of inverting the transform to obtain the solution to the original problem. Fortunately, extensive tables of transform pairs are available which can directly be utilized for the solution of many problems (Appendix L, Table L.1)

4.3.2 Temperature Example

As a demonstration of the Laplace transform method, an example of finding the transient temperature is given next.

Example 4.1: Heat Conduction in a Semi-Infinite Body with Specified Surface Temperature—*X*10*B*1*T*0-Case

Consider a semi-infinite body initially at zero temperature subjected to a constant surface temperature T_0, for times $t > 0$. There is no volume energy generation in the body. Using the Laplace transform method, find the transient temperature distribution in the body.

Solution

The differential equation and the boundary and initial conditions for this problem are given as

$$\frac{\partial^2 T(x,t)}{\partial x^2} = \frac{1}{\alpha}\frac{\partial T(x,t)}{\partial t} \tag{4.5}$$

$$T(0,t) = T_0 \tag{4.6a}$$

$$\lim T(x,t) \to 0 \qquad \text{as } x \to \infty \tag{4.6b}$$

$$T(x,0) = 0 \tag{4.6c}$$

The first step in the solution is to find the Laplace transform of the differential equation (4.5) with respect to t; that is,

$$\mathscr{L}\left[\frac{\partial^2 T(x,t)}{\partial x^2}\right] = \frac{1}{\alpha}\mathscr{L}\left[\frac{\partial T(x,t)}{\partial t}\right] \tag{4.7}$$

The use of the properties of Laplace transform yields

$$\mathscr{L}\left[\frac{\partial^2 T(x,t)}{\partial x^2}\right] = \frac{d^2\overline{T}(x,s)}{dx^2} \tag{4.8a}$$

$$\begin{aligned}\mathscr{L}\left[\frac{\partial T(x,t)}{\partial t}\right] &= s\overline{T}(x,s) - T(x,0)\\ &= s\overline{T}(x,s) \qquad \text{since } T(x,0) = 0\end{aligned} \tag{4.8b}$$

where

$$\overline{T}(x,s) = \mathscr{L}\left[T(x,t)\right] = \int_0^\infty e^{-st}T(x,t)\,dt \tag{4.8c}$$

Thus, Equation 4.7 can be written as

$$\frac{d^2\overline{T}(x,s)}{dx^2} - \frac{s}{\alpha}\overline{T}(x,s) = 0 \tag{4.9}$$

Similarly, the Laplace transform of the boundary conditions, Equation 4.6a and b, yields

$$\overline{T}(0,s) = \mathscr{L}\left[T_0\right] = \frac{T_0}{s} \tag{4.10a}$$

$$\overline{T}(x,s) \to 0 \qquad \text{as } x \to \infty \tag{4.10b}$$

Equation 4.9 is an ordinary differential equation for $\overline{T}(x,s)$ with the only independent variable being x. The solution of this equation with the boundary conditions given by Equation 4.10a and b may be written as

$$\overline{T}(x,s) = \frac{T_0}{s}e^{-x\sqrt{s/\alpha}} \tag{4.11}$$

The final step is now to transform $\overline{T}(x,s)$ back to obtain the solution for $T(x,t)$; that is,

$$T(x,t) = \mathscr{L}^{-1}\left[\frac{T_0}{s}e^{-x\sqrt{s/\alpha}}\right] \tag{4.12}$$

Equation 4.12 can be inverted simply by utilizing a table of transform pairs (Appendix L, Table L.1, number 42) to obtain

$$T(x,t) = T_0\,\mathrm{erfc}\left[\frac{x}{(4\alpha t)^{1/2}}\right] \tag{4.13}$$

This is the same solution as given by Equation 1.112 which was obtained by the GF method.

4.3.3 Derivation of Green's Functions

The short-cotime GF for many heat conduction problems are derived from the Laplace transform solution of the corresponding auxiliary equation with homogeneous boundary conditions. As discussed in Chapter 1, the auxiliary equation for a given problem is identical to the original heat conduction equation for that problem except for the heat generation term, which is replaced by a unit instantaneous heat source (a Dirac delta function). The homogeneous boundary conditions for the auxiliary problem must be of the same kind as the original problem. Determination of the GFs by the method of Laplace transform is best illustrated through the use of examples.

Example 4.2: Semi-Infinite Body with Convection—*X*30 Case

Find the GF for the problem of a semi-infinite body with the convective boundary condition at the surface.

Solution

This is the $X30$ case. The GF associated with this problem is the solution to the following auxiliary equation:

$$\frac{\partial^2 G}{\partial x^2} + \frac{1}{\alpha}\delta(x - x')\,\delta(t - 0) = \frac{1}{\alpha}\frac{\partial G}{\partial t} \qquad t \geq 0 \qquad x > 0 \tag{4.14}$$

subject to the homogeneous boundary conditions of

$$-k\frac{\partial G(0, t|x', 0)}{\partial x} + hG(0, t|x', 0) = 0 \qquad t \geq 0 \tag{4.15a}$$

$$G(\infty, t|x', 0) \text{ is bounded} \qquad t \geq 0 \tag{4.15b}$$

and initial condition

$$G(x, t|x', 0) = 0 \qquad t < 0 \tag{4.15c}$$

Notice that the second term in Equation 4.14 represents a unit instantaneous plane source at location x' released at time $\tau = 0$. Consequently, $G(x, t|x', 0)$ is the $X30$ GF for $\tau = 0$. Once the appropriate expression for $G(x, t|x', 0)$ is determined, then GF for $\tau \neq 0$ can be found by replacing t by $t - \tau$ in that expression.

In the Laplace transform approach, the auxiliary problem given by Equation 4.14 is subdivided into two problems. One gives the solution due to the instantaneous plane source at location x' and at time τ for an infinite one-dimensional body (the fundamental heat conduction solution), and the other satisfies the given initial and boundary conditions. Hence $G(x, t|x', 0)$ is written as

$$G(x, t|x', 0) = K(x - x', t - 0) + V(x, t) \tag{4.16}$$

where K is the fundamental heat conduction solution for $\tau = 0$, given by

$$K(x - x', t - 0) = (4\pi\alpha t)^{-1/2} \exp\left[-\frac{(x - x')^2}{4\alpha t}\right] \tag{4.17}$$

and $V(x, t)$ satisfies the one-dimensional heat conduction equation in the semi-infinite region; that is,

$$\frac{\partial^2 V(x,t)}{\partial x^2} = \frac{1}{\alpha}\frac{\partial V(x,t)}{\partial t} \quad \text{for } t > 0 \quad \text{and} \quad 0 < x < \infty \tag{4.18a}$$

with the initial condition of

$$V(x, 0) = 0 \tag{4.18b}$$

and V should be such that the boundary conditions, Equation 4.15a and b, is satisfied.

Next Laplace transform will be used to replace time t by transform parameter s. Taking the Laplace transform of K (use Appendix L, Table L.1, number 43) gives

$$\overline{K} = \frac{1}{2(\alpha s)^{1/2}} \exp\left[-\left(\frac{s}{\alpha}\right)^{1/2} |x - x'|\right] \tag{4.19}$$

Taking the Laplace transform of Equation 4.18a, using the same techniques discussed in Example 4.1, results in

$$\frac{d^2\overline{V}}{dx^2} - \frac{s}{\alpha}\overline{V} = 0 \tag{4.20}$$

The general solution of Equation 4.20 may be written as

$$\overline{V}(x, s) = A\exp\left[\left(\frac{s}{\alpha}\right)^{1/2} x\right] + B\exp\left[-\left(\frac{s}{\alpha}\right)^{1/2} x\right] \tag{4.21}$$

Now, taking the Laplace transform of Equation 4.16 and substituting the values for $\overline{K}$ and $\overline{V}$ from Equations 4.19 and 4.21 into the transformed equation yields

$$\begin{aligned}\overline{G}(x, s|x', 0) &= \frac{1}{2(\alpha s)^{1/2}} \exp\left[-\left(\frac{s}{\alpha}\right)^{1/2} |x - x'|\right] \\ &\quad + A\exp\left[\left(\frac{s}{\alpha}\right)^{1/2} x\right] + B\exp\left[-\left(\frac{s}{\alpha}\right)^{1/2} x\right]\end{aligned} \tag{4.22}$$

The constants A and B in Equation 4.22 are determined from the boundary conditions 4.15a, b. The Laplace transform of these equations are

$$-k\frac{\partial \overline{G}(0, s|x', 0)}{\partial x} + h\overline{G}(0, s|x', 0) = 0 \tag{4.23a}$$

$$\overline{G}(\infty, s|x', 0) \text{ is bounded} \tag{4.23b}$$

Then, by introducing the transformed conditions Equation 4.23a and b into Equation 4.22, the constants A and B are

$$A = 0 \tag{4.24a}$$

$$B = \frac{1}{2(\alpha s)^{1/2}} \frac{(s/\alpha)^{1/2} - H}{(s/\alpha)^{1/2} + H} \exp\left[-\left(\frac{s}{\alpha}\right)^{1/2} x'\right] \tag{4.24b}$$

where $H = h/k$. Substituting Equation 4.24a and b back into Equation 4.22 yields

$$\overline{G}(x, s|x', 0) = \frac{1}{2(\alpha s)^{1/2}}\left\{\exp\left[-\left(\frac{s}{\alpha}\right)^{1/2}|x - x'|\right] + \exp\left[-\left(\frac{s}{\alpha}\right)^{1/2}(x + x')\right] - \frac{2H}{[(s/\alpha)^{1/2} + H]} \times \exp\left[-\left(\frac{s}{\alpha}\right)^{1/2}(x + x')\right]\right\} \tag{4.25}$$

From the Laplace transform table (Appendix L, numbers 43 and 47), the inverse transform of Equation 4.25 gives the solution $G(x, t|x', 0)$; that is,

$$G(x, t|x', 0) = \frac{1}{2(\pi\alpha t)^{1/2}}\left\{\exp\left[-\frac{(x - x')^2}{4\alpha t}\right] + \exp\left[-\frac{(x + x')^2}{4\alpha t}\right]\right\} - \frac{h}{k}\exp\left[\frac{h}{k}(x + x') + \alpha\frac{h^2 t}{k^2}\right]\operatorname{erfc}\left[\frac{(x + x')}{(4\alpha t)^{1/2}} + \frac{h}{k}(\alpha t)^{1/2}\right] \tag{4.26}$$

which is the $X30$ GF for $\tau = 0$. The $X30$ GF for $\tau \neq 0$ can now be determined by replacing t by $t - \tau$ in Equation 4.26; that is,

$$G_{X30}(x, t|x', \tau) = \frac{1}{[4\pi\alpha(t - \tau)]^{1/2}}\left\{\exp\left[-\frac{(x - x')^2}{4\alpha(t - \tau)}\right] + \exp\left[-\frac{(x + x')^2}{4\alpha(t - \tau)}\right]\right\} - \frac{h}{k}\exp\left[\frac{h}{k}(x + x') + \alpha\frac{h^2(t - \tau)}{k^2}\right]\operatorname{erfc}\left\{\frac{x + x'}{[4\alpha(t - \tau)]^{1/2}} + \frac{h}{k}[\alpha(t - \tau)]^{1/2}\right\} \tag{4.27}$$

This equation is tabulated in Appendix X. Note that for $h \to 0$ the error function term drops out, demonstrating that the convection boundary (third kind) reduces to the insulated boundary (second kind) when convection goes to zero. That is, $G_{X30}(h \to 0) = G_{X20}$.

Example 4.3: Region outside a Spherical Cavity with Convection—*RS*30 Case

Find the GF for the infinite region outside a spherical cavity of radius a with a convective boundary condition. This is the $RS30$ case.

Solution

The GF is the solution to the auxiliary equation,

$$\frac{1}{r}\frac{\partial^2(rG)}{\partial r^2} + \frac{1}{\alpha}\delta(\mathbf{r} - \mathbf{r}')\,\delta(t - 0) = \frac{1}{\alpha}\frac{\partial G}{\partial t} \qquad a < r < \infty \quad t \geq 0 \tag{4.28}$$

Here $\delta(\mathbf{r} - \mathbf{r}')$ has units m^{-3}. The homogeneous boundary conditions are

$$-k\frac{\partial G(a, t|r', 0)}{\partial r} + hG(a, t|r', 0) = 0 \qquad t \geq 0 \tag{4.29a}$$

$$G(\infty, t|r', 0) \text{ is bounded,} \qquad t \geq 0 \tag{4.29b}$$

and the initial condition is

$$G(r,t|r',0)=0 \qquad t<0 \tag{4.29c}$$

Equations 4.28 and 4.29 represent the problem of an infinite region outside the spherical cavity of $r=a$ (initially at zero temperature) subject to a unit instantaneous spherical surface source at $r=r'$ released at time $\tau=0$ with a homogeneous convective boundary condition at $r=a$.

Again, in a manner similar to that used in the previous example, the solution for $G(r,t|r',0)$ is subdivided into two parts in the following form:

$$G(r,t|r',0)=K_s(r-r',t-0)+V(r,t) \tag{4.30}$$

where K_s is the fundamental heat conduction solution for radial flow in the spherical region; it is the GF for the *RS*00 case (see Appendix RS) and is given by

$$K_s(r-r',\ t-0)=\frac{1}{8\pi rr'(\pi\alpha t)^{1/2}}\left\{\exp\left[-\frac{(r-r')^2}{4\alpha t}\right]-\exp\left[-\frac{(r+r')^2}{4\alpha t}\right]\right\} \tag{4.31}$$

and its Laplace transform is given by (Appendix L, number 43)

$$\overline{K}_s=\frac{1}{8\pi rr'(\alpha s)^{1/2}}\left\{\exp\left[-\left(\frac{s}{\alpha}\right)^{1/2}|r-r'|\right]-\exp\left[-\left(\frac{s}{\alpha}\right)^{1/2}(r+r')\right]\right\} \tag{4.32}$$

The temperature V in this case satisfies the heat conduction equation for one-dimensional heat flow in the region outside the spherical cavity $r=a$; that is,

$$\frac{\partial^2[rV(r,t)]}{\partial r^2}=\frac{1}{\alpha}\frac{\partial[rV(r,t)]}{\partial t} \qquad \text{for } t>0 \quad \text{and} \quad a<r<\infty \tag{4.33a}$$

with the initial condition of

$$V(r,0)=0 \tag{4.33b}$$

The Laplace transform of Equation 4.33a yields

$$\frac{d^2(r\overline{V})}{dr^2}-\frac{s}{\alpha}r\overline{V}=0 \qquad \text{for } a<r<\infty \tag{4.34}$$

which has the general solution of the form

$$\overline{V}(r,s)=\frac{A}{r}\exp\left[\left(\frac{s}{\alpha}\right)^{1/2}r\right]+\frac{B}{r}\exp\left[-\left(\frac{s}{\alpha}\right)^{1/2}r\right] \tag{4.35}$$

Taking the Laplace transform of Equation 4.30 and substituting the values for $\overline{K}$ and $\overline{V}$ from Equations 4.32 and 4.35 into the result gives

$$\overline{G}(r,s|r',0)=\frac{1}{8\pi rr'(\alpha t)^{1/2}}\left\{\exp\left[-\left(\frac{s}{\alpha}\right)^{1/2}|r-r'|\right]-\exp\left[-\left(\frac{s}{\alpha}\right)^{1/2}(r+r')\right]\right\}$$
$$+\frac{A}{r}\exp\left[\left(\frac{s}{\alpha}\right)^{1/2}r\right]+\frac{B}{r}\exp\left[-\left(\frac{s}{\alpha}\right)^{1/2}r\right] \tag{4.36}$$

Equation 4.36 must satisfy the boundary conditions Equation 4.29a and b. The Laplace transforms of these equations are

$$-\frac{\partial \overline{G}(a, s|r', 0)}{\partial r} + H\overline{G}(a, s|r', 0) = 0 \tag{4.37a}$$

$$\overline{G}(\infty, s|r', 0) = 0 \tag{4.37b}$$

where h/k is denoted H. It follows from Equation 4.37b that

$$A = 0 \tag{4.38a}$$

Then, from Equation 4.37a, one can show that

$$B = \frac{1}{8\pi r'(\alpha s)^{1/2}} \left\{ \exp\left[-\left(\frac{s}{\alpha}\right)^{1/2} r' \right] - \exp\left[-\left(\frac{s}{\alpha}\right)^{1/2} (r' - 2a) \right] \right.$$
$$\left. + \frac{2(s/\alpha)^{1/2}}{(s/\alpha)^{1/2} + \frac{1}{a} + H} \exp\left[-\left(\frac{s}{\alpha}\right)^{1/2} (r' - 2a) \right] \right\} \tag{4.38b}$$

Substituting the values for A and B from Equation 4.38a and b into Equation 4.36 yields,

$$\overline{G}(r, s|r', 0) = \frac{1}{8\pi r r'(\alpha s)^{1/2}} \left\{ \exp\left[-\left(\frac{s}{\alpha}\right)^{1/2} |r - r'| \right] \right.$$
$$- \exp\left[-\left(\frac{s}{\alpha}\right)^{1/2} (r + r' - 2a) \right] + \frac{2(s/\alpha)^{1/2}}{(s/\alpha)^{1/2} + \dfrac{1}{a} + H}$$
$$\left. \times \exp\left[-\left(\frac{s}{\alpha}\right)^{1/2} (r + r' - 2a) \right] \right\} \tag{4.39}$$

which is the Laplace transform of $G(r, t|r', 0)$. Taking the inverse transform of Equation 4.39 (see Appendix L, number 43 and 47) and by replacing t by $t - \tau$ gives

$$G_{RS30}(r, t|r', \tau)$$
$$= \frac{1}{8\pi r r'[\alpha\pi(t-\tau)]^{1/2}} \left(\exp\left[-\frac{(r - r')^2}{4\alpha(t-\tau)} \right] + \exp\left[-\frac{(r + r' - 2a)^2}{4\alpha(t-\tau)} \right] \right.$$
$$- \frac{k + ah}{ak} [4\pi\alpha(t-\tau)]^{1/2} \exp\left[\alpha(t-\tau) \left(\frac{k + ah}{ak} \right)^2 + \frac{k + ah}{ak} (r + r' - 2a) \right]$$
$$\left. \times \operatorname{erfc}\left\{ \frac{(r + r' - 2a)}{2[\alpha(t-\tau)]^{1/2}} + \frac{k + ah}{ak} [\alpha(t-\tau)]^{1/2} \right\} \right) \tag{4.40}$$

which is the $RS30$ Green's function; it is included in Appendix RS.

Example 4.4: Transient Slab Body, Case X12

Use the Laplace transform method to find the transient GF in the slab with $G = 0$ at $x = 0$ and $\partial G/\partial x = 0$ at $x = L$.

Solution

The GF satisfies

$$\frac{\partial^2 G}{\partial x^2} - \frac{1}{\alpha}\frac{\partial G}{\partial u} + \frac{1}{\alpha}\delta(x - x')\delta(u) = 0 \tag{4.41}$$

$$G(x = 0, x', u) = 0 \tag{4.42}$$

$$\left.\frac{dG}{dx}\right|_{x=L} = 0 \tag{4.43}$$

$$G(x, x', u = 0) = 0 \tag{4.44}$$

where $u = t - \tau$ is the cotime. As before, the solution will be sought in the form

$$G(x, x', u) = K(x, x', u) + v(x, u) \tag{4.45}$$

where K is the fundamental heat conduction solution. Replace this form of G into the auxiliary problem for G, Equation 4.41, to find the boundary value problem for v:

$$\frac{\partial^2 v}{\partial x^2} - \frac{1}{\alpha}\frac{\partial v}{\partial u} = 0 \tag{4.46}$$

$$v(x = 0) = -K(x = 0) \tag{4.47}$$

$$\left.\frac{dv}{dx}\right|_{x=L} = -\left.\frac{dK}{dx}\right|_{x=L} \tag{4.48}$$

$$v(x, x', u = 0) = 0 \tag{4.49}$$

With this procedure the nonhomogeneous term has moved from the differential equation for G to the boundary conditions for v. Now apply the Laplace transform, to find the s-space relations for v:

$$\frac{\partial^2 \overline{v}}{\partial x^2} - \frac{s}{\alpha}\overline{v} = 0 \tag{4.50}$$

$$\overline{v}(x = 0) = -\overline{K}(x = 0) \tag{4.51}$$

$$\left.\frac{d\overline{v}}{dx}\right|_{x=L} = -\left.\frac{d\overline{K}}{dx}\right|_{x=L} \tag{4.52}$$

The general solution for $\overline{v}$ was discussed in the previous example, and it is given by

$$\overline{v}(x, x', s) = A\exp[(s/\alpha)^{1/2}x] + B\exp[-(s/\alpha)^{1/2}x] \tag{4.53}$$

Constants A and B may be found by replacing $\overline{v}$ in the boundary conditions for $\overline{v}$. The result, after some algebra, is

$$A = \frac{\exp[-(s/\alpha)^{1/2}(2L - x')] - \exp[-(s/\alpha)^{1/2}(2L + x')]}{\sqrt{4\alpha s}\,(1 + \exp[-2L(s/\alpha)^{1/2}])} \tag{4.54}$$

$$B = -\frac{\exp[-(s/\alpha)^{1/2}(2L - x')] + \exp[-(s/\alpha)^{1/2}x']}{\sqrt{4\alpha s}\,(1 + \exp[-2L(s/\alpha)^{1/2}])} \tag{4.55}$$

Substitute A and B into $\bar{v}$, and then use $\bar{G} = \bar{K} + \bar{v}$ to find the GF in Laplace transform space:

$$\bar{G}(x, x', s) = \frac{1}{\sqrt{4\alpha s}} \exp[-(s/\alpha)^{1/2}|x - x'|] \quad (4.56)$$
$$+ \frac{\exp[-(s/\alpha)^{1/2}(2L - x - x')] - \exp[-(s/\alpha)^{1/2}(2L - x + x')]}{\sqrt{4\alpha s}\,(1 + \exp[-2L(s/\alpha)^{1/2}])}$$
$$- \frac{\exp[-(s/\alpha)^{1/2}(2L + x - x')] + \exp[-(s/\alpha)^{1/2}(x + x')]}{\sqrt{4\alpha s}\,(1 + \exp[-2L(s/\alpha)^{1/2}])}$$

The next step is to inverse transform this expression. This inverse transform of the first term gives the fundamental solution K. The inverse transform of the second and third terms is more difficult, because none of the transform pairs in Table L.1 (Appendix L) contain exponentials in the denominator. Consider the binomial theorem

$$\frac{1}{1+z} = 1 - z + z^2 - z^3 + \ldots = \sum_{n=0}^{\infty}(-1)^n z^n \quad \text{for } |z| < 1 \quad (4.57)$$

The binomial theorem with $z = \exp[-2L(s/\alpha)^{1/2}]$ can be used to replace the exponential term in the denominator (Carslaw and Jaeger, 1959, p. 309), as follows:

$$\frac{1}{1 + \exp[-2L(s/\alpha)^{1/2}]} = \sum_{n=0}^{\infty}(-1)^n \exp[-2nL(s/\alpha)^{1/2}] \quad (4.58)$$

Substitute this series into the equation for $\bar{G}$, and then the inverse Laplace transform may be carried out (refer to Appendix L, Table L.1, number 43). Retaining only the $n = 0$ term from the series, the GF in the time domain is given approximately by

$$G(x, x', u) \approx \frac{1}{\sqrt{4\pi\alpha u}} \left\{ \exp\left(\frac{-(x - x')^2}{4\alpha u}\right) - \exp\left(\frac{-(x + x')^2}{4\alpha u}\right) \right.$$
$$+ \exp\left(\frac{-(2L - x - x')^2}{4\alpha u}\right) - \exp\left(\frac{-(2L - x + x')^2}{4\alpha u}\right)$$
$$\left. - \exp\left(\frac{-(2L + x - x')^2}{4\alpha u}\right) \right\} \quad (4.59)$$

This approximate expression is accurate for $\alpha u / L^2 < 0.1$. Many of the short-cotime GF given in the Appendices have been derived in this fashion. The above expression for G_{X12} also agrees with the first few terms of the series found by the image method, which is listed in Table 4.1.

In two of the example problems considered above, the inversion of the transformed solutions were obtained directly from a table of Laplace transforms. However, there are cases for which the transformed solution $\bar{G}$ does not appear in the Laplace transform tables (such as in finite bodies including plates, cylinders, and spheres). In such cases the Laplace-transform inverse is carried out with the use of a series expansion (as in Example 4.4, above) or with the inversion theorem (see Appendix L). The series expansion approach is often less complicated and more useful than the use of the inversion theorem, particularly for small times. For further discussion of the series-expansion approach see Chapter 12 of Carslaw and Jaeger (1959).

4.4 METHOD OF SEPARATION OF VARIABLES

The method of separation of variables can be used to find the GFs through the relationship between the GF and the Dirac delta function. Chapter 1 showed that GFs are proportional to the temperature rise in a body driven by a Dirac delta function initial temperature distribution. The method of separation of variables provides a straight-forward method for solving finite-body problems with arbitrary initial temperature distributions. Once the temperature $T(x,t)$ is known for an arbitrary space-variable initial temperature $F(x)$, then the GF can be found from $T(x,t)$ because an arbitrary initial temperature includes the Dirac delta function as a special case.

In this section, several one-dimensional flat plate GFs are found using the method of separation of variables. The flat plate with the temperature fixed at both sides ($X11$) is used in a full discussion of the method and the flat plate with two insulated boundaries ($X22$) is discussed in an example. A more general derivation of GFs using the separation of variables method is given by Beck (1984) for the flat plate with boundary conditions of the first, second, third, fourth, or fifth kinds.

4.4.1 Plate with Temperature Fixed at Both Sides ($X11$)

One of the simplest cases to consider using the method of separation of variables is for prescribed temperatures of zero at both boundaries of a plate. The describing partial differential equation, boundary conditions, and initial conditions are given by

$$\frac{\partial^2 T}{\partial x^2} = \frac{1}{\alpha}\frac{\partial T}{\partial t} \qquad 0 < x < L \tag{4.60}$$

$$T(0,t) = 0 \qquad T(L,t) = 0 \tag{4.61a, b}$$

$$T(x,0) = F(x) \tag{4.62}$$

Note that the boundary conditions and the partial differential equation are both homogeneous. This case has the notation $X11B00T$-.

Since the thermal diffusivity α is a constant, the differential equation can be solved by adding many solutions, each of which satisfies the differential equation. This is also called superimposing solutions. Let

$$T(x,t) = \sum_{n=1}^{\infty} T_n(x,t) \tag{4.63}$$

where the solutions $T_n(x,t)$ satisfy Equation 4.60. That is, when $T_n(x,t)$ is substituted into Equation 4.60, an identity results. In addition, each $T_n(x,t)$ solution satisfies the homogeneous boundary conditions given by Equation 4.61a and b. A $T_n(x,t)$ solution for a given n does not usually satisfy the initial condition given by Equation 4.62.

The procedure continues by assuming that

$$T_n(x,t) = \mathbf{X}(x)\,\Theta(t) \tag{4.64}$$

where $\mathbf{X}(x)$ is a function of only x, and where $\Theta(t)$ is a function of only t. In other words, $T_n(x,t)$ is chosen to be a product of two functions, one that depends only on x and the other that depends only on t. The variables have been separated in Equation 4.64, hence the name separation of variables technique. Replacing T in Equation 4.60 by T_n gives

$$\frac{\partial^2 T_n}{\partial x^2} = \frac{1}{\alpha}\frac{\partial T_n}{\partial t} \tag{4.65}$$

and substituting Equation 4.64 into Equation 4.65 gives

$$\frac{d^2\mathbf{X}}{dx^2}\Theta = \frac{\mathbf{X}}{\alpha}\frac{d\Theta}{dt} \tag{4.66}$$

Dividing Equation 4.66 by $\mathbf{X}(x)\Theta(t)$ yields

$$\frac{1}{\mathbf{X}}\frac{d^2\mathbf{X}}{dx^2} = \frac{1}{\alpha\Theta}\frac{d\Theta}{dt} \tag{4.67}$$

This equation states that a function of x is equal to a function of t. This equality can only be true if the functions are both simply the same constant. For that reason, let both sides be equal to the negative (real) quantity of $-\lambda^2$,

$$\frac{1}{\mathbf{X}}\frac{d^2\mathbf{X}}{dx^2} = \frac{1}{\alpha\Theta}\frac{d\Theta}{dt} = -\lambda^2 \tag{4.68}$$

Another choice is a positive constant λ^2, but as is shown below, a positive constant gives meaningless results. (This assumes that λ is restricted to real and not imaginary values.) In some cases, the constant may be equal to zero.

Two *ordinary* differential equations now must be solved.

$$\frac{d^2\mathbf{X}}{dx^2} + \lambda^2\mathbf{X} = 0 \tag{4.69a}$$

$$\frac{d\Theta}{dt} + \alpha\lambda^2\Theta = 0 \tag{4.69b}$$

The general solutions of these equations are

$$\mathbf{X} = C_1 \sin\lambda x + C_2 \cos\lambda x \tag{4.70a}$$

$$\Theta = C_3 e^{-\lambda^2 \alpha t} \tag{4.70b}$$

Notice that $\mathbf{X}$ is a sum of two periodic functions. Also Θ is a decaying exponential function. Note that if $-\lambda^2$ were replaced by λ^2, the solution for Θ would result in explosive growth over time—clearly not physically reasonable. (Again, if λ is allowed to be imaginary, different conclusions are possible.) For large times, the solution of the problem given by Equations 4.60 through 4.62 must tend toward zero. Consequently, the constant in Equation 4.68 must be $-\lambda^2$, where the negative sign is both necessary and important.

At this point, it has been assured that $T_n(x,t)$ satisfies the partial differential equation. Next, $T_n(x,t)$ must satisfy the two (homogeneous) boundary conditions. From the boundary condition at $x = 0$, we have

$$T_n(0,t) = \mathbf{X}(0)\,\Theta(t) = 0 \tag{4.71}$$

Since $\Theta(t)$ is an arbitrary function of time, it cannot be set equal to zero without causing $T_n(x,t)$ to be zero for all values of t; such a trivial solution clearly cannot satisfy the nonzero initial conditions which will be examined shortly. Hence, $\mathbf{X}(0) = 0$, and from Equation 4.70a it is necessary that

$$\mathbf{X}(0) = 0 = C_1 \cdot 0 + C_2 \cdot 1 \tag{4.72}$$

which yields

$$C_2 = 0 \tag{4.73}$$

Next consider the boundary condition at $x = L$ which gives

$$T_n(L,t) = \mathbf{X}(L)\,\Theta(t) = 0 \tag{4.74}$$

and again since $\Theta(t)$ cannot be always zero, the result is

$$\mathbf{X}(L) = 0 = C_1 \sin \lambda L \tag{4.75}$$

Consequently the *eigencondition* is

$$\sin \lambda_n L = 0 \tag{4.76}$$

which can occur at only certain values, namely,

$$\lambda_n L = n\pi \qquad n = \ldots, -2, -1, 0, 1, 2, \ldots$$

All of these n values are not needed, however. The negative values do not give independent eigenfunctions ($\sin \lambda_n x$ is called an eigenfunction), since

$$\sin(-\lambda_n L) = -\sin(-\lambda_n L) \tag{4.77}$$

Also the $n = 0$ value makes no contribution in this case since $\sin(0) = 0$. Hence, the eigenvalues λ_n are given by

$$\lambda_n = \frac{n\pi}{L} \qquad n = 1, 2, 3, \ldots \tag{4.78}$$

Usually the eigenvalues in this book are made dimensionless. Let the dimensionless eigenvalues be denoted β_n where for this case

$$\beta_n = n\pi \qquad n = 1, 2, 3, \ldots \tag{4.79}$$

and the eigenfunction is

$$\sin \frac{\beta_n x}{L} \tag{4.80}$$

At this point the differential equation and the two homogeneous boundary conditions for $T_n(x,t)$ have been satisfied. The next step is to bring the two parts of $T_n(x,t)$ together to find

$$T_n(x,t) = C_1 \sin \frac{\beta_n x}{L} C_3 e^{-\beta_n^2 \alpha t / L^2} = A_n \sin \frac{\beta_n x}{L} e^{-\beta_n^2 \alpha t / L^2} \tag{4.81}$$

where A_n is a constant that depends on n. Introduce this form of T_n into Equation 4.63 to get

$$T(x,t) = \sum_{n=1}^{\infty} A_n \sin \frac{\beta_n x}{L} e^{-\beta_n^2 \alpha t / L^2} \tag{4.82}$$

The remaining condition to satisfy is the initial condition, Equation 4.62. This condition is nonzero, unlike the boundary conditions. Using the value of $t = 0$ in Equation 4.82 and the value of $T(x,0) = F(x)$ gives

$$F(x) = \sum_{n=1}^{\infty} A_n \sin \frac{\beta_n x}{L} \tag{4.83}$$

The objective is now to determine values of the constants A_n, for $n = 1, 2$, etc. A result from the theory of Fourier series is that the sine functions are *orthogonal*, which can be stated as

$$\int_{x=0}^{L} \sin \frac{\beta_n x}{L} \sin \frac{\beta_m x}{L} dx = \begin{cases} \frac{L}{2} & m = n \neq 0 \\ 0 & m \neq n \end{cases} \tag{4.84}$$

for the β_n values of $n\pi$, $n = 1, 2, \ldots$. This orthogonality condition provides a very powerful tool for determining one value of A_n at a time. Multiplying both sides of Equation 4.83 by $\sin(\beta_m x / L)\, dx$ and integrating from $x = 0$ to L yields

$$\int_{x=0}^{L} F(x) \sin \frac{\beta_m x}{L} dx = \int_{x=0}^{L} \sum_{n=1}^{\infty} A_n \sin \frac{\beta_n x}{L} \sin \frac{\beta_m x}{L}\, dx \tag{4.85}$$

Now, according to the orthogonality condition, Equation 4.84, there is a nonzero term on the right-hand side of Equation 4.85 only when $m = n$. In other words, the orthogonality condition just picks out one term in the summation to give

$$\int_{x=0}^{L} F(x) \sin \frac{\beta_m x}{L} dx = \frac{A_m L}{2} \tag{4.86}$$

Another way to think of this procedure is to imagine that m is a particular value such as 2. If $m = 2$, then the right side of Equation 4.85 is

$$\int_{x=0}^{L} A_1 \sin\frac{\beta_1 x}{L} \sin\frac{\beta_2 x}{L} dx + \int_{x=0}^{L} A_2 \sin^2\frac{\beta_2 x}{L} dx + \int_{x=0}^{L} A_3 \sin\frac{\beta_3 x}{L} \sin\frac{\beta_2 x}{L} dx + \cdots$$

Only the second term (when $\beta_m = \beta_n = \beta_2$) yields a nonzero value, namely, $A_2 L/2$. See Equation 4.84. Solving Equation 4.86 for A_m yields

$$A_m = \frac{2}{L}\int_{x=0}^{L} F(x) \sin\frac{\beta_m x}{L} dx \tag{4.87}$$

where $m = 1, 2, \ldots$, and the m subscript in Equation 4.87 could be replaced by another index symbol, such as n.

Normally, the separation of variables procedure terminates at this point with the observation that A_m (with $m \to n$) in Equation 4.87 can be used to obtain the A_n values for Equation 4.82. This gives the complete solution, since the partial differential equation, the two homogeneous boundary conditions, and the initial condition are all satisfied. Since our objective is to obtain a GF, further steps are added. Introducing A_m from Equation 4.87 (with $m \to n$ and $x \to x'$) in Equation 4.82 results in

$$T(x,t) = \sum_{n=1}^{\infty} \frac{2}{L}\int_{x'=0}^{L} F(x') \sin\frac{\beta_n x'}{L} \sin\frac{\beta_n x}{L} dx' e^{-\beta_n^2 \alpha t/L^2} \tag{4.88}$$

Taking the integral outside and rearranging gives

$$T(x,t) = \int_{x'=0}^{L} \left[\frac{2}{L}\sum_{n=1}^{\infty} e^{-\beta_n^2 \alpha t/L^2} \sin\frac{\beta_n x}{L} \sin\frac{\beta_n x'}{L}\right] F(x')\, dx' \tag{4.89a}$$

$$T(x,t) = \int_{x'=0}^{L} G_{X11}(x,t|x',0)\, F(x')\, dx' \tag{4.89b}$$

Notice that the expression inside the brackets in Equation 4.89a is the $X11$ GF, evaluated at $\tau = 0$. The $X11$ GF for $\tau \neq 0$ can be found by replacing $(t-0)$ by $(t-\tau)$ inside the brackets to obtain

$$G_{X11}(x,t|x',\tau) = \frac{2}{L}\sum_{n=1}^{\infty} e^{-\beta_n^2 \alpha (t-\tau)/L^2} \sin\frac{\beta_n x}{L} \sin\frac{\beta_n x'}{L} \tag{4.90}$$

where the β_n values are

$$\beta_n = n\pi, \qquad n = 1, 2, \ldots \tag{4.91}$$

A few more comments are appropriate regarding this result. It is stated in Chapter 1 that GF can be interpreted as the temperature rise in the body caused by a Dirac delta function of unit value at position x_0 and time $t_0 = 0$. Since $F(x)$ is arbitrary, let $F(x)$ be the impulse of $T_0L\,\delta(x' - x_0)$. Then, integrating Equation 4.89b gives

$$T(x,t) = T_0L\ G_{X11}(x,t|x_0,0) \tag{4.92}$$

That is, the temperature rise is equal to the GF for the source located at x_0 and $t_0 = 0$ with strength T_0L (the units of T_0L are K-m). The symbol x_0 in Equation 4.92 could be replaced by x' to denote that the source is at x'.

The $G_{X11}(x,t|x',\tau)$ function is found by replacing t in Equation 4.90 by $t - \tau$ and limiting the time domain to $0 \leq \tau \leq t$. The $G_{X11}(\cdot)$ function satisfies the boundary conditions of $G_{X11}(0,t|x',\tau) = 0$ and $G_{X11}(L,t|x',\tau) = 0$. Also note that the $G_{X11}(\cdot)$ function given by Equation 4.90 is unchanged by interchanging x and x'. In other words, if the value of a GF at x is known for a source at x', then the same value applies to the GF at x' for a source at x; $G(\cdot)$ is symmetric in x and x'.

It is instructive to examine a plot of $G_{X11}(\cdot)$ for several values of $\alpha(t - \tau)/L^2$ and several values of x'/L. See Figure X11.1 in Appendix X. For small time values such as $\alpha(t - \tau)/L^2 < 0.025$ and x' not near the boundary, $G_{X11}(\cdot)$ is approximated by $G_{X10}(\cdot)$. See Section 4.2 and the short cotime expression given in Table 4.1. As the time $\alpha(t - \tau)/L^2$ becomes larger, the effects of the boundaries increase. The $G_{X11}(\cdot)$ function approaches zero for $\alpha(t - \tau)/L^2 > 0.5$.

The $G_{X11}(\cdot)$ expression given by Equation 4.90 and that in Table 4.1 are both exact and give the same numerical values, but the former only needs a few terms for large $\alpha(t - \tau)/L^2$ values, while the latter needs only a few terms for small $\alpha(t - \tau)/L^2$ values. In general, the large cotime expression, Equation 4.90, is easier to manipulate mathematically.

Example 4.5: Plate Insulated on Both Sides—*X*22 Case

Find the GF for a plate insulated at both $x = 0$ and at $x = L$ with the separation of variables method.

Solution

The boundary value problem for an arbitrary initial condition is given by

$$\frac{\partial^2 T}{\partial x^2} = \frac{1}{\alpha}\frac{\partial T}{\partial t} \qquad 0 < x < L \qquad t > 0 \tag{4.93}$$

$$\left.\frac{\partial T}{\partial x}\right|_{x=0} = 0 \qquad \left.\frac{\partial T}{\partial x}\right|_{x=L} = 0 \tag{4.94}$$

$$T(x,0) = F(x) \tag{4.95}$$

The solution procedure is similar to that for the *X*11 case, and Equations 4.63 through 4.70 also apply to this case. The boundary condition at $x = 0$ is different, however, and yields

$$\left.\frac{\partial T_n(x,t)}{\partial x}\right|_{x=0} = \left.\frac{d\mathbf{X}}{dx}\right|_{x=0} \Theta(t) = 0 \tag{4.96}$$

and thus $d\mathbf{X}/dx = 0$ at $x = 0$ to give, from Equation 4.70a,

$$\left.\frac{d\mathbf{X}}{dx}\right|_{x=0} = 0 = C_1\lambda\cos(0) - C_2\lambda\sin(0)$$
$$= B \cdot 1 - C \cdot 0 \tag{4.97}$$

and thus $C_1 = 0$. Repeating this procedure at $x = L$ gives

$$\left.\frac{d\mathbf{X}}{dx}\right|_{x=L} = 0 = -C_2\lambda\sin\lambda L \tag{4.98}$$

and thus the eigencondition is

$$\sin\lambda_n L = 0 \tag{4.99}$$

with eigenvalues $\beta_n = \lambda_n L = n\pi$ for $n = 0, 1, 2,$ and so on. Notice that $n = 0$ is included because the eigenfunction for this case, $\cos(\beta_n x/L)$, reduces to unity for $n = 0$. The $\mathbf{X}(x)$ function (the eigenfunction) now becomes

$$\mathbf{X}(x) = \begin{cases} C_n\cos\dfrac{\beta_n x}{L} & n = 1, 2, \ldots \\ C_0 \cdot 1 & n = 0 \end{cases} \tag{4.100}$$

where Equation 4.70 is used with Equation 4.99 and with $C_1 = 0$. At this point the partial differential equation for $T_n(x, t)$ and the two homogeneous boundary conditions are satisfied.

Using the relation that $T(x, t)$ is the sum of the $T_n(x, t)$ values gives

$$T(x, t) = \sum_{n=0}^{\infty} A_n\, e^{-\beta_n^2\alpha t/L^2}\cos\frac{\beta_n x}{L} \tag{4.101}$$

Using the initial condition, Equation 4.95, yields

$$F(x) = \sum_{n=0}^{\infty} A_n\cos\frac{\beta_n x}{L} \tag{4.102}$$

which is a Fourier cosine series. The A_n's can be found by multiplying Equation 4.102 by $\cos(\beta_m x/L)\,dx$ and integrating over the domain, which is $0 < x < L$,

$$\int_{x=0}^{L} F(x)\cos\frac{\beta_m x}{L}\,dx = \int_{x=0}^{L}\sum_{n=0}^{\infty} A_n\cos\frac{\beta_n x}{L}\cos\frac{\beta_m x}{L}\,dx \tag{4.103}$$

For the β_n values of $n\pi$, the orthogonality relation involving the cosine function is

$$\int_{x=0}^{L}\cos\frac{\beta_n x}{L}\cos\frac{\beta_m x}{L}\,dx = \begin{cases} 0 & m \neq n \\ L & m = n = 0 \\ \dfrac{L}{2} & m = n \neq 0 \end{cases} \tag{4.104}$$

Utilizing this relation in Equation 4.103 gives

$$A_0 = \frac{1}{L}\int_{x=0}^{L} F(x)\,dx \tag{4.105a}$$

$$A_m = \frac{2}{L}\int_{x=0}^{L} F(x)\cos\frac{\beta_m x}{L}\,dx \qquad m = 1, 2, \ldots \tag{4.105b}$$

As noted in connection with Equation 4.87, the subscript m in Equation 4.105b can be replaced by another index such as n.

The complete solution to this $X22$ problem posed by Equations 4.93 through 4.95 is given by Equation 4.101 with $A_m(m \to n)$ given by Equation 4.105b. However, the purpose here is to demonstrate that a GF can be derived with separation of variables theory. Hence, introduce Equation 4.105b with $m \to n$ and $x \to x'$ into Equation 4.101 to get

$$T(x,t) = \frac{1}{L}\int_{x=0}^{L} F(x')\,dx' + \sum_{n=1}^{\infty}\frac{2}{L}\int_{x=0}^{L} F(x')\cos\frac{\beta_n x'}{L}\,dx' \times e^{-\beta_n^2 \alpha t/L^2}\cos\frac{\beta_n x}{L} \tag{4.106}$$

$$T(x,t) = \int_{x=0}^{L}\left[\frac{1}{L} + \frac{2}{L}\sum_{n=1}^{\infty} e^{-\beta_n^2 \alpha t/L^2}\cos\frac{\beta_n x}{L}\cos\frac{\beta_n x'}{L}\right] F(x')\,dx' \tag{4.107}$$

in which the term in brackets is the $G_{X22}(x, t|x', \tau)$ GF evaluated at $\tau = 0$. That is, Equation 4.107 can be written as

$$T(x,t) = \int_{x=0}^{L} G_{X22}(x, t|x', 0)\,F(x')\,dx' \tag{4.108}$$

where $G_{X22}(x, t|x', \tau)$ is found from the bracketed term in Equation 4.107 by replacing $(t - 0)$ by $(t - \tau)$ for $\tau \le t$,

$$G_{X22}(x, t|x', \tau) = \frac{1}{L} + \frac{2}{L}\sum_{n=1}^{\infty} e^{-\beta_n^2 \alpha (t-\tau)/L^2}\cos\frac{\beta_n x}{L}\cos\frac{\beta_n x'}{L} \tag{4.109a}$$

$$\beta_n = n\pi \qquad n = 1, 2, \ldots \tag{4.109b}$$

Notice that the $X22$ GF in Equation 4.109a has one more explicit term than the $X11$ GF in Equation 4.90. Then $n = 0$ term is not zero in the $X22$ case because $\cos(\beta_n x/L)$ is not zero for $n = 0$. The summation terms of the $X11$ and $X22$ GFs are quite similar. Both summations contain two trigonometric functions with arguments $\beta_n x/L$ and $\beta_n x'/L$. The eigenvalues are equal to $n\pi$ for the two summations. Both summations contain the factor $\exp[-\beta_n^2 \alpha(t-\tau)/L^2]$. For "large" values of dimensionless time, such as $\alpha(t-\tau)/L^2 \ge 1$, this exponential factor causes the summations in Equations 4.90 and 4.109 to approach zero in value. That is, $G_{X11}(\cdot)$ goes to zero and $G_{X22}(\cdot)$ goes to $1/L$ for large values of $\alpha(t-\tau)/L^2$. See Figure X22.1 (Appendix X) for several plots of G_{X22}.

TABLE 4.2
Eigenfunctions for the Long-Cotime Green's Functions Given by

$$G(x,t|x',\tau) = \frac{X_0(x)}{N_0} + \sum_{m=1}^{\infty} \exp\left[-\beta_m^2 \alpha(t-\tau)/L^2\right] \frac{X_m(x)X_m(x')}{N_m}$$

Number	Eigenfunctions, $X_m(x)$	A_1	A_2
$X1J, J = 1,2,3,4,5$	$\sin \beta_m x / L$	1	0
$X2J, J = 1,2,3,4,5$	$\cos \beta_m x / L$	0	1
$X31$	$\sin \beta_m (L-x)/L$	1	0
$X32$	$\cos \beta_m (L-x)/L$	0	1
$X33, X34, X35$	$B_1 \sin(\beta_m x/L) + \beta_m \cos(\beta_m x/L)$	B_1	β_m
$X4J, J = 1,2,3,4,5$	$-C_1\beta_m \sin(\beta_m x/L) + \cos(\beta_m x/L)$	$-C_1\beta_m$	1
$X5J, J = 1,2,3,4,5$	$(B_1 - C_1\beta_m^2)\sin(\beta_m x/L) + \beta_m \cos(\beta_m x/L)$	$B_1 - C_1\beta_m^2$	β_m

Special cases:

For $X22$, $X24$, $X42$, and $X44$: $\mathbf{X}_0(x) = 1$

For all other cases $\mathbf{X}_0(x) = 0$

$$B_i = h_i L/k, C_i = (\rho c b)_i / \rho c L, i = 1,2$$

A compact list of one-dimensional GFs based on the separation of variables approach is contained in Tables 4.2 and 4.3. These are best for "large" cotimes; a complete compilation for both large and small cotimes are given in Appendix X. A brief list of eigenvalues for some flat plate geometries involving convection boundary conditions (3rd kind) are given in Table 4.4.

4.5 PRODUCT SOLUTION FOR TRANSIENT GF

The solution of certain two- and three-dimensional transient heat conduction problems can be obtained very simply as the product of one-dimensional transient solutions. In this section, certain two- and three-dimensional GFs are shown to be products of one-dimensional GFs in the rectangular and cylindrical coordinate systems. Product solutions are not permitted in the spherical coordinate system. Product solutions are not generally possible for steady heat conduction.

4.5.1 Rectangular Coordinates

In rectangular coordinates, one-dimensional transient GFs can be multiplied together to form two- and three-dimensional GFs under the following restrictions: (1) the boundary conditions are of the type 0, 1, 2, or 3 (types 4 and 5 are not permitted); (2) if boundary conditions of the third type are present, the heat transfer coefficient h_i must be a constant for a given surface s_i.

TABLE 4.3
Eigenvalues and Norms for Green's Functions Obtained Using the Method of Separation of Variables

Eigenvalues are positive roots of:

$$\tan\beta_m = \frac{\beta_m[K_1(B_2 - C_2\beta_m^2) + K_2(B_1 - C_1\beta_m^2)]}{K_1K_2\beta_m^2 - (B_1 - C_1\beta_m^2)(B_2 - C_2\beta_m^2)}$$

(K, B, and C are defined below.)

Simple Cases:
for $X11$ and $X22$, $\beta_m = m\pi, m = 1, 2, \ldots$
for $X12$ and $X21$, $\beta_m = (2m-1)\pi/2, m = 1, 2, \ldots$

Norms for $m = 1, 2, \ldots$

$$N_m = L\left(\frac{1}{2}(A_1^2 + A_2^2) + A_2^2(C_1 + C_2) + \frac{\tan\beta_m}{1+\tan^2\beta_m}\left\{\frac{1}{2\beta_m}(A_2^2 - A_1^2) + 2C_2A_1A_2 + \tan\beta_m\left[C_2(A_1^2 - A_2^2) + \frac{1}{\beta_m}A_1A_2\right]\right\}\right)$$

(A_1 and A_2 are given in Table 4.2.)

Simple cases: $N_m = L/2$ for $X11$, $X12$, $X21$, and $X22$.
Special cases: $N_0 = (1 + C_1 + C_2)L$ for $X22$, $X24$, $X42$, and $X44$ for $\beta_0 = 0$.

Use XIJ; $I, J = 1, 2, 3, 4, 5$:

I	K_1	B_1	C_1	J	K_2	B_2	C_2
1	0	1	0	1	0	1	0
2	1	0	0	2	1	0	0
3	1	B_1	0	3	1	B_2	0
4	1	0	C_1	4	1	0	C_2
5	1	B_1	C_1	5	1	B_2	C_2

and where $K_i = k_i/k$, $B_i = h_iL/k$, $C_i = (\rho cb)_i/\rho cL$, $i = 1, 2$.

The following discussion of product solutions begins with product solutions for *temperature* due to arbitrary initial conditions. Then, a particular initial condition, the Dirac delta function, is used to show that GFs also form product solutions. A two-dimensional case is demonstrated, but the procedure can be repeated to treat three-dimensional cases.

Arbitrary initial conditions. Consider first the temperature due to an arbitrary initial condition in a two-dimensional body described by rectangular coordinates. The boundary conditions are homogeneous and volume energy generation is zero. That is, consider the following heat conduction problem:

$$\frac{\partial^2 T}{\partial x^2} + \frac{\partial^2 T}{\partial y^2} = \frac{1}{\alpha}\frac{\partial T}{\partial t} \tag{4.110a}$$

$$\frac{T(x, y, t = 0)}{T_0} = F^+(x, y) \tag{4.110b}$$

TABLE 4.4
Some Eigenvalues for *X*13, *X*31, *X*23, *X*32, and *X*33 (Haji-Sheikh and Beck, 2000)

Eigenvalues	B	β_1	β_2	β_3	
Of $\tan\beta_m = -\beta_m / B$	0	1.5708	4.7124	7.8540	(also $X12$ and $X21$)
for $X13$ and $X31$	0.1	1.6320	4.7335	7.8667	
	1	2.0288	4.9132	7.9787	
	10	2.8628	5.7606	8.7083	
	100	3.1105	6.2211	9.3317	
	∞	3.1416	6.2832	9.4248	(also $X11$)
Of $\tan\beta_m = B / \beta_m$	0	0	3.1416	6.2832	(also $X22$)
for $X23$ and $X32$	0.1	0.3111	3.1731	6.2991	
	1	0.8603	3.4256	6.4373	
	10	1.4289	4.3058	7.2281	
	100	1.5552	4.6658	7.7764	
	∞	1.5708	4.7124	7.8540	(also $X12$ and $X21$)
Of $\tan\beta_m = 2\beta_m B / (\beta_m^2 - B^2)$	0	0	3.1416	6.2832	(also $X22$)
for $X33$ with $B_1 = B_2$	0.1	0.4435	3.2040	6.3149	
	1	1.3065	3.6918	6.5854	
	10	2.6277	5.3073	8.0671	
	100	3.0800	6.1601	9.2405	
	∞	3.1416	6.2832	9.4248	(also $X11$)

Source: Haji-Sheikh, A. and Beck, J.V., *Numerical Heat Transfer Part B Fundamentals*, 38, 133–156, 2000.

$$k_j \frac{\partial T}{\partial n_j} + h_j T = 0 \qquad j = 1, 2, \ldots, s \tag{4.110c}$$

where T_0 is a characteristic temperature, and s represents the number of boundary conditions ($0 \leq s \leq 4$ for the two-dimensional case). The convection heat transfer coefficient h_j must be a constant. Only boundary conditions of types 0, 1, 2, or 3 are treated.

Suppose that the dimensionless initial condition, $F^+(x, y)$, can be written as a product of two functions, one a function of x and the other a function of y:

$$F^+(x, y) = F_1^+(x)\, F_2^+(y) \tag{4.111}$$

Then, the following statement is true: the solution of the two-dimensional heat conduction problem defined by Equation 4.110a, b and c, can be written as the product of two functions

$$\frac{T(x, y, t)}{T_0} = T_1(x, t)\, T_2(y, t) \tag{4.112}$$

where T_1 and T_2 are dimensionless, and are defined by the following one-dimensional heat conduction problems:

$$x \text{ direction: } \frac{\partial^2 T_1}{\partial x^2} - \frac{1}{\alpha}\frac{\partial T_1}{\partial t} = 0 \tag{4.113a}$$

$$T_1(x, t = 0) = F_1^+(x) \tag{4.113b}$$

$$k_i \left.\frac{\partial T_1}{\partial n_i}\right|_{x=x_i} + h_i T_1|_{x=x_i} = 0 \qquad i = 1, 2 \tag{4.113c}$$

$$y \text{ direction: } \frac{\partial^2 T_2}{\partial y^2} - \frac{1}{\alpha}\frac{\partial T_2}{\partial t} = 0 \tag{4.114a}$$

$$T_2(y, t = 0) = F_2^+(y) \tag{4.114b}$$

$$k_i \left.\frac{\partial T_2}{\partial n_i}\right|_{y=y_i} + h_i T_2|_{y=y_i} = 0 \qquad i = 1, 2 \tag{4.114c}$$

Note that $i = 1, 2$ defines the two boundaries for each finite geometry. However, semi-infinite and infinite geometries are also allowed.

The above statement is proved by direct substitution of the product solution, Equation 4.112, into Equations 4.110a, b, and c. First, consider Equation 4.110a, the differential equation,

$$T_2 \frac{\partial^2 T_1}{\partial x^2} + T_1 \frac{\partial^2 T_2}{\partial y^2} - \frac{1}{\alpha}\left(T_2 \frac{\partial T_1}{\partial t} + T_1 \frac{\partial T_2}{\partial t}\right) = 0 \tag{4.115}$$

which can be written as

$$T_2 \left(\frac{\partial^2 T_1}{\partial x^2} - \frac{1}{\alpha}\frac{\partial T_1}{\partial t}\right) + T_1 \left(\frac{\partial^2 T_2}{\partial y^2} - \frac{1}{\alpha}\frac{\partial T_2}{\partial t}\right) = 0 \tag{4.116}$$

This equation is satisfied because it is the sum of the one-dimensional heat conduction Equations 4.113a and 4.114a.

Next, consider the initial condition, Equation 4.110b. Direct substitution of the product solution gives

$$T_1(x, 0)\, T_2(y, 0) = F^+(x, y) \tag{4.117}$$

and the initial condition has a product form given by Equation 4.111 to give

$$T_1(x, 0)\, T_2(y, 0) = F_1^+(x)\, F_2^+(y) \tag{4.118}$$

This equation is satisfied because it is the product of Equation 4.113b and Equation 4.114b. There are no unusual restrictions on the functions F_1^+ and F_2^+ (they may be zero, piecewise continuous functions, etc.).

Finally, consider the boundary conditions Equation 4.110c. Direct substitution of the product solution gives

$$k_j \frac{\partial (T_1 T_2)}{\partial n_j} + h_j (T_1 T_2) = 0 \tag{4.119}$$

There are two possibilities for the normal vector n_j in a two-dimensional rectangular coordinate system. The first possibility is for n_j parallel to the x-direction, in which case Equation 4.119 becomes

$$T_2\left(k_j\frac{\partial T_1}{\partial n_j}+h_jT_1\right)=0 \tag{4.120}$$

This equation is satisfied because it is Equation 4.113a multiplied by T_2. The second possibility is for n_j parallel to the y-direction, in which case Equation 4.119 is identical to Equation 4.114c multiplied by T_1. This concludes the proof of product solutions for temperature due to arbitrary initial conditions given by Equation 4.111.

Dirac delta function initial condition. Next, consider a specific initial condition, the Dirac delta function, given by

$$F^+(x,y)=L^2\delta(x-x')\,\delta(y-y') \tag{4.121}$$

where the length L may have any desired significance; it is used to make $F^+(x,y)$ dimensionless. The dimensionless initial condition, Equation 4.121, can be written as a product,

$$F^+(x,y)=L\delta(x-x')\cdot L\delta(y-y') \tag{4.122}$$

Then, the temperature $T(x,y)$ in a two-dimensional body that obeys Equation 4.110a and boundary conditions given by Equation 4.110c can also be written in product form (Equation 4.112):

$$\frac{T(x,y,t)}{T_0}=T_1(x,t)\,T_2(y,t) \tag{4.123}$$

Chapter 1 showed that the temperature, $T(r,t)$ caused by a Dirac delta function initial condition is equivalent to a GF multiplied by a constant:

$$T(r,t)=T_0L^mG(r,t|r',0) \tag{4.124}$$

where $m=1,2$, or 3 for one-, two-, or three-dimensional bodies; $G(\cdot)$ is the GF; T_0 is a characteristic temperature; and L is a characteristic length (for dimensional consistency).

Now, each of the functions $T_1(x,t)$ and $T_2(y,t)$ in Equation 4.123 can also be written in the form of GFs given in Equation 4.124,

$$T_1(x,t)=L\quad G_1(x,t|x',0) \tag{4.125a}$$

$$T_2(y,t)=L\quad G_2(y,t|y',0) \tag{4.125b}$$

Replace Equations 4.124 and 4.125 into Equation 4.123 to obtain

$$G(x,y,t|x',y',0)=G(x,t|x',0)\cdot G(y,t|y',0) \tag{4.126}$$

Finally, the time dependence of all GFs is $(t - \tau)$, so that in general, $(t - 0)$ can be replaced by $(t - \tau)$ to give

$$G(x, y, t|x', y', \tau) = G(x, t|x', \tau) \cdot G(y, t|y', \tau) \tag{4.127}$$

That is, the GF for the two-dimensional boundary value problem given in Equation 4.110 is the product of the one-dimensional GFs associated with the boundary value problems given in Equations 4.113 and 4.114.

In general, one-dimensional GFs multiply in rectangular coordinates to give two-dimensional GFs. Recall that product solutions are limited to boundary conditions of types 0, 1, 2, and 3. A repeated application of this analysis can be carried out to show the three-dimensional GF in rectangular coordinates can be found from a product of three one-dimensional GFs; that is, $G_{XYZ} = G_X \cdot G_Y \cdot G_Z$.

4.5.2 Cylindrical Coordinates

In cylindrical coordinates (r, ϕ, z), product solutions of transient GFs are allowed under the following restrictions: (1) the boundary conditions are of the type 0, 1, 2, or 3 (types 4 and 5 are not permitted); (2) if boundary conditions of the third type are present, the heat transfer coefficient h_i must be a constant for a given surface s_i; (3) a GF that depends only on the z-coordinate is multiplied by another GF that does *not* depend on the z-coordinate.

For example, let G_R, G_Φ, and G_Z represent one-dimensional GFs, let $G_{RZ}, G_{R\Phi}$, and $G_{\Phi Z}$ represent all possible two-dimensional GFs, and let $G_{R\Phi Z}$ represent the three-dimensional GF in cylindrical coordinates. Then, if the boundary conditions meet restrictions (1) and (2), the following product solutions are allowed in cylindrical coordinates:

$$G_{RZ} = G_R \cdot G_Z \tag{4.128a}$$

$$G_{\Phi Z} = G_\Phi \cdot G_Z \tag{4.128b}$$

$$G_{R\Phi Z} = G_{R\Phi} \cdot G_Z \tag{4.128c}$$

Note that the GF $G_{R\Phi}$ cannot be found by a product solution.

4.6 METHOD OF EIGENFUNCTION EXPANSIONS

We have seen that the separation of variables method, when applied to transient conduction, produces series solutions that involve eigenfunctions. In this section, eigenfunction expansions will be used directly to find the steady Green's function on finite domains. Earlier in this chapter, the eigenfunctions for the slab were found by separation of variables. In later chapters the appropriate eigenfunctions are given for the cylinder (Chapter 7) and for the sphere (Chapter 8) that can be used with this method. Eigenfunction expansions are also discussed elsewhere, for example Barton (1989, Chapter 5) and Duffy (2001, Chapter 5).

In the present discussion, we begin with the series form of the Dirac delta function. The series form of the Dirac delta function on finite domain R involves eigenfunctions ϕ_m and norm N_m, as follows (see Appendix D):

$$\sum_m \frac{\phi_m^*(r')\phi_m(r)}{N_m} = \begin{cases} \delta(r-r'); & \text{Cartesian} \\ \delta(r-r')/(2\pi r); & \text{cylindrical} \\ \delta(r-r')/(4\pi r^2); & \text{spherical-radial} \end{cases} \tag{4.129}$$

Forms for three coordinate systems are given here; note that the cylindrical and spherical coordinates include weighting factors $(2\pi r)$ and $(4\pi r^2)$, respectively*. Eigenfunctions ϕ_m satisfy the specified homogeneous boundary conditions at the boundaries of domain R, the same conditions also satisfied by the GF on R. For each combination of boundary conditions, there is a distinct GF and a distinct series form of the δ-function. For most geometries the summation begins at $m = 1$, but for bodies with all boundaries insulated, the summation begins at $m = 0$; see Section 4.7.2 for further discussion of this point.

We seek a series form of the GF identical to the δ-function series, but with an undetermined parameter (this method is also called "variation of parameters"). That is, we seek G in the form

$$G(r, r') = \sum_m C_m \frac{\phi_m^*(r')\phi_m(r)}{N_m} \tag{4.130}$$

To find unknown parameter C_m, simply replace this series into the differential equation for G. Rather than continue with a general discussion, specific examples are next given to demonstrate the procedure. The first example is a one-dimensional steady case, given only as a demonstration of the method. The second example is steady heat conduction in a two-dimensional rectangle, since the method of eigenfunction expansion is most important for two-and three-dimensional cases. The third example, a one-dimensional transient case, combines the eigenfunction expansion method and the Laplace transform method.

Example 4.6: Steady Case $X12$

Consider a 1D plate ($0 < x < L$) with boundary conditions of the first kind at $x = 0$ and of the second kind at $x = L$. Find the steady GF with eigenfunction expansions.

Solution

The Green's function satisfies the following equations:

$$\frac{d^2G}{dx^2} = -\delta(x - x') \tag{4.131}$$

$$G(0, x') = 0$$

$$\frac{dG(L, x')}{dx} = 0$$

*Other authors use weighting factors (r) and (r^2) for cylindrical and spherical coordinates, respectively.

For these boundary conditions, the eigenfunctions are $\phi_m = \sin(\beta_m x/L)$ (see Table 4.2), the norm is $N_m = L/2$, and the eigenvalues are $\beta_m = (m - 1/2)\pi$ (see Table 4.3). Next, assemble the series forms for both δ and G using Equations 4.129 and 4.130 and substitute them into Equation 4.131, the differential equation for G. The result is

$$\frac{d^2}{dx^2}\left\{\sum_{m=1}^{\infty} C_m \frac{\sin(\beta_m \frac{x'}{L})\sin(\beta_m \frac{x}{L})}{L/2}\right\} = -\sum_{m=1}^{\infty} \frac{\sin(\beta_m \frac{x'}{L})\sin(\beta_m \frac{x}{L})}{L/2} \quad (4.132)$$

Differentiate two times, and rearrange to restate the differential equation as one series:

$$\frac{2}{L}\sum_{m=1}^{\infty} \sin\left(\beta_m \frac{x'}{L}\right)\sin\left(\beta_m \frac{x}{L}\right)\left\{-C_m\left(\frac{\beta_m}{L}\right)^2 + 1\right\} = 0 \quad (4.133)$$

The above equation will be satisfied for all m if the expression in braces is zero. That is,

$$-C_m\left(\frac{\beta_m}{L}\right)^2 + 1 = 0 \quad (4.134)$$

which is satisfied by $C_m = (L/\beta_m)^2$. Replace this value for C_m into the series expansion for G to find, for steady case X12,

$$G(x, x') = \frac{2}{L}\sum_{m=1}^{\infty}\left(\frac{L}{\beta_m}\right)^2 \sin\left(\beta_m \frac{x'}{L}\right)\sin\left(\beta_m \frac{x}{L}\right) \quad (4.135)$$

In this example a series form of a steady 1D GF was found by eigenfunction expansion. Such series for 1D GF, although mathematically correct, are not recommended for numerical computation because nonseries forms for G may be found by direct integration (see Section 1.7.1). For two- or three-dimensional problems, however, the eigenfunction expansion method produces a very useful series form for the GF, as in the next example.

Example 4.7: Steady Case *X*12*Y*12

Consider a rectangle described by coordinates $(0 < x < L)$ and $(0 < y < W)$. Suppose the boundary conditions are $G = 0$ at $x = 0$ and at $y = 0$, and, $\partial G/\partial n = 0$ at $x = L$ and $y = W$. Find the steady GF.

Solution

The steady GF satisfies the following equations:

$$\frac{\partial^2 G}{\partial x^2} + \frac{\partial^2 G}{\partial y^2} = -\delta(x - x')\delta(y - y') \quad (4.136)$$

$$G(x = 0) = G(y = 0) = 0$$

$$\left.\frac{\partial G}{\partial x}\right|_{x=L} = 0; \quad \left.\frac{\partial G}{\partial y}\right|_{y=W} = 0$$

This is case X12Y12. The eigenfunction expansion for G is patterned on the series form of the Dirac delta function. The delta function appropriate for the x-direction is for case X12, which was also used in the previous example, is given by

$$\delta(x - x') = \frac{2}{L} \sum_{m=1}^{\infty} \sin\left(\beta_m \frac{x}{L}\right) \sin\left(\beta_m \frac{x'}{L}\right) \tag{4.137}$$

We seek the GF of a similar form,

$$G(x, y|x', y') = \frac{2}{L} \sum_{m=1}^{\infty} \sin\left(\beta_m \frac{x}{L}\right) \sin\left(\beta_m \frac{x'}{L}\right) P(y, y') \tag{4.138}$$

where $P(y, y')$ is an unknown *kernel function*. Replace the above series for G and $\delta(x - x')$ in Equation 4.136 to find, after some rearranging:

$$\frac{2}{L} \sum_{m=1}^{\infty} \sin\left(\beta_m \frac{x}{L}\right) \sin\left(\beta_m \frac{x'}{L}\right) \left\{-P\left(\frac{\beta_m}{L}\right)^2 + P'' + \delta(y - y')\right\} = 0 \tag{4.139}$$

This equation will be satisfied if the term in braces is zero, that is,

$$P'' - \sigma_m^2 P + \delta(y - y') = 0, \quad (0 < y < W) \tag{4.140}$$

where $\sigma_m^2 = \beta_m^2 / L^2$. We could solve for P by once again using eigenfunction expansion which would produce a *double* summation for G. However, a better-behaved solution can be found by directly integrating Equation 4.140 for P. As in Section 1.7.2, divide the domain at $y = y'$ to remove the δ-function from the differential equation. Then seek P_1 on $(0 < y < y')$ and seek P_2 on $(y' < y < W)$, that satisfy

$$P_i'' - \sigma_m^2 P_i = 0, \quad \text{for } i = 1, 2 \tag{4.141}$$

Integrate directly to find a general solution for P in the form

$$\begin{aligned} P_1 &= C_1 e^{\sigma_m y} + C_2 e^{-\sigma_m y}, \quad y < y' \\ P_2 &= C_3 e^{\sigma_m y} + C_4 e^{-\sigma_m y}, \quad y > y' \end{aligned} \tag{4.142}$$

Four conditions are needed to find the four coefficients. Two conditions on P_i come from the boundary conditions for G at $y = 0$ and $y = W$:

$$\text{(i)} \quad P_1(y = 0) = 0$$

$$\text{(ii)} \quad \left.\frac{\partial P_2}{\partial y}\right|_{y=W} = 0$$

Two more conditions at $y = y'$ are the matching condition and the jump condition (see Section 1.7.2):

$$\text{(iii)} \quad P_1(y', y') = P_2(y', y')$$

$$\text{(iv)} \quad \left.\frac{\partial P_2}{\partial y'}\right|_{y=y'} - \left.\frac{\partial P_1}{\partial y'}\right|_{y=y'} = -1$$

Using these four conditions, coefficients C_i may be found algebraically and replaced into Equation 4.142. The result is, for $(y < y')$,

$$P_1(y, y') = \frac{-e^{-\sigma_m(2W+y-y')} + e^{-\sigma_m(2W-y-y')}}{2\sigma_m(1 + e^{-2\sigma_m W})} + \frac{e^{-\sigma_m(y'-y)} - e^{-\sigma_m(y+y')}}{2\sigma_m(1 + e^{-2\sigma_m W})} \tag{4.143}$$

and for $(y > y')$,

$$P_2(y, y') = \frac{-e^{-\sigma_m(2W-y+y')} + e^{-\sigma_m(2W-y-y')}}{2\sigma_m(1 + e^{-2\sigma_m W})} + \frac{e^{-\sigma_m(y-y')} - e^{-\sigma_m(y+y')}}{2\sigma_m(1 + e^{-2\sigma_m W})} \tag{4.144}$$

Then replace P into the series for G to obtain

$$G(x, y|x', y') = \frac{2}{L}\sum_{m=1}^{\infty} \sin(\beta_m \frac{x}{L}) \sin(\beta_m \frac{x'}{L}) \left[\frac{-e^{-\sigma_m(2W+|y'-y|)} + e^{-\sigma_m(2W-y-y')}}{2\sigma_m(1 + e^{-2\sigma_m W})} + \frac{e^{-\sigma_m|y-y'|} - e^{-\sigma_m(y+y')}}{2\sigma_m(1 + e^{-2\sigma_m W})} \right] \tag{4.145}$$

Here an absolute value has been used to give P with a single expression. The above series was created by examining the δ-function along the x-direction. An alternate single-sum form for G may be found by starting with the y-direction δ function, placing eigenfunctions along the y-direction, and seeking kernel function $Q(x, x')$. Alternate forms for G are very important for checking purposes and for verification, as discussed in Chapter 5.

The above example is one of several GF that may be constructed for the rectangle. For other combinations of boundary conditions in the rectangle, see Table 4.2 for the appropriate eigenfunctions and Table X.4 (Appendix X) for the appropriate kernel functions. There is a special case for the rectangle when the series for Y22 is involved. In this case the summation begins at $m = 0$ and $\beta_0 = 0$ is an eigenvalue, and an additional kernel function P_0 is required. See Tables X.2 and X.4 in Appendix X for these kernel functions. For case X22Y22, the rectangle with all boundaries insulated, a pseudo-GF is required (see Section 4.7.2).

The eigenfunction expansion method may be used to find steady GF in any orthogonal coordinate system and for other combinations of boundary conditions. Additional examples are given elsewhere for the rectangle (Cole and Yen, 2001a), the two-dimensional slab (Cole and Yen, 2001b), the parallelepiped (Crittenden and Cole, 2002) and the cylinder (Cole, 2004).

Transient problems may be treated with the eigenfunction expansion method if the time-derivative term can first be removed. Later in Chapter 9, steady-periodic heat conduction is treated by the eigenfunction expansion method. In the following example, the Laplace transform method is combined with the eigenfunction expansion method.

Example 4.8: Transient Case *X*12

Consider the transient temperature in a one-dimensional slab ($0 < x < L$) with boundary conditions of the first kind at $x = 0$ and of the second kind at $x = L$. Find the transient GF by combining the Laplace transform method and the eigenfunction expansion method.

Solution

The GF for transient case X12 satisfies following equations:

$$\frac{\partial^2 G}{\partial x^2} - \frac{1}{\alpha}\frac{\partial G}{\partial u} + \frac{1}{\alpha}\delta(x - x')\,\delta(u) = 0 \tag{4.146a}$$

$$G(0, x', u) = 0 \tag{4.146b}$$

$$\frac{dG(L, x', u)}{dx} = 0 \tag{4.146c}$$

$$G(x, x', 0) = 0 \tag{4.146d}$$

where $u = t - \tau$ is the cotime. The GF will be sought by using the Laplace transform on the above equations, with respect to cotime, to give

$$\frac{\partial^2 \overline{G}}{\partial x^2} - \frac{s}{\alpha}\overline{G} + \frac{1}{\alpha}\delta(x - x')\cdot 1 = 0 \tag{4.147a}$$

$$\overline{G}(0, x') = 0 \tag{4.147b}$$

$$\frac{d\overline{G}(L, x')}{dx} = 0 \tag{4.147c}$$

Note that the Laplace transform of $\delta\,(u)$ is unity. Next we seek the Laplace-domain solution for $\overline{G}$ using eigenfunction expansions in the form

$$\overline{G}(x, x', s) = \sum_{m=1}^{\infty} D_m \frac{\sin(\beta_m \frac{x'}{L})\sin(\beta_m \frac{x}{L})}{L/2} \tag{4.148}$$

where D_m is an undetermined parameter and where the eigenfunctions, eigenvalues, and norm for the X12 case are taken from the from the previous example. Eigenfunctions, eigenvalues, and norms for other kinds of boundaries are given in Tables 4.2 and 4.3. Note that the above series automatically satisfies the boundary conditions at $x = 0$ and $x = L$ through the eigenfunctions and eigenvalues. We also need the series form of $\delta(x - x')$, which from the previous example is given by

$$\delta_{X12}(x - x') = \sum_{m=1}^{\infty} \frac{\sin(\beta_m \frac{x'}{L})\sin(\beta_m \frac{x}{L})}{L/2} \tag{4.149}$$

To determine parameter D_m, substitute the series expressions for $\overline{G}$ and $\delta(x - x')$ into Equation 4.147a to find:

$$\frac{2}{L}\sum_{m=1}^{\infty} \sin\left(\beta_m \frac{x'}{L}\right)\sin\left(\beta_m \frac{x}{L}\right)\left[-D_m\left(\frac{\beta_m}{L}\right)^2 - D_m\frac{s}{\alpha} + \frac{1}{\alpha}\right] = 0 \tag{4.150}$$

Here the common elements in each series have been grouped together. The above equation will be true if the expression in brackets is zero for all m. That is,

$$-D_m\left(\frac{\beta_m}{L}\right)^2 - D_m\frac{s}{\alpha} + \frac{1}{\alpha} = 0 \tag{4.151}$$

Solving for D_m, we find

$$D_m = \frac{1}{s + \alpha\beta_m^2/L^2} \tag{4.152}$$

Replace this value for D_m back into the series for $\overline{G}$, Equation 4.148, to find the solution in transform space:

$$\overline{G}(x, x', s) = \sum_{m=1}^{\infty}\left(\frac{1}{\alpha\beta_m^2/L^2 + s}\right)\frac{\sin(\beta_m\frac{x'}{L})\sin(\beta_m\frac{x}{L})}{L/2} \tag{4.153}$$

To complete the solution, this series can be inverse-Laplace-transformed term by term (using the linear property) along with the following transform pair (Appendix L, Table L.1, number 12):

$$\mathscr{L}^{-1}\left(\frac{1}{s+a}\right) = e^{-at} \tag{4.154}$$

Then the time-domain solution is given by

$$G(x, x', u) = \frac{2}{L}\sum_{m=1}^{\infty} e^{-\beta_m^2\alpha u/L^2}\sin\left(\beta_m\frac{x'}{L}\right)\sin\left(\beta_m\frac{x}{L}\right) \tag{4.155}$$

where u is the cotime. This is the large-cotime form of the GF for case X12, and this GF is also listed in Appendix X. In this example the eigenfunction expansion method has been applied to a transient problem in combination with the Laplace transform method; the result is identical to that found by the separation of variables method.

4.7 STEADY GREEN'S FUNCTIONS

Under steady-state conditions the heat conduction equation reduces to the Poisson equation. Much has been written about the Poisson equation in the fields of electrostatics, elasticity, diffusion, and heat transfer. Many books on theoretical physics contain an overview of solution methods to the Poisson equation and its special case, the Laplace equation, including Morse and Feshbach (1953), Melnikov (1999) and Duffy (2001). The method of GFs is only one of many solution methods, and we have chosen a unified treatment of GFs at the expense of completeness. Although we do not present other methods, we do not mean to imply that other methods are not important. For example, the use of complex variables and conformal transformations is a powerful method for two-dimensional problems.

In some ways the steady GFs are more difficult to apply than the transient GFs. The steady GFs behave very differently in one, two, and three dimensions. Unlike the transient GFs, the one-dimensional steady GF may not be multiplied to find two- or three-dimensional solutions; the steady GF for each geometry must be found separately.

There are sometimes two forms of the steady GF, depending on the method used to derive it. For example, in a two-dimensional rectangle, eigenfunction expansions may be carried out along x or along y to produce distinct series. These are different expansions of the same unique solution, with different convergence properties that can be used to advantage.

In this section, three topics on steady GFs are discussed. The source solutions, basic functions found by direct integration, are presented to indicate when a steady GF exists. The pseudo-GF is presented for those cases for which the (usual) steady GF does not exist. Finally, the limit method is presented to show the relationship between steady and transient GF.

4.7.1 Integration of the Auxiliary Equation: The Source Solutions

For one-dimensional cases the auxiliary equation for the steady GF can be solved directly by integration. The solution for the point source, the line source, and the plane source in the infinite body will be examined to demonstrate the method. The source solutions are important in certain numerical methods, such as the boundary element method. For the present discussion, the source solutions are useful in understanding the functional form of the steady GF before considering the added complexity of boundary conditions. The distinction between the source solutions and the GF is important: the source solution satisfies the auxiliary equation alone and may or may not satisfy homogeneous boundary conditions, but the GF satisfies a *boundary value problem* which includes the auxiliary equation and homogeneous boundary conditions.

Point source (three dimensions). The point source solution is the steady temperature induced at location $\mathbf{r}$ by a point heat source at location $\mathbf{r}'$. The point source solution depends only on the distance $(\mathbf{r} - \mathbf{r}')$, so the appropriate coordinate system is spherical polar coordinates. The point-source solution satisfies the equation

$$\frac{1}{r^2}\frac{d}{dr}\left(r^2\frac{dG}{dr}\right) = -\frac{\delta(r - r')}{4\pi r^2} \tag{4.156}$$

The solution to Equation 4.156 is:

$$G(\mathbf{r}|\mathbf{r}') = \frac{1}{4\pi|r - r'|} \tag{4.157}$$

The point-source solution is given by symbol $G(\mathbf{r}|\mathbf{r}')$ because it is also a GF: it satisfies the homogeneous boundary condition $G(r \to \infty) = 0$. The point source solution is singular at $|\mathbf{r} - \mathbf{r}'| = 0$. In rectangular coordinates the point source solution may be written

$$G(x, y, z|x', y', z') = \frac{1}{4\pi}\left[(x - x')^2 + (y - y')^2 + (z - z')^2\right]^{-1/2} \tag{4.158}$$

Derivation of the point source solution. The point source may be found by integrating the differential Equation 4.156. For the moment, let the source be located at $\mathbf{r}' = 0$ to simplify the analysis. We can translate the source back to $\mathbf{r}' \neq 0$ later. The Dirac delta function $\delta(\mathbf{r})$ is zero everywhere except at $\mathbf{r} = 0$, so except at this point, G should satisfy the Laplace equation in spherical polar coordinates

$$\frac{1}{r^2}\frac{d}{dr}\left(r^2\frac{dG}{dr}\right) = 0 \tag{4.159}$$

Integrating once:

$$r^2\frac{dG}{dr} = C_1 \qquad \frac{dG}{dr} = \frac{C_1}{r^2} \tag{4.160}$$

integrating again gives,

$$G = -\frac{C_1}{r} + C_2 \tag{4.161}$$

The constant C_2 may have any value to satisfy the Laplace equation and, if we take $C_2 = 0$, it will also satisfy the GF boundary condition $G \to 0$ at $r \to \infty$. The constant C_1 may be found to have the value $-1/(4\pi)$ by replacing G back into the differential equation 4.156 and integrating both sides of the equation over all space. The nature of the Dirac delta function allows us to equivalently integrate over a small sphere P centered at $r = 0$ with arbitrary small radius σ, because the integrand is zero for any integral that does not include the location of the Dirac delta function:

$$\int_P \nabla^2 G\, dV = -\int_P \frac{\delta(r - r')}{4\pi(r')^2}\, dV \tag{4.162}$$

Here $dV = 4\pi(r')^2 dr'$ is the differential volume. The right-hand side yields, with the sifting property of the Dirac delta function,

$$\int_P \nabla^2 G\, dV = -1 \tag{4.163}$$

The left-hand side may be simplified with the divergence theorem to give the integral over the surface of sphere P:

$$\int_{r=\sigma} d\mathbf{S} \cdot \nabla G = -1 \tag{4.164}$$

The value of ∇G in spherical coordinates evaluated at $r = \sigma$ may be substituted to give

$$\int_{r=\sigma} ds\frac{C_1}{\sigma^2} = -1 \tag{4.165}$$

Finally, the integral may be evaluated to give the surface area of the sphere,

$$\frac{4\psi^2 C_1}{\sigma^2} = -1 \tag{4.166}$$

or, $C_1 = -1/(4\pi)$, which completes the derivation for $r'=0$: $G(r|0) = 1/(4\pi r)$. Finally, the point source may be translated to arbitrary location $r' \neq 0$ by noting that $G(r|0) > 0$, and since a change of coordinate system should not change the sign of G, the vector magnitude is required: $G(\mathbf{r}|\mathbf{r}') = 1/(4\pi|\mathbf{r} - \mathbf{r}'|)$.

Line source (two dimensions). The cylindrical coordinate system is appropriate for the line source. The two-dimension differential equation for the line source in cylindrical coordinates is

$$\frac{1}{r}\frac{d}{dr}\left(r\frac{dG}{dr}\right) = -\frac{\delta(r - r')}{2\pi r'} \tag{4.167}$$

The solution to Equation 4.167 is

$$G(r|r') = \frac{-1}{2\pi} \ln |\mathbf{r} - \mathbf{r}'| \tag{4.168}$$

where $|\mathbf{r} - \mathbf{r}'|$ is a vector magnitude in cylindrical coordinates. Strictly speaking, Equation 4.168 has an error in the units because the argument of the log function should be dimensionless; however, in physical use the line source always has the form $\ln(a/|\mathbf{r} - \mathbf{r}'|)$ where a has the units of meters. In rectangular coordinates, the line source may be written

$$\begin{aligned} G(x, y|x', y') &= -\frac{1}{2\pi} \ln\left\{\left[(x - x')^2 + (y - y')^2\right]^{1/2}\right\} \\ &= -\frac{1}{4\pi} \ln\left[(x - x')^2 + (y - y')^2\right] \end{aligned} \tag{4.169}$$

Unlike the point source solution, the line source does not satisfy the homogeneous boundary condition $G \to 0$ at $r \to \infty$; the log function increases without bound as $r \to \infty$. However the heat flux approaches zero far from the line source ($k\partial G/\partial r \to 0$ as $r \to \infty$). This far-field boundary condition (second kind) is adequate for the line-source solution to be used to construct temperature solutions in the infinite body.

The line source is important in numerical methods such as the boundary element method. The boundary element method in two dimensions involves a distribution of line sources on a closed curve in the infinite body. The closed curve is broken into line segments called boundary elements, and the distribution of the line sources on the boundary elements is chosen to satisfy boundary conditions on the closed curve. The temperature in the body is evaluated by numerical summation over all the boundary elements, in effect superimposing the temperature induced by each source distribution. This is equivalent to the GF procedure of integrating over the volume to account for volume energy generation. For an introduction to the method see Brebbia and Dominquez (1992).

Plane source (one dimension). The steady plane source solution is described by the one-dimension steady-state heat equation

$$\frac{d^2 G_0}{dx^2} = -\delta(x - x') \qquad (4.170)$$

where $\delta(x - x')$ has units of $(\text{meters})^{-1}$. The solution for G_0 is

$$G_0(x|x') = -\tfrac{1}{2}|x - x'| \qquad (4.171)$$

The notation G_0 is used for the plane source solution because it is not a proper GF, because it does not satisfy homogeneous boundary conditions of the first or second kind as $x \to \infty$. The plane source solution blows up at $x \to \infty$, and in fact it blows up proportional to $|x - x'|$, which is faster than the line source solution which blows up like $\ln|\mathbf{r} - \mathbf{r}'|$. The heat flux, although not zero, is at least bounded as $x \to \infty$; this condition is sufficient for G_0 to be used for constructing temperature solutions.

The plane source solution may be derived by integrating the differential Equation 4.170 directly, but a little care is required. Since the heated plane divides the infinite body into two regions, the differential equation is integrated in two different regions and then the two solutions are linked by a jump condition at $(x - x') = 0$.

The plane source is not a GF because of a problem with the boundary conditions. The auxiliary equation always has a general solution, but the homogeneous boundary conditions cannot always be satisfied. There are several other geometries for which this problem occurs and such geometries do not have a steady GF. For example, the *X*22 geometry has no steady GF and neither do finite geometries with specified heat flux on *all* of the boundaries (boundary conditions of the second kind, also called Neumann boundary conditions).

A physical reason that some geometries do not have a steady GF function comes from the perspective of a GF as the response to a heat source. In steady heat transfer, any heat introduced inside the body must either flow out of the boundaries or flow off to infinity if the body is of infinite extent. If all the boundaries are insulated, there is nowhere for the heat to go and, consequently, there is no steady GF.

Steady *temperature* distributions can exist in bodies with no steady GF, but the usual GF method cannot be used to find the temperature. For example, the *X*22 geometry has a linear temperature distribution if the same amount of heat that flows into the body at $x = 0$ also flows out at $x = L$. In this simple case, the temperature distribution can be found by applying the *nonhomogeneous* boundary conditions to the general solution of the differential equation. The steady temperature can always be found with the transient GF solution equation (GFSE) in the limit as time becomes large ($t \to \infty$). Any questions on the existence of the steady-state temperature can be answered this way.

In the next section a pseudo-GF is discussed to deal with those geometries that do not have a steady GF because of insulated boundaries. The pseudo-GF differs from the ordinary GF by an additive constant. In physical terms, the additive constant cancels out the heat flow introduced by the heat source. A modified GFSE is then needed to calculate temperatures from the pseudo-GF.

4.7.2 Pseudo-Green's Function for Insulated Boundaries

For the special case in which all boundaries of the body are insulated, the usual steady GF does not exist and the usual steady GF solution cannot be used to find the temperature. In this section a pseudo-GF is discussed that can be used instead.

In this case the input data to the temperature problem must satisfy a constraint—the sum of the heat passing through the boundaries of the body must be equal to the (negative of the) integral of the heat introduced by volume energy generation. This is equivalent to an energy balance over the volume of the body. If there is no volume energy generation then the boundary heat fluxes must sum to zero. In addition, the solution for the temperature contains an arbitrary additive constant that must be supplied as input data.

The pseudo-GF, given the name G_{PS}, satisfies the following differential equation (Barton, 1989)

$$\nabla^2 G_{PS} = -\delta(\mathbf{r}-\mathbf{r}') + \frac{1}{V} \tag{4.172}$$

Here constant V represents the integration volume associated with $\delta(\mathbf{r}-\mathbf{r}')$. The boundary conditions (second kind) are given by

$$\frac{\partial G_{PS}}{\partial n_i} = 0 \quad \text{at boundary } i. \tag{4.173}$$

To use the pseudo-GF for finding temperature, a special form of the GF solution equation must be used:

$$T(\mathbf{r}) = \sum_i \int \frac{f_i}{k} G_{PS}(\mathbf{r},\mathbf{r}')\, dS_i' \tag{4.174}$$

$$+ \int \frac{g}{k} G_{PS}(\mathbf{r},\mathbf{r}')\, dV' + \langle T_{av} \rangle \tag{4.175}$$

where $\langle T_{av} \rangle$ is the spatial-average temperature in the body. For this solution to make sense, the boundary heating f_i and the internal heating g must satisfy an energy balance.

Two Cartesian cases are discussed below to demonstrate the pseudo-GF. For the one-dimensional slab, the pseudo-GF satisfies

$$\frac{\partial^2 G_{PS}}{\partial x^2} = -\delta(x-x') + \frac{1}{L}; \quad 0 < x < L \tag{4.176}$$

$$\left.\frac{\partial G_{PS}}{\partial x}\right|_{x=0} = \left.\frac{\partial G_{PS}}{\partial x}\right|_{x=L} = 0 \tag{4.177}$$

Note that additive constant is $1/L$ because the appropriate domain for the delta function is $(0 < x < L)$. The solution for this pseudo-GF may be found by direct integration (see Section 1.7.2):

$$G_{PS}(x,x') = \begin{cases} \left((x')^2 + x^2\right)/(2L) - x' + L/3, & x < x' \\ \left(x^2 + (x')^2\right)/(2L) - x + L/3, & x > x' \end{cases} \tag{4.178}$$

Note that this pseudo-GF contains an additive constant, $L/3$, that is needed to satisfy the differential equation but does not contribute to satisfying boundary conditions.

For the two-dimensional rectangle, case X22Y22, the pseudo-GF satisfies

$$\frac{\partial^2 G_{PS}}{\partial x^2} + \frac{\partial^2 G_{PS}}{\partial y^2} = -\delta(x - x')\,\delta(y - y') + \frac{1}{LW} \tag{4.179}$$

Here the additive constant is $1/(LW)$ because the integration domain for the 2D delta function is $(0 < x < L)$ and $(0 < y < W)$. The pseudo-GF in the rectangle has two alternate forms; using eigenfunction expansions along the x-axis,

$$G_{PS}(x, y|x', y') = \frac{P_0(y, y')}{L} + \frac{2}{L}\sum_{m=1}^{\infty} \cos(\beta_m x/L)\cos\left(\beta_m x'/L\right)\ P_m(y, y') \tag{4.180}$$

where $\beta_m = m\pi$ for $m = 0, 1, 2$, and so on. Here P_m is the usual kernel function that satisfies

$$P_m'' - \frac{\beta_m^2}{L^2} P_m + \delta(y - y') = 0 \tag{4.181}$$

Note that in this insulated-boundary case the eigenfunction expansion for $\delta(x - x')$ has the form

$$\delta(x - x') = \frac{1}{L} + \frac{2}{L}\sum_{m=1}^{\infty} \cos(\beta_m x/L)\cos\left(\beta_m x'/L\right) \tag{4.182}$$

where additive term $1/L$ is associated with the zero eigenvalue ($\beta_0 = 0$). Then, an additional kernel function, P_0, is also associated with the zero eigenvalue, which satisfies

$$P_0'' + \delta(y - y') = \frac{1}{W} \tag{4.183}$$

Kernel function $P_0(y, y')$ may be found from the 1D Cartesian pseudo-GF given above in Equation 4.178, by replacing x by y and replacing L by W.

An alternate pseudo-GF may be constructed, with eigenfunction expansions along the y-axis, in the form

$$G_{alt}(x, y|x', y') = \frac{P_0(x, x')}{W} + \frac{2}{W}\sum_{n=1}^{\infty} \cos(\gamma_n y/W)\cos\left(\gamma_n y'/W\right)\ P_n(x, x') \tag{4.184}$$

where $\gamma_n = n\pi$. Kernel functions $P_0(x, x')$ and $P_n(x, x')$ may be derived in a manner similar to that given above.

In this section the pseudo-GF has been explored for finite-domain Cartesian cases in 1D and 2D. The same principles apply for the 3D parallelepiped (case X22Y22Z22), and to insulated boundary geometries in cylindrical coordinates (cases R02, R02Z22, R02Z22ϕ22, etc.) and spherical coordinates (cases RS02, RS02ϕ22, etc.).

4.7.3 Limit Method

The steady GF can be calculated from the large-cotime transient GF by integrating over time and taking the limit as $t \to \infty$. That is,

$$G(\mathbf{r}, \mathbf{r}') = \lim_{t\to\infty} \alpha \int_{\tau=0}^{t} G(\mathbf{r}, t|\mathbf{r}', \tau)\, d\tau \tag{4.185}$$

This is the limit method. Many steady GFs can be written down immediately in integral form with the limit method. Three examples of the limit method are discussed below.

Example 4.9: Point Source in the Infinite Body

Find the steady point-source solution from the transient point source solution $G_{X00Y00Z00}$.

Solution

The point source will be located at the origin ($x' = 0,\ y' = 0,\ z' = 0$) for convenience. Later the source can be translated to any position. The limit method is given by the integral

$$G(x, y, z|0, 0, 0) = \lim_{t\to\infty} \alpha \int_{\tau=0}^{t} G_{X00Y00Z00}(x, y, z, t|0, 0, 0, \tau)\, d\tau \tag{4.186}$$

The product solution may be used for the transient $X00Y00Z00$ GF to give

$$G_{X00Y00Z00}(x, y, z, t|0, 0, 0, \tau) = G_{X00}(x, t|0, \tau)\, G_{Y00}(y, t|0, \tau)\, G_{Z00}(z, t|0, \tau)$$

where

$$G_{X00}(x, t|0, \tau) = [4\pi\alpha(t-\tau)]^{-1/2} \exp\left[\frac{-x^2}{4\alpha(t-\tau)}\right]$$

$$G_{Y00}(y, t|0, \tau) = [4\pi\alpha(t-\tau)]^{-1/2} \exp\left[\frac{-y^2}{4\alpha(t-\tau)}\right]$$

$$G_{Z00}(z, t|0, \tau) = [4\pi\alpha(t-\tau)]^{-1/2} \exp\left[\frac{-z^2}{4\alpha(t-\tau)}\right]$$

Then Equation 4.186 may be written

$$G(x, y, z|0, 0, 0) = \lim_{t\to\infty} \alpha \int_{\tau=0}^{t} [4\pi\alpha(t-\tau)]^{-3/2} \times \exp\left[\frac{-\left(x^2+y^2+z^2\right)}{4\alpha(t-\tau)}\right] d\tau \tag{4.187}$$

Note that the product of the three one-dimensional GFs is also the same as the $RS00$ GF given in Appendix RS for the case $r' = 0$. The above integral may be evaluated to give

$$G(x, y, z|0, 0, 0) = \lim_{t\to\infty} \frac{1}{4\pi r} \operatorname{erfc}\left[\frac{r}{(4\alpha t)^{1/2}}\right] = \frac{1}{4\pi|r|} \tag{4.188}$$

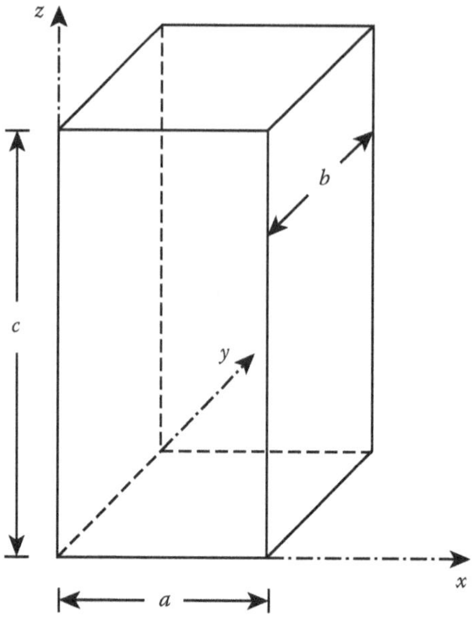

FIGURE 4.4 Parallelepiped geometry for Example 4.10.

where $r^2 = (x^2 + y^2 + z^2)$. This is the steady point-source solution located at $r' = 0$, as discussed in Section 4.7.1.

Example 4.10: Parallelepiped with Specified Surface Temperature—*X*11*Y*11*Z*11 Case

Find the steady GF in the parallelepiped with temperature boundary conditions (type 1) on all six surfaces.

Solution

The parallelepiped body is shown in Figure 4.4. The limit method integral for this case is given by

$$G(x, y, z|x', y', z') = \lim_{t\to\infty} \alpha \int_{\tau=0}^{t} G_{X11Y11Z11}(x, y, z, t|x', y', z', \tau)\, d\tau \quad (4.189)$$

The transient GF for the $X11Y11Z11$ geometry is given by the product of one-dimensional transient solutions: $G_{X11}G_{Y11}G_{Z11}$. The function G_{X11} is given in Appendix X:

$$G_{X11}(x, t|x', \tau) = \frac{2}{a}\sum_{m=1}^{\infty} e^{-m^2\pi^2\alpha(t-\tau)/a^2} \sin\frac{m\pi x}{a} \sin\frac{m\pi x'}{a} \quad (4.190)$$

where a is the length of the body in the x-direction. The functions G_{Y11} and G_{Z11} are similar; for example, G_{Y11} is given by Equation 4.190 with x and a replaced by y and b, respectively. Replace the transient GF into the integral to give

$$G(x, y, z|x', y', z') = \lim_{t\to\infty} \frac{8\alpha}{abc} \sum_{m=1}^{\infty}\sum_{n=1}^{\infty}\sum_{p=1}^{\infty} \int_{\tau=0}^{t} \exp\left[-\alpha\pi^2(t-\tau)\left(\frac{m^2}{a^2}+\frac{n^2}{b^2}+\frac{p^2}{c^2}\right)\right] d\tau$$
$$\times \sin\left(m\pi\frac{x}{a}\right)\sin\left(m\pi\frac{x'}{a}\right)\sin\left(p\pi\frac{z}{c}\right)\sin\left(p\pi\frac{z'}{c}\right)\sin\left(n\pi\frac{y}{b}\right)\sin\left(n\pi\frac{y'}{b}\right) \quad (4.191)$$

When the time integral is carried out and the limit taken, the steady GF becomes

$$G(x, y, z|x', y', z') = 8\sum_{m=1}^{\infty}\sum_{n=1}^{\infty}\sum_{p=1}^{\infty} \sin\left(m\pi\frac{x}{a}\right)\sin\left(m\pi\frac{x'}{a}\right)\sin\left(p\pi\frac{z}{c}\right)\sin\left(p\pi\frac{z'}{c}\right)$$
$$\times \sin\left(n\pi\frac{y}{b}\right)\sin\left(n\pi\frac{y'}{b}\right)\left[abc\pi^2\left(\frac{m^2}{a^2}+\frac{n^2}{b^2}+\frac{p^2}{c^2}\right)\right]^{-1} \quad (4.192)$$

Generally triple-sum series such as this converge slowly, and alternate series should be used for numerical evaluation, if possible. In the parallelepiped the eigenfunction expansion method can be used to construct three alternate double-sum series. For example, the double-sum form with kernel function along the z-direction is given by

$$G(x, y, z|x', y', z') = \sum_{m=1}^{\infty}\sum_{n=1}^{\infty} \sin\left(m\pi\frac{x}{a}\right)\sin\left(m\pi\frac{x'}{a}\right)$$
$$\times \sin\left(n\pi\frac{y}{b}\right)\sin\left(n\pi\frac{y'}{b}\right) P_{nm}(z, z') \quad (4.193)$$

where the kernel function P_{nm} is given by (Table X.4, case X11)

$$P_{nm}(z, z') = \frac{e^{-\sigma(2c+|z-z'|)} - e^{-\sigma(2c-z-z')}}{2\sigma(1-e^{-2\sigma c})} + \frac{e^{-\sigma|z-z'|} - e^{-\sigma(z+z')}}{2\sigma(1-e^{-2\sigma c})} \quad (4.194)$$

where $\sigma^2 = \pi^2(n^2 + m^2)$. Further discussion of the convergence speed of series solutions is given in Chapter 5.

Example 4.11: Two-Dimensional Slab with One Side Semi-Infinite—*X*11*Y*20 Case

Find the steady-state GF for the region $0 < x < a, y > 0$ with $G = 0$ at $x = 0$ and at $x = a$ and $\partial G/\partial y = 0$ at $y = 0$.

Solution

The limit method integral for this case is given (with $u = t - \tau$)

$$G(x, y|x', y') = \alpha \int_0^{\infty} G_{X11}(x, u|x')\, G_{Y20}(y, u|y')\, du \quad (4.195)$$

where $G_{X11}(x, u|x')$ is given by Equation 4.190 with $u = t - \tau$ and $G_{Y20}(y, u|y')$ is (see Equation X20.1, Appendix X)

$$G_{Y20}(y, u|y') = (4\pi\alpha u)^{-1/2}\left(e^{-(y-y')^2/(4\alpha u)} + e^{-(y+y')^2/(4\alpha u)}\right) \tag{4.196}$$

Integrals of the form (see integral 12 in Table I.6, Appendix I)

$$\int_0^\infty u^{-1/2} e^{-a^2 u - b^2 u^{-1}}\, du = \frac{\pi^{1/2}}{a} e^{-2ab} \tag{4.197}$$

are needed. Then, using Equations 4.190, 4.196, and 4.197 in Equation 4.195 gives

$$G(x, y|x', y') = \frac{1}{\pi}\sum_{m=1}^{\infty}\frac{1}{m}\left(e^{-m\pi|y-y'|/a} + e^{-m\pi(y+y')/a}\right) \times \sin\frac{m\pi x}{a}\sin\frac{m\pi x'}{a} \tag{4.198}$$

Observe for the point $y = y'$, $x = x'$ (with x not at 0 or a) that the value of G is unbounded, which is unlike the behavior of the one-dimensional GFs in the x-coordinate. Green's functions in the cylindrical coordinate system also have this unbounded behavior for r and r' going to zero.

PROBLEMS

Note: In many of the problems in this chapter the partial answers can be obtained by using the GFs tabulated in the appendixes. Unless otherwise requested, the reader should use Appendix I to evaluate integrals.

4.1 Using the method of images, find the transient GF for the region $0 < x < \infty, 0 < y < \infty$, with the boundary conditions of $\partial G / \partial x = 0$ at $x = 0$ and $\partial G / \partial y = 0$ at $y = 0$. Also find the GF using the product of the appropriate GFs and relate the corresponding terms.

4.2 Using the method of images, find the transient GF for the region $0 < x < L, 0 < y < \infty$, with $\partial G / \partial x = 0$ at $x = 0$ and L, and $\partial G / \partial y = 0$ at $y = 0$. Also find the GF using the product of the appropriate GFs and relate the corresponding terms.

4.3 Use the Laplace transform method to find the GF for the semi-infinite body with an insulated boundary (case $X20$).

4.4 Use the Laplace transform method to find the small-cotime form of the GF with boundaries of first kind at $x = 0$ and $x = L$ (case $X11$). Check your answer against Table 4.1.

4.5 Using a computer, evaluate $LG_{X11}(x, t|x', \tau)$ at $x / L = x' / L = 0.5$ for times $\alpha(t - \tau) / L^2 = 0.025$, 0.1, 0.5, and 1.0. Use two different expressions, one from Table 4.1 and the other from Tables 4.2 and 4.3. Determine the number of terms required for each expression for the different dimensionless times for the errors to be less than 0.0001 in value. Compare the values with those obtained from $LG_{X00}(\cdot)$.

4.6 Evaluate $LG_{X22}(x, t|x', \tau)$ at $x/L = x'/L = 0.5$ for $\alpha(t-\tau)/L^2 = 0.025$, 0.1, 0.5, and 1.0. Use two different expressions, one from Table 4.1 and the other from Tables 4.2 and 4.3. Determine the number of terms required for each expression for the different dimensionless times for the errors to be less than 0.0001. Compare the values with those obtained from $LG_{X00}(\cdot)$ and $LG_{X11}(\cdot)$.

4.7 Using expressions in Table 4.3, consider boundary conditions of the first and second kinds and also of the third kind for small values of B_1 and B_2 compared to 1 and both K_1 and K_2 not equal to zero (one K can be zero). Find an approximation of the first eigenvalue, β_1, using the approximate relation.

$$\cot x = \frac{1}{x} - \frac{x}{3}$$

in the eigencondition in Table 4.3. The use of this approximation yields a more accurate equation than a two-term approximation for $\tan x$. Why?

4.8 Show that eigenvalues calculated using the eigencondition in Table 4.3 gives $\beta_{m+1} = \beta_m + \pi$ for large β_m values.

4.9 For cases RS30 and X30, in the limit as $h \to \infty$ the following limit must be evaluated:

$$\lim_{m \to \infty} e^{m^2} \operatorname{erfc} m$$

Evaluate this limit (a) by using a series expression for the complementary error function, and (b) by using L'Hospital's rule. What kind of boundary condition results from this limit?

4.10 An instantaneous volume source from $-a$ to a in an infinite body is to be approximated by a finite number of line sources. Show that the exact solution is

$$\frac{1}{2}\left[\operatorname{erfc}\left(\frac{x-a}{\sqrt{4\alpha u}}\right) - \operatorname{erfc}\left(\frac{x+a}{\sqrt{4\alpha u}}\right)\right] \qquad \text{where } u = t - \tau$$

(The detailed derivation of this equation is not required if an appropriate integral in the book can be used.) This solution is to be approximated by a series of plane sources. Derive and evaluate the expressions for (a) a single source at $x = 0$, (b) three equally spaced, and (c) five equally spaced plane sources. Show that these approximations can be used to obtain

(a) $\operatorname{erf}(z) \simeq 2z/\pi^{1/2}$
(b) $\operatorname{erf}(z) \simeq (2z/\pi^{1/2})(1 + 2e^{-4z^2/9})/3$
(c) $\operatorname{erf}(z) \simeq (2z/\pi^{1/2})(1 + 2e^{-4z^2/25} + 2e^{-16z^2/25})/5$
Evaluate and compare these expressions with the exact values at $z = 0.05, 0.25, 1$, and 2.

4.11 Show all the steps to obtain the transient GF for case $X12$ using the separation of variables method. Check your answer with Appendix X.

4.12 Show all the steps to obtain the transient GF for case $X23$ using the separation of variables method. Check your answer with Appendix X.

4.13 Determine the transient GF for a line source at $x = x', y = y'$ for the boundary condition of the third kind at $y = 0$ and for the region of $-\infty < x < \infty, y > 0$.

4.14 Obtain the transient GF function for the case denoted $X13Z00$ using the product method.

4.15 Obtain the transient GF for the cases denoted $R02Z20$ and $R01\Phi00Z10$ using the product method.

4.16 Give the expressions for the GFs for the cases represented by $X00$, $X00Y00$, and $X00Y00Z00$. What is the physical significance for each case?

4.17 Evaluate the following integrals

$$\int_a^b G_{X00}(x,t|x',\tau)\,dx', \qquad \int_a^b G_{X00Y00}(x,y,t|x',y',\tau)\,dx',$$

and $\int_b^a G_{X00Y00Z00}(x,y,z,t|x',y',z',\tau)\,dx'$

(Perform the integration either explicitly or by using a table.) What physical situation does each integral represent? (Hint: compare to the GF solution equation.)

4.18 Compute numerical values from the series form of the steady case X12 (Example 4.6, Equation 4.135) for $x' = 0.2$ and for 50, 500, and 5000 terms of the series. Plot your numerical values over ($0 < x < L$) and discuss how well your plot agrees with the algebraic form of the this steady GF (Section 1.7.2).

4.19 Derive the steady-state GF for the $X11$ case by direct integration of the auxiliary problem

$$\frac{d^2G}{dx^2} = -\delta(x - x'); \quad G(0,x') = G(L,x') = 0$$

Compare your answer to Table X.1, Appendix X.

4.20 Derive the steady-state GF for the $X11$ case using the limit method and starting with Equation 4.190 for $G_{X11}(\cdot)$.

The answer is

$$G_{X11}(x,x') = \frac{2a}{\pi^2}\sum_{m=1}^{\infty}\frac{1}{m^2}\sin\frac{m\pi x}{a}\sin\frac{m\pi x'}{a}$$

4.21 (a) Program on a computer the expression for $G_{X11}(\cdot)/a$ given in Problem 4.20 as a function of x/a, x'/a and M; here M is the maximum number of terms used.

(b) Calculate using the computer program $G_{X11}(\cdot)/a$ as a function of the number of terms for $x/a = x'/a = 1/2$. Also tabulate the errors by using the nonseries solution of Problem 4.19. How many terms are needed to obtain accuracy within 1%? By observing the dependence the error as a function of number of terms, how many terms would be needed to obtain 0.1%?

4.22 Verify that in two-dimensional cylindrical coordinates the function

$$G\left(r|r'\right) = -\frac{1}{2\pi}\ln|r - r'|$$

satisfies the differential equation $\nabla^2 G = -\delta(r - r')/(2\pi r')$.

4.23 Show by direct integration of the energy equation that the steady GF for the *X*23 geometry is given by

$$G(x|x') = \begin{cases} L\left(\dfrac{1+B}{B} - \dfrac{x}{L}\right) & 0 \le x' \le x \\ L\left(\dfrac{1+B}{B} - \dfrac{x'}{L}\right) & x \le x' \le L \end{cases}$$

where $B = hL/k$ is the Biot number.

4.24 Use the method of eigenvalue expansions to find the steady GF in the rectangle for case X12Y11 with eigenfunctions in the x-direction. Compare your expression with Example 4.7 and comment on the similarities and/or differences.

4.25 Use the method of eigenvalue expansions to find the steady GF in the semi-infinite slab for case X11Y00. Compare your expression to Example 4.11 and comment on the similarities and/or differences.

4.26 Use the limit method to find the steady GF for case X12. Compare your answer to the result given by the eigenfunction expansion method in Example 4.6.

4.27 Use the limit method to solve for the steady-state GF for the problem denoted *X*11*Y*10.

(a) Use the *X*11 GF best for small cotimes.

(b) Use the *X*11 GF best for large cotimes.

4.28 Use the limit method to solve for the steady-state GF for the problem denoted *X*11*Y*10*Z*12. Use the *X*11 and *Z*12 GFs best for large cotimes.

REFERENCES

Arpaci, V. S., 1966, *Conduction Heat Transfer*, Addison-Wesley, Reading, Mass.

Barton, G., 1989, *Elements of Green's Functions and Propagation*, Oxford University Press, London.

Beck, J. V., 1984, Green's functions for transient heat conduction problems, *Int. J. Heat Mass Transfer*, vol. 27, no. 8, pp. 1235–1244.

Brebbia, C. A. and Dominquez, J., 1992, *Boundary Elements: An Introductory Course*, WIT Press, Southamption, U.K.

Carslaw, H. S. and Jaeger, J. C., 1959, *Conduction of Heat in Solids*, 2nd edn, Oxford University Press, New York.

Cole, K. D., 2004, Fast-converging series for steady heat conduction in the circular cylinder, *J. Eng. Math.*, vol. 49, pp. 217–232.

Cole, K. D. and Yen, D. H. Y., 2001a, Green's functions, temperature, and heat flux in the rectangle, *Int. J. Heat Mass Transfer*, vol. 44, pp. 3883–3894.

Cole, K. D. and Yen, D. H. Y., 2001b, Influence functions for the infinite and semi-infinite strip, *J. Thermophys. Heat Transfer*, vol. 15, pp. 431–438.

Crittenden, P. E. and Cole, K. D., 2002, Fast-converging steady-state heat conduction in a rectangular parallelepiped, *Int. J. Heat Mass Transfer*, vol. 45, pp. 3585–3596.

Duffy, D. G., 2001, *Green's Functions with Applications*, Chapman and Hall, Boca Raton, FL.

Haji-Sheikh, A. and Beck, J.V., 2000, An efficient method of computing eigenvalues in heat conduction, *Numerical Heat Transfer Part B Fundamentals*, vol. 38, pp. 133–156.

Luikov, A. V., 1968, *Analytical Heat Diffusion Theory*, Academic Press, New York.

Melnikov, Y. A., 1999, *Influence Functions and Matrices*, Marcel Dekker, New York.

Morse, P. M. and Feshbach, H., 1953, *Methods of Theoretical Physics*, McGraw-Hill, New York.

Ozisik, M. N., 1993, *Heat Conduction*, John Wiley, New York.

5 Improvement of Convergence and Intrinsic Verification

5.1 INTRODUCTION

For heat conduction in finite bodies, expressions for temperature often involve infinite series. This chapter is devoted to numerical issues associated with evaluation of infinite-series solutions. Slow convergence of these infinite-series expressions can make it difficult to obtain accurate numerical values because many terms must be evaluated. Slow series convergence can also contribute to lengthy computer evaluation time.

Lengthy evaluation time will always be an issue, no matter how fast computers may become, because scientists and engineers will always be able to imagine calculations that outstrip their computer resources. Lengthy evaluation times can occur in heat conduction when many temperature values are needed (at many locations in time or space), or, when very high numerical accuracy is needed.

The concept of intrinsic verification, introduced in this chapter, is the process of determining correct numerical values from an exact analytical solution, to many significant figures, in two or more independent ways. Arising as it does from the solutions themselves ("intrinsic"), this type of checking is easy to implement and provides assurance that numerical results are correct. We strongly recommend this approach.

The remainder of this section introduces the Cartesian geometries considered in this chapter, the two basic functions that arise in heat conduction for these geometries (short cotime and long cotime), and the convergence issues associated with long cotime functions. In Section 5.2 strategies are given for identifying when slow convergence is a problem. Three methods for improving convergence are discussed in Section 5.3: replacement of steady-state; the alternate Green's function (GF) solution; and, time partitioning. In Section 5.4 the concept of intrinsic verification is introduced as a means to improve one's confidence that the numerical values computed from exact solutions are correct.

5.1.1 Problems Considered in This Chapter

In this chapter some problems associated with series convergence are introduced for Cartesian bodies for purposes of illustration. The same concepts apply for other coordinate systems and multiple dimensions. Three types of problems are considered: those containing a nonzero initial temperature distribution $F(\vec{r})$; those containing an energy generation term $g(\vec{r},t)$; and, those containing one nonzero boundary-heating term f at boundary location $x = 0$. Simultaneous heating at additional boundaries can be included by superposition of additional boundary terms, evaluated at the appropriate boundary.

The describing partial differential equation for the temperature in Cartesian bodies with uniform thermal properties is

$$\frac{\partial^2 T}{\partial x^2} + \frac{\partial^2 T}{\partial y^2} + \frac{\partial^2 T}{\partial z^2} + \frac{1}{k} g(\vec{r},t) = \frac{1}{\alpha}\frac{\partial T}{\partial t} \qquad \text{in finite domain } R, \quad t > 0 \tag{5.1}$$

The finite domain R can include slabs, rectangles, and parallelepipeds. Notice that this equation contains an energy generation term, and hence is nonhomogeneous. At boundary $x = 0$ there may be a nonzero boundary-heating term of the first, second, or third kinds described by

$$T = f \qquad \text{or} \qquad \frac{\partial T}{\partial n} = f \qquad \text{or} \qquad k\frac{\partial T}{\partial n} + hT = f \tag{5.2}$$

The other boundary conditions are homogeneous ($f_i = 0$). For the nonzero initial temperature distribution one writes

$$T(\vec{r},0) = F(\vec{r}) \qquad \text{in finite domain } R \tag{5.3}$$

Analogous to Equation 1.74, the solution of the above problem using GFs for a finite body is

$$\begin{aligned} T(\vec{r},t) &= \int_R G(\vec{r},t|\vec{r}',0)F(\vec{r}')\,dv' \\ &+ \frac{\alpha}{k}\int_{\tau=0}^{t}\int_R G(\vec{r},t|\vec{r}',\tau)\,g(\vec{r}',\tau)\,dv'd\tau \\ &+ \alpha\int_{\tau=0}^{t}\int_s \left\{ \begin{array}{ll} f(r',\tau)\,\partial G/\partial x'; & \text{first kind only} \\ \frac{1}{k}\,f(r',\tau)\,G; & \text{2nd or 3rd kind} \end{array} \right\}_{x'=0} ds'\,d\tau \end{aligned} \tag{5.4}$$

For one-dimensional slab bodies the GF is given in Chapter 4 for many cases, and more extensive tables of GFs for rectangular coordinates are given in Appendix X. For 2D and 3D Cartesian bodies, the transient GF may be found by products of one-dimensional GF.

5.1.2 Two Basic Functions

The GF for one-dimensional slab bodies have the form of infinite series of basic functions. (For semi-infinite or infinite bodies, the GF is usually given as a *finite* sum of such functions.) There are two types of basic functions that occur in the expression for $G_{XIJ}(\cdot)$ for $I = 1, 2$ and $J = 0, 1$, and 2. One is the fundamental heat conduction function, $K(z + x', t - \tau)$,

$$K(z + x', t - \tau) = [4\pi\alpha(t - \tau)]^{-1/2} \exp\left[-\frac{(z + x')^2}{4\alpha(t - \tau)}\right] \tag{5.5}$$

The variable z is $2nL+x$ or $2nL-x$. See Section 4.2. In this function the variables x' and cotime $(t - \tau)$ occur in the same group as in the argument of $\exp(\cdot)$. This is a compact form, but integrations involving $K(\cdot)$ can be quite complicated and can be difficult to do analytically.

The other type of basic function involves the product of an exponential that is a function of only cotime $(t - \tau)$ and two identical eigenfunctions, one a function of x and the other of x', and a norm N_m,

$$\exp\left[-\frac{\beta_m^2 \alpha(t-\tau)}{L^2}\right]\frac{\mathbf{X}(\beta_m, x)\mathbf{X}(\beta_m, x')}{N_m} \tag{5.6}$$

The norm can be L, $L/2$, or a more complicated function. The eigenfunctions for the $X11$ and $X12$ geometries are

$$\mathbf{X}(\beta_m, x) = \sin\frac{\beta_m x}{L}$$
$$\text{and } \mathbf{X}(\beta_m, x) = \cos\frac{\beta_m x}{L}$$

The basic function given by Equation 5.6 is more convenient for mathematical manipulation than $K(\cdot)$ given by Equation 5.5, because the dependent variables x, x', and $(t - \tau)$ all occur in different terms of Equation 5.6. Thus, an integral on one variable $(x, x', \text{or } t-\tau)$ acts only on one term and does not affect integration on the other two variables. Whenever practical, the product form given by Equation 5.6 is preferred for this reason.

5.1.3 Convergence of the GF

There is an important case when the large-cotime GF has convergence difficulties for large values of $\alpha t/L^2$. It occurs when $G(\cdot)$ is integrated over the dummy time variable τ. For example, let $g(x', \tau)$ in Equation 5.4 be simply $g_0\delta(x_0 - x')$. This is a continuous (that is, constant over time) source of heat of strength g_0 located at position x_0. Then, the second integral of Equation 5.4, restated for a one-dimensional body, contains typical terms of

$$\frac{\alpha}{k}g_0\sum_{m=1}^{\infty}\int_{\tau=0}^{t}\exp\left[-\frac{\beta_m^2\alpha(t-\tau)}{L^2}\right]d\tau\frac{\mathbf{X}(\beta_m, x)\mathbf{X}(\beta_m, x_0)}{N_m} \tag{5.7}$$

[The integral over x' has been evaluated with the sifting property of the Dirac delta function, $\delta(x_0 - x')$.] Next, only the integral over τ is considered, but the upper limit is replaced by $t - \Delta t$, where Δt is discussed below. Then the τ integral can be expressed as

$$\int_{\tau=0}^{t-\Delta t}\exp\left[-\frac{\beta_m^2\alpha(t-\tau)}{L^2}\right]d\tau = \frac{L^2}{\alpha\beta_m^2}\left(e^{-\beta_m^2\alpha\Delta t/L^2} - e^{-\beta_m^2\alpha t/L^2}\right) \tag{5.8}$$

If $\Delta t = 0$ in Equation 5.8, then the term $\exp(-\beta_m^2 \alpha \Delta t / L^2)$ becomes unity, and the first term of the time integral becomes $L^2/(\alpha\beta_m^2)$. When this term is replaced back into the infinite sum in Equation 5.7, for the $X21$ and $X22$ cases at $x = 0$, the resulting term is proportional to

$$\sum_{m=1}^{\infty} \frac{1}{\beta_m^2} \tag{5.9}$$

which is part of the expression for the temperature. In many cases β_m is approximately equal to m times π

$$\beta_m \approx m\pi \tag{5.10}$$

for large values of m. For large m, the "tail" of the summation of Equation 5.9 for $m = M, M+1, M+2$, etc., is given by the Euler–Maclaurin summation formula (Abramowitz and Stegun, 1964, p. 16)

$$\sum_{m=M}^{\infty} \frac{1}{\pi^2 m^2} \approx \frac{1}{\pi^2} \int_M^{\infty} \frac{1}{m^2} dm = \frac{1}{\pi^2 M} \tag{5.11}$$

Hence, the tail of the summation is proportional to $1/M$. This means that a very large number of terms in the series is needed if accurate temperature values are desired. For example, if M is equal to 100, the error in neglecting the tail is approximately $1/(100\pi^2) \approx 0.0010$; for $M = 1000$, the error is one-tenth as large, but there is 10 times as much computation. Note that

$$\sum_{m=1}^{\infty} \frac{1}{\pi^2 m^2} = \frac{1}{6}$$

and so using $M = 100$ would result in an error of about 0.001/(1/6) or a 0.6% error. One reason that analytical solutions are used is to obtain the "exact" solution which, in practice, usually means an error of 0.01% or less. In this example, accuracy of 0.01% would require the large number of over 6000 terms in the single-sum infinite series. For double or triple series the number of terms could be much larger. However, using the methods in this book the number of terms in a given summation may be reduced to 40 or less depending upon the desired accuracy.

Suppose that the integrand for the integral is replaced by the appropriate small-cotime expression, which has terms similar to the one in Equation 5.5. Then the integral over τ in the range $(t - \Delta t < \tau < t)$ can be accurately found using only a small number of terms involving fundamental solution K. Now consider the error in the tail of the large-cotime expression with a finite number of terms. Then instead of evaluating the slowly convergent series given by Equation 5.9, it is only necessary to evaluate the sum

$$\sum_{m=1}^{M-1} \frac{1}{\beta_m^2} \exp\left[-\frac{\beta_m^2 \alpha \Delta t}{L^2}\right] \tag{5.12}$$

which requires many fewer terms for *nonzero* values of $\alpha\Delta t / L^2$. If the tail of Equation 5.12 is calculated, and $\beta_m \approx m\pi$, the result is

$$\sum_{m=M}^{\infty} \frac{1}{m^2\pi^2} \exp\left[-\frac{m^2\pi^2\alpha\Delta t}{L^2}\right] \approx \frac{1}{M\pi^{3/2}} \text{ierfc}\left[M\pi\left(\frac{\alpha\Delta t}{L^2}\right)^{1/2}\right] \tag{5.13}$$

which reduces to the Equation 5.11 result for $\Delta t = 0$. For nonzero values of $\alpha\Delta t / L^2$, the right side of Equation 5.13 decreases very rapidly as M increases. As an example, let $\alpha\Delta t / L^2$ be the small value of 0.025. Then, $\pi(\alpha\Delta t / L^2)^{1/2} = 0.497 \approx 0.5$, and then ierfc $(0.5M)$ takes on the values 400E−7, 30E−7, and 0.9E−7, for $M = 4, 5$, and 6, respectively. Hence for $\alpha\Delta t / L^2 = 0.025$ and small values of M such as 4, the error by dropping the tail of the summation is negligible. (Larger $\alpha\Delta t / L^2$ values cause the right-hand side of Equation 5.13 to decrease even more rapidly as M increases.) The contribution for the integral over τ in the range $(t - \Delta t < \tau < t)$ in Equation 5.8 is obtained using just a few terms of the small-cotime GFs.

Consequently, partitioning the time integral in Equation 5.4 has great potential to improve the computational efficiency of solutions obtained with the GF method, for two- and three-dimensional problems. It is not usually needed for one-dimensional problems. Time partitioning, discussed in Section 5.3.3, is one of several methods that can improve the convergence of a series solution.

The discussion in this section has established that large-cotime GF *may* produce slow-converging series for temperature, and that it *may* be necessary to improve the series convergence. Further discussion of improvement is premature, because first we need to determine whether or not slow convergence is actually present in the problem at hand.

5.2 IDENTIFYING CONVERGENCE PROBLEMS

Evaluating an infinite series is like using a chain saw—you can avoid serious injury if you follow the safety rules. The safely rules for evaluating an infinite series, discussed in this section, are the following: use a convergence criterion; monitor the number of terms; and, be aware that the derivative of a series converges more slowly.

5.2.1 Convergence Criterion

Every infinite series must be truncated to a finite number of terms when evaluated numerically on a computer. The number of terms sets the accuracy of the numerical result. Unfortunately, the number of terms needed for accurate evaluation can vary from place to place within the body and can vary with time. This nonuniform convergence makes it difficult to estimate beforehand how many terms of the series are needed in every circumstance. The use of a fixed number of terms, say for evaluating temperature at several locations, risks poor accuracy in some locations and risks wasting computer time in other locations. A convergence criterion is needed to choose the number of terms, at any location or time, to provide a predetermined accuracy without wasting computer cycles. Two convergence criteria are discussed here.

TABLE 5.1
Number of Terms and Truncated Sums as a Function of Convergence Criterion K_{max}*

K_{max}	$e^{-K_{max}}$	m_{max}	$\sum_{m=1}^{m_{max}} \frac{e^{-(m\pi)^2 0.01}}{(m\pi)^2}$	m_{max}	$\sum_{m=1}^{m_{max}} \frac{e^{-(m\pi)^2 0.1}}{(m\pi)^2}$
4.6	1.0E–02	7	0.1152443685	3	0.03825354276
6.9	1.0E–03	9	0.1152476499	3	0.03825354276
11.5	1.0E–05	11	0.1152477078	4	0.03825354364
23.0	1.0E–10	15	0.1152477083	5	0.03825354364

*Inaccurate digits are underlined.

Maximum exponential argument. When the series contains an exponential factor, the best convergence criterion is to specify the maximum allowable absolute value of the exponential argument. For transient heat conduction, the time-exponential is monotonically decreasing and generally dominates the convergence behavior. Tracking the value of the exponential argument is a conservative way to control the convergence. Most importantly, this convergence criterion can be applied ahead of time to choose the number of series terms needed.

Consider the series given by Equation 5.12 from the large-cotime GF:

$$\sum_{m=1}^{\infty} \frac{1}{\beta_m^2} e^{-A}, \quad \text{where } A = \beta_m^2 \alpha \Delta t / L^2 \tag{5.14}$$

The convergence criterion is to continue to add terms to the series until $A > K_{max}$ where K_{max} is the maximum allowable absolute value of the exponential argument. The value of K_{max} determines the size of the exponential factor as indicated in the first two columns of Table 5.1. For the above series, for an error of one part in 10^{10} requires that $A \leq 23$ which means that at $\alpha \Delta t / L^2 = 1$ the eigenvalue must be $\beta_m^2 = 23$ or $\beta_m \approx 4.8$. This convergence test is used later in Section 5.4.

To be more specific, consider the common case of $\beta_m = m\pi$ and the dimensionless times of 0.01 and 0.1. Table 5.1 shows results for the number of series terms and the truncated sum with the inaccurate digits underlined. The convergence criterion based on the exponential factor is shown to be conservative in each case. For example, when $\exp(-K_{max}) \approx 10^{-5}$, the truncated sum is accurate to at least seven digits.

Ratio convergence test. Unfortunately, some series do not contain an exponential factor. In this case we suggest a convergence test based on a ratio of the average of the last few terms of the series and the entire series so far. Specifically, let f_i be the ith term of the series and let S_m be the truncated series, given by

$$S_m = \sum_{i=1}^{m} f_i \tag{5.15}$$

TABLE 5.2
Number of Terms and Truncated Sums as a Function of Convergence Criterion ϵ (Equation 5.16)*

ϵ	m_{max}	$\sum_{m=1}^{m_{max}} \frac{1}{m^2\pi^2}$	m_{max}	$-\sum_{m=1}^{m_{max}} \frac{(-1)^m}{m^2\pi^2}$
1.0E–03	27	0.1629826658	24	0.0832490393
1.0E–04	81	0.1654234776	66	0.0833218794
1.0E–05	249	0.1662605703	204	0.0833321220
1.0E–06	783	0.1665373480	639	0.0833334572
1.0E–07	2469	0.1666256376	2016	0.0833333209
1.0E–08	7800	0.1666536776	6369	0.0833333346

*Inaccurate digits are underlined.

Then using the average of the last three terms, the summation is truncated when

$$\left| \frac{f_{m-2} + f_{m-1} + f_m}{3} \cdot \frac{1}{S_m} \right| < \epsilon \tag{5.16}$$

Using an average of several terms, rather than just the last term, is important because the last term of the series may not shrink in size monotonically, but may oscillate in size or repeatedly change sign because of a sine or cosine component. The absolute value is used to guard against negative values of f, which could prematurely signal truncation. An average of more than three terms could be used to test convergence, but this would require additional computer resources to little advantage.

Table 5.2 shows the results of the ratio convergence test applied to two series that contain factor $1/m^2$. The table values show that convergence criterion $\epsilon = 10^{-8}$ provides about four accurate digits for the first series and about seven accurate digits for the series with alternating signs (note that the alternating sign speeds the convergence). Clearly, the number of accurate digits given by this test varies with the convergence speed of the series. The ratio convergence test can be performed after each term is added, because the computer time needed to compute the test is generally small. However, testing every third term can be coded very simply (a simple sum rather than a moving sum) and it allows the series to establish a trend before the first test. The convergence test for the values shown in Table 5.2 was applied every third term; note that all the m_{max} values listed in the table are divisible by three.

5.2.2 Monitor the Number of Terms

Even though modern computers can rapidly saw through millions of series terms, it is important to be aware of the number of terms needed to evaluate your series. For example, more series terms are often needed near nonhomogeneous boundaries, and monitoring the number of terms can identify these problem areas. As another example, if the series is evaluated in a code that specifies the maximum number of terms, there should be a warning flag encoded if the maximum number is exceeded. In this way the code user will know that the computed value may not be accurate because the series was truncated without satisfying the convergence test.

If the number of terms needed becomes very large ($> 10^5$), then round-off errors can accumulate. Round-off error is the error introduced by the floating-point representation of each term, summed over all terms. To address round-off error, the obvious step is to increase the precision of the floating point representation, for example by changing from single-precision to double-precision. Unfortunately, this can more than double the computer evaluation time, depending on how the computer hardware processes floating-point numbers. A better approach, discussed in Section 5.3, is to find a way to improve the convergence speed of the series.

5.2.3 Slower Convergence of the Derivative

Generally the heat flux is found by differentiating the temperature according to Fourier's law. However, be careful when evaluating the heat flux from a temperature series, because differentiation degrades the convergence speed of a series. Worse, given a convergent series, there is no mathematical guarantee that its derivative will converge at all (Lanczos, 1966, p. 63).

This problem often occurs near boundaries and corners, and can be severe near boundaries of the first kind and in 2D and 3D cases. For simplicity in presentation a 1D example is given here.

Consider a nonhomogeneous boundary of the first kind, say at $x = 0$ for case X11B10T0. The GF solution has the form:

$$T(x,t) = \alpha \int_{\tau=0}^{t} T_0 \left.\frac{\partial G_{X11}}{\partial x'}\right|_{x'=0} d\tau \tag{5.17}$$

The long-cotime form of the GF is given by

$$G_{X11}(x,t|x',\tau) = \frac{2}{L}\sum_{m=1}^{\infty} \sin(m\pi x\,/\,L)\sin(m\pi x'\,/\,L)\exp[-m^2\pi^2\alpha(t-\tau)\,/\,L^2] \tag{5.18}$$

After evaluating the derivative on x' and the integral on τ, the long-cotime temperature series is given by:

$$T(x,t) = 2T_0 \sum_{m=1}^{\infty} \frac{1}{m\pi}\sin(m\pi x\,/\,L)\left(1-\exp[-m^2\pi^2\alpha t\,/\,L^2]\right) \tag{5.19}$$

This equation has a steady-state part (the sum of $\sin(m\pi x\,/\,L)\,/(m\pi)$) and an exponentially converging complementary-transient part. The steady state series converges slowly because the only factor uniformly decreasing to zero is $1\,/(m\pi)$. A series composed only of factor $1\,/(m\pi)$ will not converge; however this series contains a sine function whose positive and negative values do allow the series to converge, though very slowly. Convergence becomes slower and slower as you approach the heated boundary at $x = 0$. The slow convergence arises from the Fourier series, not from

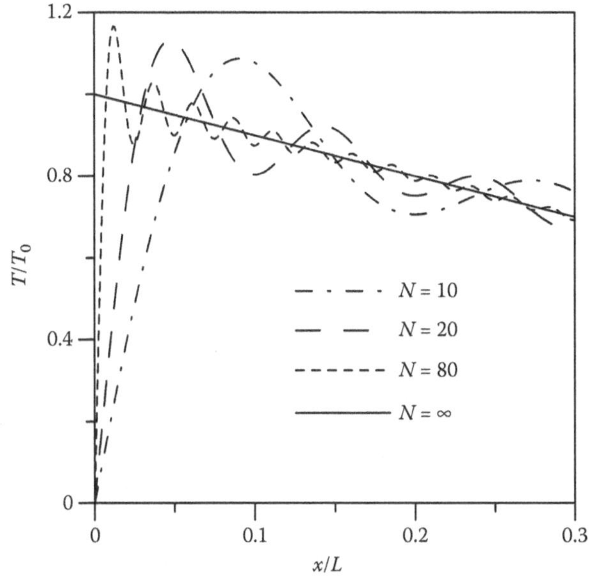

FIGURE 5.1 Series $2\sum \sin(m\pi x/L)/(m\pi)$ truncated to $N = 10, 20$ and 80 terms, demonstrating the Gibbs phenomenon for a type 1 nonhomogeneous boundary (Case $X11B10$).

the physics of heat conduction. In this case the series attempts to describe $T \neq 0$ near $x = 0$ using eigenfunctions that approach zero as $x \to 0$, requiring increasingly more series terms. This phenomenon was first explained by J. Willard Gibbs, one of America's foremost scientists*. The Gibbs phenomenon occurs whenever a truncated Fourier series is used to approximate a discontinuous function (Sommerfeld, 1949, p. 12). A demonstration of the Gibbs phenomenon is given in Figure 5.1, in a plot of the steady portion of the X11B10 temperature (Equation 5.19) where the series is truncated to $N = 10$, 20 and 80 terms. As the number of terms increases, the curve more closely approaches the exact values (straight line) except near the $x = 0$ boundary. The curve for each truncated series begins at zero at $x = 0$ and rises sharply to overshoot the exact values. Although the width of the rise-and-overshoot region shrinks as N increases, the overshoot height never vanishes. In addition, as N increases the slope at $x = 0$ becomes steeper and steeper.

Consider next the heat flux series found by term-by-term differentiation of the temperature series:

$$q = -k\frac{\partial T}{\partial x} = -\frac{2kT_0}{L}\sum_{m=1}^{\infty}\cos(m\pi x / L)\left(1 - \exp[-m^2\pi^2\alpha t / L^2]\right) \qquad (5.20)$$

Because the decreasing factor $1/(m\pi)$ has been removed by differentiation, this series for heat flux diverges for *every* value of x. That is, as you add terms to the series the

*Although renowned as a scientist, Gibbs earned the first American Ph.D. in Engineering in 1863.

numerical value increases without limit. It is important to remember that this lack of convergence is a mathematical artifact associated with the series form of the solution; the physical system is well behaved for all $t > 0$. Fortunately, for cases like this, there are alternate forms of the solution, discussed in the next section, that can be used to evaluate the heat flux (see also Beck and Cole, 2007).

5.3 STRATEGIES TO IMPROVE SERIES CONVERGENCE

When slow convergence becomes a problem, there are specific strategies that can be used to improve the convergence of series solutions. These strategies are: replacement of steady-state series; use of the alternate GF solution; and, partitioning the time integral. The first two methods are easier to implement and should be explored first. Time partitioning, important for 2D and 3D applications, may require more analytical effort.

5.3.1 Replacement of Steady-State Series

If a transient solution contains a steady-state portion, often this portion of the solution converges slowly. One strategy for improving convergence of the entire series is to replace the steady portion by a better-converging form. How to find this better form can depend on the body shape and on the number of spatial dimensions involved.

In one-dimensional transient cases, the steady-state portion of the solution can usually be found in algebraic form by direct integration. When this algebraic form is substituted for the series form, the accuracy is significantly improved and the computation time is reduced. Consider a specific one-dimensional example.

Example 5.1: Slab with Elevated Temperature on One Side—$X11B10T0$

Molten metal is suddenly poured over a plate of thickness L and an initial temperature of zero. The temperature at the back side of the plate can be considered to be fixed at zero also. A reasonable approximation for this problem is a step change in the $x = 0$ surface temperature to T_0. Assuming temperature-invariable thermal properties, model the problem and solve using GFs; replace the steady state component of the solution with a nonseries form and evaluate the heat flux.

Solution

The transient temperature satisfies the following equations:

$$\frac{\partial^2 T}{\partial x^2} = \frac{1}{\alpha}\frac{\partial T}{\partial t}; \quad 0 < x < L \tag{5.21}$$

$$T(x,0) = 0$$

$$T(0,t) = T_0$$

$$T(L,t) = 0$$

A jump in temperature of size T_0 is suddenly imposed at the $x = 0$ boundary. This geometry was discussed earlier in Section 5.2.3, and the temperature is given by (see Equation 5.19)

$$T(x,t) = 2T_0 \sum_{m=1}^{\infty} \frac{\sin(m\pi x/L)}{m\pi} - 2T_0 \sum_{m=1}^{\infty} \frac{\sin(m\pi x/L)}{m\pi} e^{-m^2\pi^2\alpha t/L^2} \tag{5.22}$$

Note that this equation has the form

$$T(x,t) = T_0[S_{ss}(x) + S_{c.t.}(x,t)] \tag{5.23}$$

where S_{ss} is the steady term and $S_{c.t.}$ is the complementary transient term. It is the steady term that converges slowly. The temperature would converge better if the slowly converging steady portion could be improved.

Next the steady-state portion is found in nonseries form. Introducing Equation 5.23 into the boundary value problem for temperature, Equation 5.21, gives

$$\frac{\partial^2 S_{ss}(x)}{\partial x^2} + \frac{\partial^2 S_{c.t.}(x,t)}{\partial x^2} = \frac{1}{\alpha}\frac{\partial S_{c.t.}(x,t)}{\partial t} \tag{5.24}$$
$$S_{ss}(x) + S_{c.t.}(x,0) = 0$$
$$S_{ss}(0) + S_{c.t.}(0,t) = 1$$
$$S_{ss}(L) + S_{c.t.}(L,t) = 0$$

For $t > 0$ we know that the two solutions S_{ss} and $S_{c.t.}$ are independent. Then we can obtain

$$\frac{d^2 S_{ss}(x)}{dx^2} = 0; \quad \frac{\partial^2 S_{c.t}(x,t)}{\partial x^2} - \frac{1}{\alpha}\frac{\partial S_{c.t.}(x,t)}{\partial t} = 0 \tag{5.25}$$

So function S_{ss} satisfies the steady heat conduction equation. Let us choose boundary conditions for S_{ss} to be

$$S_{ss}(0) = 1; \quad S_{ss}(L) = 0$$

Then function S_{ss} may be found by direct integration: the steady heat conduction equation admits a linear distribution in the form $S_{ss}(x) = ax+b$, and then constants a and b may be found from the boundary conditions. Function S_{ss} is then given by

$$S_{ss}(x) = (1 - \frac{x}{L}) = 2\sum_{m=1}^{\infty} \frac{\sin(m\pi x/L)}{m\pi} \tag{5.26}$$

Replace this algebraic form for the steady solution into the series solution above to find

$$T(x,t) = T_0\left(1 - \frac{x}{L}\right) - 2T_0 \sum_{m=1}^{\infty} \frac{\sin(m\pi x/L)}{m\pi} e^{-m^2\pi^2\alpha t/L^2} \tag{5.27}$$

This equation has much better series convergence than the previous series. At large time ($\alpha t/L^2 > 0.025$) only a few terms are needed for high accuracy.

An important aspect of the temperature expression given by Equation 5.27 is the possibility of "intrinsic verification" which is discussed in detail in Section 5.4. As shown in Figure 5.2, the temperature is nearly zero at $x/L > 0.4$ and at early time

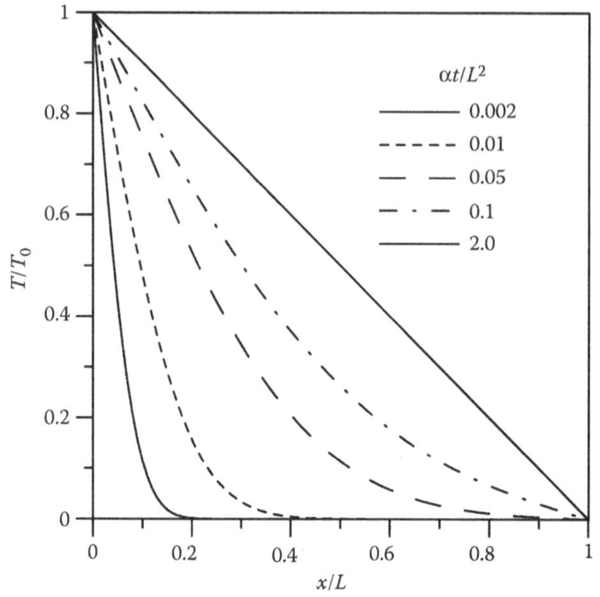

FIGURE 5.2 Case $X11B10T0$. Temperature in the plane wall initially at zero temperature, with $T = T_0$ applied at $x = 0$ for $t > 0$, and with zero temperature at $x = L$.

such that $\alpha t / x^2 < 0.01$. In this region the steady and complementary transient parts of the temperature expression must sum to zero. This is useful for checking that the series is computed accurately.

This better-converging form for the temperature, Equation 5.27, can also be differentiated term-by-term to find the heat flux, as follows:

$$q(x,t) = -k\frac{\partial T}{\partial x} = \frac{kT_0}{L}\left[1 + 2\sum_{m=1}^{\infty} \cos(m\pi x / L)e^{-m^2\pi^2\alpha t / L^2}\right] \tag{5.28}$$

This series for the heat flux, unlike Equation 5.20, converges everywhere for $t > 0$ thanks to the exponential term. At the instant $t = 0$, however, the exponential term is unity and the series diverges. The heat flux is infinite at $t = 0$ not because of some mathematical flaw, but because of the physically unrealistic boundary condition. The instantaneous jump in boundary temperature imposed at $t = 0$ results in a momentarily infinite heat flux. For small dimensionless times such as $\alpha t / L^2 = 0$ to 0.06, the finite body problem X11B10T0 is better modeled as the X10B1T0 problem. For any $t > 0$ the heat flux has a noninfinite value and continues to decay until the steady state value is reached.

In two- and three-dimensional cases, the steady portion can also be replaced by a better-converging form, but more effort may be required. For example, in rectangles, the poorly converging steady portion is a double-infinite series, which can be replaced by a single-summation series. An example is given below for a rectangle heated at a

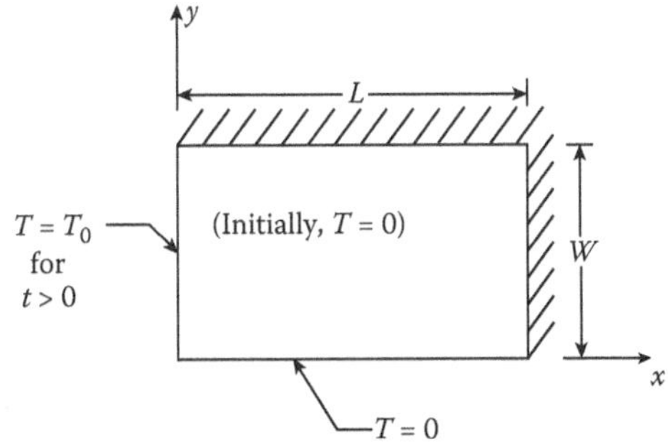

FIGURE 5.3 Geometry for the rectangle, case $X12B10Y12B00T0$.

boundary. Similar improvement in series convergence can also be obtained for steady solutions caused by internal energy generation. A large number of steady solutions, along with a discussion of the speed of series convergence, are given in the literature for the following geometries: the rectangle (Melnikov, 1999; Duffy, 2001; Cole and Yen, 2001a), the two-dimensional semi-slab and the slab (Cole and Yen, 2001b), the parallelepiped (Crittenden and Cole, 2002); and, the three-dimensional finite cylinder (Cole, 2004).

Example 5.2: Rectangle with Boundary Heating—$X12B10\ Y12B00T0$

Find the temperature in a rectangle with a suddenly applied change in temperature at $x = 0$. Replace the steady portion of the solution with a single-sum form.

Solution

The rectangle ($0 < x < L$; $0 < y < W$) with number designation X12Y12 is initially at a uniform temperature, and a sudden change in temperature is applied at the $x = 0$ boundary. The other boundaries are homogeneous as shown in Figure 5.3. The temperature satisfies the following equations:

$$\frac{\partial^2 T}{\partial x^2} + \frac{\partial^2 T}{\partial y^2} = \frac{1}{\alpha}\frac{\partial T}{\partial t} \tag{5.29}$$

$$T(0, y, t) = T_0; \quad \left.\frac{\partial T}{\partial x}\right|_{x=L} = 0; \quad T(x, 0, t) = 0; \quad \left.\frac{\partial T}{\partial y}\right|_{y=W} = 0;$$

$$T(x, y, 0) = 0$$

The GF solution is given by Equation 5.4 in the form

$$T(x, y, t) = \alpha \int_{\tau=0}^{t} \int_{y'=0}^{W} T_0 \left.\frac{\partial}{\partial x'}\right|_{x'=0} G_{X12Y12}(x, y, t|x', y', \tau)\, dy'\, d\tau \tag{5.30}$$

The transient GF is found from a product solution $G_{X12} \cdot G_{Y12}$, so the solution may be written in the form

$$T(x, y, t) = T_0\, \alpha \int_{\tau=0}^{t} \frac{\partial G_{X12}}{\partial x'}(x, t|0, \tau) \int_{y'=0}^{W} G_{Y12}(y, t|y', \tau)\, dy'\, d\tau \tag{5.31}$$

and the large cotime form of this GF is given by

$$G_{X12}G_{Y12} = \frac{2}{L}\sum_{m=1}^{\infty} \sin\frac{\beta_m x}{L} \sin\frac{\beta_m x'}{L} \exp\left(-\beta_m^2 \alpha u / L^2\right)$$
$$\times \frac{2}{W}\sum_{n=1}^{\infty} \sin\frac{\gamma_n y}{W} \sin\frac{\gamma_n y'}{W} \exp\left(-\gamma_n^2 \alpha u / W^2\right) \tag{5.32}$$

where $\beta_m = (m - 1/2)\pi; \quad \gamma_n = (n - 1/2)\pi$

and where $u = t - \tau$ is the cotime. The large-cotime form of the GF is easier to use because the derivatives and integrals in the GF solution can be carried out separately and in any order. The derivative falls on one of the sine terms, as follows:

$$\frac{\partial}{\partial x'} \sin\frac{\beta_m x'}{L}\bigg|_{x'=0} = \frac{\beta_m}{L}\cos\frac{\beta_m x'}{L}\bigg|_{x'=0} = \frac{\beta_m}{L} \tag{5.33}$$

The integral on y' falls on another sine term:

$$\int_{y'=0}^{W} \sin\frac{\gamma_n y'}{W}\,dy' = -\frac{W}{\gamma_n}\cos\frac{\gamma_n y'}{W}\bigg|_0^W = \frac{W}{\gamma_n}(1 - \cos\gamma_n) = \frac{W}{\gamma_n} \tag{5.34}$$

Finally the integral on τ falls on the exponential term

$$\int_{\tau=0}^{t} e^{-C(t-\tau)}\,d\tau = \int_{u=0}^{t} e^{-Cu}\,du = \frac{1}{C}\left(1 - e^{-Ct}\right) \tag{5.35}$$

$$\text{where} \quad C = \alpha\left(\beta_m^2/L^2 + \gamma_n^2/W^2\right) \tag{5.36}$$

Now assemble these portions into the temperature expression to find

$$T(x, y, t) = T_0 \frac{4}{L^2}\sum_{m=1}^{\infty}\sum_{n=1}^{\infty} \sin\frac{\beta_m x}{L} \sin\frac{\gamma_n y}{W} \frac{1}{\left(\frac{\gamma_n}{W}\right)^2 + \left(\frac{\beta_m}{L}\right)^2}$$
$$\times \frac{\beta_m}{\gamma_n}\left(1 - \exp\left[-\left(\gamma_n^2/W^2 + \beta_m^2/L^2\right)\alpha t\right]\right) \tag{5.37}$$

The steady-state part of the above solution, the slowly converging part, is given by

$$T_S(x, y) = T_0 \frac{4}{L^2}\sum_{m=1}^{\infty}\sum_{n=1}^{\infty} \frac{\beta_m}{\gamma_n} \frac{\sin\frac{\beta_m x}{L} \sin\frac{\gamma_n y}{W}}{\left(\gamma_n^2/W^2 + \beta_m^2/L^2\right)} \tag{5.38}$$

Because there is no exponential present to speed the convergence, the steady part converges slowly.

Next a better-converging form of the steady temperature will be sought with a single-sum GF found with the eigenfunction expansion method. As discussed in Section 4.6.3, the general form for the GF in the rectangle is given by

$$G(x, y|x', y') = \sum_{m=1}^{\infty} \frac{X_m(x)X_m(x')}{N(\beta_m)} P_m(y, y') \tag{5.39}$$

where X_m is an eigenfunction, β_m is the eigenvalue, and N_m is the norm along the x-axis in the rectangle. Function P_m is the kernel function. For the specific case under discussion, case X12Y12, the GF has previously been derived in Example 4.6, and it is given by

$$G_{X12Y12} = \sum_{m=1}^{\infty} \frac{\sin\left(\frac{\beta_m x}{L}\right)\sin\left(\frac{\beta_m x'}{L}\right)}{L/2} P_m(y, y') \tag{5.40}$$

where

$$P_m(y, y') = \frac{e^{-\sigma_m(2W-y-y')} - e^{-\sigma_m(2W+|y-y'|)} + e^{-\sigma_m|y-y'|} - e^{-\sigma_m(y+y')}}{2\sigma_m(1 + e^{-2\sigma_m W})} \tag{5.41}$$

and where $\sigma_m = \beta_m / L$. Note that the y-direction kernel function depends on the x-direction eigenvalue through parameter σ_m; this is how the two coordinate directions communicate with each other in the single-sum solution. The above GF is also given in the GF Library web site (Cole, 2000). The steady temperature solution needed here is case X12B10 Y12B00, and the steady GF solution equation for heating at $x = 0$ is given by (see Equation 5.4 or 3.46):

$$T_S(x, y) = T_0 \int_{y'=0}^{W} \left.\frac{\partial G_{X12Y12}(x, y, x', y')}{\partial x'}\right|_{x'=0} dy' \tag{5.42}$$

The derivative required was given earlier, and the integral on y' is given by

$$\int_{y'=0}^{W} P_m(y, y')dy' = \frac{1}{\sigma_m^2} - \frac{e^{-\sigma_m(2W-y)} + e^{-\sigma_m y}}{\sigma_m^2(1 + e^{-2\sigma_m W})} \tag{5.43}$$

This integral is also given in Cole and Yen (2001a). Then the single-sum steady temperature may be assembled in the form

$$T_S(x, y) = 2T_0 \sum_{m=1}^{\infty} \frac{\sin \beta_m x / L}{\beta_m} \left[1 - \frac{e^{-\sigma_m(2W-y)} + e^{-\sigma_m y}}{1 + e^{-2\sigma_m W}}\right] \tag{5.44}$$

The convergence of the above series may be further improved by recognizing that a portion of the series depends only on coordinate x. This portion of the series may be replaced by a fully summed form with the identity

$$2 \sum_{m=1}^{\infty} \frac{\sin \beta_m x / L}{\beta_m} = 1; \ \beta_m = \pi(m - 1/2) \tag{5.45}$$

(This identity is further explored in homework problem 5.6 at the end of the chapter.) Then the steady temperature is given by

$$T_S(x, y) = T_0 - 2T_0 \sum_{m=1}^{\infty} \frac{\sin \beta_m x / L}{\beta_m} \left[\frac{e^{-\sigma_m(2W-y)} + e^{-\sigma_m y}}{1 + e^{-2\sigma_m W}}\right] \tag{5.46}$$

This single-sum series for the steady temperature, converges much faster than the double-sum solution, especially near $x = 0$. A detailed discussion of the number of terms needed for convergence of this steady-temperature expression is given later in Example 5.4. The point of the present discussion is that Equation 5.46 can be used to replace the steady portion of Equation 5.37 to construct a faster-converging transient temperature in the rectangle, in this case given by

$$T(x,y,t) = T_0 - 2T_0 \sum_{m=1}^{\infty} \frac{\sin \beta_m x/L}{\beta_m} \left[\frac{e^{-\sigma_m(2W-y)} + e^{-\sigma_m y}}{1+e^{-2\sigma_m W}} \right]$$
$$- T_0 \frac{4}{L^2} \sum_{m=1}^{\infty} \sum_{n=1}^{\infty} \sin \frac{\beta_m x}{L} \sin \frac{\gamma_n y}{W}$$
$$\times \frac{\beta_m}{\gamma_n} \left[\frac{\exp\left[-(\gamma_n^2/W^2 + \beta_m^2/L^2)\alpha t \right]}{\gamma_n^2/W^2 + \beta_m^2/L^2} \right] \quad (5.47)$$

5.3.2 Alternate GF Solution Equation

The alternate GF solution equation (AGFSE), introduced in Section 3.4, is an important strategy for improving convergence of series solutions that involve a boundary-heating effect. The thrust of the AGFSE method is to replace the integral representing the boundary heating effect in the GF solution by one or more nonboundary integrals. The resulting alternate temperature series generally has better convergence behavior.

This approach works whether the causative boundary heating is steady or transient, and it works whether or not the temperature solution tends to a steady-state condition. However, if a steady-state temperature is present, applying the AGFSE method is equivalent to replacing the steady-state solution by a better-converging form (as discussed in the previous section). Next an example will be given to demonstrate the AGFSE method.

Example 5.3: Transient Boundary Heating—*X*21*B*21*T*1

The heat shield on a space vehicle entering the atmosphere experiences a heat flux which is increasing with time for a short period. Assume that this heat flux increases linearly with time at location $x = 0$ and the thermal properties are constant. There is a fixed temperature at surface $x = L$. Find the temperature with the standard GF solution and with the alternate GF solution.

Solution

The temperature satisfies the following equations:

$$\frac{\partial^2 T}{\partial x^2} = \frac{1}{\alpha}\frac{\partial T}{\partial t}; \quad 0 < x < L \quad (5.48)$$
$$T(x,0) = T_0$$
$$-k\left.\frac{\partial T}{\partial x}\right|_{x=0} = q_0 \frac{t}{t_0}$$
$$T(L,t) = T_0$$

Here q_0 and t_0 are known constants which produce a time-increasing heat flux at the boundary. The temperature solution with the GF method is given by the boundary-heating integral:

$$T(x,t) - T_0 = \frac{\alpha}{k}\int_{\tau=0}^{t} q_0 \frac{\tau}{t_0}\, G_{X21}|_{x'=0}\, d\tau \tag{5.49}$$

The long-cotime GF for this case is

$$G_{X21} = \frac{2}{L}\sum_{m=1}^{\infty} e^{-\beta_m^2 \alpha(t-\tau)/L^2} \cos\beta_m \frac{x}{L}\cos\beta_m\frac{x'}{L} \tag{5.50}$$

$$\text{where}\quad \beta_m = \pi(m - 1/2)$$

Introduce this GF into Equation 5.49 and carry out the time integral to obtain the standard temperature series:

$$\frac{T(x,t) - T_0}{\frac{q_0 L}{k}\frac{L^2}{\alpha t_0}} = 2\sum_{m=1}^{\infty}\left(\frac{t^+}{\beta_m^2} - \frac{1}{\beta_m^4} + \frac{e^{-\beta_m^2 t^+}}{\beta_m^4}\right)\cos\left(\beta_m \frac{x}{L}\right) \tag{5.51}$$

$$\text{where}\quad t^+ = \alpha t/L^2$$

The convergence of the standard series is controlled by the coefficient in parentheses, which contains three parts. The part containing the exponential converges most rapidly as m increases; it has exponential convergence. The part $1/\beta_m^4$ converges less rapidly, and finally the part containing t^+/β_m^2 converges slowly.

The alternate GF solution will now be applied to this problem. We seek to split the temperature solution as $T = T^* + T'$, where T^* satisfies the original nonhomogeneous boundary conditions and T' will therefore be freed from this responsibility. First we will find T^* that satisfies the following equations:

$$\frac{\partial^2 T^*}{\partial x^2} = 0;\quad 0 < x < L \tag{5.52}$$

$$-k\left.\frac{\partial T^*}{\partial x}\right|_{x=0} = q_0\frac{t}{t_0}$$

$$T^*(L,t) = T_0$$

The quantity T^* in this case is a quasisteady temperature. Integrate the above differential equation twice to find $T^* = ax + b$ where a and b are independent of x. Quantities a and b may be found by application of the nonhomogeneous boundary conditions at $x = 0$ and $x = L$, to give T^* in the form

$$T^*(x,t) = \frac{q_0 L}{k}\frac{t}{t_0}\left(1 - \frac{x}{L}\right) + T_0 \tag{5.53}$$

Note that if the boundary heating were constant with time, quantity T^* would be a steady-state solution. With T^* in hand we can now find the equations that define T' by replacing $T = T^* + T'$ into the original boundary value problem for T. The result is:

$$\frac{\partial^2 T'}{\partial x^2} + 0 = \frac{1}{\alpha}\frac{\partial T'}{\partial t} + \frac{1}{\alpha}\frac{q_0 L}{k}\frac{1}{t_0}\left(1 - \frac{x}{L}\right) \tag{5.54}$$

$$T_0 + T'(x,0) = T_0$$

$$q_0\frac{t}{t_0} - k\left.\frac{\partial T'}{\partial x}\right|_{x=0} = q_0\frac{t}{t_0}$$

$$T_0 + T'(L,t) = T_0$$

In the differential equation, an extra term comes from the time derivative of T^*. However, we can treat this extra term as a known source term. Rewrite the boundary value problem for T' in the form:

$$\frac{\partial^2 T'}{\partial x^2} + \frac{g^*(x)}{k} = \frac{1}{\alpha}\frac{\partial T'}{\partial t} \tag{5.55}$$
$$T'(x,0) = 0$$
$$-k\left.\frac{\partial T'}{\partial x}\right|_{x=0} = 0$$
$$T'(L,t) = 0$$

where $g^*(x)/k$ is given by

$$\frac{g^*(x)}{k} = -\frac{1}{\alpha}\frac{q_0 L}{k}\frac{1}{t_0}\left(1-\frac{x}{L}\right) \tag{5.56}$$

Because T^* satisfies the original boundary conditions, note that the boundary conditions for T' are completely homogeneous. Now the usual GF solution can be applied to find T' in the form

$$T'(x,t) = \alpha \int_{\tau=0}^{t}\int_{x'=0}^{L} \frac{g^*(x')}{k} G_{X21}|_{x'=0}\, dx'd\tau \tag{5.57}$$

Compared to the standard approach, there is an additional spatial integral needed, given by

$$\int_{x'=0}^{L}\left(1-\frac{x'}{L}\right)\cos\left(\beta_m\frac{x'}{L}\right)dx' = \frac{L}{\beta_m^2} \tag{5.58}$$

(Note that $\cos\beta_m = 0$ in this case.) Combine the spatial integral with the time integral, and assemble the solution for T' in the form

$$\frac{T'(x,t)}{(q_0 L/k)} = -2\frac{L^2}{\alpha t_0}\sum_{m=1}^{\infty}\frac{1}{\beta_m^4}\left(1-e^{-\beta_m^2\alpha t/L^2}\right)\cos\beta_m\frac{x}{L} \tag{5.59}$$

Finally, the alternate temperature series is the sum $T = T^* + T'$, which in normalized form is given by

$$\frac{T(x,t)-T_0}{\frac{q_0 L}{k}\frac{L^2}{\alpha t_0}} = t^+\left(1-\frac{x}{L}\right) - 2\sum_{m=1}^{\infty}\frac{1-e^{-\beta_m^2 t^+}}{\beta_m^4}\cos\beta_m\frac{x}{L} \tag{5.60}$$
$$\text{where} \quad t^+ = \alpha t/L^2$$

Compare the above alternate series above to the standard series (Equation 5.51) to see that the slowest-converging term from the standard series has been replaced by a fully summed form, and the remainder of the series is unaffected.

The computational advantage of the alternate solution is apparent in Table 5.3 which shows the number of terms required for evaluating temperature at several locations and times and for two values of the convergence criterion, ϵ (see Equation 5.16).

TABLE 5.3
Number of Series Terms Required for Convergence of Standard Series (Equation 5.51) and Alternate Series (Equation 5.60) for Case X21B21T1 with Convergence Criterion ϵ*

			Number of Terms, Standard Series		Number of Terms, Alternate Series	
x/L	$\alpha t/L^2$	T^+	$\epsilon = 10^{-4}$	$\epsilon = 10^{-6}$	$\epsilon = 10^{-4}$	$\epsilon = 10^{-6}$
0.01	0.01	0.659233E–03	51	351	15	36
0.2	0.01	0.289374E–04	534	5229	15	39
0.4	0.01	0.341467E–06	6219	62139	15	39
0.01	0.1	0.228144E–01	45	252	9	24
0.2	0.1	0.971327E–02	99	909	9	24
0.4	0.1	0.352408E–02	189	1764	9	24
0.01	1	0.684581E+00	39	150	9	18
0.2	1	0.511833E+00	48	399	9	15
0.4	1	0.358541E+00	66	564	9	18
0.01	10	0.956672E+01	36	150	9	18
0.2	10	0.768534E+01	39	324	9	15
0.4	10	0.573600E+01	51	444	9	18

*Convergence was tested every third term. Normalized temperature values shown were found from the alternate series with $\epsilon = 10^{-6}$.

The standard series requires many more terms than the alternate series. At small dimensionless time and far from the boundary, the standard series requires thousands of terms. Most importantly, for the alternate solution an increase in precision from $\epsilon = 10^{-4}$ to $\epsilon = 10^{-6}$ can be purchased inexpensively with about two times more series terms. The same increase in precision with the standard series requires about ten times more series terms. For more cases see Beck et al. (2008).

5.3.3 Time Partitioning

Time partitioning, first introduced by Beck and Keltner (1987), is a powerful method for improvement of series convergence. It is intended for the solution of multidimensional problems. It is not needed for one-dimensional problems. The chief motivation for time partitioning is that the convenient form of the GF given by Equation 5.6, also called the large-cotime form, cannot efficiently be used for small times. For small times, a finite body (such as a plate) behaves as if it were a semi-infinite body, since at small times each boundary condition affects only a small region near its boundary. Small times are defined by dimensionless time $\alpha t/L^2 \leq 0.06$, or $t \leq 0.06L^2/\alpha$ seconds.* Under this circumstance, the large-cotime form of the solution requires

*The number given here as 0.06 may vary between 0.025 and 0.25 depending on the circumstances. The important point is that Fourier number $\alpha t/L^2$ defines the small-time regime.

many terms of the infinite series for the GF. This is inefficient and possibly inaccurate if too few terms are used. The other form of the GF, given by Equation 5.5, also called the small-cotime form, can be evaluated accurately at small cotimes with only a few terms of its infinite series; it tends to be more difficult to evaluate analytically, however.

In the GF equation given by Equation 5.4, the $G(\cdot)$ functions for finite bodies can be of the small-cotime form (see the $X11$, $X12$, $X21$, and $X22$ GFs listed in Table 4.1 and Appendix X), or they can be of the large-cotime form (see Section 4.4 and Appendix X). The small-cotime and large-cotime forms of the GF are each solutions to the heat conduction boundary value problem, given by Equations 5.1 through 5.3. These two solutions are mathematically equivalent, as required by the uniqueness property of solutions of linear boundary value problems (Carslaw and Jaeger, 1959, pp. 35–38). The numerical values are identical for the two solutions for the same conditions. The solution of a boundary value problem is unique, but the expansion of that solution in infinite series form may not be unique. That is, the small-cotime and large-cotime solutions are different infinite-series expansions of the same solution.

Many small-cotime GFs can be derived from Laplace transform solutions of the heat conduction equation (refer to Section 4.3 for an example). For plates, small-cotime GFs take the form of an infinite series of fundamental heat conduction functions given by Equation 5.5. See also Equation 4.1 and the XIJ case of Table 4.1. For sufficiently small times, the value of a GF at any x is unaffected by the boundaries or at most by a single boundary. Hence, the GFs at sufficiently small cotimes can be described by the same GFs as for infinite or semi-infinite bodies. Consequently, the small-cotime GFs can be represented by only the few terms which emphasize the effects of a single boundary.

In contrast, many large-cotime GF expressions are derived from the separation of variables method of solution of the heat conduction equation. For slab bodies, large-cotime GFs are composed of infinite series of basic functions given by Equation 5.6. The large-cotime GFs incorporate the effect of the finite nature of the body and require only a few terms for sufficiently large times. The large-cotime GFs contain eigenvalues that are based on the finite thickness of the plate. As a consequence, the small and large cotime GFs emphasize different aspects of the physical problem in a manner so that only a few terms, in their respective infinite series, are usually needed.

Time partitioning can speed evaluation of the infinite series expressions compared to using a single form of the series. To take advantage of the different convergence properties of the small-cotime and large-cotime solutions, Equation 5.4 can be written as

$$
\begin{aligned}
T(\vec{r},t) &= \int_R G^L(\vec{r},t|\vec{r}',0)F(\vec{r}')\,dv' \\
&\quad + \frac{\alpha}{k}\int_{\tau=0}^{t-t_p}\int_R G^L(\vec{r},t|\vec{r}',\tau)g(\vec{r}',\tau)\,dv'd\tau \\
&\quad + \frac{\alpha}{k}\int_{\tau=t-t_p}^{t}\int_R G^S(\vec{r},t|\vec{r}',\tau)g(\vec{r}',\tau)\,dv'd\tau
\end{aligned}
$$

$$+\alpha\int_{\tau=0}^{t-t_p}\int_s\left\{\begin{array}{ll} f\,\partial G^L/\partial x'; & \text{first kind only} \\ (f/k)\,G^L; & \text{2nd or 3rd kind}\end{array}\right\}_{x'=0} ds'\,d\tau$$
$$+\alpha\int_{\tau=t-t_p}^{t}\int_s\left\{\begin{array}{ll} f\,\partial G^S/\partial x'; & \text{first kind only} \\ (f/k)\,G^S; & \text{2nd or 3rd kind}\end{array}\right\}_{x'=0} ds'\,d\tau \tag{5.61}$$

where $G^S(\cdot)$ and $G^L(\cdot)$ correspond to small-cotime and large-cotime GFs, respectively. The dimensionless time $\alpha t_p / L^2$ is small compared with unity.

The value of $\alpha t_p / L^2$ for time partitioning is usually chosen to be between 0.025 and 0.25. The benefit of choosing a small value of $\alpha t_p / L^2$ is that only a few terms of the series for $G^S(\cdot)$ will be needed in the last integral of Equation 5.61. An example of time partitioning for the rectangle is given later in Section 5.4.4.

5.4 INTRINSIC VERIFICATION

Intrinsic verification is the process of determining correct numerical values from an exact analytical solution, to many significant figures, in two or more independent ways. This provides assurance that the solution is correct and that the process for obtaining accurate numerical values is sound. We use the word "intrinsic" because for many exact solutions, the means of verification are contained within the solution itself.

Intrinsic verification is distinct from "code verification" or "solution verification" which the finite-element and finite-difference community use to assure that their fully numeric computer codes are sufficiently accurate (Roach, 1998). One type of code verification is to compare the fully numeric solution with an exact solution. In contrast, intrinsic verification is a comparison between two exact analytical solutions, for the purpose of assuring that values are correct, far beyond the precision generally practicable from fully numeric solutions.

Exact solutions must satisfy the governing partial differential equation and also the boundary conditions. However, analytical checks of these conditions might not reveal certain errors. For example, the eigenvalues might not be accurate or an eigenvalue might be missing. It is also possible that convergence of series may be so poor that accurate values are not obtainable, or perhaps an insufficient number of terms of the infinite series have been used. By using intrinsic verification we can quantitatively and confidently check the accuracy of numerical values generated by exact solutions. These concepts have been used in developing computer codes and have verified literally thousands of exact transient heat conduction solutions involving the parallelepiped (Beck et al., 2004).

Four different types of intrinsic verification are discussed in this section. The first type, discussed in Section 5.4.1, uses only long-cotime GF related to the method of separation of variables, and it has a parameter which can be continuously varied to demonstrate verification. This type of verification is particularly appropriate for locations removed from the heated surface where the temperature is known to be zero (or as close as desired) for sufficiently small times. The second type of verification uses limiting-case one-dimensional solutions appropriate for short times. The third type of intrinsic verification uses different eigenfunction expansions for steady-state problems; it does not have a parameter that can be continuously varied. Instead different solutions of the same problem are found and compared. Finally, the fourth type of intrinsic verification is related to time partitioning and is appropriate for 2D and 3D geometries. In this method varying a partition time changes numerical values in a minimal fashion, and these small changes indicate the accuracy of the numerical values.

5.4.1 Intrinsic Verification by Complementary Transients

Intrinsic verification by complementary transients applies to transient cases with heating (or cooling) at a surface. If the solution contains a steady component and a time-decaying component that we call the complementary transient, then intrinsic verification can be carried out at locations removed from the heated surface where negligible temperature rise occurs for sufficiently small times.

Consider a body of finite extent heated at surface $x = 0$. Using Equation 5.4 the temperature may be stated as

$$T(\vec{r},t) = \alpha \int_{u=0}^{t} \int_{s} \left\{ \begin{array}{ll} f\,\partial G^L / \partial x'; & \text{first kind only} \\ (f/k)\,G^L; & \text{2nd or 3rd kind} \end{array} \right\}_{x'=0} ds'\,d\tau \qquad (5.62)$$

where u is the cotime; note that the large-cotime form G^L is used here. Now the time-dependence of every large-cotime GF in the finite body has the form $e^{-c_n u}$ where u is the cotime. Evaluate the time integral of this exponential factor as follows:

$$\int_{u=0}^{t} e^{-c_n u}\,du = \frac{1}{c_n} - \frac{e^{-c_n t}}{c_n} \qquad (5.63)$$

Note that as $t \to \infty$ this integral gives a constant value, specifically, factor $1/c_n$. This suggests that the temperature solution can be written as the sum of a steady term and a time-decaying term, for example in the rectangle:

$$T(x,y,t) = T^L_{steady}(x,y) + T^L_{c.t.}(x,y,t) \qquad (5.64)$$

At locations removed from the heated surface, for sufficiently small times, temperature T will be essentially zero:

$$0 = T^L_{steady}(x,y) + T^L_{c.t.}(x,y,t) \quad \text{for } \alpha t / x^2 < C_0 \qquad (5.65)$$

where the value of the dimensionless cutoff time, C_0 is to be determined.

Replacing the inequality in the previous equation by an equality and solving for the steady state component gives

$$T_{steady}(x,y) = -T^L_{c.t.}(x,y,t_0 = C_0 x^2/\alpha) \qquad (5.66)$$

This equation suggests that intrinsic verification can be carried out with the complementary transient. We have the steady state expression on the left which is a function of only position while on the right side the (negative of the) complementary transient is a function of position and time. This expression can only be correct if the right side

TABLE 5.4
Intrinsic Verification for Case $X21B10$ $Y11B00T0$ Using the Complementary Transient for Interior Locations at Early Time*

x/L	y/L	$\alpha t/L^2$	K_{max}	Number of Terms	$-T_{C\cdot t}$
0.25	0.5	0.0005	4.6	450	0.1788017425
0.25	0.5	0.0010	4.6	210	0.1788031719
0.25	0.5	0.0020	4.6	105	0.1787935827
0.25	0.5	0.0040	4.6	50	0.1786858298
0.25	0.5	0.0100	4.6	18	0.1743892020
0.25	0.5	0.0005	23	2312	0.1788046222
0.25	0.5	0.0010	23	1152	0.1788046220
0.25	0.5	0.0020	23	578	0.1788035094
0.25	0.5	0.0040	23	288	0.1786662112
0.25	0.5	0.0100	23	105	0.1744284821
0.50	0.5	0.0020	11.5	288	0.0800610324
0.50	0.5	0.0040	11.5	136	0.0800610361
0.50	0.5	0.0080	11.5	72	0.0800588045
0.50	0.5	0.0160	11.5	32	0.0797857482
0.50	0.5	0.0020	23	578	0.0800610334
0.50	0.5	0.0040	23	288	0.0800610330
0.50	0.5	0.0080	23	136	0.0800588080
0.50	0.5	0.0160	23	72	0.0797857689
0.75	0.5	0.0045	11.5	128	0.0303331137
0.75	0.5	0.0090	11.5	55	0.0303331192
0.75	0.5	0.0180	11.5	32	0.0303298132
0.75	0.5	0.0360	11.5	10	0.0299547020
0.75	0.5	0.0045	23	242	0.0303331147
0.75	0.5	0.0090	23	128	0.0303331142
0.75	0.5	0.0180	23	55	0.0303298158
0.75	0.5	0.0360	23	32	0.0299548261

*Quantity K_{max} is the largest allowed absolute value of the exponential argument. Inaccurate digits are underlined.

gives the same numerical value for all acceptable times. Hence we can verify the solution by examining numerical values with times less than indicated in Equation 5.65.

Consider determination of the temperature in the rectangle with specified heat flux at the $x = 0$ boundary, cases $X21B10$ $Y11B00T0$. Table 5.4 shows results for the complementary transient component of the temperature, Equation 5.64, evaluated at $y = L/2$ and with $x = L/4$, $L/2$ and $3L/4$. The third column contains the dimensionless cotime and the fourth column contains the number of series terms needed. Note that the same numerical values are repeated in Table 5.4 for sufficiently small cotimes. This is the essence of intrinsic verification with the complementary transient, because as the cotime goes to zero the negative of the complementary transient is equal to the steady-state value.

In the upper part of Table 5.4, for $K_{max} = 4.6$, about five digits are accurate for dimensionless times $\alpha t / L^2 < 0.001$. For $\alpha t / L^2 = 0.004$ only three digits are accurate but fewer series terms are required. Further down the table, for $K_{max} = 23$, there are nine accurate digits present at $\alpha t / L^2 < 0.001$, and 1152 series terms are required. Similar observations can be made regarding the table entries for $x / L = 0.5$ and $x / L = 0.75$. See de Monte et al. (2008).

5.4.2 Complementary Transient and 1D Solution

The use of a one-dimensional semi-infinite solution at an appropriate early time provides another way to use the complementary transient for intrinsic verification. Consider the rectangle again, case X21B10Y11B00. The one-dimensional semi-infinite solution denoted X20B1T0 is a close approximation up to about $\alpha t / L^2 = 0.3$ for x small and for y away from the boundaries. In this case x can also be zero. Using Equation 5.65

$$\begin{aligned} T(x, y, t) &= T_{X20B1T0}(x, t_{1D}) \\ &= T_{steady}(x, y) + T_{c.t.}(x, y, t) \quad \text{for } t < t_{1D} \end{aligned} \tag{5.67}$$

Now solve for the steady-state result:

$$T_{steady}(x, y) = T_{X20B1T0}(x, t_{1D}) - T_{c.t.}(x, y, t_{1D}) \tag{5.68}$$

This is another expression that demonstrates intrinsic verification, with a continuously variable parameter, because there is no time dependence on the left side but time dependence on the right. Varying parameter t_{1D} over an acceptable range should give the same numerical values for the steady temperature constructed from two independent pieces. This expression has an advantage over Equation 5.66 which has only one part on the right-hand side. As a result, the above expression can be used to find an error in a multiplicative constant in the series for $T_{c.t.}$ which could not be accomplished using Equation 5.66.

Although use of a one-dimensional semi-infinite solution was discussed in this section for verification in a rectangle, the same sort of verification is possible in a three-dimensional body (the parallelepiped). It is also possible to carry out intrinsic verification on the rectangle with a two-dimensional short-cotime solution (case X20B1Y10B0 to approximate the rectangle discussed above). This would be particularly efficient near the corner $y = 0$ and $x = 0$.

5.4.3 Intrinsic Verification by Alternate Series Expansion

In this section a method of verifying steady-state solutions in two- or three-dimensional finite bodies is described. The method requires that solutions can be found with the eigenfunction expansions in more than one direction (hence the requirement of finite bodies). Next an example is given in a rectangle.

Example 5.4: Steady Rectangle—*X*12*B*10 *Y*12*B*00

Find the steady temperature in the rectangle by two different eigenfunction expansions for verification. The $x = 0$ surface of the rectangle has an elevated temperature and the other surfaces are homogeneous.

Solution

The steady solution in this rectangle was studied earlier in Example 5.2, in a discussion of improving series convergence by replacing the double-sum series with a single-sum series. The temperature satisfies the following equations:

$$\frac{\partial^2 T}{\partial x^2} + \frac{\partial^2 T}{\partial y^2} = 0 \tag{5.69}$$

$$T(0, y) = T_0; \quad \left.\frac{\partial T}{\partial x}\right|_{x=L} = 0; \quad T(x, 0) = 0; \quad \left.\frac{\partial T}{\partial y}\right|_{y=W} = 0;$$

The steady GF solution has the form

$$T_S(x, y) = T_0 \int_{y'=0}^{W} \left.\frac{\partial G}{\partial x'}\right|_{x'=0} dy' \tag{5.70}$$

The single-sum steady solution, found with eigenfunctions along the x-direction, is given by (Equation 5.46):

$$T_S(x, y) = T_0 - 2T_0 \sum_{m=1}^{\infty} \frac{\sin \beta_m x / L}{\beta_m} \left[\frac{e^{-\sigma_m(2W-y)} + e^{-\sigma_m y}}{1 + e^{-2\sigma_m W}} \right] \tag{5.71}$$

where $\sigma_m = \beta_m / L$. It should be noted that this is the preferred single-sum solution for a rectangle heated on the $x = 0$ boundary, because the eigenfunction expansions (sines in this case) are in the nonhomogeneous direction (the direction that locates the heated boundary). This arrangement provides the fastest convergence for both temperature and for heat flux series near the heated boundary (Cole and Yen, 2001a).

Next an alternate single-sum solution will be sought, for the purpose of verification, using an alternate GF with eigenfunctions along the y-direction, in the form

$$G_{X12Y12} = \frac{2}{W} \sum_{n=1}^{\infty} \sin \frac{\gamma_n y}{W} \sin \frac{\gamma_n y'}{W} P_n(x, x') \tag{5.72}$$

where kernel function P_n may be found by eigenfunction expansion (see Section 4.6)

$$P_n(x, x') = \frac{e^{-\sigma_n(2L+|x-x'|)} + e^{-\sigma_n(2L-x-x')} - e^{-\sigma_n|x-x'|} - e^{-\sigma_n(x+x')}}{2\sigma_n(1 - e^{-2\sigma_n L})} \tag{5.73}$$

where $\sigma_n = \gamma_n / W$. Now replace this GF into the GF solution, Equation 5.70, to see that one integral is needed that falls on a sine term (see Equation 5.43) and a derivative is needed that falls on the kernel function (see also Crittenden and Cole, 2002):

$$\left.\frac{\partial P_n}{\partial x'}\right|_{x'=0} = \frac{e^{\sigma_n(2L-x)} + e^{-\sigma_n x}}{\left(1 + e^{-2\sigma_n L}\right)} \tag{5.74}$$

Then the alternate temperature expression may be assembled in the form

$$\frac{T_S(x, y)}{T_0} = 2\sum_{n=1}^{\infty} \frac{\sin(\gamma_n y / W)}{\gamma_n} \left[\frac{e^{-\sigma_n(2L-x)} + e^{-\sigma_n x}}{1 + e^{-2\sigma_n L}}\right] \tag{5.75}$$

This is the result commonly given in the literature. It has eigenfunctions in the homogeneous direction, in this case the y-direction. Examination of Equation 5.75 reveals that it converges slowly at $x = 0$.

Numerical results are given in Table 5.5 for intrinsic verification based on the two series expressions for the steady temperature in the rectangle with aspect ratio $L/W = 0.5$. Convergence testing was carried out every four terms (note that the number of terms shown are multiples of four) for $\epsilon = 10^{-10}$. Ten digits of the normalized temperature are shown in Table 5.5, and the digits that do not agree between the two series expressions are underlined. The preferred series converges

TABLE 5.5

Normalized Temperature T_s/T_0 and Number of Series Terms for Steady Rectangle Case *X*12*B*10*Y*12*B*00 for the Preferred Series (Equation 5.71) and the Alternate Series (Equation 5.75)*

x/L	y/W	Alternate Series Equation 5.75	Number of Terms	Preferred Series Equation 5.71	Number of Terms
0.01	0.01	$0.7048827\underline{901}$	900	$0.7048827\underline{774}$	284
0.20	0.01	$0.064464204\underline{6}$	72	$0.064464204\underline{7}$	268
0.40	0.01	$0.033911604\underline{9}$	40	$0.033911604\underline{8}$	268
0.80	0.01	$0.020883214\underline{6}$	24	$0.020883214\underline{5}$	264
0.01	0.20	$0.985038752\underline{6}$	972	$0.985038752\underline{7}$	24
0.20	0.20	0.7240794778	64	0.7240794778	20
0.40	0.20	0.5399106059	36	0.5399106059	20
0.80	0.20	0.3878999614	20	0.3878999614	20
0.01	0.40	$0.993686161\underline{8}$	968	$0.993686161\underline{7}$	16
0.20	0.40	0.8772373586	64	0.8772373586	12
0.40	0.40	0.7732177380	36	0.7732177380	12
0.80	0.40	0.6538082660	20	0.6538082660	12
0.01	0.80	$0.997911889\underline{5}$	964	$0.997911889\underline{4}$	12
0.20	0.80	0.9589473349	60	0.9589473349	8
0.40	0.80	0.9220505381	36	0.9220505381	8
0.80	0.80	0.8743658796	20	0.8743658796	8

*The rectangle aspect ratio is $L/W = 0.5$. Convergence was tested every fourth term and with $\epsilon = 10^{-10}$.

better near $x = 0$, which is the heated surface, and of the most importance in this case. The alternate series, although requiring more terms over most of the body, does better near $y = 0$. From this perspective these two series are complementary. Neither series converges quickly close to the corner $x = y = 0$ because of the jump in temperature there.

In the above example, two different eigenfunction expansions were found in the rectangle for the steady temperature. In the parallelepiped, three different eigenfunction expansions are possible in the x-, y-, and z-directions (see Beck et al., 2006). This method has also been applied to finite cylinders with eigenfunction expansions along the r- and z-directions (Cole, 2004).

5.4.4 Time-Partitioning Intrinsic Verification

The use of time partitioning for intrinsic verification involves varying the partition time, a parameter in the solution, to construct two or more different series solutions for the same heat conduction problem. If the formulation is correct, the numerical values from the different series solutions will agree to high accuracy. This method is appropriate for 2D and 3D geometries and is best illustrated by a specific example.

Example 5.5: Time Partitioning in the Rectangle

Consider the transient heat conduction in the rectangle ($0 < x < L$; $0 < y < W$) that is heated by constant heat flux at $x = 0$ and all the other faces are held at zero temperature. The initial temperature is zero. The describing differential equation and related conditions are given by:

$$\frac{\partial^2 T}{\partial x^2} + \frac{\partial^2 T}{\partial y^2} = \frac{1}{\alpha}\frac{\partial T}{\partial t}; \quad t > 0 \tag{5.76}$$

$$-k\left.\frac{\partial T}{\partial x}\right|_{x=0} = q_0; \quad T(L, y, t) = 0; \quad T(x, 0, t) = 0; \quad T(x, W, t) = 0 \tag{5.77}$$

$$T(x, y, 0) = 0 \tag{5.78}$$

Solution

This problem is described by case X21B10Y11B00T0. This problem can be solved in several ways, including the separation of variables method which is particularly effective for large dimensionless times when fewer terms of the series are needed. The Laplace transform method is most effective (fewer terms needed and better accuracy) for small dimensionless times.

The time-partitioning method uses components of both separation of variables and Laplace transform methods, as follows. The GF solution for this problem is given by

$$T(x, y, t) = \frac{\alpha q_0}{k}\int_{u=0}^{t} G_{X21}(x, x' = 0, u)\int_{y'=0}^{W} G_{Y11}(y, y', u)\, dy'\, du \tag{5.79}$$

Here $u = t - \tau$ is the cotime. For time partitioning the solution is written in two parts, one for short cotimes and one for long cotimes, as

$$T(x, y, t) = \frac{\alpha q_0}{k} \int_{u=0}^{t_p} G_{X21}^S(x, 0, u) \int_{y'=0}^{W} G_{Y11}^S(y, y', u)\, dy'\, du$$
$$+ \frac{\alpha q_0}{k} \int_{u=t_p}^{t} G_{X21}^L(x, 0, u) \int_{y'=0}^{W} G_{Y11}^L(y, y', u)\, dy'\, du \quad (t > t_p) \tag{5.80}$$

Superscript S on the GF denotes the small-cotime form, and superscript L denotes the long-cotime form. Quantity t_p is the partition time, chosen in the range $0 < \alpha t_p / L^2 < 0.05$. In the present example the length scale L is used, however, in general the characteristic dimension for choosing the partition time should be the smallest dimension in the body. Note also that in the above equation the second integral is needed only when $t > t_p$.

The next step is to find the different GF that are needed. The small-cotime GF are given as approximations for small values of u:

$$G_{X21}^S(x, 0, u) \approx \frac{1}{\sqrt{\pi \alpha u}} e^{-x^2/(4\alpha u)} \tag{5.81}$$

(Appendix X, Equation X21.1, $n = 0$ only)

$$\int_{y'=0}^{W} G_{Y11}^S(y, y', u)\, dy' \approx erf\left(\frac{y}{\sqrt{4\alpha u}}\right) - erfc\left(\frac{W - y}{\sqrt{4\alpha u}}\right) + erfc\left(\frac{W + y}{\sqrt{4\alpha u}}\right)$$
$$+ erfc\left(\frac{2W - y}{\sqrt{4\alpha u}}\right) - erfc\left(\frac{2W + y}{\sqrt{4\alpha u}}\right) \tag{5.82}$$

(Appendix X, Equation X11.17b, $n = 1$ only)

The large-cotime GF are given exactly by

$$G_{X21}^L(x, 0, u) = \frac{2}{L} \sum_{m=1}^{\infty} e^{-\beta_m^2 \alpha u / L^2} \cos(\beta_m x / L) \tag{5.83}$$

where $\beta_m = (m - 1/2)\pi$

$$\int_{y'=0}^{W} G_{Y11}^L(y, y, u)\, dy' = 4 \sum_{n=1}^{\infty} \frac{\sin \gamma_n y / W}{\gamma_n} e^{-\gamma_n^2 \alpha u / W^2} \tag{5.84}$$

where $\gamma_n = (2n - 1)\pi$

(Appendix X, Equation X11.18)

The solution given by Equation 5.80 can be written as

$$T(x, y, t) = T^S(x, y, t_p) + T^L(x, y, u)\Big|_{u=t_p}^{t}$$
$$= T^S(x, y, t_p) - T_{c.t.}^L(x, y, t_p) + T_{c.t.}^L(x, y, t) \tag{5.85}$$

where subscript "c.t." denotes complementary transient. The short-cotime component is given by

$$T^S(x, y, t_p) = \frac{\alpha q_o}{k} \int_{u=0}^{t_p} \left[G_{X21}^S(x, 0, u) \int_{y'=0}^{W} G_{Y11}^S(y, y, u)\, dy' \right] du \tag{5.86}$$

Obtaining a general closed-form expression for the integration over u for the short-cotime components in Equation 5.80 can be difficult, while the long-cotime integration is straight forward. Generally, the simplest way to perform the short-cotime integration is numerically (McMasters et al., 2002). However, for $x = 0$ and y not too close to the boundaries, the short-cotime solution becomes semi-infinite case X20B1T0, given by

$$T^S(0, y, t_p) \approx 2\frac{q_0 L}{k}\left(\frac{\alpha t_p}{\pi L^2}\right)^{1/2} \tag{5.87}$$

The long-cotime component is found by substituting Equation 5.83 and Equation 5.84 into the second half of Equation 5.80 and integrating. The result for the two long-cotime components in the second half of Equation 5.85 is given by

$$T^L_{c.t.}(x, y, u) = 8\frac{q_0 L}{k}\sum_{m=1}^{\infty}\sum_{n=1}^{\infty}\frac{\sin(\pi(2n-1)y/W)}{\pi(2n-1)\phi_{mn}^2}\cos\left(\frac{\beta_m x}{L}\right)\frac{e^{-\phi_{mn}^2 \alpha u/L^2}}{\phi_{mn}} \tag{5.88}$$

$$\text{where } \phi_{mn}^2 = (2n-1)^2\pi^2 L^2/W^2 + (m-1/2)^2\pi^2$$

If the above expression is evaluated at $u = 0$ the series converges very slowly. The goal is to make t_p as large as possible while still obtaining the desired accuracy. The limits are that at small t_p, many term of the series for T^L are needed. At large values of t_p, the approximations in G^S are not as accurate.

As $t \to \infty$ in Equation 5.85, the last term on the right disappears and the steady solution is obtained:

$$T(x, y, \infty) = T(x, y) = T^S(x, y, t_p) - T^L_{c.t.}(x, y, t_p) \tag{5.89}$$

This equation provides for intrinsic verification, since the left side is independent of time while the right side is a function of partition time t_p. It is important to note that the two right-hand side components in Equation 5.89 are independent, since one comes from the Laplace transform solution and the other from the separation of variables method. Varying the partition cotime over the acceptable range should give precisely the same value (for a given number of significant figures), thus exhibiting intrinsic verification. Generally the steady-state component is the most difficult part of a solution to evaluate numerically, however using Equation 5.89 is very efficient. The use of time partitioning to find a rapidly converging form of the steady solution has also been discussed by Linton (1999) under the name Ewald summation.

Numerical values for intrinsic verification based on time partitioning are shown in Table 5.6. In Table 5.6 steady temperature values are computed at one location ($x/L = 0$; $y/L = 0.25$), but for several values of the partition time and for two values of convergence parameter K_{max}. In this example the simple 1D transient solution given by Equation 5.87 is used. Verification to 10 digit accuracy is demonstrated at $K_{max} = 23$ because as the partition time t_p becomes smaller, the steady temperature is unchanged to 10 digits, even though the components of the steady temperature (complementary transient $T^L_{c.t.}$ and small-cotime T^S) do vary with partition time and the number of series terms required increases. Table 5.6

TABLE 5.6
Steady-State Temperature for Case *X*21*B*10 *Y*11*B*00*T*0*

y/L	$\alpha t_p / L^2$	K_{max}	Number of Terms	$-T^L_{c.t.}(x, y, t_p)$	$T^S(x, y, t_p)$	T_{steady}
0.25	0.00050	11.5	1152	0.2788314014	0.0252313252	0.3040627266
0.25	0.00060	11.5	968	0.2764231950	0.0276395320	0.3040627270
0.25	0.00075	11.5	741	0.2731607902	0.0309019362	0.3040627263
0.25	0.00100	11.5	578	0.2683802437	0.0356824823	0.3040627260
0.25	0.00200	11.5	288	0.2536002774	0.0504626504	0.3040629278
0.25	0.00400	11.5	136	0.2327310450	0.0713649646	0.3040960096
0.25	0.00600	11.5	78	0.2168942410	0.0874038744	0.3042981154
0.25	0.01000	11.5	50	0.1926680623	0.1128379167	0.3055059790
0.25	0.00050	23	2312	0.2788314035	0.0252313252	0.3040627287
0.25	0.00060	23	1922	0.2764231967	0.0276395320	0.3040627287
0.25	0.00075	23	1485	0.2731607925	0.0309019362	0.3040627287
0.25	0.00100	23	1152	0.2683802464	0.0356824823	0.3040627287
0.25	0.00200	23	578	0.2536002814	0.0504626504	0.3040629318
0.25	0.00400	23	288	0.2327310515	0.0713649646	0.3040960161
0.25	0.00600	23	171	0.2168942541	0.0874038744	0.3042981286
0.25	0.01000	23	105	0.1926680993	0.1128379167	0.3055060161

*$x = 0$ and $y/L = 0.25$ with varying partition time for intrinsic verification. Quantity K_{max} is the largest allowed absolute value of the exponential argument. The steady temperature in the last column is the sum of the two preceding columns. Inaccurate digits are underlined.

also indicates that an adequate partition time for this case is 0.00075 because it provides a balance between high accuracy and a reasonable number of series terms. If numerical integration were used to obtain the short-cotime component for larger partition times up to $\alpha t / L^2 = 0.05$, many fewer terms of the series would be needed for the same numerical accuracy.

PROBLEMS

5.1 Evaluate the sum

$$S = 2 \sum_{m=1}^{\infty} \frac{\sin(n\pi x / L)}{n\pi} \tag{5.90}$$

at $x / L = 0.1$ by truncating the series when the average of the last three terms divided by the sum is less than 10^{-4} (see Equation 5.16). Compare your result to the exact value of $S = (1 - x / L)$ to find the number of accurate digits produced by this convergence criterion. Now repeat your calculation at $x / L = 0.01$. Does this series converge more rapidly or more slowly as $x \to 0$? Explain.

5.2 Evaluate the series

$$2\sum_{m=1}^{\infty}\frac{[1-(-1)^m]}{m\pi}\,e^{-m^2\pi^2\alpha t\,/\,L^2} \tag{5.91}$$

for $\alpha t\,/\,L^2 = 0.1$ using two convergence tests: (a) truncate the series using Equation 5.16 for $\epsilon = 10^{-5}$; and (b) truncate the series when the (absolute value of the) argument of the exponential exceeds $K_{max} = 11.5$. How many accurate digits of the series does each convergence test provide?

5.3 Evaluate the temperature for case X11B10T0 using the long-cotime GF (see Equation 5.19) at $x\,/\,L = 0.5$ and at dimensionless time $\alpha t\,/\,L^2 = 0.01$, 0.1, and 1.0. How many terms of the series are needed for four-digit accuracy?

5.4 Determine the accuracy of evaluating the series for heat flux given by Equation 5.28 at $x = 0$ and at $x = L$ by requiring that the magnitude of the exponential argument be no greater than 11.5 for dimensionless times 0.1 and 1.0. How many terms of the series are required at each time and place?

5.5 Repeat Problem 5.4 for Equation 5.27, to find the number of terms needed for the series for temperature. Compare the number of terms needed with that from the heat flux series. Note that the temperature series contains factor $m\pi$ in the denominator and the heat flux series, Equation 5.28, does not. What effect does factor $m\pi$ have in the convergence speed of the two series?

5.6 Starting with the relation inferred from Equation 5.44 which has the form

$$T_S(x, y) = T_0\left[S_x(x) + S_{xy}(x, y)\right], \tag{5.92}$$

derive the identity given by Equation 5.45 by replacing the above relation into the boundary value problem for T_S, and then solve for $S_x(x)$. What are the describing differential equation and boundary conditions for $S_{xy}(x, y)$?

5.7 Use the X11B10T0 solution given by Equation 5.27 to investigate intrinsic verification (by the method of complementary transients) at location $x\,/\,L = 0.5$. Make a table of your results at five different dimensionless times.

5.8 Compute numerical values from case X11B10T0 given by Equation 5.27 near the surface at locations $x\,/\,L = 0.01$, 0.05 and at dimensionless times $\alpha t\,/\,L^2 = 0.01$, 0.05. Verify that your numerical values are correct by comparing them with semi-infinite case X10B10T0 (see Example 1.4, Equation 1.109). Make a table of your results including values from both geometries and the percentage difference between them.

5.9 Write a computer program to evaluate temperatures from case X21B21T0 given by Equation 5.60. Use intrinsic verification (by complementary transients) at $x\,/\,L = 0.5$ to check that your calculations are correct. Make a table of your results at five different dimensionless times.

REFERENCES

Abramowitz, M. and Stegun, I. A. (Eds), 1964, *Handbook of Mathematical Functions with Formulas, Graphs, and Mathematical Tables*, Applied Mathematics Series 55, National Bureau of Standards, Washington, DC.

Beck, J. V. and Cole, K. D., 2007, Improving convergence of summations in heat conduction, *Int. J. Heat Mass Transfer*, vol. 50, pp. 257–268.

Beck, J. V. and Keltner, N. R., 1987, Green-function partitioning procedure applied to foil heat-flux gages, *J. Heat Transfer*, vol. 109, pp. 274–280.

Beck, J. V., Haji-Sheikh, A., Amos, D. E., and Yen, D. H. Y., 2004, Verification solution for partial heating of rectangular solid, *Int. J. Heat Mass Transfer*, vol. 47, pp. 4243–4255.

Beck, J. V., McMasters, R., Dowding, K. J., and Amos, D. E., 2006, Intrinsic verification methods in linear heat conduction, *Int. J. Heat Mass Transfer*, vol. 49, pp. 2984–2994.

Beck, J. V., Wright, N. T., Haji-Sheikh, A., 2008, Transient power variation in surface conditions in heat conduction for plates, *Int. J. Heat Mass Transfer*, vol. 51, pp. 2553–2565.

Carslaw, H. S. and Jaeger, J. C., 1959, *Conduction of Heat in Solids*, 2nd edn, Oxford University Press, New York.

Cole, K. D., 2000, Green's Function Library, internet site, www.greensfunction.unl.edu.

Cole, K. D., 2004, Fast converging series for heat conduction in the circular cylinder, *J. Eng. Math.*, vol. 49, pp. 217–232.

Cole, K. D. and Yen, D. H. Y., 2001a, Green's functions, temperature, and heat flux in the rectangle, *Int. J. Heat Mass Transfer*, vol. 44, pp. 3883–3894.

Cole, K. D. and Yen, D. H. Y., 2001b, Influence functions for the infinite and semi-infinite strip, *AIAA J. Thermophys. Heat Transfer*, vol. 15, no. 4, pp 431–438.

Crittenden, P. E. and Cole, K. D., 2002, Fast-converging steady-state heat conduction in a rectangular parallelepiped, *Int. J. Heat Mass Transfer*, vol. 45, pp. 3585–3596.

de Monte, F., Beck, J. V., and Amos, D.E., 2008, Diffusion of thermal disturbances in two-dimensional transient Cartesian heat conduction, *Int. J. Heat Mass Transfer*, vol. 51, pp. 5931–5941.

Duffy, D. G., 2001, *Green's Functions with Applications*, Chapman and Hall/CRC Press, Boca Rotan, FL.

Lanczos, C., 1966, *Discourse on Fourier Series*, Hafner Publishing, London.

Linton, C. M., 1999, Rapidly convergent representations for Green's functions for Laplace's equation, *Proc. Royal Soc. London A*, vol. 455, pp. 1767–1797.

McMasters, R. L., Dowding, K., Beck J. V., and Yen, D. H. Y., 2002, Methodology to generate accurate solutions for verification in transient three-dimensional heat conduction, *J. Numer. Heat Transfer B*, vol. 41, pp. 521–541.

Melnikov, Y., 1999, *Influence Functions and Matrices*, Marcel Dekker, New York.

Roach, P. J., 1998, *Verification and Validation in Computational Science and Engineering*, Hermosa, Albuquerque, NM, Chapters 3–8.

Sommerfeld, A., 1949, *Partial Differential Equations in Physics*, Academic Press, New York.

6 Rectangular Coordinates

6.1 INTRODUCTION

This chapter is concerned with heat conduction in bodies described by rectangular coordinates. Complete examples are included that demonstrate strategies for evaluating the integrals in the Green's function solution equation (GFSE).

An important feature of the Green's function (GF) solution method is the ability to simply write down the temperature in integral form. Once the problem is properly defined, one can jump to the solution and gain insight into the problem. For example, one can also immediately write down the alternative Green's function solution, and then the better form of the solution can be selected for evaluation. The student can concentrate on translating a physical heat transfer situation into a boundary value problem without getting lost in the details of the solution. There is a sense of accomplishment associated with jumping to the solution that can be a valuable part of the learning process. After the integral form is written down, the integrals can be examined. If they are familiar, the solution can be completed easily. If the integrals are unfamiliar they may be available in integral tables, approximate forms may be substituted, or finally numerical integration will always yield an answer.

The examples in this chapter are concerned with only one nonhomogeneous term at a time. The nonhomogeneous term may be the initial condition, the volume energy generation, or the boundary condition. Practical situations often involve two or more nonhomogeneous terms, but because the GF solution equation is the sum of the contributions from the various nonhomogeneous terms, the temperature resulting from initial conditions, boundary conditions, and volume energy generation can simply be added together for the complete solution.

One-dimensional geometries are emphasized in this chapter and the one-dimensional GFSE is given in Section 6.2. Semi-infinite bodies are discussed in Section 6.3. Flat plates are discussed in Section 6.4 through 6.6. Some two-dimensional cases are discussed in Sections 6.7 and 6.8, and some steady-state cases are discussed in Section 6.9.

6.2 ONE-DIMENSIONAL GREEN'S FUNCTIONS SOLUTION EQUATION

The heat conduction equation for homogeneous one-dimensional bodies in the rectangular coordinate system is

$$\frac{\partial^2 T}{\partial x^2} + \frac{1}{k} g(x,t) = \frac{1}{\alpha}\frac{\partial T}{\partial t} \tag{6.1}$$

with initial condition

$$T(x,0) = F(x) \tag{6.2}$$

and with boundary conditions

$$k_i \left.\frac{\partial T}{\partial n_i}\right|_{x_i} + h_i T|_{x_i} = f_i(t) - (\rho c b)_i \left.\frac{\partial T}{\partial t}\right|_{x_i} \tag{6.3}$$

where n_i is an outward normal from the body at the boundary, and x_i represents the two boundaries ($i = 1, 2$). Equation 6.3 represents five different kinds of boundary conditions by the choice of k_i, h_i, f_i, and b_i. These boundary conditions are discussed in detail in Chapter 2.

The solution of the temperature problem given in Equations 6.1 through 6.3 is given by the GFSE for one-dimensional rectangular coordinates (refer to Section 3.2 for a derivation)

$$\begin{aligned}
T(x,t) &= \int_{x'=0}^{L} G(x,t|x',0)F(x')\,dx' \qquad \text{(for the initial condition)} \\
&+ \alpha \sum_{i=1}^{s} \left[\frac{(\rho c b)_i}{k_i} G(x,t|x',0)F(x') \right]_{x'=x_i} \qquad \text{(for boundary conditions of the fourth and fifth kinds only)} \\
&+ \int_{\tau=0}^{t} \int_{x'=0}^{L} \frac{\alpha}{k} G(x,t|x',\tau)\, g(x',\tau)\,dx'\,d\tau \qquad \text{(for volume energy generation)} \\
&+ \alpha \int_{\tau=0}^{t} d\tau \sum_{i=1}^{2} \left[\frac{f_i(\tau)}{k_i} G(x,t|x_i,\tau) \right] \qquad \text{(for boundary conditions of the second through fifth kinds)} \\
&- \alpha \int_{\tau=0}^{t} d\tau \sum_{i=1}^{2} \left[f_i(\tau) \left.\frac{\partial G}{\partial n_i'}\right|_{x'=x_i} \right] \qquad \text{(for boundary conditions of the first kind only)}
\end{aligned} \tag{6.4}$$

where $G(x,t|x',\tau)$ is the GF. For each different set of boundary conditions there is a different GF that must be used in the GFSE.

6.3 SEMI-INFINITE ONE-DIMENSIONAL BODIES

In this section, the cases under consideration are semi-infinite bodies denoted by $XI0$, $I = 1, 2, 3, 4$. The GFs for infinite and semi-infinite bodies are listed in Table 6.1. A complete listing of rectangular-coordinate GFs, including certain derivatives, integrals, and approximations is given in Appendix X.

For the semi-infinite cases, the GFs have only one form, do not involve infinite series, and are mathematically well behaved everywhere except at the point $x - x' = 0$ and $t - \tau = 0$, where every GF approaches a Dirac delta function. The temperatures calculated by integrating these GFs are mathematically well behaved for any location x and for $t > 0$.

TABLE 6.1
GF for Infinite and Semi-Infinite Bodies

$$G_{XI0}(x,t|x',\tau) = [4\pi\alpha(t-\tau)]^{-1/2}\left\{\exp\frac{-(x-x')^2}{4\alpha(t-\tau)} + M\exp\frac{-(x+x')^2}{4\alpha(t-\tau)}\right\}$$
$$-MD_1 ER(x+x', t-\tau, D_1)$$

Number	M	$D_1 L$
$X00$	0	0
$X10$	−1	0
$X20$	1	0
$X30$	1	B_1
$X40$	−1	$(C_1)^{-1}$

where $ER(x,t,D) = \exp(Dx + D^2\alpha t)\,\text{erfc}\left[\frac{x}{(4\alpha t)^{1/2}} + D(\alpha t)^{1/2}\right]$

$$B_1 = \frac{hL}{k};\quad C_1 = \frac{\rho cb}{\rho cL}$$

and L is a reference length that cancels out.

6.3.1 Initial Conditions

For the case of *spatially uniform* initial conditions in semi-infinite bodies, the appropriate integrals in the GFSE, Equation 6.4, are known in closed form. The resulting temperature expressions for homogeneous boundary conditions are listed in Table 6.2 in compact form. The integrals that were used to create Table 6.2 are listed in the integral tables in Appendix I.

In Table 6.2, one case involves a surface film of high conductivity, numbered $X40$. The notation for the initial condition $T01$ in Table 6.2 refers to a zero initial temperature in the film and a uniform initial temperature in the body. Conversely the notation $T10$ refers to a uniform initial temperature in the film and a zero initial temperature in the body. If both the film and the body have the same uniform initial temperature, the problem can always be formulated with no contribution from the initial condition by defining a new temperature variable, $T - T_0$, where T_0 is the initial temperature.

Semi-infinite bodies with *spatially varying* initial conditions are now considered. Consider the initial temperature distribution of

$$F(x') = \begin{cases} T_0 & a < x' < b \\ 0 & \text{otherwise} \end{cases} \tag{6.5}$$

The GF for boundary conditions of type 1 or 2 can be written in terms of the fundamental heat conduction solution $K(\cdot)$,

$$G_{XI0}(x,t|x',0) = K(x-x',t) + (-1)^I K(x+x',t) \tag{6.6}$$

TABLE 6.2
Temperatures in a Semi-Infinite Body for Uniform Initial Temperature of T_i for Cases $XI0B0T1$, $I = 1,2,3$, and Cases $X40B0T10$, and $X40B0T01$

$$T(x,t) = T_i\left(1 - \frac{1}{2}(1-M)\,\mathrm{erfc}\left[\frac{x}{(4\alpha t)^{1/2}}\right] - M(D_0 - F_0)ER(x,t,D_1)\right)$$

Number	M	D_0	$D_1 L$	F_0
$X10B0T1$	−1	0	0	0
$X20B0T1$	1	0	0	0
$X30B0T1$	−1	1	B_1	0
$X40B0T10$	1	1	C_1^{-1}	1
$X40B0T01$	1	1	C_1^{-1}	0

Note: $ER(\cdot)$ is given in Table 6.1.

where $I = 1$ or 2. The solution for the temperature is obtained by substituting Equation 6.6 in Equation 6.4, the GFSE, to give (case $XI0B0T5$; $I = 1$ or 2):

$$T(x,t) = \frac{T_0}{2}\left(\mathrm{erfc}\left[\frac{-x+a}{(4\alpha t)^{1/2}}\right] - \mathrm{erfc}\left[\frac{-x+b}{(4\alpha t)^{1/2}}\right] + (-1)^I\left\{\mathrm{erfc}\left[\frac{x+a}{(4\alpha t)^{1/2}}\right] - \mathrm{erfc}\left[\frac{x+b}{(4\alpha t)^{1/2}}\right]\right\}\right) \tag{6.7}$$

Next consider the initial temperature distribution of a linear function of x' over part of the body,

$$F(x') = \begin{cases} T_0\dfrac{x'}{L} & \text{for } a < x' < b \\ 0 & \text{elsewhere} \end{cases} \tag{6.8}$$

The length L can have any desired significance; it is only present to make Equation 6.8 dimensionally consistent. The integrals in the GF equation can then be evaluated using Table I.7 (Appendix I) with z replaced by x', and $t - \tau$ replaced by t. The solution is (case $XI0B0T2$; $I = 1$ or 2)

$$\begin{aligned} T(x,t) = T_0\Bigg[&\frac{x}{2L}\left(\mathrm{erfc}\left[\frac{x-b}{(4\alpha t)^{1/2}}\right] - \mathrm{erfc}\left[\frac{x-a}{(4\alpha t)^{1/2}}\right]\right. \\ &\left.+(-1)^I\left\{\mathrm{erfc}\left[\frac{x+b}{(4\alpha t)^{1/2}}\right] - \mathrm{erfc}\left[\frac{x+a}{(4\alpha t)^{1/2}}\right]\right\}\right) \\ &+\frac{2\alpha t}{L}\{K(x-a,t) - K(x-b,t) \\ &+(-1)^I\left[K(x+a,t) - K(x+b,t)\right]\}\Bigg] \end{aligned} \tag{6.9}$$

For the boundary condition of the first kind ($I = 1$), consider the special case of $a = 0$ and $b \to \infty$, which is for a linear initial temperature $F(x') = T_0 x'/L$ over the entire body, $0 \le x \le \infty$. In this case the temperature given by Equation 6.9 reduces to

$$T(x,t) = T_0 \frac{x}{L} \tag{6.10}$$

This is a time-independent solution for the case denoted $X10B0T2$. For the boundary condition of the second kind ($I = 2$) with $a = 0$ and $b \to \infty$, Equation 6.9 gives (case $X20B0T2$)

$$T(x,t) = T_0 \left\{ \frac{x}{L} \operatorname{erfc}\left[\frac{x}{(4\alpha t)^{1/2}}\right] + \frac{4\alpha t}{L} K(x-b,t) \right\} \tag{6.11}$$

This solution is always transient and never reaches a steady state. The transient deviation from the initial straight-line temperature distribution begins at $x = 0$ and spreads to larger x values as time increases.

6.3.2 Boundary Conditions

Temperature expressions resulting from time-invariant boundary conditions are listed in Table 6.3 for four kinds of boundary conditions. These temperature expressions were found by evaluating the integrals in the GFSE. Two mathematical functions that appear in Table 6.3 are erfc and ierfc, which are the complementary error function and the integral of the complementary error function, respectively. Refer to Appendix E for more information on these functions.

TABLE 6.3
Temperatures for Semi-Infinite Bodies for Constant Source Term at $x = 0$; Case $XI0B1T0$, $I = 1, 2, 3$, and $X40B1T00$

$$T(x,t) = H_0(1+M)\left(\frac{t}{k\rho c}\right)^{1/2} \operatorname{ierfc}\left[\frac{x}{(4\alpha t)^{1/2}}\right] + \left(K_0 - \frac{MD_0}{kD_1}\right)\operatorname{erfc}\left[\frac{x}{(4\alpha t)^{1/2}}\right] + \frac{MD_0}{kD_1} ER(x,t,D_1)$$

Number	M	H_0	K_0	D_0	D_1
$X10B1T0$	-1	0	T_0	0	0
$X20B1T0$	1	q_0	0	0	0
$X30B1T0$	-1	0	0	$q_0 + hT_\infty$	h/k
$X40B1T00$	1	q_0	0	q_0	$\rho c/(\rho c b)_1$

$$ER(x,t,D) = \exp[Dx + D^2\alpha t]\operatorname{erfc}\left[\frac{x}{(4\alpha t)^{1/2}} + D(\alpha t)^{1/2}\right]$$

The following examples demonstrate the use of Table 6.1 to find the GF and demonstrate strategies for finding the integrals that occur for various nonhomogeneous boundary conditions.

Example 6.1: Semi-Infinite Body with Specified Surface Temperature—*X*10*B*-*T*0 Case

Find the temperature distribution in the semi-infinite body with specified surface temperature $f(t)$ and with zero initial condition. The volume energy generation is zero.

Solution

This is the $X10B$-$T0$ geometry. The GFSE gives the temperature as

$$T(x,t) = \alpha \int_{\tau=0}^{t} f(\tau) \left.\frac{\partial G_{X10}}{\partial x'}\right|_{x'=0} d\tau \tag{6.12}$$

The GF G_{X10} is found from Table 6.1 by choosing $M = -1$, $D_1 = 0$, and $E_1 = 0$:

$$G_{X10}(x,t|x',\tau) = \frac{1}{[4\pi\alpha(t-\tau)]^{1/2}} \left\{ \exp\left[\frac{-(x-x')^2}{4\alpha(t-\tau)}\right] - \exp\left[\frac{-(x+x')^2}{4\alpha(t-\tau)}\right] \right\} \tag{6.13}$$

The derivative of the GF with respect to x' is required here in the form $\partial/\partial x' = -\partial/\partial n_i$ at $x' = 0$. The derivative of G_{X10} is given in Appendix X as

$$\begin{aligned} \left.\frac{\partial G_{X10}}{\partial x'}\right|_{x'=0} &= \frac{x}{(4\pi)^{1/2}[\alpha(t-\tau)]^{3/2}} \exp\left[\frac{-x^2}{4\alpha(t-\tau)}\right] \\ &= \frac{x}{\alpha(t-\tau)} K(x, t-\tau) \end{aligned} \tag{6.14}$$

where $K(\cdot)$ is the fundamental heat conduction solution. The temperature solution can then be written as

$$T(x,t) = \alpha \int_{\tau=0}^{t} f(\tau) \frac{x}{\alpha(t-\tau)} K(x, t-\tau) d\tau \tag{6.15}$$

(a) *Case X10B1T0.* For the case where the boundary temperature is constant, $f(t) = T_0$, the integral in Equation 6.15 is given in Table I.8, Appendix I, as integral 3,

$$T(x,t) = T_0 \operatorname{erfc}\left[\frac{x}{(4\alpha t)^{1/2}}\right] \tag{6.16}$$

or

$$\frac{T(x,t)}{T_0} = 1 - \operatorname{erf}\left[\frac{x}{(4\alpha t)^{1/2}}\right] \tag{6.17}$$

Compare this solution to case $X10B0T1$ listed in Table 6.2 which is the temperature caused by a zero boundary temperature and uniform initial condition. The two solutions differ by a constant and a change of sign. The heat conduction equation is linear, so that multiplying a solution by (−1) gives another solution; and, adding a constant to a solution gives another solution.

(b) *Case X10B3TO.* In the case where the boundary temperature is a polynomial in $t^{n/2}$, such as

$$f(t) = a_{-1}t^{-1/2} + a_0 + a_1 t^{1/2} + a_2 t + a_3 t^{3/2} + \cdots \tag{6.18}$$

then the integral in Equation 6.15 can be written as the sum of the effects of each term in the polynomial. For the general term of such a polynomial, let $f(\tau) = T_0(\tau/t_0)^{n/2}$, where T_0 has units of temperature and t_0 is some reference time (t_0 could be 1 s). Then the integral in Equation 6.15 may be written

$$T(x,t) = \alpha \int_{\tau=0}^{t} T_0 \left(\frac{\tau}{t_0}\right)^{n/2} \frac{x}{\alpha(t-\tau)} K(x, t-\tau)\, d\tau \tag{6.19}$$

This integral is listed in Table I.8 (Appendix I) and the temperature resulting from the applied surface temperature $T_0(t/t_0)^{n/2}$ may be written

$$T(x,t) = T_0\, \Gamma\left(1 + \frac{n}{2}\right) \left(4\frac{t}{t_0}\right)^{n/2} \mathrm{i}^n\mathrm{erfc}\left[\frac{x}{(4\alpha t)^{1/2}}\right] \tag{6.20}$$

where $n = -1, 0, 1, \ldots$, and so on. The gamma function $\Gamma(1 + n/2)$ takes the values $\pi^{1/2}, 1, \pi^{1/2}/2$, and 1 for $n = -1, 0, 1$, and 2, respectively. The function $\mathrm{i}^n\mathrm{erfc}(\cdot)$ is the repeated integral of the error function plotted in Figure 6.1. The

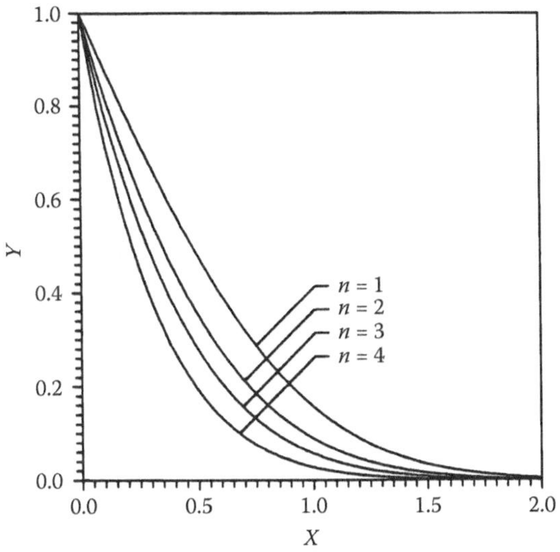

FIGURE 6.1 Repeated integrals of error function, $Y = 2^n\Gamma(n/2+1)i^n\,\mathrm{erfc}(X)$.

$\text{i}^n\text{erfc}(z)$ function is related to erfc(z) by

$$\text{i}^0\text{erfc}(z) = \text{erfc}(z) \tag{6.21a}$$

$$\text{i}^1\text{erfc}(z) = \frac{1}{\pi^{1/2}}e^{-z^2} - z\,\text{erfc}(z) \tag{6.21b}$$

$$2n\,\text{i}^n\text{erfc}(z) = \text{i}^{n-2}\text{erfc}(z) - 2z\text{i}^{n-1}\text{erfc}(z) \tag{6.21c}$$

Some numerical values of $\text{i}^1\text{erfc}(z)$ are listed in Table E.1, in Appendix E, along with other properties of the error function.

In the case where the surface temperature $f(t)$ is periodic in time, the Laplace transform technique can be used on the integral in Equation 6.15 to good advantage. Refer to Carslaw and Jaeger (1959, pp. 399–402) for a general discussion. See Chapter 9 for a discussion of the *steady periodic* portion of the temperature caused by a periodic surface temperature.

Example 6.2: Semi-Infinite Body with Specified Surface Heat Flux—*X*20*B*-*T*0 Case

Find the temperature in the semi-infinite body that has a heat flux boundary condition and zero initial condition.

Solution

This is the $X20B\text{-}T0$ geometry. The GFSE for the temperature takes the form

$$T(x,t) = \alpha \int_{\tau=0}^{t} \frac{f(\tau)}{k} G_{X20}(x,t|0,\tau)\,d\tau \tag{6.22}$$

The heat flux at the boundary is $f(t)$ with units of W/m^2. Note that the $X20$ GF is evaluated at the surface $x' = 0$. The $X20$ GF given in Table 6.1 is the sum of two fundamental heat conduction solutions, so the temperature can be written as

$$\begin{aligned} T(x,t) &= \alpha \int_{\tau=0}^{t} \frac{f(\tau)}{k}[K(x-0,t-\tau) + K(x+0,t-\tau)]\,d\tau \\ &= 2\alpha \int_{\tau=0}^{t} \frac{f(\tau)}{k} K(x,t-\tau)\,d\tau \end{aligned} \tag{6.23}$$

(a) *Case X20B1T0.* In the case where $f(t) = q_0$, a constant heat flux, the integral in Equation 6.23 is given in Table I.8 (Appendix I), and the temperature is given by

$$T(x,t) = \frac{q_0}{k}(4\alpha t)^{1/2}\,\text{ierfc}\left[\frac{x}{(4\alpha t)^{1/2}}\right] \tag{6.24}$$

This expression is also listed in Table 6.3. The temperature is plotted in Figure 6.2 in terms of $(T - T_0)/(q_0 a/k)$, where a is the reference length. Sometimes the quantity $(4\alpha t)^{1/2}$ is used as a reference length. The quantity $q_0(4\alpha t/\pi)^{1/2}/k$ is the surface temperature on the semi-infinite body resulting from the heat flux q_0 ($\text{ierfc}(0) = 1/\sqrt{\pi}$). Equation 6.24 can also be obtained from Equation 6.20 for $n = 1$; that is, a surface temperature proportional to $t^{1/2}$ produces a steady surface heat flux.

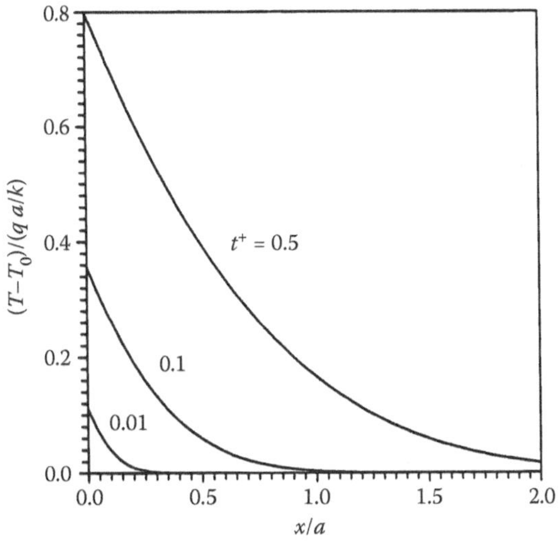

FIGURE 6.2 Temperature in semi-infinite body with constant heat flux at surface.

(b) *Case X20B3T0.* For the case where the surface heat flux is $f(t) = q_0(t/t_0)^{n/2}$, for $n = -1, 0, 1$, and so on, the integral in Equation 6.23 is also given in Table I.8 (Appendix I). After some simplification, the temperature is given by

$$T(x,t) = \frac{q_0}{k}(4\alpha t)^{1/2}\left(\frac{t}{t_0}\right)^{n/2}\Gamma\left(1+\frac{n}{2}\right)2^n$$
$$\mathrm{i}^{n+1}\operatorname{erfc}\left[\frac{x}{(4\alpha t)^{1/2}}\right] \qquad n = -1, 0, 1, \ldots \tag{6.25}$$

Note that for $n = 0$, this solution reduces to the constant heat flux case.

Example 6.3: Semi-Infinite Body with Convection—*X*30*B*1*T*0 Case

Find the temperature in a semi-infinite body due the sudden application of the convection boundary condition where both h and T_∞ are constant. The convection boundary condition is

$$-k\left.\frac{\partial T}{\partial x}\right|_{x=0} + hT|_{x=0} = hT_\infty \tag{6.26}$$

Solution

This is the $X30B1T0$ case. The temperature solution is given by the GF equation as

$$T(x,t) = \alpha\int_{\tau=0}^{t}\frac{hT_\infty}{k}G_{X30}(x,t|0,\tau)\,d\tau \tag{6.27}$$

Note that x' is evaluated at the surface, $x' = 0$. The function G_{X30} is listed in Table 6.1, and Equation 6.27 becomes

$$T(x,t) = \alpha \int_{\tau=0}^{t} \frac{hT_\infty}{k} \left(\frac{2}{[4\pi\alpha(t-\tau)]^{1/2}} \exp\left[\frac{-x^2}{4\alpha(t-\tau)}\right] \right.$$
$$\left. - \frac{h}{k} \exp\left[\frac{hx}{k} + \frac{h^2}{k^2}\alpha(t-\tau)\right] \operatorname{erfc}\left\{\frac{x}{2[\alpha(t-\tau)]^{1/2}} + \frac{h}{k}[\alpha(t-\tau)]^{1/2}\right\}\right) d\tau \tag{6.28}$$

This contains a difficult integral if $T_\infty = T_\infty(t)$. Note that if the temperature were evaluated at $x = 0$, the integral would be less difficult. Usually, the surface temperature resulting from a boundary condition is much easier to find than the temperature everywhere inside the body.

For the case where T_∞ is time invariant, the integral for any value of x given in Table I.8 (Appendix I) is used to obtain

$$T(x,t) = T_\infty \left\{ \operatorname{erfc}\left[\frac{x}{2(\alpha t)^{1/2}}\right] - \exp\left(\frac{hx}{k} + \alpha t \frac{h^2}{k^2}\right) \right.$$
$$\left. \times \operatorname{erfc}\left[\frac{x}{2(\alpha t)^{1/2}} + \frac{h}{k}(\alpha t)^{1/2}\right] \right\} \tag{6.29}$$

This temperature is plotted versus position in Figure 6.3 for several values of the (normalized) heat transfer coefficient. Note that as h increases the surface temperature approaches the fluid temperature T_∞.

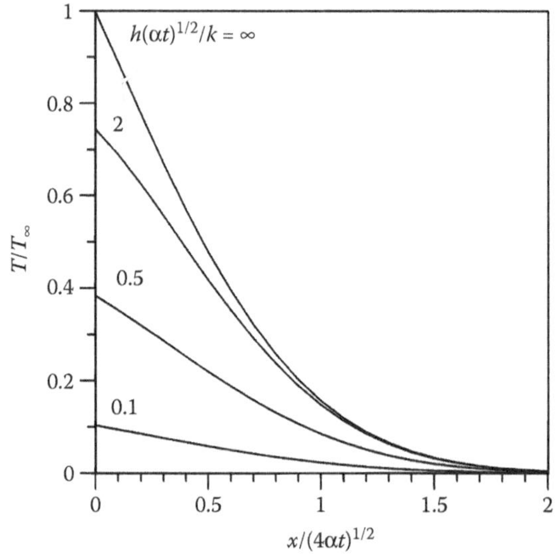

FIGURE 6.3 Temperature in semi-infinite geometry with surface convection defined by $h(\alpha t)^{1/2}/k = 0.1, 0.5, 2.0, \infty$.

6.3.3 Volume Energy Generation

Next consider the temperature in a semi-infinite body caused by volume energy generation. The boundary conditions and the initial condition are homogeneous. The temperature is given by the GFSE,

$$T(x,t) = \frac{\alpha}{k}\int_{\tau=0}^{t}\int_{x'=0}^{\infty} G(x,t|x',\tau)\, g(x',\tau)\, dx'\, d\tau \tag{6.30}$$

This expression is more complicated than the temperature resulting from a boundary condition because there are two integrals to evaluate.

Consider the case when the volume energy generation $g(x',\tau)$ is either independent of τ or a product of a function of x' and a function of τ,

$$g(x',\tau) = g_x(x')\, g_t(\tau) \tag{6.31}$$

Then the integrations over x' previously discussed can be used. For example, suppose the volume energy generation is given by one term of a polynomial in time:

$$g(x',\tau) = g_0\left(\frac{\tau}{t_0}\right)^{n/2} \qquad n = -1, 0, 1, 2, \ldots \tag{6.32}$$

where g_0 is a constant with units of W/m^3. That is, $g(x',\tau)$ is independent of x' and is proportional to $\tau^{n/2}$. The time t_0 is any convenient value and could be one unit, such as 1 s.

The solution for the temperature when $g(\cdot)$ is given by Equation 6.32 can be found for boundary conditions of the first and second kinds using the GF given by Equation 6.6. The integration over the body (x' in this case) is usually considered first, and the integrals required are listed in Table I.7 (Appendix I). Integration of Equation 6.30 over x' yields

$$T(x,t) = \frac{\alpha}{2k}\int_{\tau=0}^{t} g_0\left(\frac{\tau}{t_0}\right)^{n/2}\left(2 - \text{erfc}\left\{\frac{x}{[4\alpha(t-\tau)]^{1/2}}\right\}\right.$$
$$\left. + (-1)^I \text{erfc}\left\{\frac{x}{[4\alpha(t-\tau)]^{1/2}}\right\}\right) d\tau \tag{6.33}$$

where $I = 1$ or 2 to represent the kind of boundary condition at $x = 0$.

The remaining integral on τ in Equation 6.33 is listed in Appendix I (Table I.8, number 9) to give for $I = 1$ (boundary condition of the first kind)

$$T(x,t) = \frac{g_0\alpha t}{k}\left(\frac{1}{n/2+1}\right)\left(\frac{t}{t_0}\right)^{n/2}\left\{1 - \frac{i^{n+2}\,\text{erfc}\left[\frac{x}{(4\alpha t)^{1/2}}\right]}{i^{n+2}\,\text{erfc}(0)}\right\} \tag{6.34a}$$

for $n = -1, 0, 1, 2$, etc. Values for i^{n+2}erfc(0) are given in Appendix E (Equation E.11). This is case $X10B0T0Gt3$ and the temperature is plotted versus position in

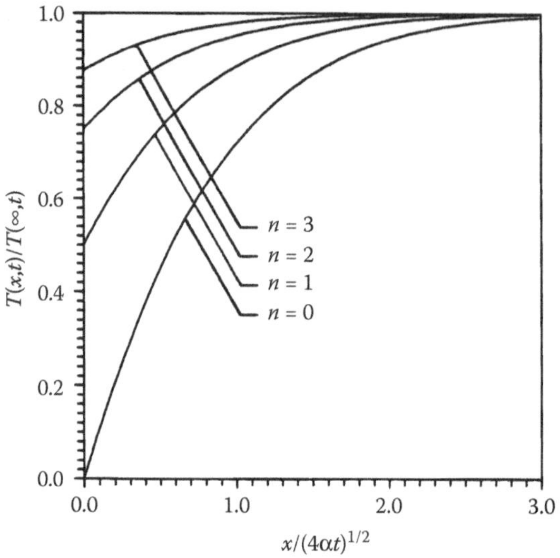

FIGURE 6.4 Temperature in the semi-infinite body with energy generation $g \simeq g_0 t^{n/2}$.

Figure 6.4. In the figure the temperature is normalized by the temperature far from the surface, which is proportional to $(g_0 t^{n/2+1})$.

In the case $I = 2$ for the boundary condition of the second kind, Equation 6.33 gives,

$$T(x,t) = \frac{g_0 \alpha t}{k} \frac{1}{n/2+1} \left(\frac{t}{t_0} \right)^{n/2} \tag{6.34b}$$

This is case $X20B0T0Gt3$, the temperature in a semi-infinite body with spatially uniform heat generation and an insulated boundary. The temperature does not depend on position because there is no heat flow in the body; the temperature increases everywhere at the same rate. The same result could have been obtained by a simple lumped capacitance description that is appropriate when the temperature is spatially uniform: $\rho c \, \partial T / \partial t = g(t)$.

6.4 FLAT PLATES: SMALL-COTIME GREEN'S FUNCTIONS

The cases discussed in this section are one-dimensional flat plates, denoted XIJ, $I, J = 1, 2, 3, 4, 5$. The small-cotime GF for these cases are infinite-series expressions or approximate truncated infinite series. The small-cotime GFs are listed in Appendix X. In general, for $\alpha(t - \tau)/L^2 < 0.05$, only three terms of the expressions for the small-cotime GF are needed for accuracy to four decimal places. Many of these expressions were derived from a Laplace transform solution of the auxiliary equation for the GF.

6.4.1 Initial Conditions

Consider the small-cotime solutions of the following equation:

$$\frac{\partial^2 T}{\partial x^2} = \frac{1}{\alpha}\frac{\partial T}{\partial t} \qquad 0 < x < L \tag{6.35}$$

The boundary conditions are homogeneous and the initial temperature distribution is

$$T(x,0) = F(x) \tag{6.36}$$

The solution using GFs is

$$T(x,t) = \int_{x'=0}^{L} G(x,t|x',0)F(x')\,dx' \tag{6.37}$$

Consider one term in a quadratic initial temperature distribution,

$$F(x') = T_0\left(\frac{x'}{L}\right)^i \qquad 0 < x' < L \qquad i = 0,1,2 \tag{6.38}$$

For boundary of the first and second kinds, the expressions for G_{XIJ} may be written in the form

$$G_{XIJ}(x,t|x',0) = \sum_{n=-\infty}^{\infty}(-1)^{(I+J)n}\left[K(2nL+x-x',t)+(-1)^I K(2nL+x+x',t)\right] \tag{6.39}$$

where I and J describe the boundary conditions types at $x = 0$ and $x = L$, respectively. Then the temperature is given by combining Equations 6.37, 6.38 and 6.39 in the form

$$T(x,t) = T_0\int_{x'=0}^{L}\sum_{n=-\infty}^{\infty}(-1)^{(I+J)n}\left[K(2nL+x-x',t)\right.$$
$$\left.+\,(-1)^I K(2nL+x+x',t)\right]\left(\frac{x'}{L}\right)^i dx' \tag{6.40}$$

This is the $XIJ\,B00T(i+1)$ case, where $i = 0,1,2$; $I = 1$ or 2; and, $J = 1$ or 2. Refer to Table I.7 (Appendix I) for closed form expressions of these integrals.

The solution given by Equation 6.40 is valid for all $t > 0$, but for small time only a few terms of the series are needed. As $\alpha t / L^2$ increases, the number of significant terms in the infinite series increases.

Example 6.4: Slab with Zero-Temperature Boundaries—*X*11*B*00*T*1 Case

Find the temperature in a slab body with zero temperature boundary conditions and with a spatially uniform initial condition $F(x) = T_0$.

Solution

This is the $X11B00T1$ case and the solution is given by Equation 6.40 where $i = 0$ and $I = J = 1$. This case involves the following integral (see Table I.7,

Appendix I):

$$D(n) = \int_{x'=0}^{L} [K(2nL + x - x', t) - K(2nL + x + x', t)]\, dx'$$
$$= \frac{1}{2}\left\{\text{erfc}\left[\frac{(2n-1)L + x}{(4\alpha t)^{1/2}}\right] - 2\,\text{erfc}\left[\frac{2nL + x}{(4\alpha t)^{1/2}}\right] + \text{erfc}\left[\frac{(2n+1)L + x}{(4\alpha t)^{1/2}}\right]\right\} \tag{6.41}$$

The major contributions to the temperature given by Equation 6.40 for small times come from the smaller values of $|n|$ such as 0 and 1. For $n = 0$ the above integral gives

$$D(0) = \frac{1}{2}\left\{2 - \text{erfc}\left[\frac{L - x}{(4\alpha t)^{1/2}}\right] - 2\,\text{erfc}\left[\frac{x}{(4\alpha t)^{1/2}}\right] + \text{erfc}\left[\frac{L + x}{(4\alpha t)^{1/2}}\right]\right\} \tag{6.42a}$$

For $n = -1$ the integral $D(n)$ gives

$$D(-1) = \frac{1}{2}\left\{-\text{erfc}\left[\frac{3L - x}{(4\alpha t)^{1/2}}\right] + 2\,\text{erfc}\left[\frac{2L - x}{(4\alpha t)^{1/2}}\right] - \text{erfc}\left[\frac{L - x}{(4\alpha t)^{1/2}}\right]\right\} \tag{6.42b}$$

Note 1. The identity $\text{erfc}(-u) = 2 - \text{erfc}(u)$ has been used in Equation 6.42 to put positive arguments in each of the terms $\text{erfc}(\cdot)$. The quantity $(x - L)$ is zero or negative since $0 \le x \le L$. Recall that $\text{erfc}(u \to +\infty) = 0$ but that $\text{erfc}(u \to -\infty) = 2$, so that positive arguments ensures that each of the $\text{erfc}(\cdot)$ terms will converge to zero as $|n| \to \infty$.

Note 2. The identity $\text{erfc}(-u) = 2 - \text{erfc}(u)$ applied to the $D(n)$ term in Equation 6.42b for $n = -1$ produced three constant terms that canceled to zero. This cancelation occurs for every $n < 0$ and it has an important numerical consequence. As you add more terms to the infinite series for the temperature to improve the accuracy, it is important to find a value for each $D(n)$ as a unit and then add that value to the temperature. This will avoid excessive loss of significant digits resulting from subtracting numbers that are very close in value.

For small values of $\alpha t/L^2$, the dominant terms in Equation 6.39 for the temperature in the $X11B00T1$ case are given by the largest terms from Equation 6.42a and b and the $n = 1$ term multiplied by the initial temperature T_0:

$$T(x,t) \approx T_0\left\{1 - \text{erfc}\left[\frac{x}{(4\alpha t)^{1/2}}\right] - \text{erfc}\left[\frac{L - x}{(4\alpha t)^{1/2}}\right] - \text{erfc}\left[\frac{L + x}{(4\alpha t)^{1/2}}\right] - \text{erfc}\left[\frac{2L - x}{(4\alpha t)^{1/2}}\right]\right\} \tag{6.43}$$

Near the boundary $x = 0$ and for $\alpha t/L^2 < 0.025$, the quantity $\text{erfc}[L/(4\alpha t)^{1/2}]$ is less than 0.0001 and the temperature is given approximately by the first two terms of Equation 6.43:

$$T(x,t) \approx T_0\left\{1 - \text{erfc}\left[\frac{x}{(4\alpha t)^{1/2}}\right]\right\}$$
$$= T_0\,\text{erf}\left[\frac{x}{(4\alpha t)^{1/2}}\right] \qquad x \ll L,\ \alpha t/L^2 \text{ small} \tag{6.44a}$$

This result is identical to the semi-infinite case $X10B0T1$. That is, near the boundary at small time, the temperature in a flat plate is given by the semi-infinite case with the same boundary condition. For $\alpha t/L^2 < 0.025$ and for $x \approx L$, the dominant terms are

$$T(x,t) \approx T_0 \left\{1 - \operatorname{erfc}\left[\frac{L-x}{(4\alpha t)^{1/2}}\right]\right\}$$
$$= T_0 \operatorname{erf}\left[\frac{L-x}{(4\alpha t)^{1/2}}\right] \qquad \begin{matrix} L - x \ll L \\ \alpha t/L^2 \text{ small} \end{matrix} \tag{6.44b}$$

6.4.2 Volume Energy Generation

Early time solutions of

$$\frac{\partial^2 T}{\partial x^2} + \frac{1}{k}g(x,t) = \frac{1}{\alpha}\frac{\partial T}{\partial t} \qquad 0 < x < L \tag{6.45}$$

are discussed in this section for homogeneous boundary conditions and zero initial condition. Actually the solutions are valid for all times but they are computationally efficient for early times.

The solution for the temperature using GFs is

$$T(x,t) = \frac{\alpha}{k}\int_{\tau=0}^{t}\int_{x'=0}^{L} G(x,t|x',\tau)\, g(x',\tau)\, dx'\, d\tau \tag{6.46}$$

The discussion will be limited to cases for which the volume energy generation $g(x',\tau)$ is the product of a function of x' and a function of τ,

$$g(x',\tau) = g_x(x')g_t(\tau) \tag{6.47}$$

The integration over x' in Equation 6.46 is similar to that for the nonzero initial temperature distribution, Equation 6.37. Integrals over time τ are given in Appendix I (Table I.8).

As an example, the case where $g(x',\tau) = g_0 L\, \delta(x' - x_0)g_t(\tau)$ is examined. This is a plane heat source located at x_0 with a time-variable source strength given by $g_t(\tau)$. Using Equation 6.46 and the small-cotime GF for geometry XIJ given by Table 4.1 gives

$$\begin{aligned} T(x,t) &= \frac{\alpha}{k}\int_{\tau=0}^{t}\int_{x'=0}^{L} G(x,t|x',\tau) g_0 L\, \delta(x'-x_0) g_t(\tau)\, dx'\, d\tau \\ &= \frac{\alpha}{k} g_0 L \int_{\tau=0}^{t} G(x,t|x_0,\tau)\, g_t(\tau)\, d\tau \\ &= \frac{\alpha}{k} g_0 L \sum_{n=-\infty}^{\infty}\int_{\tau=0}^{t} (-1)^{(I+J)n}[K(2nL + x - x_0, t-\tau) \\ &\quad + (-1)^I K(2nL + x + x_0, t - \tau)]\, g_t(\tau)\, d\tau \end{aligned} \tag{6.48}$$

where $I = 1$ or 2 and $J = 1$ or 2 determines the type of boundary conditions.

Suppose the time variation of the plane source strength $g_t(\tau)$ is given by

$$g_t(\tau) = \left(\frac{\tau}{t_0}\right)^{m/2} \qquad m = -1, 0, 1, \ldots \tag{6.49}$$

where t_0 is some convenient positive time value. Then the time integral in Equation 6.48 is given in Appendix I (Table I.8):

$$\begin{aligned} T(x,t) = {} & \frac{1}{2}\left(\frac{\alpha t_0}{L^2}\right)^{1/2} \frac{g_0 L^2}{k} \Gamma\left(\frac{m}{2}+1\right)\left(\frac{4t}{t_0}\right)^{(m+1)/2} \\ & \times \sum_{n=-\infty}^{\infty} (-1)^{-(I+J)n} \left\{ i^{m+1} \operatorname{erfc}\left[\frac{|2nL + x - x_0|}{(4\alpha t)^{1/2}}\right] \right. \\ & \left. + (-1)^I i^{m+1} \operatorname{erfc}\left[\frac{|2nL + x + x_0|}{(4\alpha t)^{1/2}}\right]\right\} \end{aligned} \tag{6.50}$$

This solution applies to geometries described by the number $XIJB00T0Gx7t3$ for $I, J = 1, 2$. The plane source at x_0 can vary with time as given by Equation 6.49 with $m = -1, 0, 1, 2$, and so on. A particularly important value of m is $m = 0$, which gives the temperature resulting from a continuous constant plane source; for $m = 0$, the t_0 values cancel in Equation 6.50.

One possible location for the plane source is at $x_0 = 0$. For this location and case XIJ with $I = 1$ (that is, geometries $X11$ and $X12$), $T(x,t)$ is equal to zero,

$$T(x,t) = 0 \qquad \text{for all } x \text{ and } t \tag{6.51}$$

while for cases XIJ with $I = 2$ (that is, geometries $X21$ and $X22$), Equation 6.50 gives

$$\begin{aligned} T(x,t) = {} & \left(\frac{\alpha t_0}{L^2}\right)^{1/2} \frac{g_0 L^2}{k} \Gamma\left(\frac{m}{2}+1\right)\left(\frac{4t}{t_0}\right)^{(m+1)/2} \\ & \times \sum_{n=-\infty}^{\infty} (-1)^{Jn} \left\{ i^{m+1} \operatorname{erfc}\left[\frac{|2nL + x|}{(4\alpha t)^{1/2}}\right]\right\} \end{aligned} \tag{6.52a}$$

By isolating the $n = 0$ term, the temperature can be written as a sum over $n = 1$ to ∞:

$$\begin{aligned} T(x,t) = {} & \left(\frac{\alpha t_0}{L^2}\right)^{1/2} \frac{g_0 L^2}{k} \Gamma\left(\frac{m}{2}+1\right)\left(\frac{4t}{t_0}\right)^{(m+1)/2} \left(i^{m+1} \operatorname{erfc}\left[\frac{x}{(4\alpha t)^{1/2}}\right] \right. \\ & \left. + \sum_{n=1}^{\infty} (-1)^{Jn} \left\{ i^{m+1} \operatorname{erfc}\left[\frac{|2nL + x|}{(4\alpha t)^{1/2}}\right] + i^{m+1} \operatorname{erfc}\left[\frac{|2nL - x|}{(4\alpha t)^{1/2}}\right]\right\}\right) \end{aligned} \tag{6.52b}$$

For small values of $\alpha t / L^2$ (such as $\alpha t / L^2 < 0.1$), only a few terms of the summation are needed.

Equation 6.52 was derived as the temperature resulting from the space- and time-varying volume energy source

$$g(x,t) = g_0 L\, \delta(x = 0) \left(\frac{t}{t_0}\right)^{m/2} \tag{6.53a}$$

which is a plane heat source located at $x = 0$, but this plane source produces an effect identical to a prescribed heat flux at $x = 0$ given by

$$-k \frac{\partial T}{\partial x}\bigg|_{x=0} = q_0 \left(\frac{t}{t_0}\right)^{m/2} \tag{6.53b}$$

Therefore, g_0 and q_0 in Equations 6.53a, b are related by

$$q_0 = g_0 L \tag{6.53c}$$

where q_0 has units of W/m^2 and g_0 has units of W/m^3. Equation 6.52 has been described as the temperature for the case $X2JB00T0Gx7t3$ (plane heat source at $x = 0$), but because the plane heat source at $x = 0$ is equivalent to a prescribed heat flux at $x = 0$, the description $X2JB30T0$ also applies to Equation 6.52.

6.5 FLAT PLATES: LARGE-COTIME GREEN'S FUNCTIONS

Large-time GFs are usually derived from a separation of variables solution of the energy equation. The separation of variables technique is discussed in Chapter 4. The large-time GFs for slab bodies have the general form

$$G(x,t|x',\tau) = \frac{\mathbf{X}_0(x)}{N_0} + \sum_{m=1}^{\infty} \exp\left[-\frac{\beta_m^2 \alpha(t-\tau)}{L^2}\right] \frac{\mathbf{X}_m(x)\mathbf{X}_m(x')}{N_m} \tag{6.54}$$

where the eigenfunctions, $\mathbf{X}_0(x)$ and $\mathbf{X}_m(x)$, and the norms N_0 and N_m are given in Tables 4.2 and 4.3. Each GF also has associated eigenvalues β_m. For cases involving only boundary conditions of kinds 1 or 2, the eigenvalues are given in Table 4.3. For cases with boundary conditions of types 3, 4, or 5, the eigenvalues must be found numerically as roots of the characteristic equation listed in Table 4.3. A complete list of large-cotime GFs with derivatives and useful approximations is given in Appendix X.

6.5.1 Initial Conditions

The temperature in a body resulting from a nonzero initial temperature distribution is discussed in this section. As an example, consider the initial temperature distribution given by

$$F(x') = \begin{cases} T_0 & a < x' < b \\ 0 & \text{otherwise} \end{cases} \tag{6.55}$$

For the specific case of a body with zero temperature at boundary $x = 0$ and with one of two possible boundary conditions at $x = L$ described by number $X1J$ where $J = 1$ or 2, the temperature distribution is found from the initial-temperature term of the GFSE with the GF given by Equation 6.54:

$$T(x,t) = 2T_0 \sum_{m=1}^{\infty} e^{-\beta_m^2 \alpha t / L^2} \sin\left(\frac{\beta_m x}{L}\right) \frac{\cos(\beta_m a / L) - \cos(\beta_m b / L)}{\beta_m} \tag{6.56}$$

The eigenvalues β_m depend on whether $J = 1$ or 2. The number for this case is $X1JB00T5$ for $J = 1$ or 2. The presence of the term $\exp(-\beta_m^2 \alpha t / L^2)$ multiplying by all the other terms in Equation 6.56 causes rapid numerical convergence of the series for large dimensionless time $\alpha t / L^2 > 0.025$.

Next the convergence criterion introduced in Section 5.2 will be used to determine the number of series terms needed for accurate evaluation of the above temperature expression. The exponential term controls the convergence speed, and the exponential term will be smaller than 0.001 when (the absolute value of) the argument of the exponential term is greater than $K_{max} = 6.9$. That is,

$$\frac{\beta_m^2 \alpha t}{L^2} > 6.9 \tag{6.57}$$

For $\alpha t / L^2 = 0.01$, and for case X11 where $\beta_m = m\pi$, the above relation gives $m_{MAX} = 8$ (refer to Table 5.1). That is, only eight terms of the infinite series are sufficient to make the exponential factor smaller than 0.001. For $\alpha t / L^2 = 0.025$, only five terms of the series are sufficient. For "large" values of $\alpha t / L^2$, such as 0.17 or larger, only one term of the series (the $m = 1$ term) is sufficient for the $X12$ case. For the $X11$ case, if $\alpha t / L^2$ is larger than 0.31, then one term of the series is sufficient to make the exponential factor less than 0.001.

Next, consider the uniform initial temperature

$$F(x') = T_0 \qquad 0 < x < L \tag{6.58}$$

applied to a body with homogeneous boundary conditions of the first kind at both $x = 0$ and L. This is case $X11B00T1$ and the temperature is given by Equation 6.56 with $a = 0$ and $b = L$:

$$T(x,t) = T_0 \frac{4}{\pi} \sum_{m=1,3,\ldots}^{\infty} e^{-m^2 \pi^2 \alpha t / L^2} \sin\left(m\pi \frac{x}{L}\right) \frac{1}{m} \tag{6.59a}$$

The first two terms of this infinite series can be used to approximate the temperature for $\alpha t / L^2$ not too small:

$$T(x,t) \approx T_0 \frac{4}{\pi} \left[e^{-\pi^2 \alpha t / L^2} \sin\left(\pi \frac{x}{L}\right) + \frac{1}{3} e^{-9\pi^2 \alpha t / L^2} \sin\left(3\pi \frac{x}{L}\right) \right] \tag{6.59b}$$

Equation 6.59b gives satisfactory accuracy for $\alpha t / L^2 \geq 0.025$. The related small-time expression, Equation 6.43, is accurate for $\alpha t / L^2 < 0.025$. The least accurate range for Equation 6.59b is in its lower limit ($\alpha t / L^2 \approx 0.025$) and the least accurate range for Equation 6.43 (the small-time form of the same problem) is near its upper limit ($\alpha t / L^2 \approx 0.025$). Hence it is instructive to evaluate both expressions for the temperature when they are least accurate at the middle of the body: at $x = L/2$ and at $\alpha t / L^2 = 0.025$. Equation 6.43 evaluated at $x = L/2$ is

$$T\left(\frac{L}{2}, t\right) \approx T_0 \left\{1 - 2\,\mathrm{erfc}\left[\frac{1}{4(\alpha t / L^2)^{1/2}}\right] + 2\,\mathrm{erfc}\left[\frac{3}{4(\alpha t / L^2)^{1/2}}\right]\right\} \tag{6.60a}$$

which has the numerical components of

$$T\left(\frac{L}{2}, 0.025\right) \approx T_0[1 - 2(0.0253473) + 2(0.197E\text{-}10)] = 0.949305\, T_0 \tag{6.60b}$$

The components of Equation 6.59b at $x = L/2$ and $\alpha t / L^2 = 0.025$ are

$$T\left(\frac{L}{2}, 0.025\right) = T_0 \frac{4}{\pi}[0.7813437 + \frac{1}{3}(0.108537)(-1)] = 0.94877\, T_0 \tag{6.60c}$$

The expression given by Equation 6.60b is slightly more accurate, but both expressions are less than 0.1% in error. Again only two terms are needed for each temperature expression near $\alpha t / L^2 = 0.025$.

6.5.2 Plane Heat Source

Consider a plane heat source located at x_0 described by

$$g(x', \tau) = g_0 L\, \delta(x_0 - x')\, g_t(\tau) \tag{6.61}$$

Then, using this expression for $g(x', \tau)$ in the GFSE gives

$$T(x, t) = \frac{\alpha}{k} \int_{\tau=0}^{t} G(x, t|x_0, \tau)\, g_0 L\, g_t(\tau)\, d\tau \tag{6.62}$$

Here the integral on x' has been evaluated using the sifting property of the Dirac delta function. When $G(\cdot)$ is given by the large-time GF from Equation 6.54, then Equation 6.62 becomes

$$\begin{aligned} T(x, t) = {} & \frac{\alpha}{k} g_0 L \int_{\tau=0}^{t} \sum_{m=1}^{\infty} e^{-\beta_m^2 \alpha (t-\tau)/L^2} \frac{\mathbf{X}(\beta_m x / L)\, \mathbf{X}(\beta_m x_0 / L)}{N_m} g_t(\tau)\, d\tau \\ & + \frac{\alpha}{k} g_0 L \int_{\tau=0}^{t} \frac{\mathbf{X}_0}{N_0} g_t(\tau)\, d\tau \end{aligned} \tag{6.63}$$

The temperature caused by a number of time-varying plane sources can be investigated with different functions $g_t(\tau)$ in Equation 6.63. One of the simplest is for $g_t(\tau) = 1$, a constant for which the time integral in Equation 6.63 may be evaluated as

$$T(x, t) = \frac{g_0 L^3}{k} \sum_{m=1}^{\infty} \left(1 - e^{-\beta_m^2 \alpha t / L^2}\right) \frac{\mathbf{X}(\beta_m x / L)\, \mathbf{X}(\beta_m x_0 / L)}{N_m \beta_m^2} + \delta_{2I} \delta_{2J} \frac{g_0 L^2}{k} \frac{\alpha t}{L^2} \tag{6.64a}$$

This solution is denoted $XIJB00T0Gx7t1$ where I and J can be 1 or 2. The symbol δ_{IJ} is called the Kronecker delta and is defined to be

$$\delta_{IJ} = \begin{cases} 1 & \text{for } I = J \\ 0 & \text{for } I \neq J \end{cases} \tag{6.64b}$$

Do not confuse δ_{IJ} with the Dirac delta function $\delta(\cdot)$ defined in Chapter 1. In Equation 6.63 there is a contribution for the $\delta_{2I}\delta_{2J}$ term only for $I = J = 2$. The term associated with $\delta_{2I}\delta_{2J}$ comes from the $m = 0$ term of the summation for $G(\cdot)$ which must be treated in a special manner when $I = J = 2$ because in this case $\beta_{m=0} = 0$ is an eigenvalue. There are two parts in this solution: a steady-state part, and a transient part. The steady-state part of Equation 6.64a can be written as

$$T(x) = \frac{g_0 L^3}{k} \sum_{m=1}^{\infty} \frac{\mathbf{X}(\beta_m x / L)\mathbf{X}(\beta_m x_0 / L)}{N_m \beta_m^2} \tag{6.65}$$

for the $X11$, $X12$, and $X21$ cases. The $X22$ case does not in general have a steady-state part. The series given by Equation 6.65 for the steady-state part converges very slowly. This slow convergence can be avoided because a simple linear function for the steady-state solution for the $X11$, $X12$, and $X21$ cases may be found with steady-state GFs (refer to Section 1.7). The steady-state solution for the X11 case is

$$T(x) = \begin{cases} \dfrac{g_0 x(L - x_0)}{k} & 0 \leq x \leq x_0 \\[2ex] \dfrac{g_0 x_0(L - x)}{k} & x_0 \leq x \leq L \end{cases} \tag{6.66}$$

Equation 6.66 is the steady-state GF multiplied by the source strength. The solution for the $X12$ case is

$$T(x) = \begin{cases} \dfrac{g_0 x L}{k} & 0 \leq x \leq x_0 \\[2ex] \dfrac{g_0 x_0 L}{k} & x_0 \leq x \leq L \end{cases} \tag{6.67}$$

Algebraic expressions such as Equations 6.66 and 6.67 are clearly much easier to evaluate than the infinite-series expression Equation 6.65. Furthermore, the simple linear dependence on x can be seen in these equations, while it is not apparent in Equation 6.65. When it is convenient to do so, the nonseries form of the steady state should be obtained.

Next, two specific temperature expressions are given that are drawn from the general expressions discussed above. The solution of the $X11B00T0Gx7t1$ problem is obtained from Equation 6.64, Tables 4.2 and 4.3, and Equation 6.66,

$$T(x,t) = \frac{g_0 L^2}{k} \frac{x}{L}\left(1 - \frac{x_0}{L}\right) - \frac{2 g_0 L^2}{k} \sum_{m=1}^{\infty} e^{-m^2\pi^2\alpha t / L^2} \frac{\sin(m\pi x / L)\sin(m\pi x_0 / L)}{m^2\pi^2} \tag{6.68}$$

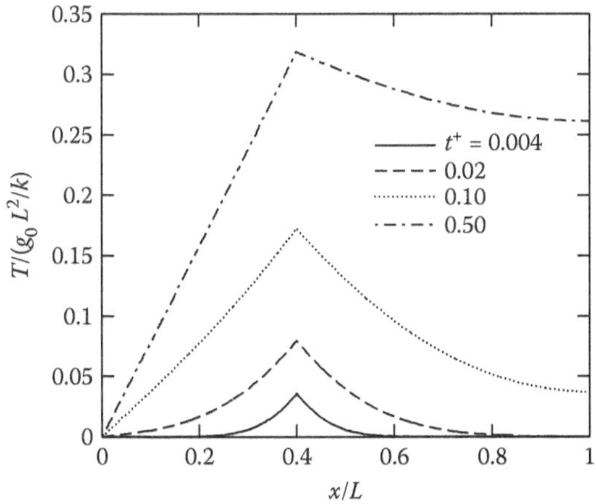

FIGURE 6.5 Temperature in a slab body, heated by a continuous plane source at $x'/L = 0.4$, and with boundary conditions $T = 0$ at $x = 0$ and $\partial T/\partial x = 0$ at $x = L$ (case $X22B00T0Gx7t1$).

for $0 \le x \le x_0$. For $x_0 \le x \le L$, the same expression applies where the x and x_0 symbols are interchanged.

For the case of $T = 0$ at $x = 0$ and $\partial T/\partial x = 0$ at $x = L$ (i.e., $X12B00T0Gx7t1$), the solution is

$$T(x,t) = \frac{g_0 L^2}{k} \min\left(\frac{x}{L}, \frac{x_0}{L}\right) - \frac{2g_0 L^2}{k} \sum_{m=1}^{\infty} e^{-\beta_m^2 \alpha t/L^2} \frac{\sin(\beta_m x/L)\sin(\beta_m x_0/L)}{\beta_m^2} \tag{6.69}$$

where $\min(x/L, x_0/L)$ means the minimum values of the choice between x/L and x_0/L, and where $\beta_m = (m - \frac{1}{2})\pi$. This temperature distribution is plotted in Figure 6.5 for the continuous plane source located at $x'/L = 0.4$. Not that some time passes before heat reaches surface $x = L$.

The two specific temperature expressions given above as Equations 6.68 and 6.69 are relatively efficient expressions for computation for $\alpha t/L^2 > 0.025$. These equations are valid for values of $\alpha t/L^2$ that are even smaller, but more computationally efficient solutions for small times can be obtained by using the small time GFs for flat plates.

Approximate solutions at small times can also be obtained from the GFs for *semi-infinite* bodies. For example, for x/L and x_0/L both less than 0.5, the temperature solution of the problems $X11B00T0Gx7t1$ and $X12B00T0Gx7t1$ can be approximated at small times by the $X10B0T0Gx7t1$ problem. In other words, for sufficiently small times, the temperature distribution is affected most by the nearest boundary. This is the nature of diffusion—the influence of any transient driving term is localized in space at early time.

6.5.3 Volume Energy Generation

Next, heating caused by volume energy generation will be addressed for finite one-dimensional cases having boundary conditions of the first and second kinds. The volume energy generation is uniform over a portion of the body. That is,

$$g(x', \tau) = \begin{cases} g_0 L\, g_t(\tau) & 0 < x' < a \\ 0 & a < x' < L \end{cases} \tag{6.70}$$

The source strength is spatially uniform from $x' = 0$ to a and is zero otherwise. For the initial temperature $F(x)$ being zero, the energy-generation term of Equation 6.4 gives

$$T(x,t) = \frac{\alpha}{k} \int_{\tau=0}^{t} \int_{x'=0}^{a} G^L(x,t|x',\tau)\, g_0 L\, g_t(\tau)\, dx'\, d\tau \tag{6.71}$$

Now, as usual, consider the integrals over x' first. The integral of $G^L(\cdot)$ over x' can be written as

$$\int_{x'=0}^{a} G^L(x,t|x',\tau)\, dx' = \sum_{m=0}^{\infty} \exp\left[-\frac{\beta_m^2 \alpha(t-\tau)}{L^2}\right] \times \frac{\mathbf{X}(\beta_m x\,/\,L)\mathbf{IX}(\beta_m a\,/\,L)}{N_m} \tag{6.72}$$

where $\mathbf{IX}(\cdot)$ is defined to be

$$\mathbf{IX}\left(\frac{\beta_m a}{L}\right) = \int_0^a \mathbf{X}\left(\frac{\beta_m x}{L}\right) dx' \tag{6.73}$$

The $\mathbf{X}(\beta_m x\,/\,L)$ functions are eigenfunctions for $G^L(\cdot)$ listed in Table 4.2 and they are either sin(·) or cos(·) depending on the boundary conditions.

The time integration of Equation 6.71 is now considered. This, in turn, requires a choice of the form of $g_t(\tau)$. Two cases are considered here:

$$g_t(\tau) = 1 \qquad \text{and} \qquad g_t(\tau) = \frac{\tau}{t_0} \tag{6.74a, b}$$

For the first of these, $g_t(\tau) = 1$, integration over τ in Equation 6.71 yields,

$$T(x,t) = \frac{g_0 L^2}{k}\left(L \sum_{m=1}^{\infty}\left(1 - e^{-\beta_m^2 \alpha t\,/\,L^2}\right)\frac{\mathbf{X}(\beta_m x\,/\,L)\mathbf{IX}(\beta_m a\,/\,L)}{N_m} + \delta_{2I}\delta_{2J}\frac{\alpha t}{L^2}\right) \tag{6.75}$$

The cases covered by Equation 6.75 are denoted $XIJB00T0Gx5t1$ for $I, J = 1, 2$. Equation 6.75 can also be broken into steady state and transient parts and the speed of convergence could be improved by replacing the steady series by a nonseries form.

Now consider the linear time variation of the volume energy generation, $g_t(\tau) = \tau / t_0$. The solution is

$$T(x,t) = \frac{g_0 L^2}{k}\left[\frac{L^2}{\alpha t_0} L \sum_{m=1}^{\infty}\left\{[\beta_m^2 t^+ - 1] + e^{-\beta_m^2 t^+}\right\}\right.$$
$$\left.\times \frac{\mathbf{X}(\beta_m x / L)\mathbf{IX}(\beta_m a / L)}{N_m \beta_m^4} + \frac{1}{2}\delta_{2I}\delta_{2J}\frac{t}{t_0}\frac{\alpha t}{L^2}\right] \tag{6.76}$$

where $t^+ = \alpha t / L^2$. Equation 6.76 has the notation of $XIJB00T0Gx5t2$, with $I, J = 1, 2$. Notice that for the case $I = J = 2, (X22)$, the Kronecker delta terms gives $\delta_{2I}\delta_{2J} = 1$ and the temperature increases like t^2; hence, there is no steady-state portion in Equation 6.76. The standard separation of variables procedure does not work for this problem because the source term is not a constant. Note that the above series expression contains a slowly converging part, proportional to $(\beta_m^2 t^+ - 1)$, because this portion does not include an exponential term. To improve the series convergence, the slowly converging part could be replaced by a nonseries expression through use of the alternate GF solution method (see Section 3.4).

6.6 FLAT PLATES: THE NONHOMOGENEOUS BOUNDARY

In this section, temperature caused by heating effects at a boundary is explored for flat plates. Recall that the general boundary condition for temperature has the form

$$k_i \left.\frac{\partial T}{\partial n_i}\right|_{x_i} + h_i T\Big|_{x_i} + (\rho c b)_i \left.\frac{\partial T}{\partial t}\right|_{x_i} = f_i(t) \tag{6.77}$$

When $f_i \neq 0$ we say that the boundary condition is nonhomogeneous. Recall that the associated GF must satisfy homogeneous boundary conditions ($f_i = 0$) of the same type at this location. This difference between the boundary conditions for the GF and the temperature, although required by the GF method, sometimes produces a poorly converging temperature solution, compared to a solution caused by initial conditions or internal-heating solutions.

As discussed earlier in Section 5.3, there are several ways to improve the convergence of a solution caused by a nonhomogeneous boundary. Sometimes it is possible to transform a nonhomogeneous boundary, through a suitable variable normalization, into a homogeneous boundary. Such a transformation will concurrently shift the causative heating effect into a nonzero initial condition. Another approach is the Alternative GF Solution method, discussed in Section 3.4, which is a formal, step-by-step method to remove the causative heating effect from the boundary; both the initial condition and the internal-generation term may be affected. Finally, the method of time partitioning can be used to improve the convergence properties of a solution. Next some examples are given for heat conduction caused by nonhomogeneous boundaries.

Example 6.5: Slab with One Side Heated, One Side at Fixed Temperature—*X*21*B*10*T*0 Case

Find the temperature in a flat plate suddenly heated by a constant heat flux q_0 at $x = 0$ and with a fixed zero temperature at $x = L$. The initial temperature is zero and there is no volume heat generation. Find expressions that are numerically efficient for all values of time.

Solution

The GF solution is given by

$$T(x,t) = \alpha \int_{\tau=0}^{t} \frac{q_0}{k} G_{X21}(x,t|0,\tau)\,d\tau \tag{6.78}$$

(a) *Small-time solution.* The temperature for small values of time ($\alpha t/L^2 < 0.025$) is most efficiently found from the GF equation involving the small-cotime GF. The small-cotime GF is given in Appendix X as an infinite series. Substituting $G^S_{X21}(\cdot)$ into Equation 6.78 gives

$$T(x,t) = \frac{\alpha q_0}{k} \int_{\tau=0}^{t} \frac{2}{[4\pi\alpha(t-\tau)]^{1/2}} \sum_{n=-\infty}^{\infty} (-1)^n \exp\left[\frac{-(2nL+x)^2}{4\alpha(t-\tau)}\right] d\tau \tag{6.79}$$

This integral can be stated in terms of the fundamental heat conduction solution, $K(w_n, t-\tau)$, as

$$T(x,t) = 2\frac{\alpha q_0}{k} \sum_{n=-\infty}^{\infty} (-1)^n \int_{\tau=0}^{t} K(w_n, t-\tau)\,d\tau \tag{6.80}$$

where $w_n = 2nL + x$. The integral in Equation 6.80 is given in Appendix I (Table I.8, number 1) as

$$T(x,t) = 2\frac{q_0 L}{k} \sum_{n=-\infty}^{\infty} (-1)^n \left(\frac{\alpha t}{L^2}\right)^{1/2} \text{ierfc}\left[\frac{|2n + x/L|}{2(\alpha t/L^2)^{1/2}}\right] \tag{6.81}$$

This expression applies for any $t > 0$, however it is numerically efficient for small times. For $\alpha t/L^2 < 0.025$ only three terms of the series ($n = 0, 1, -1$) are sufficient to give a temperature that is exact to over 13 digits. (This can be shown by evaluating the "tail" of the series, $n = \pm 2, \pm 3$, etc.)

(b) *Large-time solution.* The temperature expression that is best for large time ($\alpha t/L^2 > 0.025$) involves the large-cotime GF. The large time GF for the X21 case is given in Appendix X (see also Tables 4.2 and 4.3). Using the large-time GF, evaluated at $x' = 0$, Equation 6.78 may be written

$$T(x,t) = \alpha \int_{\tau=0}^{t} \frac{q_0}{k} \frac{2}{L} \sum_{m=1}^{\infty} e^{-\beta_m^2 \alpha(t-\tau)/L^2} \cos\left(\beta_m \frac{x}{L}\right) d\tau \tag{6.82}$$

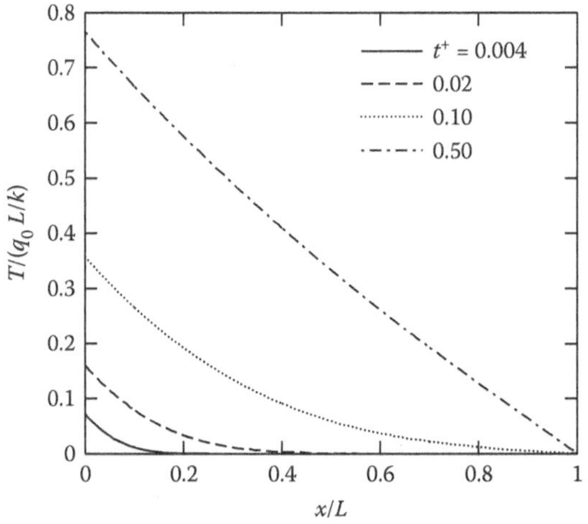

FIGURE 6.6 Temperature in a slab body with a constant heat flux at $x = 0$ and $T = 0$ at $x = L$ (case $X21B10T0$).

where $\beta_m = \pi(m - 1/2)$. The integral on time may be evaluated to give

$$T(x,t) = 2\frac{q_0 L}{k}\sum_{m=1}^{\infty}\frac{\cos\left(\beta_m \frac{x}{L}\right)}{\beta_m^2}\left(1 - e^{-\beta_m^2\alpha(t-\tau)/L^2}\right) \tag{6.83}$$

As noted in earlier examples, the steady-state portion of the series for the temperature converges very slowly. The convergence speed of the transient temperature can be greatly improved by replaced the steady series with a nonseries form. For this one-dimensional case, the steady temperature may be found by direct integration (see Section 1.7). The steady temperature is given by

$$T_{steady}(x) = \frac{q_0}{k}\left(1 - \frac{x}{L}\right) \tag{6.84}$$

Using this form of the steady temperature, the large-time form of the transient temperature, Equation 6.83, may be written

$$T(x,t) = \frac{q_0 L}{k}\left(1 - \frac{x}{L}\right) - 2\frac{q_0 L}{k}\sum_{m=1}^{\infty} e^{-\beta_m^2\alpha t/L^2}\frac{\cos(\beta_m x/L)}{\beta_m^2} \tag{6.85}$$

This expression can be evaluated for any $t > 0$ and it converges rapidly at large times ($\alpha t/L^2 > 0.025$). The temperature for this example is plotted in Figure 6.6. In this figure, at time $\alpha t/L^2 = 0.5$ the temperature is approaching steady state (a straight line).

Example 6.6: Slab with One Side Heated, One Side Insulated—*X22B10T0* Case

Consider the flat plate insulated on one side and heated by a steady heat flux on the other side. Find the temperature using the standard and alternative GFSEs. The boundary value problem is given by

$$\frac{\partial^2 T}{\partial x^2} = \frac{1}{\alpha}\frac{\partial T}{\partial t} \tag{6.86a}$$

$$-k\left.\frac{\partial T}{\partial x}\right|_{x=0} = q_0 \qquad \left.\frac{\partial T}{\partial x}\right|_{x=L} = 0 \tag{6.86b}$$

$$T(x, 0) = 0 \tag{6.86c}$$

(a) *Standard solution.* The standard GF solution is given by Equation 6.4 where the only nonhomogeneous term (the only driving term) is the heat flux at $x = 0$. This is the $X22B10T0$ case. Then using Equation 6.4 and the $X22$ GF from Appendix X gives

$$\begin{aligned} T(x, t) &= \frac{\alpha}{k}\int_{\tau=0}^{t} q_0 G_{X22}(x, t|0, \tau)\, d\tau \\ &= \frac{\alpha q_0}{k}\int_{\tau=0}^{t}\frac{1}{L}\left(1 + 2\sum_{m=1}^{\infty} e^{-m^2\pi^2\alpha(t-\tau)/L^2}\cos\left(m\pi\frac{x}{L}\right)\right) d\tau \\ &= \frac{q_0 L}{k}\left[\frac{\alpha t}{L^2} + \frac{2}{\pi^2}\sum_{m=1}^{\infty}\frac{1}{m^2}\cos\left(m\pi\frac{x}{L}\right)\left(1 - e^{-m^2\pi^2\alpha t/L^2}\right)\right] \end{aligned} \tag{6.87}$$

This expression has three main parts. The first part is proportional to time and thus increases without limit over time. The last part contains an exponential factor that decays with time. The middle part that does not depend on time is given by

$$\frac{q_0 L}{k}\frac{2}{\pi^2}\sum_{m=1}^{\infty}\frac{\cos(m\pi x/L)}{m^2} \tag{6.88}$$

This part of the temperature expression converges very slowly, that is, many terms of the infinite series must be evaluated for accurate numerical values, particularly for small values of x/L.

Next, another temperature expression with better convergence properties will be found with the alternative GFSE equation.

(b) *Alternative solution.* The alternative GFSE (AGFSE) involves a known solution T^* that satisfies the boundary conditions but does not need to satisfy the initial condition. Since Equation 6.87 contains a term proportional to time that dominates the temperature for large times, the T^* solution should display that behavior. The T^* solution for this problem is

$$T^*(x, t) = f(x) + \frac{q_0 L}{k}\frac{\alpha t}{L^2} \tag{6.89}$$

where $f(x)$ must be chosen to satisfy the boundary conditions. Substitute T^* into the energy equation:

$$\frac{\partial^2 T^*}{\partial x^2} = \frac{1}{\alpha}\frac{\partial T^*}{\partial t} \quad \text{or} \quad \frac{d^2 f}{dx^2} = \frac{q_0}{kL} \tag{6.90}$$

Solve Equation 6.90 for $f(x)$ (by integrating twice) and then substitute $f(x)$ back into Equation 6.89 to give

$$T^*(x,t) = \frac{q_0}{kL}\frac{x^2}{2} + C_1 x + C_2 + \frac{q_0 L}{k}\frac{\alpha t}{L^2} \tag{6.91}$$

Using the boundary conditions at $x = 0$ and L given by Equations 6.86b, c allows C_1 to be found as

$$C_1 = -\frac{q_0}{k} \tag{6.92}$$

Since C_2 cannot be found using these boundary conditions, it is set equal to zero. The choice of C_2 is arbitrary because both boundary conditions for T^* are gradient conditions and a constant can be subtracted from T^* without changing the properties of the solution. Then T^* is given by

$$T^*(x,t) = \frac{q_0 L}{k}\left[\frac{1}{2}\left(\frac{x}{L}\right)^2 - \frac{x}{L} + \frac{\alpha t}{L^2}\right] \tag{6.93}$$

Now T^* will be used in the alternative GFSE. The only nonzero integral in the alternative GFSE, Equation 3.66, is the one corresponding to the initial condition. (Why does the last integral drop out in the case?) The alternative GFSE gives

$$\begin{aligned} T(x,t) &= T^*(x,t) + \int_{x'=0}^{L} G_{X22}(x,t|x',0)\,[-T^*(x',0)]\,dx' \\ &= T^*(x,t) - \frac{q_0}{k}\int_{x'=0}^{L}\left[1 + 2\sum_{m=1}^{\infty} e^{-m^2\pi^2\alpha t/L^2}\right. \\ &\quad \left.\times \cos\left(m\pi\frac{x'}{L}\right)\cos\left(m\pi\frac{x}{L}\right)\right]\left[\frac{1}{2}\left(\frac{x'}{L}\right)^2 - \frac{x'}{L} + 0\right] dx' \\ T(x,t) &= \frac{q_0 L}{k}\left[\frac{\alpha t}{L^2} + \frac{1}{2}\left(\frac{x}{L}\right)^2 - \frac{x}{L} + \frac{1}{3} - \frac{2}{\pi^2}\sum_{m=1}^{\infty}\frac{1}{m^2}\right. \\ &\quad \left.\times \cos\left(m\pi\frac{x}{L}\right) e^{-m^2\pi^2\alpha t/L^2}\right] \end{aligned} \tag{6.94}$$

This temperature is plotted in Figure 6.7 at several dimensionless times. Only after $\alpha t/L^2 > 0.1$ does the temperature at $x = L$ begin to rise above the initial value; only then is the insulated boundary evident. Equation 6.94 is valid for any time value but it has good convergence properties for $\alpha t/L^2 > 0.025$. For $\alpha t/L^2 < 0.025$ the temperature may be found approximately from the semi-infinite body solution with the same boundary heat flux (the $X20B1T0$ case) for five-digit numerical accuracy near $x = 0$ and with lesser accuracy near $x = L$.

It is interesting to equate the two expressions for the temperature found from the standard and alternate GFSE, Equations 6.87 and 6.94. Setting them equal and canceling identical terms leaves the equality

$$\frac{2}{\pi^2}\sum_{m=1}^{\infty}\frac{1}{m^2}\cos\left(m\pi\frac{x}{L}\right) = \frac{1}{2}\left(\frac{x}{L}\right)^2 - \frac{x}{L} + \frac{1}{3} \tag{6.95}$$

In effect, we have found the exact value of the infinite sum.

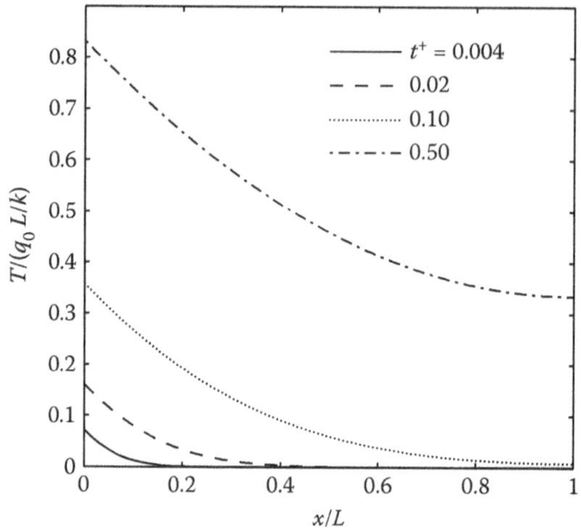

FIGURE 6.7 Temperature in a slab body with a constant heat flux at $x = 0$ and $\partial T / \partial x = 0$ at $x = L$ (case $X22B10T0$).

Example 6.7: Slab with Convection on Both Sides

A large flat plate of thickness $2L$, initially at temperature T_0, is quenched in a large tank of fluid at temperature T_∞. The heat transfer coefficient for the quenching process is h, a constant. Find the temperature distribution $T(x, t)$.

Solution

The geometry of the quenching problem is shown in Figure 6.8a. This problem is modeled as the $X32$ geometry shown in Figure 6.8b. The centerline of the plate is a plane of symmetry, which is modeled as an insulated boundary. The initial condition can be made homogeneous by defining a new variable $T = T - T_0$, and the fluid temperature becomes $(T_\infty - T_0)$. This is the $X32B10T0$ case.

The GF solution using time partitioning is given by the boundary-heating integral of the GF solution equation,

$$T(x, t) - T_0 = \alpha \int_{\tau=0}^{t} \frac{h(T_\infty - T_0)}{k} G^L_{X32}(x, t|0, \tau)\, d\tau \tag{6.96}$$

The large-cotime GF for case X32 is listed in Tables 4.2 and 4.3 (also Appendix X), and upon substitution into Equation 6.96, the result is

$$T(x, t) - T_0 = \alpha \int_{\tau=0}^{t} d\tau \frac{h(T_\infty - T_0)}{k} \frac{2}{L} \sum_{m=1}^{\infty} e^{-\beta_m^2 \alpha (t-\tau)/L^2}$$
$$\times \frac{\beta_m^2 + B^2}{\beta_m^2 + B^2 + B} \cos\left[\beta_m\left(1 - \frac{x}{L}\right)\right] \cos \beta_m \tag{6.97}$$

where eigenvalues β_m are roots of the equation $\beta_m \tan \beta_m = hL/k$. The time-integral in Equation 6.97 operates only on the exponential term, and the result is given by

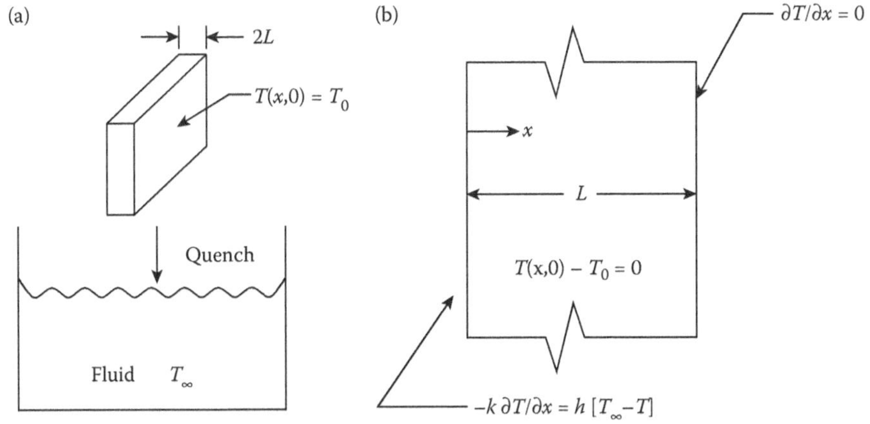

FIGURE 6.8 (a) Quenching of large plate of thickness $2L$. (b) One-dimensional model using $X32$ geometry on $0 \le x \le L$.

$$T(x,t) - T_0 = \frac{2h(T_\infty - T_0)L}{k} \sum_{m=1}^{\infty} \frac{1}{\beta_m^2} \left(1 - e^{-\beta_m^2 \alpha t / L^2}\right)$$
$$\times \frac{\beta_m^2 + B^2}{\beta_m^2 + B^2 + B} \cos\left[\beta_m\left(1 - \frac{x}{L}\right)\right] \cos\beta_m \tag{6.98}$$

This form of the series solution converges slowly because of the steady-state term. Two ways to improve the convergence are given here.

(a) *Replace steady-state term.* The steady-state part of the solution can be found in a nonseries form, as follows. The steady-state portion of the solutions satisfies:

$$\frac{d^2T}{dx^2} = 0 \tag{6.99}$$

$$\text{at } x = 0, \quad -k\frac{dT}{dx} = h(T_\infty - T) \tag{6.100}$$

$$\text{at } x = L, \quad \frac{dT}{dx} = 0 \tag{6.101}$$

Note the sign of the convection boundary condition; heat flux will be in the $+x$-direction for $T_\infty > T$. The general steady solution, found by integrating twice, is given by

$$T(x) = ax + b \tag{6.102}$$

and constants a and b may be found by applying the boundary conditions

$$\text{at } x = L: \quad a = 0 \tag{6.103}$$

$$\text{at } x = 0: \quad 0 = h(T_\infty - (0 + b)) \;\rightarrow\; b = T_\infty \tag{6.104}$$

Then the steady solution is simply $T(x) = T_\infty$ which makes sense because eventually the body takes on the temperature of the fluid.

Now replace the (uniform) steady solution into the series solution to obtain the improved-convergence solution:

$$T(x,t) - T_0 = (T_\infty - T_0) - 2(T_\infty - T_0)\frac{hL}{k}\sum_{m=1}^{\infty}\frac{1}{\beta_m^2}e^{-\beta_m^2\alpha t/L^2}$$
$$\times \frac{\beta_m^2 + B^2}{\beta_m^2 + B^2 + B}\cos\left[\beta_m\left(1-\frac{x}{L}\right)\right]\cos\beta_m \tag{6.105}$$

(b) *Convert to homogeneous boundary.* In the original transient problem, there are two nonzero temperatures given, T_0 for the initial condition, and T_∞ for the fluid temperature. In the solution discussed above, the initial condition was set to zero with normalization $T - T_0$. Here, the nonhomogeneous boundary will be set to zero (made homogeneous) by normalization $\theta = T - T_\infty$. Then the original transient problem becomes

$$\frac{\partial^2\theta}{\partial x^2} = \frac{1}{\alpha}\frac{\partial\theta}{\partial t} \tag{6.106}$$

$$\text{at } x = 0, \quad -k\frac{\partial\theta}{\partial x} + h\theta = 0 \tag{6.107}$$

$$\text{at } x = L, \quad \frac{d\theta}{dx} = 0 \tag{6.108}$$

$$\text{at } t = 0, \quad \theta(x,t) = T_0 - T_\infty \tag{6.109}$$

This is case X32B00T1. Now only the initial-condition integral is needed from the GF solution, Equation 6.4, as follows:

$$T(x,t) - T_\infty = \int_{x'=0}^{L}(T_0 - T_\infty)\,G_{X23}^L(x,t|x',0)\,dx' \tag{6.110}$$

Using the same GF as before, but evaluated at $\tau = 0$, the temperature is

$$T(x,t) - T_\infty = (T_0 - T_\infty)\frac{2}{L}\sum_{m=1}^{\infty}e^{-\beta_m^2\alpha t/L^2}\frac{\beta_m^2 + B^2}{\beta_m^2 + B^2 + B}\cos\left[\beta_m\left(1-\frac{x}{L}\right)\right]$$
$$\times\int_{x'=0}^{L}\cos\left[\beta_m\left(1-\frac{x'}{L}\right)\right]dx' \tag{6.111}$$

After evaluating the integral on x', the temperature is given by

$$T(x,t) - T_\infty = 2(T_0 - T_\infty)\sum_{m=1}^{\infty}e^{-\beta_m^2\alpha t/L^2}$$
$$\times\frac{\beta_m^2 + B^2}{\beta_m^2 + B^2 + B}\cos\left[\beta_m\left(1-\frac{x}{L}\right)\right]\frac{\sin\beta_m}{\beta_m} \tag{6.112}$$

At first glance this solution appears to be different than that found earlier by replacing the steady-state part. However, by rearranging the eigencondition into the form

$$\frac{\sin\beta_m}{\beta_m} = \frac{hL}{k}\frac{\cos\beta_m}{\beta_m^2} \tag{6.113}$$

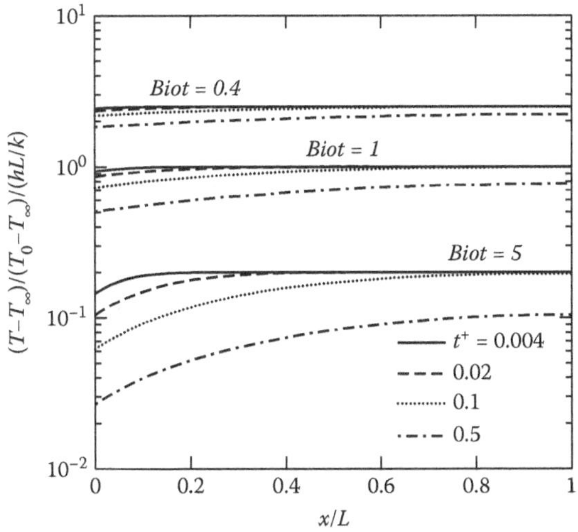

FIGURE 6.9 Normalized temperature in a slab body initially at T_0 and cooled by convection, fluid temperature T_∞, at surface $x = 0$. The $x = L$ boundary is insulated. Three levels of convection are shown for $hL/k = 0.4, 1.0$, and 5.0.

then Equation 6.112 takes the form

$$T(x,t) - T_\infty = 2(T_0 - T_\infty)\frac{hL}{k}\sum_{m=1}^{\infty} e^{-\beta_m^2 \alpha t/L^2}\frac{\beta_m^2 + B^2}{\beta_m^2 + B^2 + B} \times \cos\left[\beta_m\left(1 - \frac{x}{L}\right)\right]\frac{\cos\beta_m}{\beta_m^2} \tag{6.114}$$

which is equivalent to Equation 6.105. Note that this approach provides a rapidly converging solution in a single step. The point of this example is that when a nonhomogenous boundary is present, if it is possible to do so, convert the nonhomogeneous boundary into a homogeneous boundary.

The temperature in the convectively cooled slab wall is plotted in Figure 6.9 for several dimensionless times and for several values of the Biot number hL/k. When the Biot number is small, the temperature is nearly uniform across the body. For Biot < 0.1 (not shown) the temperature is uniform within a few percent and a lumped-capacitance model may be used to describe the temperature as a function of time alone (Ozisik, 1993, p. 27).

6.7 TWO-DIMENSIONAL RECTANGULAR BODIES

Transient temperatures in two-dimensional rectangular bodies are discussed in this section. The transient GF for two-dimensional cases can be found by multiplying one-dimensional GF together for boundary conditions of type 0, 1, 2, and 3. Thus for many cases the temperature solution can be written down immediately in integral form.

Multidimension cases are often more difficult than one-dimensional cases because the integrals in the GFSE are more difficult. Often the spatial integrals can be evaluated, but sometimes the time integral cannot be evaluated in closed form. In this event, numerical methods may be required to get accurate numbers for the temperature.

Some two-dimension rectangle cases are solved in the literature. Ozisik (1993, Chapter 2) gives two examples of separation of variables applied to transient temperature in the rectangle. Carslaw and Jaeger (1959, Chapter 5) discuss several examples of steady and unsteady temperature in rectangles. Solutions for the rectangle also appear in recent papers on improving series convergence (Beck and Cole, 2007) and on intrinsic verification (Beck et al., 2004); see also Sections 5.3 and 5.4 of this book. In this section two examples are discussed for boundary conditions of type 1 and 2.

Example 6.8: Rectangular Body with Several Different Boundary Conditions—*X*21*B*10*Y*21*B*01 Case

Consider a rectangle with zero initial temperature, with one side uniformly heated, one side at a fixed temperature, T_0, one side at a fixed temperature of zero, and one side insulated. Find the temperature by using large cotime GFs.

Solution

This is the $X21B10Y21B01$ case and the geometry is shown in Figure 6.10. The boundary value problem is given by

$$\frac{\partial^2 T}{\partial x^2} + \frac{\partial^2 T}{\partial y^2} = \frac{1}{\alpha}\frac{\partial T}{\partial t} \tag{6.115}$$

$$T(x, y, 0) = 0 \tag{6.116a}$$

$$-k\left.\frac{\partial T}{\partial x}\right|_{x=0} = q_0 = \text{constant} \tag{6.116b}$$

$$T(a, y, t) = 0 \tag{6.116c}$$

$$\left.\frac{\partial T}{\partial y}\right|_{y=0} = 0 \tag{6.116d}$$

$$T(x, b, t) = T_0 \tag{6.116e}$$

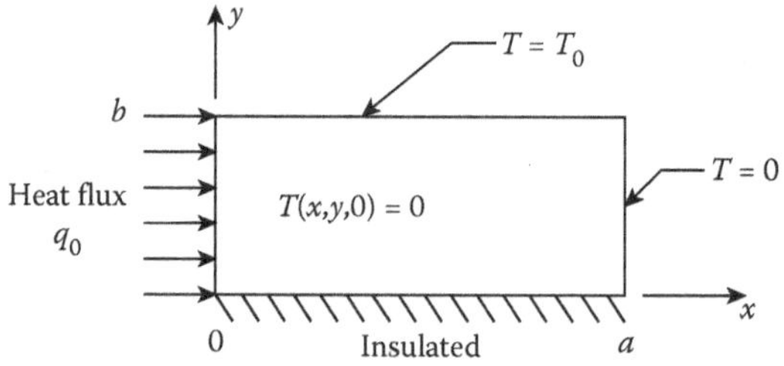

FIGURE 6.10 Geometry for rectangular body in Example 6.8.

The integral expression for the temperature can be written down immediately from the GFSE. There are two terms to account for the heating at $x = 0$ and the nonzero temperature at $y = b$:

$$T(x, y, t) = \alpha \int_{\tau=0}^{t} d\tau \int_{y'=0}^{b} \frac{q_0}{k} G_{X21Y21}(x, y, t|0, y', \tau)\, dy' - \alpha \int_{\tau=0}^{t} d\tau \int_{x'=0}^{a} T_0 \frac{\partial G_{X21Y21}}{\partial y'}\bigg|_{y'=b} dx' \tag{6.117}$$

The GF is formed by multiplying two one-dimensional GFs together. That is,

$$G_{X21Y21}(x, y, t|x', y', \tau) = G_{X21}(x, t|x', \tau)\, G_{Y21}(y, t|y', \tau) \tag{6.118}$$

where G_{X21} and G_{Y21} can be readily obtained from Appendix X as

$$G_{X21}(x, t|x', \tau) = \frac{2}{a} \sum_{m=1}^{\infty} e^{-\beta_m^2 \alpha(t-\tau)/a^2} \cos\left(\frac{\beta_m x}{a}\right) \cos \frac{\beta_m x'}{a} \tag{6.119a}$$

$$G_{Y21}(y, t|y', \tau) = \frac{2}{b} \sum_{n=1}^{\infty} e^{-\beta_n^2 \alpha(t-\tau)/b^2} \cos\left(\frac{\beta_n y}{b}\right) \cos \frac{\beta_n y'}{b} \tag{6.119b}$$

where

$$\beta_m = \pi(m - \tfrac{1}{2}) \qquad \beta_n = \pi(n - \tfrac{1}{2}) \tag{6.120}$$

The spatial integrals in Equation 6.117 operate only on the cosine terms. The time integral can be carried out independently on the product of the exponentials:

$$\int_{\tau=0}^{t} e^{-\beta_m^2 \alpha(t-\tau)/a^2} e^{-\beta_n^2 \alpha(t-\tau)/b^2}\, d\tau = \frac{1}{\alpha C}(1 - e^{-\alpha t C}) \tag{6.121}$$

where $C = (\beta_m/a)^2 + (\beta_n/b)^2$. Then the temperature is given by Equation 6.117 with Equations 6.118 and 6.119 (Beck, 1984),

$$T(x, y, t) = 4 \sum_{m=1}^{\infty} \sum_{n=1}^{\infty} (1 - e^{-\alpha t C}) \cos\left(\frac{\beta_m x}{a}\right) \cos\left(\frac{\beta_n y}{b}\right) \times (-1)^n \left\{ \frac{q_0 a}{k} \frac{1}{\beta_n[\beta_m^2 + \beta_n^2 (a/b)^2]} + T_0 \frac{\beta_n (-1)^m}{\beta_m[\beta_n^2 + \beta_m^2 (b/a)^2]} \right\} \tag{6.122}$$

where $C = (\beta_m/a)^2 + (\beta_n/b)^2$. There are two difficulties with this solution. First, this is the large-time solution that converges rapidly only for $\alpha t/b^2$ and $\alpha t/a^2$ large (greater than 0.05, say). Second, the most difficult part of the solution to evaluate directly is the steady-state part for $T_0 \neq 0$ and $q_0 = 0$:

$$T(x, y) = 4T_0 \sum_{m=1}^{\infty} \sum_{n=1}^{\infty} \cos\left(\frac{\beta_m x}{a}\right) \cos\left(\frac{\beta_n y}{b}\right) \frac{\beta_n (-1)^{m+n}}{\beta_m[\beta_n^2 + \beta_m^2 (b/a)^2]} \tag{6.123}$$

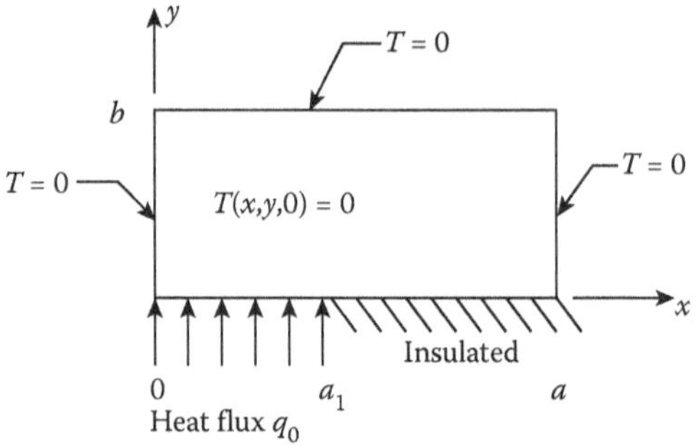

FIGURE 6.11 Geometry for rectangular body heated over part of one face.

This part of the solution converges slowly because for $m \gg 1$ and $n \gg 1$, the series converges something like $n(-1)^{m+n}/(mn^2 + m^3)$ which is painfully close to the slowly converging series $1/n^2$. This double-summation form of the steady temperature can be replaced by a better-converging single-sum form with a GF based on eigenfunction expansions; see Section 4.6. See Beck et al. (2004) for further discussion of improving the convergence of series expressions for temperature in the rectangle.

Example 6.9: Rectangular Body Heated over Part of One Face

Consider a rectangle heated over part of one face. The other faces are held at a fixed temperature of zero and the initial temperature is also zero. The geometry is shown in Figure 6.11. The boundary value problem is given by

$$\frac{\partial^2 T}{\partial x^2} + \frac{\partial^2 T}{\partial y^2} = \frac{1}{\alpha}\frac{\partial T}{\partial t} \qquad 0 < x < a \qquad 0 < y < b \qquad t > 0 \tag{6.124a}$$

$$T(0, y, t) = T(a, y, t) = T(x, b, t) = T(x, y, 0) = 0 \tag{6.124b}$$

$$-k\left.\frac{\partial T}{\partial y}\right|_{y=0} = \begin{cases} q_0 & 0 < x < a_1 \\ 0 & a_1 < x < a \end{cases} \tag{6.124c}$$

(a) Solve the problem using the large-cotime GFs.

(b) Solve the problem using small-cotime GFs and retain only the terms needed for small times near $x = a_1$, and near $y = 0$.

Solution

The number for this case is $X11B00Y21B(x5)0T0$. The GFSE for this problem is

$$T(x, y, t) = \frac{\alpha q_0}{k}\int_{x'=0}^{a_1}\int_{\tau=0}^{t} G_{X11}(x, t|x', \tau)G_{Y21}(y, t|0, \tau)\, dx'\, d\tau \tag{6.125}$$

(a) *Large-time solution.* The large-cotime forms of the GFs are

$$G_{X11}^{L}(x, t|x', \tau) = \frac{2}{a}\sum_{m=1}^{\infty} e^{-m^2\pi^2\alpha(t-\tau)/a^2}\sin\left(m\pi\frac{x}{a}\right)\sin\left(m\pi\frac{x'}{a}\right) \tag{6.126a}$$

$$G^L_{Y21}(y,t|y',\tau) = \frac{2}{b}\sum_{n=1}^{\infty} e^{-\beta_n^2\alpha(t-\tau)/b^2}\cos\left(\beta_n\frac{y}{b}\right)\cos\left(\beta_n\frac{y'}{b}\right) \tag{6.126b}$$

where $\beta_n = \pi(n-1/2)$. Solving the problem using the GFSE and $G^L(\cdot)$ requires the integrals

$$\int_{x'=0}^{a_1} \sin\left(m\pi\frac{x'}{a}\right)dx' = \frac{a}{m\pi}\left[1-\cos\left(m\pi\frac{a_1}{a}\right)\right] \tag{6.127a}$$

$$\int_{\tau=0}^{t} e^{-D\alpha(t-\tau)}\,d\tau = \frac{1}{D\alpha}(1-e^{-D\alpha t}) \tag{6.127b}$$

where D is equal to

$$D = \left(\frac{m^2\pi^2}{a^2}+\frac{\beta_n^2}{b^2}\right) \tag{6.128}$$

Using these integrals in Equation 6.125 gives

$$T(x,y,t) = \frac{4q_0a^2}{\pi kb}\sum_{m=1}^{\infty}\sum_{n=1}^{\infty}(1-e^{-D\alpha t})\frac{1-\cos(m\pi a_1/a)}{m[m^2\pi^2+(a^2/b^2)\beta_n^2]} \times \sin\left(m\pi\frac{x}{a}\right)\cos\left(\beta_n\frac{y}{b}\right) \tag{6.129}$$

There are two parts to this solution: steady-state and transient. The steady-state part converges something like $1/m^3$, which is faster than the steady state in the previous example but which may still require many terms of the series for accurate evaluation. The double-summation steady temperature can be replaced by a better-converging single-summation form with GF based on eigenvalue expansions; this particular case is discussed later in Example 6.10, Section 6.9.

(b) *Small-time solution.* At early times, any temperature changes occur near the heated boundary $y = 0$, and elsewhere, the temperature remains zero. The small-cotime GFs useful for the early time solution are given in Appendix X in the form of infinite series. Near the point $x = a_1$ and $y = 0$, however, just the dominant terms of the series may be used. An equivalent point of view at early time is to replace the rectangle by the quarter-infinite body described by number $X10B0Y20(x5)T0$. The appropriate GFs are

$$G^S_{X11}(x,t|x',\tau) \simeq G_{X10}(x,t|x',\tau) = \frac{1}{[4\pi\alpha(t-\tau)]^{1/2}}\left(e^{-(x-x')^2/[4\alpha(t-\tau)]} - e^{-(x+x')^2/[4\alpha(t-\tau)]}\right) \tag{6.130}$$

$$G^S_{Y21}(y,t|0,\tau) \simeq G_{Y20}(y,t|0,\tau) = \frac{2}{[4\pi\alpha(t-\tau)]^{1/2}}e^{-y^2/[4\alpha(t-\tau)]} \tag{6.131}$$

Next, replace these GFs into the temperature expression given by Equation 6.125. The integral over x' should be familiar; by focusing on the area of interest near

$x = a_1$ the integral over x' may be written as

$$\int_{x'=0}^{a_1} G_{X10}^S(x, t|x', \tau)\, dx'$$
$$= \frac{1}{2}\left(\text{erfc}\left\{\frac{x - a_1}{[4\alpha(t-\tau)]^{1/2}}\right\} - 2\,\text{erfc}\left\{\frac{x}{[4\alpha(t-\tau)]^{1/2}}\right\}\right.$$
$$\left. + \text{erfc}\left\{\frac{x + a_1}{[4\alpha(t-\tau)]^{1/2}}\right\}\right)$$
$$\simeq \frac{1}{2}\text{erfc}\left\{\frac{x - a_1}{[4\alpha(t-\tau)]^{1/2}}\right\} \tag{6.132}$$

for x near a_1. Then the solution for small y values and for x near a_1 becomes

$$T(x, y, t) = \frac{\alpha q_0}{k}\int_{\tau=0}^{t} \frac{1}{2}\,\text{erfc}\left\{\frac{x - a}{[4\alpha(t-\tau)]^{1/2}}\right\} \frac{1}{[\pi\alpha(t-\tau)]^{1/2}} e^{-y^2/[4\alpha(t-\tau)]}\, d\tau \tag{6.133}$$

This is a difficult integral and it will be evaluated below with an approximate integrand. This integral is evaluated exactly in Section 6.8 in the form of an infinite series.

The integral in Equation 6.133 may be evaluated in closed form if an approximation for the complementary error function is used. The erfc(z) function for "small" values of z can be approximated by

$$\text{erfc}(z) = \begin{cases} 1 - Az, & -A^{-1} < z < A^{-1} \\ 0 & z > A^{-1} \\ 2 & z < -A^{-1} \end{cases} \tag{6.134}$$

where $A = 2/\pi^{1/2}$. Using this approximation then gives, for the temperature for small y and near $x = a_1$,

$$T(x, y, t) = T_0 + \frac{\alpha q_0}{k}\int_{u=u_m}^{t} \frac{1}{2}\left[1 - A\frac{x - a_1}{(4\alpha u)^{1/2}}\right]$$
$$\times \frac{1}{(\pi\alpha u)^{1/2}} e^{-y^2/[4\alpha u]}\, du \qquad t > u_m \tag{6.135}$$

where $u_m = A^2(x - a_1)^2/\alpha$. Note that the region of $u = 0$ to u_m (which corresponds to $\tau = t$ to $t - u_m$) has no contribution to the temperature using the above approximation for erfc(z). This equation also implies that the region under which the approximation for erfc(z) is useful is given by coordinate x in the range

$$a_1 - A^{-1}(4\alpha t)^{1/2} < x < a_1 + A^{-1}(4\alpha t)^{1/2} \tag{6.136}$$

This equation defines what the phrase "x near a_1 at early time" means in describing the range of application of Equation 6.135.

For smaller x values (but not near $x = 0$), the temperature distribution is given by

$$T(x,y,t) = \frac{\alpha q_0}{k} \int_{u=0} \frac{1}{(\pi\alpha u)^{1/2}} e^{-y^2/4\alpha u}\, du$$
$$= 2q_0 \left(\frac{t}{k\rho c}\right)^{1/2} \operatorname{ierfc}\left[\frac{y}{(4\alpha t)^{1/2}}\right] \qquad (6.137)$$

This result is exactly the same as for a semi-infinite body that is uniformly heated over its entire surface.

For x values larger than a_1, the surface at $y = 0$ is insulated. Sufficiently far from a_1 indicated by

$$x > a_1 + A^{-1}(4\alpha u)^{1/2} \qquad (6.138)$$

the temperature near the surface $y = 0$ is simply zero.

(c) *Surface temperature.* The temperature on the heated surface can be found directly by substituting $y = 0$ into the temperature expression at any point in the derivation. Often the surface temperature is easier to find than interior temperatures. The surface temperature is given by

$$T(x,0,t) \simeq \frac{\alpha q_0}{k} \int_{u_m}^{t} \frac{1}{2}\left[1 - \frac{x-a_1}{(4\alpha u)^{1/2}}\right] \frac{1}{(\pi\alpha u)^{1/2}}\, du$$
$$= \frac{\alpha q_0}{k}\left[\frac{1}{(\pi\alpha)^{1/2}}(t^{1/2} - u_m^{1/2}) - A\frac{x-a_1}{2\pi^{1/2}\alpha}\ln\left(\frac{t}{u_m}\right)\right]$$
$$= q_0\left[\left(\frac{t}{\pi k\rho c}\right)^{1/2} - \frac{A(x-a_1)}{k\sqrt{\pi}}\right] - \frac{q_0(x-a_1)}{k}\frac{A}{2\sqrt{\pi}}\ln\frac{\alpha t}{A^2(x-a_1)^2} \qquad (6.139)$$

for $-A^{-1}(4\alpha t)^{1/2} < x - a_1 < A^{-1}(4\alpha t)^{1/2}$ and $x \neq a_1$.

For larger values of x in the range $a_1 < x < a$, such that

$$x > a_1 + A^{-1}(4\alpha t)^{1/2}$$

the surface temperature is simply zero, and for smaller values of x in the range $0 < x < a_1$ such that

$$x < a_1 - A^{-1}(4\alpha t)^{1/2} \qquad (6.140a)$$

the surface temperature is given by

$$T(x,0,t) = 2q_0\left(\frac{t}{\pi k\rho c}\right)^{1/2} \qquad (6.140b)$$

which is the same as the surface temperature for a uniformly heated semi-infinite body.

6.8 TWO-DIMENSIONAL SEMI-INFINITE BODIES

The temperature in a semi-infinite body heated over half of the surface and insulated over the other half is treated in this section. This is a basic solution of two-dimensional heat conduction because it serves as a building block for other solutions and it is a kernel function for the unsteady surface element method discussed in Chapter 12.

The temperature is presented first in integral form, and then two series expressions for the integral are presented to evaluate the temperature efficiently at any location in the body and at any value of time.

6.8.1 Integral Expression for the Temperature

The geometry for the semi-infinite body heated over the half-plane is shown in Figure 6.12. The initial temperature is zero and the spatially uniform heat flux q_0 begins at time zero. This is the $X00Y20B5T0$ case. The temperature is given by the GF equation in the form

$$T(x, y, t) = \frac{\alpha q_0}{k} \int_{\tau=0}^{t} \int_{x'=-\infty}^{0} G_{X00Y20}(x, y, t|x', 0, \tau)\, dx'\, d\tau \tag{6.141}$$

where q_0 is a constant. Note that the GF is evaluated at the surface $y' = 0$, and that the integral over surface extends over only the heated half plane $-\infty < x' < 0$. The GF is given by a product solution of two familiar one-dimensional GFs, $G_{X00Y20} = G_{X00}G_{Y20}$.

The integral on x' in Equation 6.141 falls only on G_{X00}, and this integral should be familiar, so Equation 6.141 can be written

$$T(x, y, t) = \frac{1}{2}\frac{\alpha q_0}{k} \int_{\tau=0}^{t} G_{Y20}(y, t|0, \tau)\, \text{erfc}\left\{\frac{x}{[4\alpha(t-\tau)]^{1/2}}\right\} d\tau \tag{6.142}$$

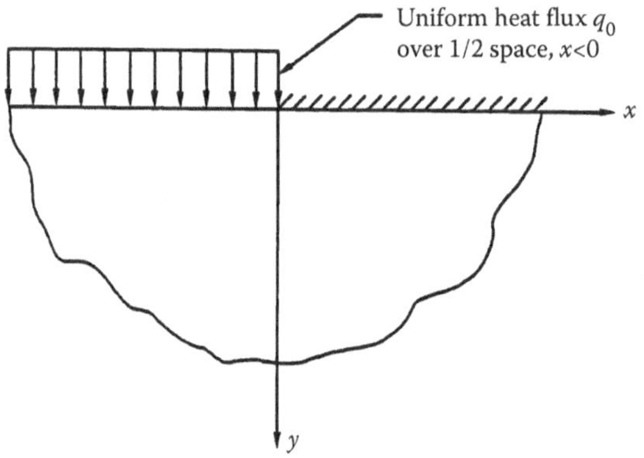

FIGURE 6.12 Geometry for semi-infinite region with uniform heat flux q_0 over half-space $\infty < x < 0$ and $y = 0$.

The function G_{Y20} is listed in Appendix X (replace x by y wherever it appears in the listing for G_{X20}), and the general expression for the temperature can be written

$$T(x,y,t) = \frac{1}{2}\frac{\alpha q_0}{k}\left(\frac{1}{\pi}\right)^{1/2}\int_{\tau=0}^{t}\frac{2}{[4\alpha(t-\tau)]^{1/2}} \times \exp\left[\frac{-y^2}{4\alpha(t-\tau)}\right]\operatorname{erfc}\left\{\frac{x}{[4\alpha(t-\tau)]^{1/2}}\right\}d\tau \quad (6.143)$$

This expression is valid for all locations in the body ($-\infty < x < \infty, y \geq 0$) and for any time $t \geq 0$.

6.8.2 Special Cases

The time integral in Equation 6.143 can be evaluated in closed form in two special cases.

Surface temperature. For the special case of $y = 0$, the temperature on the surface is given by (Carslaw and Jaeger, 1959, p. 264)

$$T(x,0,t) = \frac{q_0}{k}\left(\frac{\alpha t}{\pi}\right)^{1/2}\left\{\operatorname{erfc}\left[\frac{x}{2(\alpha t)^{1/2}}\right] - \frac{x}{2(\pi\alpha t)^{1/2}}E_1\left(\frac{x^2}{4\alpha t}\right)\right\} \quad (6.144)$$

The function $E_1(\cdot)$ is the exponential integral, defined by

$$E_1(z) = \int_z^{\infty}\frac{e^{-u}}{u}du \quad (6.145)$$

It is tabulated in Abramowitz and Stegun (1964) and it is available in computer libraries. See also Appendix I, Table I.1, for some expressions involving function E_1.

Centerline temperature. For the special case of $x = 0$, the temperature at the centerline is given by

$$T(0,y,t) = \frac{q_0}{k}(\alpha t)^{1/2}\operatorname{ierfc}\left[\frac{y}{2(\alpha t)^{1/2}}\right] \quad (6.146)$$

which is exactly one-half of the solution for a semi-infinite body heated over the entire $y = 0$ surface.

6.8.3 Series Expression for the Temperature

The time integral for the temperature, Equation 6.143 is evaluated in this section with series expressions. To begin, the time integral is written with a change of variables using

$$u = \frac{y}{2[\alpha(t-\tau)]^{1/2}} \quad (6.147)$$

and Equation 6.143 can be written as

$$T(x,y,t) = \frac{q_0 y}{2k\pi^{1/2}}\int_{y/(4\alpha t)^{1/2}}^{\infty}\frac{du}{u^2}e^{-u^2}\operatorname{erfc}\left(\frac{xu}{y}\right) \quad (6.148)$$

Further, a set of dimensionless variables will be used to present the temperature results:

$$X = \frac{x}{2(\alpha t)^{1/2}} \qquad Y = \frac{y}{2(\alpha t)^{1/2}} \tag{6.149a, b}$$

$$p = \frac{y}{x} = \frac{Y}{X} \qquad \Theta = \frac{T}{(q_0 / k)(\alpha t / \pi)^{1/2}} \tag{6.149c, d}$$

Notice that the variable p is independent of time. With these new variables, Equation 6.148 can be written

$$\Theta(p, Y) = Y \int_Y^\infty \frac{du}{u^2} e^{-u^2} \operatorname{erfc}\left(\frac{u}{p}\right) \tag{6.150}$$

The number of independent variables has been reduced from three (x, y, t) in Equation 6.148 to two dimensionless variables (p, Y) in Equation 6.150. The time dependence of the temperature has been absorbed into the coordinates and into the dimensionless temperature by normalizing them by the "length" $\surd(\alpha t)$.

This type of coordinate transformation is called a similarity transformation, and the variables are called similarity variables. Heat conduction problems can be solved this way where the solution depends on a penetration depth $\surd(\alpha t)$, usually because the geometry has no intrinsic length scale. Certain fluid flow problems may also be solved with similarity transformations. Equation 6.150 can be integrated by parts to give (Litkouhi, 1982)

$$\begin{aligned}\Theta(X, Y) = \pi^{1/2} \operatorname{ierfc}(Y) - e^{-Y^2} \operatorname{erf}(X) - \frac{X}{\pi^{1/2}} E_1(X^2 + Y^2) \\ + 2pY \int_X^\infty e^{-p^2u^2} \operatorname{erfc}(u)\, du\end{aligned} \tag{6.151}$$

Here u is a dummy variable.

The integral in the last term of Equation 6.151 can be represented by a function H defined as

$$H(X, Y) = \frac{2p}{\pi^{1/2}} \int_X^\infty e^{-p^2u^2} \operatorname{erf}(u)\, du \tag{6.152}$$

Recall that $p = Y / X$. Then the general temperature solution for a constant heat flux over the half plane can be written

$$\begin{aligned}\Theta(X, Y) = \pi^{1/2} \operatorname{ierfc}(Y) - e^{-Y^2} \operatorname{erf}(X) \\ - \frac{X}{\pi^{1/2}} E_1(X^2 + Y^2) + \pi^{1/2} Y\, H(X, Y)\end{aligned} \tag{6.153}$$

The general solution given by Equation 6.153 is valid for all times and any location in the body. However, Equation 6.153 is recommended onlyfor $X > 0$. For $X < 0$, a complementary expression is recommended:

$$\Theta(X < 0, Y) = Y\pi^{1/2} \operatorname{ierfc}(Y) - \Theta(X > 0, Y) \tag{6.154}$$

where the first term on the right-hand side of Equation 6.154 is the solution to the same problem if the entire surface was heated by a constant heat flux.

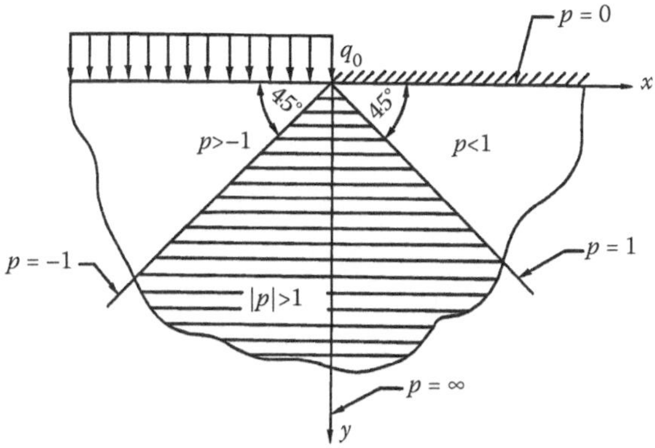

FIGURE 6.13 Geometry showing various regions $|p| < 1$, $|p| = 1$, and $|p| > 1$.

The function $H(X, Y) = H(X, p)$ can be represented in a series form for the three different regions indicated in Figure 6.13.

Region $|p| > 1$. The region $|p| > 1$ represents the region closest to the surface of the semi-infinite body. In this region, $H(X, p)$ is given by

$$H(X, Y) = H(X, p) = \frac{2}{\pi} \sum_{n=0}^{\infty} \frac{(-1)^n \Gamma(n+1, p^2X^2)}{p^{2n+1}(2n+1)n!} \tag{6.155}$$

where the truncated exponential function is defined in Abramowitz and Stegun (1964)

$$\Gamma(n, u) = \int_u^{\infty} e^{-t} t^{n-1}\, dt \tag{6.156}$$

Region $|p| < 1$. For the region $|p| < 1$, the expression 6.155 cannot be used for $H(X, Y)$ since the term p^{2n+1} appearing in the denominator causes the summation to diverge. In this case the following expression is provided:

$$H(X, p) = 1 - \operatorname{erf}(X)\operatorname{erf}(pX) - \frac{2}{\pi} \sum_{n=0}^{\infty} \frac{(-1)^n p^{2n+1} \Gamma(n+1, X^2)}{(2n+1)n!} \tag{6.157}$$

Region $|p| = 1$. On the line $|p| = 1$, it can be shown that $H(X, p)$ is given by

$$H(X, 1) = -H(X, -1) = \frac{1 - [\operatorname{erf}(X)]^2}{2} \tag{6.158}$$

Next some numerical results are presented. Figure 6.14 is a plot of function $H(X, p)$ versus X as calculated from the series expressions. [Numerical results for $H(X, p)$ to six decimal places are tabulated in Litkouhi, 1982.] Dimensionless temperature in the semi-infinite body is plotted versus X in Figure 6.15. Recall that $\Theta(X, Y)$ is normalized by the time, so time does not explicitly appear in the figure.

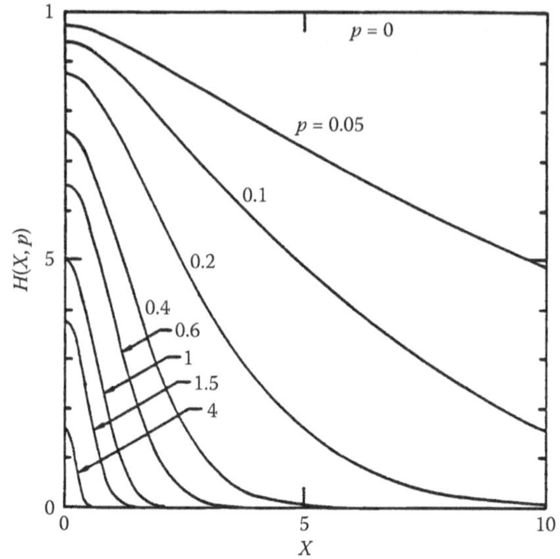

FIGURE 6.14 Function $H(X, p)$ versus X for different values of p.

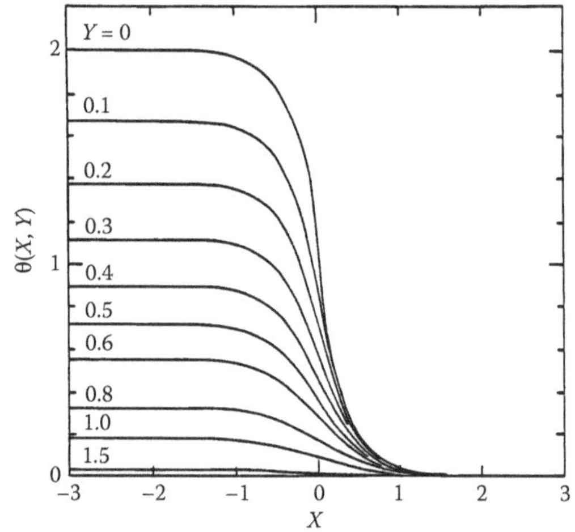

FIGURE 6.15 Dimensionless temperature $\Theta(X, Y)$ versus X for different values of Y in semi-infinite body with uniform heat flux over half-space $x < 0$ and $y = 0$.

6.8.4 Application to the Strip Heat Source

Other boundary conditions can be obtained by using the half-plane solution and superposition, and Figure 6.16 shows several geometries that are possible. One case of interest is the semi-infinite body heated by a constant heat flux over an infinite strip of width $2a$ and insulated elsewhere as shown in Figure 6.16a. This solution can be

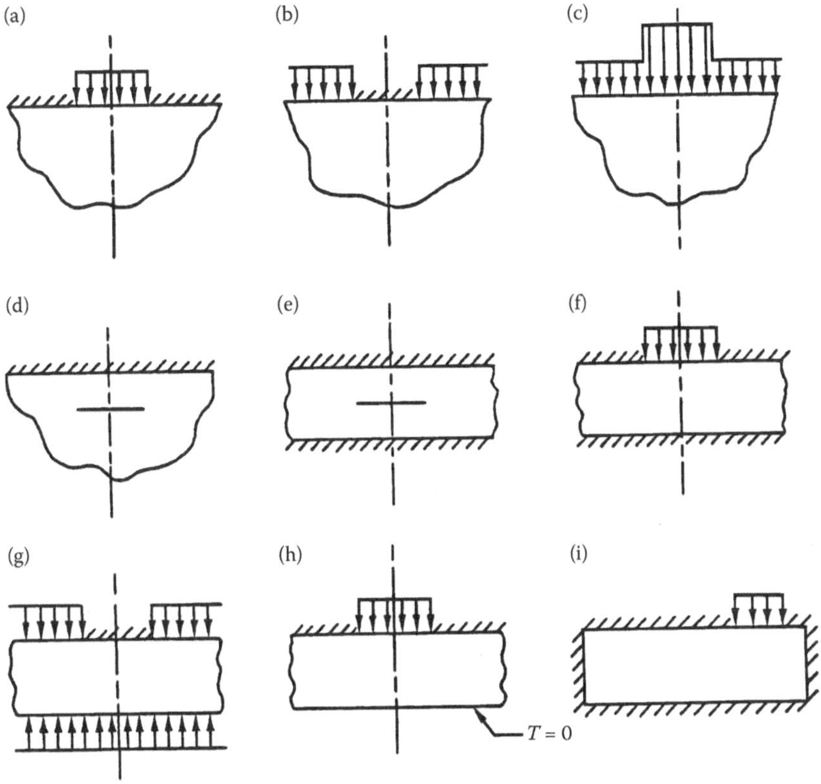

FIGURE 6.16 Various possible cases that can be treated using solution given in Figure 6.15 as a building block.

found from the superposition of two half-plane solutions: one half-plane is located at $x - a = 0$ with a positive heat flux, and the other half-plane is located at $x + a = 0$ with a negative heat flux. The resulting temperature is given by (Litkouhi, 1982)

$$\begin{aligned}\Theta(x^+, y^+, t^+) = {} & e^{-(y^{+2}/4t^+)}\left\{-\mathrm{erf}\left[\frac{x^+ - 1}{(4t^+)^{1/2}}\right] + \mathrm{erf}\left[\frac{x^+ + 1}{(4t^+)^{1/2}}\right]\right\} \\ & - \left(\frac{x^+ - 1}{(4\pi t^+)^{1/2}}\right) E_1\left[\frac{(x^+ - 1)^2 + (y^+)^2}{4t^+}\right] \\ & + \left[\frac{x^+ + 1}{(4\pi t^+)^{1/2}}\right] E_1\left[\frac{(x^+ + 1)^2 + (y^+)^2}{4t^+}\right] \\ & + \frac{\pi^{1/2} y^+}{(4t^+)^{1/2}} H\left[\frac{x^+ - 1}{(4t^+)^{1/2}}, p\right] - H\left[\frac{x^+ + 1}{(4t^+)^{1/2}}, p\right]\end{aligned} \tag{6.159}$$

where now the coordinates are normalized by a, the characteristic length:

$$x^+ = \frac{x}{a} \qquad y^+ = \frac{y}{a} \tag{6.160a, b}$$

$$t^+ = \frac{\alpha t}{a^2} \qquad p = \frac{y}{x} = \frac{y^+}{x^+} \tag{6.160c, d}$$

and $\Theta = T/[(q_0/k)(\alpha t/\pi)^{1/2}]$ as before. Note that the definition of parameter p has not changed from when it was introduced in Equation 6.149.

Surface temperature. For the special case of $y^+ = 0$, the surface of the semi-infinite body, the temperature due to the heated strip is (Carslaw and Jaeger, 1959)

$$\Theta(x^+, 0, t^+) = \text{erf}\left[\frac{x^+ + 1}{(4t^+)^{1/2}}\right] - \text{erf}\left[\frac{x^+ - 1}{(4t^+)^{1/2}}\right] + \left[\frac{x^+ + 1}{(4\pi t^+)^{1/2}}\right] E_1\left[\frac{(x^+ + 1)^2}{4t^+}\right]$$
$$- \left[\frac{x^+ - 1}{(4\pi t^+)^{1/2}}\right] E_1\left[\frac{(x^+ - 1)^2}{4t^+}\right] \tag{6.161}$$

6.8.5 Discussion

Round-off error. The expressions for the temperature in the strip heater case are recommended only for $x^+ > 0$ due to the possibility of computer round-off error. The geometry is symmetric about the x-axis so that the temperature for $x^+ < 0$ can easily be found from $T^+(x^+ < 0, y^+, t^+) = T^+(x^+ > 0, y^+, t^+)$.

Round-off error comes from subtracting two numbers that are close in value. For the strip heater problem, round-off error can come from the two superposed half-plane solutions. The temperature due to the heated strip can be written as

$$\Theta_{strip}(x^+) = \Theta_{half\text{-}plane}(x^+ - 1) - \Theta_{half\text{-}plane}(x^+ + 1) \tag{6.162}$$

(For the moment, the dependence on y^+ and t^+ has been left out.) Now, the physics of the heat transfer problem requires that sufficiently far from the heated strip, the temperature must approach zero. As $x^+ \to +\infty$, the two superposed solutions each approach zero (within the computation limits of the computer) because the half-plane solution is heated on the *left half* of the plane. There is no round-off error associated with the temperature at $x^+ > 0$. As $x^+ \to -\infty$, however, the two half-plane solutions are evaluated near their heated regions and the half-plane temperatures can be very large (especially near the surface $y^+ = 0$); the strip-heater temperature is near zero due to cancellation of the nearly equal half-plane temperatures. This process of canceling when $x^+ < 0$ can be demonstrated by a numerical example.

Suppose the temperature is evaluated directly at $x^+ = -3$, $y^+ = 0$, and $t^+ = 0.5$ for the heated strip located over $(-1 \leq x^+ \leq 1)$. The numerical value will be calculated with seven-digit accuracy using floating-point notation appropriate for a computer. Using Equation 6.162 with $x^+ = -3$,

$$\begin{aligned}\Theta_{strip}(x^+ = -3) &= \Theta_{half\text{-}plane}(-3 - 1) - \Theta_{half\text{-}plane}(-3 + 1)\\ &= 0.2000000\text{E} + 01 - 0.1999587\text{E} + 01\\ &= 0.413\text{E} - 03\end{aligned}$$

Note that the two half-plane temperatures are nearly equal, so the subtraction problem has reduced the accuracy from seven digits to three digits. Loss of accuracy is only part of the error, however, because a computer working with seven-digit accuracy will usually give the answer in seven digits, such as 0.4132662E–03 where the last four digits of the mantissa are computer-generated gibberish (round-off error). Most computers won't tell you when this type of error occurs. Again, for the strip heater problem, this type of error can be avoided by evaluating the temperature only at $x^+ > 0$ and using symmetry to find the temperature at $x^+ < 0$.

Lack of a steady state. The heated half-plane temperature, $T(x, y, t)$, has no steady state. As $t \rightarrow \infty$, the temperature increases without limit. In Equation 6.153, this dependence on time is hidden by the normalized temperature Θ which results in a dimensionless temperature expression that does not explicitly depend on time; however, the actual temperature in degrees kelvin represented by Equation 6.153 and Figure 6.15 does depend on time and there is no steady state.

It is not always clear if a semi-infinite body with heat flux boundary conditions has a steady-state temperature. In general, a semi-infinite body will have a steady-state temperature if a *finite* amount of heat (joules) is added to the body. There are at least three ways that a finite amount of heat can be added to a semi-infinite body: through a heated region that is finite in spatial extent, through a short duration of heating, or through a net zero heat flow into a body (sources and sinks of heat that balance out). For example, the heated strip solution discussed in this section is infinite in extent in the z-direction and an infinite amount of heat enters the body per unit time; consequently, there is no steady state. As a counterexample, a semi-infinite body heated over its surface for a short period and insulated thereafter always has a steady-state temperature of zero if you wait long enough after the heating has ended; in the limit of an infinitesimally short heating period, the temperature is similar to the GF G_{X20}, which goes to zero as $t - \tau$ goes to infinity.

6.9 STEADY STATE

Steady-state solutions have already been touched on in connection with the alternative GF solution method in Examples 6.5 and 6.6. In this section, three examples of steady heat conduction in rectangular coordinates are presented for two- and three-dimensional geometries. For one-dimensional steady cases in rectangular coordinates the GFs are listed in Appendix X, Tables X.1 through X.4.

Example 6.10: Rectangle Heated over Part of the $y = 0$ Boundary

In the rectangle ($0 < x < a$; $0 < y < b$), the $y = 0$ surface has a uniform heat flux over $0 < x < a_1$ and zero heat flux (insulated condition) over $a_1 < x < a$. The other three boundaries of the rectangle are at zero temperature. Find the steady temperature.

Solution

This case X11B00Y21B(x5)0. The geometry is shown in Figure 6.11 and the transient temperature for this rectangle was discussed in Example 6.9. The steady temperature is given by

$$T(x, y) = \frac{1}{k} \int_{x'=0}^{a_1} q_o\, G_{X11Y21}(x, y, |x', y' = 0)\, dx' \tag{6.163}$$

Here the steady GF will be constructed from eigenfunction expansions (see Section 4.6). There are two alternate forms of the steady GF in the rectangle. Using eigenfunctions along the x-direction (X11) gives eigenfunctions which are sines (see Table 4.2), so the GF has the form

$$G_{X11Y21} = \frac{2}{a} \sum_{m=1}^{\infty} \sin\left(\beta_m \frac{x}{a}\right) \sin\left(\beta_m \frac{x'}{a}\right) P_m(y, y') \tag{6.164}$$

where the eigenvalues are $\beta_m = m\pi$. Kernel function $P_m(y, y')$ is denoted case Y21 and is given by (see Table X.4, Appendix X)

$$P_m(y, y') = \frac{-e^{-\sigma(2b-|y-y'|)} - e^{-\sigma(2b-y-y')}}{2\sigma(1 + e^{-2\sigma b})} + \frac{e^{-\sigma(|y-y'|)} + e^{-\sigma(y+y')}}{2\sigma(1 + e^{-2\sigma b})} \tag{6.165}$$

where $\sigma = \beta_m / a$. Replace the GF into the temperature expression, Equation 6.163, and carry out the integral to find

$$T(x, y) = \frac{q_0 a}{k} \sum_{m=1}^{\infty} \sin\left(\frac{m\pi x}{a}\right) \frac{\left[1 - \cos\left(\frac{m\pi a_1}{a}\right)\right]}{m\pi} \left(\frac{e^{-\sigma y} - e^{-\sigma(2b-y)}}{\sigma a(1 + e^{-2\sigma b})}\right) \tag{6.166}$$

An alternate temperature expression, useful for intrinsic verification, can be constructed with eigenfunctions in the y-direction (Y21) and a kernel function in the x-direction (X11).

Example 6.11: Two-Dimensional Slab Heated over a Small Region

Find the steady temperature in a two-dimensional slab caused by a uniform heat flux q_0 over a small region $-a \leq x' \leq a$ and insulated elsewhere on one side of the slab, and fixed temperature T_0 on the other side. The region is very large in the x-direction and has thickness L in the y-direction. This geometry is related to the study of surface-mounted heated films.

Solution

This is the $X00Y21$ geometry, and the temperature distribution in the body is driven by heating on the surface $y = 0$. The geometry is shown in Figure 6.16h. The steady temperature is given by the surface heating term of the GFSE:

$$T(x, y) - T_0 = \int_{x'=-a}^{a} \frac{q_0}{k} G_{X00Y21}(x, y|x', y' = 0)\, dx' \tag{6.167}$$

The steady GF may be found from the method of limits and by the product of two one-dimensional transient GFs:

$$G_{X00Y21}(x, y|x', y') = \lim_{t\to\infty} \alpha \int_{\tau=0}^{t} G_{X00}(x, t|x', \tau)\, G_{Y21}(y, t|y', \tau)\, d\tau \tag{6.168}$$

The transient GFs are available in Appendix X, and Equation 6.168 may be written

$$G_{X00Y21}(x, y|x', y') = \lim_{t\to\infty} \alpha \int_{\tau=0}^{t} \frac{d\tau}{\sqrt{4\pi\alpha(t-\tau)}} \exp\left[-\frac{(x-x')^2}{4\alpha(t-\tau)}\right] \times \frac{2}{L}\sum_{m=1}^{\infty} \exp\left[-\frac{\beta_m^2\alpha(t-\tau)}{L^2}\right] \cos\left[\frac{\beta_m y'}{L}\right] \cos\left[\frac{\beta_m y}{L}\right] \tag{6.169}$$

where $\beta_m = \pi(m - 1/2)$. Note that the large-time form of the function G_{Y21} is used. The time integral in the above equation involves the error function and is given in Appendix I (Table I.6, number 12). After the limit is taken, the result is

$$G_{X00Y21}(x, y|x', y') = \sum_{m=1}^{\infty} \frac{1}{\beta_m} \exp\left[\frac{-\beta_m|x-x'|}{L}\right] \cos\left[\frac{\beta_m y'}{L}\right] \cos\left[\frac{\beta_m y}{L}\right] \tag{6.170}$$

The absolute value $|x - x'|$ is introduced by the time integral, and it reflects the symmetry of the GF about $(x - x') = 0$. It also guarantees that the exponential term dies away as $|x - x'|$ increases. The same form of the GF may also be found by the method of eigenfunction expansion (Section 4.6).

Now that the GF has been found, the temperature caused by heating the body over a small region may be found from Equation 6.167:

$$T(x, y) - T_0 = \frac{q_0}{k} \int_{x'=-a}^{a} \sum_{m=1}^{\infty} \frac{dx'}{\beta_m} \exp\left[\frac{-\beta_m|x-x'|}{L}\right] \cos\left[\frac{\beta_m y}{L}\right] \tag{6.171}$$

The absolute value must be treated carefully by examining $(x - x') > 0$ separately from $(x - x') < 0$. The result is two expressions for the temperature depending on the region:

$$\frac{T(x, y) - T_0}{q_0 L/k} = \sum_{m=1}^{\infty} \frac{\cos\left[\frac{\beta_m y}{L}\right]}{\beta_m^2} \times \left\{ \begin{array}{ll} e^{-\beta_m(|x|-a)/L} - e^{-\beta_m(|x|+a)/L}; & |x| > a \\ 2 - e^{-\beta_m(|x|-a)/L} - e^{-\beta_m(|x|+a)/L}; & |x| < a \end{array} \right\} \tag{6.172}$$

The convergence of this infinite series for $|x| > a$ is controlled by the exponential terms whose values rapidly go to zero with increasing m. On the heated region, $|x| \le a$, there is a portion of the series that does not contain an exponential, and this term causes slow convergence. The slow-converging portion of the series may be replaced with the following identity (Beck and Cole, 2007)

$$\sum_{m=1}^{\infty} \frac{\cos\left[\frac{\beta_m y}{L}\right]}{\beta_m^2} = 1 - \frac{y}{L} \tag{6.173}$$

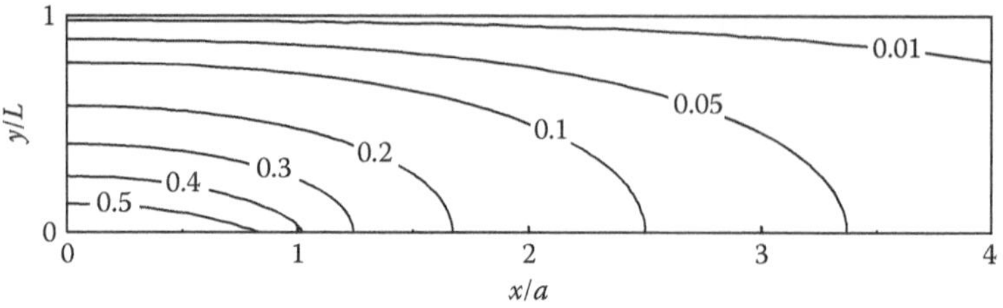

FIGURE 6.17 Contour plot of (normalized) temperature for $X00Y21$ geometry with heating over a small region at $y = 0$. Region $x < 0$ may be inferred from symmetry.

This identity may alternately be deduced by recognizing that the temperature solution in the range $|x| < a$ has the form

$$\frac{T(x,y) - T_0}{q_0 L / k} = S_y(y) + S_{xy}(x,y) \tag{6.174}$$

Upon replacing the above expression into the original boundary value problem for the two-dimensional temperature, the solution for S_y may be found by direct integration (see homework problem 5.6).

Figure 16.17 is a contour plot of the steady temperature for this case, from Cole and Yen (2001). The thickness is $L = a$ and the unheated surface of the slab at $y = L$ is held at temperature T_0. The heated region is located at $y = 0$ over $-a < x < a$. The temperature is normalized as $(T - T_0)/(q_0 L/k)$ where q_0 is the heat flux on the surface. The boundary heat flux is proportional to the slope of the contour lines where they meet the $y = 0$ boundary. For example, the contours are perpendicular to the $y = 0$ surface for $x/a > 1$ which indicates the zero-flux conditions there.

Example 6.12: Parallelepiped with Specified Surface Temperature—*X*11*Z*11*Y*11 Case

Find the steady temperature in the parallelepiped with five faces at zero temperature and one face (at $x = 0$) maintained at temperature T_0.

Solution

The GF for this geometry was treated in Example 4.10 and the parallelepiped body is shown in Figure 4.4. The triple-sum GF for this case is given by

$$G(x,y,z|x',y',z') = 8\sum_{m=1}^{\infty}\sum_{n=1}^{\infty}\sum_{p=1}^{\infty} \sin\left(m\pi\frac{x}{a}\right)\sin\left(m\pi\frac{x'}{a}\right)\sin\left(p\pi\frac{z}{c}\right) \times \sin\left(p\pi\frac{z'}{c}\right)\sin\left(n\pi\frac{y}{b}\right)\sin\left(n\pi\frac{y'}{b}\right)\left[abc\pi^2\left(\frac{m^2}{a^2}+\frac{n^2}{b^2}+\frac{p^2}{c^2}\right)\right]^{-1} \tag{6.175}$$

The temperature for this case is given by the boundary term of the steady GFSE,

$$T(x,y,z) = -\int_{y'=0}^{b} dy' \int_{z'=0}^{c} dz'\, T_0 \left.\frac{\partial G}{\partial n'}\right|_{x'=0} \tag{6.176}$$

where the surface integral is carried out over the $x = 0$ face of the parallelepiped. The required derivative and integrals are elementary, and the temperature is

$$T(x,y,z) = 8T_0 \sum_{m=1}^{\infty}\sum_{n=1}^{\infty}\sum_{p=1}^{\infty}[1-(-1)^p][1-(-1)^n] \\ \times \sin\left(m\pi\frac{x}{a}\right)\sin\left(p\pi\frac{z}{c}\right)\sin\left(n\pi\frac{y}{b}\right)\left[a^2np\pi^3\left(\frac{m^2}{a^2}+\frac{n^2}{b^2}+\frac{p^2}{c^2}\right)\right]^{-1} \tag{6.177}$$

This triple summation expression is not recommended for numerical evaluation because it converges very slowly. A better-converging temperature expression can be found using a double-summation GF; one such GF was discussed in Example 4.10. The double-summation GF with the kernel function along the z-direction, is given by

$$G(x,y,z|x',y',z') = \sum_{m=1}^{\infty}\sum_{n=1}^{\infty}\sin\left(m\pi\,\frac{x}{a}\right)\sin\left(m\pi\,\frac{x'}{a}\right) \\ \times \sin\left(n\pi\,\frac{y}{b}\right)\sin\left(n\pi\,\frac{y'}{b}\right)\, P_{nm}(z,z') \tag{6.178}$$

where the kernel function P_{nm} is given by (Table X.4, case X11)

$$P_{nm}(z,z') = \frac{e^{-\sigma(2c+|z-z'|)} - e^{-\sigma(2c-z-z')}}{2\sigma(1-e^{-2\sigma c})} + \frac{e^{-\sigma|z-z'|} - e^{-\sigma(z+z')}}{2\sigma(1-e^{-2\sigma c})} \tag{6.179}$$

where $\sigma^2 = \left(\frac{n^2\pi^2}{b^2} + \frac{m^2\pi}{a^2}\right)$

Using this GF in the GF solution, Equation 6.176, the double-summation temperature is given by

$$T(x,y,z) = T_0 \sum_{m=1}^{\infty}\sum_{n=1}^{\infty}\sin\left(m\pi\frac{x}{a}\right)\sin\left(n\pi\frac{y}{b}\right)\left(\frac{m\pi}{a}\right)\frac{b}{n\pi}[1-\cos(n\pi)] \\ \times\left(\frac{1}{\sigma^2} - \frac{e^{-\sigma(c+z)} - e^{-\sigma(c-z')}}{\sigma^2(1-e^{-2\sigma c})}\right) \tag{6.180}$$

The double-summation form converges much faster than the triple summation form. Further convergence improvements are possible if the solution is written in the form

$$T(x,y,z) = T_0[S_{xy}(x,y) + S_{xyz}(x,y,z)] \tag{6.181}$$

The slower-converging term in the above equation is S_{xy} because it does not contain any z-exponentials. The convergence speed of term $S_{xy}(x, y)$ can be improved by recognizing that it is the solution to a certain two-dimensional heat conduction problem in a rectangle. The double-sum form of $S_{xy}(x, y)$ in the above expression can be replaced by a better-converging *single sum* form using an appropriate GF based on eigenfunction expansions (see Crittenden and Cole, 2002).

PROBLEMS

6.1 Find the temperature in a semi-infinite body resulting from the following surface temperature.

$$T(x = 0, t) = \begin{cases} T_1 & 0 < t \leq t_1 \\ T_2 & t > t_1 \end{cases}$$

The initial temperature is zero. What is the number of this case?

6.2 Find the temperature in a semi-infinite body heated at the surface by a square pulse of heat:

$$q(t) = \begin{cases} q_0 & 0 < t < t_1 \\ 0 & t > t_1 \end{cases}$$

Find the steady-state temperature as $t \to \infty$.

6.3 Suppose the surface temperature on a semi-infinite solid due to surface heating is given by

$$T(t) - T_0 = a\sqrt{\frac{t}{t_0}} + b\left(\frac{t}{t_0}\right)$$

where a, b, and t_0 are constants. Find the surface heat flux that caused the temperature to rise.

6.4 Find the prescribed surface temperature, $f(t)$, such that when applied to the semi-infinite solid with zero initial condition ($X10B$-$T0$ case), the surface heat flux is given by

$$-k\frac{\partial T}{\partial x}\bigg|_{x=0} = q_0\left(\frac{t}{t_0}\right)^{n/2} \qquad \text{for } n = 0, 1, 2, \ldots$$

6.5 Consider a semi-infinite solid with a thin, high conductivity film at $-\delta \leq x \leq 0$. Let $x > 0$ be the semi-infinite body. Find the temperature at $x = 0$ for the following heating condition:

$$-k\frac{\partial T}{\partial x}(0, t) = q_0$$
$$T(x, 0) = 0$$

This is case $X40B1T0$. Compare your answer to the $X20B1T0$ case. What is a dimensionless parameter that describes the added effect of the thin surface film on the heated semi-infinite body at early times after heating begins?

6.6 Consider the same geometry as in the previous problem, but now the surface of the thin film is suddenly heated by a convection process. The initial temperature is zero. Find the transient temperature and compare it to the $X30B1T0$ case.

6.7 Find the small-time form of the temperature for a one-dimensional slab geometry with one surface heated with a constant heat flux, one surface insulated, and zero initial conditions ($X22B10T0$ case). Compare your result to the (semi-infinite) $X20B1T0$ solution listed in Table 6.3, and comment on the differences.

6.8 Derive the following expression:

$$\frac{\partial T_{X2JB10T0}}{\partial t}(x,t) = \frac{\alpha}{k} q_o G_{X2J}(x,0|t,0), \quad \text{for } J = 0,1,2,3.$$

6.9 Based on Problem 6.8, show that for $\alpha t / L^2 < 0.1$ to an accuracy of 1 part in 10^4, that

$$\left. \frac{\partial T_{X2JB10T0}}{\partial t}(0,t) \right|_{\alpha t / L^2 < 0.1} \approx \frac{\alpha}{k} q_o \frac{1}{\sqrt{\pi \alpha t}} \quad \text{for } J = 0,1,2,3.$$

Verify numerically by using the exact solution for case X22B10T0.

6.10 Exponential heating is sometimes used to model runaway heating of nuclear fuel rods. Write down the integral form of the temperature for the following problem with exponential heating and convection cooling. Assume that the GF has the name $G_{X33}(x,t|x',\tau)$. Do not evaluate the GF. Do not evaluate the integrals.

$$\frac{\partial^2 T}{\partial x^2} + \frac{1}{k} g(t) = \frac{1}{\alpha}\frac{\partial T}{\partial t}$$

$$T(x,0) = 0$$

$$-k \left. \frac{\partial T}{\partial n} \right|_{x_i} = h(T|_{x_i} - T_\infty), \ i = 1,2.$$

$$g(t) = g_0 e^{at}$$

6.11 Use the standard Green's function solution equation (GFSE) to obtain the temperature distribution for the problem

$$\frac{\partial^2 T}{\partial x^2} + \frac{g(x)}{k} = \frac{1}{\alpha}\frac{\partial T}{\partial t} \qquad 0 < x < L \qquad t > 0$$

$$T(0,t) = T_0 \qquad T(L,t) = T_0 \qquad T(x,0) = T_0$$

where $g(x) = g_0 = \text{constant for } 0 < x < L_1 < L$
$= 0$ otherwise

Use the large cotime GF.

6.12 Restate Problem 6.11 in dimensionless form with new variables $\xi = x / L$, $\eta = \alpha t / L^2$ and $\theta = (T - T_0)/(g_0 L^2 / k)$. Do not solve for θ.

6.13 Solve, using the GFSE, the problem

$$\frac{\partial^2 T}{\partial x^2} = \frac{1}{\alpha}\frac{\partial T}{\partial t} \qquad 0 < x < L \qquad t > 0$$

$$-k\frac{\partial T}{\partial x}\bigg|_{x=0} = q_0 \qquad T(L,t) = T_0 \qquad T(x,0) = T_0$$

Use the large cotime GF.

6.14 Solve Problem 6.13 using the Alternate Green's Function Solution Equation (AGFSE).

6.15 Solve Problem 6.13 using the GFSE with the small cotime GF.

6.16 Consider the following one-dimensional problem.

$$\frac{\partial^2 T}{\partial x^2} = \frac{1}{\alpha}\frac{\partial T}{\partial t} \qquad -k\frac{\partial T}{\partial x}\bigg|_{x=0} = q_0$$

$$-k\frac{\partial T}{\partial x}\bigg|_{x=L} = h(T|_{x=L} - T_\infty)$$

and initial condition $T(x,0) = 0$.

(a) Using $T(x,t) = T^*(x) + T_1(x,t)$, write down an alternative boundary value problem for $T_1(x,t)$, where $T^*(x)$ is the solution to the following steady problem.

$$\frac{\partial^2 T^*}{\partial x^2} = 0$$

$$-k\frac{\partial T^*}{\partial x}\bigg|_0 = q_0$$

$$-k\,\frac{\partial T^*}{\partial x}\bigg|_{x=L} = h(T^*|_{x=L} - T_\infty)$$

(b) Carry out the transient solution using the *large-cotime* form of the GF and the AGFSE. Write your answer in terms of dimensionless parameters hL/k, x/L, and dimensionless temperature $(T - T_\infty)/(q_0 L/k)$.

6.17 Write down the GF solution equation for the following two-dimensional case. Do not derive the GF; do not solve the integrals. However, use the correct form of dV and ds_i. Use the name G_{X22Y11} $(x, y, t|x', y', \tau)$ in your expression.

$$\frac{\partial^2 T}{\partial x^2} + \frac{\partial^2 T}{\partial y^2} + \frac{g_0}{k} = \frac{1}{\alpha}\frac{\partial T}{\partial t} \qquad g_0 \text{ is constant}$$

$$T(x,y,0) = ax + by + c$$

$$\frac{\partial T}{\partial x}(x = 0, y, t) = 0$$

$$\frac{\partial T}{\partial x}(x = L_x, y, t) = 0$$

$$T(x, y = 0, t) = T_0$$

$$T(x, y = L_y, t) = 0$$

6.18 Consider the surface temperature on a semi-infinite body heated over two infinite strips of equal size, case $X00T0Y20B(x5)$.

(a) Find the surface temperature resulting from the following anti-symmetric heat flux distribution:

$$q(x,t) = \begin{cases} 0 & t = 0 \\ -q_0 & t > 0 \quad -b < x < -a \\ +q_0 & t > 0 \quad a < x < b \\ 0 & t > 0 \quad \text{otherwise} \end{cases}$$

(b) Where does the maximum temperature occur?

(c) Plot the steady state surface temperature $T(x, y = 0)$.

6.19 Find the surface temperature $T(x, y, t)$ on a semi-infinite body heated by a line source located at $x = 0$, $y = 0$. The surface heating is given by:

$$q(x,t) = \begin{cases} 0 & t = 0 \\ q_0\delta(x) & t > 0 \end{cases}$$

6.20 Find the steady-state temperature at $y = 0$ for the $X00T0Y21B(x5)0$ case (strip heat source) as follows:

(a) First find the integral form of the transient temperature with the large-time form of the GF. The boundary conditions are the following:

$$T(x, y = D, t) = 0$$

$$-k\frac{\partial T(x,0,t)}{\partial y} = \begin{cases} q_0 & -a < x < a \\ 0 & \text{elsewhere} \end{cases}$$

$$T(x \to -\infty, y, t) = 0$$

$$T(x \to +\infty, y, t) = 0$$

$$T(x, y, 0) = 0$$

(b) Evaluate the temperature at $y = 0$ and evaluate the integrals.

(c) Suggest one method to improve the convergence of the series expression.

6.21 A rectangle $0 \leq x \leq L, 0 \leq y \leq L$ is initially at temperature zero. Surfaces $y = 0$, $y = L$, and $x = L$ are insulated. Surface $x = 0$ is heated by constant heat flux q_0 over $0 < y < L/2$ and is insulated over $L/2 < y < L$. Find the temperature at location $x = 0$, $y = 0$ as a function of time.

(a) Using large-cotime GFs.

(b) Using time-partitioning at $\alpha\Delta t / L^2 = 0.005$.

6.22 Solve the following problem of two-dimensional heat flow.

$$\frac{\partial^2 T}{\partial x^2} + \frac{\partial^2 T}{\partial y^2} = \frac{1}{\alpha}\frac{\partial T}{\partial t}$$

$$T(0, y, t) = T(a, y, t) = T(x, b, t) = T_0, \; T(x, y, 0) = T_0$$

$$T(x, 0, t) = \begin{cases} T_1 \neq T_0 & \text{for } 0 < x < a_1 < a \\ T_0 & \text{for } a_1 < x < a \end{cases}$$

(a) Use the GFSE with large-cotime GFs.

(b) Use the GFSE with small-cotime GFs.

6.23 Starting with the transient GF for case X31 for the body with convection at $x = 0$ and a zero temperature at $x = L$, show that (a) in the limit as $h \to \infty$ the GF reduces to case X11, and (b) in the limit as $h \to 0$ the GF reduces to case X21.

6.24 Starting with the steady-fin GF for case X13 (Table X.4, Appendix X), show that (a) in the limit as $h_2 \to \infty$ the GF reduces to (steady fin) case X11, and (b) in the limit as $h_2 \to 0$ the GF reduces to (steady fin) case X12.

6.25 Consider the steady temperature in a rectangular fin which satisfies

$$\frac{d^2T}{dx^2} - m^2(T - T_\infty) = 0 \tag{6.182}$$

where $m^2 = 2h/(dk)$, h is a heat transfer coefficient (W/m^2/K), d is the fin thickness, k is the fin conductivity (W/m/K), and T_∞ is the fluid temperature. The boundary at $x = 0$ has $T = T_0$ and the boundary at $x = L$ has $dT/dx = 0$. Find the temperature in the fin using the steady-fin GF given in Table X.4, Appendix X. Compare your result for $T(x)$ to that given in a heat transfer text for the fin with an insulated tip and comment on the differences. (Hint: you will need cosh and sinh.)

6.26 Find the steady temperature in the rectangle with one side at elevated temperature and the other three sides at zero temperature, case X11B10Y11B00.

6.27 (a) Find an integral expression for a semi-infinite body heated at the surface over a rectangular area (three-dimensional problem). This is the $X00Y00Z20B(x5y5)T0$ case. The surface heating is given by

$$-k\frac{\partial T(x,y,z=0,t)}{\partial y} = \begin{cases} q_0 & -a < y < a \\ & -b < x < b \qquad t > 0 \\ 0 & \text{elsewhere on surface} \end{cases}$$

Initially the temperature is zero.

(b) Find the average temperature on the rectangle in the form of an integral on τ (evaluate spatial integrals).

6.28 Solve, using the GFSE, the problem

$$\frac{\partial^2 T}{\partial x^2} + \frac{g_0 e^{-x/x_0}}{k} = \frac{1}{\alpha}\frac{\partial T}{\partial t} + \frac{1}{\alpha}U_0\frac{\partial T}{\partial x} \qquad 0 < x < \infty$$

$$T(0,t) = T_0, T(x,0) = 0$$

The quantities, g_0, U_0, and x_0, are constants. What is the number of this case?

REFERENCES

Abramowitz, M. and Stegun, I. A. (Eds), 1964, *Handbook of Mathematical Functions with Formulas, Graphs, and Mathematical Tables*, Applied Mathematics Series 55, National Bureau of Standards, Washington, D.C.

Beck, J. V., 1984, Green's function solution for transient heat conduction problems, *Int. J. Heat Mass Transfer*, vol. 27, pp. 1135–1244.

Beck, J. V. and Cole, K. D., 2007, Improving convergence of summations in heat conduction, *Int. J. Heat Mass Transfer*, vol. 50, pp. 257–268.

Beck, J. V., Haji-Sheikh, A., Amos, D. E., and Yen, D., 2004, Verification solution for partial heating of rectangular solids, *Int. J. Heat Mass Transfer*, vol. 47, pp. 4243–4255.

Carslaw, H. S. and Jaeger, J. C., 1959, *Conduction of Heat in Solids*, 2nd edn, Oxford University Press, New York.

Cole, K. D. and Yen, D. H. Y., 2001, Influence functions for the infinite and semi-infinite strip, *AIAA J. Thermophys. Heat Transfer*, vol. 15, no. 4, pp 431–438.

Crittenden, P. E. and Cole, K. D., 2002, Fast-converging steady-state heat conduction in a rectangular parallelepiped, *Int. J. Heat Mass Transfer*, vol. 45, pp. 3585–3596.

Litkouhi, B., 1982, Surface element method in transient heat conduction problems, Ph.D. thesis, Michigan State University Mechanical Engineering Department, East Lansing, Michigan.

Ozisik, M. N., 1993, *Heat Conduction*, John Wiley, New York.

7 Cylindrical Coordinates

7.1 INTRODUCTION

Heat conduction in geometries described by the cylindrical coordinate system (r, ϕ, z) are discussed in this chapter. Transient radial heat flow is covered in Sections 7.2 through 7.7, for the infinite body, the long cylinder, and the infinite body with a cylindrical hole. The thin shell is discussed in Section 7.8. The use of limiting cases for two- and three-dimensional bodies is discussed in Section 7.9. Two-dimensional transient heat transfer is discussed in Sections 7.10 through 7.12 for finite cylinders and for a disk heat source on a semi-infinite body. Several steady-state cases are given in Section 7.13.

7.2 RELATIONS FOR RADIAL HEAT FLOW

Temperature and Green's functions (GFs) for radial flow of heat in the cylindrical coordinate system (r, ϕ, z) are discussed in this section. For radial flow of heat, the temperature depends on position r and time t, and the heat conduction equation has the form

$$\frac{1}{r}\frac{\partial}{\partial r}\left[r\frac{\partial T}{\partial r}\right] + \frac{1}{k}\,g(r,t) = \frac{1}{\alpha}\frac{\partial T}{\partial t} \tag{7.1}$$

That is, the temperature does not depend on ϕ or z. The radial GF equation is given by

$$\begin{aligned}
T(r,t) &= \int_{r'} G(r,t|r',0)\,F(r')\,2\pi r' dr' + \alpha \sum_{i=1}^{s} \frac{(\rho c b)_i}{k_i} \\
&\quad \times G(r,t|r_i,0)\,F(r_i)\,2\pi r_i \qquad \text{(for boundary conditions of fourth and fifth kinds only)} \\
&\quad + \int_{\tau=0}^{t} \int_{r'} \frac{\alpha}{k} G(r,t|r',\tau) g(r',\tau)\,2\pi r' dr'\,d\tau \\
&\quad + \alpha \int_{\tau=0}^{t} \sum_{i=1}^{s} \frac{f_i(r_i,t)}{k_i} \qquad \text{(for boundary conditions of the second through fifth kinds)} \\
&\quad \times G(r,t|r_i,\tau)\,2\pi r_i\,d\tau \\
&\quad - \alpha \int_{\tau=0}^{t} \sum_{j=1}^{s} f_j(r_j,t) \\
&\quad \times \left.\frac{\partial G}{\partial n'_j}\right|_{r'=r_j} 2\pi r_j\,d\tau \qquad \text{(for boundary condition of the first kind only)}
\end{aligned} \tag{7.2}$$

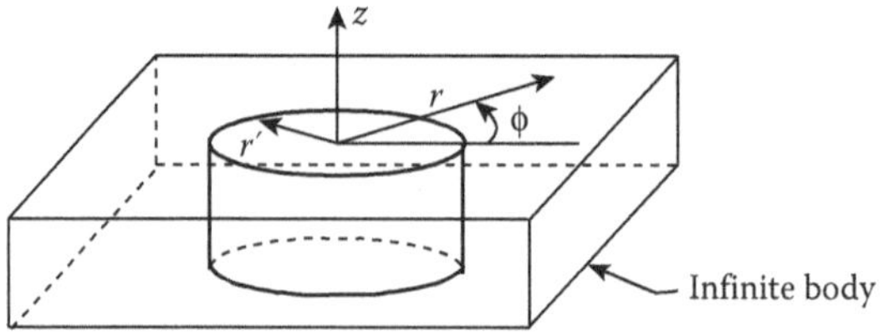

FIGURE 7.1 Cylindrical surface heat source located at r'.

Note that $dV' = 2\pi r' dr'$, and the integrals over boundary surface s_i have been replaced by $2\pi r_i$, the area per unit length. Equation 7.2 may be applied to bodies with boundary conditions of type 0 through 5. However, the radial heat flow GF actually listed in this book (Appendix R) are denoted $G_{RIJ}(\cdot)$, where $I, J = 0, 1, 2$, and 3.

7.3 INFINITE BODY

7.3.1 The *R*00 Green's Function

The GF for the radial flow of heat in the infinite body is denoted $G_{R00}(r, t|r', \tau)$. This GF can be interpreted as the response to a cylindrical surface heat source located at radius r' (refer to Figure 7.1), and it is given by

$$G_{R00}(r, t|r', \tau) = \frac{1}{4\pi\alpha(t-\tau)} \exp\left[\frac{-(r^2 + r'^2)}{4\alpha(t-\tau)}\right] I_0\left[\frac{rr'}{2\alpha(t-\tau)}\right] \tag{7.3}$$

for $0 \le r \le \infty$ and $0 \le r' \le \infty$. The function $I_0(\cdot)$ is the modified Bessel function of the first kind of order zero [$I_0(0) = 1$ and $I_0(z \to \infty) = \infty$]. Refer to Appendix B for more information on the Bessel functions. The units of G_{R00} are (meters)$^{-2}$. Note that the reciprocity relation holds for this GF because r and r' can be reversed and the function is unchanged.

In the special case where $r' = 0$, the cylindrical source that generates the function $G_{R00}(\cdot)$ collapses into a line source located at $r' = 0$, given by

$$G_{R00}(r, t|0, \tau) = \frac{1}{4\pi\alpha(t-\tau)} \exp\left[\frac{-r^2}{4\alpha(t-\tau)}\right] \tag{7.4a}$$

Recall that a line source can also be represented by the product of two plane sources, and that $r' = 0$ corresponds to the point $x' = 0$, $y' = 0$. Thus, the identity is

$$G_{R00}(r, t|0, \tau) = G_{X00}(x, t|0, \tau)\, G_{Y00}(y, t|0, \tau) \tag{7.4b}$$

This product solution also demonstrates that the units of $G_{R00}(\cdot)$ are $m^{-1}m^{-1} = m^{-2}$.

7.3.2 Derivation of the *R*00 Green's Function

There are several ways to derive the *R*00 GF from first principles (Ozisik, 1993, p. 107; Carslaw and Jaeger, 1959, p. 259). The following derivation involves an infinite body with heat generation in rectangular coordinates. The GFSE in two-dimensional rectangular coordinates is given by

$$T(x,y,t) = \frac{\alpha}{k}\int_{\tau=0}^{t}\int_{x'}\int_{y'} G_{X00Y00}(x,y,t|x',y',\tau)\, g(x',y',\tau)\, dy'\, dx'\, d\tau \quad (7.5)$$

The appropriate heat generation term is an instantaneous cylindrical surface heat source shown in Figure 7.1 and given by

$$g(x',y',\tau) = g_0 \frac{\delta(\tau-\tau_0)\,\delta(r'-r_0)}{2\pi r_0} \quad (7.6)$$

Parameter r_0 is the radius of the cylindrical surface heat source that introduces heat at time τ_0 and g_0 (J/m) is the strength of the heat source per unit length of the cylindrical surface. [The strength of the heat source *per unit area* is $g_0/(2\pi r_0)$].

The appropriate two-dimensional GF in rectangular coordinates is given by

$$G_{X00Y00}(x,y,t|x',y',\tau) = \frac{1}{4\pi\alpha(t-\tau)}\exp\left[-\frac{(x-x')^2+(y-y')^2}{4\alpha(t-\tau)}\right] \quad (7.7)$$

Recall that G_{X00Y00} represents the response to an instantaneous line heat source located at (x', y', τ), and that $G_{X00Y00} = G_{X00}\, G_{Y00}$.

To evaluate the temperature in Equation 7.5, the integral over the infinite body must be changed to cylindrical coordinates. First, the distance between points (x, y) and (x', y') that appears in the expression for G_{X00Y00} must be converted to cylindrical coordinates. If the cylindrical coordinates of points (x, y) and (x', y') are (r, ϕ) and (r', ϕ'), respectively, then the distance between these points is given by

$$R^2 = (x-x')^2 + (y-y')^2 = r^2 + (r')^2 - 2rr'\cos(\phi-\phi') \quad (7.8)$$

Second, the spatial integrals in Equation 7.5 that extend over the entire (x', y') plane, where $dA = dx'dy'$ must be converted to equivalent integrals in the (r', ϕ') coordinate system over $0 \le r' < \infty$ and $0 \le \phi' \le 2\pi$ with $dA = r'\,dr'\,d\phi'$. Then Equation 7.5 can be combined with Equations 7.6 through 7.8 to give

$$T(r,\phi,t) = \frac{\alpha}{k}\int_{\tau=0}^{t}\int_{r'=0}^{\infty}\int_{\phi'=0}^{2\pi}\left\{\frac{1}{4\pi\alpha(t-\tau)}\exp\left[-\frac{r^2+(r')^2-2rr'\cos(\phi-\phi')}{4\alpha(t-\tau)}\right]\right.$$
$$\left.\times\, g_0\frac{\delta(\tau-\tau_0)\,\delta(r'-r_0)}{2\pi r_0}\right\} d\tau\, r'\,dr'\,d\phi' \quad (7.9)$$

The integrals over r' and τ can be evaluated easily with the sifting property of the Dirac delta functions:

$$T(r,\phi,t) = \frac{\alpha}{k}\int_{\phi'=0}^{2\pi} d\phi'\left\{\frac{r_0}{4\pi\alpha(t-\tau_0)}\exp\left[-\frac{r^2+r_0^2-2rr_0\cos(\phi-\phi')}{4\alpha(t-\tau_0)}\right]\frac{g_0}{2\pi r_0}\right\}$$
$$= \frac{\alpha}{k}\frac{g_0/(2\pi)}{4\pi\alpha(t-\tau_0)}\exp\left[\frac{-(r^2+r_0^2)}{4\alpha(t-\tau_0)}\right]\int_{\phi'=0}^{2\pi}\exp\left[\frac{rr_0\cos(\phi-\phi')}{2\alpha(t-\tau_0)}\right]d\phi' \quad (7.10)$$

TABLE 7.1
Approximate Expressions for $G_R00\ (r, t\,|\,r', \tau)$ Listed in Appendix R

Range of Application	Error (%)	Equation Number in Appendix R
$\frac{\alpha(t-\tau)}{rr'} < 0.25$	0.016	R00.4
$\frac{\alpha(t-\tau)}{rr'} > 0.33$	−0.012	R00.5
$\frac{\alpha(t-\tau)}{r^2}$ large and $\frac{\alpha(t-\tau)}{(r')^2}$ large		R00.6

The final integral on ϕ' is given by Watson (1944). The GF is given by the temperature divided by the source strength, or

$$G_{R00}(r, t|r_0, \tau_0) = \frac{T(r, \phi, t)}{\alpha g_0 / k}$$

$$= \frac{1}{4\pi\alpha(t-\tau_0)} \exp\left[\frac{-(r^2 + r_0^2)}{4\alpha(t-\tau_0)}\right] I_0\left[\frac{rr_0}{2\alpha(t-\tau_0)}\right] \tag{7.11}$$

Note that the result does not depend on angle ϕ. Finally, the GF is usually written with the heat source located at (r', τ) instead of at (r_0, τ_0), to give the same result as in Equation 7.3.

7.3.3 Approximations for the *R*00 Green's Function

The $R00$ GF usually must be integrated to find the temperature, but it is not an easy function to integrate. Most integrals of function G_{R00} must be evaluated numerically unless a simple approximate expression can be found. A few approximate expressions for G_{R00} are listed in Appendix R, and Table 7.1 is a reference list of these approximations. These approximations are composed of exponentials and powers and they are generally easier to manipulate than the exact expression for G_{R00}. Table 7.1 lists the region of application, the maximum error, and the location in Appendix R of several approximate expressions for G_{R00}.

7.3.4 Temperatures from Initial Conditions

The temperature in an infinite body resulting from a nonuniform initial condition is given by the Green's function solution equation (GFSE) as

$$T(r, t) = \int_{r'} G_{R00}(r, t|r', 0)\, F(r')\, 2\pi r'\, dr' \tag{7.12}$$

In this section, the above integral is discussed for the specific case of a uniform initial temperature near the origin and zero temperature elsewhere:

$$F(r') = \begin{cases} T_0 & 0 \le r' \le a \\ 0 & r' > a \end{cases} \tag{7.13}$$

This is the $R00T5$ case. The transient temperature is given by

$$\begin{aligned} T(r,t) &= T_0 \int_{r'=0}^{a} G_{R00}(r,t|r',0)\, 2\pi r'\, dr' \\ &= \frac{T_0}{4\pi\alpha t} \int_{r'=0}^{a} \exp\left[\frac{-(r^2+r'^2)}{4\alpha t}\right] I_0\left[\frac{rr'}{2\alpha t}\right] 2\pi r'\, dr' \end{aligned} \tag{7.14}$$

Note that the integral is written over $0 \le r' \le a$ because $F(r')$ is zero elsewhere. In general, this integral must be evaluated numerically, and some numerical values of this integral are listed in Table R00.1 in Appendix R.

Over the region $0 \le r < a/2$, the temperature given by Equation 7.14 remains within 0.03% of T_0 for small values of the time parameter ($\alpha t/a^2 < 0.01$). That is, $T(r = a/2,\ \alpha t/a^2 = 0.01) = 0.9997\, T_0$.

Several approximate closed form expressions are also available for the integral given by Equation 7.14, and these expressions are listed in Appendix R. For example, for $\alpha t/a^2 < 0.25$ and at $r/a = 1.0$, the temperature resulting from initial temperature T_0 over $0 \le r' \le a$ is given approximately by Equation R00.9, Appendix R:

$$\frac{T(r,t)}{T_0} = \frac{1}{2}\left[1 - \left(\frac{u}{\pi}\right)^{1/2} - \frac{1}{4\sqrt{\pi}} u^{3/2}\right] \qquad \text{where } u = \frac{\alpha t}{a^2} \tag{7.15}$$

Equation 7.15 and several other approximate expressions for the integral given by Equation 7.14 are summarized in Table 7.2 with their region of application, maximum error, and location in Appendix R. Some of the expressions referenced in Table 7.2 have been found by integration of the expressions for G_{R00} referenced in Table 7.1.

TABLE 7.2
Approximate Closed-Form Expressions for $\int_{r'=0}^{a} G_{R00}(r,t|r',\tau)\, 2\pi r'\, dr'$

Range of Application	Error (%)	Equation Number in Appendix R
$u < 0.1, r/a \ge 1$		R00.7
$u < 0.25, r/a = 1$	1.3	R00.9
$u < 0.01, 0.5 < r/a < 1$	0.03	R00.10
$u \ge 0.25, (r/a)^2/(4u)$ small	−0.016	R00.11

Note: (1) $u = \alpha(t-\tau)/a^2$. (2) As $a \to \infty$, the integral approaches the value 1.0.

In the special case $r = 0$, the temperature in the infinite body in Equation 7.14 may be found in closed form. This temperature is given by Equation 7.14 evaluated at $r = 0$:

$$T(r = 0, t) = \frac{T_0}{4\pi\alpha t} \int_{r'=0}^{a} \exp\left[\frac{-r'^2}{4\alpha t}\right] 2\pi r'\, dr' \tag{7.16a}$$

Note that $I_0(0) = 1$. This integral can be evaluated by a change of variables to $z = r'/(4\alpha t)^{1/2}$ to give

$$T(0, t) = 2T_0 \int_{z=0}^{a/(4\alpha t)^{1/2}} e^{-z^2} z\, dz = T_0 \left(1 - \exp\frac{-a^2}{4\alpha t}\right) \tag{7.16b}$$

This expression is exact for all t. Thus, the temperature at $r = 0$ decays with time as $(1 - e^{-1/4u})$, where $u = \alpha t / a^2$, the time parameter.

7.4 SEPARATION OF VARIABLES FOR RADIAL HEAT FLOW

In this section the separation of variables method is used to show how the Bessel functions arise for cylindrical geometries. For the geometries *RIJ*, $I = 0, 1, 2, 3$, and $J = 1, 2, 3$, the large-time GFs can be derived by this method. For further discussion of the separation of variables method for cylinders, refer to Ozisik (1993, Chapter 3). It is important to note that Ozisik's notation for GFs in cylindrical and spherical coordinates differs from this book by a factor of (2π); that is, G (Ozisik, 1993)/$2\pi = G$ (this volume).

In this section the separation of variables technique will be demonstrated with the *R*01 GF (solid cylinder with temperature boundary conditions), but the method also applies to hollow cylinders. Consider the following initial-value problem for a solid cylinder:

$$\frac{1}{r}\frac{\partial}{\partial r}\left[r\frac{\partial T}{\partial r}\right] = \frac{1}{\alpha}\frac{\partial T}{\partial t} \tag{7.17}$$

$$T(b, t) = 0 \tag{7.18}$$

$$T(0, t) < M \quad \text{where } M \text{ is a finite constant} \tag{7.19}$$

$$T(r, 0) = F(r) \tag{7.20}$$

The initial condition is an arbitrary function of position. There is no energy generation and the boundary condition at $r = b$ is homogeneous. An equivalent boundary condition at $r = 0$ is that the temperature is symmetric, $\partial T / \partial r = 0$. The same solution can be derived with either condition.

The separation of variables technique produces a series solution of the form

$$T(r, t) = \sum_{n=1}^{\infty} T_n(r, t) \tag{7.21}$$

where $T_n(r, t)$ satisfies the differential equation and the boundary conditions. Individually the $T_n(r, t)$ solutions do not satisfy the initial condition given by Equation 7.20, and the series form is used precisely to satisfy the initial condition. The issue of convergence raised by the infinite series in Equation 7.21 is an important one, but for the purpose of this book, the solution converges for heat conduction problems and the results are physically meaningful.

The separation of variables method assumes that the solutions $T_n(r,t)$ have the form

$$T_n(r,t) = \mathbf{R}(r)\,\theta(t) \tag{7.22}$$

That is, the dependence on r and t has been separated into a product of a function of position and a function of time. The function T_n must satisfy the differential equation

$$\frac{\partial^2 T_n}{\partial r^2} + \frac{1}{r}\frac{\partial T_n}{\partial r} = \frac{1}{\alpha}\frac{\partial T_n}{\partial t} \tag{7.23}$$

Substitute Equation 7.22 in Equation 7.23 to give, after some rearrangement,

$$\frac{1}{\mathbf{R}}\left[\frac{\partial^2 \mathbf{R}}{\partial r^2} + \frac{1}{r}\frac{\partial \mathbf{R}}{\partial r}\right] = \frac{1}{\alpha\theta}\frac{\partial \theta}{\partial t} = -\lambda^2 \tag{7.24}$$

The negative constant $-\lambda^2$ is introduced because (a) a function of r set equal to a function of t must both be equal to a constant function, and (b) the negative value is required to give physically meaningful results for $\theta(t)$. Equation 7.24 represents two ordinary differential equations. The equation for $\mathbf{R}$ is

$$\frac{d^2\mathbf{R}}{dr^2} + \frac{1}{r}\frac{d\mathbf{R}}{dr} + \lambda^2\mathbf{R} = 0 \tag{7.25a}$$

This is the Bessel equation of order zero, and the elementary solutions are

$$\mathbf{R}(r) = A\ J_0(\lambda r) + B\ Y_0(\lambda r) \tag{7.25b}$$

where $J_0(\cdot)$ and $Y_0(\cdot)$ are Bessel functions of order zero and A and B are constants. A graph of these functions in shown in Figure 7.2.

The differential equation for $\theta(t)$ is

$$\frac{d\theta}{dt} + \lambda^2\alpha\theta = 0 \tag{7.26a}$$

and the elementary solution is

$$\theta(t) = C\ e^{-\lambda^2\alpha t} \tag{7.26b}$$

where C is a constant. Thus, the solution $T_n(r,t)$ is given by Equation 7.23, with Equations 7.25b and 7.26b:

$$T_n(r,t) = e^{-\lambda_n^2\alpha t}[A_n J_0(\lambda_n r) + B_n Y_0(\lambda_n r)] \tag{7.27}$$

Here new names have been given to the constants λ_n, A_n, and B_n, which must be determined from the boundary conditions and the initial condition for each geometry. Up to this point the analysis applies to both solid and hollow cylinders.

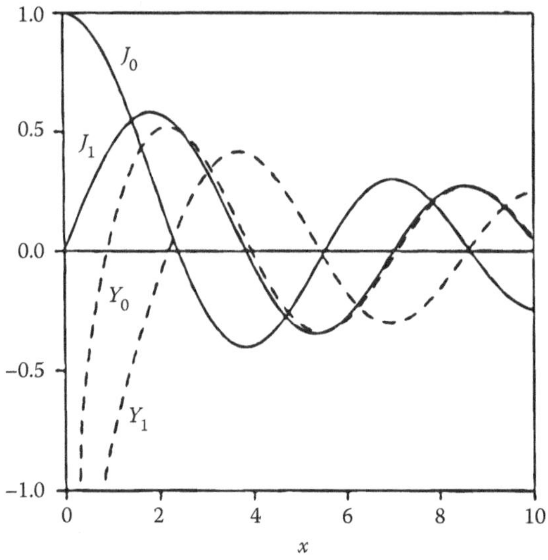

FIGURE 7.2 Bessel functions $J_0(x)$, $Y_0(x)$, $J_1(x)$, $Y_1(x)$.

Next, the boundary conditions are applied to the general solution given by Equation 7.27, so the following analysis applies only to the $R01$ geometry. At $r = 0$, the natural boundary condition given by Equation 7.19 yields

$$\lim_{r \to 0} e^{-\lambda_n^2 \alpha t} [A_n J_0(\lambda_n r) + B_n Y_0(\lambda_n r)] \neq \infty \tag{7.28}$$

In the limit as $r \to 0$, the function $J_0(\lambda_n r)$ approaches one (1), but the function $Y_0(\lambda_n r)$ becomes infinite. The term containing $Y_0(\lambda_n r)$ does not belong in the solution, and Equation 7.28 can be satisfied only by

$$B_n = 0 \tag{7.29}$$

Next the temperature boundary condition at $r = b$ given by Equation 7.18 is applied to the general solution to give

$$T_n(b, t) = 0 = e^{-\lambda_n^2 \alpha t} A_n J_0(\lambda_n b) \qquad \text{or} \qquad 0 = A_n J_0(\lambda_n b) \tag{7.30}$$

The exponential is never zero, so it may be canceled out. The constant A_n cannot be zero or the entire solution will be identically zero, a trivial result. Equation 7.30 is satisfied by choosing

$$J_0(\beta_n) = 0 \tag{7.31}$$

where $\beta_n = \lambda_n b$ are the dimensionless eigenvalues for $n = 1, 2$, and so on. There are an infinite number of eigenvalues that are distinct for each cylinder geometry. The first few eigenvalues are listed in Appendix B for the cylinder cases $R01$, $R02$, $R03$, $R11$, $R12$, and $R22$.

Next, the initial condition must be satisfied. So far, the complete solution has the form

$$T(r,t) = \sum_{n=1}^{\infty} e^{-\beta_n^2 \alpha t / b^2} A_n J_0\left(\frac{\beta_n r}{b}\right) \tag{7.32}$$

The initial condition requires that

$$T(r,0) = F(r) = \sum_{n=1}^{\infty} A_n J_0\left(\frac{\beta_n r}{b}\right) \tag{7.33}$$

The initial condition can be satisfied if an arbitrary function $F(r)$ can be expressed as an infinite series of Bessel functions. In Chapter 4, expansions of arbitrary functions in terms of Fourier sine and cosine series arose from one-dimensional plate cases. Fourier series are a special case of the general theory of orthogonal functions (Wylie and Barrett, 1995). Bessel functions are simply another class of functions for which infinite-series expansions are possible, and the infinite series expansion is needed to satisfy the initial condition.

The orthogonality condition for $J_0(\cdot)$ on $0 \leq r \leq b$ is (Appendix B)

$$\int_0^b J_0\left(\frac{\beta_m r}{b}\right) J_0\left(\frac{\beta_n r}{b}\right) 2\pi r\, dr = \begin{cases} 0 & m \neq n \\ \pi b^2 J_1^2(\beta_n) & m = n \end{cases} \tag{7.34a}$$

To apply the orthogonality condition to find A_n, multiply Equation 7.33 by $J_0(\beta_m r / b)$ and integrate over the volume of the cylinder ($0 \leq r \leq b$):

$$\int_0^b J_0\left(\frac{\beta_m r}{b}\right) F(r)\, 2\pi r\, dr = \int_0^b \sum_{n=1}^{\infty} A_n J_0\left(\frac{\beta_n r}{b}\right) J_0\left(\frac{\beta_m r}{b}\right) 2\pi r\, dr \tag{7.34b}$$

The orthogonality condition applied to the right-hand side of Equation 7.34b gives exactly one nonzero term from the infinite series, at $m = n$. Solving for A_n gives

$$A_n = \frac{1}{\pi b^2 J_1^2(\beta_n)} \int_0^b J_0\left(\frac{\beta_n r'}{b}\right) F(r') 2\pi r'\, dr' \tag{7.35}$$

Note that the subscript n is really a dummy subscript, and any letter could be substituted. Also, the variable of integration has been written as r', as it too is a dummy variable.

Next, replace A_n into the solution given by Equation 7.32 to give the particular solution to the initial-value problem (case *R*01*B*0*T*-),

$$T(r,t) = \sum_{n=1}^{\infty} e^{-\beta_n^2 \alpha t / b^2} \int_{r'=0}^{b} \frac{J_0(\beta_n r' / b) J_0(\beta_n r / b)}{\pi b^2 J_1^2(\beta_n)} F(r') 2\pi r'\, dr' \tag{7.36}$$

After some rearrangement, this solution can be written

$$T(r,t) = \int_{r'=0}^{b} F(r') \left[\frac{1}{\pi b^2} \sum_{n=1}^{\infty} e^{-\beta_n^2 \alpha t / b^2} \frac{J_0(\beta_n r' / b) J_0(\beta_n r / b)}{J_1^2(\beta_n)}\right] 2\pi r'\, dr' \tag{7.37}$$

This is the separation of variables result for the temperature resulting from an arbitrary initial condition on the $R01$ geometry.

Finally, the GF can be deduced from the separation of variables solution by also solving the initial value problem (Equations 7.17 through 7.20) with the GFSE to give

$$T(r,t) = \int_{r'=0}^{b} F(r')\left[G_{R01}(r,t|r',0)\right] 2\pi r'\,dr' \tag{7.38}$$

Equations 7.37 and 7.38 are solutions to the same boundary value problem, and since a boundary value problem has only one unique solution, the expression in brackets in Equation 7.37 must be identically $G_{R01}(r,t|r',0)$, the GF evaluated at $\tau = 0$:

$$G_{R01}(r,t|r',0) = \frac{1}{\pi b^2}\sum_{n=1}^{\infty} e^{-\beta_n^2 \alpha t/b^2}\,\frac{J_0(\beta_n r'/b)\,J_0(\beta_n r/b)}{J_1^2(\beta_n)} \tag{7.39}$$

The last step in finding the GF from the separation of variables solution is to replace $(t-0)$ in Equation 7.39 by $(t-\tau)$. Recall that the time dependence of all GFs is in the form $(t-\tau)$. Then,

$$G_{R01}(r,t|r',\tau) = \frac{1}{\pi b^2}\sum_{n=1}^{\infty} e^{-\beta_n^2 \alpha (t-\tau)/b^2}\,\frac{J_0(\beta_n r'/b)\,J_0(\beta_n r/b)}{J_1^2(\beta_n)} \tag{7.40}$$

This GF is also listed in Appendix R.

This method for finding the large-time GFs can be used on all of the solid cylinder and hollow cylinder cases, denoted G_{RIJ} for which $I = 0,1,2,3$ and $J = 1,2,3$. It is not necessary to derive these GFs, however, since they are listed in Appendix R.

7.5 LONG SOLID CYLINDER

Some worked examples are next discussed for the temperature in long solid cylinders. Time partitioning is introduced on a case-by-case basis because the choice of an appropriate small-time GF depends on time, on geometry, and on location in the cylinder.

7.5.1 INITIAL CONDITIONS

Example 7.1: Solid Cylinder with Zero Surface Temperature—*R01B0T*-Case

Find the temperature in a solid cylinder, $0 \le r \le b$, with initial temperature $F(r)$ and a boundary temperature fixed at $T = 0$.

Solution

This is the *R01B0T*- case and it was examined in Section 7.4. The temperature is given by Equation 7.37, where the expression in brackets is the GF $G_{R01}(r,t|r',0)$. The eigenvalues β_n are defined by the eigencondition $J_0(\beta_n) = 0$, and the first 10 values of β_n are listed in Appendix B. The integral on r' acts on just a portion of Equation 7.37:

$$T(r,t) = \frac{1}{\pi b^2} \sum_{n=1}^{\infty} e^{-\beta_n^2 \alpha t / b^2} \frac{J_0(\beta_n r / b)}{J_1^2(\beta_n)} \int_{r'=0}^{b} F(r')\, J_0\left(\frac{\beta_n r'}{b}\right) 2\pi r'\, dr' \tag{7.41}$$

(a) *Case R01B0T1.* If the initial temperature is uniform, $F(r) = T_0$, then the above integral is given by

$$\int_{r'=0}^{b} J_0\left(\frac{\beta_n r'}{b}\right) 2\pi r' dr' = \frac{2\pi b^2 J_1(\beta_n)}{\beta_n} - 0 \tag{7.42}$$

and the temperature resulting from initial temperature T_0 becomes (case $R01B0T1$)

$$T(r,t) = 2T_0 \sum_{n=1}^{\infty} e^{-\beta_n^2 \alpha t / b^2} \frac{J_0(\beta_n r / b)}{\beta_n J_1(\beta_n)} \tag{7.43}$$

For $\alpha t / b^2$ small, the temperature near the center of the cylinder (at $r = 0$) remains at T_0, because the effect of the surface temperature has not yet penetrated to the center of the cylinder.

(b) *Case R01B0T5.* For the initial condition

$$F(r) = \begin{cases} T_0 & 0 \le r \le a \\ 0 & a < r \le b \end{cases}$$

the temperature is given by

$$T(r,t) = 2\,T_0 \frac{a}{b} \sum_{n=1}^{\infty} e^{-\beta_n^2 \alpha t / b^2} \frac{J_1(\beta_n a / b)\, J_0(\beta_n r / b)}{\beta_n J_1^2(\beta_n)} \tag{7.44}$$

This solution converges efficiently for large values of time. Small time expressions for the temperature for the case when $a / b \neq 1$ can be found by approximating the cylinder as an infinite body. Initially, heat diffuses outward from the point $r = a$, and it takes a little time before the diffusion is influenced by the zero-temperature boundary at $r = b$. During this small time period, the temperature distribution is identical to that in an infinite body with the same initial condition. Thus, the appropriate early-time GF is G_{R00}, and the expressions referenced in Table 7.2 (integral of G_{R00}) may be used to find the temperature at small times. The criterion for small time is $\alpha t / (b - a)^2$ small (<0.01) because it is the distance between the initial temperature region and the boundary, $(b - a)$, that determines the time span of infinite-body behavior.

Example 7.2: Solid Cylinder with Surface Convection—*R*03*B*0*T*1 Case

Find the transient temperature in a cylinder initially at temperature T_0 that is suddenly quenched in a large tank of fluid at temperature T_∞ with heat transfer coefficient h.

Solution

The boundary and initial conditions are given by

$$T \text{ is finite as } r \to 0$$

$$-k \frac{\partial T(b,t)}{\partial r} = h(T(b,t) - T_\infty)$$

$$T(r,0) = T_0 \tag{7.45}$$

This boundary value problem has two nonhomogeneous conditions resulting from the two temperatures T_0 and T_∞. Two integrals from the GFSE are needed to directly describe this problem; however, one nonhomogeneous condition can be removed by defining a new variable $(T - T_\infty)$. The new boundary and initial conditions are given by

$$(T - T_\infty) \text{ is finite as } r \to 0$$

$$-k\frac{\partial(T - T_\infty)}{\partial r}\bigg|_{r=b} - h(T|_{r=b} - T_\infty) = 0$$

$$T(r,0) - T_\infty = (T_0 - T_\infty) \tag{7.46}$$

Note that the boundary condition at $r = b$ is now homogeneous in terms of variable $(T - T_\infty)$. Variable $T - T_0$ could have been chosen, but it would result in a form of the solution less well suited to numerical evaluation at small values of dimensionless time. The temperature in the cylinder is now given by the initial condition term of the GFSE:

$$T(r,t) - T_\infty = \int_{r'=0}^{b} (T_0 - T_\infty)\, G_{R03}(r,t|b,0)\, 2\pi r'\, dr' \tag{7.47}$$

Using the $R03$ GF listed in Appendix R gives

$$T(r,t) - T_\infty = (T_0 - T_\infty)\int_{r'=0}^{b} \sum_{n=1}^{\infty} e^{-\beta_n^2\alpha t/b^2}$$

$$\times \frac{\beta_n^2\, J_0(\beta_n r/b)\, J_0(\beta_n r'/b)}{\pi b^2(B^2 + \beta_n^2)J_0^2(\beta_n)} 2\pi r'\, dr' \tag{7.48}$$

where $B = hb/k$ (the Biot number) and eigenvalues β_n are the roots of

$$-\beta_n J_1(\beta_n) + B\, J_0(\beta_n) = 0 \tag{7.49}$$

Values of β_n for several values of B are given in Carslaw and Jaeger (1959).

The integral on r' in Equation 7.48 was given earlier in Example 7.1, so the temperature in the cylinder is given by

$$T(r,t) - T_\infty = 2(T_0 - T_\infty)\sum_{n=1}^{\infty} e^{-\beta_n^2\alpha t/b^2} \frac{\beta_n J_1(\beta_n)\, J_0(\beta_n r/b)}{(B^2 + \beta_n^2)\, J_0^2(\beta_n)} \tag{7.50}$$

This temperature is plotted in Figure 7.3 for specific case $hb/k = 5$. Note that the slope at $r = b$ varies with time according to the temperature there.

7.5.2 Boundary Conditions

Example 7.3: Solid Cylinder with Elevated Surface Temperature—*R*01*B*1*T*0 Case

Find the temperature in a solid cylinder, $0 \le r \le b$, that has zero initial condition and has temperature T_0 suddenly applied at boundary $r = b$.

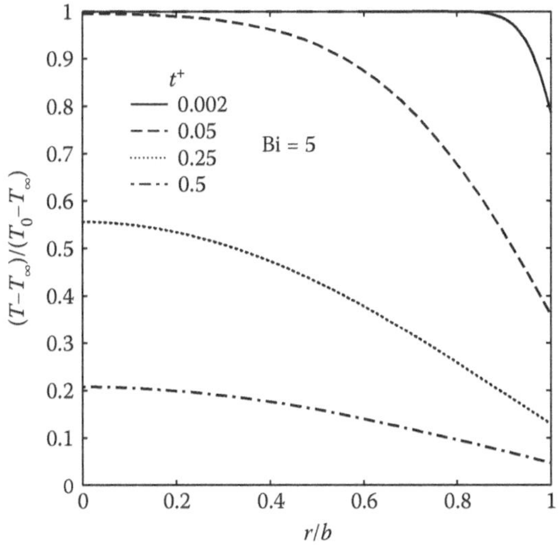

FIGURE 7.3 Solid cylinder initially at T_0 and cooled by surface convection with $hb/k = 5$ and fluid temperature T_∞. This is case $R03B0T1$ discussed in Example 7.2.

Standard Solution

The temperature resulting from a boundary temperature is given by the last term of Equation 7.2 with $r_j = b$:

$$T(r,t) = -\alpha \int_{\tau=0}^{t} T_0 \left.\frac{\partial G}{\partial n'}\right|_{r'=b} 2\pi b\, d\tau \tag{7.51}$$

The required $R01$ GF and its derivative $\partial G_{R01}/\partial n'$ is given in Appendix R, so the integral in Equation 7.51 is given by

$$\alpha T_0 \int_{\tau=0}^{t} \frac{2}{b^2} \sum_{n=1}^{\infty} e^{-\beta_n^2 \alpha (t-\tau)/b^2} \frac{\beta_n J_0(\beta_n r/b)}{J_1(\beta_n)}\, d\tau \tag{7.52}$$

and the eigenvalues are given by $J_0(\beta_m) = 0$. The integral on τ is easily evaluated to give

$$T(r,t) = 2T_0 \sum_{n=1}^{\infty} (1 - e^{-\beta_n^2 \alpha t/b^2}) \frac{J_0(\beta_n r/b)}{\beta_n J_1(\beta_n)} \tag{7.53}$$

This solution suffers from poor numerical convergence which can be made clear by writing the solution as the sum of two series,

$$T(r,t) = 2T_0 \sum_{n=1}^{\infty} \frac{J_0(\beta_n r/b)}{\beta_n J_1(\beta_n)} - 2T_0 \sum_{n=1}^{\infty} e^{-\beta_n^2 \alpha t/b^2} \frac{J_0(\beta_n r/b)}{\beta_n J_1(\beta_n)} \tag{7.54}$$

The first series converges slowly and it does not depend on the dimensionless time. Time partitioning could be used to find a temperature expression that converges more efficiently, but in this case there is a simple alternative solution.

Alternative Solution

The alternative solution method discussed in Chapter 3 is useful for improving the numerical convergence of the temperature driven by nonhomogeneous boundary conditions. In this case, the steady-state solution is simply $T(r,\ t \to \infty) = T_0$. Let the known solution be $T^*(r,t) = T_0$, and let the unknown temperature be given by $T(r,t) = T^*(r,t) + T'(r,t)$. Temperature T^* is of course also a solution to the transient energy equation. A new solution is now sought for the temperature $T'(r,t) = T(r,t) - T^*(r,t)$ subject to the following boundary value problem:

$$\begin{aligned}&\frac{\partial^2 T'}{\partial r^2} + \frac{1}{r}\frac{\partial T'}{\partial r} = \frac{1}{\alpha}\frac{\partial T'}{\partial t}\\ &T'(b,t) = T(b,t) - T^*(b,t) = T_0 - T_0 = 0\\ &T'(r,0) = T(r,0) - T^*(r,0) = 0 - T_0\end{aligned} \tag{7.55}$$

Then the alternative solution is given by Equation 3.66:

$$T(r,t) = T_0 + \int_{r'=0}^{b} (-T_0) \times \left[\sum_{n=1}^{\infty} e^{-\alpha\beta_n^2 t/b^2} \frac{J_0(\beta_n r'/b)\, J_0(\beta_n r/b)}{\pi b^2 J_1^2(\beta_n)}\right] 2\pi r'\, dr' \tag{7.56}$$

Effectively, the boundary heating problem has been transformed into an initial heating problem, and this integral has been solved previously as in Example 7.1:

$$T(r,t) = T_0 - T_0 \sum_{n=1}^{\infty} e^{-\alpha\beta_n^2 t/b^2} \frac{J_0(\beta_n r/b)}{\beta_n J_1(\beta_n)} \tag{7.57}$$

This expression converges better than Equation 7.54 for all values of $\alpha t/b^2$.

Example 7.4: Solid Cylinder with Heating at the Surface—*R02B1T0* Case

A cylinder whose initial temperature is zero is heated by a suddenly applied surface heat flux q_0. Find (a) the surface temperature on the cylinder at early time, and (b) the spatial average temperature in the cylinder at any time.

Solution

The boundary conditions are given by

$$\begin{aligned}&\frac{\partial T(0,t)}{\partial r} = 0 \qquad \text{(symmetry condition)}\\ &-k\frac{\partial T(b,t)}{\partial r} = q_0 \qquad T(r,0) = 0\end{aligned} \tag{7.58}$$

(a) *Surface temperature at early time.* The surface temperature is given by the GFSE evaluated at $r = b$:

$$T(b,t) = \alpha \int_{\tau=0}^{t} \frac{q_0}{k} G_{R02}(b,t|b,\tau)\, 2\pi b\, d\tau \tag{7.59}$$

For small times, only the small-cotime form of the GF is needed to find the temperature. For the G_{R02} small-cotime form, Equation R02.5 from Appendix R is appropriate:

$$G_{R02}(b, t|b, \tau) \approx \frac{1}{2\pi b^2}\left[\frac{1}{\sqrt{\pi}}(u)^{-1/2} + \frac{1}{2} + \frac{3}{4\sqrt{\pi}}(u)^{1/2} + \frac{3}{8}u\right] \tag{7.60}$$

where $u = \alpha(t-\tau)/b^2 < 0.1$. Then the surface temperature can be written in terms of the integral given by Equation 7.59 with a change of variable to $u = \alpha(t-\tau)/b^2$:

$$T(b, t) \cong \int_{u=0}^{\alpha t/b^2} \frac{q_0 b}{k}\left[\frac{1}{\sqrt{\pi}}(u)^{-1/2} + \frac{1}{2} + \frac{3}{4\sqrt{\pi}}(u)^{1/2} + \frac{3}{8}u\right] du \tag{7.61}$$

and the time integral can be evaluated to give

$$T(b, t) \cong \frac{q_0 b}{k}\left[\frac{2}{\sqrt{\pi}}\left(\frac{\alpha t}{b^2}\right)^{1/2} + \frac{1}{2}\left(\frac{\alpha t}{b^2}\right) + \frac{1}{2\sqrt{\pi}}\left(\frac{\alpha t}{b^2}\right)^{3/2} + \frac{3}{16}\left(\frac{\alpha t}{b^2}\right)^2\right] \tag{7.62}$$

for $\alpha t/b^2 < 0.1$. In the above expression, the first term inside the brackets

$$\frac{2}{\sqrt{\pi}}\left(\frac{\alpha t}{b^2}\right)^{1/2}$$

is the same as the temperature on a plane wall caused by a suddenly applied heat flux. From this perspective, the next term $(\alpha t/2b^2)$ is the first correction term for the curvature of the cylinder wall (Beck et al., 1985).

(b) *Spatial average temperature.* The spatial average temperature in the cylinder may be found from an overall energy balance on the cylinder

$$q_{in} = q_{storage} \quad \text{or} \quad q_0 = \rho c \pi b^2 \frac{\partial T_{av}}{\partial t} \tag{7.63}$$

This may be integrated from the initial temperature of zero to find

$$T_{av}(t) = \frac{q_0 t}{\rho c \pi b^2} \tag{7.64}$$

Note that the spatial average temperature increases linearly with time. The same behavior occurs in a body with uniform energy generation if the boundary is insulated ($R02B0T0G1$); in both cases there is heating specified and no heat loss.

7.5.3 Volume Energy Generation

In this section two examples are given of long solid cylinders with volume energy generation.

Example 7.5: Solid Cylinder with Uniform Energy Generation—*R01B0T0G1* Case

A cylinder is initially at zero temperature and the boundary at $r = b$ is maintained at $T = 0$. Find the temperature in the cylinder resulting from a uniform volume energy generation g_0 (W/m^3).

Solution

The temperature is given by the GFSE:

$$T(r,t) = \int_{\tau=0}^{t} \int_{r'=0}^{b} \frac{\alpha}{k} g_0 \, G_{R01}(r,t|r',\tau) \, 2\pi r' \, dr' \, d\tau \tag{7.65}$$

Using the large-cotime form of G_{R01} from Appendix R, the temperature is given by

$$T(r,t) = 2\frac{g_0 b^2}{k} \sum_{n=1}^{\infty} \left(1 - e^{-\beta_n^2 \alpha t / b^2}\right) \frac{J_0(\beta_n r / b)}{\beta_n^3 J_1(\beta_n)} \tag{7.66}$$

where the eigenvalues are roots of the equation $J_0(\beta_n) = 0$.

At large time values the convergence of Equation 7.66 is controlled by a term that does not depend on time. This is the steady solution, and it can be found independently (by solving the steady boundary value problem) to give a better expression for numerical evaluation. Equation 7.66 can be written

$$T(r,t) = \frac{g_0 b^2}{4k}\left[1 - \left(\frac{r}{b}\right)^2\right] - 2\frac{g_0 b^2}{k} \sum_{n=1}^{\infty} e^{-\beta_n^2 \alpha t / b^2} \frac{J_0(\beta_n r / b)}{\beta_n^3 J_1(\beta_n)} \tag{7.67}$$

The same result can also be found from the alternate GFSE.

The above series expression converges rapidly for large time ($\alpha t / b^2 > 0.025$). At early time, the interior of the cylinder will behave like an infinite body and the zero-temperature boundary at $r = b$ will have only a local influence. To find a form of the temperature that is numerically efficient at small time, refer to Appendix R for a suitable small-cotime form of G_{R01}.

Example 7.6: Solid Cylinder with Nonuniform Energy Generation—*R02B0T0Gr5* Case

Consider the solid cylinder $0 \leq r \leq b$ initially at zero temperature with an insulated boundary. The cylinder is heated by volume energy generation

$$g(r',\tau) = \begin{cases} 0 & 0 \leq r \leq a \\ g_0 & a \leq r \leq b \end{cases} \tag{7.68}$$

where g_0 (W/m^3) is the energy generation rate. The energy generation is zero deep inside the cylinder and it has the value g_0 near the surface of the cylinder. This geometry approximately describes microwave heating of food or nuclear radiation heating of reactor control rods (approximately, because actual radiation heating is attenuated inside the body). Find the temperature after a long period.

Solution

This is the $R02$ geometry with energy generation. The GF equation for this case is given by the second term of Equation 7.2, and G_{R02} is listed in Appendix R. The temperature in the cylinder is

$$T(r,t) = \frac{\alpha}{k} \int_{\tau=0}^{t} \int_{r'=a}^{b} \frac{g_0}{\pi b^2} \left[1 + \sum_{n=1}^{\infty} e^{-\beta_n^2 \alpha (t-\tau)/b^2} \frac{J_0(\beta_n r / b) \, J_0(\beta_n r' / b)}{J_0^2(\beta_n)}\right] 2\pi r' dr' d\tau \tag{7.69}$$

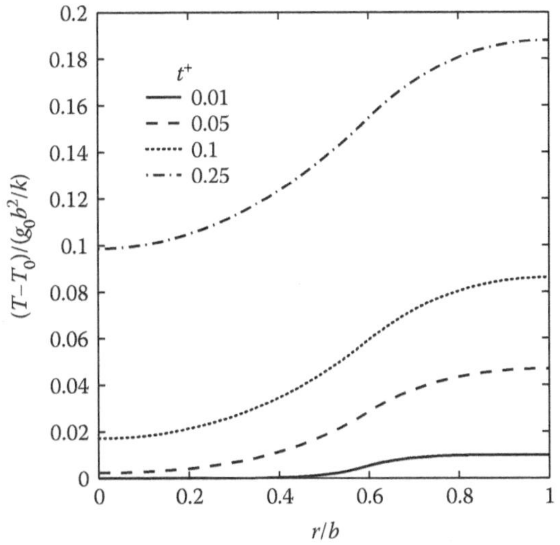

FIGURE 7.4 Solid cylinder with internal energy generation which is nonzero only in the region $a < r < b$ and with $a/b = 0.6$. This is case $R02B0T0G(r5)$ discussed in Example 7.6.

where the eigenvalues β_n are defined by $J_1(\beta_n) = 0$. Note that the GF contains an additive constant that may be interpreted as the $n = 0$ term of the series. The integrals can be evaluated to give

$$T(r,t) = \frac{g_0 b^2}{k}\left[1 - \left(\frac{a}{b}\right)^2\right]\frac{\alpha t}{b^2} + 2\frac{g_0 ab}{k}\sum_{n=1}^{\infty} e^{-\beta_n^2 \alpha t/b^2}\frac{J_0(\beta_n r/b)\,J_1(\beta_n a/b)}{\beta_n^3 J_0^2(\beta_n)}$$
$$- 2\frac{g_0 ab}{k}\sum_{n=1}^{\infty}\frac{J_0(\beta_n r/b)\,J_1(\beta_n a/b)}{\beta_n^3 J_0^2(\beta_n)} \tag{7.70}$$

This temperature is plotted in Figure 7.4 for the specific case $a/b = 0.6$.

The above solution contains three pieces, one of which does not depend on time. At $t = 0$, the temperature is zero as required because the second and third pieces cancel out, not because each piece is zero. The first piece of the above solution is the spatial average temperature in the cylinder, $T_{av}(t)$. This can be demonstrated by integrating Equation 7.70 over the volume of the cylinder:

$$T_{av}(t) \equiv \frac{1}{\pi b^2}\int_{r=0}^{b} T(r,t)\,2\pi r\,dr = \frac{g_0 b^2}{k}\left[1 - \left(\frac{a}{b}\right)^2\right]\frac{\alpha t}{b^2} \tag{7.71}$$

The average temperature increases with time because the heat that is added has no place to go (the boundary is insulated). There is no steady-state temperature.

In the limit as $a/b \to 0$, the cylinder will be heated uniformly over its volume. In this case the temperature given by Equation 7.70 reduces to the spatial average temperature given by Equation 7.71. No heat can escape at the boundary, and every point in the cylinder is heated equally.

The second piece of the solution given by Equation 7.70 is

$$2\frac{g_0 ab}{k}\sum_{n=1}^{\infty} e^{-\beta_n^2 \alpha t/b^2} \frac{J_0(\beta_n r/b)\, J_1(\beta_n a/b)}{\beta_n^3 J_0^2(\beta_n)} \tag{7.72}$$

This series decreases exponentially over time. For $\alpha t/b^2 > 0.3$ or so, this series becomes negligible and the spatial distribution of the temperature stops changing. The temperature is said to be quasisteady, because although the shape of the temperature distribution is fixed, the magnitude of the temperature distribution increases linearly with time according to the average temperature term.

The third piece of the solution given by Equation 7.70 is the quasisteady temperature distribution which does not depend on time,

$$-2\frac{g_0 ab}{k}\sum_{n=1}^{\infty} \frac{J_0(\beta_n r/b)\, J_1(\beta_n a/b)}{\beta_n^3 J_0^2(\beta_n)} \tag{7.73}$$

For large time (say $\alpha t/b^2 > 0.3$), this piece of the solution describes the shape of the temperature distribution in the form of deviations from the average temperature given by Equation 7.71. That is, the temperature is above average near $r = b$ (the heated region), and the temperature is below average near $r = 0$ (the unheated region).

7.6 HOLLOW CYLINDER

Two examples are given for the temperature in hollow cylinders. Compared to solid cylinders, hollow cylinders have one more physical boundary and consequently the GF and the eigenconditions are more complex; however, all of the analytical techniques for solid cylinders also apply to hollow cylinders. Another approach for cylinders is the Galerkin-based GF developed in Chapters 10 and 11. For hollow cylinders, it may be possible to obtain numerical results more easily from the Galerkin-based GFs than from the analytical GFs discussed in this section.

Example 7.7: Hollow Cylinder with Zero Surface Temperature—*R*11*B*00*T*1 Case

Consider the hollow cylinder $a \leq r \leq b$ with uniform initial temperature T_0. Find the temperature for $t > 0$ for the boundaries fixed at zero temperature.

Solution

The temperature due to an initial condition is given by the GFSE:

$$T(r,t) = \int_{r'=a}^{b} G_{R11}(r,t|r',0)\, T_0\, 2\pi r'\, dr' \tag{7.74}$$

Note that the integral is evaluated over the hollow cylinder, $a \leq r' \leq b$. The large-time GF is listed in Appendix R to give

$$T(r,t) = T_0 \int_{r'=a}^{b} \frac{\pi}{4a^2} \sum_{m=1}^{\infty} e^{-\beta_m^2 \alpha t/a^2} \frac{\beta_m^2 J_0^2(\beta_m)}{J_0^2(\beta_m) - J_0^2(\beta_m b/a)} \mathbf{R}(r)\mathbf{R}(r')\, 2\pi r'\, dr' \tag{7.75}$$

TABLE 7.3
Bessel Function Integral over r' of $\int J_0\left(\beta_m \frac{r'}{b}\right) F(r')\, r'\, dr'$

$F(r')$	Integral $\left(\text{let } x = \beta_m \frac{r'}{b}\right)$
1	$\left(\frac{b}{\beta_m}\right)^2 x J_1(x)$
$\frac{r'}{b}$	$\frac{b^2}{\beta_m^3}[x^2 J_1(x) + x J_0(x) - \int J_0(x)\, dx]$
$\left(\frac{r'}{b}\right)^2$	$\frac{b^2}{\beta_m^4}\left[(x^3 - 4x)\, J_1(x) + 2x^2\, J_0(x)\right]$
$\ln \frac{r'}{b}$	$\frac{b^2}{\beta_m^2}\left[J_0(x) + x \ln\left(\frac{r'}{b}\right) J_1(x)\right]$

TABLE 7.4
Bessel Function Integral over r' of $\int Y_0\left(\beta_m \frac{r'}{b}\right) F(r')\, r'\, dr'$

$F(r')$	Integral $\left(\text{let } x = \beta_m \frac{r'}{b}\right)$
1	$\left(\frac{b}{\beta_m}\right)^2 x Y_1(x)$
$\frac{r'}{b}$	$\frac{b^2}{\beta_m^3}[x^2 Y_1(x) + x Y_0(x) - \int Y_0(x)\, dx]$
$\left(\frac{r'}{b}\right)^2$	$\frac{b^2}{\beta_m^4}\left[(x^3 - 4x)\, Y_1(x) + 2x^2\, Y_0(x)\right]$
$\ln \frac{r'}{b}$	$\frac{b^2}{\beta_m^2}\left[Y_0(x) + x \ln\left(\frac{r'}{b}\right) Y_1(x)\right]$

where

$$\mathbf{R}(r) = J_0\left(\frac{\beta_m r}{a}\right) Y_0\left(\frac{\beta_m b}{a}\right) - J_0\left(\frac{\beta_m b}{a}\right) Y_0\left(\frac{\beta_m r}{a}\right) \tag{7.76}$$

and where the eigenvalues β_m satisfy

$$J_0(\beta_m) Y_0\left(\frac{\beta_m b}{a}\right) - J_0\left(\frac{\beta_m b}{a}\right) Y_0(\beta_m) = 0 \tag{7.77}$$

The first five eigenvalues are listed in Appendix B for various values of b/a (various cylinder geometries). The integral on r' operates only on $J_0(\beta_m r'/a)$ and $Y_0(\beta_m r'/a)$, and the integral can be carried out with the first integrals from Tables 7.3 and 7.4

to give (see also Appendix B)

$$T(r,t) = T_0 \frac{\pi^2}{2a} \sum_{m=1}^{\infty} e^{-\beta_m^2 \alpha t/a^2} \frac{\beta_m J_0^2(\beta_m)}{J_0^2(\beta_m) - J_0^2(\beta_m b/a)} \mathbf{R}(r)$$
$$\times \left[bJ_1\left(\beta_m \frac{b}{a}\right) Y_0\left(\beta_m \frac{b}{a}\right) - aJ_1(\beta_m) Y_0\left(\beta_m \frac{b}{a}\right) \right.$$
$$\left. - bJ_0\left(\beta_m \frac{b}{a}\right) Y_1\left(\beta_m \frac{b}{a}\right) + aJ_0\left(\beta_m \frac{b}{a}\right) Y_1(\beta_m) \right] \quad (7.78)$$

This is the large-time form of the temperature in the hollow cylinder where $\mathbf{R}(r)$ is given by Equation 7.76. There are four Bessel functions involved: J_0, Y_0, J_1, and Y_1. For large values of $\alpha t/a^2$ only a few terms of the series are needed for accurate numerical values.

Example 7.8: Hollow Cylinder Insulated Inside—*R*21*B*00*T*- Case

Consider the hollow cylinder $a \leq r \leq b$ has a steady temperature distribution T_i due to steady heating at the boundary $r = a$ and a zero temperature $r = b$. That is,

$$-k \frac{\partial T_i(a)}{\partial r} = q_0 \quad (7.79)$$

$$T_i(b) = 0 \quad (7.80)$$

Suppose that for $t > 0$, the heat flux at $r = a$ suddenly becomes zero (the boundary becomes insulated). Find (a) the initial temperature distribution, and (b) the transient temperature due to the change in the heating at the boundary $r = a$.

Solution

(a) *Initial temperature.* The initial temperature may be found from the steady GFSE, Equation 3.94, in radial cylindrical coordinates:

$$T_i(r) = \frac{q_0}{k} G(r|r' = a)\, 2\pi a \quad (7.81)$$

The steady GF is given by Table R.1 in Appendix R:

$$G_{R21}(r|r') = \begin{cases} \dfrac{1}{2\pi} \ln\left(\dfrac{b}{r'}\right) & r < r' \\ \dfrac{1}{2\pi} \ln \dfrac{b}{r} & r > r' \end{cases} \quad (7.82)$$

When G_{R21} is substituted into Equation 7.81, the steady temperature is given by

$$T_i(r) = \frac{q_0 a}{k} \ln \frac{b}{r} = \frac{q_0 a}{k} \left(\ln \frac{b}{a} - \ln \frac{r}{a} \right) \quad (7.83)$$

This form of the steady temperature is convenient for part (b) discussed below. The steady temperature may also be found by direct integration of the steady energy equation

$$\frac{1}{r} \left[\frac{d}{dr} \left(r \frac{dT_i}{dr} \right) \right] = 0 \quad (7.84)$$

with boundary conditions given by Equations 7.79 and 7.80.

(b) *Transient temperature.* The transient temperature is given by the initial condition form of the GF equation, Equation 7.2, because the boundary conditions are homogeneous:

$$T(r,t) = \int_{r'=a}^{b} G_{R21}(r,t|r',0)\, T_i(r')\, 2\pi r' dr' \tag{7.85}$$

The large-time form of the GF is given in Appendix R, and the initial temperature is given by Equation 7.84 to give,

$$T(r,t) = \int_{r'=a}^{b} \frac{\pi}{4a^2} \sum_{m=1}^{\infty} e^{-\beta_m^2 \alpha t/a^2} \frac{\beta_m^2 J_1^2(\beta_m)}{J_1^2(\beta_m) - J_0^2(\beta_m b/a)} \mathbf{R}(r)\mathbf{R}(r')$$
$$\times \frac{q_0 a}{k}\left(\ln\frac{a}{b} - \ln\frac{r'}{a}\right) 2\pi r'\, dr' \tag{7.86}$$

where

$$\mathbf{R}(r) = J_0\left(\frac{\beta_m r}{a}\right) Y_0\left(\frac{\beta_m b}{a}\right) - J_0\left(\frac{\beta_m b}{a}\right) Y_0\left(\frac{\beta_m r}{a}\right) \tag{7.87}$$

and where the eigenvalues β_m satisfy

$$J_1(\beta_m)\, Y_0\left(\frac{\beta_m b}{a}\right) - J_0\left(\frac{\beta_m b}{a}\right) Y_1(\beta_m) = 0 \tag{7.88}$$

The integral in Equation 7.86 contains two basic forms:

$$\int W_0\left(\frac{\beta_m r'}{a}\right) r'\, dr' \quad \text{and} \quad \int W_0\left(\frac{\beta_m r'}{a}\right)\left(\ln\frac{r'}{a}\right) r'\, dr'$$

where $W_0(\cdot)$ is either J_0 or Y_0. These integrals are listed in Table 7.3 and Table 7.4 and can also be written as

$$\int W_0\left(\beta_m \frac{r'}{a}\right) r'\, dr' = \frac{a^2}{\beta_m^2}\left(\beta_m \frac{r'}{a}\right) W_1\left(\beta_m \frac{r'}{a}\right) \tag{7.89}$$

$$\int W_0\left(\beta_m \frac{r'}{a}\right)\left(\ln\frac{r'}{a}\right) r'\, dr' = \frac{a^2}{\beta_m^2}\left[W_0\left(\beta_m \frac{r'}{a}\right) + \beta_m \frac{r'}{a}\left(\ln\frac{r'}{a}\right)\right.$$
$$\left.\times\, W_1\left(\beta_m \frac{r'}{a}\right)\right] \tag{7.90}$$

where W_1 is either J_1 or Y_1. After some simplification involving the eigencondition Equation 7.88, to cancel some terms, Equation 7.86 may be written

$$T(r,t) = \pi^2 \frac{q_0 b}{k} \sum_{m=1}^{\infty} e^{-\beta_m^2 \alpha t/a^2} \frac{\beta_m J_1^2(\beta_m)}{J_1^2(\beta_m) - J_0^2(\beta_m b/a)} \mathbf{R}(r)$$
$$\times \ln\frac{a}{b}\left[J_1\left(\beta_m \frac{b}{a}\right) Y_0\left(\beta_m \frac{b}{a}\right) - J_0\left(\beta_m \frac{b}{a}\right) Y_1\left(\beta_m \frac{b}{a}\right)\right] \tag{7.91}$$

where $\mathbf{R}(r)$ is given by Equation 7.87. This expression involves four Bessel functions, J_0, Y_0, J_1, and Y_1. Only a few terms of the series are needed for $\alpha t/a^2$ large. For $\alpha t/a^2$ very small, the analysis can be repeated with an approximate small-time GF, such as $(1/2\pi a)G_{X20}$ for $r \approx a$ (refer to Example 7.3).

7.7 INFINITE BODY WITH A CIRCULAR HOLE

The radial heat flow in an infinite body containing a circular hole is discussed in this section. Some of the applications for this heat transfer geometry are buried pipes, oil wells, and a heated wire in a quiescent fluid at early time. The GFs for cases numbered *R*10, *R*20, and *R*30 are available in Appendix R. Do not confuse these numbers with the solid cylinder numbers *R*01, *R*02, and *R*03.

The GFs discussed below are derived from Laplace transformation methods (Carslaw and Jaeger, 1959, p. 334). The GFs for the infinite body with a hole involve integrals over a continuous range of eigenvalues instead of a series over discrete eigenvalues. Although the GFs are more complex, they are used to find the temperature as any other cylindrical GF.

Example 7.9: Infinite Body with a Circular Hole and Specified Surface Temperature—*R*10*B*1*T*0 Case

An infinite body bounded internally by the circular hole $r = a$ has an initial temperature of zero. At $t > 0$ the surface $r = a$ has a fixed temperature T_0. Find the temperature in the body for $t > 0$.

Solution

The GF equation for radial flow of heat, Equation 7.2, applies to this case as

$$T(r, t) = -\alpha \int_{\tau=0}^{t} T_0 \frac{\partial G_{R10}}{\partial n'}(r, t|a, \tau)\, 2\pi a\, d\tau \tag{7.92}$$

The derivative $\partial G_{R10} / \partial n'$ is given in Appendix R as

$$-\left.\frac{\partial G_{R10}}{\partial n'}\right|_{r'=a} = -\frac{1}{\pi^2 a^3} \int_{\beta=0}^{\infty} e^{-\beta^2 \alpha(t-\tau)/a^2} \times \frac{\beta\left[J_0\left(\beta\frac{r}{a}\right) Y_0(\beta) - Y_0\left(\beta\frac{r}{a}\right) J_0(\beta)\right]}{J_0^2(\beta) + Y_0^2(\beta)}\, d\beta \tag{7.93}$$

Then, replace Equation 7.93 into Equation 7.92 to find the temperature:

$$T(r, t) = T_0 \frac{2\alpha}{\pi a^2} \int_{\tau=0}^{t} \int_{\beta=0}^{\infty} e^{-\beta^2 \alpha(t-\tau)a^2} \times \frac{\beta\left[J_0\left(\beta\frac{r}{a}\right) Y_0(\beta) - Y_0\left(\beta\frac{r}{a}\right) J_0(\beta)\right]}{J_0^2(\beta) + Y_0^2(\beta)}\, d\beta\, d\tau \tag{7.94}$$

The time integral may be evaluated to give

$$T(r, t) = T_0 \frac{2}{\pi} \int_{\beta=0}^{\infty} \left[1 - e^{-\beta^2 \alpha t / a^2}\right] \times \frac{\left[J_0\left(\beta\frac{r}{a}\right) Y_0(\beta) - Y_0\left(\beta\frac{r}{a}\right) J_0(\beta)\right]}{\beta[J_0^2(\beta) + Y_0^2(\beta)]}\, d\beta \tag{7.95}$$

The integral on β must be evaluated numerically, but the temperature is bounded by T_0 at $r = a$ and the temperature decays to zero as $r \to \infty$ (as $\alpha t/r^2 \to 0$). At steady state defined by $\alpha t/r^2 \to \infty$, the temperature approaches T_0 everywhere. A plot of the temperature given by Equation 7.95 is given by Carslaw and Jaeger (1959, p. 337). An approximate small-time form of this solution is also listed by Carslaw and Jaeger (1959) on p. 336; another approximate small-time solution is $G_{R10} \approx 1/(2\pi a)\, G_{X10}$ (refer to Example 7.3).

Example 7.10: Infinite Body with a Circular Hole and Specified Surface Heat Flux—*R20B-T0* Case

An infinite body bounded internally by the circular hole $r = a$ has a zero initial temperature. At $t \geq 0$ the surface $r = a$ sees an instantaneous pulse of heat given by $q'\delta(t)$, where $\delta(t)$ is the Dirac delta function and q' has units of J/m^2. Find the surface temperature $T(a, t)$ due to this heat pulse.

Solution

The temperature is given by the GF equation (7.2), for a boundary condition of the second kind:

$$T(r, t) = \alpha \int_{\tau=0}^{t} \frac{q'\delta(\tau)}{k} G_{R20}(r, t|a, \tau)\, 2\pi a\, d\tau \tag{7.96}$$

To evaluate the temperature at $r = a$ apply the sifting property of the Dirac delta function to the time integral to give

$$T(a, t) = 2\pi\alpha \frac{q' a}{k} G_{R20}(a, t|a, 0) = 2\pi \frac{q' a}{\rho c} G_{R20}(a, t|a, 0) \tag{7.97}$$

Note that $q'a/(\rho c)$ has units of Km^2 and that G_{R20} has units of m^{-2}, as expected. The GF G_{R20} is listed in Appendix R as

$$G_{R20}(a, t|a, 0) = \frac{2}{\pi^2 a^2} \int_{\beta=0}^{\infty} \frac{e^{-\beta\alpha t/a^2}}{\beta\left[J_1^2(\beta) + Y_1^2(\beta)\right]} d\beta \tag{7.98}$$

In general, this integral must be evaluated numerically. However, several approximate expressions for $G_{R20}(a, t|a, \tau)$ are listed in Appendix R, and some numerical values of $G_{R20}(a, t|a, \tau)$ are listed in Table R20. For small values of time $\alpha t/a^2$, the surface temperature is approximately

$$\begin{aligned} T(a, t) \approx \frac{q'}{a\rho c} \Big[& (\pi t^+)^{-1/2} - 0.5 + 0.413434\,(t^+)^{1/2} - 0.299877\, t^+ \\ & + 0.154483\,(t^+)^{3/2} - 0.045263\,(t^+)^2 + 0.005484\,(t^+)^{5/2} \Big] \end{aligned} \tag{7.99}$$

where $t^+ = \alpha t/a^2 < 6$. The first term inside the square brackets in Equation 7.99 is the same as the temperature in a plane wall, and the second term is the first correction for the curvature of the cylindrical hole. For large values of time $\alpha t/a^2 > 6$, the surface temperature is approximately

$$T(a, t) \approx \frac{q'}{a\rho c} \frac{1}{t^+} \left\{ 1 - \frac{1}{2t^+} L \left[1 + \frac{3}{4t^+}(1 - L) \right] - (\pi^2 + 4) \frac{C}{(4t^+)^2} \right\} \tag{7.100}$$

where $L = \ln(4t^+) - \gamma$
$\gamma = 0.57722$ (Euler's constant)
$C = 0.5$

7.8 THIN SHELLS, $T = T(\phi, t)$

Thin shells are bodies for which the coordinates (ϕ, t) completely describe the heat transfer. These are closely related to the one-dimensional rectangular cases, although there are some differences. A thin shell has radius a, thickness δ, and angle ϕ_0. The shell is thin if the temperature at $r = a$ and at $r = a + \delta$ are approximately equal. If not, then the variable r must be included in the analysis, and the body must be analyzed with the variables (r, ϕ, t).

Example 7.11: Thin Shell Heated at One Point and Cooled by Convection

A transient experiment was carried out in a nonrotating railroad roller bearing to determine how heat moves from a single heated roller through the outer bearing race. Treat the bearing race as a thin shell heated at one point and with heat loss, both internal and external, described by $q_{loss} = h(T(\phi, t) - T_0)$ where h is the heat transfer coefficient (W/m^2 K), and T_0 is the temperature of the surroundings. A schematic of the geometry is shown in Figure 7.5. Find (a) the transient temperature in the thin shell and (b) the steady temperature.

Solution

The energy equation for the thin shell with heat losses from the sides is given by

$$\frac{1}{a^2}\frac{\partial^2 T}{\partial \phi^2} - m^2(T - T_0) = \frac{1}{\alpha}\frac{\partial T}{\partial t} \qquad (7.101a)$$

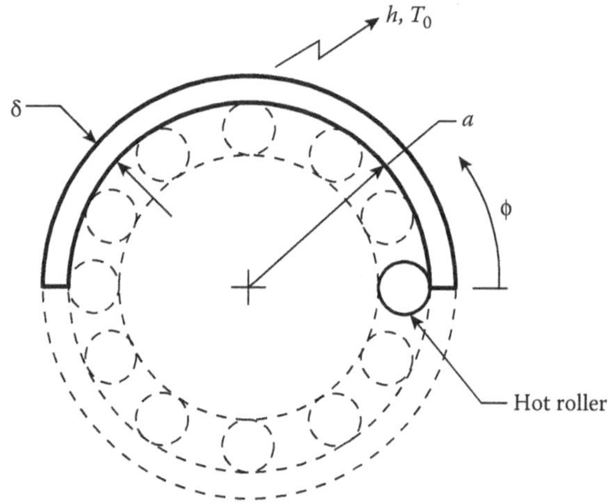

FIGURE 7.5 Schematic of a thin-shell model of the outer race of a railroad roller bearing heated by one hot roller and with side losses described by convective coefficient h.

Parameter $m^2 = h/(k\delta)$ describes heat loss from the sides of the thin shell. As there is symmetry in the heating geometry, we choose to treat the thin shell as geometry $\Phi 22$ on $(0 < \phi < \pi)$. The boundary conditions are

$$-\frac{k}{a}\frac{\partial T}{\partial \phi}\bigg|_{\phi=0} = q_0 \qquad (7.101\text{b})$$

$$\frac{\partial T}{\partial \phi}\bigg|_{\phi=\pi} = 0 \text{ by symmetry} \qquad (7.101\text{c})$$

If the heat flow from the roller to the whole shell is Q_0 (watts), then the heat flux into half of the shell is $q_0 = Q_0/(2w\delta)$ where w is the roller length out-of-plane. The initial condition is:

$$T(\phi, t - 0) = T_0 \qquad (7.101\text{d})$$

(a) *Transient temperature.* The GF solution for the transient temperature involves a variable transformation to eliminate the fin term. Let a new temperature variable be defined

$$W(\phi, t) = [T(\phi, t) - T_0]\, e^{m^2 \alpha t} \qquad (7.102)$$

Refer to Section 3.5 for a complete discussion of this procedure. Then, Equations 7.101a through d can be written with the new temperature variable as

$$\frac{1}{a^2}\frac{\partial^2 W}{\partial \phi^2} = \frac{1}{\alpha}\frac{\partial W}{\partial t} \qquad (7.103)$$

$$-\frac{k}{a}\frac{\partial W}{\partial \phi}\bigg|_{\phi=0} = q_0 e^{m^2 \alpha t} \qquad (7.104)$$

$$\frac{\partial W}{\partial \phi}\bigg|_{\phi=\pi} = 0 \qquad (7.105)$$

$$W(\phi, t = 0) = 0 \qquad (7.106)$$

Note that the fin term is gone and the initial condition is homogeneous, but the boundary condition is more complicated.

The GF solution to this transformed equation is given by a single integral for heating at the $\phi = 0$ boundary boundary:

$$W(\phi, t) = \frac{\alpha}{k}\int_{\tau=0}^{t} q_0 e^{m^2 \alpha \tau} G_{\Phi 22}(\phi, t|\phi' = 0, \tau)\, \delta\, d\tau \qquad (7.107)$$

The GF, described in Appendix Φ, is given by

$$G_{\Phi 22}(\phi, t|0, \tau) = \frac{1}{\phi_0 a\delta} + \frac{2}{\phi_0 a\delta}\sum_{n=1}^{\infty} e^{-n^2\pi^2\alpha(t-\tau)/(a^2\phi_0^2)} \cos(n\pi\phi/\phi_0) \qquad (7.108)$$

where in this case $\phi_0 = \pi$. The time integral may be carried out to give

$$W(\phi, t) = \frac{q_0 a}{k\phi_0}\left\{\frac{1}{m^2 a^2}\left(e^{m^2\alpha t} - 1\right)\right.$$
$$\left. + 2\sum_{n=1}^{\infty}\frac{e^{m^2\alpha t} - e^{-n^2\pi^2\alpha t/(a^2\phi_0^2)}}{m^2 a^2 + n^2\pi^2/\phi_0^2}\cos(n\pi\phi/\phi_0)\right\} \qquad (7.109)$$

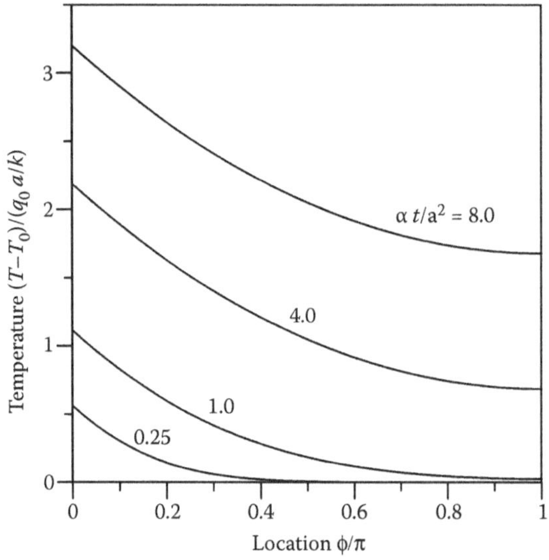

FIGURE 7.6 Normalized temperature along the thin shell heated at $\phi = 0$ for convective losses described by $ma = 0.2$ at times $\alpha t / a^2 = 0.25$, 1.0, 4.0, and 8.0.

In evaluating the integral, do not confuse the fin parameter m^2 with quantity n^2 which comes from the eigenvalue and is part of the infinite series.

Finally, to find the temperature in the original problem, convert back according to the transformation $T - T_0 = W(\phi, t)\, e^{-m^2 \alpha t}$:

$$T(\phi, t) - T_0 = \frac{q_0 a}{k\phi_0}\left\{\frac{1}{m^2 a^2}\left(1 - e^{-m^2 \alpha t}\right)\right.$$
$$\left. + 2\sum_{n=1}^{\infty} \frac{1 - \exp[-(m^2 a^2 + n^2\pi^2/\phi_0^2)\alpha t / a^2]}{m^2 a^2 + n^2 \pi^2 / \phi_0^2} \cos(n\pi\phi/\phi_0)\right\} \tag{7.110}$$

A plot of the spatial distribution of the (normalized) temperature in the thin shell heated at $\phi = 0$ is given in Figure 7.6 for convection condition $ma = 0.2$ at several dimensionless times. For more information on this thermal model and its use in determining h from transient experiments on railroad roller bearings, see (Cole et al., 2009).

(b) *Steady temperature.* The steady-state temperature is given by the limit as $t \to \infty$, or,

$$T_{steady}(\phi) - T_0 = \frac{q_0 a}{k\phi_0}\left\{\frac{1}{m^2 a^2} + 2\sum_{n=1}^{\infty} \frac{\cos(n\pi\phi/\phi_0)}{m^2 a^2 + n^2\pi^2/\phi_0^2}\right\} \tag{7.111a}$$

If slow series convergence is a problem, the steady-state series can be replaced by a nonseries form, constructed from the steady-fin GF (adapted from Table X.4, Appendix X). The result is

$$T_{steady}(\phi) - T_0 = \frac{q_0 a}{k}\left[\frac{e^{-2ma\phi_0}e^{ma\phi} + e^{-ma\phi}}{ma(1 - e^{-2ma\phi_0})}\right] \tag{7.111b}$$

where for the half-shell the maximum angle is $\phi_0 = \pi$.

7.9 LIMITING CASES FOR 2D AND 3D GEOMETRIES

The ability to analyze two- and three-dimensional heat transfer geometries is an important feature of the GF method. Multidimensional geometries can be so challenging that finding ways to verify the results, including analysis of limiting cases, is usually an important step in the solution process. Limiting cases can improve one's insight and contribute to a better understanding of the whole problem.

One-dimensional limiting cases can be important for checking the analysis of two- or three-dimensional temperature expressions and for checking the numerical results. Under the limiting conditions, the multidimension expression for the temperature should reduce to the limiting-case expression and the multidimension computer program should give numerical values that agree with the limiting case. Numerical values for simple one-dimensional cases are sometimes tabulated in books such as this one, whereas numerical values for two-dimensional cases are rarely available. Comparison with more than one limiting case should be used whenever possible.

7.9.1 FOURIER NUMBER

All of the transient cases discussed in this chapter depend upon a Fourier number $\alpha t / L^2$, where t is the characteristic time, α is the thermal diffusivity, and L is a characteristic length. The trick to constructing a limiting case based on the Fourier number is to use the significant characteristic length. The characteristic length can depend on (1) time (early, middle, or late); (2) body shape (slab, cylinder, etc.); (3) location of the driving force for the transient heat transfer (at the surface or internally); or (4) location of the temperature of interest.

For example, in a solid cylinder heated at the boundary (case $R01B1T0$), the Fourier number is $\alpha t / b^2$, where b is cylinder radius. For sufficiently small values of this Fourier number, the temperature near $r = b$ is given approximately by the semi-infinite case $X10$. The $X10$ geometry is a limiting case for small time because the surface heating penetrates the cylinder so slightly that the curvature of the cylinder may be neglected.

For energy generation inside a body and for small dimensionless times, the characteristic length depends on the heating location. For example, in a cylinder heated by a cylindrical-shell heat source, $R01B0T0Gr7$, the significant Fourier number is $\alpha t /(b - r_0)^2$ where b is the cylinder radius, r_0 is the location of the cylindrical-shell heat source, and $r_0 / b < 0.5$ (this last condition ensures that the boundary is far enough from the heat source). Then, for $\alpha t /(b - r_0)^2$ sufficiently small, the temperature is given by an infinite region ($R00$) heated by a cylindrical-shell heat source. The characteristic length is $b - r_0$, the distance from the heating location to the boundary.

7.9.2 Aspect Ratio

Limiting cases may be found by changing the aspect ratio of the body, the ratio of the width to the length of the body. For solid cylinders, the aspect ratio is b / L, where b is the cylinder radius and L is the cylinder length. In a solid cylinder there are two limiting cases based on variations of the aspect ratio. First, consider the cylinder with aspect ratio $b / L > 5$ shown in Figure 7.7a. This cylinder is more like a flat disk, and depending how it is heated, the limiting case may be the one-dimensional slab of thickness L. Second, for the solid cylinder with aspect ratio $b / L < 1 / 10$ shown in Figure 7.7b, the limiting case is the infinite cylinder of radius b for which $T = T(r,t)$.

Three-dimensional bodies may have two aspect ratios. For hollow cylinders, an additional aspect ratio is $(b-a)/b$, where $(b-a)$ is the thickness of the cylinder wall.

7.9.3 Nonuniform Heating

When a body is heated nonuniformly over position or over time, the limiting case of uniform heating is useful for checking purposes. The uniformly heated cases are generally easier to analyze. For example, a cylinder heated over part of its surface is shown in Figure 7.8, and it is described by the number $R01B(z5)Z11B11T0$. The

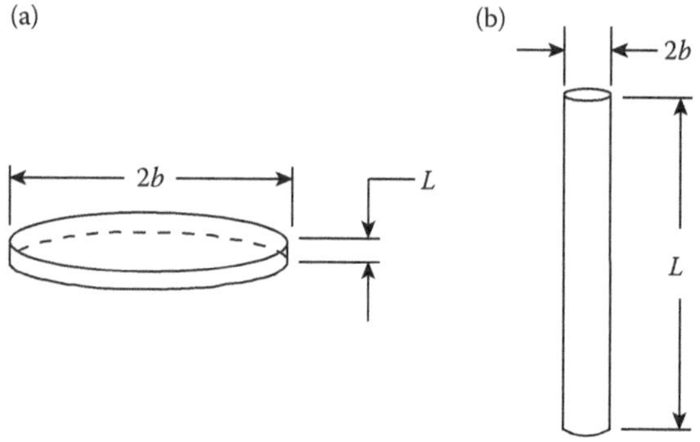

FIGURE 7.7 (a) Cylinder with aspect ratio $b / L > 5$. (b) Cylinder with aspect ratio $b / L < 0.1$.

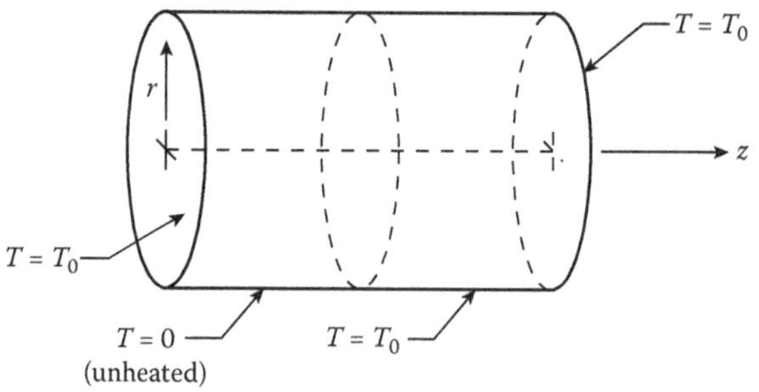

FIGURE 7.8 Cylinder with specified temperature over part of its surface.

limiting case of a uniformly heated cylinder, number $R01B1Z11B1T0$, is particularly easy to find by multiplying two one-dimensional temperature solutions, as discussed in the next section.

7.10 CYLINDERS WITH $T = T(r, z, t)$

In this section cylinders are discussed for which the temperature depends on coordinates r and z. The GFs for these cases can be constructed by multiplying two one-dimensional GFs. That is,

$$G_{RZ} = (G_R)(G_Z) \tag{7.112}$$

The boundary conditions of types 0, 1, 2, and 3, may be treated. Two examples are given to illustrate the method.

Example 7.12: Finite Cylinder with Specified Surface Temperature—*R*01*Z*11 Geometry

A finite cylinder of length L and radius b has a uniform initial temperature T_0. For $t > 0$, the entire surface of the cylinder is suddenly set to temperature T_1. Find the temperature in the cylinder for large times.

Solution

The cylinder is shown in Figure 7.9. A detailed statement of the boundary and initial conditions of this example are

$$T(r = b, z, t) = T_1 \tag{7.113a}$$

$$T(r, z = 0, t) = T_1 \tag{7.113b}$$

$$T(r, z = L, t) = T_1 \tag{7.113c}$$

$$T(r, z, t = 0) = T_0 \tag{7.113d}$$

The heat conduction numbering system for this case is $R01B1T1Z11B11$.

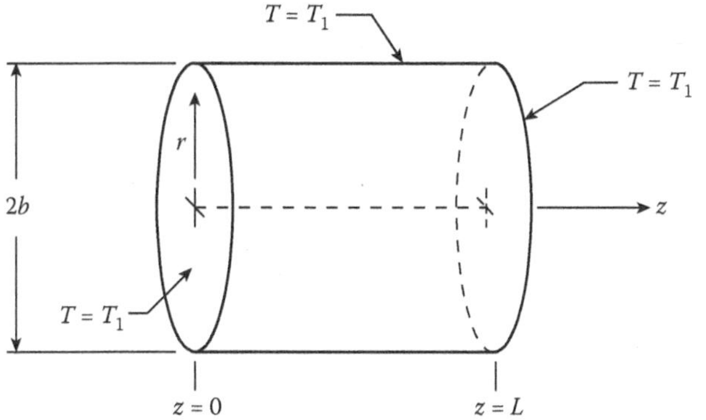

FIGURE 7.9 Solid cylinder with temperature boundary conditions, $R01B1T1Z11B11$.

The best solution for this problem using GFs is to cast the problem as an initial-temperature case, by defining a new temperature $T = T - T_1$. Then, the boundary temperature is zero, and the initial temperature is $T_0 - T_1$. This is equivalent to the alternative GF solution method with $T^* = T_1$. The solution is given by

$$T(r, z, t) - T_1 = \int_0^b 2\pi r' dr' \int_0^L dz' \, (T_0 - T_1) \, G_{R01Z11}(r, z, t|r', z', 0) \tag{7.114}$$

The GF G_{R01Z11} may be found from a product solution as shown in Equation 7.112, and substituted into the spatial integral in Equation 7.114 to give

$$T(r, z, t) - T_1 = (T_0 - T_1)\left[\int_0^b G_{R01}(r, t|r', 0) \, 2\pi r' \, dr'\right] \times \left[\int_0^L G_{Z11}(z, t|z', 0) \, dz'\right] \tag{7.115}$$

The product of the two integrals in Equation 7.115 can be interpreted as the product of two dimensionless temperatures, one for an infinite cylinder $R01B0T1$, and one for an infinite slab $X11B0T1$. A product solution for temperature is possible only for certain initial conditions, but a product solution for the GF is always possible for coordinates r and z. Refer to Section 4.6 for a discussion of this point.

Function G_{R01} comes from Appendix R and function G_{Z11} comes from Appendix X (with x and x' replaced by z and z'). The result for the temperature is

$$T(r, z, t) - T_1 = (T_0 - T_1)\left[2 \sum_{m=1}^{\infty} e^{-\beta_m^2 \alpha t / b^2} \frac{J_0(\beta_m r / b)}{\beta_m J_1^2(\beta_m)}\right] \times \left\{2 \sum_{n=1}^{\infty} e^{-n^2\pi^2 \alpha t / L^2} \sin\left(n\pi \frac{z}{L}\right) \frac{[1 - (-1)^n]}{n\pi}\right\} \tag{7.116}$$

where the eigenvalues β_m are the roots of $J_0(\beta_m) = 0$; some values for β_m are given in Appendix B. The expression $[1 - (-1)^n]$ comes from evaluating the integral of the sine to give $\cos(0) - \cos(n\pi)$. This expression gives zero for all the odd terms in the second summation ($n = 1, 3, 5, \ldots$), and the summation on n could be rewritten in a form to represent just the nonzero terms:

$$\sum_{n=1}^{\infty} f(n)[1 - (-1)^n] = 2 \sum_{k=0}^{\infty} f(2k + 1) \tag{7.117}$$

Although the solution given by Equation 7.116 contains two summations, the summations converge very rapidly for large times $\alpha t / b^2 \gg 1$.

Example 7.13: Cylinder Heated over Half of Its Surface—*R02Z00* Case

An infinite cylinder initially at zero temperature is suddenly heated over half of its surface ($z < 0$) with heat flux q_0. The other half of the cylinder is insulated ($z > 0$). Refer to Figure 7.10. Find the temperature on the surface of the cylinder soon after the heating begins.

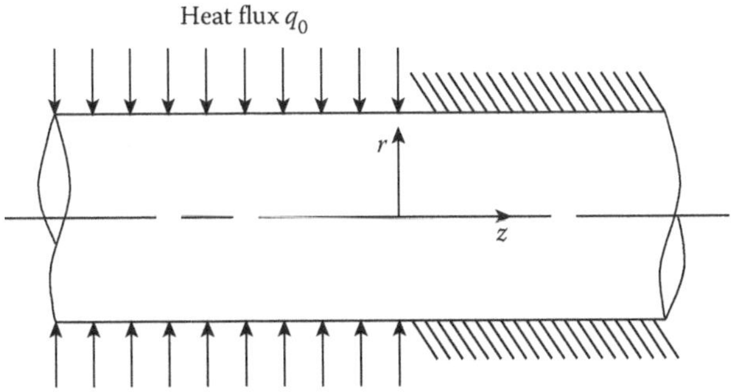

FIGURE 7.10 Infinite cylinder heated over half of the surface and insulated elsewhere, case $R02B5T0Z00$.

Solution

This is the $R02B(z5)T0Z00$ geometry with heating at the surface and with no initial temperature or internal heat generation. The maximum temperature occurs on the surface of the cylinder and the surface temperature is given by the surface heating term of the GFSE:

$$T(b, z, t) = \frac{q_0 \alpha}{k} \int_{\tau=0}^{t} \int_{z'=-\infty}^{0} G_{R20Z00}(b, z, t | b, z', \tau)\, 2\pi b dz'\, d\tau \qquad (7.118)$$

The GF is evaluated at $r' = b$ where the heating occurs, and the temperature is evaluated at $r = b$. Note that the spatial integral over the surface involves differential area $ds = 2\pi b dz'$ and the integral limits are $-\infty < z' < 0$. The GF is given by the product $(G_{R02})(G_{Z00})$. To find the temperature soon after heating begins only the small-time forms of the GFs are needed. An approximate form of G_{R02} for small times evaluated at the surface $r = b$ is given in Appendix R:

$$G_{R02}(b, t | b', \tau) \approx \frac{1}{2\pi b^2} \left\{ \frac{b}{[\pi\alpha(t-\tau)]^{1/2}} + \frac{1}{2} + \frac{3}{4\sqrt{\pi}} \left[\frac{\alpha(t-\tau)}{b^2} \right]^{1/2} \right\} \qquad (7.119)$$

The GF G_{Z00} is given in Appendix X:

$$G_{Z00}(z, t | z', \tau) = \frac{1}{[4\pi\alpha(t-\tau)]^{1/2}} \exp\left[\frac{-(z-z')^2}{4\alpha(t-\tau)} \right] \qquad (7.120)$$

It is generally better to evaluate spatial integrals first, and the integral on z' in Equation 7.118 may be written

$$T(b, z, t) = \frac{q_0 \alpha}{k} \frac{1}{2b} \int_{\tau=0}^{t} d\tau \left\{ \frac{b}{[\pi\alpha(t-\tau)]^{1/2}} + \frac{1}{2} + \frac{3}{4\sqrt{\pi}} \left[\frac{\alpha(t-\tau)}{b^2} \right]^{1/2} \right\} \operatorname{erfc}\left\{ \frac{z}{[4\alpha(t-\tau)]^{1/2}} \right\} \qquad (7.121)$$

The integral on τ may be evaluated in three terms and the final result for the surface temperature may be written (see integral Table I.8, Appendix I)

$$
\begin{aligned}
T(b,z,t) = \frac{q_0 b}{k}\frac{1}{4\pi}\Bigg(& 2(t^+)^{1/2} \\
& \times \left\{ \operatorname{erfc}\left[\frac{z}{(4\alpha t)^{1/2}}\right] - \frac{z}{(4\pi\alpha t)^{1/2}} E_1\left(\frac{z^2}{4\alpha t}\right) \right\} \\
& + 2(t^+)\, \mathrm{i}^2 \operatorname{erfc}\left[\frac{z}{(4\alpha t)^{1/2}}\right] + \frac{12}{\sqrt{\pi}}(t^+)^2 \\
& \times \left\{ \mathrm{i}^2 \operatorname{erfc}\left[\frac{z}{(4\alpha t)^{1/2}}\right] + \mathrm{i}^4 \operatorname{erfc}\left[\frac{z}{(4\alpha t)^{1/2}}\right] \right\} \Bigg)
\end{aligned}
\quad (7.122)
$$

where $t^+ = \alpha t / b^2$.

Discussion. One part of the given temperature expression is multiplied by $(t^+)^{1/2}$. This part is identical to the temperature in a semi-infinite plane body heated over half of its surface which was studied in Section 6.8. For early times, the surface of the cylinder displays behavior similar to a plane body. From this perspective the other terms in the temperature expression Equation 7.122 are corrections to account for the curvature in the surface of the cylinder.

The temperature expression contains factors like $\sqrt{t^+}$, t^+, and $(t^+)^2$, which indicate that the temperature increases over time without limit; there is no steady-state solution since all the heat that enters the cylinder remains in the cylinder. The surface temperature is the largest temperature on the cylinder at any given time.

On the heated region of the cylinder ($z < 0$) and far away from the point $z = 0$, the temperature is described by one-dimensional radial heat conduction, $T = T(r,t)$. Here "far" is determined by $z^2/\alpha t \gg 1$, because the correct Fourier number along the z-axis is $\alpha t / z^2$. On the nonheated end of the cylinder and for $z^2/\alpha t \gg 1$, the temperature is identically zero.

7.11 DISK HEAT SOURCE ON A SEMI-INFINITE BODY

In this section, the cylindrical GFs are applied to a semi-infinite body heated at the surface by a disk heat source. Over the disk heat source, the heat flux is constant with position and with time, while outside the disk, the surface is insulated. This case is a basic building block in transient heat conduction and in the surface element method discussed in Chapter 12. Applications of the disk heat source solution include constriction resistance, the intrinsic thermocouple, and laser heating of a flat surface.

The GF solution yields an exact solution in the form of an integral with limits of zero and infinity, and the integrand involves error functions and Bessel functions. The integral is difficult to evaluate numerically because the domain is infinite and because of the sinusoidal behavior of the Bessel functions. Though this integral represents a solution valid for any position (r, z), accurate numerical values are difficult to obtain directly except along the centerline ($r = 0$).

The purpose of this section is to present the exact solution with the GF method, to present closed-form expressions for some special cases, and to present series expressions for the surface temperature that are accurate and easy to evaluate numerically. Expressions for interior temperatures ($z > 0$) are given by Beck (1980, 1981).

7.11.1 Integral Expression for the Temperature

The geometry for the disk heat source problem is shown in Figure 7.11. The surface of the semi-infinite body is insulated except for the disk $0 < r < a$. The initial temperature is zero. A mathematical statement of the energy equation and boundary conditions is given by

$$\frac{1}{r}\frac{\partial}{\partial r}\left(r\frac{\partial T}{\partial r}\right) + \frac{\partial^2 T}{\partial z^2} = \frac{1}{\alpha}\frac{\partial T}{\partial t} \tag{7.123}$$

$$-k\frac{\partial T(r,0,t)}{\partial z} = \begin{cases} q_0 & \text{for } 0 < r < a \\ 0 & \text{for } r > a \end{cases} \tag{7.124a}$$

$$T(r,z,t) \to 0 \qquad \text{for } r \to \infty \qquad \text{and } z \to \infty \tag{7.124b}$$

$$T(r,z,0) = 0 \tag{7.124c}$$

The GFSE is

$$T(r,z,t) = \frac{\alpha}{k}\int_{\tau=0}^{t}\int_{r'=0}^{a} q_0\, G_{R00Z20}(r,z,t|r',0,\tau)2\pi r'\,dr'\,d\tau \tag{7.125}$$

The integral over the surface involves the area element $dA = 2\pi r' dr'$, and the GF is evaluated at the surface $z' = 0$.

The GF is given by the multiplication of two one-dimensional functions, $G_{R00Z20} = (G_{R00})(G_{Z20})$, where

$$G_{Z20}(z,t|z'=0,\tau) = \frac{2}{[4\pi\alpha(t-\tau]^{1/2}}\exp\left[\frac{-z^2}{4\alpha(t-\tau)}\right] \tag{7.126}$$

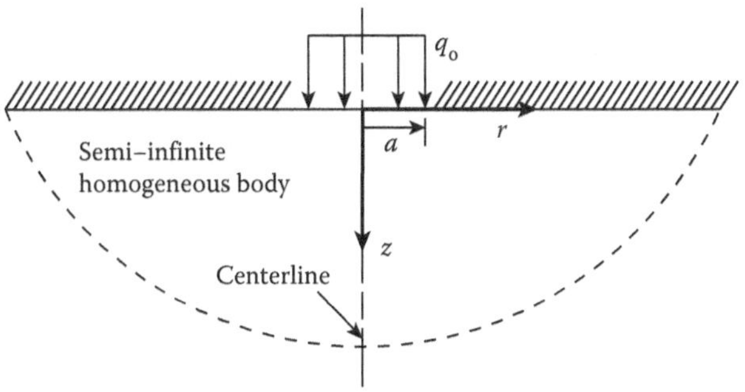

FIGURE 7.11 Semi-infinite body heated over a disk-shaped region centered at $r = 0$ and $z = 0$ and insulated elsewhere at $z = 0$.

from Appendix X, and where

$$G_{R00}(r,t|r',\tau) = \frac{1}{2\pi a^2}\int_{\beta=0}^{\infty} e^{-\beta^2\alpha(t-\tau)/a^2}\beta\, J_0\left(\frac{\beta r}{a}\right) J_0\left(\frac{\beta r'}{a}\right) d\beta \qquad (7.127)$$

from Appendix R. Note that β is a dimensionless eigenvalue, and that G_{R00} has units of meters^{-2}. This form of G_{R00} is difficult to evaluate for reasons discussed above; but, because the integrand depends on r' and τ in *separate terms*, the GF equation (7.125) can be integrated separately over r' and τ leaving only the integral on β. The other form of G_{R00} listed in Appendix R contains the term $I_0[(rr')/(2\alpha(t-\tau)]$, and the integrals on r' and τ cannot be evaluated separately.

The GF G_{R00} and G_{Z20} can now be substituted into the expression for the temperature, and the integrals on r' and τ can be evaluated. The integral on r' acts only on the term $r'J_0(\beta r'/a)$, so the integral is given by (Appendix B)

$$\int_{r'=0}^{a} J_0\left[\frac{\beta r'}{a}\right] 2\pi r'\, dr' = \frac{2\pi a^2}{\beta} J_1(\beta) \qquad (7.128)$$

Combine the integral on r' with the temperature expression, Equation 7.125, to get

$$T(r,z,t) = \frac{\alpha}{k}\int_{\tau=0}^{t} q_0 \frac{1}{[\pi\alpha(t-\tau)]^{1/2}} \exp\left[\frac{-z^2}{4\alpha(t-\tau)}\right] \times \int_{\beta=0}^{\infty} e^{-\beta^2\alpha(t-\tau)/a^2} J_0\left(\frac{\beta r}{a}\right) J_1(\beta)\, d\beta\, d\tau \qquad (7.129)$$

The integral on τ can now be identified as (Appendix I, Table I.6),

$$\int_{\tau=0}^{t} \frac{1}{[\alpha(t-\tau)]^{1/2}} \exp\left[-\frac{\beta^2\alpha(t-\tau)}{a^2} - \frac{z^2}{4\alpha(t-\tau)}\right] d\tau = \frac{2\pi^{1/2}}{2\beta\alpha}\left(e^{-\beta z/a}\left\{1+\operatorname{erf}\left[\frac{\beta(\alpha t)^{1/2}}{a} - \frac{z}{2(\alpha t)^{1/2}}\right]\right\} - e^{\beta z/a}\operatorname{erfc}\left[\frac{\beta(\alpha t)^{1/2}}{a} + \frac{z}{2(\alpha t)^{1/2}}\right]\right) \qquad (7.130)$$

Then, Equation 7.129 can be written as

$$T(r,z,t) = \frac{1}{2}\frac{q_0 a}{k}\int_{\beta=0}^{\infty} J_0\left(\frac{\beta r}{a}\right) J_1(\beta) \times \left(e^{-\beta z/a}\left\{1+\operatorname{erf}\left[\frac{\beta(\alpha t)^{1/2}}{a} - \frac{z}{2(\alpha t)^{1/2}}\right]\right\} - e^{\beta z/a}\operatorname{erfc}\left[\frac{\beta(\alpha t)^{1/2}}{a} + \frac{z}{2(\alpha t)^{1/2}}\right]\right)\frac{d\beta}{\beta} \qquad (7.131)$$

This is the exact solution in integral form, valid for all values $r > 0, z > 0, t > 0$ (Carslaw and Jaeger, 1959).

At the surface where $z = 0$ the exact solution reduces to

$$\frac{T(r,0,t)}{q_0 a / k} = \int_{\beta=0}^{\infty} \text{erf}\left[\frac{\beta(\alpha t)^{1/2}}{a}\right] J_0\left(\frac{\beta r}{a}\right) J_1(\beta) \frac{d\beta}{\beta} \tag{7.132}$$

This expression still contains the difficult integral on β.

7.11.2 Closed-Form Expressions for the Temperature

In general, the infinite integral on β in Equation 7.132 cannot be evaluated in closed form. In restricted cases, however, convenient temperature expressions may be found.

Steady-state temperature. Thomas (1957) derived an exact steady solution for the surface temperature ($z = 0$) in terms of known functions. The steady surface temperature for $0 < r < a$ is given by

$$\frac{T(r,0,\infty)}{q_0 a / k} = \frac{2}{\pi} E\left(\frac{r}{a}\right) \tag{7.133}$$

and, for $r > a$,

$$\frac{T(r,0,\infty)}{q_0 a / k} = \frac{2r}{\pi}\left[E\left(\frac{a}{r}\right) - (1 - r^{-2}) K\left(\frac{a}{r}\right)\right] \tag{7.134a}$$

The functions $K(\cdot)$ and $E(\cdot)$ are the complete elliptic integrals of the first and second kinds;

$$K(\epsilon) = \int_0^{\pi/2} (1 - \epsilon^2 \sin^2\theta)^{-1/2} d\theta \tag{7.134b}$$

$$E(\epsilon) = \int_0^{\pi/2} (1 - \epsilon^2 \sin^2\theta)^{1/2}\, d\theta \tag{7.134c}$$

These functions are tabulated in Abramowitz and Stegun (1964) and are available in computer libraries.

At $r \approx 0$, $[T(r,0,\infty) - T_0]/(q_0 a / k) = 2/\pi$. For large values of r/a, Equation 7.134a can be approximated by

$$\frac{T(r,0,\infty)}{q_0 a / k} = \frac{1}{2r}\left[1 + \frac{1}{2(2r/a)^2} + \frac{1}{2^2(2r/a)^4} + \cdots\right] \tag{7.135}$$

The leading term of Equation 7.135 is proportional to $1/(r)$, which is the same as a steady point heat source on the surface.

Centerline temperature. At the centerline of the body at $r = 0$, the exact solution is given by (Carslaw and Jaeger, 1959)

$$\frac{T(0,z,t)}{q_0 a / k} = 2\frac{\sqrt{\alpha t}}{a}\left\{\text{ierfc}\left[\frac{z}{2(\alpha t)^{1/2}}\right] - \text{ierfc}\left[\frac{(z^2+a^2)^{1/2}}{2(\alpha t)^{1/2}}\right]\right\} \tag{7.136}$$

where ierfc(·) is the integral of the complementary error function (see Appendix E).

The centerline temperature in Equation 7.136 can be derived using Equation 7.125 and using the following form of the $R00$ GF listed in Appendix R:

$$G_{R00}(r,t|r',\tau) = [4\pi\alpha(t-\tau)]^{-1}\exp\left[\frac{-(r^2+r'^2)}{4\alpha(t-\tau)}\right] I_0\left[\frac{rr'}{2\alpha(t-\tau)}\right]$$

At $r = 0$, the modified Bessel function drops out ($I_0(0) = 1$), and the integrals on r' and τ in Equation 7.125 produce function ierfc (·).

Surface temperature far from the disk source. The surface temperature far from the disk heat source behaves as if the heat is introduced by a point source and the temperature is given approximately by

$$\frac{T(r,0,t)}{q_0 a / k} = \frac{1}{2(r/a)}\,\text{erfc}\left[\frac{r}{2(\alpha t)^{1/2}}\right] \tag{7.137}$$

In the limit as $t \to \infty$, the steady-state surface temperature goes like $(1/r)$. At steady-state, Equation 7.137 gives for $r/a = 8$ the value of 0.0625 while the exact value is 0.062623 which is 0.2% higher. For larger r/a, the error in using Equation 7.137 is less, but the percent error for a given r/a tends to become larger as $\alpha t/a^2$ is reduced.

7.11.3 Series Expression for the Surface Temperature at Large Times

By using the relation erf = 1 − erfc, Equation 7.132 for the surface temperature is given by

$$\frac{T(r,0,t)}{q_0 a / k} = \int_{\beta=0}^{\infty} J_0\left(\frac{\beta r}{a}\right) J_1(\beta)\frac{d\beta}{\beta} - \int_{\beta=0}^{\infty}\text{erfc}\left[\frac{\beta(\alpha t)^{1/2}}{a}\right] J_0\left(\frac{\beta r}{a}\right) J_1(\beta)\frac{d\beta}{\beta} \tag{7.138}$$

Notice that the first integral is a steady-state term and the second integral goes to zero as $t \to \infty$. Hence, the first integral is equal to the steady-state temperature given either by Equation 7.133 or 7.134 depending on the range of r.

Consider now the second integral in Equation 7.138. Using the dimensionless variables $r^+ = r/a$ and $t^+ = \alpha t/a^2$, an exact series expression for this integral is given by (Beck, 1981),

$$I_2 = -\frac{1}{2\sqrt{\pi\, t^+}}\sum_{k=1}^{\infty}\frac{(-1)^k}{C_{k-1}(t^+)^{k-1}}\sum_{j=1}^{k}\frac{k-j+1}{k}U_{kj}^2 \tag{7.139}$$

where

$$C_k = 4^k(2k+1)[(k+1)!] \tag{7.140}$$

$$U_{k1} = 1 \tag{7.141a}$$

$$U_{kj} = U_{k,j-1}\frac{(k-j+2)r^+}{(j-1)} \qquad k = 1, 2, \ldots \qquad j = 2, 3, \ldots, k \tag{7.141b}$$

where Equation 7.141b is a recursion relation.

In summary for $0 < r^+ < 1$, a series expression for T at $z = 0$ is

$$\frac{T(r^+, 0, t)}{q_0 a / k} = \frac{2}{\pi}E(r^+) - I_2(r^+, t^+) \tag{7.142a}$$

where the fundamental dependence of I_2 is noted. For $r^+ > 1$, the temperature is given by

$$\frac{T(r^+, 0, t)}{q_0 a / k} = \frac{2r^+}{\pi}\left[E\left(\frac{1}{r^+}\right) - [1 - (r^+)^{-2}]K\left(\frac{1}{r^+}\right)\right] - I_2(r^+, t^+) \tag{7.142b}$$

The function $I_2(r^+, t^+)$ is calculated using Equations 7.139 through 7.141. These exact expressions are very efficient for "large" times because the infinite summation in I_2 can be approximated with just a few terms.

In order to display clearly the nature of the summation in I_2, several terms are now given.

$$\begin{aligned} T(r^+, 0, t^+) = T(r^+, 0, \infty) &- \frac{1}{2\sqrt{\pi t^+}}\left\{1 - \frac{1 + 2(r^+)^2}{24t^+} + \frac{1}{480(t^+)^2}\right. \\ &\times [1 + 6(r^+)^2 + 3(r^+)^4] - \frac{1}{10752(t^+)^3}[1 + 12(r^+)^2 \\ &\left. + 18(r^+)^4 + 4(r^+)^6] + \cdots\right\} \end{aligned} \tag{7.143}$$

Note that the denominators 24, 480, etc., are the C_k values given by Equation 7.140. The number of terms required in the series for I_2 increases quite rapidly as the dimensionless times become small. Fortunately, for a large range of t^+, the required number of terms is quite modest, that is, less than 7 for $r^+ = 0$ and for $t^+ > 1$ to obtain eight-significant-figure accuracy. Also the number of additional terms required to go from three to eight significant figures is not large. The series solution, however, is not appropriate for very small dimensionless times. The limiting appropriate dimensionless times are about $t^+ = 0.01, 0.05$, and 0.1 for $r^+ = 0, 1$ and 2, respectively. For $r^+ \geq 1$, a convenient limiting time expression is

$$\frac{t^+}{(r^+)^2} \geq 0.05 \tag{7.144}$$

Temperatures for $r^+ = 0, 0.25, 0.5, 0.75, 0.9$, and 1.0 are plotted in Figure 7.12. For the small dimensionless time values at $r^+ = 0$, temperatures were calculated

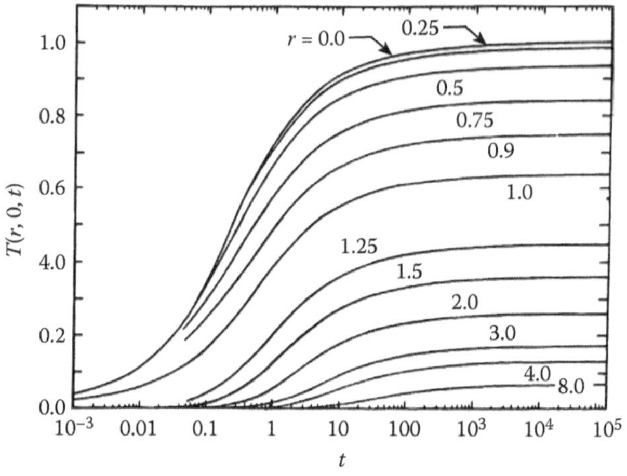

FIGURE 7.12 Local temperature versus time at $z = 0$ on semi-infinite body heated over a circular disk.

utilizing Equation 7.136. The $r^+ = 1$ curve for small dimensionless time was found using

$$T(1,0,t^+) \approx \left(\frac{t^+}{\pi}\right)^{1/2} - \frac{t^+}{2\pi}\left[1 + \frac{t^+}{8} + \frac{9(t^+)^2}{96}\right] \tag{7.145}$$

where $t^+ = \alpha t / a^2$. This expression is accurate to five significant figures for $t^+ < 0.1$. For very small t^+ values (about 10^{-4}) the T given by Equation 7.145 is one-half the center value given by Equation 7.136.

7.11.4 Average Temperature

The temperature averaged over position is of interest for determining the contact conductance and for other purposes. For the average temperature between $r^+ = 0$ and $r^+ = c$ (where c is an arbitrary dimensionless radial location), one can multiply $T(r, 0, t)$ by $2\pi r dr$, integrate from $r / a = 0$ to c, and divide by πc^2. The result is

$$\overline{T}(c,0,t) = \overline{T}(c,0,\infty) - \overline{I}_2(c,t^+) \tag{7.146}$$

where $\overline{I}_2(c, t^+)$ is exactly the same expression as given by Equation 7.139 except the inner summation has kj in the denominator instead of simply k and in Equation 7.141b, r^+ is replaced by c. The term $\overline{T}(c, 0, \infty)$ in Equation 7.146 for $0 < c \leq 1$ is given by

$$\overline{T}(c,0,\infty) = \frac{4}{3\pi c^2}[(1+c^2)E(c) - (1-c^2)K(c)] \tag{7.147a}$$

and for $1 \leq c \leq \infty$

$$\overline{T}(c,0,\infty) = \frac{4}{3\pi c^2}[(1+c^2)E(c^{-1}) - (1-c^2)K(c^{-1})] \tag{7.147b}$$

where $E(\cdot)$ and $K(\cdot)$ are elliptic integrals defined in Equation 7.134. At $c = 0$, $\overline{T}(0, 0, \infty) = 1$, and at $c = 1, \overline{T}(1, 0, \infty) = 8/3\pi$. For large c values, Equation 7.147b can be approximated by

$$\overline{T}(c, 0, \infty) \approx \frac{1}{c}\left(1 - \frac{1}{8c^2} - \frac{1}{64c^2}\right) \tag{7.147c}$$

An expanded form of Equation 7.146 for a few terms is

$$\begin{aligned}\overline{T}(c, 0, t) = \overline{T}(c, 0, \infty) &- \frac{1}{2\sqrt{\pi t^+}} \\ &\times \left[1 - \frac{1 + c^2}{24t^+} + \frac{1}{480(t^+)^2}(1 + 3c^2 + c^4)\right. \\ &\left. - \frac{1}{10752(t^+)^3}(1 + 6c^2 + 6c^4 + 6c^6) + \cdots\right]\end{aligned} \tag{7.148}$$

For small t, the average temperature from $r^+ = 0$ to 1 can be approximated by (Beck, 1980)

$$\overline{T}(c, 0, t) \approx 2\left(\frac{t^+}{\pi}\right)^{1/2} - \frac{t^+}{\pi}\left[2 - \frac{t^+}{4} - \frac{(t^+/4)^2}{4} - \frac{15(t^+/4)^3}{4}\right] \tag{7.149}$$

which is accurate to five significant digits for $0 < t^+ < 0.1$.

The average temperatures are plotted in Figure 7.13 (Beck, 1981). The curve of $\overline{T}$ for small t^+ and for $r^+ > 1$ shown in Figure 7.13 can be obtained by using

$$\overline{T}(c, 0, t) \approx \frac{2}{c^2}\left(\frac{t^+}{\pi}\right)^{1/2} \tag{7.150}$$

This expression becomes more accurate as $t \to 0$ and as c becomes larger. For $c = 1.5$ and $t^+ = 0.2$, it gives a number that is 5% too large, but for $c = 8$ and $t^+ = 4$, the value given by Equation 7.150 is only 0.2% large.

A comparison of Figures 7.12 and 7.13 shows that they have the same general shape, but the average curves start to rise sooner and reach larger steady-state values. This is true for all curves except for $c = r^+ = 0$ for which the curves are identical.

7.12 BODIES WITH $T = T(r, \phi, t)$

When the temperature depends on coordinates r and ϕ, the GFs cannot be found by multiplying one-dimensional GFs. Consequently, these GFs are tabulated separately in Appendix RΦ for boundary conditions of type 0, 1, 2, and 3. In this section, two examples are given of cylinders with angular dependence of the temperature.

In the full cylinder for which $0 < \phi < 2\pi$, the GFs contain Bessel functions $J_n(\cdot)$ and $Y_n(\cdot)$ where n is an integer. The GFs for full cylinders are numbered $RIJ\Phi00$ where $I = 0, 1, 2$, or 3 and $J = 1, 2$, or 3. These GFs are derived from separation of variable methods (Ozisik, 1993).

In the sector of a cylinder for which $0 < \phi < \phi_0$, the GFs contain Bessel functions of fractional order $J_\nu(\cdot)$ and $Y_\nu(\cdot)$ where ν is a rational number. Some other names for

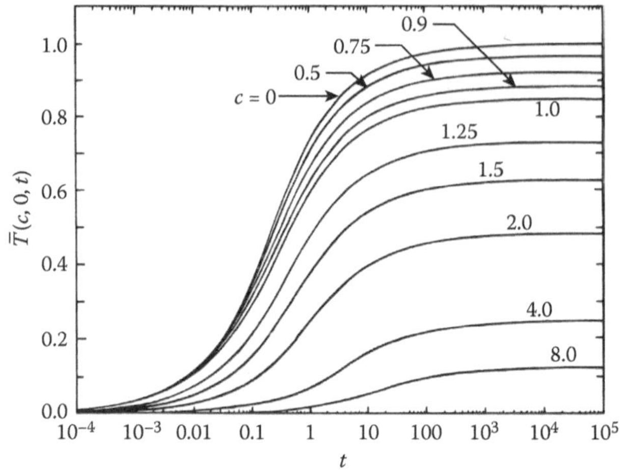

FIGURE 7.13 Average temperature versus time at $z = 0$ on semi-infinite body heated over a circular disk.

the sector of a cylinder are the wedge, the partial cylindrical shell, and the cylinder with a radial slot. The temperature in these bodies is described by the GFs numbered $RIJ\Phi KL$ where J, K, and L are not zero. These functions are listed in table form in Appendix RΦ in Tables RΦ.1 through RΦ.4.

These Bessel functions can be difficult to work with and there are few closed-form solutions that result from the GF method. An attractive alternative to the Bessel functions is the use of Galerkin-based GFs discussed in Chapters 10 and 11. Galerkin-based GFs apply with equal ease to any coordinate system because the GFs are constructed numerically. Since, with the Bessel functions, numerical integration is often needed to find the temperature distribution, the Bessel function (exact) form of the GF has little advantage of accuracy over the Galerkin-based form.

Example 7.14: Cylinder with Initial Temperature Varying with Angle—*R*01*B*0*T*-Φ00 Case

Find the temperature in the full cylinder with initial condition $F(r, \phi)$ and with specified zero temperature on the boundary $r = a$.

Solution

This is the *R*01*B*0*T*-Φ00 case. The GF is listed in Appendix *R*Φ as

$$G_{R01\Phi00}(r, \phi, t|r', \phi', \tau)$$

$$= \frac{2}{2\pi a^2}\sum_{m=1}^{\infty} e^{-\beta_{m0}^2\alpha(t-\tau)/a^2}\frac{J_0(\beta_{m0}r/a)J_0(\beta_{m0}r'/a)}{[J_0'(\beta_{m0})]^2}$$

$$+\frac{2}{\pi a^2}\sum_{n=1}^{\infty}\cos n(\phi-\phi')\sum_{m=1}^{\infty} e^{-\beta_{mn}^2\alpha(t-\tau)/a^2}\frac{J_n(\beta_{mn}r/a)J_n(\beta_{mn}r'/a)}{[J_n'(\beta_{mn})]^2} \tag{7.151}$$

The index on n is for the eigenfunctions $J_n(\cdot)$ and the index on m is for the eigenvalues β_{mn} associated with each eigenfunction. The eigenvalues β_{mn} are the zeroes of

$$\begin{aligned} J_0(\beta_{m0}) &= 0 \qquad \text{for } n = 0 \\ J_1(\beta_{m1}) &= 0 \qquad \text{for } n = 1 \\ J_2(\beta_{m2}) &= 0 \qquad \text{for } n = 2, \text{and so on} \end{aligned} \tag{7.152}$$

The first 10 eigenvalues for $n=0$ through $n=5$ are listed in Ozisik (1993, p. 679). The convergence of the double infinite series in Equation 7.151 is determined primarily by the exponential term so that for small values of $(t-\tau)$, many terms and many eigenvalues are required; finding the eigenvalues can require significant effort.

The temperature in the cylinder is given by the initial condition term of the GFSE:

$$T(r,\phi,t) = \int_{r'=0}^{a} \int_{\phi'=0}^{2\pi} F(r',\phi') \times G_{R01\Phi00}(r,\phi,t|r',\phi,0)\, r'\, d\phi'\, dr' \tag{7.153}$$

Note that $dV' = r'\, d\phi'\, dr'$ for this integral. Replace the GF into Equation 7.153 to get the temperature:

$$\begin{aligned} T(r,\phi,t) = \int_{r'=0}^{a} \int_{\phi'=0}^{2\pi} F(r',\phi') \Bigg\{ &\frac{2}{2\pi a^2} \sum_{m=1}^{\infty} \\ &\times e^{-\beta_{m0}^2 \alpha t/a^2} \frac{J_0(\beta_{m0} r/a) J_0(\beta_{m0} r'/a)}{[J_0'(\beta_{m0})]^2} \\ &+ \frac{2}{\pi a^2} \sum_{n=1}^{\infty} \cos n(\phi - \phi') \sum_{m=1}^{\infty} \\ &\times e^{-\beta_{mn}^2 \alpha t/a^2} \frac{J_n(\beta_{mn} r/a) J_n(\beta_{mn} r'/a)}{[J_n'(\beta_{mn})]^2} \Bigg\} r'\, d\phi'\, dr' \end{aligned} \tag{7.154}$$

The integral on r' cannot be evaluated in closed form even if $F(r,\phi) = F(\phi)$, because the integral

$$\int_{r'=0}^{a} J_n\left(\frac{\beta_{mn} r'}{a}\right) r'\, dr'$$

is not available in closed form.

Example 7.15: Cylindrical Sector (Wedge) Heated over the Curved Surface and Insulated Elsewhere—*R*02*B*–*T*0Φ22 Case

Find the temperature in a sector of a cylinder that is heated over the surface at $r = b$ by a heat flux $q(\phi)$, a steady heat flux that varies with position. The faces of the sector at $\phi = 0$ and $\phi = \phi_0$ are insulated as shown in Figure 7.14. The initial temperature is zero.

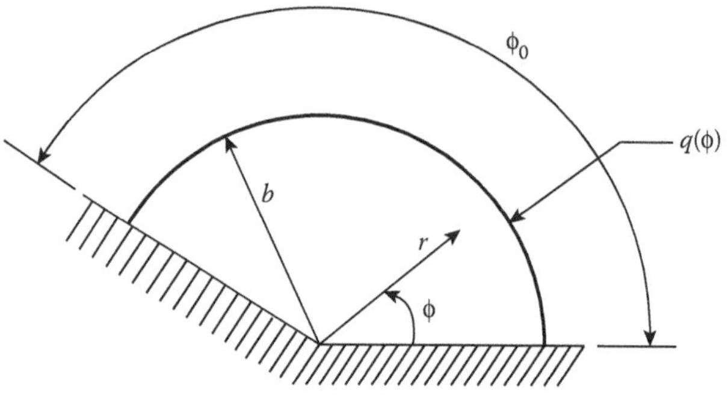

FIGURE 7.14 Sector of a cylinder heated at $r = b$ and insulated elsewhere, case $R02\phi22$.

Solution

This is the $R02B$-$T0\Phi22$ geometry. The temperature in the sector due to heat flux $q(\phi)$ at the surface is given by the GFSE

$$T(r, \phi, t) = \frac{\alpha}{k} \int_{\tau=0}^{t} d\tau \int_{\phi'=0}^{2\pi} q(\phi')\, G_{R02\Phi22}(r, \phi, t|b, \phi, \tau)\, bd\phi' \tag{7.155}$$

The spatial integral extends over the surface at $r = b$. The GF for this case is found from Table RΦ.1 in Appendix RΦ,

$$G(r, \phi, t|r', \phi', \tau) = \frac{R_0(\beta_{00}, r/b)}{N(\beta_{00})N(\nu = 0)} + \sum_{m=1}^{\infty} \sum_{\nu}^{\infty} e^{\beta_{m\nu}^2 \alpha(t-\tau)/a^2} \times \frac{R_\nu(\beta_{m\nu}, r/b)\, R_\nu(\beta_{m\nu}, r'/b)}{N(\beta_{m\nu})} \frac{\Phi(\nu, \phi)\, \Phi(\nu, \phi')}{N(\nu)} \tag{7.156}$$

The GF is constructed from the particular R_ν, Φ, N, and eigenconditions listed in Tables RΦ.1 through RΦ.4 in Appendix RΦ. In the $R02\Phi22$ case,

$$R_\nu(\beta_{m\nu}, r/b) = \begin{cases} 1 & m = 0 \\ J_\nu\left(\dfrac{\beta_{m\nu} r}{b}\right) & \text{for } m \neq 0 \text{ and } \nu \neq 0 \end{cases} \tag{7.157a}$$

$$\frac{1}{N(\beta_{m\nu})} = \begin{cases} \dfrac{2}{b^2} & \text{for } m = 0; \\ \dfrac{2\,\beta_{m\nu}^2}{b^2 J_\nu(\beta_{m\nu})(\beta_{m\nu}^2 - \nu^2)} & \text{for } m \geq 0 \end{cases} \tag{7.157b}$$

$$\Phi(\nu, \phi) = \begin{cases} 1 & \text{for } \nu = 0 \\ \cos \nu\phi & \text{for } \nu \geq 1 \end{cases} \tag{7.157c}$$

$$\frac{1}{N(\nu)} = \begin{cases} \dfrac{1}{\phi_0} & \text{for } \nu = 0 \\ \dfrac{2}{\phi_0} & \text{for } \nu \geq 1 \end{cases} \quad (7.157\text{d})$$

and the eigenvalues are given by the roots of

$$J'_\nu(\beta_{m\nu}) = 0 \quad (7.158\text{a})$$

$$\sin(\nu\phi_0) = 0 \qquad \left(\text{that is, } \nu = \frac{n\pi}{\phi_0};\ n = 0, 1, 2, \ldots\right) \quad (7.158\text{b})$$

The GF may now be assembled by substituting the pieces (7.157) into the general expression (7.156). In the following expression the term for $\nu = 0$ has been separated from the other summation terms:

$$G(r, \phi, t|r', \phi', \tau)$$
$$= \frac{2}{b^2\phi_0} + \sum_{m=1}^{\infty} 2\, e^{-\beta_{m0}^2\alpha(t-\tau)/b^2} \frac{J_0(\beta_{m0} r/b)\, J_0(\beta_{m0} r'/b)}{b^2\phi_0}$$
$$+ \sum_{m=1}^{\infty}\sum_{\nu}^{\infty} 4\, e^{-\beta_{m\nu}^2\alpha(t-\tau)/b^2} \frac{\beta_{m\nu}^2 J_\nu(\beta_{m\nu} r/b)\, J_\nu(\beta_{m\nu} r'/b)}{b^2 J_\nu^2(\beta_{m\nu})\,(\beta_{m\nu}^2 - \nu^2)} \frac{\cos(\nu\phi)\cos(\nu\phi')}{\phi_0} \quad (7.159)$$

where now $\nu = n\pi/\phi_0$ for $n = 1, 2, 3, \ldots$.

The GF may be replaced into Equation 7.155 to give the temperature expression. After the time integral is evaluated, the temperature is given by

$$T(r, \phi, t) = \frac{\alpha t}{b^2} \int_{\phi'=0}^{2\pi} \frac{2\, q(\phi')}{k\,\phi_0} b\, d\phi'$$
$$+ \frac{\phi b}{k} \int_{\phi'=0}^{2\pi} q(\phi') \left[\sum_{m=1}^{\infty} 2\left(1 - e^{-\beta_{m0}^2\alpha t/a^2}\right) \right.$$
$$\left. \times \frac{J_0(\beta_{m0} r/a)\, J_0(\beta_{m0} r'/a)}{\beta_{m0}^2\phi_0} \right] d\phi'$$
$$+ \frac{\phi b}{k} \int_{\phi'=0}^{2\pi} q(\phi') \left[\sum_{m=1}^{\infty}\sum_{\nu}^{\infty} 4\left(1 - e^{-\beta_{m\nu}^2\alpha t/a^2}\right) \right.$$
$$\left. \times \frac{J_\nu(\beta_{m\nu} r/b)\, J_\nu(\beta_{m\nu} r'/b)\cos(\nu\phi)\cos(\nu\phi')}{J_\nu^2(\beta_{m\nu})(\beta_{m\nu}^2 - \nu^2)\,\phi_0} \right] d\phi' \quad (7.160)$$

The units of each term are qb/k which gives temperature as required. The first term in the temperature is the quasisteady term. As $t \to \infty$, the summation terms drop out and the quasisteady term causes the temperature to increase linearly with time.

In the special case $\phi_0 = \pi, \pi/2, \pi/3, \ldots$, where the sector is an even fraction of a full cylinder, then the temperature may be found by the method of images on the full cylinder with case $R02B$-$\Phi00$. The method of images on the cylinder is discussed in Carslaw and Jaeger (1959), but in brief, the method involves a fictitious full cylinder with a surface heating pattern composed of "images" of the original heating pattern on $0 < \phi < \phi_0$ so that the surface $\phi = 0$ and $\phi_0 = 0$ satisfy $\partial T / \partial\phi = 0$ (insulated condition). The temperature in the fictitious full cylinder is found with $R02\Phi00$ analysis and then the desired temperature can be found in the region $0 < \phi < \phi_0$.

7.13 STEADY STATE

Three examples of steady heat transfer in cylindrical coordinates are given in this section. Included are a long cylinder, a finite cylinder with axisymmetry, and a long cylinder with angular effects.

For cylindrical-radial cases, several steady GFs are given in Appendix R, Table R.1. Several 2D GF for cylinders are given elsewhere including Barton (1989, pp. 149–150), Melnikov (1999, Section 5.2), and Duffy (2001, Section 5.2). Several steady-temperature examples for cylinders are given by Carslaw and Jaeger (1959, Sections 8.2 and 8.3) and by Ozisik (1993, Section 3.7).

Example 7.16: Solid Cylinder with Internal Energy Generation—*R03B0G-* Case

Find the steady temperature in a solid cylinder with internal heating. The surface of the cylinder is cooled by convection heat transfer and T_∞ is the fluid temperature.

Solution

If the temperature is evaluated in the form $(T - T_\infty)$, then the convection boundary condition is homogeneous and this is the $R03B0G$- case. The energy generation term of the GFSE may be used to find the temperature as

$$T(r) - T_\infty = \int_{r'=0}^{b} \frac{g(r')}{k} G_{R03}(r|r')\, 2\pi r'\, dr' \tag{7.161}$$

where $g(r')$ is the volume energy generation. The steady GF is given in Appendix R, Table R.1 as

$$G(r|r') = \begin{cases} \dfrac{\ln(b/r') + 1/B_2}{2\pi} & r < r' \\[2ex] \dfrac{\ln(b/r) + 1/B_2}{2\pi} & r > r' \end{cases} \tag{7.162}$$

where B_2 is hb/k, and h is the heat transfer coefficient. Because the GF is piecewise continuous, the spatial integral in Equation 7.161 must be carried out in two pieces:

$$\begin{aligned} T(r) - T_\infty = &\int_{r'=0}^{r} \frac{g(r')}{k}\left[\ln\left(\frac{b}{r}\right) + \frac{1}{B_2}\right] r'dr' \\ &+ \int_{r'=r}^{b} \frac{g(r')}{k}\left[\ln\left(\frac{b}{r'}\right) + \frac{1}{B_2}\right] r'dr' \end{aligned} \tag{7.163}$$

(a) *Case R03B0G1*. For uniform heat generation $g(r) = g_0$ the integrals in Equation 7.163 may be evaluated to give

$$T(r) - T_\infty = \frac{g_0 b^2}{k}\left[\frac{1}{2B_2} + \frac{1-(r/b)^2}{4}\right] \tag{7.164}$$

(b) *Case R03B0G5*. For piecewise constant energy generation

$$g(r') = \begin{cases} 0 & 0 < r' < a \\ g_0 & a < r' < b \end{cases} \tag{7.165}$$

the temperature given by Equation 7.164 must be carried out in two parts depending on the location of the observation point. For $0 < r < a$, the temperature does not depend on location r. Only the first term of Equation 7.163 is used with limits $a < r' < b$ to give

$$\begin{aligned} T(r<a) - T_\infty = \frac{g_0 b^2}{k}&\left\{\frac{1}{2B_2}\left[1-\left(\frac{a}{b}\right)^2\right]\right. \\ &\left.+\frac{1}{4}\left[1-\left(\frac{a}{b}\right)^2\right]+\frac{1}{2}\left(\frac{a}{b}\right)^2 \ln\left(\frac{a}{b}\right)\right\} \end{aligned} \tag{7.166a}$$

For $a < r < b$ both terms of Equation 7.163 are needed and the temperature is

$$\begin{aligned} T(r>a) - T_\infty = \frac{g_0 b^2}{k}&\left\{\frac{1}{2B_2}\left[1-\left(\frac{a}{b}\right)^2\right]\right. \\ &\left.+\frac{1}{4}\left[1-\left(\frac{r}{b}\right)^2\right]+\frac{1}{2}\left(\frac{a}{b}\right)^2 \ln\left(\frac{r}{b}\right)\right\} \end{aligned} \tag{7.166b}$$

Note that the piecewise continuous temperature distributions are equal at $r = a$. Also, as B_2 increases, the surface temperature at $r = b$ approaches T_∞.

Example 7.17: Finite Cylinder with Arbitrary Surface Temperature on the Curved Surface—*R*01*B*-*Z*11*B*00 Case

On a finite cylinder of length L find the steady temperature due a specified temperature $f(z)$ over surface $r = b$ and zero temperature at the ends $z = 0$ and $z = L$.

Solution

This is the *R*01*B*-*Z*11*B*00 geometry. The temperature is given by the steady GFSE equation as

$$T(r, z) = -\int_{z'=0}^{L} f(z') \left.\frac{\partial G_{R01Z11}}{\partial n}\right|_{r'=b} 2\pi b \, dz' \tag{7.167}$$

One form of the steady GF is given by the method of limits combined with the multiplicative property of transient GFs:

$$G(r, z|r', z') = \lim_{t\to\infty} \alpha \int_{t=0}^{t} G_{R01}(r, t|r', \tau)\, G_{Z11}(z, t|z', \tau)\, d\tau \tag{7.168}$$

The transient GFs are given in Appendixes R and X. The time integral and the limit may be evaluated to give the steady GF:

$$G(r, z|r', z') = \sum_{m=1}^{\infty} \sum_{n=1}^{\infty} \frac{2}{\pi b^2 L} \left[\left(\frac{\beta_m}{b} \right)^2 + \left(\frac{n\pi}{L} \right)^2 \right]^{-1} \times \frac{J_0(\beta_m r / b) J_0(\beta_m r' / b)}{J_1^2(\beta_m)} \sin \frac{n\pi z}{L} \sin \frac{n\pi z'}{L} \quad (7.169)$$

The steady GF may be substituted in Equation 7.167 to find the temperature. The derivative on r' is elementary, and the temperature due to surface temperature distribution $f(z)$ is given by

$$T(r, z) = \sum_{m=1}^{\infty} \sum_{n=1}^{\infty} \frac{4}{b^2 L} \left[\left(\frac{\beta_m}{b} \right)^2 + \left(\frac{n\pi}{L} \right)^2 \right]^{-1} \frac{\beta_m J_0(\beta_m r / b)}{J_1(\beta_m)} \sin \frac{n\pi z}{L} \times \int_{z'=0}^{L} f(z') \sin \frac{n\pi z'}{L} \, dz' \quad (7.170)$$

The integral on z' can be found in closed form for many functions $f(z')$. In the case of uniform surface temperature, $f(z) = T_0$, the integral on z' may be evaluated to give

$$T(r, z) = T_0 \sum_{m=1}^{\infty} \sum_{n=1}^{\infty} \frac{4}{n\pi} \frac{1 - (-1)^n}{\beta_m^2 + (n\pi b / L)^2} \frac{\beta_m J_0(\beta_m r / b)}{J_1(\beta_m)} \sin \frac{n\pi z}{L} \quad (7.171)$$

The double summation in Equation 7.171 converges somewhat slowly. Next an alternate solution is given based on a single-sum GF.

Alternative Solution

A single-sum GF for geometry R01Z11 can be found with the method of eigenfunction expansion (Section 4.6). Using eigenfunctions in the z-direction, the GF has the form

$$G(r, z|r', z') = \frac{2}{L} \sum_{n=1}^{\infty} \sin(\beta_n z) \sin(\beta_n z') Q_n(r, r') \quad (7.172)$$

with $\beta_n = n\pi / L$ appropriate for the Z11 geometry. The defining equation for kernel function Q_n may be found by substituting the above series for the GF, along with the series form for $\delta(z - z')$ (see Appendix D), into the defining auxiliary equation for the GF. Then Q_n satisfies

$$\frac{\partial^2 Q_n}{\partial r^2} + \frac{1}{r} \frac{\partial Q_n}{\partial r} - \beta_n^2 Q_n + \frac{\delta(r - r')}{2\pi r'} = 0 \quad (7.173)$$

along with homogeneous boundary conditions for case R01: $Q_n(0)$ is bounded; and, $Q_n(b) = 0$. The above equation for Q_n is a modified Bessel equation (see Appendix B) and the kernel function is given by

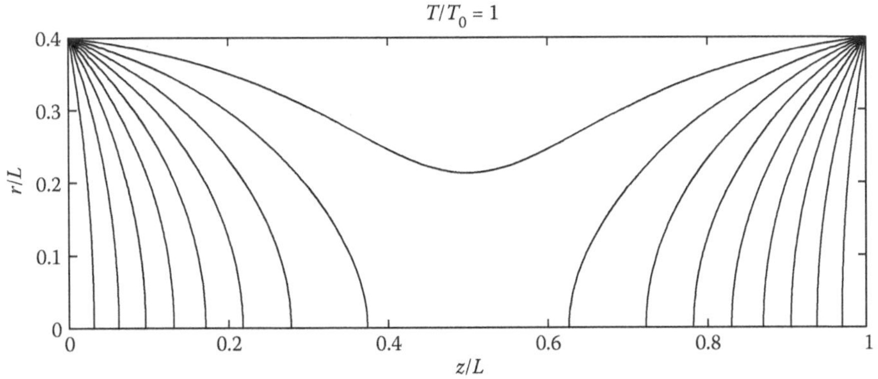

FIGURE 7.15 Contour plot of temperature T/T_0 in the finite cylinder, aspect ratio $b/L = 0.4$, with $T/T_0 = 1$ at the outer radius and $T/T_0 = 0$ at ends $z = 0$ and $z = L$. Contours are at even intervals $T/T_0 = 0.1, 0.2$, etc.

$$Q_n(r, r') = \frac{1}{2\pi}\begin{Bmatrix} [A_2 I_0(\beta_n r') + K_0(\beta_n r')]\, I_0(\beta_n r); & r < r' \\ [A_2 I_0(\beta_n r) + K_0(\beta_n r)] I_0(\beta_n r'); & r > r' \end{Bmatrix} \tag{7.174}$$

$$\text{where} \quad A_2 = -K_0(\beta_n b)/\, I_0(\beta_n b). \tag{7.175}$$

(See also Section 9.3.2 for kernel functions RIJ derived for *steady-periodic* conditions.) With the above kernel function, the single-sum GF may be replaced into the temperature integral, Equation 7.167, to find the alternate single-sum expression for the temperature caused by $f(z) = T_0$:

$$T(r, z) = T_0 \frac{2b}{L} \sum_{n=1}^{\infty} \sin(\beta_n z)[1 - (-1)^n] \left[\frac{K_0(\beta_n b)}{I_0(\beta_n b)} I_1(\beta_n b) + K_1(\beta_n b) \right] I_0(\beta_n r) \tag{7.176}$$

This series converges rapidly everywhere except near $r = b$ (such behavior is common for nonhomogeneous type 1 boundaries). A contour plot of temperature T/T_0 for this case is given in Figure 7.15. Note that the centerline $r = 0$ is a line of symmetry and the plane $z/L = 0.5$ is a plane of symmetry.

Example 7.18: Long Cylinder with Specified Surface Temperature—*R*01*B*–Φ00 Case

A long cylinder has a piecewise constant temperature imposed on its surface. The temperature satisfies

$$\frac{\partial^2 T}{\partial r^2} + \frac{1}{r}\frac{\partial T}{\partial r} + \frac{1}{r^2}\frac{\partial^2 T}{\partial \phi^2} = 0 \tag{7.177}$$

$$T(b, \phi) = \begin{cases} T_0, & 0 < \phi < \phi_0 \\ 0, & \phi_0 < \phi < 2\pi \end{cases} \tag{7.178}$$

Find the steady temperature.

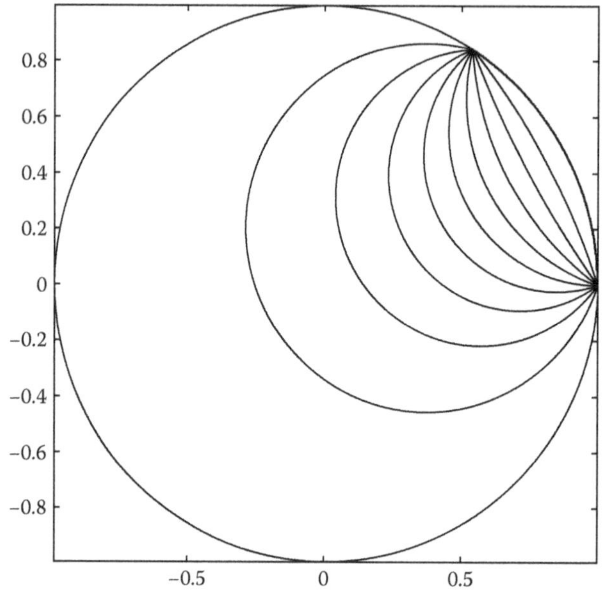

FIGURE 7.16 Contour plot of temperature in the cylinder with elevated boundary temperature over $0 < \phi < \pi/4$ and zero temperature elsewhere on the boundary. Contours are at even intervals $T/T_0 = 0.1, 0.2$, etc.

Solution

The steady temperature in the cylinder is given by the GF solution equation which contains an integral over the surface of the cylinder

$$T(r,\phi) = -\int_{\phi'=0}^{\phi_0} T_0 \left.\frac{\partial G}{\partial r'}\right|_{r'=b} b\, d\phi' \tag{7.179}$$

The appropriate GF satisfies

$$\frac{\partial^2 G}{\partial r^2} + \frac{1}{r}\frac{\partial G}{\partial r} + \frac{1}{r^2}\frac{\partial^2 G}{\partial \phi^2} = -\frac{\delta(r-r')}{2\pi r'}\delta(\phi-\phi') \tag{7.180}$$

A single-sum form of the GF may be developed by eigenfunction expansion with eigenfunctions in the ϕ-direction. The GF is given by (Melnikov, 1999, p. 223)

$$G(r,\phi|r',\phi') = -\frac{1}{2\pi}\begin{bmatrix} \ln(r'/b); & r<r' \\ \ln(r/b); & r>r' \end{bmatrix} + \sum_{n=1}^{\infty} \frac{\cos[n(\phi-\phi')]}{2\pi n} \begin{bmatrix} (r/r')^n - \left(\frac{rr'}{b^2}\right)^n; & r<r' \\ (r'/r)^n - \left(\frac{rr'}{b^2}\right)^n; & r>r' \end{bmatrix} \tag{7.181}$$

Note that the $n = 0$ term is treated separately. Only the $r < r'$ portion of the GF is needed in the temperature integral, Equation 7.179. After evaluating the derivative on r' and the integral on ϕ', the temperature is given by

$$T(r, \phi) = T_0 \left[\frac{\phi_0}{2\pi} + \sum_{n=1}^{\infty} \frac{[\sin n\phi - \sin n(\phi - \phi_0)]}{n\pi} \left(\frac{r}{b}\right)^n \right] \quad (7.182)$$

Temperature contours computed from the above temperature expression are plotted in Figure 7.16 for case $\phi_0 = \pi/4$. That is, the surface temperature is $T_0 = 1$ over $(0 < \phi < \pi/4)$ and is zero elsewhere on the surface. Note that all of the temperature contours begin and end where there are jumps in the surface temperature. A double-sum form of the GF may also be found with the method of limits, but it is not recommended for numerical computation.

PROBLEMS

7.1 Derive the relation of $\alpha(t - \tau)/r^2 = 0.25$ for the time of maximum $G_{R00}(r, t|0, \tau)$ for a given r (not equal to zero).

7.2 Plot $b^2 G_{R00}(r, t|0, \tau)$ versus $\alpha(t - \tau)/b^2$ for $\alpha(t - \tau)/b^2$ values from 0.1 to 2 for $r/b = 0$ and 1.

7.3 Derive the dimensionless distance for the values for the GF to drop to 1% of the $r = 0$ value for a given value of $\alpha(t - \tau)$. (Answer: $r^2/\alpha(t - \tau) = 18.42$ or $\alpha(t - \tau)/r^2 = 0.054$.)

7.4 Plot $b^2 G_{R00}(r, t|0, \tau)$ versus r/b for $\alpha t/b^2 = 0.01$, 1, and 10. (Three separate plots are to be done.)

7.5 Derive the first few terms of Equation R00.4.

7.6 Under what conditions does $G_{X00}(r, t|r', \tau)$ approximate $G_{R00}(r, t| r', \tau)\, 2\pi r'$?

7.7 Derive Equation R00.5.

7.8 Derive the approximate expression below for $G_{R00}(r, t|r', \tau)$ by approximating the circular source by four line sources.

$$G_{R00}(r, t|r', \tau)$$
$$\simeq \frac{1}{16\pi\alpha(t - \tau)} \left\{ \exp\left[-\frac{(r - r')^2}{4\alpha(t - \tau)} \right] + \exp\left[-\frac{(r + r')^2}{4\alpha(t - \tau)} \right] \right.$$
$$\left. + 2\exp\left[-\frac{r^2 + r'^2}{4\alpha(t - \tau)} \right] \right\}$$

Show that the ratio of this approximate expression to the exact one is

$$\frac{G_{R00,\,\text{app}}}{G_{R00,\,\text{exact}}} = \frac{\cosh\{rr'/[2\alpha(t - \tau)]\} + 1}{2I_0\{rr'/[2\alpha(t - \tau)]\}}$$

Calculate values of this ratio, showing that the errors are less than 0.5% for $\alpha(t - \tau)/rr'$ greater than 0.5.

7.9 Compare the numerical values of $G_{R00}(a, t|0, \tau)$ with the average GF over r and r' from the center to $r = a$, which is denoted $\overline{G}_{R00}(t, \tau)$, Equation R00.15. Plot the values from $\alpha(t - \tau)/a^2 = 0$ to 2.

7.10 A line source is frequently used to measure the thermal conductivity. It is made of a thin wire which has an electric current flowing through it. The temperature of the wire is measured and its asymptotic response is used to measure k. Derive an expression using GFs for the temperature distribution in an infinite solid with a line source. The initial temperature is zero. The source is to simulate a wire of radius a and volume energy generation of g_0 in W/m^3.

7.11 Find an expression for the temperature at $r = 0$ in infinite body with the following initial temperature distribution:

$$F(r) = \begin{cases} 0 & r' < a \\ T_0 & a \le r' \le b \\ 0 & r' > b \end{cases}$$

7.12 Find an expression for the temperature everywhere in an infinite body with the following initial temperature:

$$F(r') = \begin{cases} 0 & r' < a \\ T_0\left(\frac{r_0}{r'}\right) & a < r' < b \ (r_0 \text{ is a constant}) \\ 0 & r' > b \end{cases}$$

7.13 Using the series definition of the Bessel function $J_\nu(z)$:

$$J_\nu(z) = \left(\frac{1}{2}z\right)^\nu \sum_{k=0}^{\infty} (-1)^k \frac{(z/2)^{2k}}{k!\Gamma(\nu + k + 1)}$$

show that

$$\frac{d}{dz}\left[J_0(z)\right] = -J_1(z).$$

7.14 Find a small-time temperature for the *R01B0T0G1* case (Example 7.5).

7.15 Find the small-time temperature for the case *R03B0T1* which represents quenching of a hot cylinder in a cold fluid.

7.16 Find the small-time temperature for the case *R02B0T0Gr5* (Example 7.6).

7.17 In hot-wire anemometry, a heated wire is cooled by a fluid flow. An important issue is the time constant of the wire, which is the time for the heated wire to come to steady state. This problem is a simple model of the time constant of the wire alone (without supports).

(a) Find the spatial average, time-varying temperature in the wire heated uniformly by energy generation g_0 (case *R03B0G1T0*).

(b) As an estimate of the time constant, find the time it takes for the *average* temperature in the wire to reach 90% of the steady temperature. (See Example 7.16 for the steady temperature.)

7.18 Consider the *R21B10T0* case:

(a) Write down the GF solution equation for this case.

(b) Find the transient temperature in *integral form* using the GF from Appendix R. (Hint: use $R(r)$ $R(r')$ as a shorthand notation for the eigenfunctions, and refer to Equation 7.87.)

(c) Carry out the integral on τ to find the temperature in closed form.

7.19 A steel rod 25 mm in diameter is heated to a temperature of 1000°C, then quenched in a liquid bath. The temperature of the bath remains constant and equal to 50°C. If the heat transfer coefficient is 10,000 W/m^{2}K, calculate the time required for the center temperature to reach 500°F. What is the surface temperature at the calculated time? The thermophysical properties are $k = 32$ W/m K, $c_p = 700$ J/kg K, and $\rho = 7800$ kg/m^3.

7.20 An electrical cable with a 1-cm diameter copper wire ($k = 400$ W/m K) and 0.5-cm thick electrical insulation ($k = 0.5$ W/m K) carries electricity. The current is 300 A and resistance is 0.006 ohm/m. When the ambient temperature is 25°C, use a steady-state solution to calculate the surface heat transfer coefficient if the wire temperature is not to exceed 100°C. For the same heat transfer coefficient, calculate the temperature variations as a function of time at the center of wire. The line frequency is 60 cycles per second.

7.21 Find the steady-state temperature in a thin-walled tube that is cooled by steady uniform convection inside and heated by incident solar radiation on the outside. Assume that all of the incident radiation is absorbed and that the incident radiation is described by $q = q_0 \cos\phi$ for $-\pi/2 < \phi < \pi/2$, and $q = 0$ otherwise. Model the tube as a fin (see Example 7.11) and treat the surface heat flux as energy generation $g(\phi) = q(\phi)/\delta$ where δ is the tube-wall thickness.

7.22 Show that $\int_{\phi'=0}^{\phi_0} G_{RIJ\Phi22}(\cdot)\, d\phi' = G_{RIJ}(\cdot)$.

7.23 Show that $\int_{\phi'=0}^{2\pi} G_{RIJ\Phi00}(\cdot)\, d\phi' = G_{RIJ}(\cdot)$.

7.24 Does $G_{RIJ\Phi11}(\cdot)$ for $G = 0$ at $\phi = \phi_0$ equal $G_{RIJ\Phi12}(\cdot)$ for $\partial G/\partial\phi = 0$ at $\phi = \phi_0/2$? Examine both physically and mathematically.

7.25 Does $G_{RIJ\Phi22}(\cdot)|_{\phi_0=2\pi} = G_{RIJ\Phi00}(\cdot)$? Examine both physically and mathematically.

7.26 Derive the centerline temperature ($r = 0$) for the disk heat source on a semi-infinite body given by Equation 7.136. Use the form of G_{R00} that contains the modified Bessel function $I_0(r\, r'/[4\alpha(t-\tau)])$.

7.27 Find an integral expression for the surface temperature caused by a short laser pulse of duration δt on a large flat surface. The surface heating by the laser may be modelled as a uniform disk heat source of radius a. Find the average surface temperature over the laser-heated region as a function of time for $t > \delta t$.

7.28 A more realistic model of laser beam absorption involves a distribution of energy across the beam. Find an integral expression for the transient temperature on the surface of an opaque semi-infinite solid caused by a short laser pulse of duration δt where the incident energy has a Gaussian distribution:

$$q(r,t) = q_0\, e^{-2(r/a)^2} \qquad \text{for } 0 < t < \delta t$$

where now a is the Gaussian beam radius. The initial temperature is zero. Find the maximum temperature on the surface and the time when it occurs.

7.29 Find an integral expression for the temperature in a half-cylinder (sector $0 < \phi < \pi$) with initial temperature T_0. The surface $r = b$ is insulated and the flat surface ($\phi = 0$ and $\phi = \pi$) is held at a fixed temperature T_0.

7.30 Induction heating is a rapid, highly localized heating method that is used to harden bearing surfaces on crankshafts. If the crankshaft may be modelled as a solid cylinder and the induction heating may be modelled as surface heating, find the transient temperature in the cylinder suddenly heated over a small portion of its length $-a < z < 0$. The remainder of the cylinder surface is insulated. Initially the cylinder has zero temperature.

7.31 A pin fin is a cylinder with a fixed elevated temperature of T_0 at $z = 0$ and convection cooling by a fluid at T_∞ over the other surfaces.

(a) Find an exact expression for the steady two-dimensional heat flow into the pin fin at $z = 0$ (in watts) by analyzing the geometry $R03B0Z11B10$ (assume the temperature at $z = L$ is T_∞).

(b) Find an approximate expression for the steady heat flow into the pin fin at $z = 0$ by analyzing the fin equation $\nabla^2 T - m^2(T - T_\infty) = 0$ for geometry $Z11$.

(c) Compare the numerical answer from parts (*a*) and (*b*) in the specific cases of a pin fin with length/radius of $L/a = 3, 10$, and comment on the conditions for which the fin approximation is useful.

7.32 Find the transient temperature $T(r, \phi, t)$ in a long circular cylinder initially at zero temperature and with uniform heat flux on a sector of its surface $0 < \phi < \phi_0$. The remainder of the surface is insulated. This is a model of a split-film anemometer sensor formed from a platinum heater bonded to a quartz cylinder. Write the temperature as the sum of three terms: the spatial-average (or lumped) term proportional to time, the transient term that dies away as $t \to \infty$, and the quasisteady term that does not depend on time. If the cylinder properties are $k = 1.4\,\text{W/(m K)}$ and $\alpha = 8.3\text{E–}07\,\text{m}^2/\text{s}$, the cylinder radius is 2.5E–05 m, and $\phi_0 = \pi$, find the time for the transient term to die away to 10% of its initial value; this is a measure of the response time of the split-film anemometer sensor.

REFERENCES

Abramowitz, M. and Stegun, I. A. (Eds), 1964, *Handbook of Mathematical Functions with Formulas, Graphs, and Mathematical Tables*, Applied Mathematics Series 55, National Bureau of Standards, Washington, DC.

Barton, G., 1989, *Elements of Green's Functions and Propagation*, Oxford University Press, London.

Beck, J. V., 1980, Average transient temperature within a body heated by a disk heat source, in *Heat Transfer, Thermal Control and Heat Pipes*, ed. W. B. Olstad, vol. 70, Progress in Astronautics and Aeronautics, American Institute of Aeronautics and Astronautics (AIAA) pp. 3–24.

Beck, J. V., 1981, Large time solutions for temperatures in a semi-infinite body with a disk heat source, *Int. J. Heat Mass Transfer*, vol. 24, pp. 155–164.

Beck, J. V., Keltner, N. R., and Schisler, I. P., 1985, Influence functions for the unsteady surface element method, *AIAA J.*, vol. 23, pp. 1978–1982.

Carslaw, H. S. and Jaeger, J. C., 1959, *Conduction of Heat in Solids*, 2nd edn, Oxford University Press, New York.

Cole, K. D., 2004, Computer software for fins and slab bodies with Green's functions, *Computer Appl. Eng. Education*, vol. 12, no. 3, pp. 189–197.

Cole, K. D., Tarawneh, C., and Wilson, B., 2009, Analysis of flux-based fins for estimation of heat transfer coefficient, *Int. J. Heat Mass Transfer*, vol. 52, pp. 92–99.

Duffy, D. G., 2001, *Green's Functions with Applications*, Chapman and Hall/CRC, Boca Rotan, FL.

Melnikov, Y. A., 1999, *Influence Functions and Matrices*, Marcel Dekker, New York.

Ozisik, M. N., 1993, *Heat Conduction*, John Wiley, New York.

Thomas, P. H., 1957, Some conduction problems in the heating of small areas on large solids, *Q. J. Mech. Appl. Math.*, vol. 10, pp. 482–493.

Watson, G. N., 1944, *A Treatise on the Theory of Bessel Functions*, 2nd edn, Cambridge University Press, London.

Wylie, C. R. and Barrett, L. C., 1995, *Advanced Engineering Mathematics*, 6th edn, McGraw-Hill, New York.

8 Radial Heat Flow in Spherical Coordinates

8.1 INTRODUCTION

The applications of the Green's function (GF) solution approach to the problems posed in the spherical coordinate system are discussed in this chapter. The general heat conduction equation for linear flow of heat in spherical polar coordinates has the form

$$\frac{1}{r^2}\frac{\partial}{\partial r}\left(r^2\frac{\partial T}{\partial r}\right)+\frac{1}{r^2\sin\theta}\frac{\partial}{\partial\theta}\left(\sin\theta\frac{\partial T}{\partial\theta}\right)+\frac{1}{r^2\sin^2\theta}\frac{\partial^2 T}{\partial\phi^2}+\frac{1}{k}g(r,\theta,\phi,t)=\frac{1}{\alpha}\frac{\partial T}{\partial t} \tag{8.1}$$

where $T = T(r,\theta,\phi,t)$, g represents the generation rate per unit volume (W/m^3) within the spherical region, and k is constant.

As was mentioned in the previous chapter, the applications of the GF solution method to multidimensional problems involve cumbersome analytical work. In addition, for spherical coordinates, unlike the rectangular and cylindrical coordinates, the two- and three-dimensional GFs cannot be obtained from the product of the one-dimensional solutions. Because of these problems and because many heat conduction problems in spherical coordinates involve spherical symmetry (i.e., the temperature does not depend on θ and ϕ) this chapter emphasizes problems with temperature distributions that are functions only of time t and radius r (radial flow of heat). For radial flow of heat, Equation 8.1 reduces to:

$$\frac{1}{r^2}\frac{\partial}{\partial r}\left(r^2\frac{\partial T}{\partial r}\right)+\frac{g(r,t)}{k}=\frac{1}{r}\frac{\partial^2(rT)}{\partial r^2}+\frac{1}{k}g(r,t)=\frac{1}{\alpha}\frac{\partial T}{\partial t} \tag{8.2}$$

Later in this chapter, we will show how this equation can further be simplified and put into the rectangular form by introducing a new temperature $U(r,t) = rT(r,t)$. Topics covered in the remainder of this chapter include the Green's function solution equation (GFSE) for radial flow of heat in spherical coordinates (Section 8.2), the infinite body with radial flow of heat (Section 8.3), methods for obtaining the related GFs (Section 8.4), and how the GF solution method can be used to solve a number of important problems for radial flow of heat in the geometries of solid and hollow spheres and in the region outside a spherical cavity (Sections 8.6 through 8.8).

8.2 GREEN'S FUNCTION EQUATION FOR RADIAL SPHERICAL HEAT FLOW

From the general GF equation for heat conduction given by Equation 3.46, one can write down the GF equation for the radial heat flow in spherical coordinates with the exclusion of the term associated with the boundary conditions of the fourth and fifth kinds, as

$$T(r,t) = \int_{r'} G(r,t|r',0)\,F(r')4\pi r'^2 dr'$$

(for the initial condition)

$$+ \int_{\tau=0}^{t} \int_{r'} \frac{\alpha}{k} G(r,t|r',\tau) g(r',\tau) 4\pi r'^2 dr'\, d\tau$$

(for volume energy generation)

$$+ \alpha \int_{\tau=0}^{t} \sum_{i=1}^{S} \frac{f_i(r_i,\tau)}{k_i} G(r,t|r_i,\tau) 4\pi r_i^2 d\tau$$

(for boundary conditions of the second and third kinds)

$$- \alpha \int_{\tau=0}^{t} \sum_{j=1}^{S} f_j(r_j,\tau) \frac{\partial G}{\partial n'_j}\bigg|_{r'=r_j} 4\pi r_j^2 d\tau$$

(for boundary condition of the first kind only) (8.3)

Note that $dv' = 4\pi r'^2 dr'$, and that the integrals over boundary surface s_i have been replaced by $4\pi r_i^2$.

The GFs associated with different set of boundary conditions for radial spherical heat flow are denoted by $G_{RSIJ}(\cdot)$, where subscript RS stands for radial spherical according to the heat conduction numbering system. Only boundary conditions of the zeroth through the third kinds are considered here ($I, J = 0, 1, 2, 3$). A listing of the available GFs for radial spherical heat flow is provided in Appendix RS.

8.3 INFINITE BODY

The GF for spherical radial flow of heat in an infinite body is denoted by $G_{RS00}(r,t|r',\tau)$. It is called the fundamental heat conduction solution for spherical radial heat flow and is given by

$$G_{RS00}(r,t|r',\tau) = \frac{1}{8\pi r r'[\pi\alpha(t-\tau)]^{1/2}} \times \left\{ \exp\left[-\frac{(r-r')^2}{4\alpha(t-\tau)}\right] - \exp\left[-\frac{(r+r')^2}{4\alpha(t-\tau)}\right] \right\} \quad (8.4)$$

This GF represents the temperature response due to a unit instantaneous spherical surface source of radius r' at time τ in an infinite body with zero initial condition.

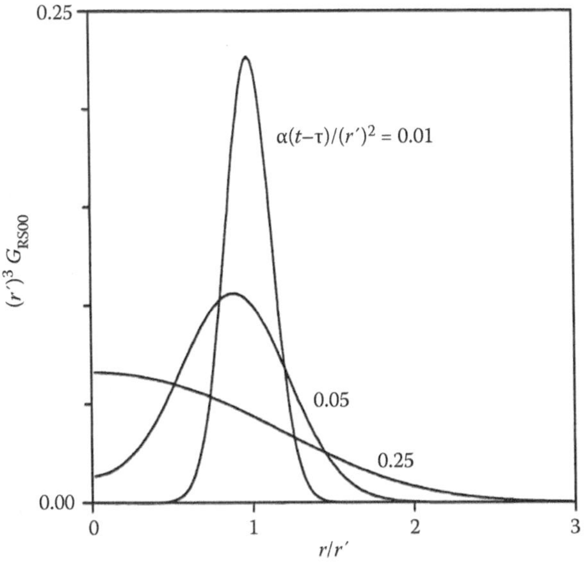

FIGURE 8.1 The $RS00$ GF.

Do not confuse the spherical-surface source with the point source discussed in Section 4.7.1. This GF satisfies

$$\frac{1}{r}\frac{\partial^2(rG)}{\partial r^2} + \frac{1}{\alpha}\delta(\mathbf{r}-\mathbf{r}')\,\delta(t-\tau) = \frac{1}{\alpha}\frac{\partial G}{\partial t} \tag{8.5a}$$

$$\frac{\partial G}{\partial r}(0, t|r', \tau) = 0 \tag{8.5b}$$

$$G(\infty, t|r', \tau) = 0 \tag{8.5c}$$

$$G(r, 0|r', \tau > 0) = 0 \tag{8.5d}$$

Figure 8.1 shows $r'^3 G_{RS00}(\cdot)$ versus $r^+ = r/r'$ for various values of $t^+ = \alpha(t-\tau)/r'^2$. Note that $G_{RS00}(\cdot)$ is unaffected by the axisymmetric condition of $\partial G/\partial r = 0$ at $r = 0$ for $t^+ < 0.03$ and approaches the Dirac delta function as t^+ goes to zero. For larger values of t^+, the position of the maximum G moves to smaller r^+ values.

It is interesting to note that for the special case where $r' \to 0$, the $G_{RS00}(\cdot)$ becomes:

$$G_{RS00}(r, t|0, \tau) = \frac{1}{[4\pi\alpha(t-\tau)]^{3/2}} \exp\left[-\frac{r^2}{4\alpha(t-\tau)}\right] \tag{8.6}$$

which represents the response due to an instantaneous point source at the origin. It can also be shown that for the case where r' is fixed and $r \to 0$, a similar equation to (8.6) is obtained; that is,

$$G_{RS00}(0,t|r',\tau) = \frac{1}{[4\pi\alpha(t-\tau)]^{3/2}} \exp\left[-\frac{r'^2}{4\alpha(t-\tau)}\right] \tag{8.7}$$

which gives the response at the origin due to an instantaneous spherical surface source at r'. From Equations 8.6 and 8.7, it is obvious that the reciprocity relation holds for this GF.

It is also interesting to note that the point source solution, Equation 8.6, can be represented by the product of three plane heat sources for the x-, y-, and z-directions; that is,

$$G_{RS00}(r,t|0,\tau) = G_{X00}(x,t|0,\tau)\, G_{Y00}(y,t|0,\tau)\, G_{Z00}(z,t|0,\tau) \tag{8.8a}$$

which also shows that the unit of $G_{RS00}(\cdot)$ is $\text{m}^{-1}\text{m}^{-1}\text{m}^{-1} = \text{m}^{-3}$. Note that for this case where the source is at the origin ($r' = 0$), the distance between the impulse and response points is given by

$$r = (x^2 + y^2 + z^2)^{1/2} \tag{8.8b}$$

Similarly, for the case where r' is fixed and $r = 0$, we can write

$$G_{RS00}(0,t|r',\tau) = G_{X00}(0,t|x',\tau)\, G_{Y00}(0,t|y',\tau)\, G_{Z00}(0,t|z',\tau) \tag{8.9a}$$

where

$$r' = (x'^2 + y'^2 + z'^2)^{1/2} \tag{8.9b}$$

represents the distance between the impulse and the response points for this case.

The *RS*00 GF given by Equation 8.4 may be employed to obtain GFs for other cases of radial flow of heat in spherical geometry with different types of boundary conditions. For instance, in Section 4.3, we saw how $G_{RS00}(\cdot)$ was used in the Laplace transform approach to obtain G_{RS30} which is the GF for the infinite region outside the spherical cavity $r = a$ with convective boundary condition at $r = a$.

8.3.1 Derivation of the *RS*00 Green's Function

The derivation given here is based on the physical interpretation that $G_{RS00}(\mathbf{r},t|\mathbf{r}',\tau)$ is equal to the temperature rise due to an instantaneous spherical surface source at time τ and location $\mathbf{r} = \mathbf{r}'$ divided by the strength of the source and multiplied by ρc. Having this in mind, we start with the general form of the GF solution equation (3.46). From this equation, the temperature due to a distributed energy source is an infinite body with zero initial condition is given by

$$T(\mathbf{r},t) = \int_{\tau=0}^{t} \int_R \frac{\alpha}{k} G(\mathbf{r},t|\mathbf{r}',\tau)\, g(\mathbf{r}',\tau) dV' \, d\tau \tag{8.10}$$

where $G(\mathbf{r},t|\mathbf{r}',\tau)$ represents the temperature response at point $\mathbf{r}$ and time t in an infinite body due to an instantaneous impulse at point $\mathbf{r}'$ and time τ. It is given by

$$G(\mathbf{r},t|\mathbf{r}',t) = \frac{1}{[4\pi\alpha(t-\tau)]^{3/2}} \exp\left[-\frac{R^2}{4\alpha(t-\tau)}\right] \tag{8.11}$$

where R represents the distance between the points $\mathbf{r}$ and $\mathbf{r}'$. In rectangular coordinate system, this distance is given by

$$R^2 = (x - x')^2 + (y - y')^2 + (z - z')^2 \tag{8.12}$$

where (x, y, z) and (x', y', z') are rectangular coordinates of points $\mathbf{r}$ and $\mathbf{r}'$, respectively. The rectangular coordinates (x, y, z) and (x', y', z') can be transformed into the spherical coordinates (r, θ, ϕ) and (r', θ', ϕ') through the following relations:

$$x = r\ \sin\theta\ \cos\phi \tag{8.13a}$$

$$y = r\ \sin\theta\ \sin\phi \tag{8.13b}$$

$$z = r\ \cos\theta \tag{8.13c}$$

Then the distance R in spherical coordinates may be presented by

$$R^2 = r^2 + r'^2 - 2rr'[\sin\theta\ \sin\theta' \cos(\phi - \phi') + \cos\theta \cos\theta'] \tag{8.14a}$$

For the case of only radial flow of heat in spherical system, there is no temperature variation with θ and ϕ. This implies that the temperature at any point $\mathbf{r}$ over a spherical surface which is at an arbitrary distance from the spherical surface source (at $r = r'$) is the same regardless of the values of θ and ϕ. Accordingly, for simplicity, we choose the spherical coordinates of point $\mathbf{r}$ to be $(0, 0, r)$. Then Equation 8.14a simplifies to

$$R^2 = r^2 + r'^2 - 2rr' \cos\theta' \tag{8.14b}$$

The generation term in Equation 8.10 represents a *continuous distributed volumetric source* and has the unit of W/m^3. However, since we are seeking the temperature solution due to an *instantaneous spherical surface source*, $g(\mathbf{r}', \tau)$, in Equation 8.10 is replaced by

$$g(\mathbf{r}', \tau) = \frac{\delta(\tau - \tau_0)\delta(r' - r_0)g_0}{4\pi r_0^2} \tag{8.15}$$

where r_0 is the radius of the spherical surface source that pulses at time τ_0, and g_0 (Joule) represents the strength of the source. (The strength per unit area is given by $g_0 / 4\pi r_0^2$ and function $\delta(r - r_0)$ has units of m^{-1}).

Now by substituting the values of $G(\mathbf{r}, t|\mathbf{r}', \tau)$, R, and $g(\mathbf{r}', \tau)$ from Equations 8.11, 8.14b, and 8.15 into Equation 8.10 and integrating over the appropriate ranges for $r'(0 \to \infty)$, $\phi'(0 \to 2\pi)$, and $\theta'(0 \to \pi)$, one can write

$$T(r, t) = \frac{\alpha}{k} \int_{\tau=0}^{t} \int_{r'=0}^{\infty} \int_{\theta'=0}^{\pi} \int_{\phi'=0}^{2\pi} \times \left\{ \frac{1}{[4\pi\alpha(t - \tau)]^{3/2}} \exp\left[-\frac{r^2 + r'^2 - 2rr' \cos\theta'}{4\alpha(t - \tau)} \right] \times \frac{\delta(\tau - \tau_0)\delta(r' - r_0)g_0}{4\pi r_0^2} \right\} r'^2 \sin\theta'\, dr'\, d\theta'\, d\phi'\, d\tau \tag{8.16}$$

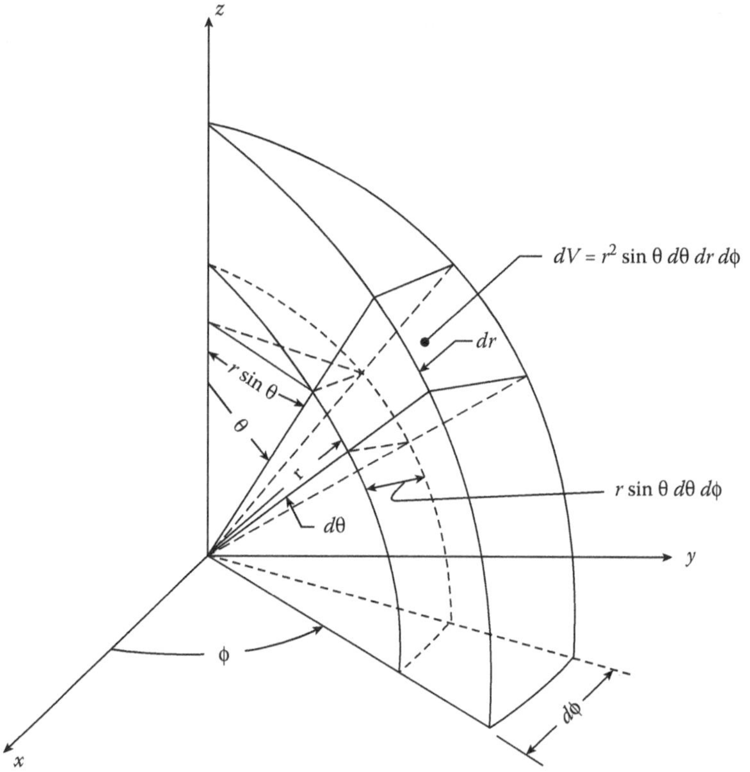

FIGURE 8.2 Spherical polar coordinates.

where dV' in Equation 8.10 has been replaced by $r'^2 \sin\theta'\, dr'\, d\theta'\, d\phi'$ (see Figure 8.2). Note that Equation 8.16 gives the temperature due to an instantaneous spherical surface source at radius r_0 and time τ_0. The integrals over r' and τ can be evaluated easily with the sifting property of the Dirac delta functions. Then Equation 8.16 reduces to

$$T(r,t) = \frac{\alpha}{k} \frac{g_0/(4\pi)}{[4\pi\alpha(t-\tau_0)]^{3/2}} \exp\left[-\frac{(r^2+r_0^2)}{4\alpha(t-\tau_0)}\right] \int_{\phi'=0}^{2\pi} d\phi' \int_{\theta'=0}^{\pi} \times \exp\left[-\frac{rr_0\cos\theta'}{2\alpha(t-\tau_0)}\right] \sin\theta'\, d\theta' \quad (8.17)$$

The integral over ϕ' is equal to 2π and the integral over θ' can be evaluated easily by choosing a new variable $\mu = \cos\theta'$ to give

$$T(r,t) = \frac{g_0}{\rho c} \frac{1}{8\pi r r_0[\alpha\pi(t-\tau_0)]^{1/2}} \left\{\exp\left[-\frac{(r-r_0)^2}{4\alpha(t-\tau_0)}\right] - \exp\left[-\frac{(r+r_0)^2}{4\alpha(t-\tau_0)}\right]\right\} \quad (8.18)$$

Then, the GF is given by the temperature divided by the source strength and multiplied by ρc.

$$G_{RS00}(r,t|r_0,\tau_0) = \frac{T(r,t)}{g_0/\rho c} = \frac{1}{8\pi r r_0[\alpha\pi(t-\tau_0)]^{1/2}} \times \left\{\exp\left[-\frac{(r-r_0)^2}{4\alpha(t-\tau_0)}\right] - \exp\left[-\frac{(r+r_0)^2}{4\alpha(t-\tau_0)}\right]\right\} \tag{8.19}$$

Finally, by considering the conventional form, that is, the heat source being at (r', τ) instead of at (r_0, τ_0), the same result as in Equation 8.4 is obtained.

8.4 SEPARATION OF VARIABLES FOR RADIAL HEAT FLOW IN SPHERES

In the previous chapters, we saw that the separation of variables method provides an easy and straightforward approach for obtaining the GFs for finite-body problems posed in the Cartesian and cylindrical coordinate systems with arbitrary initial temperature distributions provided that the differential equations and the boundary conditions are homogeneous. This method can also be applied to the radial spherical heat flow problems to obtain the appropriate GFs. However, for radial flow of heat in spheres, there is an alternative approach which is more convenient and involves less analytical work than the separation of variables method; that is, the *RSIJ* GFs can be obtained from the *XIJ* GFs through a simple transformation of the variables. Since the method of separation of variables has already been discussed and demonstrated in detail for the problems posed in the Cartesian and cylindrical coordinate systems, in this section we demonstrate only the second approach by considering various examples.

The heat conduction equation for linear radial flow of heat in spherical coordinates is given by Equation 8.2 as

$$\frac{1}{r}\frac{\partial^2(rT)}{\partial r^2} + \frac{1}{k}g(r,t) = \frac{1}{\alpha}\frac{\partial T}{\partial t} \tag{8.20}$$

This equation can be put into the rectangular form by introducing a new temperature U (the dependent variable) as

$$U(r,t) = rT(r,t) \tag{8.21}$$

Then Equation 8.20 becomes

$$\frac{\partial^2 U}{\partial r^2} + \frac{1}{k}g^*(r,t) = \frac{1}{\alpha}\frac{\partial U}{\partial t} \tag{8.22}$$

where

$$g^*(r,t) = rg(r,t) \tag{8.23}$$

Note that $U(r,t)$ in Equation 8.22 is similar to $T(x,t)$ in Equation 1.113 for the Cartesian coordinate system. The above transformation should also be applied to the boundary conditions and the initial condition of the problem under consideration. Once the problem is completely transformed into the Cartesian coordinate system, the appropriate *XIJ* GFs can be found from the separation of variable method or more conveniently from the available tables given in Appendix X. Then, the *RSIJ* GFs are obtained by transforming the results back into the spherical coordinate system. Note that only the homogeneous part of Equation 8.22 is considered for derivation of the GFs. The relation between the G_{RSIJ} and $G_{XIJ}(I,\ J = 0, 1, 2, 3)$ may vary from case to case depending on the geometry and the type of the boundary conditions. This is best illustrated through the following examples.

Example 8.1: Derivation of G_{RS03}—Solid Sphere with Convective Boundary Condition

The *RS*03 GF is obtained from the solution to the following homogeneous problem with an arbitrary initial condition described by

$$\frac{1}{r}\frac{\partial^2(rT)}{\partial r^2} = \frac{1}{\alpha}\frac{\partial T}{\partial t} \qquad 0 \le r \le b \quad t > 0 \tag{8.24}$$

$$T \text{ is finite at } r = 0 \qquad t > 0 \tag{8.25a}$$

$$\frac{\partial T(b,t)}{\partial r} + HT(b,t) = 0 \qquad t > 0 \tag{8.25b}$$

where

$$H \equiv \frac{h}{k} \tag{8.25c}$$

$$T(r,0) = F(r) \qquad 0 \le r \le b \tag{8.26}$$

By introducing the new variable U as

$$U(r,t) = r\,T(r,t) \tag{8.27}$$

Equations 8.24 through 8.26 become

$$\frac{\partial^2 U(r,t)}{\partial r^2} = \frac{1}{\alpha}\frac{\partial U}{\partial t} \qquad 0 \le r \le b \qquad t > 0 \tag{8.28}$$

$$U(0,t) = 0 \qquad t > 0 \tag{8.29a}$$

$$\frac{\partial U(b,t)}{\partial r} + H^* U(b,t) = 0 \qquad t > 0 \tag{8.29b}$$

where

$$H^* = H - \frac{1}{b} \tag{8.29c}$$

$$U(r,0) = rF(r) = F^*(r) \qquad 0 \le r \le b \tag{8.30}$$

The transformed problem described by Equations 8.28 through 8.30 represents a flat plate problem with an arbitrary initial condition and the homogeneous boundary conditions of the first kind on one side and the third kind on the other side ($X13$). See Figures 8.3a and b.

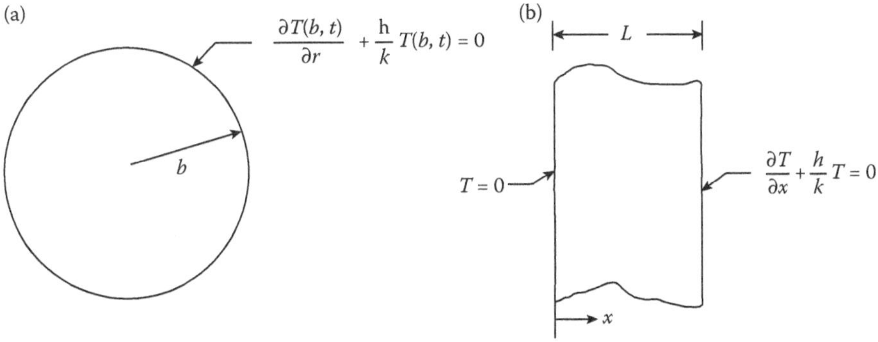

FIGURE 8.3 (a) Solid sphere with convection boundary condition ($RS03$). (b) Flat plate with boundary conditions of the first and third kinds ($X13$).

From the separation of variables method or the available tables in Appendix X, the GF for the $X13$ geometry for $\tau = 0$ is given by

$$G_{X13}(x,t|x',0) = \frac{2}{L}\sum_{m=1}^{\infty} \exp\left(-\beta_m^2 \frac{\alpha t}{L^2}\right) \frac{(\beta_m^2 + B^2)\sin(\beta_m x/L)\sin(\beta_m x'/L)}{\beta_m^2 + B^2 + B} \tag{8.31}$$

where

$$\beta_m \cot\beta_m = -B \qquad B = HL = \frac{hL}{k} \tag{8.32a,b}$$

and L represents the thickness of the plate.

Then with $L \to b$, $x \to r$, $x' \to r'$, and $H \to H^*$, one can write,

$$G_{X13}(r,t|r',0) = \frac{2}{b}\sum_{m=1}^{\infty} \exp\left(-\beta_m^2 \frac{\alpha t}{b^2}\right) \frac{(\beta_m^2 + B^2)\sin(\beta_m r/b)\sin(\beta_m r'/b)}{\beta_m^2 + B^2 + B} \tag{8.33}$$

where

$$\beta_m \cot\beta_m = -B \qquad B = H^*b = Hb - 1 \tag{8.34a,b}$$

From the first term of Equation 3.16 which gives the temperature in a flat plate due to a nonuniform initial condition, one can write

$$U(r,t) = \int_{r'=0}^{b} F^*(r')G_{X13}(r,t|r',0)dr' \tag{8.35}$$

Replacing for $F^*(r') = r'F(r')$, and transforming $U(r,t)$ back into $T(r,t)$, gives

$$T(r,t) = \frac{1}{r}U(r,t) = \int_{r'=0}^{b} \frac{r'}{r}F(r')G_{X13}(r,t|r',0)dr' \tag{8.36}$$

Equation 8.36 can be rearranged to give

$$T(r,t) = \int_{r'=0}^{b} \left[\frac{1}{4\pi r r'} G_{X13}(r,t|r',0) \right] F(r') 4\pi r'^2 \, dr' \tag{8.37}$$

$T(r, t)$ can also be obtained by solving the initial value problem (Equations 8.24 through 8.26) with the GF equation for radial spherical heat flow (Equation 8.3) to give

$$T(r,t) = \int_{r'=0}^{b} G_{RS03}(r,t|r',0) F(r') 4\pi r'^2 \, dr' \tag{8.38}$$

Now, by comparing Equations 8.37 and 8.38, which both represent the same solution, one can conclude that the term in the brackets in Equation 8.37 must be the *RS*03 GF evaluated at $\tau = 0$; that is,

$$G_{RS03}(r,t|r',0) = \frac{1}{4\pi r r'} G_{X13}(r,t|r',0) \tag{8.39}$$

Finally by substitution of $G_{X13}(r,t|r',0)$ from Equation 8.33 and by replacement of $(t-0)$ by $(t-\tau)$, one can write

$$G_{RS03}(r,t|r',\tau) = \frac{1}{2\pi b r r'} \sum_{m=1}^{\infty} \exp\left[-\beta_m^2 \frac{\alpha(t-\tau)}{b^2} \right] \times \frac{(\beta_m^2 + B^2)\sin(\beta_m r/b)\sin(\beta_m r'/b)}{\beta_m^2 + B^2 + B} \tag{8.40}$$

where

$$\beta_m \cot\beta_m = -B \qquad B = Hb - 1 \tag{8.41a,b}$$

This GF is also listed in Appendix RS.

Example 8.2: Derivation of G_{RS33}—Hollow Sphere with Convective Boundary Conditions

Consider the following homogeneous initial-value problem for a hollow sphere as shown in Figure 8.4a:

$$\frac{1}{r}\frac{\partial^2(rT)}{\partial r^2} = \frac{1}{\alpha}\frac{\partial T}{\partial t} \qquad a \le r \le b \quad t > 0 \tag{8.42}$$

$$\frac{\partial T(a,t)}{\partial r} - H_1 T(a,t) = 0 \qquad t > 0 \tag{8.43a}$$

$$\frac{\partial T(b,t)}{\partial r} + H_2 T(b,t) = 0 \qquad t > 0 \tag{8.43b}$$

where

$$H_1 = \frac{h_1}{k} \qquad H_2 = \frac{h_2}{k} \tag{8.43c}$$

$$T(r,0) = F(r) \qquad a \le r \le b \tag{8.44}$$

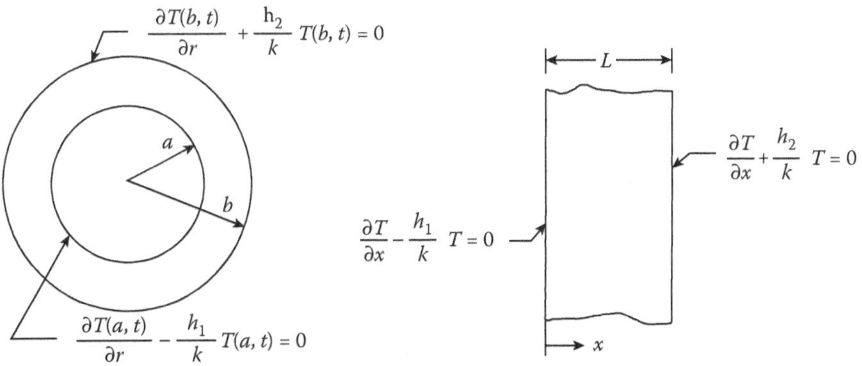

FIGURE 8.4 (a) Hollow sphere with convection boundary conditions (b) Flat plate with convection boundary conditions ($X33$).

The solution procedure is the same as that used for Example 8.1. By introducing the new dependent variable $U(r, t) = rT(r, t)$, we get a similar differential equation, in terms of U, as that in the previous example. However, the boundary conditions are different from the previous case. For this case the transformation of the variables yields

$$\frac{\partial^2 U}{\partial r^2} = \frac{1}{\alpha}\frac{\partial U}{\partial t} \qquad a \le r \le b \quad t > 0 \tag{8.45}$$

$$\frac{\partial U(a,t)}{\partial r} - H_1^* U(a,t) = 0 \qquad t > 0 \tag{8.46a}$$

$$\frac{\partial U(b,t)}{\partial r} + H_2^* U(b,t) = 0 \qquad t > 0 \tag{8.46b}$$

where

$$H_1^* = H_1 + \frac{1}{a} \qquad H_2^* = H_2 - \frac{1}{b} \tag{8.46c}$$

$$U(r,0) = r\,F(r) = F^*(r) \tag{8.47}$$

The transformed equations 8.45 through 8.47 represent a flat plate problem with an arbitrary initial condition and the homogeneous convective boundary conditions on both sides ($X33$) (see Figure 8.4b). From Appendix X, Equation X33.2, the GF for this case, $X33$ geometry, for $\tau = 0$ is given by

$$G_{X33}(x,t|x',0) = \frac{2}{L}\sum_{m=1}^{\infty} \exp\left(-\beta_m^2 \frac{\alpha t}{L^2}\right) \times \frac{[\beta_m \cos(\beta_m x/L) + B_1 \sin(\beta_m x/L)] \times [\beta_m \cos(\beta_m x'/L) + B_1 \sin(\beta_m x'/L)]}{(\beta_m^2 + B_1^2)[1 + B_2/(\beta_m^2 + B_2^2)] + B_1} \tag{8.48}$$

where the β_m values are the positive eigenvalues of

$$\tan\beta_m = \frac{\beta_m(B_1+B_2)}{\beta_m^2 - B_1B_2} \qquad B_1 = H_1L = \frac{h_1L}{k} \qquad B_2 = H_2L = \frac{h_2L}{k} \tag{8.49a, b, c}$$

and L represents the plate thickness. Again in a similar manner to that of the previous example, one can show that

$$T(r,t) = \frac{1}{r}U(r,t) = \int_{r'=a}^{b}\left[\frac{1}{4\pi rr'}G_{X33}(r,t|r',0)\right]F(r')4\pi r'^2\,dr' \tag{8.50}$$

which yields

$$G_{RS33}(r,t|r',0) = \frac{1}{4\pi rr'}G_{X33}(r,t|r',0) \tag{8.51}$$

However, it should be noted that, for this case, the transformation of the Cartesian variables (x, x', L, etc.) to the spherical variables (r, r', b, a, etc.) is not the same as that for the previous example. From Figures 8.4a and b, one can see that for this case,

$$L \to (b-a) \qquad x \to (r-a) \qquad x' \to (r'-a) \tag{8.52a,b,c}$$

$$B_1 \to H_1^*(b-a) = \left(H_1 + \frac{1}{a}\right)(b-a) \tag{8.53a}$$

$$B_2 \to H_2^*(b-a) = \left(H_2 - \frac{1}{b}\right)(b-a) \tag{8.53b}$$

Finally by substituting for L, x, x', B_1, and B_2 from Equation 8.52a, b, c and 8.53a, b into Equation 8.48, and replacing $(t-0)$ by $(t-\tau)$, one can write

$$\begin{aligned} G_{RS33}(r,t|r',\tau) &= \frac{1}{2\pi rr'(b-a)}\sum_{m=1}^{\infty}\exp\left[-\beta_m^2\frac{\alpha(t-\tau)}{(b-a)^2}\right] \\ &\quad \{\beta_m\cos[\beta_m(r-a)/(b-a)] + B_1\sin[\beta_m(r-a)/(b-a)]\} \\ &\quad \times \frac{\times\{\beta_m\cos[\beta_m(r'-a)/(b-a)] + B_1\sin[\beta_m(r'-a)/(b-a)]\}}{(\beta_m^2+B_1^2)[1+B_2/(\beta_m^2+B_2^2)]+B_1} \end{aligned} \tag{8.54}$$

where

$$B_1 = \left(\frac{a\,h_1}{k}+1\right)\left(\frac{b}{a}-1\right) \tag{8.55a}$$

$$B_2 = \left(\frac{b\,h_2}{k}-1\right)\left(1-\frac{a}{b}\right) \tag{8.55b}$$

with

$$\tan \beta_m = \frac{\beta_m(B_1 + B_2)}{\beta_m^2 - B_1 B_2} \tag{8.55c}$$

This GF is also listed in Appendix RS.

The procedure demonstrated in the two previous examples can also be used to obtain G_{RSIJ} from G_{XIJ} for other types of boundary conditions (I, $J = 0, 1, 2, 3$). Table 8.1 gives a summary of how $RSIJ$ GFs are obtained from XIJ GFs for different values of I, $J = 0, 1, 2, 3$ with the appropriate variable transformations.

8.5 TEMPERATURE IN SOLID SPHERES

In this section, we demonstrate the application of the GF solution method to the solid sphere problems with radial flow of heat numbered by $RS0J$ where $J = 1, 2$, and 3. Three groups of problems are considered: those with a nonzero initial temperature distributions $F(r)$; those with nonhomogeneous boundary conditions; and those containing an energy generation term $g(r,t)$.

The describing partial differential equation for these groups of problems is

$$\frac{1}{r}\frac{\partial^2(rT)}{\partial r^2} + \frac{1}{k}g(r,t) = \frac{1}{\alpha}\frac{\partial T}{\partial t} \qquad 0 \le r \le b \qquad t > 0 \tag{8.56}$$

where $g(r,t)$ represents an energy generation term that makes this equation nonhomogeneous. The boundary conditions at the center of the sphere ($r = 0$) are homogeneous and are given by

$$\frac{\partial T(0,t)}{\partial r} = 0 \qquad \text{or} \qquad T(0,t) \ne \infty \qquad t > 0 \tag{8.57}$$

The condition at the surface of the sphere ($r = b$), can be of the first, second, or third kinds depending on the values of J, that is,

$$\text{for } J = 1: \qquad T(b,t) = T_b(t) \qquad t > 0 \tag{8.58a}$$

$$\text{for } J = 2: \qquad k\frac{\partial T(b,t)}{\partial r} = q_b(t) \qquad t > 0 \tag{8.58b}$$

$$\text{and for } J = 3: \qquad k\frac{\partial T(b,t)}{\partial r} + hT(b,t) = hT_\infty(t) \qquad t > 0 \tag{8.58c}$$

Here q_b is heat flow *into* the sphere. The initial condition is considered to be an arbitrary function of time, given by

$$T(r,0) = F(r) \qquad 0 \le r \le b \tag{8.59}$$

From the GFSE for radial flow of heat in spheres, Equation 8.3, the temperature solution is

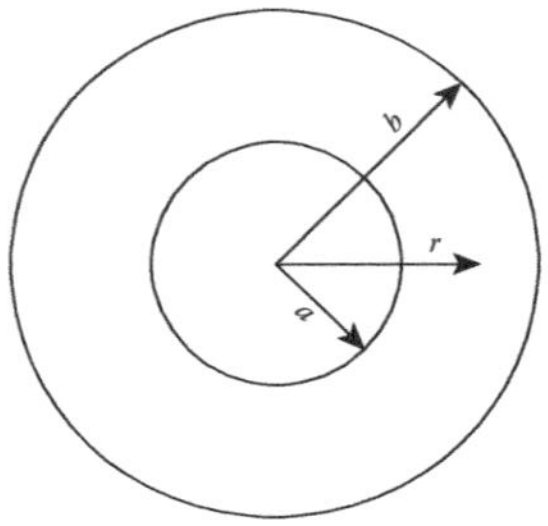

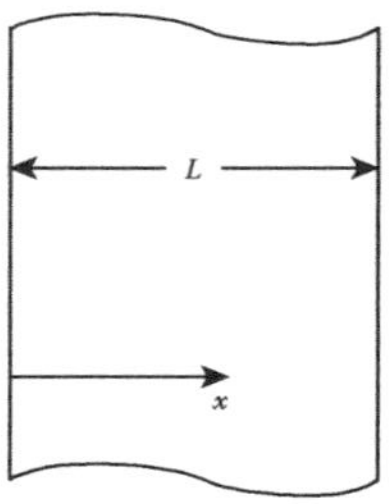

TABLE 8.1 **Conversion Components for Derivation of the *RSIJ* Green's Function from the *XI′J′* Green's Function for $I, J = 0,1,2,3$ $G_{RSIJ}(r, t|r', \tau) = \frac{1}{4\pi rr'} G_{XI'J'}(x, t|x', \tau)$**

No.	*IJ*	*I′J′*	*L*	*x*	*x′*	*H*	*B*	B_1	B_2	Characteristic Equation
1	00	10	–	r	r'	–	–	–	–	–
2	01	11	b	r	r'	–	–	–	–	–
3*	02	13	b	r	r'	$-\frac{1}{b}$	-1	–	–	$\beta_m \cot \beta_m = 1$
4	03	13	b	r	r'	$\frac{h}{k} - \frac{1}{b}$	$\frac{hb}{k} - 1$	–	–	$\beta_m \cot \beta_m = -B$
5	10	10	–	$r - a$	$r' - a$	–	–	–	–	–
6	11	11	$b - a$	$r - a$	$r' - a$	–	–	–	–	–
7	12	13	$b - a$	$r - a$	$r' - a$	$-\frac{1}{b}$	$\frac{a}{b} - 1$	–	–	$\beta_m \cot \beta_m = -B$
8	13	13	$b - a$	$r - a$	$r' - a$	$\frac{h}{k} - \frac{1}{b}$	$\left(\frac{hb}{k} - 1\right)\left(1 - \frac{a}{b}\right)$	–	–	$\beta_m \cot \beta_m = -B$

9	20	30	–	$r-a$	$r'-a$	$\frac{1}{a}$	–	–	–	–
10	21	31	$b-a$	$r-a$	$r'-a$	$\frac{1}{a}$	$\frac{b}{a}-1$	–	–	$\beta_m \cot \beta_m = -B$
11[†]	22	33	$b-a$	$r-a$	$r'-a$	–	–	$\frac{b}{a}-1$	$\frac{a}{b}-1$	$\tan \beta_m = \frac{\beta_m(B_1+B_2)}{\beta_m^2 - B_1 B_2}$
12	23	33	$b-a$	$r-a$	$r'-a$	–	–	$\frac{b}{a}-1$	$\left(\frac{bh}{k}-1\right)\left(1-\frac{a}{b}\right)$	$\tan \beta_m = \frac{\beta_m(B_1+B_2)}{\beta_m^2 - B_1 B_2}$
13	30	30	–	$r-a$	$r'-a$	$\frac{h}{k}+\frac{1}{a}$	–	–	–	–
14	31	31	$b-a$	$r-a$	$r'-a$	$\frac{h}{k}+\frac{1}{a}$	$\left(\frac{ha}{k}+1\right)\left(\frac{b}{a}-1\right)$	–	–	$\beta_m \cot \beta_m = -B$
15	32	33	$b-a$	$r-a$	$r'-a$	$\frac{h}{k}+\frac{1}{a}$	–	$\left(\frac{ha}{k}+1\right)\left(\frac{b}{a}-1\right)$	$\frac{a}{b}-1$	$\tan \beta_m = \frac{\beta_m(B_1+B_2)}{\beta_m^2 - B_1 B_2}$
16	33	33	$b-a$	$r-a$	$r'-a$	–	–	$\left(\frac{ah_1}{k}+1\right)\left(\frac{b}{a}-1\right)$	$\left(\frac{bh_2}{k}-1\right)\left(1-\frac{a}{b}\right)$	$\tan \beta_m = \frac{\beta_m(B_1+B_2)}{\beta_m^2 - B_1 B_2}$

*For this case, the characteristic equation has a zero root; consequently, a term $3/(4b^3)$ has to be added to the value of G_{RS02} obtained from G_{X13}. See Example 8.4.

[†]A term $3/[4(b^3-a^3)]$ has to be added to the value of G_{RS22} obtained from G_{X33}.

$$\begin{aligned}T(r,t) = &\int_{r'=0}^{b} F(r')G_{RS0J}(r,t|r',0)4\pi r'^2\,dr'\\&+\frac{\alpha}{k}\int_{\tau=0}^{t}\int_{r'=0}^{b} G_{RS0J}(r,t|r',\tau)\,g(r',t)\,4\pi r'^2\,dr'\,d\tau\\&-\alpha\int_{\tau=0}^{t} T_b(\tau)\frac{\partial G_{RS01}(r,t|b,\tau)}{\partial r}4\pi b^2\,d\tau \qquad (\text{for } J=1 \text{ only})\\&+\frac{\alpha}{k}\int_{\tau=0}^{t} q_b(\tau)G_{RS02}(r,t|b,\tau)\,4\pi b^2\,d\tau \qquad (\text{for } J=2 \text{ only})\\&+\frac{\alpha}{k}\int_{\tau=0}^{t} hT_\infty(\tau)G_{RS03}(r,t|b,\tau)\,4\pi b^2\,d\tau \qquad (\text{for } J=3 \text{ only}) \quad (8.60)\end{aligned}$$

Note that since the boundary condition at the center of the solid sphere is homogeneous, the last two integrals, associated with boundary conditions, are evaluated only at the sphere's outer surface at radius b. The $RS0J$ GFs for the cases of $J = 1,\ 2,\ 3$ are given in Appendix RS or can be obtained from XIJ GFs through the appropriate transformations provided in Table 8.1. Some example problems are discussed next.

Example 8.3: Solid Sphere with Arbitrary Initial Temperature—*RS01 B0T*-Case

A solid sphere, $0 \le r \le b$, has a known initial temperature distribution $F(r)$. The surface temperature is kept at $T = 0$. Find the transient temperature distribution in the sphere.

Solution

The solution can be obtained from Equation 8.60 by considering only the first integral on the right-hand side which is due to the nonzero initial temperature distribution. The second through the last integrals vanish since there is no volume energy generation in the above problem, $g(r, t) = 0$, and the boundary conditions are homogeneous. The required GF for this case is $G_{RS01}(r, t|r', 0)$ which is equivalent to G_{X11} in Cartesian coordinates. Following the procedure explained in Section 8.4, from Table 8.1, with $x \to r$, $x' \to r'$, and $L \to b$, one can write

$$\begin{aligned}G_{RS01}(r,t|r',0) &= \frac{1}{4\pi rr'}G_{X11}(r,t|r',0)\\&= \frac{1}{2\pi brr'}\sum_{m=1}^{\infty}\exp\left[-\frac{m^2\pi^2\alpha t}{b^2}\right]\sin\left(m\pi\frac{r}{b}\right)\sin\left(m\pi\frac{r'}{b}\right) \quad (8.61)\end{aligned}$$

Note that the expression for $X11$ GF used in the above equation is suitable for large times. Substituting for G_{RS01} from Equation 8.61 into 8.60 yields

$$\begin{aligned}T(r,t) = &\int_{r'=0}^{b} F(r')\\&\times\left[\frac{1}{2\pi brr'}\sum_{m=1}^{\infty}\exp\left(-\frac{m^2\pi^2\alpha t}{b^2}\right)\sin\left(m\pi\frac{r}{b}\right)\sin\left(m\pi\frac{r'}{b}\right)\right]4\pi r'^2\,dr'\end{aligned} \quad (8.62)$$

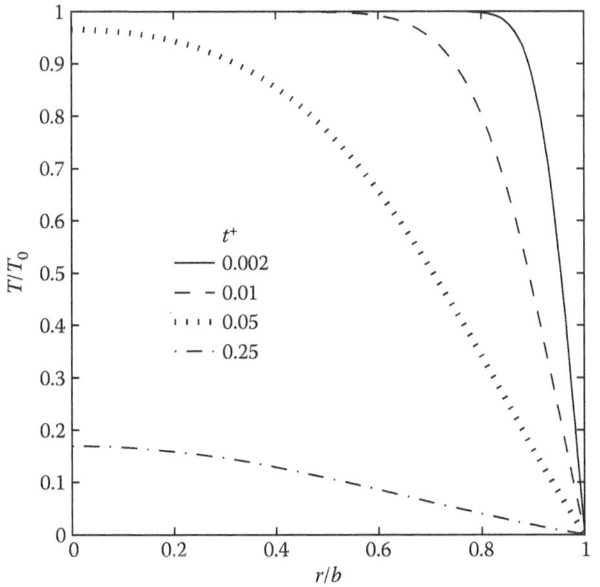

FIGURE 8.5 Temperature in a solid sphere initially at uniform temperature T_0 and with surface temperature set to zero at $t = 0$. Case $RS01B0T1$.

which can be simplified and rearranged to give

$$T(r,t) = \frac{2}{br} \sum_{m=1}^{\infty} \exp\left(-\frac{m^2\pi^2\alpha t}{b^2}\right) \times \sin\left(m\pi\frac{r}{b}\right) \int_{r'=0}^{b} F(r') \sin\left(m\pi\frac{r'}{b}\right) r'\,dr' \tag{8.63}$$

For the special case ($RS01B0T1$) where there is a uniform initial temperature distribution, $T(r,0) = F(r) = T_0$, the integral in solution (8.63) can easily be evaluated as

$$\int_{r'=0}^{b} F(r') \sin\left(m\pi\frac{r'}{b}\right) r'\,dr' = T_0 \int_{r'=0}^{b} \sin\left(m\pi\frac{r'}{b}\right) r'\,dr' = -\frac{T_0 b^2}{m\pi}\cos(m\pi) = -\frac{T_0 b^2}{m\pi}(-1)^m \tag{8.64}$$

since $\int x \sin x\,dx = \sin x - x\cos x$ and $\cos(m\pi) = (-1)^m$. Then the solution becomes

$$T(r,t) = 2T_0 \sum_{m=1}^{\infty} (-1)^{m+1} \exp[-m^2\pi^2\alpha t/b^2] \frac{\sin(m\pi r/b)}{(m\pi r/b)} \tag{8.65}$$

This temperature is plotted in Figure 8.5. The above solution represents a rapidly converging series for large times ($\alpha t/b^2$) since the exponential term rapidly decreases as m increases. However, for small times, it takes a large number of terms for convergence. For small values of $\alpha t/b^2$, it is more computationally efficient to use the small-cotime GF. For small times, the $RS01$ GF can be obtained from the small-cotime expression for $X11$ GF given by Equation X11.1 in Appendix X. That is,

$$G^S_{RS01}(r,t|r',0) = (4\pi r r')^{-1}(4\pi\alpha t)^{-(1/2)} \sum_{n=-\infty}^{+\infty} \times \left\{ \exp\left[-\frac{(2nb+r-r')^2}{4\alpha t}\right] - \exp\left[-\frac{(2nb+r+r')^2}{4\alpha t}\right] \right\} \tag{8.66}$$

In a similar manner to that used for the large-time solution, the small-time solution, for $F(r) = T_0$, is obtained by substituting Equation 8.66 into Equation 8.60 and integrating from $r' = 0$ to $r' = b$ to give

$$T(r,t) = T_0 - \frac{bT_0}{r} \sum_{n=-\infty}^{\infty} \left\{ \operatorname{erfc}\left[\frac{(2n+1)b-r}{(4\alpha t)^{1/2}}\right] - \operatorname{erfc}\left[\frac{(2n+1)b+r}{(4\alpha t)^{1/2}}\right] \right\} \tag{8.67}$$

Note that since the complementary error function erfc(·), decreases rapidly with an increase in its argument, for small times (such as $\alpha t/b^2 < 0.4$), the major contribution to the temperature in the above solution is due to the first two terms of the series ($n = 0, 1$). For smaller times, say $\alpha t/b^2 < 0.1$, even one term in the series is sufficient to give accurate results.

The large- and small-time solutions given by Equations 8.65 and 8.67 are applicable for $r > 0$. For the temperature at the center of the sphere, these solutions approach the following expressions as $r \to 0$ at the limit:

$$T(0,t) = 2T_0 \sum_{m=1}^{\infty} (-1)^{m+1} \exp\left[-\frac{m^2\pi^2\alpha t}{b^2}\right] \quad \text{(for large times)} \tag{8.68}$$

$$T(0,t) = T_0 - \frac{bT_0}{(\pi\alpha t)^{1/2}} \sum_{n=-\infty}^{\infty} \exp\left[-\frac{(2n+1)^2 b^2}{4\alpha t}\right] \quad \text{(for small times)} \tag{8.69}$$

In the above problem, the boundary condition at the surface $r = b$ was considered to be homogeneous, and consequently, in the derivation of the solution, we did not have to consider the contributions of the last three integrals in Equation 8.60.

If the boundary condition at $r = b$ is nonhomogeneous but constant at T_b, the problem can be cast as one with homogeneous boundary condition, by defining a new temperature variable $T - T_b$. Therefore, the solutions given by Equations 8.65 and 8.67 through 8.69 can still be used by replacing T and T_0 by $T - T_b$ and $T_0 - T_b$, respectively, in these solutions. For the case where the boundary condition at $r = b$ is not constant, the corresponding integral in Equation 8.60 must be included in the solution. This is best illustrated in the following example.

Example 8.4: Solid Sphere Heated at Surface—*RS*02*B*-*T*- Case

A solid sphere, $0 \le r \le b$, has a known initial temperature distribution $F(r)$. The surface of the sphere is heated uniformly by a known heat flux as a function of time, $q_b(t)$. Find the temperature distribution in the sphere for large times.

Solution

The partial differential equation, the initial condition, and the boundary condition at the center of the sphere $r=0$ for this case are the same as those for Example 8.3. The boundary condition at the surface ($r=b$) is of second kind ($J=2$) and nonhomogeneous, given by Equation 8.58b as

$$k\frac{\partial T(b,t)}{\partial r} = q_b(t) \tag{8.70}$$

From the Equation 8.60, the solution is

$$\begin{aligned} T(r,t) = &\int_{r'=0}^{b} G_{RS02}(r,t|r',0)\, F(r') 4\pi r'^2\, dr' \\ &+ \frac{\alpha}{k}\int_0^t q_b(\tau)\, G_{RS02}(r,t|b,\tau) 4\pi b^2\, d\tau \end{aligned} \tag{8.71}$$

Note that since there is no energy generation in the sphere, the second integral in Equation 8.60 is not included in the solution. The required GF for this case, $G_{RS02}(r,t|r',\tau)$, can be obtained from the $X13$ GF through the appropriate transformation of the variables given in Table 8.1. It also can be obtained from the *RS*03 GF given by Equation 8.40 by setting $H=0$. However, note that when $H=0$, the corresponding eigenfunction, Equation 8.41, has a zero root, and consequently, a term $3/(4\pi b^3)$ has to be added to the value of G_{RS03} given by Equation 8.40 with $H=0$. Therefore, one can write

$$\begin{aligned} G_{RS02}(r,t|r',\tau) = &\frac{1}{2\pi b r r'}\sum_{m=1}^{\infty} \exp\left[-\frac{\beta_m^2 \alpha(t-\tau)}{b^2}\right] \\ &\times \frac{(\beta_m^2+1)\sin\left(\beta_m \frac{r}{b}\right)\sin\left(\beta_m\frac{r'}{b}\right)}{\beta_m^2} + \frac{3}{4\pi b^3} \end{aligned} \tag{8.72}$$

where β_m are the roots of the eigenfunction

$$\beta_m \cot \beta_m = 1 \tag{8.73}$$

The first five values of β_m are 4.4934, 7.7253, 10.9041, 14.0662, and 17.2208, respectively. (Note: Some sources count ($\beta = 0$) as the first eigenvalue, however in our GF expression, Equation 8.72, the contribution of the zero eigenvalue is included as the additive term outside the series.)

Note that the *RS*02 GF given here (also listed in Appendix RS) is valid for any time but is best for "large" values of $\alpha(t-\tau)/b^2$. Substituting Equation 8.72 into

Equation 8.71 yields

$$T(r,t) = \frac{3}{b^3}\int_{r'=0}^{b} r'^2 F(r')dr'$$
$$+\frac{2}{br}\sum_{m=1}^{\infty}\exp\left(-\frac{\beta_m^2\alpha t}{b^2}\right)\frac{\beta_m^2+1}{\beta_m^2}\sin\left(\beta_m\frac{r}{b}\right)\int_{r'=0}^{b} r'F(r')$$
$$\times\sin\left(\beta_m\frac{r'}{b}\right)dr' + \frac{3\alpha}{bk}\int_0^t q_b(\tau)d\tau$$
$$+\frac{2\alpha}{rk}\sum_{m=1}^{\infty}\frac{\beta_m^2+1}{\beta_m^2}\sin(\beta_m)\sin\left(\beta_m\frac{r}{b}\right)\int_0^t q_b(\tau)$$
$$\times\exp\left[-\frac{\beta_m^2\alpha(t-\tau)}{b^2}\right]d\tau \tag{8.74}$$

Some special cases are considered next.

(a) Case $RS02B1T1$. The initial temperature and the surface heat flux are constant and given by

$$T(r,0) = F(r) = T_0 \qquad q_b(t) = q_0 \tag{8.75a, b}$$

For this case, the last integral for Equation 8.74 is given in Table I.6 (Appendix I) and the solution becomes

$$T(r,t) = T_0 + \frac{3\alpha q_0 t}{bk} + \frac{q_0(5r^2-3b^2)}{10kb}$$
$$-\frac{2q_0b^2}{kr}\sum_{m=1}^{\infty}\frac{\sin(\beta_m r/b)}{\beta_m^2\sin\beta_m}\exp\left(-\frac{\beta_m^2\alpha t}{b^2}\right) \tag{8.76}$$

Note that the second integral in Equation 8.74 vanishes for this case since from the characteristic equation 8.73, we have $\sin\beta_m - \beta_m\cos\beta_m = 0$. The third term on the right-hand side of Equation 8.76 is the nonseries form of the quasisteady temperature for this case. It can be found by applying the alternative solution method presented in Example 6.6. The temperature computed from the above series is plotted in Figure 8.6.

(b) Case $RS02B2T0$. Zero initial temperature and the surface heat flux is a linear function of time, that is,

$$T(r,0) = F(r) = 0 \qquad q_b(t) = q_0 t \tag{8.77}$$

since the initial condition is zero, there is no contribution to the solution due to the initial condition and consequently the first two terms in solution (8.74) vanish. The time integral in the last term can be evaluated using integral number 18 from Table I.6. Then the solution becomes;

$$T(r,t) = \frac{3\alpha q_0}{2bk}t^2 + \frac{2q_0b^4}{kr}\sum_{m=1}^{\infty}\frac{\sin(\beta_m r/b)}{\beta_m^4\sin\beta_m}$$
$$\times\left[\exp\left(-\frac{\beta_m^2\alpha t}{b^2}\right)+\beta_m\frac{\alpha t}{b^2}-1\right] \tag{8.78}$$

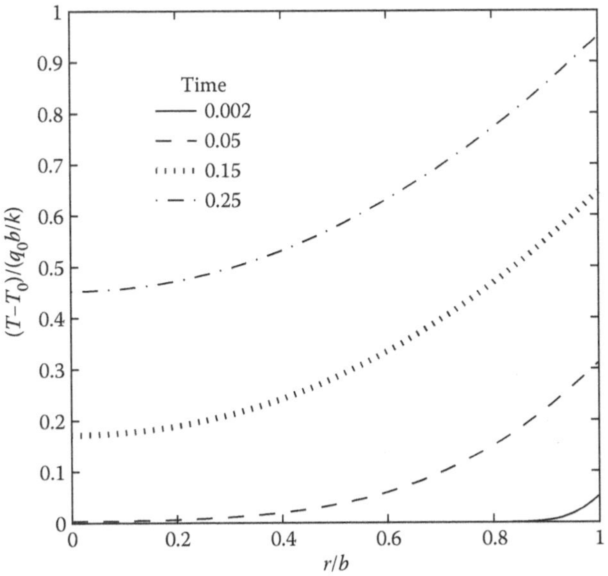

FIGURE 8.6 Temperature in a solid sphere initially at uniform temperature T_0 with a suddenly-applied surface heat flux. Case $RS02B1T1$.

Note that the solutions presented in this example are most efficient for large values of $\alpha t / b^2$. In the next example, time partitioning, introduced in Chapter 5, is used to find a solution that is numerically efficient for both small and large times.

Example 8.5: Solid Sphere with Convective Boundary Condition—*RS03B1T0* Case

A solid sphere, $0 \leq r \leq b$, initially at temperature T_0, is suddenly immersed in a fluid at a constant temperature T_∞. The heat transfer coefficient for this process is h, a constant. Find separate expressions for temperature distribution, $T(r, t)$, that are numerically efficient for small time and for large time.

Solution

For very small times, the change in temperature is limited to a small region near the surface of the sphere. Using normalized temperature $T - T_0$, the solution due to the convection boundary condition at the surface of the sphere ($r = b$) is given by the last term of Equation 8.60,

$$T(r, t) - T_0 = \frac{\alpha}{k} \int_{\tau=0}^{t} h(T_\infty - T_0) G^S_{RS03}(r, t|b, \tau)\, 4\pi b^2\, d\tau \tag{8.79}$$

For small-time behavior, the small-cotime GF, G^S_{RS03}, is needed in this integral, and it can be obtained from G^S_{X13} given by Equation X13.1 in Appendix X. Following the procedure explained in Section 8.4 with the appropriate transformation of the variables from Table 8.1, one can get

$$G^S_{RS03}(r,t|r',\tau) = [4\pi r r']^{-1}[4\pi\alpha(t-\tau)]^{-1/2}\left\{\exp\left[-\frac{(r-r')^2}{4\alpha(t-\tau)}\right]\right.$$
$$-\exp\left[-\frac{(r+r')^2}{4\alpha(t-\tau)}\right]+\exp\left[-\frac{(2b-r-r')^2}{4\alpha(t-\tau)}\right]\Bigg\}$$
$$-(4\pi r r')^{-1}H^*\exp[H^*(2b-r-r')$$
$$+H^{*2}\alpha(t-\tau)]\text{erfc}\left\{\frac{(2b-r-r')}{[4\alpha(t-\tau)]^{1/2}}\right.$$
$$\left.+H^*[\alpha(t-\tau)]^{1/2}\right\} \tag{8.80a}$$

where

$$H^* = H - \frac{1}{b} \qquad H = \frac{h}{k} \tag{8.80b, c}$$

Evaluating G^S_{RS03} at $r' = b$ and substituting the result into Equation 8.79 with $(t-\tau)$ replaced by cotime u yields

$$T^S(r,t) = \frac{\alpha b h T_\infty}{kr}\int_{u=0}^{t}\Bigg((4\pi\alpha u)^{-1/2}$$
$$\times\left\{2\exp\left[-\frac{(b-r)^2}{4\alpha u}\right]-\exp\left[-\frac{(r+b)^2}{4\alpha u}\right]\right\}$$
$$-H^*\exp[H^*(b-r)+H^{*2}\alpha u]$$
$$\times\,\text{erfc}\left[\frac{(b-r)}{(4\alpha u)^{1/2}}+H^*(\alpha u)^{1/2}\right]\Bigg)du, \quad \text{for } \alpha t/b^2 < 0.022 \tag{8.81}$$

The above integral can be evaluated using the Laplace transform method. This method has already been demonstrated in Examples 4.1 and 4.2, where similar integrals have been solved for the $X10$ and $X30$ geometries, respectively. Then the small-time solution becomes

$$T^S(r,t) = \frac{T_\infty bh}{kr}\left\{\frac{1}{H^*}\text{erfc}\left[\frac{(b-r)}{2(\alpha t)^{1/2}}\right]-\frac{1}{H^*}\right.$$
$$\times\exp[(b-r)H^*+\alpha t H^{*2}]\,\text{erfc}\left[\frac{(b-r)}{2(\alpha t)^{1/2}}+H^*(\alpha t)^{1/2}\right]$$
$$\left.-(\alpha t)^{1/2}\,\text{ierfc}\left[\frac{r+b}{2(\alpha t)^{1/2}}\right]\right\}, \quad \text{for } \alpha t/b^2 < 0.022 \tag{8.82a}$$

where

$$H^* = \frac{h}{k} - \frac{1}{b} \tag{8.82b}$$

Note that this expression is good for r away from the center where the temperature remains unchanged for small values of $\alpha t/b^2$.

Next an expression will be found for the temperature that is rapidly convergent at larger times and suitable for all times. The required large-cotime GF is found in Appendix RS, Equation RS03.1. Still using normalized temperature $(T(r,t) - T_0)$ and Equation 8.79, the temperature is given by

$$T(r,t) - T_0 = \frac{\alpha h(T_\infty - T_0)}{k} \int_{\tau=0}^{t} d\tau \frac{2}{r} \sum_{m=1}^{\infty} \times e^{-\beta_m^2 \alpha(t-\tau)/b^2} \frac{(\beta_m^2 + B^2)\sin(\beta_m r/b)\sin\beta_m}{\beta_m^2 + B^2 + B} \tag{8.83a}$$

where

$$\beta_m \cot\beta_m = -B \qquad \text{and} \qquad B = \frac{hb}{k} - 1 \tag{8.83b}$$

The integral on τ involves only the exponential term and can easily be evaluated to give

$$T(r,t) - T_0 = \frac{2h(T_\infty - T_0)b^2}{kr} \sum_{m=1}^{\infty} \frac{1}{\beta_m^2}(1 - e^{-\beta_m^2 \alpha t/b^2}) \times \frac{(\beta_m^2 + B^2)\sin(\beta_m r/b)\sin\beta_m}{\beta_m^2 + B^2 + B} \tag{8.84}$$

This series can be split into a time-varying series and a steady series. The steady series, which represents the steady-state temperature, converges slowly. As discussed in Chapter 5, it is usually helpful to replace the steady-state series by a nonseries form if it can be found. In the present case, the steady temperature in the sphere takes on the temperature of the fluid. That is, the steady series has the value $(T_\infty - T_0)$. Making this substitution yields a rapidly converging form of the series:

$$T(r,t) - T_0 = (T_\infty - T_0) - 2\frac{hb}{k}\frac{b}{r}(T_\infty - T_0) \sum_{m=1}^{\infty} \frac{1}{\beta_m^2} e^{-\beta_m^2 \alpha t/b^2} \times \frac{(\beta_m^2 + B^2)}{\beta_m^2 + B^2 + B} \sin(\beta_m r/b)\sin\beta_m \tag{8.85}$$

Note that term T_0 appears on both side of the equal sign and could be cancelled. The above series is suitable for all time and converges rapidly at large times, with only a few terms of the series needed for $\alpha t/b^2 > 0.022$.

Alternate Solution

An alternate derivation of the large-time temperature may be found using the normalized temperature $(T - T_\infty)$, which makes the boundary condition homogeneous and moves the causative effect to the initial-condition term of the GF solution. That is, using the initial condition term of Equation 8.60, the temperature is given by

$$T(r,t) - T_\infty = \int_{r'=0}^{b} h(T_0 - T_\infty) G_{RS03}(r,t|r',\tau=0)\, 4\pi(r')^2\, dr' \tag{8.86a}$$

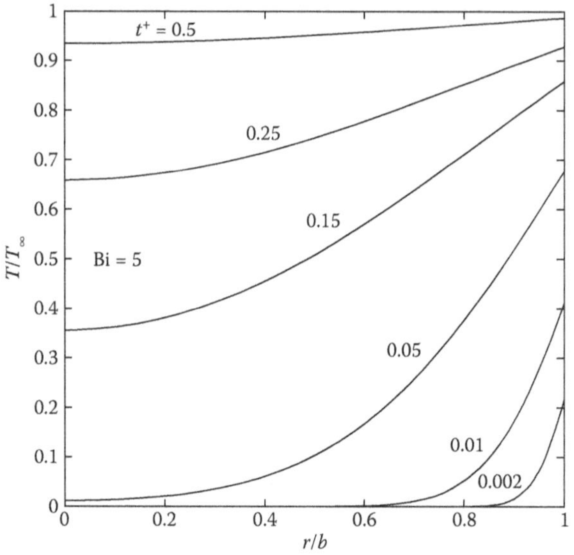

FIGURE 8.7 Solid sphere with convection boundary condition with $hb/k = 5$.

Upon substituting the large-cotime GF given in Appendix RS, Equation RS03.1 and carrying out the spatial integral, the alternate solution is given by

$$T(r,t) - T_\infty = 2\frac{b}{r}(T_0 - T_\infty)\sum_{m=1}^{\infty}\frac{1}{\beta_m^2}\,e^{-\beta_m^2\alpha t/b^2}$$

$$\times\frac{(\beta_m^2 + B^2)}{\beta_m^2 + B^2 + B}\sin(\beta_m r/b)[\sin\beta_m - \beta_m\cos\beta_m] \qquad (8.86b)$$

At first glance this alternate series appears to be quite different than that found by the direct solution given in Equation 8.85; however, the alternate series can be shown to be identical with Equation 8.85 by use of the eigencondition identity, Equation 8.83b. See Figure 8.7 for a plot of normalized temperature versus position for the specific condition ($hb/k = 5$).

Example 8.6: Solid Sphere with Internal Energy Generation and Insulated Surface—*RS*02*B*0*T*0*G*- Case

A solid sphere, $0 \le r \le b$, is initially at zero temperature. For times $t > 0$, heat is produced within the sphere at the rate $g(r, t)$ per unit time per unit volume while the surface boundary is kept insulated. See Figure 8.8. Find the temperature distribution within the sphere, $T(r, t)$ for large times.

Solution

The temperature due to the heat generation within the sphere is given by the second integral on the right-hand side of Equation 8.60,

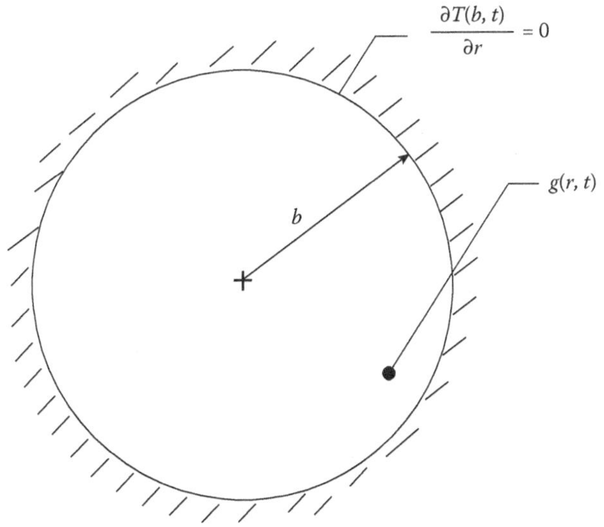

FIGURE 8.8 Solid sphere with insulated surface and energy generation.

$$T(r,t) = \frac{\alpha}{k} \int_{\tau=0}^{t} \int_{r'=0}^{b} G_{RS02}(r,t|r',\tau)\, g(r',\tau)\, 4\pi (r')^2 dr'\, d\tau \tag{8.87}$$

Note that since the initial temperature is zero and the boundary conditions are homogeneous, the other integrals in Equation 8.60 have no contribution in $T(r, t)$. The large-time form of G_{RS02} is found in Appendix RS and is also given by Equation 8.72. Substituting G_{RS02} from Equation 8.72 into Equation 8.87 yields

$$T(r,t) = \frac{\alpha}{k} \int_{\tau=0}^{t} \int_{r'=0}^{b} \left\{ \frac{1}{2\pi b r r'} \sum_{m=1}^{\infty} \exp\left[-\frac{\beta_m^2 \alpha (t-\tau)}{b^2} \right] \times \frac{(\beta_m^2 + 1)\sin(\beta_m r/b)\sin(\beta_m r'/b)}{\beta_m^2} + \frac{3}{4\pi b^3} \right\} g(r',\tau) 4\pi (r')^2 dr'\, d\tau \tag{8.88}$$

which can be simplified and rearranged into two different terms as

$$T(r,t) = T_1(r,t) + T_2(r,t) \tag{8.89}$$

where

$$T_1(r,t) = \frac{3\alpha}{kb^3} \int_{\tau=0}^{t} d\tau \int_{r'=0}^{b} g(r',\tau)\,(r')^2 dr' \tag{8.90a}$$

$$T_2(r,t) = \frac{2\alpha}{kbr} \sum_{m=1}^{\infty} \frac{(\beta_m^2+1)\sin(\beta_m r/b)}{\beta_m^2} \int_{\tau=0}^{t} \times \exp\left[-\frac{\beta_m^2 \alpha (t-\tau)}{b^2} \right] d\tau \int_{r'=0}^{b} \sin\left(\beta_m \frac{r'}{b} \right) g(r',\tau)\, r' dr' \tag{8.90b}$$

and the eigenvalues β_m are defined by $\beta_m \cot \beta_m - 1 = 0$.

In the above equations, usually, it is more convenient to carry out the integrals over r' first and then over τ. The solution given by Equations 8.89 and 8.90 is now examined for some special cases.

(a) Case RS*02*B*0*T*0*G*1*. Heat is generated within the sphere at a constant rate, $g(r, t) = g_0 =$ constant. For this case the integrals in Equation 8.90 are easily evaluated and the solution becomes

$$T(r, t) = T_1(r, t) = \frac{\alpha g_0 t}{k} \tag{8.91}$$

Note that $T_2(r, t)$ is equal to zero in this case since the integration over r' in Equation 8.90b results in the term $(\sin\beta_m - \beta_m \cos\beta_m) = 0$.

(b) Case RS*02*B*0*T*0*Gr*2*. Heat is generated within the sphere as a linear function of radius given by

$$g(r, t) = \frac{g_0(b - r)}{b} \tag{8.92}$$

The integrals in $T_1(r, t)$ are easily evaluated to give

$$T_1(r, t) = \frac{\alpha t g_0}{4k} \tag{8.93a}$$

The integrals over r' and τ in the $T_2(r, t)$ solution are evaluated to give

$$T_2(r, t) = \frac{2g_0 b^3}{kr} \sum_{m=1}^{\infty} \frac{[2 - (2 + \beta_m^2)\cos\beta_m](\beta_m^2 + 1)\sin(\beta_m r/b)}{\beta_m^5} \times \left[1 - \exp\left(-\frac{\beta_m^2 \alpha t}{b^2}\right)\right] \tag{8.93b}$$

Then the solution becomes

$$T(r, t) = \frac{g_0 b^2}{4k} \left\{ \frac{\alpha t}{b^2} + \frac{8b}{r} \sum_{m=1}^{\infty} \frac{(\beta_m^2 + 1)[2 - (2 + \beta_m^2)\cos\beta_m]\sin(\beta_m r/b)}{\beta_m^5} \times \left[1 - \exp\left(-\frac{\beta_m^2 \alpha t}{b^2}\right)\right]\right\} \tag{8.94}$$

It is interesting to note that $T_1(r, t)$ solution is not a function of r and changes linearly with t. This term represents the volume-average temperature in the sphere, defined by

$$T_{av}(t) = \left(\frac{4}{3}\pi b^3\right)^{-1} \int_{r=0}^{b} T(r, t)\, 4\pi r^2\, dr \tag{8.95a}$$

and can be verified by substituting for $T(r, t)$ from Equation 8.89 and carrying out the integration; that is,

$$T_{av}(t) = \left(\frac{4}{3}\pi b^3\right)^{-1} \int_{r=0}^{b} [T_1(r, t) + T_2(r, t)]\, 4\pi r^2\, dr \tag{8.95b}$$

The integration over $T_2(r, t)$ becomes zero, since it involves the eigencondition $\beta_m \cot\beta_m - 1 = 0$. Then, since T_1 is not a function of r, one can write

$$T_{av}(t) = \left(\frac{4}{3}\pi b^3\right)^{-1} T_1(t) \int_{r=0}^{b} 4\pi r^2\, dr = T_1(t) \tag{8.96a}$$

or

$$T_{av}(t) = \frac{\alpha t g_0}{4k} \tag{8.96b}$$

It is also interesting to note that the $T_2(r, t)$ solution contains two terms: one is the transient term which decreases exponentially to zero over time, while the other term represents the quasisteady temperature distribution which does not depend on time. After the transient term becomes zero, the shape of the temperature distribution remains unchanged due to the latter term. Note that even though the transient term in T_2 solution dies out with time, since T_1 solution is a function of time, there is no steady-state temperature for this problem. In other words, the average temperature of the sphere increases with time since, due to the insulated surface condition, the heat that is generated has no place to go.

(c) *Case* RS02B0T0Gr4. Heat is generated within the sphere as an exponential function of radius given by

$$g(r, t) = g_0\, e^{-\gamma r/b} \tag{8.97}$$

The $T_1(r, t)$ solution for this case is given by

$$T_1(r, t) = \frac{3\alpha t g_0}{kb^3} \int_{r'=0}^{b} e^{-\gamma r'/b} r'^2\, dr' \tag{8.98}$$

The integral over r' in Equation 8.98 can be evaluated by parts to give

$$T_1(r, t) = \frac{3\alpha t g_0}{k\gamma^3}[2 - e^{-\gamma}(\gamma^2 + 2\gamma + 2)] \tag{8.99}$$

After integrating over time, the $T_2(r, t)$ solution becomes

$$T_2(r, t) = \frac{2g_0 b^3}{kr} \sum_{m=1}^{\infty} \frac{(\beta_m^2 + 1)\sin(\beta_m r/b)}{\beta_m^4} \times \left[1 - \exp\left(-\frac{\beta_m^2 \alpha t}{b^2}\right)\right] \int_0^1 \sin\left(\beta_m \frac{r'}{b}\right) \times e^{-\gamma r'/b} \left(\frac{r'}{b}\right) d\left(\frac{r'}{b}\right) \tag{8.100}$$

The integral over r' in Equation 8.100 can be evaluated by using the relation

$$\int x\, e^{Ax} \sin(Bx)\, dx = \frac{xe^{Ax}}{A^2 + B^2} \times (A \sin Bx - B\cos Bx) - \frac{e^{Ax}}{(A^2 + B^2)^2} \times [(A^2 - B^2)\sin Bx - 2AB\cos Bx] \tag{8.101}$$

Then, one can write

$$T_2(r,t) = \frac{2g_0 b^3\gamma}{kr}\sum_{m=1}^{\infty} \times \frac{(1+\beta_m^2)[2\beta_m - e^{-\gamma}\sin(\beta_m)(2+2\gamma+\gamma^2+\beta_m^2)]}{\beta_m^4(\gamma^2+\beta_m^2)^2} \times \sin\left(\beta_m \frac{r}{b}\right)\left[1-\exp\left(-\frac{\beta_m^2\alpha t}{b^2}\right)\right] \tag{8.102}$$

Finally, the solution for $T(r,t)$ becomes

$$T(r,t) = \frac{3g_0 b^2}{k}\left\{\frac{\alpha t}{b^2}\left[\frac{2-e^{-\gamma}(\gamma^2+2\gamma+2)}{\gamma^3}\right] + \frac{2\gamma b}{3r}\sum_{m=1}^{\infty}\frac{(\beta_m^2+1)}{\beta_m^4} \times \frac{[2\beta_m - e^{-\gamma}\sin(\beta_m)(2+2\gamma+\gamma^2+\beta_m^2)]}{(\gamma^2+\beta_m^2)^2} \times \sin\left(\beta_m\frac{r}{b}\right)\left[1-\exp\left(-\frac{\beta_m^2\alpha t}{b^2}\right)\right]\right\} \tag{8.103}$$

Note that for this case, similar to the previous case, the T_1 solution is not a function of position, and the T_2 solution contains a transient decaying term and a quasisteady term.

(d) *Case* RS02B0T0Gt4. Heat is generated within the sphere as an exponential function of time given by

$$g(r,t) = g_0 e^{-\lambda t} \tag{8.104}$$

Similar to case 1, since g is not a function of r, $T_2(r,t)$ becomes zero. Then, the solution is given by

$$T(r,t) = T_1(r,t) = \frac{\alpha g_0}{k\lambda}(1-e^{-\lambda t}) \tag{8.105}$$

Note that, there is a steady-state temperature for this case given by $\alpha g_0 / k\lambda$.

(e) *Case* RS02B0T0Gr6. Heat is generated within the sphere with generation rate given by

$$g(r,t) = \frac{g_0}{r}\sin\frac{\pi r}{b} \tag{8.106}$$

The integrals in the T_1 solution, for this case, can easily be evaluated to give,

$$T_1(t) = \frac{3\alpha g_0 t}{\pi k b} \tag{8.107}$$

After integrating over time, the T_2 solution becomes

$$T_2(r,t) = \frac{2g_0 b^2}{kr} \sum_{m=1}^{\infty} \times \frac{(\beta_m^2+1)\sin(\beta_m r/b)}{\beta_m^4}\left[1-\exp\left(-\frac{\beta_m^2 \alpha t}{b^2}\right)\right] \times \int_{r'=0}^{1} \sin\left(\beta_m \frac{r'}{b}\right)\sin\left(\pi\frac{r'}{b}\right) d\left(\frac{r'}{b}\right) \tag{8.108}$$

The integral over r' in Equation 8.108 can be evaluated by using the relation

$$\sin A \sin B = ½[\cos(A-B) - \cos(A+B)] \tag{8.109}$$

Then, the solution for $T(r, t)$ becomes

$$T(r,t) = \frac{3g_0 b}{\pi k}\left\{\frac{\alpha t}{b^2} + \frac{2b\pi^2}{3r}\sum_{m=1}^{\infty}\frac{\sin(\beta_m r/b)}{\beta_m^2(\pi^2-\beta_m^2)} \times \left[1-\exp\left(-\frac{\beta_m^2\alpha t}{b^2}\right)\right]\right\} \tag{8.110}$$

Again, there is no steady-state temperature for this case.

8.6 TEMPERATURE IN HOLLOW SPHERES

In this section, we demonstrate, with examples, the application of the GF solution method to the hollow sphere problems with radial flow of heat, denoted by *RSIJ*, for $I, J = 1, 2, 3$. The describing equations and the analytical techniques used for solid spheres in the previous section are also applicable to hollow spheres. However, hollow spheres have one more physical boundary ($I = 1$, 2, or 3) at the inner surface $r = a$ as compared to solid spheres with no physical boundary ($I = 0$) at the center $r = 0$. Accordingly, for hollow spheres, we may have one of the following boundary conditions at the inner surface $r = a$,

for $I = 1$,
$$T(a,t) = T_a(t) \qquad t > 0 \tag{8.111a}$$

for $I = 2$,
$$-k\frac{\partial T(a,t)}{\partial r} = q_a(t) \qquad t > 0 \tag{8.111b}$$

and for $I = 3$,
$$-k\frac{\partial T(a,t)}{\partial r} + h_1 T(a,t) = h_1 T_\infty(t) \qquad t > 0 \tag{8.111c}$$

The appropriate GF for the hollow sphere problems with radial flow of heat can be obtained from the *XIJ* GFs by following the procedure explained in Section 8.4. See Example 8.2. The required transformations are provided in Table 8.1.

Example 8.7: Hollow Sphere Heated on the Inside Surface—*RS*21*B*10*T*0 Case

A hollow sphere, $a \le r \le b$, is initially at zero temperature. For time $t > 0$, the inner surface of the hollow sphere is heated by a constant heat flux q_0, while the outer surface is kept at zero temperature. Find the temperature distribution in the hollow sphere for large times.

Solution

This is the *RS*21 geometry with no heat generation and zero initial temperature. The temperature solution is only due to the boundary conditions and is given by

$$T(r,t) = \alpha \int_{\tau=0}^{t} \frac{q_0}{k} G_{RS21}(r,t|a,\tau) 4\pi a^2 \, d\tau \tag{8.112}$$

Note that since the boundary condition at $r = b$ is homogeneous, the *RS*21 GF in Equation 8.112 is evaluated only at the inner surface boundary, $r = a$, where the heat flux is located. The GF for this case is obtained from the large-time form of G_{X31} with the appropriate transformations given in Table 8.1.

$$G_{RS21}(r,t|r',\tau) = \frac{1}{2\pi(b-a)rr'} \sum_{m=1}^{\infty} \exp\left[-\beta_m^2 \frac{\alpha(t-\tau)}{(b-a)^2}\right] \times \frac{[\beta_m^2 + B^2] \sin\{\beta_m[1-(r-a)/(b-a)]\} \sin\{\beta_m[1-(r'-a)/(b-a)]\}}{\beta_m^2 + B^2 + B} \tag{8.113}$$

where $B = (b-a)/a$ and the eigenvalues β_m are the positive roots of the characteristic equation

$$\beta_m \cot \beta_m = -B \tag{8.114}$$

Evaluating $G_{RS21}(r,t|r',\tau)$ at $r' = a$ and substituting the result into Equation 8.112 gives

$$T(r,t) = \frac{2\alpha q_0 a}{k(b-a)r} \sum_{m=1}^{\infty} \frac{(\beta_m^2 + B^2) \sin[\beta_m(b-r)/(b-a)] \sin \beta_m}{\beta_m^2 + B^2 + B} \times \int_{\tau=0}^{t} \exp\left[-\beta_m^2 \frac{\alpha(t-\tau)}{(b-a)^2}\right] d\tau \tag{8.115}$$

The time integral in the above equation can be carried out to give

$$T(r,t) = \frac{2q_0 a(b-a)}{kr} \sum_{m=1}^{\infty} \frac{(\beta_m^2 + B^2) \sin[\beta_m(b-r)/(b-a)] \sin \beta_m}{\beta_m^2(\beta_m^2 + B^2 + B)} \times \left\{1 - \exp\left[-\frac{\beta_m \alpha t}{(b-a)^2}\right]\right\} \tag{8.116}$$

See Figure 8.9 for a plot of this temperature. Note that the steady-state part of the temperature in Equation 8.116 is given in a series form. The nonseries form of this part can be found by solving the above problem under the steady-state conditions to give

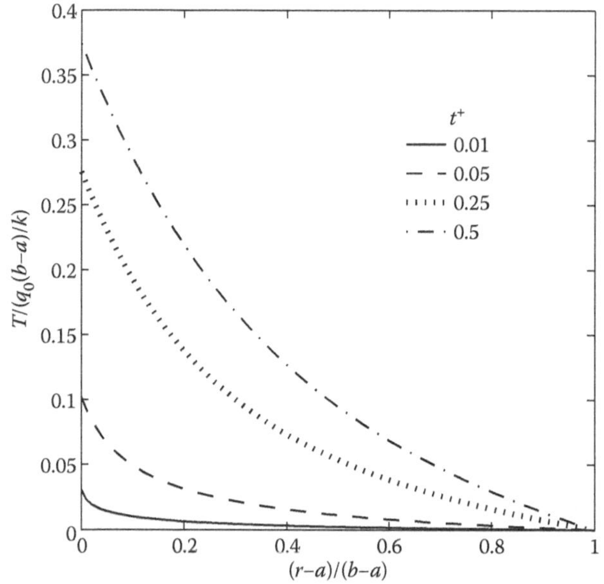

FIGURE 8.9 Hollow sphere with suddenly-applied heat flux at the inner boundary and outer boundary at zero temperature. Here $t^+ = \alpha t/(b-a)^2$.

$$T_s(r) = \frac{q_0 a^2 (b-a)}{kbr} \tag{8.117}$$

Then the alternative form of the solution (8.116) is given by

$$T(r,t) = \frac{q_0 a^2 (b-a)}{kbr} - \frac{2q_0 a(b-a)}{kr} \sum_{m=1}^{\infty} \frac{(\beta_m^2 + B^2)}{\beta_m^2(\beta_m^2 + B^2 + B)} \times \sin\left[\beta_m \frac{(b-r)}{(b-a)}\right] \sin\beta_m \exp\left[-\frac{\beta_m \alpha t}{(b-a)^2}\right] \tag{8.118}$$

Example 8.8: Hollow Sphere Exposed to Convection with Large Heat Transfer Coefficient at the Inside Surface—*RS*11*B*10*T*0 Case

A hollow sphere, $a \leq r \leq b$, initially at zero temperature is suddenly exposed to a fluid at a constant temperature T_∞ at its inner surface. The outer surface temperature remains constant at its initial value $T = 0$. The heat transfer coefficient between the fluid and the inner surface is very large. Find the transient temperature distribution within the hollow sphere.

Solution

In this problem, the inner surface boundary is exposed to a fluid with a very large heat transfer coefficient, which is equivalent to the case where there is a sudden step change in the surface temperature to the fluid's temperature T_∞. Possible examples might be when the fluid is a liquid metal or is changing phase since these processes usually have very large heat transfer coefficients. Therefore, this is the $RS11$ geometry with a homogeneous boundary condition at the outer surface of the hollow sphere. Since there is no energy generation within the body, and the initial temperature is zero, the GF solution is only due to the boundary condition at the inner surface and is given by

$$T(r,t) = \alpha \int_{\tau=0}^{t} T_\infty(\tau) \frac{\partial G_{RS11}(r,t|a,\tau)}{\partial r} 4\pi a^2 \, d\tau \tag{8.119}$$

Note that the above integral is evaluated only at the inner surface with $r' = a$. The derivative of G_{RS11} is listed in Appendix RS, Equation RS11.5 as

$$\frac{\partial G_{RS11}(r,t|a,\tau)}{\partial r} = \frac{1}{2(b-a)^2 ra} \sum_{m=1}^{\infty} m \sin\left(m\pi \frac{r-a}{b-a}\right) \exp\left[-\frac{m^2\pi^2\alpha(t-\tau)}{(b-a)^2}\right] \tag{8.120}$$

Substituting for $\partial G_{RS11} / \partial r$ from Equation 8.120 into Equation 8.119 gives

$$T(r,t) = \frac{2\pi\alpha a T_\infty}{(b-a)^2 r} \sum_{m=1}^{\infty} m \sin\left(m\pi \frac{r-a}{b-a}\right) \times \int_0^t \exp\left[-\frac{m^2\pi^2\alpha(t-\tau)}{(b-a)^2}\right] d\tau \tag{8.121}$$

The time integral in Equation 8.121 can easily be evaluated to give

$$T(r,t) = \frac{2aT_\infty}{\pi r} \sum_{m=1}^{\infty} \frac{\sin[m\pi(r-a)/(b-a)]}{m} \left\{1 - \exp\left[-\frac{m^2\pi^2\alpha t}{(b-a)^2}\right]\right\} \tag{8.122}$$

See Figure 8.10 for a plot of this temperature. Note that the steady-state part of the above solution is given in a series form. The nonseries form of this part is obtained by solving the problem under the steady-state conditions. Then the solution becomes

$$T(r,t) = \frac{aT_\infty}{r} \left\{1 - \frac{r-a}{b-a} - \frac{2}{\pi} \sum_{m=1}^{\infty} \frac{\sin[m\pi(r-a)/(b-a)]}{m} \times \exp\left[-\frac{m^2\pi^2\alpha t}{(b-a)^2}\right]\right\} \tag{8.123}$$

8.7 TEMPERATURE IN AN INFINITE REGION OUTSIDE A SPHERICAL CAVITY

In this section, the GF solution method is applied to two example problems with radial flow of heat in an infinite region outside a spherical cavity, denoted by $RSI0$, $I = 1, 2, 3$.

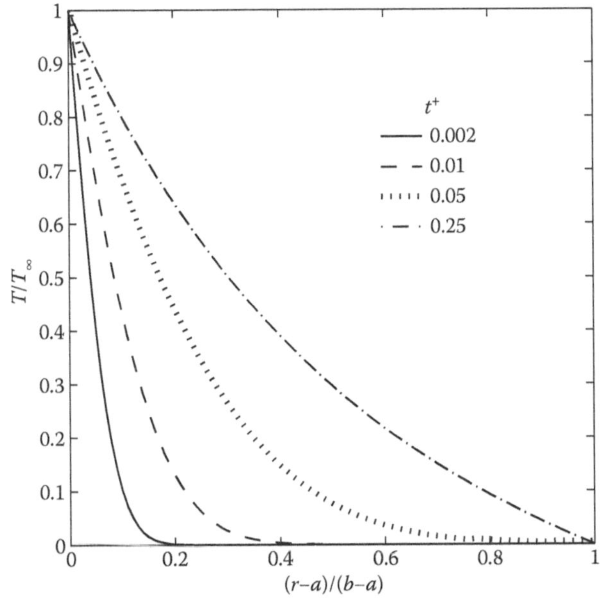

FIGURE 8.10 Hollow sphere with sudden increase of temperature at the inner boundary, case $RS11B10T0$. Here $t^+ = \alpha t/(b-a)^2$.

There is no physical boundary for this geometry at $r = \infty$. The possible boundary conditions at the inner surface ($r = a$) are similar to those given by Equations 8.111a, b, c for $I = 1, 2, 3$, respectively. The GFs for $RSI0$ cases can be obtained from the $XI0$ GFs with the appropriate transformations given in Table 8.1. Note that these GFs do not involve infinite series; consequently, the solutions are mathematically well behaved, for all values of time.

Example 8.9: Infinite Body Heated at the Surface of a Spherical Cavity—*RS*20*B*1*T*0 Case

An infinite body bounded internally by the spherical cavity $r = a$, is initially at zero temperature. For $t > 0$, the surface of the body ($r = a$) is heated uniformly by a known heat flux as a function of time, $q_a(t)$. Find the transient temperature distribution within the body.

Solution

This is the $RS20$ geometry with an arbitrary heat flux boundary condition at $r = a$, given by Equation 8.111b. The temperature is given by

$$T(r, t) = \frac{\alpha}{k} \int_{\tau=0}^{t} q_a(\tau) G_{RS20}(r, t|a, \tau) 4\pi a^2 \, d\tau \tag{8.124}$$

The GF function for this case is obtained from the $X30$ GF with the appropriate transformations given in Table 8.1.

$$
\begin{aligned}
G_{RS20}(r,t|r',\tau) &= (4\pi r r')^{-1}[4\pi\alpha(t-\tau)]^{-1/2} \\
&\quad \times \left\{ \exp\left[-\frac{(r-r')^2}{4\alpha(t-\tau)}\right] + \exp\left[-\frac{(r+r'-2a)^2}{4\alpha(t-\tau)}\right]\right\} \\
&\quad - (4\pi r r' a)^{-1} \exp\left[\frac{\alpha(t-\tau)}{a^2} + \frac{1}{a}(r+r'-2a)\right] \\
&\quad \times \operatorname{erfc}\left\{\frac{(r+r'-2a)}{[4\alpha(t-\tau)]^{1/2}} + \frac{1}{a}[\alpha(t-\tau)]^{1/2}\right\} \qquad (8.125)
\end{aligned}
$$

Evaluating $G_{RS20}(r,t|r',\tau)$ at $r' = a$ and substituting the result into Equation 8.124 gives

$$
\begin{aligned}
T(r,t) &= \frac{\alpha a}{kr}\int_0^t q_a(\tau)\Bigg(2[4\pi\alpha(t-\tau)]^{-1/2} \\
&\quad \times \exp\left[-\frac{(r-r')^2}{4\alpha(t-\tau)}\right] - \frac{1}{a}\exp\left[\frac{\alpha(t-\tau)}{a^2} + \frac{r-a}{a}\right] \\
&\quad \times \operatorname{erfc}\left\{\frac{r-a}{[4\alpha(t-\tau)]^{1/2}} + \frac{1}{a}[\alpha(t-\tau)]^{1/2}\right\}\Bigg)\, d\tau \qquad (8.126)
\end{aligned}
$$

Depending on the functional form of $q_a(t)$, different solutions can be obtained from Equation 8.126. For special case where $q_a(t) = q_0 =$ constant, the solution becomes

$$
\begin{aligned}
T(r,t) &= \frac{q_0 a^2}{kr}\left\{ \operatorname{erfc}\left[\frac{r-a}{(4\alpha t)^{1/2}}\right] - \exp\left(\frac{r-a}{a} + \frac{\alpha t}{a^2}\right)\right. \\
&\quad \left. \times \operatorname{erfc}\left[\frac{r-a}{(4\alpha t)^{1/2}} + \frac{(\alpha t)^{1/2}}{a}\right]\right\} \qquad (8.127)
\end{aligned}
$$

See Figure 8.11 for a plot of this temperature at several dimensionless times.

Example 8.10: Infinite Body with a Fixed-Temperature Spherical Cavity with Internal Energy Generation—*RS10B00T0Gr5* Case

An infinite body bounded internally by the spherical cavity $r = a$ is initially at zero temperature. For time $t > 0$, the body is heated by a volume energy source given by

$$g(r,t) = g_0 \qquad \text{for } a \le r \le b \qquad (8.128a)$$

$$g(r,t) = 0 \qquad \text{for } r > b \qquad (8.128b)$$

and the surface temperature at $r = a$ is kept at its initial value $T = 0$. Find the transient temperature distribution within the body.

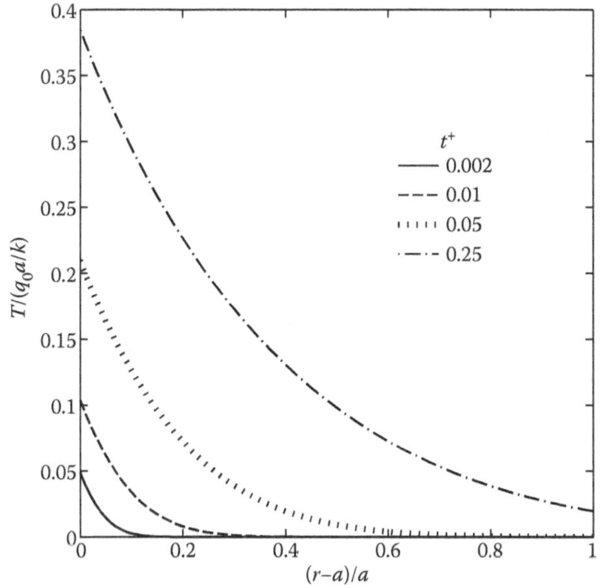

FIGURE 8.11 Infinite region outside the spherical cavity with heat flux suddenly applied at boundary $r = a$, case $RS20B1T0$.

Solution

This is the $RS10$ geometry with energy generation shown in Figure 8.12. The GF solution for this case is given by

$$T(r, t) = \frac{\alpha}{k} \int_{\tau=0}^{t} \int_{r'=a}^{b} G_{RS10}(r, t|r', \tau) g_0 4\pi r'^2 \, dr' \, d\tau \tag{8.129}$$

Note that the integral on r' is evaluated from a to b since the generation is zero for $r > b$. The Green's function is given in Appendix RS as

$$\begin{aligned} G_{RS10}(r, t|r', \tau) &= (4\pi rr')^{-1} [4\pi\alpha(t-\tau)]^{-1/2} \\ &\times \left\{ \exp\left[-\frac{(r-r')^2}{4\alpha(t-\tau)} \right] - \exp\left[-\frac{(r+r'-2a)^2}{4\alpha(t-\tau)} \right] \right\} \end{aligned} \tag{8.130}$$

Substituting for G_{RS10} into Equation 8.129 gives

$$\begin{aligned} T(r, t) &= \frac{\alpha g_0}{kr} \int_{\tau=0}^{t} d\tau \int_{r'=a}^{b} [4\pi\alpha(t-\tau)]^{-1/2} \\ &\times \left\{ \exp\left[-\frac{(r-r')^2}{4\alpha(t-\tau)} \right] - \exp\left[-\frac{(r+r'-2a)^2}{4\alpha(t-\tau)} \right] \right\} r' \, dr' \end{aligned} \tag{8.131}$$

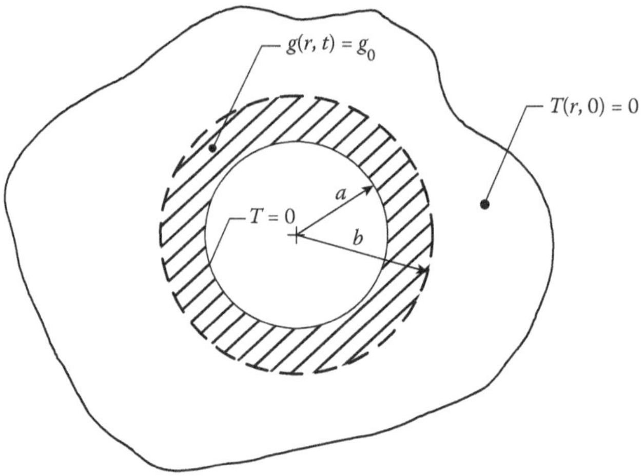

FIGURE 8.12 Infinite region outside the spherical cavity with temperature specified at boundary $r = a$ and with non-uniform energy generation.

The integral over r' is carried out first with the use of Table I.7 (Appendix I) to give

$$T(r,t) = \frac{\alpha g_0}{kr} \int_{\tau=0}^{t} \left(\frac{r}{2} \left\{ \operatorname{erfc} \frac{r-b}{[4\alpha(t-\tau)]^{1/2}} - \operatorname{erfc} \frac{r+b-2a}{[4\alpha(t-\tau)]^{1/2}} \right\} \right.$$
$$\left. + a \left\{ \operatorname{erfc} \frac{r+b-2a}{[4\alpha(t-\tau)^{1/2}]} - \operatorname{erfc} \frac{r-a}{[4\alpha(t-\tau)]^{1/2}} \right\} \right) d\tau \tag{8.132}$$

Next, the integral over time can be evaluated with Table I.8 (Appendix I) to give

$$T(r,t) = \frac{2\alpha g_0 t}{k} \left[\mathrm{i}^2\operatorname{erfc} \frac{r-b}{(4\alpha t)^{1/2}} - \left(1 - \frac{2a}{r}\right) \right.$$
$$\left. \times \mathrm{i}^2\operatorname{erfc} \frac{r+b-2a}{(4\alpha t)^{1/2}} - \frac{2a}{r} \mathrm{i}^2\operatorname{erfc} \frac{r-a}{(4\alpha t)^{1/2}} \right] \tag{8.133}$$

8.8 STEADY STATE

In this section, three examples of steady heat conduction in radial spherical coordinates are presented. The GFs are listed in Appendix RS, Table RS.1. For two- and three-dimensional heat conduction, the steady GF must be found on a case-by-case basis.

Example 8.11: Hollow Sphere Heated on the Inside Surface—*RS*21*B*10 Case

Find the steady temperature in the geometry of Example 8.7, the hollow sphere heated at the inside surface ($r = a$) by heat flux q_0. The outside surface ($r = b$) is maintained at zero temperature.

Solution

This is the $RS21B10$ geometry. The steady temperature is given by the boundary condition term of the steady GFSE, Equation 3.94,

$$T(r) = \frac{q_0}{k} G_{RS21}(r|r' = a) \tag{8.134}$$

The steady GF given in Appendix RS Table RS.1 is a piecewise continuous function,

$$G_{RS21}(r|r') = \begin{cases} \dfrac{1/r' - 1/b}{4\pi} & r \le r' \\ \dfrac{1/r - 1/b}{4\pi} & r \ge r' \end{cases} \tag{8.135}$$

Substitute the above GF evaluated at $r' = a$ into Equation 8.134 by using the $r \ge r'$ portion of the function to find the temperature,

$$T(r) = \frac{q_0}{k}\left(\frac{1}{r} - \frac{1}{b}\right) a^2 \tag{8.136}$$

Example 8.12: Hollow Sphere with Temperature Fixed on Both Surfaces—*RS11B10* Case

Find the steady temperature in the geometry of Example 8.8, the hollow sphere with zero temperature on the outside surface ($r = b$) and with temperature T_∞ maintained at the inside surface ($r = a$).

Solution

This is the $RS11$ geometry. The steady temperature is driven by the boundary condition at $r = a$ and the steady GFSE, Equation 3.94, gives

$$T(r) = -T_\infty \frac{\partial G_{RS11}}{\partial n'}\bigg|_{r'=a} 4\pi a^2 \tag{8.137}$$

The steady GF is given in Appendix RS Table RS.1 as a piecewise continuous function,

$$G_{RS11}(r) = \begin{cases} \dfrac{(b - r')(1 - a/r)}{4\pi r'(b - a)} & r \le r' \\ \dfrac{(b - r)(1 - a/r')}{4\pi r(b - a)} & r \ge r' \end{cases} \tag{8.138}$$

The derivative in Equation 8.137 is evaluated at the surface $r' = a$ so that the $r \ge r'$ portion of the GF is used:

$$-\frac{\partial G_{RS11}}{\partial n'}\bigg|_{r'=a} = \frac{\partial G_{RS11}}{\partial r'}\bigg|_{r'=a} = \frac{(b - r)a/(r')^2}{4\pi r(b - a)}\bigg|_{r'=a} = \frac{b - r}{4\pi a r(b - a)} \tag{8.139}$$

Then the temperature is given by Equation 8.137

$$T(r) = T_\infty \frac{a(b-r)}{r(b-a)} = T_\infty \frac{a}{r}\left(1 - \frac{r-a}{b-a}\right) \tag{8.140}$$

Example 8.13: Solid Sphere with Internal Energy Generation and Convective Boundary Condition

Find the steady temperature in a solid sphere heated by internal energy generation $g(r)$ and cooled by convection from the surface. The heat transfer coefficient is h and the fluid temperature is T_∞.

Solution

This geometry is number *RS03B0G-* if the temperature is normalized in the form $T(r) - T_\infty$. The temperature is given by the energy generation term of the steady GFSE, Equation 3.94,

$$T(r) - T_\infty = \frac{1}{k}\int_{r'=0}^{b} g(r')G_{RS03}(r|r')4\pi(r')^2\, dr' \tag{8.141}$$

The steady GF is given by

$$G_{RS03}(r|r') = \begin{cases} \dfrac{1/r' + (1/B_2 - 1)/b}{4\pi} & r \le r' \\[2ex] \dfrac{1/r + (1/B_2 - 1)/b}{4\pi} & r \ge r' \end{cases} \tag{8.142}$$

where $B_2 = hb/k$, the Biot number. Because G_{RS03} is piecewise continuous, the integral in Equation 8.141 must be carried out in two pieces:

$$\begin{aligned} T(r) - T_\infty &= \frac{1}{k}\int_{r'=0}^{r} g(r')\left[\frac{1}{r} + \frac{1/B_2 - 1}{b}\right](r')^2\, dr' \\ &\quad + \frac{1}{k}\int_{r'=r}^{b} g(r')\left[\frac{1}{r'} + \frac{1/B_2 - 1}{b}\right](r')^2\, dr' \end{aligned} \tag{8.143}$$

A symbolic mathematics computer program is very helpful in finding the correct solution to these integrals. When the GF is piecewise continuous, it is particularly important to get all the signs correct because there are usually terms with opposite sign that cancel out.

(a) *Case* RS03B0G1. In the simple case when the internal energy generation is constant, $g(r) = g_0$, the temperature in Equation 8.143 is given by a second-order polynomial:

$$T(r) - T_\infty = \frac{g_0 b^2}{3k}\left[\frac{1}{2} + \frac{1}{B_2} - \frac{(r/b)^2}{2}\right] \tag{8.144}$$

Note that Equation 8.143 contains six polynomial terms but the solution (8.144) contains only three terms. Two terms canceled and two terms were summed together.

(b) *Case* RS03B0G4. If the energy generation is a maximum at the surface $r = b$ and is attenuated exponentially inside the body as in microwave heating, then the energy generation term may be written

$$g(r) = g_0 e^{-c(1-r/b)} \tag{8.145}$$

where c is the attenuation parameter (dimensionless) and g_0 is the maximum energy generation (W/m^3). The steady temperature is given by Equations 8.143 and 8.145 as

$$T(r) - T_\infty = \frac{g_0 b^2}{kc^3}\left[\left(\frac{2b}{r} - c\right) e^{-c(1-r/b)} + \left(1 - \frac{b}{r} - \frac{1}{B_2}\right)\right.$$
$$\left. \times e^{-c} + c - 2 + \frac{c^2 - 2c + 2}{B_2}\right] \tag{8.146}$$

In the limiting case as $B_2 \to \infty$, the temperature at the surface is $T(r = b) = T_\infty$.

PROBLEMS

8.1 A solid sphere, $0 \le r \le b$, is initially at a uniform temperature T_0 when its surface temperature is suddenly changed to T_b and maintained at this value for times $t > 0$. Using the GF method, find the transient temperature distribution in the sphere for small and large times.

8.2 A solid sphere, $0 \le r \le b$, is initially at a uniform temperature of T_0. For times $t > 0$, its surface temperature changes linearly with time as $T(b, t) = ct$. Using the GF method, find the temperature distribution in the sphere for large times.

8.3 Using the method described in Section 8.4, derive the small-time GF for a solid sphere, $0 \le r \le b$, with a heat flux boundary condition (G_{RS02}).

8.4 A solid sphere, $0 \le r \le b$, has an initial temperature distribution given by

$$T(r, 0) = \begin{cases} T_0 & \text{for } 0 \le r \le a \\ 0 & \text{for } a \le r \le b \end{cases}$$

For times $t > 0$, the surface temperature is kept at its initial value. Using the GF method, find the transient temperature distribution in the sphere.

8.5 A solid sphere, $0 \le r \le b$, is initially at zero temperature. For times $t > 0$, heat is generated uniformly within the sphere with a constant rate of g_0 W/m^3, while the surface temperature is kept at its initial value. Using the GF method, find the transient temperature distribution in the sphere.

8.6 Using the method described in Section 8.4, derive the small-time and the large-time GFs for a hollow sphere, $a \le r \le b$, with a prescribed temperature boundary condition at $r = a$ and a convection boundary condition at $r = b (G_{RS13})$.

8.7 A hollow sphere, $a \leq r \leq b$, is initially at temperature T_0. For times $t > 0$, the boundary at $r = a$ is kept at zero temperature while the boundary at $r = b$ is dissipating heat by convection into a medium at zero temperature with a constant heat transfer coefficient h. Using the GF method, find the transient temperature distribution in the hollow sphere.

8.8 A spherical capsule contains a fluid with volumetric heat capacity $\rho_f c_f$. The inside and outside radii of the capsule wall are r_i and r_o. The boundary conditions are

$$-\frac{k\partial T}{\partial r} = h(T - T_\infty) \qquad \text{at } r = r_0$$

$$-\left(\frac{3k}{r_i}\right)\frac{\partial T}{\partial r} = \rho_f c_f \frac{\partial T}{\partial t} \qquad \text{at } r = r_i$$

$$T = T_i \qquad \text{at } t = 0$$

Find an expression for the temperature at $r = r_i$. If the encapsulated fluid is an ideal gas at a quasi-uniform temperature and the initial pressure is p_i, find p / p_i as a function of time where p is the instantaneous pressure within the capsule.

8.9 Lead shot is sometimes manufactured in shot towers, where molten lead falls through the air to solidify and then is quenched in a liquid to cool.

(a) Suppose the molten lead droplet of radius a starts falling at the solidification temperature T_s, and the latent heat of fusion is f_0 (J / m^3). If the heat transfer coefficient is h_a and the air temperature is T_a, find an expression for the distance the droplet must fall to solidify. (Hint: use a lumped analysis on the droplet; neglect air friction.)

(b) If $T_s = 327°C$, $T_a = 30°C$, and $h_a = 50$ W/(m^2 K), what is the largest size shot that can be dropped in a 50-m tower? [$f_0 = 23$ kJ/(kg K); $\rho c = 1330$ kJ/(m^{3}K).]

(c) Now the shot is quenched in a liquid at T_∞ with heat transfer coefficient h. Find an expression for the transient temperature at the surface of the shot.

8.10 In a pulverized-coal furnace coal particles are blown in with preheated air at a temperature T_i. A new coal particle does not begin to burn until its surface temperature reaches the combustion temperature T_c. If the primarily radiant heat transfer in the furnace may be modeled as a uniform heat flux q_0, on the surface of an approximately spherical coal particle of radius a,

(a) Find a small-time expression for the surface temperature of the particle.

(b) Find an approximate expression for the time it takes the coal particle to begin burning. (Hint: use only one term of the series for the temperature.)

8.11 One technique for handling radioactive waste is encapsulating it in ceramic. Suppose a sphere of the radioactive material of radius a is covered by ceramic to form an encapsulated sphere of radius $b > a$. If the radioactive material produces heat at the rate g_0 W/m^3 and if the outer surface of the sphere (at $r = b$) is cooled by convection (temperature T_∞, heat transfer coefficient h),

(a) Find the steady-state temperature in the ceramic shell, $a < r < b$.

(b) Find an expression for the maximum temperature in the radioactive material.

8.12 In food processing of prepared foods like soup, the heating must continue until every part of the largest chunk in the soup exceeds a specified temperature (to kill bacteria).

(a) Find an expression for the center temperature of a (spherical) dumpling initially at T_i and suddenly exposed to heated liquid at T_∞. (Assume a heat transfer coefficient h.)

(b) Find an approximate expression for the time it takes to raise the center temperature of the dumpling to specified temperature T_f. (Hint: use one term of the large-time form of the temperature.)

9 Steady-Periodic Heat Conduction

9.1 INTRODUCTION

In this chapter transient conduction is treated for which the causative heating is periodic and has continued for a long time. Steady-periodic heating is important in reciprocating machinery, in manufacturing processes, and in naturally occurring temperature cycles such as day/night and summer/winter. Steady-periodic heat transfer is also important in thermal property measurements. Generally thermal properties are measured indirectly, with the thermal properties deduced from a systematic comparison between measured temperatures and a thermal model. Thermal modeling is the focus of this chapter; analysis of experimental data (sometimes called parameter estimation) is beyond the scope of this book. For examples of thermal-property measurements involving steady-periodic heating, see for example: Haji-Shiekh et al. (1998); Hu et al. (1999); Naziev (2001); and Wang et al. (2004).

There are two analytic approaches to steady-periodic heat transfer. In the time-domain approach, the standard transient Green's function approach is used (see Section 3.4 for an example). A particular simplification is possible if the time-history of the heating has a simple wave shape (such as on–off, saw-tooth, etc.). In these cases the time-integral can be evaluated in closed form and the time-dependence reduces to a series of decaying time-exponentials (Carslaw and Jaeger, 1959, p. 108).

The frequency-domain approach, discussed in this chapter, is appropriate if the heating history is sinusoidal, or if the heating is simply periodic and a phase-locked amplifier is used to select the response at the periodic frequency. In the frequency-domain approach the temperature is interpreted as a function of frequency, rather than of time.

The chapter is divided into Green's function in Cartesian, cylindrical, and spherical geometries. The remaining sections cover temperatures constructed from Green's functions in one-dimensional geometries, in layered bodies, and in two- and three-dimensional geometries.

9.2 STEADY-PERIODIC RELATIONS

In this section the relations for transient heat conduction are restated under steady-periodic conditions to emphasize the effect of frequency. The heat equation, the GF solution equation, and the auxiliary equation will each be discussed.

The usual boundary value problem for heat conduction can be written

$$\nabla^2 T - \frac{1}{\alpha}\frac{\partial T}{\partial t} = -\frac{1}{k} g(\mathbf{r}, t) \tag{9.1}$$

$$k_i \frac{\partial T}{\partial n_i} + h_i T + (\rho c b)_i \frac{\partial T}{\partial t} = f_i(\mathbf{r}_i, t) \quad \text{at boundary } i. \tag{9.2}$$

Note that five kinds of boundary conditions are represented by this general boundary condition which may include a high-conductivity layer of thickness b_i (see Chapter 2 for the kinds of boundary conditions). Since in this chapter solutions are sought for which the heating terms are periodic and have continued for a long time, we take the heating terms g and f and the resulting temperature T to be steady periodic at a single frequency. That is, let

$$\begin{aligned} g(\mathbf{r}, t) &= \text{Real}[\tilde{g}(\mathbf{r}, \omega) e^{j\omega t}] \\ f_i(\mathbf{r}_i, t) &= \text{Real}[\tilde{f}_i(\mathbf{r}_i, \omega) e^{j\omega t}] \\ T(\mathbf{r}, t) &= \text{Real}[\tilde{T}(\mathbf{r}, \omega) e^{j\omega t}] \end{aligned} \tag{9.3}$$

where $j = \sqrt{-1}$ is a complex number. Now in Equations 9.1 and 9.2 replace g, f_i, and T by $\tilde{g}$, $\tilde{f}_i$, and $\tilde{T}$, respectively, to find the steady periodic heat conduction equation:

$$\nabla^2 \tilde{T} - \sigma^2 \tilde{T} = -\frac{1}{k} \tilde{g}(\mathbf{r}, \omega); \quad \text{in domain } \Omega \tag{9.4}$$

$$k_i \frac{\partial \tilde{T}}{\partial n_i} + [h_i + j\omega(\rho c b)_i] \tilde{T} = \tilde{f}_i(\mathbf{r}_i, \omega); \quad \text{at boundary } i \tag{9.5}$$

where $\sigma^2 = j\omega/\alpha$. In this chapter complex-valued $\tilde{T}(\mathbf{r}, \omega)$ is interpreted as the steady-periodic temperature (kelvin) at a single frequency ω. For further discussion of this point see Mandelis (2001, pp. 2–3). Later in the chapter complex-valued temperature results will be presented in the form of amplitude and phase, which are defined

$$\begin{aligned} \text{amp} &= [\tilde{T} \cdot \tilde{T}^*]^{1/2} \\ \text{phase} &= \tan^{-1}[\text{Imag}(\tilde{T}) / \text{Real}(\tilde{T})] \end{aligned}$$

where $\tilde{T}^*$ is the complex conjugate of the temperature.

The differential equation for $\tilde{T}$, Equation 9.4, has the character of a steady heat conduction equation with an additional term $(-\sigma^2 \tilde{T})$. This term is similar to the fin term discussed in Chapter 3, except that the coefficient in front of $\tilde{T}$ is now a complex number. Thus, many steady-periodic solutions may be found from steady-fin solutions, with the generalization that σ^2 is complex. The boundary condition for $\tilde{T}$, Equation 9.5, contains term $j\omega(\rho c b)_i \tilde{T}$ which represents heat storage in a surface film, which is important for boundary conditions of the fourth and fifth kinds. Boundary conditions of the fourth and fifth kind, under steady-periodic conditions, have the form of boundary conditions of the third kind, except a complex quantity is added to the heat transfer coefficient.

Since every solution discussed in this chapter will be steady periodic, from this point on the notation will be simplified by dropping the tilde notation. For example, replace $\tilde{T}$ by T. Steady-periodic quantities will be identified by their dependence on frequency ω, for example $T(\omega)$ and $G(\omega)$.

Assume for the moment that the Green's function G is known, then the steady-periodic temperature is given by the following integral equation:

$$T(\mathbf{r},\omega) = \frac{\alpha}{k}\int g(\mathbf{r}',\omega)G(\mathbf{r},\mathbf{r}',\omega)dv' \text{ (for volume heating)} \qquad (9.6)$$

$$+\alpha\sum_i \int f_i(\mathbf{r}_i',\omega) \times \begin{bmatrix} -\partial G / \partial n' & \text{(first kind only)} \\ \frac{1}{k}G(\mathbf{r},\mathbf{r}_i',\omega) & \text{(2nd–5th kind)} \end{bmatrix} ds_i'$$

The first integral is the effect of internal heat generation and the second integral is the effect of each of the nonhomogeneous boundary terms f_i. Note that the same GF appears in each integral but it is evaluated at locations appropriate for each integral.

The GF associated with Equations 9.4 and 9.5 is the response at $\mathbf{r}$ to a steady-periodic heat source located at $\mathbf{r}'$, and the GF satisfies

$$\nabla^2 G - \sigma^2 G = -\frac{1}{\alpha}\delta(\mathbf{r}-\mathbf{r}') \qquad (9.7)$$

$$k_i\frac{\partial G}{\partial n_i} + [h_i + j\omega(\rho c b)_i]G = 0; \quad \text{on boundary } i \qquad (9.8)$$

Here $\sigma^2 = j\omega / \alpha$ and $\delta(\mathbf{r}-\mathbf{r}')$ is the Dirac delta function in the appropriate coordinate system. The coefficient $1/\alpha$ preceding the delta function in Equation 9.7 provides for units of the steady-periodic Green's function that are consistent with the time-domain Green's functions discussed in earlier chapters.

9.3 ONE-DIMENSIONAL GF

In this section one-dimensional steady-periodic GF are given for geometries in the Cartesian, cylindrical, and spherical coordinate systems. In each coordinate system the defining differential equation for the GF is different, but the basic approach to constructing the GF is the same. In a later section examples are given on finding temperature with these GF.

9.3.1 ONE-DIMENSIONAL GF IN CARTESIAN COORDINATES

The one-dimensional steady-periodic GF in Cartesian coordinates, appropriate for slab bodies, semi-infinite bodies, and infinite bodies, satisfies the following equations:

$$\frac{d^2G}{dx^2} - \sigma^2 G = -\frac{1}{\alpha}\delta(x-x') \qquad (9.9)$$

$$k_i\frac{dG}{dn_i} + \lambda_i G = 0; \quad i = 1,2 \qquad (9.10)$$

Here $\sigma^2 = j\omega/\alpha$ and $\lambda_i = h_i + j\omega(\rho cb)_i$. The GF that satisfies Equations 9.9 and 9.10 is given by (see Appendix X, Table X.4 or Crittenden and Cole, 2002)

$$G_X(x,x',\sigma) = \frac{S_2^-(S_1^- e^{-\sigma(2L-|x-x'|)} + S_1^+ e^{-\sigma(2L-x-x')})}{2\alpha\sigma(S_1^+ S_2^+ - S_1^- S_2^- e^{-2\sigma L})} + \frac{S_2^+(S_1^+ e^{-\sigma(|x-x'|)} + S_1^- e^{-\sigma(x+x')})}{2\alpha\sigma(S_1^+ S_2^+ - S_1^- S_2^- e^{-2\sigma L})} \tag{9.11}$$

where the subscripts 1 and 2 represent the two boundaries at the smallest and largest x-values, respectively. Coefficients S_M^+ and S_M^- depend on the boundary conditions on side M and are given by

$$S_M^+ = \begin{cases} 1 & \text{if side } M \text{ is kind 0, kind 1, or kind 2} \\ k\sigma + \lambda_M & \text{if side } M \text{ is kind 3, 4, or 5} \end{cases} \tag{9.12a}$$

$$S_M^- = \begin{cases} 0 & \text{if side } M \text{ is kind 0} \\ -1 & \text{if side } M \text{ is kind 1} \\ 1 & \text{if side } M \text{ is kind 2} \\ k\sigma - \lambda_M & \text{if side } M \text{ is kind 3, 4, or 5} \end{cases} \tag{9.12b}$$

A boundary of kind 0 designates a far-away boundary, as in a semi-infinite body. The derivation of the steady-periodic GF in Equation 9.11 parallels that for steady-state GF given in Section 1.3.2; however in the present case quantity σ contains an imaginary component and λ_i may include effects of surface convection and a thin surface film.

The above GF can also be expressed in terms of hyperbolic functions sinh and cosh. However we recommend the form given in Equation 9.11 because it contains exponentials whose arguments are always negative or zero and therefore this form is particularly well-behaved for machine computation (Cole and Yen, 2001). More importantly, temperature expressions based on these GF are similarly well-behaved for any thickness L and for any frequency.

9.3.2 One-Dimensional GF in Cylindrical Coordinates

The steady-periodic GF for the cylinder is treated in this section. Consider a cylindrical annulus with inner radius a and outer radius b. The GF satisfies the following equations:

$$\frac{1}{r}\frac{d}{dr}\left(r\frac{dG}{dr}\right) - \sigma^2 G = -\frac{1}{\alpha}\frac{\delta(r-r')}{2\pi r'}; \quad a < r < b \tag{9.13}$$

$$k\frac{dG}{dn_i} + \lambda_i G = 0; \quad \text{at boundary } i \tag{9.14}$$

Here, $\sigma^2 = j\omega/\alpha$ and $\lambda_i = h_i + j\omega(\rho cb)_i$ where $b_i \ll r_i$ is assumed (thin surface layer). Be careful not to confuse outer radius b with surface-layer thickness b_i. Note that the radial-cylindrical form of the Dirac delta function is used here. This Green's function represents the response at location r to a steady-periodic, cylindrical-shell heat source located at r'. The GF is constructed in two pieces, that is, function G_1 on $(a < r < r')$ and function G_2 on $(r' < r < b)$. Then the defining equations for G can be written:

$$\frac{1}{r}\frac{d}{dr}\left(r\frac{dG_m}{dr}\right) - \sigma^2 G_m = 0; \quad m = 1,\, 2 \tag{9.15}$$

$$-k\frac{dG_1}{dr} + \lambda_1 G_1 = 0; \quad \text{at } r = a \tag{9.16}$$

$$k\frac{dG_2}{dr} + \lambda_2 G_2 = 0; \quad \text{at } r = b \tag{9.17}$$

Note that the Dirac delta function has been removed from the differential equation. Two matching conditions must be introduced at $r = r'$, which are

$$G_1(r', r') = G_2(r', r') \tag{9.18}$$

$$\left.\frac{dG_2}{dr}\right|_{r'} - \left.\frac{dG_1}{dr}\right|_{r'} = -\frac{1}{2\pi\alpha r'} \tag{9.19}$$

This last condition, the jump condition, comes from integrating the original differential equation, Equation 9.13, over $(r' - \epsilon, r' + \epsilon)$ and then taking the limit as $\epsilon \to 0$.

The general solution of the differential equation for G has the form

$$G = \begin{cases} C_1 I_0(\sigma r) + C_2 K_0(\sigma r); & r < r' \\ C_3 I_0(\sigma r) + C_4 K_0(\sigma r); & r > r' \end{cases} \tag{9.20}$$

where I_0 and K_0 are modified Bessel functions of order zero and where C_i are four undetermined coefficients. The four coefficients may be found by substituting the general form of G into the four conditions, Equations 9.16 through 9.19. After considerable algebra, the GF may be written (Cole and Crittenden, 2009):

$$G_R(r|r') = \frac{1}{2\pi\alpha(1 - S_1 S_2)} \times \begin{Bmatrix} [S_2 I_0(\sigma r') + K_0(\sigma r')][I_0(\sigma r) + S_1 K_0(\sigma r)], & r < r' \\ [S_2 I_0(\sigma r) + K_0(\sigma r)][I_0(\sigma r') + S_1 K_0(\sigma r')], & r > r' \end{Bmatrix} \tag{9.21}$$

Subscript R denotes radial cylindrical. Coefficients S_1 and S_2 depend on the kind of boundary conditions at a and b, as follows:

$$S_1 = \begin{cases} 0 & \text{kind 0 for } a \to 0 \\ -I_0(\sigma a)/K_0(\sigma a) & \text{kind 1 at } r = a \\ I_1(\sigma a)/K_1(\sigma a) & \text{kind 2 at } r = a \\ \left[\dfrac{k\sigma I_1(\sigma a) - \lambda_1 I_0(\sigma a)}{k\sigma K_1(\sigma a) + \lambda_1 K_0(\sigma a)}\right] & \text{kind 3, 4, or 5 at } r = a \end{cases} \tag{9.22a}$$

$$S_2 = \begin{cases} 0 & \text{kind 0 at } b \to \infty \\ -K_0(\sigma b)\,/\,I_0(\sigma b) & \text{kind 1 at } r = b \\ K_1(\sigma b)\,/\,I_1(\sigma b) & \text{kind 2 at } r = b \\ \left[\dfrac{k\sigma K_1(\sigma b) - \lambda_2 K_0(\sigma b)}{k\sigma I_1(\sigma b) + \lambda_2 I_0(\sigma b)}\right] & \text{kind 3, 4, or 5 at } r = b \end{cases} \tag{9.22b}$$

To obtain this form of the GF for the cylinder, the Wronskian identity has been used (Abramowitz and Stegun, 1964, p. 375):

$$K_0(z)I_1(z) + K_1(z)I_0(z) = 1\,/\,z \tag{9.23}$$

The above GF was constructed from case R55, and then conditions of the first or second kind were deduced by taking $\lambda_i \to \infty$ or $\lambda_i \to 0$, respectively, at each boundary. The boundary condition of kind 0 is also included here to describe the solid cylinder (cases R0J for J = 1, 2, 3, 4, or 5), the body surrounding a cylindrical hole (cases RI0 for I = 1, 2, 3, 4, or 5), and the infinite one-dimensional body (case R00).

9.3.3 One-Dimensional GF in Spherical Coordinates

The steady-periodic GF for the sphere is treated in this section. Consider a hollow sphere with inner radius a and outer radius b. The GF satisfies the following equations:

$$\frac{1}{r^2}\frac{d}{dr}\left(r^2\frac{dG}{dr}\right) - \sigma^2 G = -\frac{1}{\alpha}\frac{\delta(r - r')}{4\pi(r')^2}; \quad a < r < b \tag{9.24}$$

$$k\frac{dG}{dn_i} + \lambda_i G = 0; \quad \text{at boundary } i \tag{9.25}$$

As before, $\sigma^2 = j\omega\,/\,\alpha$ and $\lambda_i = h_i + j\omega(\rho cb)_i$, where $b_i \ll r_i$ is assumed. As in the cylinder case, do not confuse radius b with surface-layer thickness b_i. Note that the radial-spherical form of the Dirac delta function is used here. This Green's function represents the response at location r to a steady-periodic spherical-shell heat source located at r'. The GF is constructed in two pieces, that is, function G_1 on $(a < r < r')$ and function G_2 on $(r' < r < b)$. Then the above equations for G can be written:

$$\frac{1}{r^2}\frac{d}{dr}\left(r^2\frac{dG_m}{dr}\right) - \sigma^2 G_m = 0; \quad m = 1,\ 2 \tag{9.26}$$

$$-k\frac{dG_1}{dr} + \lambda_1 G_1 = 0; \quad \text{at } r = a \tag{9.27}$$

$$k\frac{dG_2}{dr} + \lambda_2 G_2 = 0; \quad \text{at } r = b \tag{9.28}$$

Note that the Dirac delta function has been removed from the differential equation. Two matching conditions must be introduced at $r = r'$, which are

$$G_1|_{r'} = G_2|_{r'} \tag{9.29}$$

$$\left.\frac{dG_2}{dr}\right|_{r'} - \left.\frac{dG_1}{dr}\right|_{r'} = -\frac{1}{4\pi\alpha(r')^2} \tag{9.30}$$

This last condition, the jump condition, comes from integrating the original differential equation over $(r' - \epsilon, r' + \epsilon)$ and then taking the limit as $\epsilon \to 0$.

Under these conditions the differential equation for G (Equation 9.26), has a general solution of the form

$$G = \begin{cases} \frac{1}{r}(C_1 e^{\sigma r} + C_2 e^{-\sigma r}); & r < r' \\ \frac{1}{r}(C_3 e^{\sigma r} + C_4 e^{-\sigma r}); & r > r' \end{cases} \tag{9.31}$$

There are four undetermined coefficients which may be found by substituting the general form of G into the four conditions, Equations 9.27 through 9.30. After considerable algebra, the GF may be written:

$$\begin{aligned} G_{RS}(r, r', \sigma) &= \frac{S_1^- S_2^- e^{-\sigma(b-a-|r-r'|)} + S_1^+ S_2^- e^{-\sigma(b+a-(r+r'))}}{8\pi\alpha\sigma r r'(S_1^+ S_2^+ - S_1^- S_2^- e^{-\sigma(b-a)})} \\ &+ \frac{S_1^- S_2^+ e^{-\sigma(r+r'-2a)} + S_1^+ S_2^+ e^{-\sigma|r-r'|}}{8\pi\alpha\sigma r r'(S_1^+ S_2^+ - S_1^- S_2^- e^{-\sigma(b-a)})} \end{aligned} \tag{9.32}$$

Here subscript RS denotes radial-spherical. Notation $|r - r'|$ is used to provide a single expression for the GF which was constructed from two functions. This single expression makes it easy to see that the GF satisfies reciprocity, that is, $G_{RS}(r, r') = G_{RS}(r', r)$. Coefficients S_1^+, S_1^-, S_2^+, and S_2^- in the expression for the GF depend on the kind of boundary conditions at a and b, as follows:

$$S_1^+ = \begin{cases} 1 & \text{kind 0 with } a \to 0 \\ 1 & \text{kind 1 at } r = a \\ \sigma a + 1 & \text{kind 2 at } r = a \\ \sigma a + 1 + \lambda_1 a / k & \text{kind 3, 4, or 5 at } r = a \end{cases} \tag{9.33a}$$

$$S_1^- = \begin{cases} 0 & \text{kind 0 with } a \to 0 \\ -1 & \text{kind 1 at } r = a \\ \sigma a - 1 & \text{kind 2 at } r = a \\ \sigma a - 1 - \lambda_1 a / k & \text{kind 3, 4, or 5 at } r = a \end{cases} \tag{9.33b}$$

$$S_2^+ = \begin{cases} 1 & \text{kind 0 with } b \to \infty \\ 1 & \text{kind 1 at } r = b \\ \sigma b - 1 & \text{kind 2 at } r = b \\ \sigma b - 1 + \lambda_2 b / k & \text{kind 3, 4, or 5 at } r = b \end{cases} \tag{9.33c}$$

$$S_2^- = \begin{cases} 0 & \text{kind 0 with } b \to \infty \\ -1 & \text{kind 1 at } r = b \\ \sigma b + 1 & \text{kind 2 at } r = b \\ \sigma b + 1 - \lambda_2 b / k & \text{kind 3, 4, or 5 at } r = b \end{cases} \tag{9.33d}$$

The boundary condition of kind 0 is included to describe the solid sphere (cases RS0J for J = 1, 2, 3, 4, or 5), the body surrounding a spherical void (cases RSI0 for

I = 1, 2, 3, 4, or 5), and the infinite one-dimensional body with spherical symmetry (case RS00).

9.4 ONE-DIMENSIONAL TEMPERATURE

In this section several examples are given for steady-periodic temperature expressions constructed from one-dimensional GF in the Cartesian, cylindrical, and spherical coordinate systems.

Example 9.1: Slab Heated on One Side, X23B60

Consider a slab wall heated steady-periodically at $x = 0$ and cooled by convection on the other side. The temperature satisfies

$$\frac{d^2T}{dx^2} - \sigma^2 T = 0 \tag{9.34}$$

$$-k\frac{dT}{dx}\bigg|_{x=0} = q_0(\omega) \tag{9.35}$$

$$-k\frac{dT}{dx}\bigg|_{x=L} = h(T-0) \tag{9.36}$$

The temperature expression is given by Equation 9.6 applied to a 1D Cartesian body

$$T(x,\omega) = \frac{\alpha}{k}\, q_0\, G_{X23}(x, x' = 0, \omega) \tag{9.37}$$

where q_0 is magnitude of the specified steady-periodic heat flux at $x = 0$. The required GF is case X23 which may be found from Equation 9.11 with $S_1^+ = 1$, $S_1^- = 1$, $S_2^+ = k\sigma + h_2$, and $S_2^- = k\sigma - h_2$. That is,

$$G_{X23}(x, x', \omega) = \frac{R_2\left(e^{-\sigma(2L-|x-x'|)} + e^{-\sigma(2L-x-x')}\right)}{2\alpha\sigma(1 - R_2 e^{-2\sigma L})} + \frac{e^{-\sigma|x-x'|} + e^{-\sigma(x-x')}}{2\alpha\sigma(1 - R_2 e^{-2\sigma L})} \tag{9.38}$$

$$\text{where} \quad R_2 = \frac{k\sigma - h_2}{k\sigma + h_2}$$

Replace this GF into the above temperature expression to find the temperature, which can be stated in dimensionless form by dividing by $q_0 L/k$:

$$\frac{T(x,\omega)}{q_0 L/k} = \frac{\frac{\sigma^+ - B_2}{\sigma^+ + B_2}\left(e^{-\sigma^+(2-x/L)}\right) + e^{-\sigma^+ x/L}}{\sigma^+(1 - \frac{\sigma^+ - B_2}{\sigma^+ + B_2} e^{-2\sigma^+})} \tag{9.39}$$

$$\text{where} \quad \sigma^+ = \frac{(1+j)}{\sqrt{2}}\sqrt{\omega L^2/\alpha}$$

Here $B_2 = h_2 L/k$ is a Biot number which describes the level of convective cooling at $x = L$. Plots of the amplitude and phase of the temperature are shown versus dimensionless frequency $(\omega L^2/\alpha)$ in Figure 9.1.

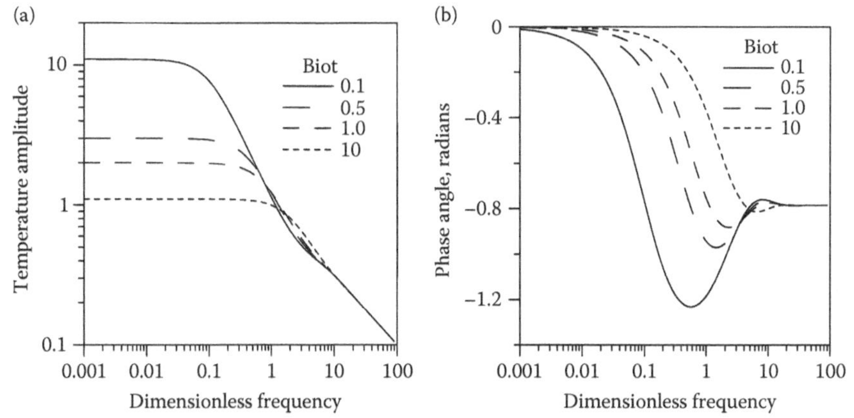

FIGURE 9.1 Amplitide and phase of the temperature on the heated surface (at $x = 0$) of the slab body, case X23B10, as a function of steady-periodic heating frequency. The Biot number determines the amount of convection at $x = L$.

Example 9.2: Semi-Infinite Body with Internal Heating

Consider a one-dimensional semi-infinite region which is heated internally and cooled by convection at the $x = 0$ surface. The steady-periodic internal heating has a spatial distribution described by

$$g(x, \omega) = g_0(\omega)e^{-x/a} \tag{9.40}$$

which is typical of microwave absorption (or optical absorption) where a is the energy penetration depth. This is case X30B0G(x4t6).

The temperature may be formally stated with the GF solution equation:

$$T(x, \omega) = \frac{\alpha}{k}\int_{x'=0}^{\infty} g(x')\, G_{X30}(x, x', \omega)\, dx' \tag{9.41}$$

The GF is given by Equation 9.11 with, $S_1^+ = k\sigma + h_1$, $S_1^- = k\sigma - h_1$, $S_2^+ = 1$, and $S_2^- = 0$. Then the GF may be written

$$G_{X30}(x, x', \omega) = \frac{e^{-\sigma|x-x'|} + R_1 e^{-\sigma(x-x')}}{2\alpha\sigma}$$

$$\text{where} \quad R_1 = \frac{k\sigma - h_1}{k\sigma + h_1} \tag{9.42}$$

Replace this GF into the temperature expression to find

$$T(x, \omega) = \frac{g_0}{2k\sigma}\int_{x'=0}^{\infty} e^{-x'/a}\left(e^{-\sigma|x-x'|} + R_1 e^{-\sigma(x-x')}\right) dx' \tag{9.43}$$

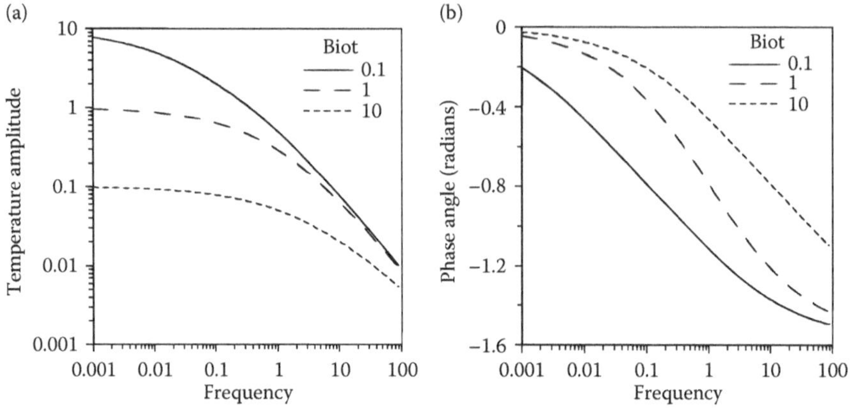

FIGURE 9.2 Amplitide and phase of the temperature at the convectively cooled surface (at $x = 0$) of a semi-finite body heated internally, case X30B0G(x4t6). The results are normalized with length scale a, which is the length scale of the exponentially-decaying internal heating.

The portion of the integrand containing the absolute value must be evaluated in two pieces, that is,

$$T(x, \omega) = \frac{g_0}{2k\sigma}\left[\int_{x'=0}^{x} e^{-x'/a}e^{-\sigma(x-x')}dx' + \int_{x'=x}^{\infty} e^{-x'/a}e^{-\sigma(x'-x)}dx'\right]$$
$$+ \frac{g_0}{2k\sigma}\int_{x'=0}^{\infty} e^{-x'/a}R_1 e^{-\sigma(x-x')}\,dx' \qquad (9.44)$$

All the integrals can be evaluated to obtain

$$\frac{T(x, \omega)}{g_0a^2/k} = \frac{1}{2\sigma a}\left[\frac{e^{-x/a} - e^{-\sigma x}}{\sigma a - 1} + \frac{e^{-x/a} + \left(\frac{\sigma a - B_1}{\sigma a + B_1}\right)e^{-\sigma x}}{\sigma a + 1}\right] \qquad (9.45)$$

where $B_1 = h_1a/k$ is a Biot number. The dimensionless temperature depends on Biot number B_1, dimensionless location x/a, and dimensionless frequency $\omega a^2/\alpha$. Phase and amplitude of dimensionless temperature at the surface ($x = 0$) are plotted versus dimensionless frequency in Figure 9.2 for several values of the Biot number.

Example 9.3: Cylinder with Internal Heating and Convective Cooling

Consider a long cylinder with steady-periodic internal heating and with convective cooling at the surface. This could represent an electric wire carrying alternating current at frequency ω which produces Joule heating. Suppose that the internal heating, $g_0(\omega)$, is spatially uniform, and the convective environment around the cylinder is characterized by heat transfer coefficient h_2 and fluid temperature T_∞. This is case R03B0G(t6x1). Then the temperature in the cylinder is given by

$$T(r, \omega) - T_\infty = \frac{\alpha}{k}\int_{r'=0}^{b} g_0(\omega)\; G_{R03}(r, r', \omega)2\pi r'dr' \qquad (9.46)$$

Note that the differential volume for the 1D cylinder is $2\pi r' dr'$. As the GF is composed of two pieces, the integral expression for the temperature must be evaluated in two pieces also. That is,

$$T(r,\omega) - T_\infty = \frac{2\pi\alpha g_0}{k}\left[\int_{r'=0}^{r} G_{R03}|_{r>r'}\, r' dr' + \int_{r'=r}^{b} G_{R03}|_{r<r'}\, r' dr'\right] \tag{9.47}$$

The required GF is given by Equation 9.21 with $S_1 = 0$ (appropriate for the solid cylinder). Replace the GF into the above integral expression to obtain

$$T(r,\omega) - T_\infty = \frac{g_0}{k\sigma}\int_{r'=0}^{r} [S_2 I_0(\sigma r') I_0(\sigma r) + I_0(\sigma r') K_0(\sigma r)] r' dr'$$
$$+ \frac{g_0}{k\sigma}\int_{r'=r}^{b} I_0(\sigma r)[S_2 I_0(\sigma r') + K_0(\sigma r')] r' dr' \tag{9.48}$$

The integrals can be evaluated and simplified with the following identities (see Appendix B):

$$\int I_0(\sigma r') r' dr' = \frac{\sigma}{r'} I_1(\sigma r')$$
$$\int K_0(\sigma r') r' dr' = -\frac{\sigma}{r'} K_1(\sigma r')$$
$$K_0(z) I_1(z) + I_0(z) K_1(z) = z^{-1} \quad \text{(Wronskian)}$$

Then the temperature may be written

$$\frac{T(r,\omega) - T_\infty}{g_0 b^2/k} = \frac{1}{\sigma b}[S_2 I_1(\sigma b) - K_1(\sigma b)] I_0(\sigma r) + \frac{1}{\sigma^2 b^2} \tag{9.49}$$
$$\text{where} \quad S_2 = \frac{\sigma b K_1(\sigma b) - B_2 K_0(\sigma b)}{\sigma b I_1(\sigma b) + B_2 I_0(\sigma b)}$$

and where Biot number $B_2 = h_2 b/k$ describes the level of external convection. Note that the temperature distribution has two terms. The first term, which depends on radius r in the form $I_0(\sigma r)$, is the effect of convection on the temperature distribution. If the convection is turned off, $h_2 \to 0$, then the surface of the cylinder is insulated, and the contribution to convection disappears (note $S_2 \to K_1(\sigma b)/I_1(\sigma b)$ for $h_2 \to 0$). The second term, $1/(\sigma^2 b^2)$, represents the steady-periodic (normalized) temperature in the absence of convection cooling. This term is independent of radius because the heat is introduced uniformly. Amplitude and phase of the normalized temperature are plotted versus frequency in Figure 9.3 for several Biot values.

Example 9.4: Large Region Heated by a High-Conductivity Sphere

Consider a large, low-conductivity region containing a sphere of high conductivity which is heated periodically at rate $Q_0(\omega)$ (units are watts). This is case RS40B6 and the geometry is shown in Figure 9.4. For low frequency heating, the sphere will

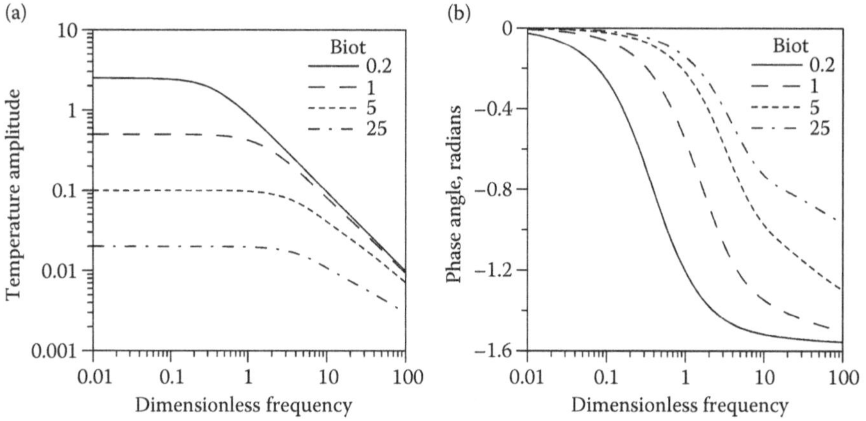

FIGURE 9.3 Amplitide and phase of the temperature at the convectively cooled surface (at $r = b$) of a cylinder heated internally. The heating is spatially uniform and time-periodic.

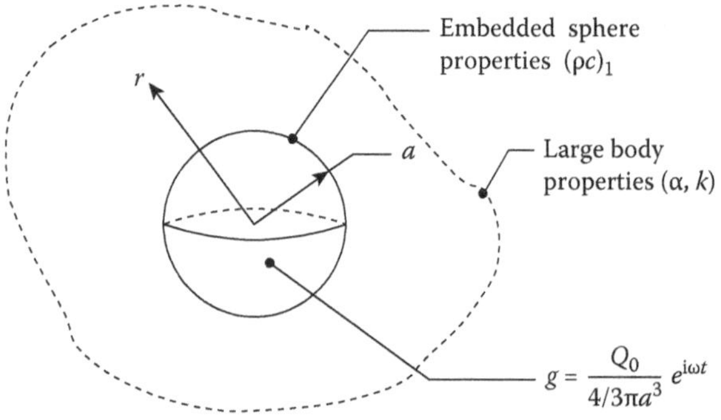

FIGURE 9.4 Geometry for a steady-periodic heating of a small high-conductivity sphere which is embedded in a large, low-conductivity body. Case $RS40B6$.

behave as a lumped body on the boundary of the large body. One application for this case is thermal property measurement in the large body in which the sphere is embedded. The temperature satisfies

$$\frac{1}{r^2}\frac{d}{dr}\left(r^2\frac{dT}{dr}\right) - \frac{j\omega}{\alpha}T = 0; \quad r > a \tag{9.50}$$

The boundary condition at $r = a$ can be found from an energy balance on the sphere in the form $Q_{in} - Q_{out} = Q_{stored}$, for steady-periodic heating:

$$Q_0(\omega) - q_{out}A_1 = (\rho c)_1 V_1 j\omega\, T|_{r=a} \tag{9.51}$$

where $(\rho c)_1$ is specific heat (per unit volume) of the sphere, and A_1 and V_1 are the surface area and volume of the sphere, respectively. Divide this relation by A_1, replace q_{out} by the heat conduction into the surrounding body, and rearrange to find

$$-k\frac{dT}{dr}\bigg|_{r=a} + j\omega(\rho c)_1\frac{a}{3}T|_{r=a} = \frac{Q_0}{4\pi a^2} \tag{9.52}$$

This boundary condition is similar to the thin-layer boundary condition given in Equation 9.5, with surface-film thickness b_1 replaced by length scale $V_1/A_1 = a/3$.

The temperature is given by the Green's function solution, Equation 9.6, for this case:

$$T(r,\omega) = \frac{\alpha}{k}\frac{Q_0}{4\pi a^2}G_{RS40}(r, r' = a, \sigma)4\pi a^2 \tag{9.53}$$

where properties of the body surrounding the sphere are k and α. Note that the boundary surface area must be included. The Green's function for this case is given by Equation 9.32 with $S_2^+ = 1$, and $S_2^- = 0$. Then

$$G_{RS40}(r, r', \sigma) = \frac{R \cdot e^{-\sigma(r+r'-2a)} + e^{-\sigma|r+r'|}}{8\pi\alpha\sigma r r'} \tag{9.54}$$
$$\text{where} \quad R = \frac{\sigma a - 1 - \lambda_1 a/k}{\sigma a + 1 + \lambda_1 a/k}$$

and where $\lambda_1 = j\omega(\rho c)_1 a/3$. Evaluate the GF at $r' = a$ and replace into the temperature expression to obtain

$$\frac{T(r,\omega)}{Q_0/(ak)} = \frac{1}{8\pi}\left[1 + \frac{\sigma a - 1 - (\sigma a)^2 C_r/3}{\sigma a + 1 + (\sigma a)^2 C_r/3}\right]\frac{e^{-\sigma(r-a)}}{\sigma r} \tag{9.55}$$

Note that $\lambda_1 a/k$ has been restated as $(\sigma a)^2 C_r/3$ where $C_r = (\rho c)_1/(\rho c)$ to show that the results depend on the (volumetric) specific-heat ratio. The above temperature expression applies to low frequency where the high-conductivity sphere remains lumped (temperature spatially uniform in the sphere). As frequency increases, eventually this assumption will break down, approximately where $\omega a^2/\alpha_1 > 1$ for $\alpha_1 = k_1/(\rho c)_1$. See Figure 9.5 for plots of amplitude and phase of this temperature for several values of the specific-heat ratio.

9.5 LAYERED BODIES

One-dimensional steady-periodic heat conduction in a layered body is discussed in this section. Each layer may have different thermal properties, and the layers may be in perfect or imperfect contact with one another. Internal heating may be present in one layer or in all of the layers. The temperature in the layered body will be found by assigning a GF within each layer and assigning unknown heat fluxes at the interfaces linking the layers. A matrix solution is used to find the unknown heat fluxes from which the temperature may be found. This technique for finding temperature has application in the measurement of thermal properties.

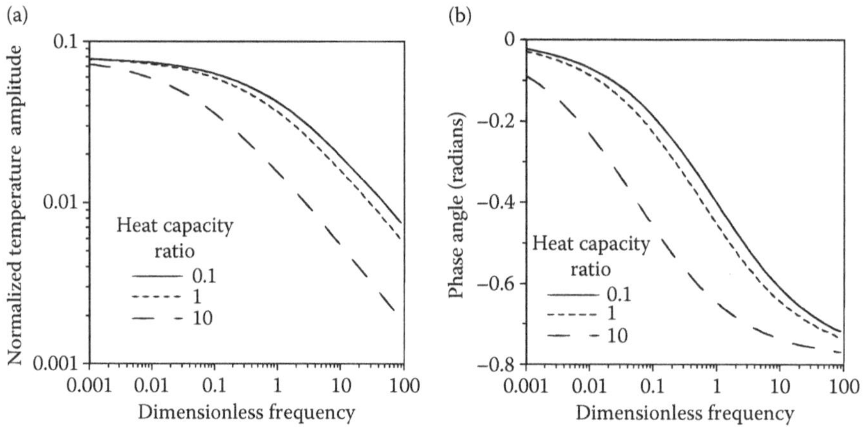

FIGURE 9.5 Amplitide and phase of the (spatially lumped) temperature in the high conductivity sphere embedded in a small-conductivity body. Case $RS40B6$.

It is possible to define one GF to describe a multilayered body, but such an approach is limited to two or three layers because the complexity of the GF increases very rapidly as layers are added (Mandelis 2001, Section 1.7). In contrast, the matrix method presented here involves a simple GF in each layer. Adding internal heating is straightforward, and adding layers simply increases the size of the matrix solution. The matrix method has been used with up to 50 layers to simulate a functionally graded material (Cole, 2004a).

The geometry to be studied is shown in Figure 9.6. There are $N + 2$ layers, numbered from 0 to $N + 1$, with $N + 1$ interfaces between the layers. Layer i has thickness L_i and thermal properties k_i and α_i. Within layer i, the interfaces are located at coordinates $x_i = 0$ and $x_i = L_i$. At the interfaces between the layers, let q_{nm} represent the heat flux leaving layer n and entering layer m. In the formulation given below, heating is caused by internal energy generation within any layer. This could describe a semitransparent material heated by a periodically modulated laser beam. Although heating at the boundaries is not shown, it could easily be added to layers 0 and $N + 1$.

Consider first the temperature in layer 0 evaluated at its interface with layer 1:

$$T_0(L_0, \omega) = \frac{\alpha_0}{k_0} G_0(L_0, L_0, \omega) q_{10} + E_0(L_0) \tag{9.56}$$

In layer i; $i = 1, 2, \ldots, N$: the interface temperatures are:

$$T_i(0, \omega) = \frac{\alpha_i}{k_i} G_i(0, 0, \omega) q_{i-1,i} + \frac{\alpha_i}{k_i} G_i(0, L_i, \omega) q_{i+1,i} + E_i(0) \tag{9.57}$$

$$T_i(L_i, \omega) = \frac{\alpha_i}{k_i} G_i(L_i, 0, \omega) q_{i-1,i} + \frac{\alpha_i}{k_i} G_i(L_i, L_i, \omega) q_{i+1,i} + E_i(L_i) \tag{9.58}$$

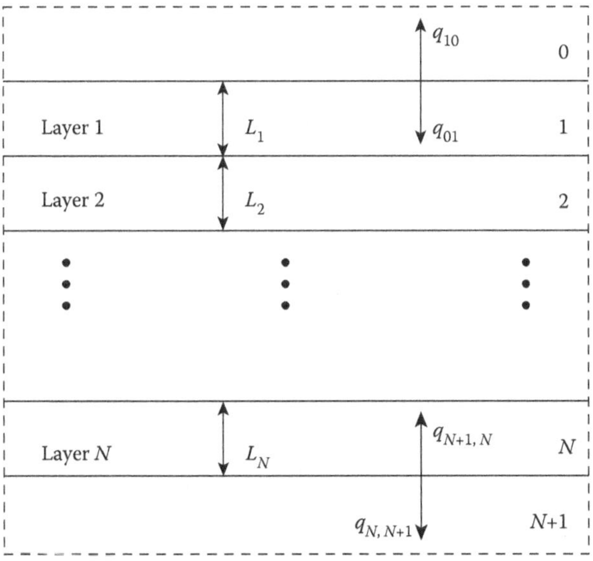

FIGURE 9.6 Geometry for heat conduction in a body composed of many plane layers.

In the last layer (substrate) the temperature at interface $N + 1$ is:

$$T_{N+1}(0, \omega) = \frac{\alpha_{N+1}}{k_{N+1}} G_{N+1}(0, 0, \omega) q_{N,N+1} + E_{N+1}(0) \tag{9.59}$$

In the above expressions, symbol E_i has been used for the volume-heating integral term from the GF solution equation, Equation 9.6, specifically,

$$E_i(x) = \frac{\alpha_i}{k_i} \int_{x'} g(x', \omega)\, G_i(x, x', \omega)\, dx' \tag{9.60}$$

In the case of laser heating, quantity g is the laser energy absorbed in the layer per unit volume; this can be determined without approximation from the optical properties of the layers (McGahan and Cole, 1992).

In the above temperature expressions, all of the interface heat fluxes are initially unknown. The heat flux leaving one layer enters the adjacent layer, $q_{i-1,i} = -q_{i,i-1}$ and the temperature difference between adjacent layers is related to heat flux through a contact resistance at each interface:

$$q_{i-1,i}\, R_i = T_i(0, \omega) - T_{i-1}(L_{i-1}, \omega); \quad i = 1, 2, \ldots, N + 1 \tag{9.61}$$

The contact resistance R_i describes the size of the temperature jump across the interface. Perfect contact is described by $R_i = 0$. Next Equations 9.56 through 9.59 are combined with Equation 9.61 to eliminate temperature. The result is a set of $N + 1$ linear algebraic equations for the unknown heat fluxes, which may be stated in matrix form:

$$\begin{bmatrix} U_0+U_1+R_1 & -V_1 & 0 & \cdots & 0 \\ -V_1 & U_1+U_2+R_2 & -V_2 & \cdots & 0 \\ 0 & -V_2 & U_2+U_3+R_3 & \cdots & 0 \\ \cdots & \cdots & \cdots & \ddots & -V_N \\ 0 & 0 & \cdots & -V_N & U_N+U_{N+1}+R_{N+1} \end{bmatrix}$$

$$\times \begin{bmatrix} q_{10} \\ q_{21} \\ q_{32} \\ \cdots \\ q_{N+1,N} \end{bmatrix} = \begin{bmatrix} E_1(0)-E_0(L_0) \\ E_2(0)-E_1(L_1) \\ E_3(0)-E_2(L_2) \\ \cdots \\ E_{N+1}(0)-E_N(L_N) \end{bmatrix} \tag{9.62}$$

Symbols U_i and V_i used in the above expression are given below:

$$U_i = \frac{\alpha_i}{k_i} G_i(0,0,\omega) = \frac{\alpha_i}{k_i} G_i(L_i, L_i, \omega) \tag{9.63}$$

$$V_i = \frac{\alpha_i}{k_i} G_i(0, L_i, \omega) = \frac{\alpha_i}{k_i} G_i(L_i, 0, \omega) \tag{9.64}$$

For any multilayered system, it is now possible to calculate the $N+1$ unknown heat fluxes (q_{ij}) through all interfaces in the system. The above result is *exact*, and Cramer's rule may be used to solve for the $q's$ for a sample composed of two or three layers. For a sample with three or more layers, a numerical solution is best, and the well-known tridiagonal algorithm may be used (Press et al., 1992, p. 42). Once the heat fluxes are found, the temperature at any interface is given by Equations 9.56 through 9.59.

Several different GF may be used in a layered material. For layers $i = 1, 2, \ldots, N$ the GF needed are type X22 (specified boundary heat flux). The GF for the outermost layers depends on the heat transfer environment there. For example, an outer layer exposed to a fluid could be described by GF number X23, and a thick substrate could be described by GF number X20.

The above discussion is for plane layers, however layered cylinders or layered spheres can be treated in a similar manner. First, substitute the appropriate cylindrical or spherical GF and use the appropriate surface area and differential volume in the temperature expressions in each layer. For example, in cylindrical layer i located at $(r_{i-1} < r < r_i)$, the temperature at the inner radius is given by

$$T_i(r_{i-1}, \omega) = \frac{\alpha_i}{k_i} G_i(r_{i-1}, r_{i-1}, \omega) q_{i-1,i} 2\pi r_{i-1} + \frac{\alpha_i}{k_i} G_i(r_{i-1}, r_i, \omega) q_{i+1,i} 2\pi r_i + E_i(r_{i-1})$$

$$\text{where} \quad E_i(r_{i-1}) = \frac{\alpha_i}{k_i} \int g(r', \omega) G_i(r_{i-1}, r', \omega) 2\pi r' dr' \tag{9.65}$$

Second, for cylinders and spheres there are two U-terms and two V-terms needed in each layer, because $G_i(r_{i-1}, r_{i-1}, \omega) \neq G_i(r_i, r_i, \omega)$ and $G_i(r_{i-1}, r_i, \omega) \neq G_i(r_i, r_{i-1}, \omega)$.

9.6 TWO- AND THREE-DIMENSIONAL CARTESIAN BODIES

In this section several two- and three-dimensional geometries are discussed for steady-periodic heat conduction in the Cartesian coordinate system.

9.6.1 Rectangles and Slabs

The GF for steady-periodic heat transfer in rectangles (cases XIJYKL) and two-dimensional slabs (cases X00YKL) may be found with the method of eigenfunction expansion (see Chapter 4). The GFs for these body shapes that satisfy Equations 9.7 and 9.8 may be found in the form of a single sum (Cole, 2006):

$$G(x, y, \omega \,|\, x', y') = \sum_{n=0}^{\infty} \frac{Y_n(y) Y_n^*(y')}{N_y(\gamma_n)} P(x, x', \nu) \tag{9.66}$$

where eigenfunction Y_n, norm N_y, and eigenvalues γ_n are identical to those discussed earlier in Chapter 4, except that here the eigenfunctions lie along the y-axis. The $n = 0$ term of the series is needed only when there are boundaries of the second kind at $y = 0$ and $y = W$ (when Y22 is part of the GF number). Kernel function P satisfies

$$\frac{d^2 P}{dx^2} - \nu^2 P = -\delta(x - x') \tag{9.67}$$

along with appropriate homogeneous boundary conditions at $x = 0$ and $x = L$. Kernel function P is similar to the one-dimensional GF given in Equation 9.11 except that σ is replaced by $\nu = \sqrt{\sigma^2 + \gamma_n^2}$. That is, $P(x, x', \nu) = G_X(x, x', \nu)$.

The above series expression for Cartesian GF also applies to boundary conditions of the fourth or fifth kind, which include a thin surface film. However, these boundary conditions require special care because the eigenvalues are complex numbers and the eigenfunctions contain complex-valued sine and/or cosine. Complex-valued eigenvalues have been previously shown to occur for heat conduction in multilayer, multidimensional bodies (Haji-Shiekh and Beck, 2002).

Although the GF is unique, for many geometries there exist alternate forms for the same unique GF. These alternate forms have a very important role in numerical evaluation of the GF and the temperatures constructed from them. Specifically, the alternate GF can be used to verify that numerical values are correctly computed.

Alternate GF for rectangles. In the rectangle an alternate series for the GF may be found by placing the kernel functions in the y-direction and the eigenfunctions in the x-direction. The alternate GF is important because at a point in the rectangle where one series converges slowly, the other series converges rapidly, and vice versa. In previous work with steady temperature (at $\omega = 0$), we have shown that there are locations in the domain at which the slowly converging series requires thousands of times more terms than the rapidly converging series (Cole and Yen, 2001; Crittenden and Cole, 2002). A double-sum form of the GF may also be found from Fourier expansions along both x and y, however it generally converges very slowly and should not be used when a single-sum form is available.

Alternate GF for the 2D slab. An alternate GF for two-dimensional slab bodies may be found with a spatial Fourier transform. Consider slab bodies described by cases X00YIJ for which I, J = 1, 2, 3, 4, or 5. The solution will be found with a spatial Fourier transform defined by the following transform pair (Carslaw and Jaeger, 1959, p. 57):

$$\overline{G}(\beta) = \int_{-\infty}^{\infty} G(x)e^{-j\beta x}\,dx \tag{9.68}$$

$$G(x) = \frac{1}{2\pi}\int_{-\infty}^{\infty} \overline{G}(\beta)e^{j\beta x}\,d\beta \tag{9.69}$$

Note that variable x' has been suppressed by a change of variable, replacing $(x - x')$ by x, which is allowed under the differential equation which defines G. Apply the above transform to Equations 9.7 and 9.8 to obtain

$$\frac{d\overline{G}^2}{dy^2} - \nu^2\overline{G} = -\frac{1}{\alpha}\delta(y - y') \tag{9.70}$$

$$k_i\frac{d\overline{G}}{dn_i} + \lambda_i\overline{G} = 0 \ \text{ at boundary } i \tag{9.71}$$

$$\text{where } \nu^2 = \beta^2 + j\omega/\alpha \tag{9.72}$$

$$\lambda_i = h_i + j\omega(\rho cb)_i \tag{9.73}$$

Equation 9.70 is similar to Equation 9.9 which defines the one-dimensional GF, so $\overline{G}$ is given by Equation 9.11 with parameter σ replaced by ν and x replaced by y. That is, $\overline{G}(y, y') = G_X(y, y', \nu)$. Finally, the GF in the slab may be formally stated in x-space by use of the inverse transform:

$$G(x, y, \omega|x', y') = \frac{1}{2\pi}\int_{-\infty}^{\infty} G_X(y, y', \nu)\, e^{j\beta(x-x')}d\beta \tag{9.74}$$

Here variable x' has been recovered by reversing the earlier change of variable and replacing x by $(x - x')$. In general the inverse-transform integral must be evaluated numerically, which is possible because the integrand approaches zero as $\beta \to \pm\infty$.

Example 9.5: Slab Heated over a Small Region

In this example a two-dimensional slab body is heated over a small region and cooled by convection at $y = W$. The body is insulated at $y = 0$. This geometry is a model of a thermal conductivity sensor. The number for this case is X00Y23B0(x5t6) and the geometry is shown in Figure 9.7. The temperature is formally given by Equation 9.6 with volume heating:

$$T(x, y, \omega) = \frac{\alpha}{k}\int\int g(x', y')\, G_{X00Y23}(x, x', y, y', \omega)\, dx'dy' \tag{9.75}$$

The heated region is of infinitesimal thickness along y and is piecewise constant along x, described by $g(x', y') = q(x')\delta(y' - W)$ where

$$q(x') = \begin{cases} q_0; & |x'| < a \\ 0; & |x'| > a \end{cases}$$

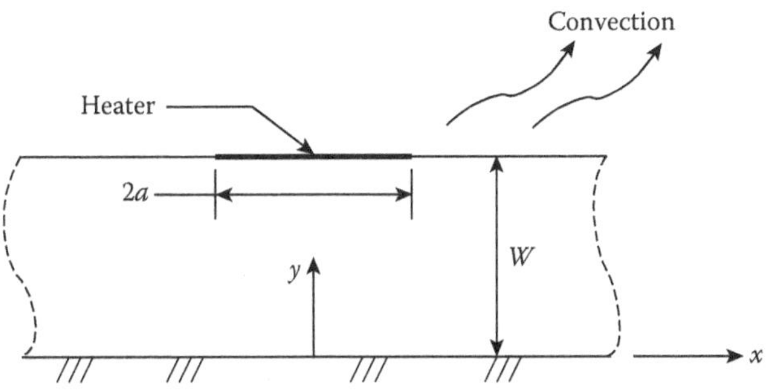

FIGURE 9.7 Two-dimensional slab heated over a small area and cooled by convection on one side, case $X00Y23B00G(x5y7t6)$.

This heating function could represent a thin metal film which is heated electrically. Substitute this heating function into the above temperature expression to obtain

$$T(x, y, \omega) = \frac{\alpha}{k}\int_{-a}^{a} q_0 G_{X00Y23}(x, x', y, y' = W, \omega)\, dx' \qquad (9.76)$$

Note that the integral on y' has been stripped away by the Dirac delta function. The series form of the GF, Equation 9.66, will be used to find the temperature. Substitute the appropriate eigenfunction from Table 4.2, norm from Table 4.3, and kernel function (1D planar GF in this case) from Equation 9.11 to obtain

$$T(x, y, \omega) = \frac{\alpha q_0}{k}\sum_{n=0}^{\infty}\frac{\cos(\gamma_n W)\cos(\gamma_n y)}{N_y}\int_{-a}^{a}\frac{e^{-\nu|x-x'|}}{2\nu\alpha}dx' \qquad (9.77)$$

$$\text{where}\quad N_y^{-1} = \frac{2}{W}\frac{\gamma_n^2 W^2 + (B_2 W/a)^2}{\gamma_n^2 W^2 + (B_2 W/a)^2 + B_2 W/a} \qquad (9.78)$$

and where $\nu^2 = \sigma^2 + \gamma_n^2$. Note that the integral on x' may be carried out in closed form.

Figure 9.8 shows the temperature and amplitude and phase on the heated surface of the slab at $B_2 = 1$ for three frequency frequency values defined by $\omega^+ = \omega a^2/\alpha = 0.1, 1.0$, and 10. At low frequency, the amplitude on the heater is large, the phase is small, and the spatial influence of the heater extends far beyond the heated region ($-1 < x/a < 1$). As the frequency increases, the amplitude is smaller on the heater and the spatial extent of the temperature is limited to the immediate vicinity of the heater.

The alternate GF may also used to find the temperature in this case. The alternate GF is given by Equation 9.74 with the necessary 1D GF given by Equation 9.11 with $S_1^+ = S_1^- = 1$, $S_2^+ = k\nu + h_2$, and $S_2^- = k\nu - h_2$ as appropriate for case X00Y23. Then the GF, evaluated at $y' = y = W$, is given by

$$G_{X00Y23}(x, 0|x', 0, \omega) = \frac{1}{2\pi}\int_{-\infty}^{\infty}\frac{e^{-j\beta(x-x')}W(1 - e^{-2\nu W})\; d\beta}{\alpha[\nu W + B_2 - (\nu W - B_2)e^{-2\nu W}]} \qquad (9.79)$$

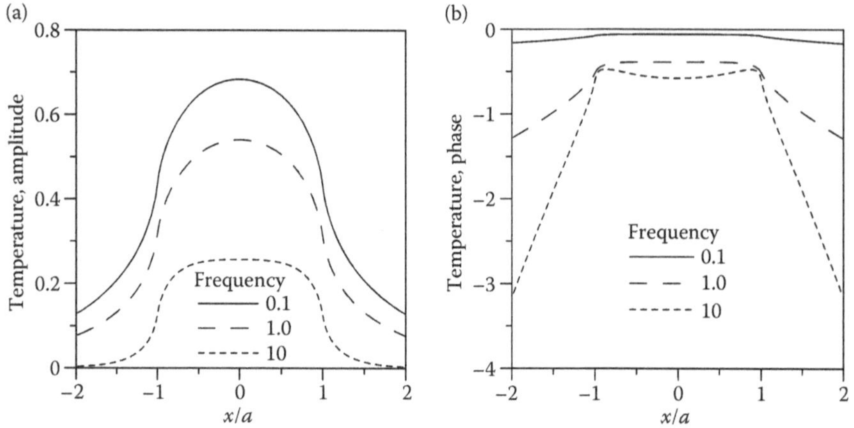

FIGURE 9.8 Amplitide (a) and phase (b) of the surface temperature on a slab heated over $(-1 < x/a < 1)$ for three heating frequencies. The surface convection is $ha/k = 1$ and the slab thickness is $W/a = 1$. Case $X00Y23B0G(x5t6)$.

where in this expression $\nu^2 = \sigma^2 + \beta^2$. The temperature is found by replacing the above GF into Equation 9.76, and evaluating the integral over x':

$$T(x, y, \omega) = \frac{q_0 W}{k} \int_{-\infty}^{\infty} \frac{e^{-j\beta(x-a)} - e^{-j\beta(x+a)}}{j\beta} \times \frac{(1 - e^{-2\nu W})\, d\beta}{[\nu W + B_2 - (\nu W - B_2)e^{-2\nu W}]} \tag{9.80}$$

Here the integral over x' has been carried out in closed form. The remaining integral on β must be carried out numerically, however the integrand rapidly approaches zero at large $|\beta|$, so the infinite limits can be truncated while providing accurate numerical values (Cole, 2006).

9.6.2 Infinite and Semi-Infinite Bodies

The GF for infinite and semi-infinite bodies are found in the same manner as the alternate GF discussed above for the two-dimensional slab body. For cases X00YI0 for I = 0, 1, 2, 3, 4 or 5, the 1D GF identified in Equation 9.74 may be simplified by taking $S_2^- = 0$ and $S_2^+ = 1$. Then the GF for infinite and semi-infinite bodies may be written:

$$G(x, y, \omega | x', y') = \frac{1}{2\pi} \int_{-\infty}^{\infty} \frac{e^{j\beta(x-x')}}{2\alpha\nu} [e^{-\nu|y-y'|} + De^{-\nu(y+y')}] d\beta \tag{9.81}$$

$$\text{where} \quad D = \begin{cases} 0 & \text{(infinite body)} \\ -1 & \text{(first kind)} \\ +1 & \text{(second kind)} \\ (k\nu - \lambda)/(k\nu + \lambda) & \text{(3rd, 4th, or 5th kind)} \end{cases}$$

For some cases the β-integral is known in closed form. For cases X00Y00, X00Y10, and X00Y20, the GF may be written (Mandelis, 2001, p. 231)

$$G(x, y, \omega|x', y') = \frac{1}{2\pi\alpha}\left[K_0\left(\nu\sqrt{(x-x')^2+(y-y')^2}\right)\right.$$
$$\left. + DK_0\left(\nu\sqrt{(x+x')^2+(y+y')^2}\right)\right] \quad (9.82)$$

where K_0 is the modified Bessel function of order zero (with complex argument) and $D = 0, \; -1,$ or 1 for cases X00Y00, X00Y10, and X00Y20, respectively.

9.6.3 Rectangular Parallelepiped

Consider the rectangular parallelepiped on domain ($0 < x < L_1$, $0 < y < L_2$, $0 < z < L_3$). The GF for steady-periodic conduction for this case satisfies the following equations:

$$\frac{\partial^2 G}{\partial x^2} + \frac{\partial^2 G}{\partial y^2} + \frac{\partial^2 G}{\partial z^2} - \sigma^2 G = -\frac{1}{\alpha}\delta(x-x')\delta(y-y')\delta(z-z') \quad (9.83)$$

$$k_i \frac{\partial G}{\partial n_i} + [h_i + j\omega(\rho c b)_i]G = 0; \quad \text{on boundary } i \quad (9.84)$$

This GF may be found in the form of a double summation with the method of eigenfunction expansion applied along two directions. For eigenfunctions X_m, norm N_x, and eigenvalues β_m along the x-direction, and, eigenfunctions Y_n, norm N_y, and eigenvalues γ_n along the y-direction, the GF is given by

$$G(x, y, z, \omega \,|\, x', y', z') = \sum_{n=0}^{\infty}\sum_{m=0}^{\infty} \frac{X_m(x)X_m^*(x')Y_n(y)Y_n^*(y')}{N_x(\beta_m)N_y(\gamma_n)} P(z, z', \nu_{nm}) \quad (9.85)$$

The eigenfunctions satisfy homogeneous boundary conditions at the appropriate limits of the x and y domains. The $m = 0$ term of the series is needed only when there are boundaries of the second kind at both $x = 0$ and $x = L_1$ (X22), and the $n = 0$ term is needed only when there are boundary conditions of the second kind at both $y = 0$ and $y = L_2$ (Y22).

Kernel function P satisfies

$$\frac{d^2 P}{dx^2} - \nu^2 P = -\delta(z - z') \quad (9.86)$$

along with appropriate homogeneous boundary conditions at $z = 0$ and $z = L_3$. The kernel function is similar to the 1D (slab) Green's function discussed earlier in Equation 9.11, except that here σ is replaced by $\nu_{nm} = \sqrt{\sigma^2 + \beta_m^2 + \gamma_n^2}$. That is,

$$P(z, z', \nu_{nm}) = G_X(z, z', \nu_{nm})$$

There are two alternate series expressions for this GF. With kernel functions along the y-direction, eigenfunctions along the x- and z-directions, and $\nu_{mp}^2 = \sigma^2 + \beta_m^2 + \eta_p^2$, one alternate GF is

$$G(x, y, z, \omega \,|\, x', y', z') = \sum_{p=0}^{\infty} \sum_{m=0}^{\infty} \frac{X_m(x) X_m^*(x') Z_p(z) Z_p^*(z')}{N_x(\beta_m) N_z(\eta_p)} P(y, y', \nu_{mp}) \quad (9.87)$$

With kernel functions along the x-direction, eigenfunctions along the y- and z-directions, and $\nu_{np}^2 = \sigma^2 + \gamma_n^2 + \eta_p^2$, another alternate GF is

$$G(x, y, z, \omega \,|\, x', y', z') = \sum_{n=0}^{\infty} \sum_{p=0}^{\infty} \frac{Y_n(y) Y_n^*(y') Z_p(z) Z_p^*(z')}{N_y(\gamma_n) N_z(\eta_p)} P(x, x', \nu_{np}) \quad (9.88)$$

The alternate GF can be used to construct alternate temperature expressions which are important for verification. The alternate series expressions are also important if slow series convergence becomes a problem, because they have complementary convergence behavior. That is, at a location where one series converges slowly (often at a boundary), an alternate series may be found that converges rapidly at that location. A triple-sum form of the GF may also be constructed from eigenfunction expansions along all three axes, but it is not recommended for numerical computation. See Crittenden and Cole (2002) for further discussion of the same issues for *steady* heat conduction (not steady-periodic) in the parallelepiped.

9.7 TWO-DIMENSIONAL BODIES IN CYLINDRICAL COORDINATES

In this section steady-periodic heat conduction is treated in geometries in cylindrical coordinates with axisymmetry, that is, described by coordinates (r, z). The GF is defined by

$$\frac{1}{r}\frac{\partial}{\partial r}\left(r\frac{\partial G}{\partial r}\right) + \frac{\partial^2 G}{\partial z^2} - \sigma^2 G = -\frac{1}{\alpha}\frac{\delta(r - r')}{2\pi r'}\delta(z - z'); \quad (9.89)$$

$$k\frac{\partial G}{\partial n_i} + \lambda_i G = 0; \quad \text{at boundary } i \quad (9.90)$$

where $\sigma^2 = j\omega/\alpha$ and $\lambda_i = h_i + j\omega(\rho c b_i)$ to describe five kinds of boundary conditions. Earlier the method of eigenfunction expansion was applied to rectangular geometries. The following development is similar, with eigenfunctions and kernel functions appropriate for cylinders.

9.7.1 GF WITH EIGENFUNCTIONS ALONG r

Consider steady-periodic conduction in the right circular cylinder, on domain $(0 < r < b)$, The GF may be constructed with eigenfunctions along the r-direction in the form

TABLE 9.1
Eigenfunction, Norm, and Eigencondition for Solid Cylinders. Note $B_2 = \lambda_2 b/k$.

Case	R_m	$\frac{2\pi}{N_r(\beta_m)}$	Eigencondition
R01	$J_0(\beta_m r/b)$	$\frac{2}{b^2 J_1(\beta_m)}$	$J_0(\beta_m) = 0$
R02*	$J_0(\beta_m r/b)$	$\frac{2}{b^2 J_0^2(\beta_m)}$	$J_0'(\beta_m) = 0$
R03	$J_0(\beta_m r/b)$	$\frac{2}{b^2 J_0^2(\beta_m)} \frac{b^2\beta_m^2}{(B_2^2+b^2\beta_m^2)}$	$\beta_m J_0'(\beta_m) + B_2 J_0(\beta_m) = 0$

*For this case $\beta_0 = 0$ is also an eigenvalue; the corresponding eigenfunction is $R_0 = 1$ and the norm is $2\pi / N_r(\beta_0) = 2/b^2$.

$$G(r, z, \omega \,|\, r', z') = \sum_{m=0}^{\infty} \frac{R_m(r) R_m^*(r')}{N_r(\beta_m)} P(z, z', \nu_m) \tag{9.91}$$

Eigenfunctions R_m satisfy the Bessel equation of order zero,

$$R_m'' + \frac{1}{r} R_m' + \beta_m^2 R_m = 0 \tag{9.92}$$

The eigenfunctions, norms and eigenconditions are given in Table 9.1. The differential equation for the kernel functions P may be found by replacing the series for the GF, Equation 9.91, into the auxiliary equation for G, Equation 9.89, along with the series expansion for $\delta(r - r')$, given by

$$\frac{\delta(r - r')}{2\pi r'} = \sum_{m=0}^{\infty} \frac{R_m(r) R_m^*(r')}{N_r(\beta_m)} \tag{9.93}$$

The factor of 2π here is used for consistency with the cylindrical GF defined in Section 7.4. After some algebra, Equation 9.89 takes the form

$$\sum_{m=0}^{\infty} \frac{R_m(r) R_m^*(r')}{N_r(\beta_m)} \left\{ -\beta_m^2 P + P'' - \sigma^2 P + \frac{1}{\alpha} \delta(z - z') \right\} = 0 \tag{9.94}$$

This equation will be satisfied if kernel function P satisfies

$$\frac{d^2 P}{dz^2} - \nu_m^2 P = -\frac{1}{\alpha} \delta(z - z') \tag{9.95}$$

where $\nu_m^2 = \beta_m^2 + \sigma^2$. The kernel function also satisfies homogeneous boundary conditions at $z = 0$ and $z = L$. Kernel function P is given by Equation 9.11 with σ replaced by ν_m. The special case of $\omega = 0$ (steady) requires special care for boundaries of the second kind for which $\beta_m = 0$ is an eigenvalue (Cole, 2004b).

In this section the discussion has focused on the solid cylinder ($0 < r < b$). The GF for the hollow cylinder has the same form, except that the eigenfunctions contain both J_0 and Y_0. The hollow-cylinder eigenfunctions may be deduced from the transient cylinder cases given in Appendix R, and they are also tabulated by Ozisik (1993, Chapter 3). It is important to note, however, that the norms for cylinders given by Ozisik differ from this book by a factor of (2π), as discussed in Section 7.4. For the finite cylinder ($0 < z < L$), an alternate series exists that contains eigenfunctions along the z-direction, as discussed in the next section.

9.7.2 GF WITH EIGENFUNCTIONS ALONG Z

Consider the region bounded by two planes (at $z = 0$ and $z = L$) described by spatial coordinates (r, z). For this geometry the steady-periodic GF can be constructed by an eigenfunction expansion along the z-direction and a kernel function along the r-direction, in the form

$$G(r, z, \sigma \,|\, r', z') = \sum_{p=0}^{\infty} \frac{Z(z) Z_p^*(z')}{N_z(\eta_p)} Q(r, r', \nu_p) \tag{9.96}$$

where $\sigma^2 = j\omega/\alpha$ and $\nu_p^2 = \eta_p^2 + \sigma^2$. Eigenfunction Z_p, eigenvalues η_p, and norm N_z are identical to the Cartesian functions discussed earlier for the rectangle and the parallelepiped except here they are renamed for the z-direction (see also Tables 4.1 through 4.3).

Kernel function Q satisfies

$$\frac{1}{r}\frac{d}{dr}\left(r\frac{dQ}{dr}\right) - \nu_p^2 Q = -\frac{1}{\alpha}\frac{\delta(r - r')}{2\pi r'} \tag{9.97}$$

The kernel function along the r-direction is similar to the 1D cylindrical radial GF given earlier in Equation 9.21, except here σ is replaced by $\nu_p = \sqrt{\eta_p^2 + \sigma^2}$. That is,

$$Q(r, r', \nu_p) = G_R(r, r', \nu_p)$$

The GF given by Equation 9.96 applies to the following geometries: the infinite slab ($0 < z < L, r > 0$); the infinite slab with a cylindrical hole ($0 < z < L, a < r < \infty$); the finite hollow cylinder ($0 < z < L$, $a < r < b$); and, the finite solid cylinder ($0 < z < L, 0 < r < b$).

Example 9.6: Pin Fin with Heat Flux at Base

Steady-periodic heat transfer in fins has been studied several times; see Kraus et al. (2001, Chapter 17) for a thorough literature review. Generally a fin is long and thin and the temperature varies down the long axis, however this example is concerned with a short cylindrical fin in which two-dimensional heat transfer is present. The base of the fin is uniformly heated by a steady-periodic heat flux, and the other surfaces are cooled by convection. The geometry for this case, shown in Figure 9.9, is described by heat conduction number R03Z23. The temperature satisfies the following equations:

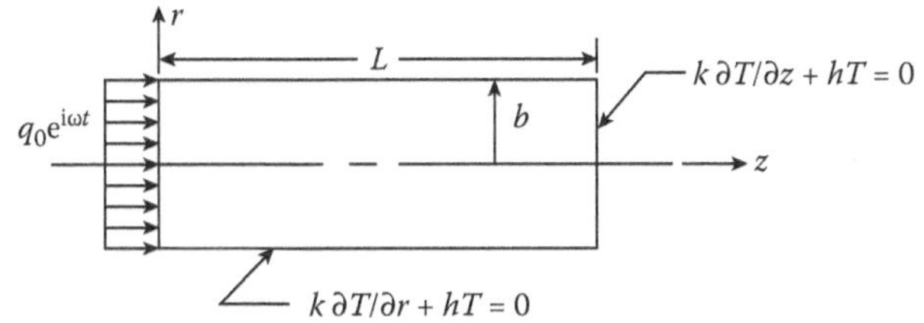

FIGURE 9.9 Solid cylinder heated steady-periodically over one end and cooled by convection over the other surfaces, case *R*03*Z*23. This is a model of a short pin fin.

$$\frac{1}{r}\frac{\partial}{\partial r}\left(r\frac{\partial T}{\partial r}\right)+\frac{\partial^2 T}{\partial z^2}-\sigma^2 T=0 \tag{9.98}$$

$$\text{at } z=0, \quad -k\frac{\partial T}{\partial r}=q_0(\omega) \tag{9.99}$$

$$\text{at } z=L, \quad k\frac{\partial T}{\partial z}+hT=T_\infty \tag{9.100}$$

$$\text{at } r=b, \quad k\frac{\partial T}{\partial r}+hT=T_\infty \tag{9.101}$$

The temperature may be expressed as an integral involving the appropriate Green's function, as follows:

$$T(r,z,\omega)-T_\infty=\frac{\alpha}{k}\int_{r'=0}^{b} q_0\; G(r,z,\omega|r',z'=0)\,2\pi r'\,dr' \tag{9.102}$$

Mathematically, the temperature has a unique solution. However, there are two series forms of the GF that can provide two distinct series expressions for the temperature.

Eigenfunctions along z**.** With eigenfunctions along the z-direction, the GF is given by Equation 9.96 where the eigenfunction and norm are given by

$$Z_p(z)=\cos(\eta_p z) \tag{9.103}$$

$$\frac{1}{N_z(\eta_p)}=\frac{2}{L}\frac{(\eta_p L)^2+B_2^2}{(\eta_p L)^2+B_2^2+B_2} \tag{9.104}$$

where $B_2=hL/k$. Eigenvalues η_p satisfy $\eta_p L\tan(\eta_p L)=hL/k$. The kernel function is given by

$$Q_p(r,r',\nu)=\frac{1}{2\pi\alpha}\begin{Bmatrix}[A_2 I_0(\nu_p r')+K_0(\nu_p r')]I_0(\nu_p r), & r<r'\\ [A_2 I_0(\nu_p r)+K_0(\nu_p r)]I_0(\nu_p r'), & r>r'\end{Bmatrix} \tag{9.105}$$

where A_2 is given by

$$A_2=\frac{\nu_p L K_1(\nu_p b)-B_2 K_0(\nu_p b)}{\nu_p L I_1(\nu_p b)+B_2 I_0(\nu_p b)} \tag{9.106}$$

and where $\nu_p^2=\beta_p^2+\sigma^2$

Replace this GF into the temperature integral, Equation 9.102, and evaluate the integral over r' to find:

$$\frac{T(r,z,\omega)-T_\infty}{(q_0 b/k)} = \sum_{p=1}^{\infty} \cos(\eta_p z)\frac{2b}{L}\frac{(\eta_p L)^2 + B_2^2}{(\eta_p L)^2 + B_2^2 + B_2} \times \left\{ \frac{1}{\nu_p b}[A_2 I_1(\nu_p b) - K_1(\nu_p b)] I_0(\nu_p r) + \frac{1}{\nu_p^2 b^2} \right\} \tag{9.107}$$

Some care is required when combining eigenfunctions and kernel functions, because each (in general) depends on separate Biot numbers with separate length scales and separate λ-values. In this example $\lambda = h$ is the same in each direction, but the length scales are different. Hence the z-direction Biot number is hL/k and the r-direction Biot number is hb/k. In the above temperature expression we have chosen to normalize the temperature by length scale b.

Eigenfunctions along *r*. An alternate form of the GF has eigenfunctions along the r-direction, and is given by Equation 9.91 with $z' = 0$,

$$G(r,z|r',z'=0,\omega) = \sum_{m=1}^{\infty} \frac{R_m(r)R_m(r')}{N_r(\beta_m)} P_m(z,z'=0,\nu_m) \tag{9.108}$$

The eigenfunction, norm, and eigencondition are given by (Table 9.1)

$$R_m(r) = J_0(\beta_m r) \tag{9.109}$$

$$\frac{2\pi}{N_r} = \frac{2}{J_0^2(\beta_m b)} \frac{\beta_m^2}{[(hb/k)^2 + b^2\beta_m^2)]} \tag{9.110}$$

$$0 = \beta_m b J_0'(\beta_m b) + (hb/k) J_0(\beta_m b). \tag{9.111}$$

Kernel function P is given by Equation 9.11 for a type 2 boundary at $z = 0$ and a type 3 boundary at $z = L$ (case Z23):

$$P(z,z'=0,\nu_m) = \frac{S_2^- e^{-\nu_m(2L-z)} + S_2^+ e^{-\nu_m z}}{\nu_m(S_2^+ - S_2^- e^{-2\nu_m L})} \tag{9.112}$$

where $\nu_m^2 = \beta_m^2 + \sigma^2$, $S_2^+ = k\nu_m + h$, and $S_2^- = k\nu_m - h$. This form of the GF may be substituted into the temperature integral to find an alternate series expression for the temperature:

$$\frac{T(r,z,\omega)-T_\infty}{(q_0 b/k)} = \sum_{m=1}^{\infty} \frac{J_0(\beta_m r)J_1(\beta_m b)}{\pi\beta_m b} \frac{\beta_m^2 b^2}{J_0^2(\beta_m b)[(hb/k)^2 + b^2\beta_m^2]} \times \frac{(\nu_m b - hb/k)e^{-\nu_m(2L-z)} + (\nu_m b + hb/k)e^{-\nu_m z}}{\nu_m b[(\nu_m b + hb/k) - (\nu_m b - hb/k)e^{-2\nu_m L}]} \tag{9.113}$$

Numerical values for the temperature in the pin fin were computed using both temperature series, Equations 9.107 and 9.113, providing a very strong check on the correctness of the results. In Figure 9.10 contour plots of the amplitude and phase of the temperature are given for a fin of aspect ratio $b/L = 0.5$. In Figure 9.10 the

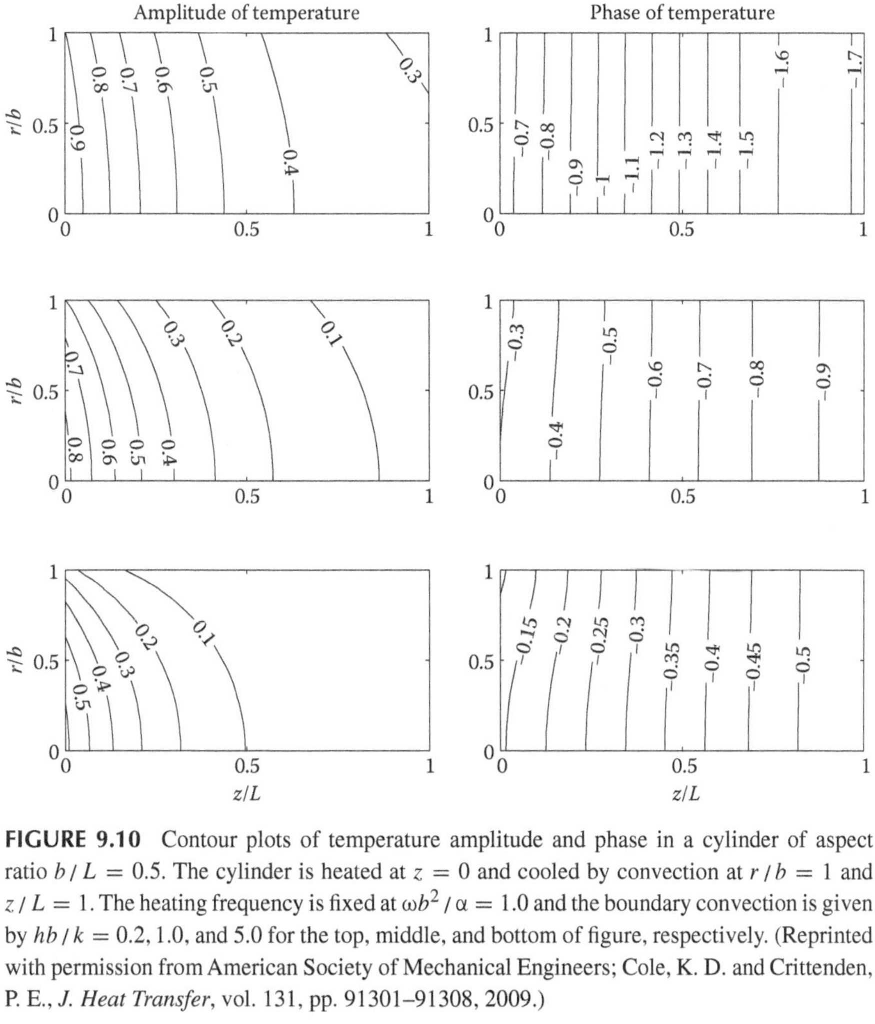

FIGURE 9.10 Contour plots of temperature amplitude and phase in a cylinder of aspect ratio $b / L = 0.5$. The cylinder is heated at $z = 0$ and cooled by convection at $r / b = 1$ and $z / L = 1$. The heating frequency is fixed at $\omega b^2 / \alpha = 1.0$ and the boundary convection is given by $hb / k = 0.2$, 1.0, and 5.0 for the top, middle, and bottom of figure, respectively. (Reprinted with permission from American Society of Mechanical Engineers; Cole, K. D. and Crittenden, P. E., *J. Heat Transfer*, vol. 131, pp. 91301–91308, 2009.)

frequency is fixed at $\omega b^2 / \alpha = 1.0$ and the results for Biot number $hb / k = 0.2$, 1.0, and 5.0 are shown at the top, middle and bottom of the figure, respectively. The amplitude of the temperature is largest where the heat is added ($z = 0$) and decreases along the length of the fin. For the smallest Biot number (at the top of the figure), the change in phase along the fin is most pronounced, and as the Biot number increases there is less change in phase along the fin.

9.7.3 Axisymmetric Half-Space

Consider the half-space with axisymmetry on domain ($r > 0$, $z > 0$). The steady-periodic GF will be sought with the Hankel transform defined as follows (Carslaw and Jaeger, p. 458):

$$\overline{G}(\beta) = \int_{r=0}^{\infty} G(r)\, J_0(\beta r)\, 2\pi r\, dr \tag{9.114}$$

$$G(r) = \frac{1}{2\pi}\int_{\beta=0}^{\infty} \overline{G}(\beta)\, J_0(\beta r)\, \beta\, d\beta \tag{9.115}$$

To apply this transform, multiply the partial differential equation for G, Equation 9.89, by $J_0(\beta r) 2\pi r$ and integrate over $(0 < r < \infty)$, to find:

$$-\beta^2\overline{G} + \frac{d^2\overline{G}}{dz^2} - \sigma^2\overline{G} = \frac{1}{\alpha}\delta(z - z')J_0(\beta r') \tag{9.116}$$

Note that the Hankel transform eliminates the r-direction derivative. Also, the sifting property is used to evaluate the integral of the δ-function term. To solve the above ordinary differential equation, define new variable G^* such that $\overline{G} = G^* \cdot J_0(\beta r')$, and replace into the above equation:

$$\left[\frac{d^2G^*}{dz^2} - \nu^2 G^* = \frac{1}{\alpha}\delta(z - z')\right] J_0(\beta r') \tag{9.117}$$

where $\nu^2 = \beta^2 + \sigma^2$. The transformed boundary condition at $z = 0$ has the same form as the original boundary condition. In this form, function G^* is similar to the 1D Cartesian Green's function given in Equation 9.11. For the present discussion in the half-space $(z > 0)$, function G^* may be written

$$G^* = \frac{e^{-\nu|z-z'|} + R \cdot e^{-\nu(z+z')}}{2\alpha\nu} \tag{9.118}$$

$$\text{where } R = \begin{cases} 0; & \text{infinite body} \\ -1; & \text{case Z10} \\ 1; & \text{case Z20} \\ \frac{k\nu-\lambda_1}{k\nu+\lambda_1}; & \text{case Z30, Z40, Z50} \end{cases} \tag{9.119}$$

where λ_1 comes from the boundary condition at $z = 0$. Finally, use the definition of G^* given above, and the inverse Hankel transform (Equation 9.115) to obtain the r-space Green's function:

$$G(r, z, \sigma|r', z') = \frac{1}{2\pi}\int_{\beta=0}^{\infty} \frac{e^{-\nu|z-z'|} + R \cdot e^{-\nu(z+z')}}{2\alpha\nu} J_0(\beta r')J_o(\beta r)\,\beta\, d\beta \tag{9.120}$$

Although the integrand of this integral approaches zero at $\beta \to \infty$, it does so slowly at $z = z' = 0$. Fortunately, temperature expressions constructed by integrating this GF generally contain a faster-decaying integrand which can be evaluated numerically, as in the following example.

Example 9.7: Half-Space with Heating over a Circular Region

Consider the temperature in a half-space heated over a circle $(0 < r < a)$ and insulated elsewhere on the $z = 0$ surface. This is case R00Z20B(t6r5) and the

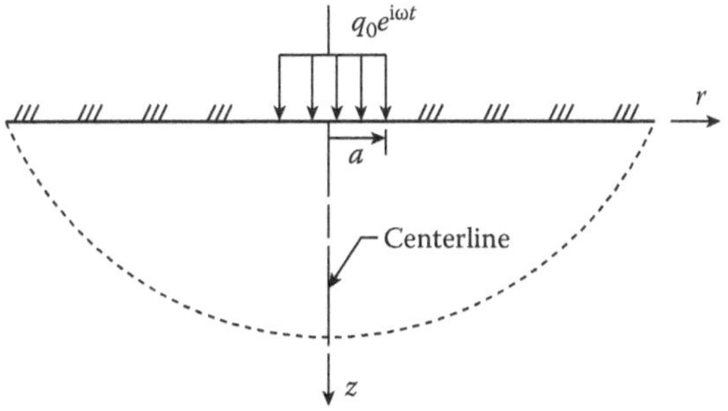

FIGURE 9.11 Geometry for a half space heated over a small circular region.

geometry is shown in Figure 9.11. The heating condition is described by

$$-k\frac{\partial T}{\partial r} = \begin{cases} q_0(\omega), & 0 < r < a \\ 0, & r > a \end{cases} \tag{9.121}$$

The GF solution for this problem involves an integration over the $z = 0$ surface of the body, given by

$$T(r, z, \omega) = \frac{1}{k}\int_{r'=0}^{a} q_0 G_{R00Z20}(r, z, \sigma|r', z' = 0)2\pi r'\, dr' \tag{9.122}$$

The GF is given by Equation 9.120 with $R = 1$. Replacing this GF into the temperature expression, and evaluating the integral over r' gives

$$T(r, z, \omega) = \frac{q_0 a}{k}\int_{\beta=0}^{\infty} \frac{e^{-z(\beta^2+\sigma^2)^{1/2}}}{\sqrt{\beta^2+\sigma^2}} J_0(\beta r) J_1(\beta a)\, d\beta \tag{9.123}$$

This improper integral is easily evaluated for all $z > 0$ because the exponential term rapidly vanishes as β increases. At the surface $z = 0$, the rate of decrease of the integrand is controlled by the Bessel functions, which decrease in size as $J_n \sim 1/\sqrt{\beta}$ (for $n = 0, 1$). Thus for $z = 0$, the integrand vanishes like $1/\beta^2$, rapidly enough to allow accurate numerical evaluation of the surface temperature with a truncated domain of integration.

9.8 CYLINDER WITH $T = T(r, \phi, z, \omega)$

In this section the cylinder with three-dimensional steady-periodic heat conduction is treated. That is, temperature depends on spatial coordinates (r, ϕ, z) and frequency ω.

The associated GF for 3D steady-periodic heat conduction in the cylinder satisfies

$$\frac{1}{r}\frac{\partial}{\partial r}\left(r\frac{\partial G}{\partial r}\right) + \frac{1}{r^2}\frac{\partial^2 G}{\partial \phi^2} + \frac{\partial^2 G}{\partial z^2} - \sigma^2 G = -\frac{1}{\alpha}\frac{\delta(r - r')}{2\pi r}\delta(z - z')\delta(\phi - \phi') \tag{9.124}$$

and at the boundaries

$$k_i\frac{\partial G}{\partial n_i} + \lambda_i G = 0 \tag{9.125}$$

The set of GF represented by Equations 9.124 and 9.125 represent 1296 combinations of boundary conditions (36 along r and 36 along z), denoted by GF number RIJZKLΦ00. Here $I, J, K, L = 0, 1, \ldots, 5$ to denote the types of boundary conditions present and Φ00 denotes the angular dependence for the full cylinder.

There are two forms of the double-sum GF, one with eigenfunctions along the z-direction and the other with eigenfunction along the r-direction. Both are important, as one can be used to check the other, and where one converges slowly the other generally converges rapidly.

9.8.1 GF with Eigenfunctions along z

The steady-periodic GF with eigenfunctions along the z-direction, appropriate for the finite length cylinder ($0 < z < L$), has the form

$$G(r, \phi, z|r', \phi', z', \omega) = \sum_{n=0}^{\infty}\sum_{p=0}^{\infty} \frac{Z_p(z)Z_p(z')}{N_z(\eta_p)} \frac{\cos[n(\phi - \phi')]}{N_\phi} Q_{np}(r, r', \nu_p) \quad (9.126)$$

This eigenfunction expansion has been chosen so that the boundary conditions at $z = 0$ and $z = L$ are satisfied by the eigenfunctions in the z-direction, and the conditions at $\phi = 0$ and $\phi = 2\pi$ are satisfied by the eigenfunctions in the ϕ-direction. Norm N_ϕ is equal to π for $n = 0$ and 2π for $n \geq 1$.

Kernel function Q_{np} satisfies

$$Q''_{np} + \frac{1}{r}Q'_{np} - \left(\nu_p^2 + \frac{n^2}{r}\right) Q_{np} = -\frac{1}{\alpha}\frac{\delta(r - r')}{2\pi r} \quad (9.127)$$

where $\nu_p^2 = \eta_p^2 + \sigma^2$. This is the modified Bessel equation of order n, with general solution K_n and I_n. The particular solution may be found with a development similar to the one-dimensional radial GF discussed earlier, with the Bessel function of order n replacing that of order zero and with ν_p replacing σ. Applying the boundary conditions the solution can be written in the same form as before

$$Q_{np}(r, r', \nu_p) = \frac{1}{2\pi\alpha(1 - A_1 A_2)} \times \begin{Bmatrix} [A_2 I_n(\nu_p r') + K_n(\nu_p r')][I_n(\nu_p r) + A_1 K_n(\nu_p r)], & r < r' \\ [A_2 I_n(\nu_p r) + K_n(\nu_p r)][I_n(\nu_p r') + A_1 K_n(\nu_p r')], & r > r' \end{Bmatrix} \quad (9.128)$$

except here

$$A_1 = \frac{[\nu_p a I_{n+1}(\nu_p a) + n I_n(\nu_p a))] - B_1 I_n(\nu_p a)}{[\nu_p a K_{n+1}(\nu_p a) - n K_n(\nu_p a)] + B_1 K_n(\nu_p a)} \quad (9.129)$$

and

$$A_2 = \frac{[\nu_p b K_{n+1}(\nu_p b) - n K_n(\nu_p b)] - B_2 K_n(\nu_p b)}{[\nu_p b I_{n+1}(\nu_p b) + n I_n(\nu_p b)] + B_2 I_n(\nu_p b)} \quad (9.130)$$

The quantities $B_1 = \lambda_1 a/k_1$ and $B_2 = \lambda_2 b/k_2$ are modified Biot numbers at the inner and outer radius, respectively. The above values for A_1 and A_2 are for the most general boundary condition (fifth kind). Values for other kinds of boundaries can be found by analogy with Equation 9.22.

9.8.2 GF WITH EIGENFUNCTIONS ALONG r

An alternate GF that satisfies Equation 9.124 may also be constructed using eigenfunctions along the r-direction for finite-radius cylinders ($r < b$). If the r-direction eigenfunctions are denoted $R_{nm}(r)$, then the alternate double-sum GF may be written

$$G(r, \phi, z|r', \phi', z', \omega) = \sum_{n=0}^{\infty}\sum_{m=0}^{\infty} \frac{R_{nm}(r)R_{nm}(r')}{N_r(\beta_{nm})} \frac{\cos[n(\phi - \phi')]}{N_\phi} P_m(z, z', \nu_m) \tag{9.131}$$

The $m = 0$ term of the series is needed only when zero is an eigenvalue (for cases R02, R22). The series on n involves the functions $cos[n(\phi - \phi')]$ which satisfy periodic boundary conditions at $\phi = 0$ and $\phi = 2\pi$. Eigenfunctions R_{nm} satisfy

$$\frac{1}{r}\frac{\partial}{\partial r}\left(r\frac{\partial R_{nm}}{\partial r}\right) = -\left(\beta_{nm}^2 - \frac{n^2}{r^2}\right) R_{nm} \tag{9.132}$$

along with boundary conditions at $r = a$ and $r = b$. For solid cylinders, eigenfunctions R_{nm} have the form of Bessel functions of order n and may be deduced from Table 9.1 for $n \neq 0$. For hollow cylinders the eigenfunctions may be deduced from the transient GF listed in Appendix R.

Kernel function P_m satisfies

$$P_m'' - \nu_m^2 P_m = -\frac{1}{\alpha}\delta(z - z') \tag{9.133}$$

where $\nu_m^2 = \beta_{nm}^2 + \sigma^2$. This kernel function is identical to G_X discussed earlier (Equation 9.11) with x and σ replaced by z and ν_m, respectively.

Example 9.8: Solid Cylinder Heated over a Sector of Its Surface and Cooled by Convection

Consider a solid cylinder with steady-periodic heating over an angular sector of the curved surface, parallel to the cylinder axis, and cooled by convection over the entire curved surface. The flat ends of the cylinder are fixed at the fluid temperature. This geometry is an approximate thermal model of a hot-film sensor used to measure fluid flow. The temperature satisfies the following equations:

$$\frac{1}{r}\frac{\partial}{\partial r}\left(r\frac{\partial T}{\partial r}\right) + \frac{1}{r^2}\frac{\partial^2 T}{\partial \phi^2} + \frac{\partial^2 T}{\partial z^2} - \sigma^2 T + \frac{g(r, \phi, z)}{k} = 0 \tag{9.134}$$

$$\text{at } z = 0,\ T = T_\infty \tag{9.135}$$

$$\text{at } z = L,\ T = T_\infty \tag{9.136}$$

$$\text{at } r = b,\ k\frac{\partial T}{\partial r} + hT = hT_\infty \tag{9.137}$$

The heating function is given by

$$g(r, \phi, z) = \begin{cases} q_0 \delta(r - b) & 0 < \phi < \phi_0 \\ 0 & \phi_0 < \phi < 2\pi \end{cases} \tag{9.138}$$

Note that the heat is introduced at surface $r = b$. This is geometry R03Z11Φ00 in the heat conduction numbering system. The temperature may be stated in the form of an integral with the GF, as follows:

$$T(r, \phi, z, \omega) - T_\infty = \frac{\alpha}{k} \int_{r'=0}^{b} \int_{\phi'=0}^{\phi_0} \int_{z'=0}^{L} q_0 \, \delta(r' - b) \, G(\cdot) \, dz' \, d\phi' \, r' \, dr' \tag{9.139}$$

The integral on r' may be evaluated with the sifting property of δ:

$$T(r, \phi, z, \omega) - T_\infty = \frac{\alpha}{k} \int_{\phi'=0}^{\phi_0} \int_{z'=0}^{L} q_0 \, G(r, \phi, z, \omega | b, \phi', z') \, dz' \, d\phi' \, b \tag{9.140}$$

There are two forms of the GF that allow for two distinct series expressions for the temperature.

Eigenfunctions along *z*. With eigenfunctions along the z-direction, the GF is given by Equation 9.126 with the eigenfunction and norm given by (case Z11)

$$Z_p(z) = \sin(\eta_p z), \text{ where } \eta_p = p\pi / L \tag{9.141}$$

$$1 / N_z = 2 / L \tag{9.142}$$

The kernel function is given by Equation 9.128, (case R03):

$$Q_{np}(r, b, \nu_p) = \frac{1}{2\pi\alpha} \left\{ [A_2 I_n(\nu_p b) + K_n(\nu_p b)][I_n(\nu_p r)] \right\} \tag{9.143}$$

where $\nu_p^2 = \eta_p^2 + \sigma^2$ and where $A_2(n)$ is given by

$$A_2(n) = \frac{[\nu_p b K_{n+1}(\nu_p b) - n K_n(\nu_p b)] - B_2 K_n(\nu_p b)}{[\nu_p b I_{n+1}(\nu_p b) + n I_n(\nu_p b)] + B_2 I_n(\nu_p b)} \tag{9.144}$$

Here $B_2 = hb / k$. Replace this GF into the temperature integral, Equation 9.140, and evaluate the integrals on ϕ' and z':

$$\begin{aligned} \frac{T(r, \phi, z) - T_\infty}{q_0 b / k} &= \sum_{p=1}^{\infty} \sum_{n=0}^{\infty} \frac{2 \sin(p\pi z / L)[1 - (-1)^p]}{p\pi} \\ &\quad \times C_n [A_2(n) \, I_n(\nu_p b) + K_n(\nu_p b)] \frac{I_n(\nu_p r)}{2\pi} \end{aligned} \tag{9.145}$$

$$\text{where} \quad C_n = \begin{cases} \phi_0 / \pi; & n = 0 \\ \{\sin(n\phi) - \sin n(\phi - \phi_0)\} / (2\pi n); & n \neq 0 \end{cases}$$

Note that the integral over ϕ' must be treated separately when $n = 0$.

Eigenfunctions along *r*. An alternate GF, with eigenfunctions along the r-direction, is given by Equation 9.131 with eigenfunction and norm given by

$$R_{nm}(r) = J_n(\beta_{nm} r) \tag{9.146}$$

$$\frac{2\pi}{N_r} = \frac{2}{b^2 J_n^2(\beta_{nm} b)} \frac{b^2 \beta_{nm}^2}{((hb / k)^2 + b^2 \beta_{nm}^2 - n^2)} \tag{9.147}$$

The eigenvalue β_{nm} satisfies

$$\beta_{nm} b J_n'(\beta_{nm} b) + (hb/k) J_n(\beta_{nm} b) = 0, \quad m = 1, 2, 3, \ldots \tag{9.148}$$

The kernel function P_m is given by Equation 9.11, with $\nu^2 = \beta_{nm}^2 + \sigma^2$ (case Z11):

$$P(z, z', \nu) = \frac{e^{-\nu(2L-|z-z'|)} - e^{-\nu(2L-z-z')}}{2\alpha\nu(1 - e^{-2\nu L})} + \frac{e^{-\nu(|z-z'|)} - e^{-\nu(z+z')}}{2\alpha\nu(1 - e^{-2\nu L})} \tag{9.149}$$

Now replace the alternate GF into the integral expression for the GF, and evaluate the integrals over ϕ' and z', to find the an alternate series expression for the temperature (Cole and Crittenden, 2009):

$$\frac{T(r, \phi, z) - T_\infty}{q_0 b/k} = \frac{1}{\pi} \sum_{m=1}^{\infty} \sum_{n=0}^{\infty} J_n(\beta_{nm} r) \frac{\beta_{nm}^2}{J_n(\beta_{nm} b)[(hb/k)^2 + b^2\beta_{nm}^2 - n^2]}$$
$$\times C_n \left[\frac{1}{\nu^2} + \frac{e^{-\nu(2L-z)} - e^{-\nu(L-z)} - e^{-2\nu L} - e^{-\nu z}}{\nu^2(1 - e^{-2\nu L})} \right] \tag{9.150}$$

where C_n is given in Equation 9.145. Note that additive term $1/\nu^2$, from integration on z' of the kernel function, may cause slow series convergence, because this portion of the series does not contain a convergence-promoting exponential function. The series containing this additive term can be replaced by a faster-converging single-sum form (see Crittenden and Cole, 2002).

Figure 9.12 shows amplitude and phase of the dimensionless temperature computed on the cylinder surface $r = b$ and at the midpoint $z = L/2$. The heated strip is located on $0 < \phi < 0.2$ and the aspect ratio of the cylinder is $b/L = 0.2$. The

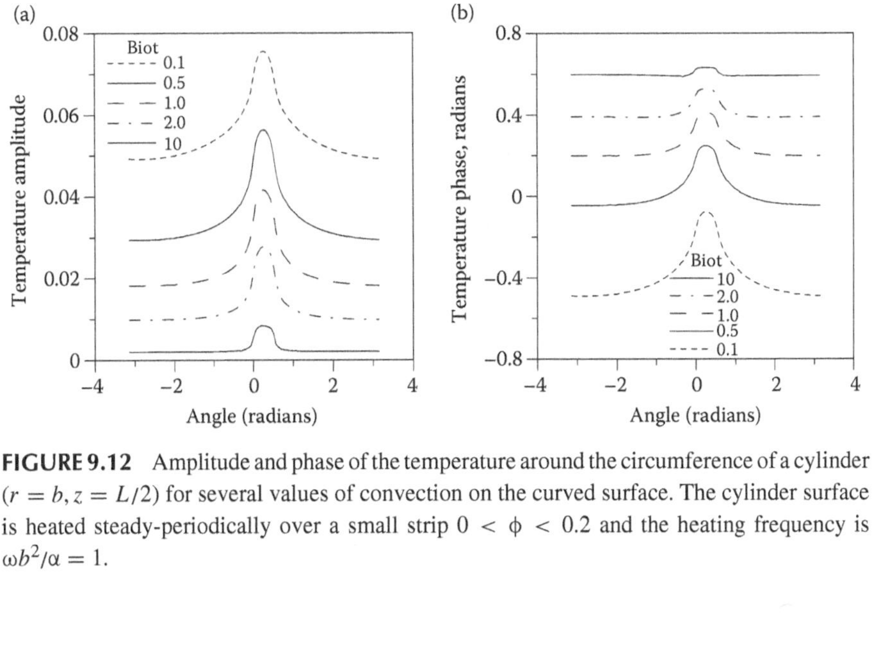

FIGURE 9.12 Amplitude and phase of the temperature around the circumference of a cylinder ($r = b, z = L/2$) for several values of convection on the curved surface. The cylinder surface is heated steady-periodically over a small strip $0 < \phi < 0.2$ and the heating frequency is $\omega b^2/\alpha = 1$.

figure shows the dimensionless temperature at heating frequency $\omega b^2/\alpha = 1$ for several values of the Biot number $B_2 = hb/k$, which could represent different cross-flow velocities. The figure shows that at larger Biot numbers, the temperature amplitude is localized to the heater location and the phase is nearly uniform. At smaller Biot numbers, the temperature amplitude is distributed further around the cylinder, and the phase difference between the heated and unheated surface is more pronounced.

PROBLEMS

9.1 The Fourier transform is defined by the following equations

$$\tilde{T}(\mathbf{r},\omega) = \int_{-\infty}^{\infty} T(\mathbf{r},t)e^{j\omega t}\,dt$$

$$T(\mathbf{r},t) = \frac{1}{2\pi}\int_{-\infty}^{\infty} \tilde{T}(\mathbf{r},\omega)e^{-j\omega t}\,d\omega$$

Apply the Fourier transform to the heat conduction Equations 9.1 and 9.2 for an alternate derivation of the steady-periodic heat equation, (9.4 and 9.5. Show all the steps in your derivation. What are the units of $\tilde{T}$?

9.2 Derive the steady-periodic GF for the following cases in the Cartesian coordinate system by solving Equation 9.9 directly. Check your answers against Equation 9.11 with appropriate values for the coefficients.

(a) Case X10, a semi-infinite body with boundary condition of the first kind at $x = 0$.

(b) Case X13, a finite body with boundary condition of the first kind at $x = 0$ and of the third kind at $x = L$.

9.3 Using direct integration of Equation 9.13, derive the steady-periodic GF for the following cases in the cylindrical coordinate system. Check your answers against Equation 9.21.

(a) Case R02, a solid cylinder with specified heat flux at $r = b$.

(b) Case R11, a hollow cylinder with boundary conditions of the first kind at both $r = a$ and $r = b$.

9.4 Using direct integration of Equation 9.24, derive the steady-periodic GF for the following cases in the spherical coordinate system. Check your answers against Equation 9.32 with appropriate values for the coefficients.

(a) Case RS10, a large body with a spherical cavity with boundary condition of first kind at $r = a$.

(b) Case RS03, a solid sphere with boundary condition of third kind at $r = a$.

9.5 A spacecraft in earth orbit is slowly rotating to produce periodic heating by absorbed sunlight. Model the spacecraft as a thin shell with wall thickness d, and use lumped capacitance theory:

$$q_0 \cos \omega t - hT = \rho c d \frac{\partial T}{\partial t}$$

(a) Find the steady-periodic form of the lumped capacitance equation.
(b) If q_0 is the solar constant (W/m^2) and h describes heat loss, find the steady-periodic temperature excursion as a function of rotation frequency, heat loss rate, etc.
(c) Suggest two improvements that could be made to make this model more realistic.

9.6 Find the steady-periodic temperature in deep soil caused by the day/night cycles of the air temperature. Assume the soil has uniform properties and the heat transfer coefficient between the air and soil is constant. What is the heat conduction number of this case?

9.7 Find the steady-periodic temperature in deep soil with a steady-periodic temperature imposed at the surface. Using this solution to describe the response of the soil to *yearly* temperature variation at the surface, find the soil depth at which the amplitude of temperature variation is 1% of the surface temperature variation.

9.8 Find the steady-periodic temperature in a slab with a specified heat flux at $x = 0$ and zero temperature at $x = L$, case X21B60. Compare your answer to Example 9.1 in the limit for $B_2 \to \infty$.

9.9 Find the temperature in a slab for steady periodic heating at $x = 0$ and for an insulated condition at $x = L$, case X22B10. Show that at high frequency the solution reduces to the semi-infinite case X20B1, and discuss this limiting case on the basis of the physics involved.

9.10 Find the steady-periodic temperature in a solid sphere caused by surface convection, case RS03B1, where the time-variation of the surface heating is caused by a time-varying fluid temperature.

9.11 Consider a 2D solid cylinder with spatially uniform heat flux at $z = 0$, and with zero temperature at $z = L$ and $r = b$. This is case R01B0Z21B60. Find the steady-periodic temperature (a) with eigenfunctions along the z-direction, and (b) with eigenfunctions along the r-direction.

9.12 Apply the Hankel transform given by Equation 9.114 to the heat conduction equation on (r, z) coordinates, Equation 9.89, to verify Equation 9.116. Show all your steps.

9.13 Find an integral expression for the temperature in the axisymmetric half-space ($r > 0$, $z > 0$) heated by laser absorption at the surface, $g(r, z = 0)$, and cooled by surface convection. What is the number of this case?

REFERENCES

Abramowitz, M. and Stegun, I. A. (Eds), 1964, *Handbook of Mathematical Functions with Formulas, Graphs, and Mathematical Tables*, Applied Mathematics Series 55, National Bureau of Standards, Washington, DC.

Carslaw, H. S. and Jaeger, J. C., 1959, *Conduction of Heat in Solids*, Oxford University Press, New York.

Cole, K. D., 2004a, Analysis of photothermal characterization of layered materials—Design of optimal experiments, *Int. Journal Thermophys.*, vol. 25, pp. 1567–1584.

Cole, K. D., 2004b, Fast converging series for heat conduction in the circular cylinder, *J. Eng. Math.*, vol. 49, p. 217–232.

Cole, K. D., 2006, Steady-periodic Green's functions and thermal measurement applications in Cartesian coordinates, *J. Heat Transfer*, vol. 128, pp. 709–716.

Cole, K. D. and Crittenden, P. E., 2009, Steady-periodic heating of a cylinder, *J. Heat Transfer*, vol. 131, pp. 91301–91308.

Cole, K. D. and Yen, D. H. Y., 2001, Green's functions, temperature, and heat flux in the rectangle, *Int. J. Heat Mass Transfer*, vol. 44, pp. 3883–3894.

Crittenden, P. E. and Cole, K. D., 2002, Fast-converging steady-state heat conduction in a rectangular parallelepiped, *Int. J. Heat Mass Transfer*, vol. 45, pp. 3585–3596.

Haji-Shiekh, A. and Beck, J. V., 2002, Temperature solution in multi-dimensional multi-layer bodies, *Int. J. Heat Mass Transfer*, vol. 45, pp. 1865–1877.

Haji-Sheikh, A., Hong, Y. S., You, S. M., and Beck, J. V., 1998, Sensitivity analysis for thermophysical property measurements using the periodic method, *J. Heat Transfer*, vol. 120, pp. 568–576.

Hu, H., Wang, X., and Xu, X., 1999, Generalized theory of the photoacoustic effect in a layered material, *J. Appl. Phys.* vol. 86, pp. 3953–3958.

Kraus, A. D., Aziz, A., and Welty, J., 2001, *Extended Surface Heat Transfer*, John Wiley, New York.

Mandelis, A., 2001, *Diffusion-Wave Fields, Mathematical Methods and Green's Functions*, Springer, New York.

McGahan, W. A. and Cole, K. D., 1992, Solution of the heat conduction equation in multilayers for photothermal deflection experiments, *J. Appl. Phys.*, vol. 72, pp. 1362–1373.

Naziev, Y. M, 2001, The effect of variability characteristics of substances during determination of their thermophysical properties by the method of periodic heating, *High Temperature*, vol. 39, no. 6, pp. 856–859.

Ozisik, N. M., 1993, *Heat Conduction*, John Wiley, New York.

Press, W. H., Teukolsky, S. A., Vetterling, W. T., and Flannery, B. P., 1992, *Numerical Recipes: The Art of Scientific Computing*, Cambridge University Press, New York.

Wang, C., Mandelis, A., and Liu, Y., 2004, Photothermal radiometry with solid cylindrical samples, *J. Appl. Phys.*, vol. 96, no. 7, pp. 3756–3762.

10 Galerkin-Based Green's Functions and Solutions

10.1 INTRODUCTION

The Green's functions (GFs) for regularly shaped bodies, such as plates, cylinders, and spheres, can be obtained by classical methods. These regularly shaped bodies shall be called orthogonal bodies. A normal at any point on the boundary of an orthogonal body is parallel to the direction of a coordinate axis. The solution methods and the derivation of the GFs for various orthogonal bodies are discussed elsewhere in this book. The formulation of the GFs for nonorthogonal bodies, as investigated by Haji-Sheikh and Lakshminarayanan (1987) and Haji-Sheikh and Beck (1988), are in this chapter.

The objective is to provide a methodology for solving the diffusion equation in various orthogonal and nonorthogonal bodies. The orthogonal bodies include plates, solid cylinders, hollow cylinders, solid spheres, and hollow spheres. The examples in this chapter consist of one-dimensional conduction in isotropic media problems that have exact solutions. The procedure is described, convenient and appropriate expressions are provided, and the accuracy of the results is compared with the exact solution. The study of multidimensional conduction in orthogonal and nonorthogonal bodies and related examples are included in Chapter 11. Also, the utility of this method when applied to conduction in heterogeneous problems is demonstrated.

The solution method discussed in this chapter is a Galerkin-based integral method and it is referred to as the Galerkin-based integral (GBI) method. The range of its usefulness encompasses thermal conduction problems with homogeneous or nonhomogeneous boundary conditions. The diffusion equation, Equation 3.28, can be written in a generalized form

$$\nabla \cdot [k(\mathbf{r})\nabla T] + g(\mathbf{r}, t) - m(\mathbf{r})^2 T = \rho(\mathbf{r})c_p(\mathbf{r})u(\mathbf{r})\frac{\partial T}{\partial t} \tag{10.1}$$

where $T = T(\mathbf{r}, t)$ is temperature, $\mathbf{r}$ is position vector, and t is time. The thermophysical properties $\rho(\mathbf{r})$, $c_p(\mathbf{r})$, and $k(\mathbf{r})$ are position-dependent density, specific heat, and thermal conductivity, respectively. The term $m(\mathbf{r})^2 T$ is the fin convection effect. The function $u(\mathbf{r})$ is designated as the velocity function. When dealing with pure conduction, the function $u(\mathbf{r})$ is equal to 1. Section 11.5 deals with flow in ducts where the functional values of $u(\mathbf{r})$ are considered and t is replaced by the axial coordinate. The nonhomogeneous boundary conditions are accommodated by using the GF solution method.

An outline of the remainder of this chapter follows. First, in Section 10.2, the standard derivation of the Green's function solution equation (GFSE) given in Chapter 3 is modified to account for the position-dependent thermophysical properties. Section 10.3 presents an alternative derivation of the GFSE which, when available, provides a rapidly converging temperature solution. This method uses a set of basis functions that need not be orthogonal. In Section 10.4, we demonstrate that, unlike the exact solutions, the functional form of the basis functions for one-dimensional solutions remains unchanged in different coordinates; however, the boundary conditions affect the values of each basis function.

Examples 10.1 and 10.2 are presented mainly to elaborate on the mathematical steps and to introduce the numerical steps in the integral method. Example 10.3 shows the use of the alternative GF solution when the surface temperature is prescribed. Example 10.4 uses a unified solution and compares the accuracy of different GF solutions.

Extensions of this method to deal with multidimensional conduction problems in heterogeneous materials and steady-state conduction are in Chapter 11. A study of heat transfer in the thermal entrance region of ducts is also included in Chapter 11.

10.2 GREEN'S FUNCTIONS AND GREEN'S FUNCTION SOLUTION METHOD

The GF method permits the solution of diffusion problems with nonhomogeneous boundary conditions. The GF solution method described in Chapter 3 is modified. The modifications allow the properties ρ, c_p, and k to be position-dependent, and the results are useful for the study of conduction of heat in homogeneous as well as heterogeneous bodies. The GF for a body with given boundary conditions describes the temperature effect at point $\mathbf{r}$ at time t if there is an impulsive point energy source of strength unity located at point $\mathbf{r}'$ and released at time τ. The GFs become the solutions of Equation 10.1 if the term $g(\mathbf{r}, t)$ in Equation 10.1 is replaced by a point energy source mathematically described by the following delta functions

$$g(\mathbf{r}, t) = \rho(\mathbf{r})c_p(\mathbf{r})\delta(\mathbf{r} - \mathbf{r}')\delta(t - \tau) \tag{10.2}$$

Accordingly, the GF is defined so that it satisfies homogeneous boundary conditions and it is the solution of the following auxiliary equation:

$$\begin{aligned} \nabla \cdot [k(\mathbf{r})\nabla G(\mathbf{r}, t|\mathbf{r}', \tau)] + C(\mathbf{r})\delta(\mathbf{r} - \mathbf{r}')\delta(t - \tau) \\ - m(\mathbf{r})^2 G(\mathbf{r}, t|\mathbf{r}', \tau) = C(\mathbf{r})u(\mathbf{r})\frac{\partial G(\mathbf{r}, t|\mathbf{r}', \tau)}{\partial t} \end{aligned} \tag{10.3}$$

where

$$C(\mathbf{r}) = \rho(\mathbf{r})c_p(\mathbf{r}) \tag{10.4}$$

where $\rho(\mathbf{r}), c_p(\mathbf{r}), k(\mathbf{r}), u(\mathbf{r})$, and $m(\mathbf{r})^2$ are position-dependent density, specific heat, thermal conductivity, velocity function, and fin effect as described for Equation 10.1. Based on the above-mentioned descriptions of the GF, Equation 10.1 for temperature and Equation 10.3 for the GF are the same, except that in Equation 10.3 the functional value of $g(\mathbf{r}, t)$ is specified. The function $G(\mathbf{r}, t|\mathbf{r}', \tau)$ is called the GF (Ozisik, 1993).

A formal solution of Equation 10.3 based on the Galerkin method yields the GF and is presented in this section. Following the discussion of the properties of the GFs, the GF solution method is presented.

10.2.1 Galerkin-Based Integral Method

The solution of the diffusion equation in a relatively general form, Equation 10.1, or the auxiliary equation for the GFs, Equation 10.3, is derived using a GBI method. It is assumed that the thermal conductivity, density, specific heat, velocity function, and fin effect are independent of temperature; however, no other restriction as to spatial variation of these thermophysical properties is needed. As described earlier, when the boundary conditions are homogeneous, $T(\mathbf{r}, 0) = 0$, and $g(\mathbf{r}, t) = C(\mathbf{r})\delta(\mathbf{r} - \mathbf{r}')\delta(t - \tau)$, the function $T(\mathbf{r}, t)$ is equal to the GF $G(\mathbf{r}, t|\mathbf{r}', \tau)$. Therefore, the value of the GF is readily available after a generalized solution of Equation 10.1 is accomplished. The GBI solution described here was used by LeCroy and Eraslan (1969) in the study of temperature development in the entrance region of an MHD parallel plate channel.

To solve a differential equation with a nonhomogeneous term, the solution is frequently broken into two parts, complementary and particular. The complementary form of Equation 10.1, that is, in essence, the diffusion equation in the absence of energy generation, is (Haji-Sheikh and Mashena, 1987)

$$\nabla \cdot (k\nabla\Theta) - m(\mathbf{r})^2\Theta = \rho(\mathbf{r})c_p(\mathbf{r})u(\mathbf{r})\frac{\partial\Theta}{\partial t} \tag{10.5}$$

The boundary conditions for Equation 10.5 are the same as those for Equations 10.1 or 10.3 and must be homogeneous. They are of the first kind (prescribed temperature), the second kind (prescribed heat flux), and the third kind (convective). It is also permissible for different parts of the boundary to have different kinds of boundary conditions. A solution to Equation 10.5 can be written as

$$\Theta = \sum_{n=1}^{N} c_n\psi_n(\mathbf{r})\exp(-\gamma_n t) \tag{10.6}$$

where γ_n is the nth eigenvalue and is independent of $\mathbf{r}$, and c_n is a constant to be evaluated. For convenience, assume that the body has finite dimensions. Because Θ is the complementary solution, it is not necessary to specify the initial condition at this time. The function $\psi_n(\mathbf{r})$ is selected so that (1) the homogeneous boundary conditions are satisfied, and (2) Equation 10.6 is a solution of Equation 10.5. The former condition is exactly satisfied if $\psi_n(\mathbf{r})$ satisfies the boundary conditions. The latter is accommodated if Equation 10.6 is substituted in Equation 10.5, resulting in

$$\nabla \cdot [k\nabla\psi_n(\mathbf{r})] - m(\mathbf{r})^2\psi_n + \rho(\mathbf{r})c_p(\mathbf{r})u(\mathbf{r})\gamma_n\psi_n(\mathbf{r}) = 0 \tag{10.7}$$

for every n value. The diffusion equation now becomes an eigenvalue problem and the function $\psi_n(\mathbf{r})$ is the eigenfunction.

When an exact solution does not exist or a simpler approximate solution is preferred, Equation 10.7 will be approximately satisfied. A function $\psi_n(\mathbf{r})$ is to be constructed as a linear combination of a properly selected set of basis functions. A properly selected set of basis functions is a complete set, its members are linearly independent, each member satisfies exactly the same homogeneous boundary conditions as those given for Θ, and not all members become zero at any interior point. The function $\psi_n(\mathbf{r})$, for $n = 1, 2, \ldots, N$, is chosen to be a linear combination of N basis functions,

$$\psi_n(\mathbf{r}) = \sum_{j=1}^{N} d_{nj} f_j(\mathbf{r}) \tag{10.8}$$

where $f_j(\mathbf{r})$ is an element of a set of basis functions and the d_{nj}'s are constants to be evaluated.

The Galerkin procedure (Kantorovich and Krylov, 1960) is now used; that is, both sides of Equation 10.7 are multiplied by $f_i dV$ and integrated over the volume V to get

$$\int_V f_i \nabla \cdot (k \nabla \psi_n) dV - \int_V m(\mathbf{r})^2 f_i \psi_n dV + \gamma_n \int_V \rho(\mathbf{r}) c_p(\mathbf{r}) u(\mathbf{r}) f_i \psi_n dV = 0 \tag{10.9}$$

Substituting ψ_n from Equation 10.8 into Equation 10.9 yields

$$\sum_{j=1}^{N} d_{nj} \left[\int_V f_i \nabla \cdot (k \nabla f_j) dV - \int_V m(\mathbf{r})^2 f_i f_j dV + \gamma_n \int_V \rho(\mathbf{r}) c_p(\mathbf{r}) u(\mathbf{r}) f_i f_j dV \right] = 0 \tag{10.10}$$

in which $i = 1, 2, \ldots, N$. The matrix form of Equation 10.10 is

$$(\mathbf{A} + \gamma_n \mathbf{B}) \mathbf{d}_n = \mathbf{0} \tag{10.11}$$

where $\mathbf{A}$ and $\mathbf{B}$ are square matrices of size N with the elements

$$a_{ij} = \int_V f_i \nabla \cdot (k \nabla f_j)\, dV - \int_V m(\mathbf{r})^2 f_i f_j\, dV \tag{10.12}$$

and

$$b_{ij} = \int_V \rho(\mathbf{r}) c_p(\mathbf{r}) u(\mathbf{r}) f_i f_j\, dV \tag{10.13}$$

The coefficients $d_{n1}, d_{n2}, \ldots, d_{nN}$ in Equation 10.8 are the member elements of the vector $\mathbf{d}_n$ in Equation 10.11. The second integral in Equation 10.12 vanishes in the absence of the fin effect. The fin effect, $m(\mathbf{r})^2$, influences only the elements of matrix $\mathbf{A}$.

An examination of Equation 10.13 reveals that matrix $\mathbf{B}$ is symmetric; that is, $b_{ij} = b_{ji}$. When i and j are switched, the second integral on the right side of Equation 10.12 will not be affected. Matrix $\mathbf{A}$ is also symmetric if the first term on the right side of Equation 10.12 is symmetric. This is accomplished by using the identities 1, 2, and 3 in Note 1 at the end of this chapter to show that

$$\begin{aligned}\int_V f_i \nabla \cdot (k \nabla f_j) dV &= \int_V \nabla \cdot (k f_i \nabla f_j) dV - \int_V k \nabla f_i \cdot \nabla f_j dV \\ &= \int_S k f_i \nabla f_j \cdot \mathbf{n} dS - \int_V k \nabla f_i \cdot \nabla f_j dV \\ &= \int_S k f_i \left(\frac{\partial f_j}{\partial n} \right) dS - \int_V k \nabla f_i \cdot \nabla f_j dV \qquad (10.14)\end{aligned}$$

When dealing with homogeneous boundary conditions of the first kind (prescribed temperature $f_j = 0$) or the second kind (prescribed heat flux $\partial f_j / \partial n = 0$), the first term on the right side of Equation 10.14 is zero while the second term is always symmetric. For homogeneous boundary conditions of the third kind (convective, $-k \partial f_j / \partial n = h f_j$), the first term on the right side of Equation 10.14 becomes

$$\int_S k f_i \, (\partial f_j / \partial n) dS = - \int_S h f_i f_j dS \qquad (10.15)$$

which is also symmetric when i and j are switched. In as much as the boundary conditions for f_i and f_j are always homogeneous, matrix $\mathbf{A}$ is always symmetric.

The calculation procedure for temperature distribution is summarized below:

(a) It is important to select a complete set of basis functions that are linearly independent. A complete set requires that all contributing members of the set be included. The members of a set are linearly independent if no member of the set is a linear combination of the other members.

(b) The computations of the values of a_{ij} and b_{ij} in Equations 10.12 and 10.13 are the major analytical or numerical computational tasks. For $N = 1$ and 2 and simple geometries, the computations are not difficult. For some complex geometries, it is convenient to utilize a symbolic software to carry out the analytical integrations which result in more accurate values and often require less computation time. When exact integration is not possible, numerical integrations can be used.

(c) The next step is to calculate the eigenvalues and eigenvectors of Equation 10.11 to be used in Equation 10.6. When $N = 1$ or $N = 2$, the procedure is discussed in Example 10.2. The details, when N is large, are given after Example 10.2.

(d) Following the calculation of eigenvalues and eigenvectors, the eigenfunctions, Equation 10.8, are known. The solution for temperature is complete after calculation of c_n in Equation 10.6. The initial temperature distribution is used to calculate the c_n values.

Examples 10.1 and 10.2 demonstrate the steps itemized above. Notice that the boundary conditions are homogeneous. The nonhomogeneous boundary conditions will be included using the Green's function solution method in Section 10.2.3.

Example 10.1:

Consider an infinite homogeneous plate with thickness L and having boundary conditions $T(0, t) = T(L, t) = 0$ when $t > 0$. The thermal properties have constant values, $u(\mathbf{r}) = 1$, and $m(\mathbf{r}) = 0$. Furthermore, the initial temperature distribution is $F(x) = T_0$. Find the temperature distribution using orthogonal basis functions.

Solution

The number for this case is $X11B00T1$. A mathematical statement of this problem is

$$\alpha \frac{\partial^2 T}{\partial x^2} = \frac{\partial T}{\partial t} \qquad \text{for } 0 < x < L \text{ and } t > 0 \tag{10.16}$$

$$T(0, t) = 0 \quad T(L, t) = 0 \quad \text{and} \quad T(x, 0) = T_0$$

If the set of orthogonal basis functions has the two members, $f_1 = \sin(\pi x / L)$, and $f_2 = \sin(2\pi x / L)$, the function ψ_n, using Equation 10.8, is

$$\psi_n = d_{n1} \sin\left(\frac{\pi x}{L}\right) + d_{n2} \sin\left(\frac{2\pi x}{L}\right) \tag{10.17}$$

Both $f_1 = \sin(\pi x / L)$ and $f_2 = \sin(2\pi x / L)$ functions satisfy the homogeneous boundary conditions ($f_1 = f_2 = 0$ at $x = 0$ and L). Here, the energy equation in its integral form, Equation 10.10, must be satisfied instead of Equation 10.7.

It is convenient to designate $f_i = \sin(i\pi x / L)$, for $i = 1, 2$, and $f_j = \sin(j\pi x / L)$, for $j = 1, 2$. Then, when k is a constant, $\nabla \cdot (k\nabla f_j) = k\nabla^2 f_j = -k(j\pi / L)^2 \sin(j\pi x / L)$. The elements of matrix **A**, using Equation 10.12, become

$$a_{ij} = -k\left(\frac{j\pi}{L}\right)^2 \int_0^L \sin\left(\frac{j\pi x}{L}\right) \sin\left(\frac{i\pi x}{L}\right) dx$$

$$\text{for } i = 1, 2, \text{ and } j = 1, 2 \tag{10.18}$$

For off-diagonal elements where i and j are not the same, this equation yields $a_{12} = a_{21} = 0$. When $i = j$, $a_{jj} = -k(j\pi)^2 / 2L$, resulting in $a_{11} = -k\pi^2 / 2L$ and $a_{22} = -2k\pi^2 / L$. Similarly, the elements of matrix **B** using Equation 10.13 are

$$b_{ij} = \rho c_p \int_0^L \sin\left(\frac{j\pi x}{L}\right) \sin\left(\frac{i\pi x}{L}\right) dx$$

$$\text{for } i = 1, 2, \text{ and } j = 1, 2 \tag{10.19}$$

Because the basis functions are orthogonal, only the diagonal terms have nonzero values; for example, $b_{12} = b_{21} = 0$. However, the diagonal terms are $b_{11} = b_{22} = \rho c_p L / 2$. Then, Equation 10.11 takes the following dimensionless form for $n = 1$ and $n = 2$:

$$\begin{bmatrix} -\frac{\pi^2}{2} + \frac{L^2\gamma}{\alpha}\frac{1}{2} & 0 \\ 0 & -(2\pi^2) + \frac{L^2\gamma}{\alpha}\frac{1}{2} \end{bmatrix} \begin{bmatrix} d_{n1} \\ d_{n2} \end{bmatrix} = 0 \tag{10.20}$$

where $\alpha = k / \rho c_p$ is the thermal diffusivity.

The eigenvalues γ_1 and γ_2 are chosen to make the determinant of this matrix equal to zero. Because all the off-diagonal terms are zero, the determinant is the product of the diagonal terms; the eigenvalues are obtained by setting each diagonal element equal to zero, $\gamma_1 = \pi^2\alpha/L^2$ and $\gamma_2 = 4\pi^2\alpha/L^2$. Since the simultaneous equations resulting from Equation 10.20 are homogeneous, one of the coefficients, d_{n1} or d_{n2}, can be arbitrarily selected. By choosing $d_{11} = d_{22} = 1$, for both $n = 1$ and 2, the other unknowns become $d_{12} = d_{21} = 0$. In as much as the differential equation for T is the same as that for Θ, then $T = \Theta$ and the solution using Equation 10.6 is

$$T = c_1 \sin\left(\frac{\pi x}{L}\right)\exp\left(-\frac{\pi^2\alpha t}{L^2}\right) + c_2 \sin\left(\frac{2\pi x}{L}\right)\exp\left(-\frac{4\pi^2\alpha t}{L^2}\right) \tag{10.21}$$

Substitute $t=0$ and the initial temperature $T(x,0)=T_0$ into Equation 10.21, to obtain

$$T_0 = c_1 \sin\left(\frac{\pi x}{L}\right) + c_2 \sin\left(\frac{2\pi x}{L}\right) \tag{10.22}$$

Analogous to the exact solution and the Fourier series expansion, both sides of this equation are multiplied by $\sin(\pi x/L)$, and then integrated over x from 0 to L to yield $c_1 = 4T_0/\pi$. Repeating the calculation but using $\sin(2\pi x/L)$ produces $c_2 = 0$. The final temperature solution is

$$\frac{T}{T_0} = \frac{4}{\pi}\sin\left(\frac{\pi x}{L}\right)\exp\left(-\frac{\pi^2\alpha t}{L^2}\right) \tag{10.23}$$

The generalization of this procedure is discussed in Section 10.2.3 and later verified in Section 10.4.

Equation 10.23 is identical to the first two terms of the exact solution. The procedure used to approximately satisfy the initial condition is not required when calculating the GF. However, it is used in Haji-Sheikh and Mashena (1987) in the integral solution as a standard procedure of dealing with the initial condition. It is used here to show the equivalence of the GF solution method and the Galerkin-based integral solution as they deal with the initial temperature distribution.

Example 10.2:

Repeat the procedure used in Example 10.1 and use nonorthogonal basis functions.

Solution

Because the boundaries of the slab are at $x = 0$ and $L - x = 0$ surfaces, the function $(L - x)x$ will vanish on both surfaces. Also, the product of $(L - x)x$ and a member of a polynomial series (e.g., $1, x, x^2, \ldots$) will vanish on $x = 0$ and $x = L$ surfaces. For this two-term solution, both $(L - x)x$ and $(L - x)x^2$ functions satisfy the boundary conditions. These functions will be designated as the basis functions. More details concerning the method of selecting these basis functions are given in Section 10.4. Then, in the dimensionless form, one may write $f_1 = (1 - x/L)(x/L)$

and $f_2 = (1 - x/L)(x/L)^2$. The eigenfunction ψ_n becomes

$$\psi_n = d_{n1}\left(1 - \frac{x}{L}\right)\frac{x}{L} + d_{n2}\left(1 - \frac{x}{L}\right)\left(\frac{x}{L}\right)^2 \qquad \text{for } n = 1 \text{ and } 2 \tag{10.24}$$

Equation 10.12 is used to compute a_{ij} using $f_i = (1 - x/L)(x/L)^i$, for $i = 1$ and 2, and $f_j = (1 - x/L)(x/L)^j$, for $j = 1$ and 2. When $m(\mathbf{r}) = 0$ and $k =$ constant, Equation 10.12 for a one-dimensional Cartesian system is

$$a_{ij} = k\int_0^L f_i\left(\frac{d^2 f_j}{dx^2}\right)dx \tag{10.25a}$$

in which

$$\frac{d^2 f_j}{dx^2} = \frac{j(j-1)(x/L)^{j-2} - (j+1)j(x/L)^{j-1}}{L^2} \tag{10.25b}$$

resulting in

$$\begin{aligned} a_{ij} &= -k\left(\frac{1}{L}\right)^2 \int_0^L \left(1 - \frac{x}{L}\right)\left(\frac{x}{L}\right)^i \\ &\quad \times \left[j(j-1)\left(\frac{x}{L}\right)^{j-2} - (j+1)j\left(\frac{x}{L}\right)^{j-1}\right]dx \\ &= \frac{k}{L}\left[\frac{j(j-1)}{i+j-1} - \frac{j(j-1)}{i+j} - \frac{(j+1)j}{i+j} \right. \\ &\quad \left. + \frac{(j+1)j}{i+j+1}\right] \qquad \text{for } i = 1, 2, \text{ and } j = 1, 2 \end{aligned} \tag{10.25c}$$

Substituting for i and j results in $a_{11} = -k/3L$, $a_{12} = a_{21} = -k/6L$, $a_{22} = -2k/15L$. Similarly, the substitution of f_i and f_j in Equation 10.13 produces (set $u = 1$)

$$\begin{aligned} b_{ij} &= \int_0^L \rho c_p \left[\left(1 - \frac{x}{L}\right)\left(\frac{x}{L}\right)^i\right] \times \left[\left(1 - \frac{x}{L}\right)\left(\frac{x}{L}\right)^j\right]dx \\ &= \rho c_p L\left[\frac{1}{i+j+1} - \frac{2}{i+j+2} + \frac{1}{i+j+3}\right] \\ &\quad \text{for } i = 1, 2, \text{ and } j = 1, 2 \end{aligned} \tag{10.26}$$

which results in $b_{11} = \rho c_p L/30$, $b_{12} = b_{21} = \rho c_p L/60$, $b_{22} = \rho c_p L/105$. After substituting matrices **A** and **B** in Equation 10.11 and putting parameters in the dimensionless form, one obtains

$$\begin{bmatrix} -\frac{1}{3} + \frac{\gamma L^2}{\alpha}\frac{1}{30} & -\frac{1}{6} + \frac{\gamma L^2}{\alpha}\frac{1}{60} \\ -\frac{1}{6} + \frac{\gamma L^2}{\alpha}\frac{1}{60} & -\frac{2}{15} + \frac{\gamma L^2}{\alpha}\frac{1}{105} \end{bmatrix} \begin{bmatrix} d_{n1} \\ d_{n2} \end{bmatrix} = 0 \tag{10.27}$$

Notice that this square matrix is symmetric, as discussed in the derivation of Equations 10.14 and 10.16. Unlike Example 10.1, the off-diagonal elements are not equal to zero; therefore, the basis functions f_1 and f_2 are not orthogonal. However, it is easy to show that ψ_n's are orthogonal; see Problem 10.7.

Since the two equations described by Equation 10.27 are homogeneous, the values of d_{n1} and d_{n2} exist if the determinant of their coefficients is zero, that is,

$$\left(-\frac{1}{3}+\frac{\gamma L^2/\alpha}{30}\right)\left(-\frac{2}{15}+\frac{\gamma L^2/\alpha}{105}\right)-\left(-\frac{1}{6}+\frac{\gamma L^2/\alpha}{60}\right)^2$$

$$=\left(\frac{1}{3150}-\frac{1}{3600}\right)\left(\frac{\gamma L^2}{\alpha}\right)^2+\left(\frac{1}{180}-\frac{1}{315}-\frac{1}{225}\right)$$

$$\times\left(\frac{\gamma L^2}{\alpha}\right)+\left(\frac{2}{45}-\frac{2}{36}\right)=0 \tag{10.28}$$

The solution of this quadratic equation yields $\gamma_1 L^2/\alpha = 10, \gamma_2 L^2/\alpha = 42$. When $n = 1$, the value of γ_1 is substituted in Equation 10.27. Note that Equation 10.27 is homogeneous and one of the d's can be selected arbitrarily. After selecting $d_{11} = 1$, either one of two equations yields $d_{12} = 0$. Repeating this process, but using $n = 2$ and $d_{22} = 1$, gives $d_{21} = -½$. The eigenfunctions ψ_1 and ψ_2, using Equation 10.24 are plotted on Figure 10.1. For comparison, the corresponding eigenfunctions of the exact solution are plotted on the same figure. Except for a scale factor, the shape of the ψ_1 and $\sin(\pi x/L)$ are similar. Notice that ψ_2 and $\sin(2\pi x/L)$ have opposite signs which will be accounted for when calculating the coefficient c_2. The function $T = \Theta$ using Equation 10.24 in Equation 10.6 becomes

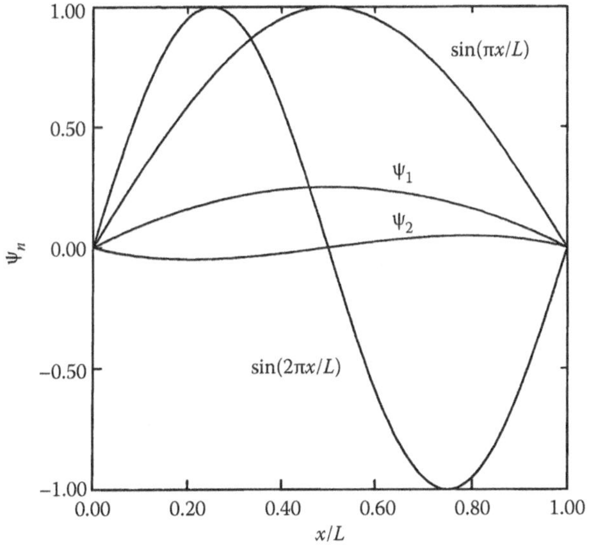

FIGURE 10.1 Eigenfunctions for Example 10.2 and exact eigen functions.

$$\begin{aligned} T &= \sum_{n=1}^{N} c_n \psi \exp(-\gamma_n t) \\ &= c_1(d_{11} f_1 + d_{12} f_2) \exp(-\gamma_1 t) + c_2(d_{21} f_1 + d_{22} f_2) \exp(-\gamma_2 t) \\ &= c_1(1)\left(1 - \frac{x}{L}\right)\frac{x}{L} \exp\left(-\frac{10\alpha t}{L_2}\right) + c_2\left[-\frac{1}{2}\left(1 - \frac{x}{L}\right)\frac{x}{L} \right. \\ &\quad \left. + (1)\left(1 - \frac{x}{L}\right)\left(\frac{x}{L}\right)^2\right] \exp\left(-\frac{42\alpha t}{L^2}\right) \end{aligned} \tag{10.29}$$

The solution is complete, except for the evaluation of c_1 and c_2 which are found from the initial condition.

Applying the initial condition, $T(0, x) = T_0$ at $t = 0$, to Equation 10.29 and multiplying the resulting relation by $f_1 = (1 - x/L)(x/L)$ and then integrating over x from 0 to L results in one equation. Repeating this process but using $f_2 = (1 - x/L)(x/L)^2$ produces a second equation. The simultaneous solution of these two equations yields $c_1 = 5T_0$ and $c_2 = 0$. Because ψ_1 is smaller than $\sin(\pi x/L)$ in Example 10.1, the calculated value of c_1 is larger than the corresponding value of $4t_0/\pi$ obtained in Example 10.1. The solution in this example, as well as in Example 10.1, is for $T(0, x) = T_0$, which gives $c_2 = 0$. Therefore, the resulting solution

$$\frac{T}{T_0} = 5\left(1 - \frac{x}{L}\right)\left(\frac{x}{L}\right) \exp\left(-\frac{10\alpha t}{L^2}\right) \tag{10.30}$$

is also a one-term solution.

Table 10.1 provides a comparison of the temperatures at $x = 0.5L$ obtained from the one-term solutions given in Examples 10.1 an 10.2 with the exact solution. Also, the polynomial-based solutions for $N = 3$ and $N = 5$ are recorded. The one-term solution using the GBI method is usually less accurate when $\alpha t/L^2 > 0.06$ than the results of the first term of the exact solution. However, when $N = 3$

TABLE 10.1
Comparison of $T(0.5L, t)$ Using GBI Solution and Exact Solution for a Slab in Examples 10.1 and 10.2

$\frac{\alpha t}{L^2}$	GBI Solution $N = 1$	GBI Solution $N = 3$	GBI Solution $N = 5$	Exact Solution (one term)	Exact Solution
0.02	1.0234	0.9927	0.97605	1.0452	0.97516
0.04	0.8379	0.8505	0.84629	0.8579	0.84580
0.06	0.6860	0.7028	0.70229	0.7043	0.70220
0.08	0.5617	0.5775	0.57777	0.5781	0.57775
0.10	0.4598	0.4741	0.47449	0.4745	0.47449
0.15	0.2789	0.2894	0.28971	0.2897	0.28971
0.20	0.1692	0.1767	0.17687	0.1769	0.17687
0.25	0.1026	0.1079	0.10798	0.1080	0.10798
0.30	0.0622	0.0659	0.06592	0.0659	0.06592

(actually a two-term solution), the accuracy of the GBI solution is substantially improved (within 0.1% when $\alpha t/L^2 \geq 0.06$). When $N = 5$, the GBI solution exhibits extremely good accuracy (within 0.013% at $\alpha t/L^2 \geq 0.06$).

10.2.2 Numerical Calculation of Eigenvalues

As a generalized and formal procedure, the computation of the temperature T and subsequent determination of the GF can be accomplished by algebraic manipulation of $N \times N$ square matrices **A** and **B**. The next step is the evaluation of the needed eigenvalues. The eigenvalues $\gamma_1, \gamma_2, \ldots, \gamma_N$ can be obtained analytically if N is small; otherwise, numerical steps may become necessary. In Example 10.2, the method of calculating eigenvalues and eigenvectors when $N = 2$ was discussed. However, when N is larger than four, the eigenvalues must be computed numerically. There are many numerical methods available in the literature (Carnahan et al., 1969) with various degrees of efficiency. In order to utilize these eigenvalue-solving routines, Equation 10.11 should be reduced to the following form:

$$(\overline{\mathbf{A}} + \gamma_n \mathbf{I}) \cdot \overline{\mathrm{d}}_n = 0 \tag{10.31}$$

where **I** is the identity matrix and $\overline{\mathrm{d}}_n$ is a column vector with N elements.

The symmetric nature of matrices **A** and **B** permits accurate and fast numerical computation of eigenvalues and eigenvectors, e.g., by the Jacobi method (Carnahan et al., 1969). An accurate method is to use the Cholesky decomposition (Forsyth and Moler, 1967) to decompose matrix **B** into $\mathbf{L} \cdot \mathbf{L}^T$, where **L** is a lower triangular matrix and $\mathbf{L}^T$ is its transposed matrix. This decomposition of matrix **B** is instrumental in reducing Equation 10.11, following some elementary matrix algebra, to

$$(\mathbf{L}^{-1} \cdot \mathbf{A} \cdot \mathbf{L}^{-T} + \gamma_n \mathbf{I}) \cdot \overline{\mathbf{d}}_n = 0 \tag{10.32}$$

where $\mathbf{L}^{-1}$ and $\mathbf{L}^{-T}$ are the inverses of **L** and $\mathbf{L}^T$, respectively. Now, Equation 10.32 has an acceptable form of Equation 10.31 for all available eigenvalue solvers. Since matrix $\overline{\mathbf{A}} = \mathbf{L}^{-1} \cdot \mathbf{A} \cdot \mathbf{L}^{-T}$ is symmetric, the computationally efficient Jacobi method (Carnahan et al., 1969), can be used to find the eigenvalues and eigenvectors of Equation 10.32. The eigenvectors $\overline{\mathbf{d}}_n$, computed using Equation 10.32, are different from but related to the eigenvectors of Equation 10.11 through the relation

$$\mathbf{d}_n = \mathbf{L}^{-T} \cdot \overline{\mathbf{d}}_n \tag{10.33}$$

A Mathematica (Wolfram, 2005) program that uses Cholesky decomposition and determines eigenvalues, eigenvectors, and temperature for Example 10.2 is given in Note 2.

Once the eigenvalues are found, the values of the coefficient d_{nn}, for $n = 1, 2, \ldots, N$, may be selected equal to unity without any loss of generality. For convenience of analysis, matrix **D** is defined so that its nth row has the components of the eigenvector $\mathbf{d}_n$; the components are $d_{n1}, d_{n2}, \ldots, d_{nN}$

$$\mathbf{D} = \begin{bmatrix} \mathbf{d}_1^T \\ \mathbf{d}_2^T \\ \vdots \\ \mathbf{d}_N^T \end{bmatrix} = \begin{bmatrix} d_{11} & d_{12} & \ldots & d_{1N} \\ d_{21} & d_{22} & \ldots & d_{2N} \\ \ldots & \ldots & \ldots & \ldots \\ \ldots & \ldots & \ldots & \ldots \\ d_{N1} & d_{N2} & \ldots & d_{NN} \end{bmatrix} \tag{10.34}$$

Often, standard subroutine packages place eigenvectors in the columns of a matrix which must be transposed to obtain matrix **D**.

10.2.3 Nonhomogeneous Solution

The objective of the following derivation is to solve Equation 10.3 which yields an expression for the GF. Equation 10.3 is essentially the same as Equation 10.1, except the volume energy source term in Equation 10.1 is specified in Equation 10.3. A solution for the nonhomogeneous equation, Equation 10.1, is now proposed by considering c_n in Equation 10.6 to be time dependent. The variation of parameters method is used to solve the nonhomogeneous, first-order, ordinary differential equations. Now, a general solution is considered as

$$T = \sum_{n=1}^{N} c_n(t)\, \psi_n\, (\mathbf{r}) e^{-\gamma_n t} \tag{10.35}$$

Equation 10.35 is an acceptable solution if it can satisfy the basic differential equation 10.1. The substitution of Equation 10.35 into Equation 10.1, followed by multiplying both sides of the resulting equation by f_i, for $i = 1, 2, \ldots, N$, and then integrating over the volume yields

$$\begin{aligned} &\sum_{n=1}^{N} c_n \left[\int_V \nabla \cdot (k \nabla \psi_n) f_i dV - \int_V m(\mathbf{r})^2 f_i\, \psi_n\, dV \right. \\ &\left. + \gamma_n \int_V \rho(\mathbf{r}) c_p(\mathbf{r}) u(\mathbf{r}) \psi_n\, f_i dV \right] e^{-\gamma_n t} \\ &+ \int_V f_i g(\mathbf{r}, t) dV - \sum_{n=1}^{N} \left[\frac{dc_n(t)}{dt} \right] e^{-\gamma_n t} \\ &\times \int_V \rho(\mathbf{r}) c_p(\mathbf{r}) u(\mathbf{r}) \psi_n\, f_i dV = 0 \end{aligned} \tag{10.36}$$

The above procedure is called the Galerkin method (Kantorovich and Krylov, 1960). The first summation term is zero for any value of n because of Equation 10.9. The remaining two terms constitute a system of N ordinary differential equations

$$\sum_{n=1}^{N} \left[\frac{dc_n(t)}{dt} \right] e^{-\gamma_n t} \int_V \rho(\mathbf{r}) c_p(\mathbf{r}) u(\mathbf{r}) \psi_n\, f_i dV = g_i^* \tag{10.37}$$

where

$$g_i^* = \int_V [g(\mathbf{r},t)]\, f_i(\mathbf{r}) dV \tag{10.38}$$

for $i = 1, 2, \ldots, N$. Note that the homogeneous partial differential equation, Equation 10.5 and the nonhomogeneous partial differential equation, Equation 10.1, are approximated by the Galerkin integral procedure. Also, the solution for T satisfies homogeneous boundary conditions.

Once the expression for $\psi_n(\mathbf{r})$ from Equation 10.8 is substituted into Equation 10.37, the result will be (see Problem 10.8)

$$\sum_{n=1}^{N} e_{in} \left[\frac{dc_n(t)}{dt}\right] e^{-\gamma_n t} = g_i^*(t) \tag{10.39}$$

where

$$e_{in} = \sum_{j=1}^{N} d_{nj} b_{ji} \tag{10.40}$$

and $i = 1, 2, \ldots, N$. Therefore, e_{in} in Equation 10.40 is an element of the square of matrix **E**. Matrix **E** is also obtained if **D**, Equation 10.34, is multiplied by **B**, whose elements are defined by Equation 10.13, and the resulting matrix transposed

$$\mathbf{E} = (\mathbf{DB})^T \tag{10.41}$$

Let

$$\chi_n = \left[\frac{dc_n(t)}{dt}\right] \exp(-\gamma_n t) \tag{10.42a}$$

in Equation 10.39; then the following set of N simultaneous equations

$$\sum_{n=1}^{N} e_{in} \chi_n = g_i^* \qquad \text{for } i = 1, 2, \ldots, N \tag{10.42b}$$

are obtained. They can be presented in matrix form as

$$\mathbf{E} \cdot \{X\} = \{\mathbf{g}^*\} \tag{10.43}$$

The notation $\{\cdot\}$ indicates that the arrays X and $\mathbf{g}^*$ in Equation 10.43 are column vectors.

Since the elements of the vector $\mathbf{g}^* = \{g_1^*, g_2^*, \ldots, g_N^*\}$ are known, the elements of the array X can be calculated if both sides of Equation 10.43 are premultiplied by $\mathbf{E}^{-1}$, the inverse of matrix **E**, to obtain

$$\{X\} = \mathbf{E}^{-1} \cdot \{\mathbf{g}^*\} \tag{10.44}$$

The matrix $\mathbf{E}^{-1}$ may be designated as $\mathbf{P}$

$$\mathbf{P} = \mathbf{E}^{-1} \tag{10.45}$$

with elements p_{ni}. Then the elements of array X given by Equation 10.42a are determined by Equation 10.44 as

$$\chi_n = \left[\frac{dc_n(t)}{dt}\right] e^{-\gamma_n t} = \sum_{i=1}^{N} p_{ni}\, g_i^*(t) \qquad n = 1, 2, \ldots, N \tag{10.46}$$

which can be solved for $dc_n(t)/dt$ to obtain

$$\frac{dc_n(t)}{dt} = \sum_{i=1}^{N} p_{ni}\, g_i^*(t) e^{\gamma_n t} \qquad n = 1, 2, \ldots, N \tag{10.47}$$

in which p_{ni}'s are constants given by Equation 10.45.

The integration of Equation 10.47 yields the function $c_n(t)$ as

$$c_n(t) = A_n + \sum_{i=1}^{N} p_{ni} \int_0^t g_i^*(t') e^{\gamma_n t'} dt' \qquad n = 1, 2, \ldots, N \tag{10.48}$$

where A_n represents the constant of integration. The expression for $c_n(t)$ is now known. Substitution of Equation 10.48 into Equation 10.35 yields the final form of the solution

$$T = \sum_{n=1}^{N} \psi_n(\mathbf{r}) e^{-\gamma_n t} \left[A_n + \sum_{i=1}^{N} p_{ni} \int_0^t g_i^*(t') \exp(\gamma_n t')\, dt' \right] \tag{10.49}$$

The second term in the square bracket represents the contribution of the internal energy source. The solution presented by Equation 10.49 is completed after A_n is evaluated. The initial condition (the $t = 0$ temperature distribution), $F(\mathbf{r})$, can be utilized to compute the constant A_n. When $t = 0$, the integral in Equation 10.49 vanishes and the resulting equation is

$$F(\mathbf{r}) = \sum_{n=1}^{N} \psi_n(\mathbf{r}) A_n \tag{10.50}$$

When calculating the GF, the initial temperature $F(\mathbf{r}) = 0$; hence, $A_n = 0$. However, to show the equivalence between the GF solution and the GBI solution when boundary conditions are homogeneous, the calculation of A_n is necessary.

The calculation of A_n when $F(\mathbf{r})$ is nonzero can be carried out by the GBI method (Section 10.2.6 shows that the following procedure agrees with the GFSE). The procedure to determine A_n is to multiply both sides of Equation 10.50 by $\rho c_p u(\mathbf{r}) f_i(\mathbf{r}) dV$

and to integrate over the volume. Then, using Equation 10.8 for $\psi_n(\mathbf{r})$ results in a set of N linear algebraic equations for evaluating $A_1, A_2, \ldots, A_N$,

$$\sum_{n=1}^{N} A_n e_{in} = \lambda_i \qquad \text{for } i = 1, 2, \ldots, N \tag{10.51}$$

where

$$\lambda_i = \int_V \rho(\mathbf{r}) c_p(\mathbf{r}) u(\mathbf{r}) F(\mathbf{r}) f_i(\mathbf{r})\, dV \tag{10.52}$$

in which e_{in} is the element of matrix $\mathbf{E}$ and defined in Equation 10.40. The inverse of matrix $\mathbf{E}$ is given as $\mathbf{P}$ by Equation 10.45. The coefficients A_1, A_2, A_N, are obtained when matrix $\mathbf{P}$ is multiplied by a column vector whose elements are $\lambda_1, \lambda_2, \ldots, \lambda_N$ as

$$\begin{aligned} A_n &= \sum_{i=1}^{N} p_{ni} \lambda_i \\ &= \sum_{i=1}^{N} p_{ni} \int_V \rho(\mathbf{r}) c_p(\mathbf{r}) u(\mathbf{r}) F(\mathbf{r}) f_i(\mathbf{r}) dV \qquad \text{for } n = 1, 2, \ldots, N \end{aligned} \tag{10.53}$$

The coefficients $A_1, A_2, \ldots, A_N$ are analogous to the Fourier coefficients in the exact solutions.

Equation 10.49, following the substitution of g_i^* from Equation 10.38 and A_n from Equation 10.53, becomes

$$\begin{aligned} T &= \sum_{n=1}^{N} \sum_{i=1}^{N} p_{ni} \psi_n(\mathbf{r}) \int_V e^{-\gamma_n t} \rho(\mathbf{r}^*) c_p(\mathbf{r}^*) u(\mathbf{r}^*) F(\mathbf{r}^*)\ f_i(\mathbf{r}^*)\, dV^* \\ &\quad + \sum_{n=1}^{N} \sum_{i=1}^{N} p_{ni} \psi_n(\mathbf{r}) \int_0^t \int_V e^{-\gamma_n (t-t')} g(\mathbf{r}^*, t')\ f_i(\mathbf{r}^*)\, dV^*\, dt' \end{aligned} \tag{10.54}$$

where $\mathbf{r}^*$ and t' are dummy variables of integration, dV^* is the volume element in $\mathbf{r}^*$ space, and $\psi_n(\mathbf{r})$ is obtained from Equation 10.8.

10.2.4 Green's Functions Expression

It is now possible to obtain an expression for the GF. Equation 10.54 is the solution of Equation 10.1 when the boundary conditions are homogeneous and the initial temperature distribution is $F(\mathbf{r})$. The temperature T in Equation 10.54 is identical to $G(\mathbf{r}, t|\mathbf{r}', \tau)$ if $F(\mathbf{r}) = 0$ and $g(\mathbf{r}, t) = \rho(\mathbf{r}) c_p(\mathbf{r}) \delta(\mathbf{r} - \mathbf{r}') \delta(t - \tau)$; see Equation 10.2. Because $F(\mathbf{r}) = 0$, the first term on the right side of Equation 10.54 is zero. The next step is to replace the variable $\mathbf{r}$ and t in $g(\mathbf{r}, t)$ by $\mathbf{r}^*$ and t', and insert $g(\mathbf{r}^*, t') = \rho(\mathbf{r}^*) c_p(\mathbf{r}^*) \delta(\mathbf{r}^* - \mathbf{r}') \delta(t' - \tau)$ in Equation 10.54. After performing the integration over

$\mathbf{r}^*$ and t' and using the Identity 6 in Note 1, the GF becomes

$$G(\mathbf{r},t|\mathbf{r}',\tau) = C(\mathbf{r}') \sum_{n=1}^{N} \sum_{j=1}^{N} \sum_{i=1}^{N} d_{nj}\, p_{ni} \exp[-\gamma_n(t-\tau)] f_j(\mathbf{r}) f_i(\mathbf{r}') \tag{10.55}$$

where $C(\mathbf{r}') = \rho(\mathbf{r}')c_p(\mathbf{r}')$, and d_{nj} and p_{ni} are numbers.

10.2.5 Properties of Green's Functions

The GF defined by Equation 10.55 has the following three properties:

1. If t is replaced by $-\tau$ and τ by $-t$, the following GF property applies:

$$G(\mathbf{r},t|\mathbf{r}',\tau) = G(\mathbf{r},-\tau|\mathbf{r}',-t) \tag{10.56}$$

   This can readily be proved by replacing t by $-\tau$ and τ by $-t$ in Equation 10.55.
2. The GF remains the same if $\mathbf{r}$ is changed to $\mathbf{r}'$ and $\mathbf{r}'$ to $\mathbf{r}$, provided $C(\mathbf{r})$ or $\rho(\mathbf{r})c_p(\mathbf{r})$ is constant,

$$G(\mathbf{r},t|\mathbf{r}',\tau) = G(\mathbf{r}',t|\mathbf{r},\tau) \tag{10.57}$$

3. It is also possible to derive the following GF relation when $C(\mathbf{r})$ is variable.

$$\frac{G(\mathbf{r},t|\mathbf{r}',\tau)}{C(\mathbf{r}')} = \frac{G(\mathbf{r}',t|\mathbf{r},\tau)}{C(\mathbf{r})} \tag{10.58}$$

The above GF properties are useful in the derivation of the GF solution discussed in Section 10.2.6.

The derivation of the second and third properties of the GF is accomplished by considering that the temperature at point $\mathbf{r}'$ is caused by an energy source located at the point $\mathbf{r}$. The temperature distribution is the solution of the equation

$$\begin{aligned} &\nabla_0 \cdot [k(\mathbf{r}')\nabla_0 G(\mathbf{r}',t|\mathbf{r},\tau)] + C(\mathbf{r}')\delta(\mathbf{r}'-\mathbf{r})\delta(t-\tau) - m(\mathbf{r}')^2 G(\mathbf{r}',t|\mathbf{r},\tau) \\ &\quad = C(\mathbf{r}')u(\mathbf{r}')\frac{\partial G(\mathbf{r}',t|\mathbf{r},\tau)}{\partial t} \end{aligned} \tag{10.59}$$

which is Equation 10.3 with $\mathbf{r}$ and $\mathbf{r}'$ interchanged; the del operator ∇_0 uses the components of the $\mathbf{r}'$ position vector. The solution of Equation 10.59 is identical to that of Equation 10.3, except $\mathbf{r}$ and $\mathbf{r}'$ have switched places. Repeating the same algebraic steps that led to the derivation of Equation 10.55 yields

$$G(\mathbf{r}',t|\mathbf{r},\tau) = C(\mathbf{r}) \sum_{n=1}^{N} \sum_{j=1}^{N} \sum_{i=1}^{N} d_{nj}\, p_{ni} \exp[-\gamma_n(t-\tau)]\, f_j(\mathbf{r}') f_i(\mathbf{r}) \tag{10.60}$$

Because the volume integrals and other algebraic operations used to compute p_{ni} and d_{nj} are not affected by switching $\mathbf{r}$ and $\mathbf{r}'$, one can conclude from a comparison of Equations 10.55 and 10.60 that Equation 10.58 is valid. Equation 10.58 implies that the GF remains the same when $\mathbf{r}$ and $\mathbf{r}'$ are switched if $C(\mathbf{r})$ is a constant and Equation 10.57 is valid.

In the above analysis, it is possible to modify the source term, Equation 10.2, [by omitting $C(\mathbf{r})$] so that the GF remains symmetric in $\mathbf{r}$ and $\mathbf{r}'$ even when the properties are variable. However, the GF is not modified here in order to adhere to the existing technical literature. The three properties of the GF described above are essential when deriving the GFSE.

10.2.6 Green's Function Solution Equation

The purpose of this section is to derive an equation for the temperature distribution in terms of the GF. The solution will consider the effects of nonzero initial conditions, distributed volumetric energy source, and nonhomogeneous boundary conditions of the first, second, and third kinds. The body may be nonhomogeneous (i.e., composed of several different materials with different k and ρc_p values) and have irregular shapes. As usual, the solution is restricted to linear problems which means that k and ρc_p cannot be functions of temperature.

The del operator ∇ in Equation 10.3 uses the components of $\mathbf{r}$ (not $\mathbf{r}'$). The del operator ∇_0 is defined earlier that uses the components of $\mathbf{r}'$. If $\mathbf{r}$ is now replaced by $\mathbf{r}'$, $\mathbf{r}'$ by $\mathbf{r}$, t by $-\tau$, and τ by $-t$, Equation 10.3 becomes

$$\nabla_0 \cdot [k(\mathbf{r}')\nabla_0 G(\mathbf{r}', -\tau|\mathbf{r}, -t)] + C(\mathbf{r}')\delta(\mathbf{r}' - \mathbf{r})\delta(\tau - t) - m(\mathbf{r}')^2 G(\mathbf{r}', -\tau|\mathbf{r}, -t)$$
$$= -C(\mathbf{r}')u(\mathbf{r}')\frac{\partial G(\mathbf{r}', -\tau|\mathbf{r}, -t)}{\partial \tau} \tag{10.61}$$

The diffusion equation, Equation 10.1, using $\mathbf{r}'$ and τ as the independent variables can be written as

$$\nabla_0 \cdot [k(\mathbf{r}')\nabla_0 T(\mathbf{r}', \tau)] + g(\mathbf{r}', \tau) - m(\mathbf{r}')^2 T(\mathbf{r}', \tau) = C(\mathbf{r}')u(\mathbf{r}')\frac{\partial T(\mathbf{r}', \tau)}{\partial \tau} \tag{10.62}$$

where $g(\mathbf{r}', \tau)$ is the contribution of a distributed volumetric energy source. To shorten the equations, the function $G(\mathbf{r}', -\tau|\mathbf{r}, -t)$ will be designated as G.

Equation 10.62 is now multiplied by G, and Equation 10.61 multiplied by T. The resulting equations are then subtracted from each other to produce

$$T\nabla_0 \cdot [k(\mathbf{r}')\nabla_0 G] - G\nabla_0 \cdot [k(\mathbf{r}')\nabla_0 T] + C(\mathbf{r}')T\delta(\mathbf{r}' - \mathbf{r})\delta(\tau - t) - Gg(\mathbf{r}', \tau)$$
$$= -C(\mathbf{r}')u(\mathbf{r}')\frac{\partial (TG)}{\partial \tau} \tag{10.63}$$

The following two relations, derived using Identity 1 in Note 1,

$$\nabla_0 \cdot [Tk(\mathbf{r}')\nabla_0 G] = T\nabla_0 \cdot [k(\mathbf{r}')\nabla_0 G] + k(\mathbf{r}')\nabla_0 T \cdot \nabla_0 G \tag{10.64}$$
$$\nabla_0 \cdot [Gk(\mathbf{r}')\nabla_0 T] = G\nabla_0 \cdot [k(\mathbf{r}')\nabla_0 T] + k(\mathbf{r}')\nabla_0 G \cdot \nabla_0 T \tag{10.65}$$

provide the expressions for the first two terms on the left side of Equation 10.63 when Equation 10.64 is subtracted from Equation 10.65. Upon substituting the results in Equation 10.63, integrating in $\mathbf{r}'$ space over the volume V, and over τ from 0 to $t^* = t + \epsilon$ where ϵ has a small but positive value, one obtains

$$\begin{aligned}
&\int_{\tau=0}^{t^*} \int_V \{\nabla_0 \cdot [k(\mathbf{r}')T\nabla_0 G] - \nabla_0 \cdot [k(\mathbf{r}')G\nabla_0 T]\} dV'\, d\tau \\
&\quad + \int_{\tau=0}^{t^*} \int_V C(\mathbf{r}')T\delta(\mathbf{r}' - \mathbf{r})\delta(\tau - t) dV' d\tau \\
&\quad - \int_{\tau=0}^{t^*} \int_V Gg(\mathbf{r}', \tau) dV'\, d\tau \\
&= - \int_{\tau=0}^{t^*} \int_V \times \left[C(\mathbf{r}')u(\mathbf{r}') \frac{\partial(TG)}{\partial\tau} \right] d\tau\, dV'
\end{aligned} \tag{10.66}$$

Various terms in Equation 10.66 are now considered. Green's theorem, Identity 3 in Note 1, can be used to reduce the first volume integral on the left side of Equation 10.66 to a surface integral. In addition, Identity 6 in Note 1 reduces the second term on the left side of Equation 10.66 to become $C(\mathbf{r})T$. Furthermore, the term on the right side of Equation 10.66 can be readily integrated over τ. Note that

$$\int_{\tau=0}^{t^*} \left[\frac{\partial(GT)}{\partial\tau} \right] d\tau = G(\mathbf{r}', -t^*|\mathbf{r}, -t)T(\mathbf{r}', t^*) - G|_{\tau=0}\, T(\mathbf{r}', 0) \tag{10.67a}$$

and the value of the GF, G, at the upper limit when $\tau = t^*$ is (see Equations 10.56 and 10.58)

$$G(\mathbf{r}', -t^*|\mathbf{r}, -t) = \frac{G(\mathbf{r}, t|\mathbf{r}', t^*)C(\mathbf{r})}{C(\mathbf{r}')} = 0 \tag{10.67b}$$

which is the value of temperature at time t when a pulse appears at a later time, $t^* = t + \epsilon$; hence, the first term on the right side of Equation 10.67a is zero. Equation 10.66 then becomes

$$\begin{aligned}
C(\mathbf{r})T(\mathbf{r}, t) = &\int_V C(\mathbf{r}')u(\mathbf{r}')G|_{\tau=0} F(\mathbf{r}') dV' \\
&+ \int_{\tau=0}^{t} d\tau \int_V g(\mathbf{r}', \tau)G\, dV' + \int_{\tau=0}^{t} d\tau \int_S \\
&\times k(S') \left[G\left(\frac{\partial T}{\partial n}\right) - T\left(\frac{\partial G}{\partial n}\right) \right]_{S'} dS'
\end{aligned} \tag{10.68}$$

where $F(\mathbf{r}') = T(\mathbf{r}', 0)$, $C(\mathbf{r}) = \rho(\mathbf{r})c_p(\mathbf{r})$ and $G = G(\mathbf{r}', -\tau|\mathbf{r}, -t)$. Equation 10.68 is the basic GFSE for heterogeneous and homogeneous materials.

The operator $\partial/\partial n$ designates differentiation along the outer normal to the external surface and $F(\mathbf{r}')$ is the initial temperature distribution. The first term on the right side of Equation 10.68 is the contribution of the initial temperature distribution. The influence of the volumetric energy source is included in the second term. The boundary conditions for G in Equation 10.68 are homogeneous. The third term on the right side describes the boundary condition effects. When the surface temperature $T|_{S'}$ is prescribed (boundary condition of the first kind), then $G|_{S'} = 0$. When the heat flux is given (boundary condition of the second kind), $\partial T/\partial n = -(q/k)|_{S'}$, then $\partial G/\partial n|_{S'} = 0$. For convective boundary conditions (boundary conditions of the third kind), the boundary conditions are

$$-k\frac{\partial T}{\partial n} = h(T - T_\infty) \qquad \text{on } S' \tag{10.69}$$

and

$$-k\frac{\partial G}{\partial n} = hG \qquad \text{on } S' \tag{10.70}$$

where k and h may vary with position. When Equation 10.69 is multiplied by $G|_{S,}$ and Equation 10.70 by $T|_S$, and the resulting equations are subtracted from each other, the following relation is obtained

$$\left[G\left(\frac{\partial T}{\partial n}\right) - T\left(\frac{\partial G}{\partial n}\right)\right]\Big|_{S'} = \left(\frac{h}{k}\right) GT_\infty|_{S'} \tag{10.71}$$

The right side of Equation 10.71 then replaces the term in square brackets in Equation 10.68.

In the derivation of Equation 10.54 the boundary conditions were considered to be homogeneous. It is of interest to compare Equations 10.68 and 10.54. For homogeneous boundary conditions, the third term on the right side of the GFSE, Equation 10.68, is equal to 0. Then, taking the Green's function, $G = G(\mathbf{r}', -\tau|\mathbf{r}, -t) = G(\mathbf{r}', t|\mathbf{r}, \tau)$, from Equation 10.60 and substituting it into Equation 10.68 yields Equation 10.54. This indicates the procedure used to include the initial conditions in Equation 10.54 and in Examples 10.1 and 10.2 are consistent with the derivation of the GFSE.

The last term in the GFSE, Equation 10.68, contains the contribution of nonhomogeneity of the boundary conditions. Boundary conditions of the first and (or) second kinds are nonhomogeneous if the surface temperature and (or) the surface heat flux are nonzero. Boundary conditions of the third kind (convective) are nonhomogeneous if the ambient temperature is nonzero. The convergence of Equation 10.68, in some cases, is slow. For instance, when the surface temperature is prescribed, the term $f_j(\mathbf{r})$ in the GF, Equation 10.60, takes the value of zero after $\partial G/\partial n$ is computed over S'. The temperature solution at the surface becomes singular (cannot be computed) and is inaccurate in the vicinity of the surface. Similar situations also exist for other boundary conditions. A GF expression that behaves more favorably and converges more rapidly for nonhomogeneous boundary conditions is derived in the next section.

10.3 ALTERNATIVE FORM OF THE GREEN'S FUNCTION SOLUTION

As discussed in the previous section, when the temperature is prescribed on the external surface, there is a singularity associated with using Equation 10.68. Equation 10.68 yields a value of zero for the surface temperature because $f_j(\mathbf{r})$ has a zero value at the

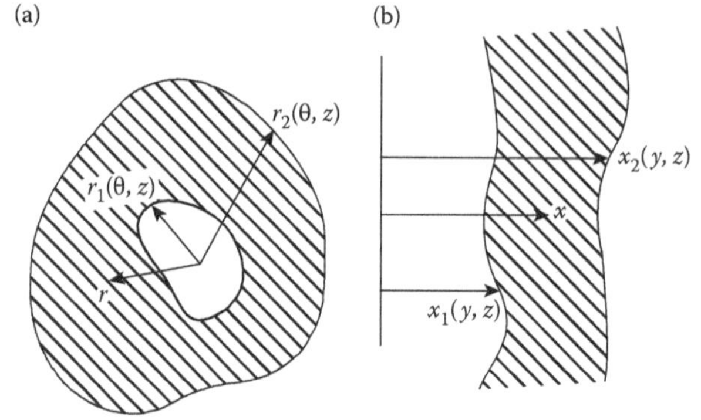

FIGURE 10.2 (a) Geometry for Equation 10.72b, and (b) geometry for Equation 10.72c.

surface. The implication is that Equation 10.68 may provide inaccurate temperature values in the vicinity of the wall and erroneous heat flux at the wall. When the wall heat flux is prescribed, the convergence at the boundaries can be very slow.

The following procedure removes this singularity and improves the convergence of the GF solution for the temperature distribution (Haji-Sheikh, 1988; Haji-Sheikh and Beck, 1988). It begins by defining a differentiable temperature function that satisfies the boundary conditions used in Equation 10.68. This new function is designated as T^*. It is usually possible to find a function T^* such as

$$T^* = c_1 u_p + c_2 \tag{10.72a}$$

For example, consider a body bounded by two surfaces as shown in Figure 10.2a. When the surface temperature is prescribed, the steady-state temperature is approximated by

$$T^* = (T_2 - T_1)\frac{\ln(r/r_1)}{\ln(r_2/r_1)} + T_1 \tag{10.72b}$$

where $r_1 = r_1(\theta, z)$ and $r_2 = r_2(\theta, z)$ are coordinates of two arbitrarily selected inner and outer surfaces whose respective temperatures are $T_1 = T_1(\theta, z, t)$ and $T_2 = T_2(\theta, z, t)$. For a different geometry, shown in Figure 10.2b, the following form is sometimes preferred:

$$T^* = (T_2 - T_1)\frac{x - x_1(y, z)}{x_2(y, z) - x_1(y, z)} + T_1 \tag{10.72c}$$

The function T^* is called the quasisteady solution if it satisfies the Laplace equation and the prescribed boundary conditions. In this part of the analyses, any internal source can be ignored. However, T^* given by Equation 10.72b or c does not always satisfy the Laplace equation, but it will satisfy the Laplace equation in cylindrical coordinates if r_1 and r_2 are constants. Also, T^* given by Equation 10.72c will satisfy the Laplace equation in Cartesian coordinates if x_1 and x_2 are constants. In one-dimensional coordinates, except when dealing with prescribed heat flux at both surfaces, it is possible to use Equation 10.72a to derive an equation for T^* that satisfies the boundary conditions. The function u_p takes the value of x in Cartesian coordinates and $\ln r$ or $-1/r$ in the radial cylindrical or spherical coordinates, respectively.

Except when the heat flux is prescribed on both surfaces, the constants c_1 and c_2 can be determined by applying the appropriate boundary conditions. The calculation of c_1 and c_2, for nonhomogeneous boundary conditions of the second and third kinds, is included in Examples 10.5 and 10.6. However, when the heat flux on both surfaces is given, the constant c_2 in Equation 10.72a should be replaced by $c_2 r^2$ before calculating c_1 and c_2; see Example 10.7. Although there are numerous conduction problems for which a T^* can be computed, it is sometimes impossible or cumbersome to find this function for many problems, e.g., locally varying heat flux and heat transfer coefficients in multidimensional bodies. In the absence of a suitable T^*, the time partitioning of the GF discussed in Chapter 5 is a logical approach.

Whenever an auxiliary function T^* is available, a function $f^*(\mathbf{r}', \tau)$ is defined (Haji-Sheikh and Beck, 1988) so that

$$\nabla_0 \cdot [k\nabla_0 T^*(\mathbf{r}', \tau)] = f^*(\mathbf{r}', \tau) \tag{10.73}$$

The function $f^*(\mathbf{r}', \tau)$ defined by Equation 10.73 is unrelated to the basis functions $f_j(\mathbf{r})$. When Equation 10.73 is multiplied by G and Equation 10.3 is multiplied by $T^* = T^*(\mathbf{r}', \tau)$, then, after subtracting the former from the latter, the following equation is obtained

$$\begin{aligned} &T^*\nabla_0 \cdot (k\nabla_0 G) - G\nabla_0 \cdot (k\nabla_0 T^*) + T^* C(\mathbf{r}')\,\delta(\mathbf{r}' - \mathbf{r})\,\delta(\tau - t) \\ &\quad = -Gf^* - T^* C(\mathbf{r}')u(\mathbf{r}')\frac{\partial G}{\partial \tau} \end{aligned} \tag{10.74}$$

Integration of Equation 10.74 over τ is carried out between the limits of 0 and $t^* = t + \epsilon$, where ϵ is a small positive number. Then integration with respect to $\mathbf{r}'$ over the entire volume, application of the Green's theorem (see Note 1), and reduction of some algebraic terms, results in

$$\begin{aligned} &\int_{\tau=0}^{t^*} d\tau \int_V k\left(G\frac{\partial T^*}{\partial n} - T^*\frac{\partial G}{\partial n}\right)\bigg|_{S'} dS' \\ &\quad = C(\mathbf{r})T^* + \int_V C(\mathbf{r}')u(\mathbf{r}')\left[\int_{\tau=0}^{t^*} T^*\left(\frac{\partial G}{\partial \tau}\right) d\tau\right] dV' \\ &\qquad + \int_{\tau=0}^{t^*} d\tau \int_V Gf^*\, dV' \end{aligned} \tag{10.75}$$

Integrating by parts and then letting ϵ go to zero, the term in the square bracket in Equation 10.75 becomes

$$\begin{aligned} \lim_{\epsilon\to 0}\int_{\tau=0}^{t^*} T^*\left(\frac{\partial G}{\partial \tau}\right) d\tau &= \lim_{\epsilon\to 0}\left[GT^*\big|_{\tau=0}^{t^*} - \int_{\tau=0}^{t^*} G\left(\frac{\partial T^*}{\partial \tau}\right) d\tau\right] \\ &= -G|_{\tau=0}T^*(\mathbf{r}, 0) - \int_{\tau=0}^{t} G\left(\frac{\partial T^*}{\partial \tau}\right) d\tau \end{aligned} \tag{10.76}$$

The substitution of Equation 10.76 into Equation 10.75 followed by the substitution of the resulting equation into Equation 10.68, the basic GFSE for heterogeneous materials, and some minor algebraic simplifications produces the alternative form of the GF solution (Haji-Sheikh and Beck, 1988) for $T(\mathbf{r},t)$,

$$
\begin{aligned}
C(\mathbf{r})T(\mathbf{r},t) = C(\mathbf{r})T^*(\mathbf{r},t) &+ \int_{\tau=0}^{t} d\tau \int_V G\left[g(\mathbf{r}',\tau) - C(\mathbf{r}')u(\mathbf{r}')\,\frac{\partial T^*(\mathbf{r}',\tau)}{\partial\tau}\right] dV' \\
&+ \int_V C(\mathbf{r}')u(\mathbf{r}')G|_{\tau=0}[F(\mathbf{r}') - T^*(\mathbf{r}',0)]dV' \\
&+ \int_{\tau=0}^{t} d\tau \int_V Gf^*\, dV'
\end{aligned}
\tag{10.77}
$$

An expression for the function $G = G(\mathbf{r}', -\tau|\mathbf{r}, -t) = G(\mathbf{r}', t|\mathbf{r}, \tau)$ is given by Equation 10.60. The function T^* contains the contribution of nonhomogeneous boundary conditions. If $f^*(r',\tau)$ is zero, then $T^*(\mathbf{r}',\tau)$ satisfies the Laplace equation and it is the quasisteady solution; accordingly, the term that contains f^* in Equation 10.77 is equal to zero.

As a special case, when the temperature of the entire surface T_s has a constant value and is different from $F(\mathbf{r}) = T_0 =$ constant and $g = 0$, then $T^* = T_s$, $f^* = 0$, and Equation 10.77 reduces to

$$
C(\mathbf{r})\left[\frac{T(\mathbf{r},\tau) - T_s}{T_0 - T_s}\right] = \int_V C(\mathbf{r}')u(\mathbf{r})G\Big|_{\tau=0} dV'
\tag{10.78}
$$

Example 10.3:

Consider a slab of isotropic and homogeneous material with thickness L and which is initially at zero temperature, $F(x) = 0$. Assume thermophysical properties are constants and there is no volumetric energy source, $g = 0$. The boundary conditions are $T(0,t) = 0$ and $T(L,t) = T_L\sin(\omega t)$. Use Equation 10.77 to calculate temperature at $x = L/2$ and compare the results with the exact solution.

Solution

A function T^*, using Equation 10.72a, that satisfies both boundary conditions is $T^* = T_L(x/L)\sin(\omega t)$. In this example, the value of $\nabla^2 T^* = d^2T^*/dx^2 = f^*/k = 0$; see Equation 10.73. The last term in Equation 10.77 is zero and the function T^* is called the quasisteady solution. The solution using Equation 10.77 is

$$
T = T^* - \int_0^t d\tau \int_0^L G\left(\frac{\partial T^*}{\partial \tau}\right) dx'
\tag{10.79}
$$

since $g = 0$ and $F(x) - T^*(x,0) = 0$. The basis functions, $X11$ case, are presented in Example 10.2 as $f_j = (1 - x/L)(x/L)^j$. The elements of matrices **A** and **B** are given by Equations 10.25 and 10.26. The eigenvalues are calculated as discussed in

Example 10.2. For $N = 3$, the eigenvalues are $\gamma_1 = 9.86975\alpha/L^2, \gamma_2 = 42\alpha/L^2$, and $\gamma_3 = 102.13\alpha/L^2$. Using Equations 10.11 matrix **D** is

$$\mathbf{D} = \begin{bmatrix} 1 & 1.1331 & -1.1331 \\ -0.5 & 1 & 0 \\ 0.21584 & -1 & 1 \end{bmatrix} \tag{10.80}$$

Equations 10.41 and 10.45 are used to calculate the elements of matrix **P**:

$$\mathbf{P} = \begin{bmatrix} 19.395 & 21.977 & -21.977 \\ -420 & 840 & 0 \\ 3802 & -17615 & 17615 \end{bmatrix} \tag{10.81}$$

The GF is obtained from Equation 10.60 as

$$G(x', -\tau|x, -t) = C(\mathbf{r}) \sum_{n=1}^{N} \sum_{j=1}^{N} \sum_{i=1}^{N} \times d_{nj} p_{ni} \exp[-\gamma_n(t-\tau)]\, f_j(x')\, f_i(x) \tag{10.82}$$

where $f_j = (1 - x/L)(x/L)^j$. The final solution is obtained by substituting the GF from Equation 10.82 into Equation 10.79 as

$$\frac{T}{T_L} = \frac{x}{L}\sin(\omega t) - \omega \sum_{n=1}^{N} \times \left[\sum_{j=1}^{N} d_{nj}\left(1 - \frac{x}{L}\right)\left(\frac{x}{L}\right)^j\right]\left[\sum_{i=1}^{N} p_{ni}\left(\frac{1}{i+2} - \frac{1}{i+3}\right)\right] \times \frac{\gamma_n \cos(\omega t) + \omega \sin(\omega t) - \gamma_n \exp(-\gamma_n t)}{\gamma_n^2 + \omega^2} \tag{10.83}$$

For $N = 2$ and $N = 3$, the results obtained by this equation are given in Table 10.2. The exact solution is (Ozisik, 1993, Equation 5-50, p. 203)

$$\frac{T}{T_L} = \frac{x}{L}\sin(\omega t) + \frac{2\omega}{\pi} \sum_{n=1}^{\infty} \frac{(-1)^n}{n} \sin\left(\frac{n\pi x}{L}\right) \times \frac{\alpha(n\pi/L)^2\{\cos(\omega t) - \exp[-(n\pi/L)^2 \alpha t]\} + \omega \sin(\omega t)}{\alpha^2(n\pi/L)^4 + \omega^2} \tag{10.84}$$

which is used to check the accuracy of the alternative GF solution. The entries in Table 10.2 are for $x = 0.5L$ and $L^2\omega/\alpha = \pi/10$. The results for $N=2$ and $N=3$ are quite accurate. Table 10.2 shows that a three-term solution yields results comparable to 10 terms of the exact solution. Even for $N = 2$, the solution closely agrees with the exact solution when the dimensionless time is larger than 0.2. A significantly better agreement with the exact solution is attributed to the lack of step change in the surface temperature. In addition to removing the singularity at the surface associated with Equation 10.68, the alternative form of the GF solution, Equation 10.77, has another advantage; it provides a faster converging solution than Equation 10.68 when t becomes large. For further discussion, see Example 10.4.

TABLE 10.2
Results for Example 10.3 for $L^2\omega/\alpha = \pi/10$. Comparison of $T(0.5L, t)$ Using the Alternative Green's Function Solution and Exact Solution for a Slab

$\frac{\alpha t}{L^2}$	AGFS[a], Equation 10.83		Exact Solution, Equation 10.84	
	$N = 2$	$N = 3$	$N = 10$	$N = 30$
0.1	0.00330	0.00362	0.00362	0.00365
0.2	0.01443	0.01459	0.01458	0.01459
0.5	0.05888	0.05889	0.05888	0.05889
1	0.13566	0.13566	0.13565	0.13566
2	0.27766	0.27765	0.27764	0.27765
3	0.39248	0.39246	0.39245	0.39246
4	0.46888	0.46886	0.46885	0.46886
5	0.49938	0.49936	0.49936	0.49936
6	0.48098	0.48098	0.48098	0.48098

[a]Alternative GF solution, Equation 10.77.

10.4 BASIS FUNCTIONS AND SIMPLE MATRIX OPERATIONS

A major step in obtaining an integral solution is to construct a set of basis functions. The set must contain linearly independent elements, and each element must satisfy all homogeneous boundary conditions. If the boundary conditions are nonhomogeneous, the basis functions must be homogeneous and of the same type. Consideration is given to two types of problems. First, the basis functions for one-dimensional and regular geometries are presented. It is shown that a unified solution procedure is possible for regular-shaped bodies. Then, the method of finding basis functions for some irregular-shaped bodies is presented; an irregular-shaped body refers to a nonorthogonal body. Although obtaining the basis functions for many irregular-shaped bodies is a simple task, for many others it can become cumbersome. The reason is that each irregular-shaped body must be treated differently. After the basis functions are determined, Equations 10.12 and 10.13 yield matrices **A** and **B**. Next, Equation 10.11 yields the eigenvalues and eigenvectors. The computation of matrix **P** completes the variables needed for calculation of the GF using Equation 10.60.

10.4.1 One-Dimensional Bodies

The method for establishing the basis functions for one-dimensional problems is an interesting feature of the GBI method. When these basis functions are established, the remaining steps for finding temperature solutions follow the same procedures as discussed in Examples 10.1 through 10.3. Moreover, the basis functions for multidimensional regular geometries can be constructed as a product of one-dimensional basis functions. The product method of finding the basis functions is valid even when the GF cannot be obtained using a product of the corresponding one-dimensional GF; see

Example 10.8. This subsection describes a method of obtaining the basis functions for one-dimensional bodies subject to boundary conditions of the first, second, and third kinds. The basis functions must satisfy homogeneous boundary conditions whether the actual boundary conditions are homogeneous or nonhomogeneous. For example, if a boundary condition for temperature is nonhomogeneous, the basis functions must satisfy the homogeneous boundary condition of the same kind. The variable z used in this derivation stands for axial, radial, or angular coordinates.

A generalized set of basis functions that satisfies the homogeneous boundary conditions $k_1 df_j/dz = h_1 f_j$ at $z = a$ and $-k_2 df_j/dz = h_2 f_j$ at $z = b$ (where $b > a$) is

$$f_j = (\delta_j z^2 + \beta_j z + \eta_j) z^{j-1} \qquad \text{for } j = 1, 2, \ldots, N \tag{10.85}$$

The variable z stands for the specific coordinate system; for example, z is x in Cartesian coordinates, XIJ, or r in cylindrical, RIJ, and spherical coordinates, $RSIJ$.

Two equations for determining the three coefficients, δ_j, β_j, and η_j, are obtained by evaluating Equation 10.85 at the two boundaries. Since one of the coefficients δ_j, β_j, or η_j can be selected arbitrarily, the coefficient δ_j is set equal to the determinant of the coefficients in the two equations. The resulting expressions for δ_j, β_j, and η_j are

$$\delta_j = a(j - aB_1)(j - 1 + bB_2) - b(j + bB_2)(j - 1 - aB_1) \tag{10.86a}$$

$$\beta_j = a^2(aB_1 - j - 1)(j - 1 + bB_2) + b^2(bB_2 + j + 1)(j - 1 - aB_1) \tag{10.86b}$$

$$\eta_j = -ab^2(j - aB_1)(bB_2 + j + 1) - ba^2(j + bB_2)(aB_1 - j - 1) \tag{10.86c}$$

for $j = 1, 2, 3, \ldots, N$.

The parameters B_1 and B_2 appearing in Equations 10.86a through c are h_1/k_1 and h_2/k_2, respectively. The parameters B_1 and B_2 are finite for $X33$, $R33$, and $RS33$ cases. If the surface $z = a$ is insulated, then $B_1 = 0$; $X23$, $R23$, or $RS23$. Similarly, $B_2 = 0$ if the $z = b$ surface is insulated; $X32$, $R32$, or $RS32$. For $X22$, $R22$, and $RS22$ problems, B_1 and B_2 are set equal to 0 in Equations 10.86a through c. Equation 10.85 holds for any one-dimensional conduction problem in a finite domain. Modifications are necessary when B_1, B_2, or both, are infinite; boundary conditions of the first kind, see Table 10.3a. In special cases when $a = 0$ or both a and B_1 are equal to zero, the coefficients δ_j, β_j, and η_j also must be modified. The values of δ_j, β_j, and η_j for special cases are found in Table 10.3b.

10.4.2 Matrices A and B for One-Dimensional Problems

After expressions for δ_j, β_j, and η_j are available, Equations 10.12 and 10.13 yield the values of a_{ij} and b_{ij}. Next, two indefinite integrals useful for calculating the values of a_{ij} and b_{ij} are presented.

TABLE 10.3a
Coefficients δ_j, β_j, and η_j in Equation 10.85 When B_1 or B_2 Are Infinite

$B_1 = \infty$; prescribed T at $z = a$, i.e., boundary conditions of the first kind at $z = a$; $X13^a$

$$\delta_j = -a(j - 1 + bB_2) + b(j + bB_2) \quad (10.87a)$$
$$\beta_j = a^2(j - 1 + bB_2) - b^2(bB_2 + j + 1) \quad (10.87b)$$
$$\eta_j = ab^2(bB_2 + j + 1) - ba^2(j + bB_2) \quad (10.87c)$$

for $j = 1, 2, 3, \ldots, N$

$B_2 = \infty$; prescribed T at $z = b$, i.e., boundary conditions of the first kind at $z = b$; $X31^a$

$$\delta_j = a(j - aB_1) - b(j - 1 - aB_1) \quad (10.88a)$$
$$\beta_j = a^2(aB_1 - j - 1) + b^2(j - 1 - aB_1) \quad (10.88b)$$
$$\eta_j = -ab^2(j - aB_1) - ba^2(aB_1 - j - 1) \quad (10.88c)$$

for $j = 1, 2, 3, \ldots, N$

$B_1 = \infty$ and $B_2 = \infty$; boundary conditions of the first kind at $z = a$ and $z = b$; $X11^a$

$$\delta_j = 1 \quad (10.89a)$$
$$\beta_j = -(a + b) \quad (10.89b)$$
$$\eta_j = ab \quad (10.89c)$$

for $j = 1, 2, 3, \ldots, N$

[a] Also for *RIJ* and *RSIJ* cases.

Matrix A

When calculating the elements of matrix **A**, and the thermal conductivity is constant, the following integral can be used as a computational aid:

$$I_a(z) = \int f_i \left(\nabla^2 f_j\right) z^p \, dz = \sum_{k=1}^{5} P_k \frac{z^{i+j+p+2-k}}{i + j + p + 2 - k} \quad (10.94)$$

where

$$P_1 = \delta_i \delta_j (j + 1)(j + p) \quad (10.95a)$$
$$P_2 = \beta_i \delta_j (j + 1)(j + p) + \beta_j \delta_i (j + p - 1) j \quad (10.95b)$$
$$P_3 = \eta_i \delta_j (j + 1)(j + p) + \beta_i \beta_j (j + p - 1) j + \eta_j \delta_i (j + p - 2)(j - 1) \quad (10.95c)$$
$$P_4 = \eta_i \beta_j (j + p - 1) j + \eta_j \beta_i (j - 1)(j + p - 2) \quad (10.95d)$$
$$P_5 = \eta_i \eta_j (j - 1)(j + p - 2) \quad (10.95e)$$

TABLE 10.3b
Coefficients δ_j, β_j, and η_j in Equation 10.85 for Special Cases When $a = 0$

$a = 0$, B_1 and B_2 finite, and $B_1 > 0$; $X33$[a]

$$\delta_1 = B_1 + B_2 + bB_1B_2 \quad (10.90a)$$

$$\eta_1 = -2b - b^2B_2 \quad (10.90b)$$

$$\beta_1 = -2bB_1 - b^2B_1B_2 \quad (10.90c)$$

and for $j > 1$

$$\delta_j = j + bB_2 \quad (10.90d)$$

$$\eta_j = 0 \quad (10.90e)$$

$$\beta_j = -b(j+1) - b^2B_2 \quad (10.90f)$$

for $j = 2, 3, \ldots, N$

$a = 0$, $B_1 = 0$, and B_2 finite; $X23$[a]

$$\delta_j = j - 1 + bB_2 \quad (10.91a)$$

$$\beta_j = 0 \quad (10.91b)$$

$$\eta_j = -b^2(j+1) - b^3B_2 \quad (10.91c)$$

for $j = 1, 3, 5, \ldots, N$

$a = 0$, $B_1 = 0$, and $B_2 = \infty$; $X21$[a]

$$\delta_j = 1 \quad (10.92a)$$

$$\beta_j = 0 \quad (10.92b)$$

$$\eta_j = -b^2 \quad (10.92c)$$

for $j = 1, 3, 5, \ldots, N$

$a = 0$ and $B_1 = \infty$; $X13$[a]

$$\delta_j = j + bB_2 \quad (10.93a)$$

$$\beta_j = -b^2B_2 - (j+1)b \quad (10.93b)$$

$$\eta_j = 0 \quad (10.93c)$$

for $j = 1, 2, 3, \ldots, N$

[a] Also for *RIJ* and *RSIJ* cases.

Note that when $p = 0$ and $j = 1$, the term containing P_4 is zero. Also, when $j = 1$ or $j + p = 2$, the term containing P_5 is zero. When there is a fin effect, additional terms are necessary (see Equation 10.12).

Matrix B

The following integral can be used as a computational aid to calculate the elements of matrix **B** when $\rho c_p u(\mathbf{r})$ is constant.

$$I_b(z) = \int f_i f_j z^p dz = \sum_{k=1}^{5} Q_k \frac{z^{i+j+p+4-k}}{i+j+p+4-k} \quad (10.96)$$

where

$$Q_1 = \delta_i\delta_j \tag{10.97a}$$
$$Q_2 = \beta_i\delta_j + \beta_j\delta_i \tag{10.97b}$$
$$Q_3 = \eta_i\delta_j + \beta_i\beta_j + \eta_j\delta_i \tag{10.97c}$$
$$Q_4 = \beta_i\eta_j + \beta_j\eta_i \tag{10.97d}$$
$$Q_5 = \eta_i\eta_j \tag{10.97e}$$

Also this integral provides the contribution of the fin effect for the elements of matrix **A** when $m(\mathbf{r}) = \text{constant}$.

10.4.3 Matrix Operations When $N = 1$ and $N = 2$

Following the computation of the values of the components of matrices **A** and **B**, the eigenvalues and the eigenvectors are computed using Equation 10.11. After the computation of p_{ni}'s, using Equations 10.41 and 10.45, the GF is obtained using Equation 10.60. The mathematical procedure is to solve Equation 10.11 when $N = 1$ to obtain $\gamma_1 = -a_{11}/b_{11}, d_{11} = 1$, and then Equations 10.41 and 10.45 yield $p_{11} = 1/b_{11}$. The one-term GF is

$$G(z', -\tau|z, -t) = \rho c_p d_{11} p_{11} \exp[-\gamma_1(t-\tau)]\, f_1(z') f_1(z) \tag{10.98}$$

Expressions for finding γ_n, d_{nj} and p_{ni}, when $N = 2$, follow the procedure presented in Example 10.2. The eigenvalues are the roots of a quadratic equation

$$\text{Det}|\mathbf{A} + \gamma\mathbf{B}| = D_1\gamma^2 + D_2\gamma + D_3 = 0 \tag{10.99}$$

where

$$D_1 = \text{Det}(\mathbf{B}) = b_{11}b_{22} - b_{12}b_{21} \tag{10.100a}$$
$$D_2 = (a_{11}b_{22} - a_{12}b_{21}) + (b_{11}a_{22} - b_{12}a_{21}) \tag{10.100b}$$
$$D_2 = \text{Det}(\mathbf{A}) = a_{11}a_{22} - a_{12}a_{21} \tag{10.100c}$$

The elements of matrix **D** are computed after arbitrarily selecting $d_{11} = d_{22} = 1$. The reason for the arbitrary choice of d_{11} and d_{22} is that c_n in Equation 10.6 is yet to be determined and, at this stage, c_n can be multiplied or divided by a constant. The other elements are

$$d_{12} = -\frac{a_{12} + \gamma_1 b_{12}}{a_{22} + \gamma_1 b_{22}} \tag{10.101a}$$

$$d_{21} = -\frac{a_{12} + \gamma_2 b_{12}}{a_{11} + \gamma_2 b_{11}} \tag{10.101b}$$

Since the transpose of matrix **BD** is matrix **E**, Equation 10.41, and matrix **P** is the inverse of matrix **E**, Equation 10.45, the elements of matrix **P** are for $N = 2$

$$p_{11} = \frac{e_{22}}{\det(\mathbf{E})} \tag{10.102a}$$

$$p_{12} = -\frac{e_{12}}{\det(\mathbf{E})} \tag{10.102b}$$

$$p_{21} = -\frac{e_{21}}{\det(\mathbf{E})} \tag{10.102c}$$

$$p_{22} = \frac{e_{11}}{\det(\mathbf{E})} \tag{10.102d}$$

The GF is obtained when $\mathbf{r}$ and $\mathbf{r}'$ are replaced by z and z', and N is set equal to 2 in Equation 10.60.

Example 10.4:

A homogeneous hollow cylinder (Figure 10.3), with inner radius a and outer radius $b = 2a$, is considered with the boundary conditions $kdT/dr = h_1 T$ at $r = a$ and $-kdT/dr = q(t)$. Furthermore, it is assumed that the initial temperature distribution is zero, $T(r, 0) = 0$. The heat flux $q(t)$ at $r = b$ is given by the relation $q(t) = q_0 t$ so that q varies linearly with t. The $h_1 a/k$ ratio is selected to be 1. Compare the alternative GF solution with the exact solution.

Solution

The notation for this case is $R32B02T0$. The temperature at $r = b$ is to be found as a function of time. The step-by-step procedure for obtaining the GF is presented in this example for $N = 2$. The basic procedure, except for the method of obtaining the basis functions, is applicable to all transient, one-dimensional conduction problems.

1. It is necessary to introduce a set of basis functions, f_j. A general set of basis functions that satisfies the homogeneous convective conditions $kdf_j/dr = h_1 f_j$ at $r = a$ and $kdf_j/dr = 0$ at $r = b$ are obtained from Equation 10.85 when z is replaced by r. Using $j = 1$ and 2, the coefficients δ_j, β_j, and η_j in Equation 10.85 are available from Equations 10.86a through c for $b = 2a$ as $\delta_1 = 1$,

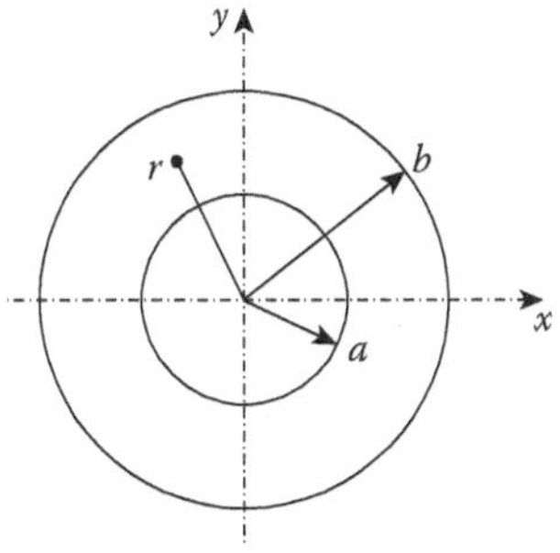

FIGURE 10.3 Hollow cylinder in Example 10.4; boundary conditions denoted $R32$.

$\beta_1 = -4, \eta_1 = 1, \delta_2 = 1, \beta_2 = -2$, and $\eta_2 = -4$. Because the basis functions are not unique, the function f_j as given by Equation 10.85 is divided by δ_j and then made dimensionless. The basis functions f_1 and f_2, with $B_1 = 1$ and $B_2 = 0$ are

$$f_1 = \left(\frac{r}{a}\right)^2 - 4\frac{r}{a} + 1 \tag{10.103a}$$

$$f_2 = \left(\frac{r}{a}\right)^3 - 2\left(\frac{r}{a}\right)^2 - 4\frac{r}{a} \tag{10.103b}$$

2. Once the basis functions are available, Equation 10.103a and b, the elements of matrices **A** and **B** are calculated using Equations 10.12 and 10.13. Because the conduction is one-dimensional and there is no fin effect, the definite integrals can be evaluated using Equations 10.94 through 10.97 to yield

$$a_{11} = -\frac{34\pi}{3a} \tag{10.104a}$$

$$a_{12} = a_{21} = -\frac{446\pi}{15a} \tag{10.104b}$$

$$a_{22} = -\frac{1181\pi}{15a} \tag{10.104c}$$

and

$$b_{11} = \frac{337\pi a}{15} \tag{10.105a}$$

$$b_{12} = b_{21} = \frac{1231\pi a}{21} \tag{10.105b}$$

$$b_{22} = \frac{21421\pi a}{140} \tag{10.105c}$$

3. The eigenvalues and the eigenvectors are computed using Equation 10.11. Equation 10.99 is used (for $N = 2$) to obtain

$$\gamma_1 = \frac{0.50276\alpha}{a^2} \tag{10.106a}$$

$$\gamma_2 = \frac{11.9830\alpha}{a^2} \tag{10.106b}$$

4. The elements of matrix **D** are computed after arbitrarily selecting

$$d_{11} = d_{22} = 1 \tag{10.107a}$$

The other elements according to Equations 10.101a and b are

$$d_{12} = -\frac{a_{12} + \gamma_1 b_{12}}{a_{22} + \gamma_1 b_{22}} = -0.14495 \tag{10.107b}$$

$$d_{21} = -\frac{a_{12} + \gamma_2 b_{12}}{a_{11} + \gamma_2 b_{11}} = -2.6085 \tag{10.107c}$$

5. Following computation of matrix **E** from Equation 10.40 or Equation 10.41, matrix **P** is the inverse of matrix **E**. When $N = 2$, Equations 10.102a through d provide the elements of matrix **P** as

$$p_{11} = \frac{e_{22}}{\det(\mathbf{E})} = 3.6638 \tag{10.108a}$$

$$p_{12} = -\frac{e_{12}}{\det(\mathbf{E})} = -0.0053107 \tag{10.108b}$$

$$p_{21} = -\frac{e_{21}}{\det(\mathbf{E})} = -13.637 \tag{10.108c}$$

$$p_{22} = \frac{e_{11}}{\det(\mathbf{E})} = 5.2279 \tag{10.108d}$$

6. All the parameters needed to calculate the GF relation are now available and Equation 10.60 becomes

$$G(r', -\tau | r, -t) = C(r) \sum_{n=1}^{2} \sum_{j=1}^{2} \sum_{i=1}^{2} d_{nj} p_{ni} \exp[-\gamma_n (t - \tau)] f_j(r') f_i(r) \tag{10.109}$$

The temperature distribution is given by Equation 10.68 or Equation 10.77. When using Equation 10.77, T^* is not a unique function. A function such as $T^* = -bq_0 t[k/(ha) + \ln(r/a)]/k$ would satisfy the nonhomogeneous boundary conditions. For this case, the value of f^* is zero; see Equation 10.73. Therefore, the analysis is simpler. The temperature solution when $ha/k = 1$ is obtained using Equation 10.77:

$$\frac{kT}{q_0} = -bt\left(1 + \ln\frac{r}{a}\right) + \int_0^t d\tau \int_a^b 4\pi G \left(1 + \ln\frac{r'}{a}\right) r' dr' \tag{10.110}$$

Since $b = 2a$ and G is given by Equation 10.109, Equation 10.77 becomes

$$\begin{aligned}
\frac{kT}{a^3 q_0/\alpha} = {} & \frac{-2\alpha t}{a^2}\left(1 + \ln\frac{r}{a}\right) + 4\pi \sum_{n=1}^{N} \\
& \times \left\{ \sum_{j=1}^{N} d_{nj} \left[\delta_j \left(\frac{r}{a}\right)^2 + \beta_j \frac{r}{a} + \eta_j \right] \left(\frac{r}{a}\right)^{j-1} \right\} \\
& \times \left\{ \sum_{i=1}^{N} p_{ni} \left[\delta_j \frac{2^{i+3}(1 + \ln 2) - 1}{i+3} + \beta_i \frac{2^{i+2}(1 + \ln 2) - 1}{i+2} \right.\right. \\
& + \eta_i \frac{2^{i+1}(1 + \ln 2) - 1}{i+1} - \delta_i \frac{2^{i+3} - 1}{(i+3)^2} \\
& \left.\left. - \beta_i \frac{2^{i+2} - 1}{(i+2)^2} - \eta_i \frac{2^{i+1} - 1}{(i+1)^2} \right] \right\} \frac{1 - \exp(-\gamma_n t)}{a^2 \gamma_n/\alpha}
\end{aligned} \tag{10.111}$$

Notice that $\partial T^*/\partial t = -bq_0[k/(ha) + \ln(r/a)]/k$ is multiplied by $G = G(r', -\tau | r, -t)$ and $dA = 2\pi r\, dr$ and then integrated between a and $b = 2a$.

Table 10.4 shows the numerical values of the GFs. The values of the GFs using the integral method agree well with the exact values when $N \geq 5$ and the $\alpha(t - \tau)/(b - a)^2 > 0.04$. Table 10.4 suggests using the small-time solution when

TABLE 10.4

Values of the Green's Functions $G(b, t|b, \tau)$ for the Integral Method, Exact, and Small-Time Asymptotic Solutions ($R32$ Case with $B_1 = 1$)

$\frac{\alpha(t-\tau)}{(b-a)^2}$	Galerkin-Based Integral Method				Small-Time Solution	Exact Solution
	$N = 2$	$N = 3$	$N = 5$	$N = 7$		
0.01	0.26457	0.38672	0.49460	0.47371	0.46974	0.46975
0.02	0.24802	0.32655	0.34709	0.33841	0.33862	0.33863
0.05	0.20838	0.22820	0.22280	0.22275	0.22275	0.22277
0.1	0.16600	0.16691	0.16495	0.16498	0.16491	0.16499
0.2	0.12670	0.12632	0.12588	0.12588	0.12480	0.12588
0.5	0.096912	0.097070	0.097042	0.097042	0.091206	0.097042
1	0.075063	0.075199	0.075185	0.075184	0.076939	0.075184
2	0.045402	0.045485	0.045477	0.045477	–	0.045477
5	0.010047	0.010066	0.010064	0.010064	–	0.010064

TABLE 10.5

Partitioned and Alternative Green's Function Solution for Dimension-Less Surface Temperature $-kT/(a^3 q_0/\alpha)$ at $r/a = 2$

Time	$N = 2$		$N = 7$			Exact Series (first 14,000 terms)
$\alpha t/a^2$	AGFS[a]	PGFS[b]	AGFS[a]	GFS[c]	PGFS[b]	
0.01	0.001372	0.000765	0.000765	0.000608	0.000765	0.000765
0.02	0.003072	0.002179	0.002179	0.001870	0.002179	0.002179
0.05	0.009945	0.008740	0.008740	0.007973	0.008740	0.008739
0.1	0.026438	0.025135	0.025136	0.023607	0.025136	0.025134
0.2	0.074303	0.072923	0.072872	0.069817	0.072870	0.072868
0.5	0.30721	0.30574	0.30518	0.29755	0.30518	0.30517
1	0.93291	0.93117	0.93001	0.91475	0.93000	0.92999
2	2.8434	2.8406	2.8393	2.8088	2.8393	2.8392
5	11.243	11.235	11.237	11.161	11.237	11.237
10	27.715	27.696	27.709	27.557	27.709	27.709

[a] Alternative Green's function solution, Equation 10.111.
[b] Partitioned Green's function solution, see Chapter 5.
[c] Green's function solution, Equation 10.68.

$\alpha(t-\tau)/(b-a)^2 < 0.12$. This latter number is used for the time partitioning of the integral solution results appearing in Table 10.5 when $N = 2$.

The dimensionless temperature solution, $kT/(a^3 q_0/\alpha)$, at $r = 2a$ and for $N = 2$ and 7 is presented in Table 10.5. The results for the alternative GF solution, which uses Equation 10.77, agree closely with the exact solution. However, the standard GF solution using Equation 10.68 is much less accurate for $N = 7$. When $N < 7$ (not shown in this table), the accuracy further decreases. Notice the remarkable accuracy of the solution using the time-partitioned GF when $N = 2$. The eigenvalues in this example were computed using Cholesky's decomposition of matrix **B** and applying the Jacobi method (Carnahan et al., 1969, p. 250; see p. 255 for FORTRAN subroutine) to Equation 10.32.

As discussed earlier, the alternative GF solution requires an auxiliary equation which is usually available for one-dimensional geometries. However, in multidimensional and complex geometries, this auxiliary equation either is unavailable or difficult to obtain. The partitioning of the Green's function, as given in Chapter 5, Equation 5.17, is an attractive option. Equation R02.5 in Appendix R provides the small-time GF while Equation 10.60 is used to obtain the large-time GF. As shown in Table 10.5, partitioning of the GF and alternative GF solutions exhibit good accuracy. Even when $N = 2$, the time-partitioned solution produces accurate results.

Example 10.5:

Derive Equation 10.72a and calculate c_1 and c_2 when the boundary conditions are

$$\frac{\partial T}{\partial z} = \frac{h_1}{k_1}(T - T_{\infty 1}) \qquad \text{at } z = a \tag{10.112a}$$

$$\frac{\partial T}{\partial z} = -\frac{h_2}{k_2}(T - T_{\infty 2}) \qquad \text{at } z = b \tag{10.112b}$$

Solution

The generalized form of the Laplace equation in one-dimensional bodies is

$$\frac{1}{z^p}\frac{d}{dz}\left(z^p \frac{dT^*}{dz}\right) = 0 \tag{10.113}$$

where $p = 0$ in Cartesian coordinates
$p = 1$ in cylindrical coordinates
$p = 2$ in spherical coordinates

Integrate twice to obtain

$$T^* = c_1 \int \frac{dz}{z^p} + c_2 = c_1 u_p(z) + c_2 \tag{10.114}$$

This equation assumes different forms in different coordinate systems:

In Cartesian coordinates, $z = x, p = 0$, and $u_p(z) = x$
In cylindrical coordinates, $z = r, p = 1$, and $u_p(z) = \ln r$
In spherical coordinates, $z = r, p = 2$, and $u_p(z) = -1/r$

The first term on the right side of Equation 10.77 contains T^*, which must satisfy the nonhomogeneous boundary conditions because the remaining terms in Equation 10.77 only satisfy the homogeneous boundary conditions. Then, the boundary conditions for T^* are

$$\frac{\partial T^*}{\partial z} = \frac{h_1}{k_1}(T^* - T_{\infty 1}) \qquad \text{at } z = a \tag{10.115a}$$

$$\frac{\partial T^*}{\partial z} = -\frac{h_2}{k_2}(T^* - T_{\infty 2}) \qquad \text{at } z = b \tag{10.115b}$$

Introducing T^* from Equation 10.114 in Equation 10.115a and b results in the following two simultaneous equations:

$$\left[u_p(a) - \frac{k_1}{h_1 a^p}\right] c_1 + c_2 = T_{\infty 1} \tag{10.116a}$$

$$\left[u_p(b) + \frac{k_2}{h_2 b^p}\right] c_1 + c_2 = T_{\infty 2} \tag{10.116b}$$

The solutions for c_1 and c_2 are

$$c_1 = \frac{T_{\infty 2} - T_{\infty 1}}{u_p(b) - u_p(a) + k_1/(h_1 a^p) + k_2/(h_2 b^p)} \tag{10.117a}$$

$$c_2 = \frac{T_{\infty 2}[u_p(b) + k_2/(h_2 b^p)] - T_{\infty 1}[u_p(a) - k_1/(h_1 a^p)]}{u_p(b) - u_p(a) + k_1/(h_1 a^p) + k_2/(h_2 b^p)} \tag{10.117b}$$

The above choice of T^* forces f^* to become equal to zero. The basis functions, as usual, must satisfy homogeneous boundary conditions, and they are given by Equations 10.85 and 10.86 when h_1 or h_2 are nonzero. The case when $h_1 = 0$ or $h_2 = 0$ is trivial. When $h_1 = 0$, $T^* = T_{\infty 2}$, and when $h_2 = 0$, $T^* = T_{\infty 1}$. Equations 10.177a and b can be used to calculate c_1 and c_2 when either h_1 or h_2, or both are infinite.

Example 10.6:

In this example, T^*, using Equation 10.72a, is to be calculated when heat flux is prescribed on one surface and the other surface is exposed to a convective boundary condition.

Solution

First consider the following boundary conditions:

$$\frac{\partial T^*}{\partial z} = \frac{q_1}{k_1} \qquad \text{at } z = a \tag{10.118a}$$

$$\frac{\partial T^*}{\partial z} = -\frac{h_2}{k_2}(T^* - T_{\infty 2}) \qquad \text{at } z = b \tag{10.118b}$$

Using these boundary conditions, the following two simultaneous equations are obtained

$$\frac{c_1}{a^p} = +\frac{q_1}{k} \tag{10.119a}$$

$$\left[u_p(b) + \frac{k_2}{h_2 b^p}\right] c_1 + c_2 = T_{\infty 2} \tag{10.119b}$$

The solutions for c_1 and c_2 are

$$c_1 = +\frac{q_1 a^p}{k_1} \tag{10.120a}$$

and

$$c_2 = -\frac{\left[k_2/(h_2 b^p) + u_p(b)\right] q_1 a^p}{k_1} + T_{\infty 2} \tag{10.120b}$$

when the boundary conditions at $z = a$ and $z = b$ are switched, that is,

$$\frac{\partial T^*}{\partial z} = \frac{h_1}{k_1}(T^* - T_{\infty 1}) \qquad \text{at } z = a \tag{10.121a}$$

$$\frac{\partial T^*}{\partial z} = -\frac{q_2}{k_2} \qquad \text{at } z = b \tag{10.121b}$$

The two simultaneous equations are

$$c_1 = -\frac{q_2 b^p}{k_2} \tag{10.122a}$$

$$c_2 = -\frac{[k_1/(h_1 a^p) - u_p(a)] q_2 b^p}{k_2} + T_{\infty 1} \tag{10.122b}$$

In this example, similar to Example 10.5, $f^* = 0$ because the Laplace equation is satisfied.

Example 10.7:

Consider a one-dimensional conduction problem when heat flux is prescribed on both surfaces. The goal is to calculate the values of T^* and f^*.

Solution

As before, the nonhomogeneous boundary conditions are assigned to T^*. The boundary conditions are

$$\frac{\partial T^*}{\partial z} = \frac{q_1}{k_1} \qquad \text{at} \quad z = a \tag{10.123a}$$

$$\frac{\partial T^*}{\partial z} = -\frac{q_2}{k_2} \qquad \text{at} \quad z = b \tag{10.123b}$$

The proposed auxiliary solution T^* is

$$T^* = c_1 u_p(z) + c_2 z^2 \tag{10.124}$$

This equation must satisfy the boundary conditions given by Equation 10.123a and b. After substituting T^* from Equation 10.124 in Equation 10.123a and b, the following two simultaneous relations are obtained:

$$\frac{k_1 u_p(a)}{a^p} c_1 + 2ac_2 = q_1 \tag{10.125a}$$

$$\frac{k_2 u_p(b)}{b^p} c_1 + 2bc_2 = -q_2 \tag{10.125b}$$

The constants c_1 and c_2 are calculated as

$$c_1 = \frac{q_2 a + q_1 b}{k_1 b a^{-p} u_p(a) - k_2 a b^{-p} u_p(b)} \tag{10.126a}$$

and

$$c_2 = -\frac{q_2 + k_2 b^{-p} u_p(b) c_1}{2b} \tag{10.126b}$$

The value of f^* is obtained from the relation

$$\begin{aligned} f^* &= \frac{1}{z^p}\frac{d}{dz}\left(z^p \frac{dT}{dz}\right) = \frac{1}{z^p}\frac{d}{dz}\left[z^p\left(\frac{c_1}{z^p} + 2c_2 z\right)\right] \\ &= 2c_2(p+1) \end{aligned} \tag{10.127}$$

Using Equation 10.77, the term that contains f^* behaves as a uniform energy source that liberates $2c_2(p+1)$ units of energy per unit time and per unit volume. Indeed, as a general rule, the function $f^*(\mathbf{r}', \tau)$ can be lumped together with $g(\mathbf{r}', \tau)$.

The basis functions needed for calculating the GF are given by Equation 10.127 for which the values of δ_j, β_j, and η_j are given by Equations 10.86a through c as

$$\delta_j = j(j-1)(a-b) \tag{10.128a}$$

$$\beta_j = (j^2-1)(b^2-a^2) \tag{10.128b}$$

$$\eta_j = abj(j+1)(a^2-b^2) \tag{10.128c}$$

For a special case when $a = 0$, Equations 10.91a through c yield

$$\delta_j = j-1 \qquad \beta_j = 0 \qquad \text{and} \qquad \eta_j = -b^2(j+1) \tag{10.129}$$

Example 10.8:

Consider a finite cylinder with boundary conditions

$$\frac{\partial T}{\partial x} = 0 \qquad \text{at } x = 0 \tag{10.130a}$$

$$\frac{\partial T}{\partial x} = -\frac{h_2}{k}(T - T_\infty) \qquad \text{at } x = L \tag{10.130b}$$

$$\frac{\partial T}{\partial r} = 0 \qquad \text{at } r = 0 \tag{10.130c}$$

$$T = T_\infty \qquad \text{at } r = r_0 \tag{10.130d}$$

and find a set of basis functions.

Solution

The basis functions must satisfy homogeneous boundary conditions of the same types as the boundary conditions on temperature

$$\frac{\partial f_j}{\partial x} = 0 \qquad \text{at } x = 0 \tag{10.131a}$$

$$\frac{\partial f_j}{\partial x} = -\frac{h_2}{k} f_j \qquad \text{at } x = L \tag{10.131b}$$

$$\frac{\partial f_j}{\partial r} = 0 \qquad \text{at } r = 0 \tag{10.131c}$$

$$f_j = 0 \qquad \text{at } r = r_0 \tag{10.131d}$$

The contribution of the x-direction to the basis functions is obtained using Equation 10.85 for which δ_j, β_j, and η_j coefficients are given by Equation 10.91a through c. The contribution of the r-direction to the basis functions is computed, in a similar manner, using Equation 10.85. Setting $B_2 = h_2/k$, the basis functions become

$$f_j = [(m_j - 1 + LB_2)x^2 - L^2(m_j + 1) - L^3 B_2](r^2 - r_0^2)x^{m_j-1} r^{n_j-1} \tag{10.132}$$

The variables m_j and n_j replaced j in Equation 10.85 to account for all combinations of $m_j = 1, 2, 3, \ldots$, and $n_j = 1, 2, 3, \ldots$, for example

$j = 1$	$m_j = 1$	and	$n_j = 1$
$j = 2$	$m_j = 2$	and	$n_j = 1$
$j = 3$	$m_j = 1$	and	$n_j = 2$
$j = 4$	$m_j = 3$	and	$n_j = 1$
$j = 5$	$m_j = 2$	and	$n_j = 2$
$j = 6$	$m_j = 1$	and	$n_j = 3$
.....	....	....	....

To include all relevant powers of x and r, use $m_j = 1$ and $n_j = 1$ for a one-term solution. For more accuracy, $N = 3$ should be used for which the polynomial coefficients are $x^0 r^0, x^1 r^0, x^0 r^1$. The accuracy can be further improved by including higher order polynomial coefficients $x^2 r^0$, $x^1 r^1$, and $x^0 r^2$ with N taking the value of 6. The next higher level of accuracy is achieved when $N = 10$, 15, and so on.

Instead of finding the set of two-dimensional basis functions and then calculating the GF, the two-dimensional GF can be computed as a product of two one-dimensional GFs. However, the two-dimensional basis functions, Equation 10.132, can be used to calculate the temperature field in a finite cylinder if it also contains inclusions with different thermophysical properties; see Section 11.3. In this latter case, the GF is not a product of two one-dimensional GFs.

The product method can be used for all multidimensional regular bodies with a few exceptions. The noted exception is the case when the regular geometry is cylindrical or spherical and there is a surface that convects heat in angular directions. However, boundary conditions of the first and second kinds in an angular direction can be accommodated by the product method.

10.5 FINS AND FIN EFFECT

The GF solution, Equation 10.77, is modified to solve temperature and heat flux in bodies with fin effect (Haji-Sheikh et al., 1991). The bodies can be single layer or multilayers, although bodies with multilayers are treated in Chapter 11. For multidimensional conduction with fin effect, Equation 10.77 is valid. For quasi-one-dimensional conduction Equation 10.77 is written as

$$T(r,t) = T^*(r,t) + \frac{1}{\rho(r)c_p(r)}\left\{\int_{\tau=0}^{t} d\tau \int_V \times G\left[f^* + g(r',\tau) - \rho(r')c_p(r')\frac{\partial T^*(r',\tau)}{\partial \tau}\right] dV' + \int_V \rho(r')c_p(r')G|_{\tau=0}[F(r') - T^*(r',0)]\,dV'\right\} \quad (10.133)$$

where r is the axial coordinate.

The auxiliary function T^* must satisfy the nonhomogeneous boundary conditions but it is not necessarily the steady-state or quasisteady-state solution. The function T^* contains only the contribution of nonhomogeneous boundary conditions. The function f^*, appearing as a source term in Equation 10.133, compensates for the arbitrary nature of T^* and is given by the modified form of Equation 10.73 as

$$f^*(r',\tau) = \nabla_0 \cdot [k\nabla_0 T^*(r',\tau)] - m(r)^2 T^* \quad (10.134)$$

where ∇_0 implies the derivatives are in r' space. If $f^*(r',\tau) = 0$ and $m(r)^2 = 0$, then $T^*(r',\tau)$ satisfies the Laplace equation and it is the quasisteady solution. The function T^* is chosen in the same manner as discussed in the examples with no fin effect.

Example 10.9:

Calculate the fin efficiency in a straight cylindrical fin. The boundary conditions are: $T = T_b = 1$ at $r = r_1$ and $q = 0$ at $r = r_2$. Finally, compare the results with the exact values.

Solution

The selection of the basis functions for this problem is exactly the same as earlier examples with no fin effect. Also, the function T^* is selected in a similar manner using Equation 10.72a mainly to satisfy the boundary conditions; here, $T^* = 1$ and $f^* = -m^2$ where m^2 is $2h/\delta^*$, and δ^* is the fin thickness. Equation 10.1 in cylindrical coordinates, when $T = T(r,t), g(\mathbf{r}) = 0, u(\mathbf{r}) = 1$ and thermophysical properties are constant, is

$$\frac{1}{r}\frac{\partial}{\partial r}\left(r\frac{\partial T}{\partial r}\right) - \frac{2h}{k\delta^*}T = \frac{1}{\alpha}\frac{\partial T}{\partial t} \quad (10.135)$$

The basis functions are given by Equation 10.85 and the coefficients δ_j, β_j, and η_j by Equation 10.89a through c. The fin efficiency is defined as the ratio of heat transfer from an actual fin to the heat transfer from an isothermal fin at $T = T_b$. Figure 10.4 shows the efficiency, when $r_2 / r_1 = 2$, as a function of dimensionless time, $\alpha t / r_1^2$, for different values of $m^* = r_1(2h/k\delta^*)^{0.5}$.

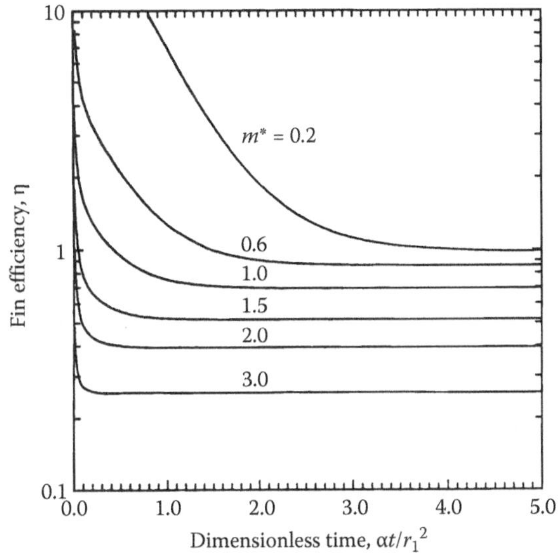

FIGURE 10.4 Fin efficiency versus time when $m^* = 2$ and $r_2/r_1 = 2$.

TABLE 10.6
Fin Efficiency Using GBI Method and Comparison with the Exact Solution for Cylindrical Fins

	$r_2/r_1 = 1.2$		$r_2/r_1 = 2.0$		$r_2/r_1 = 3.0$	
m^*	GBI	Exact	GBI	Exact	GBI	Exact
0.0	1.00000	1.00000	1.00000	1.00000	1.00000	1.00000
0.4	0.99766	0.99767	0.93025	0.93024	0.73696	0.73700
0.8	0.99074	0.99075	0.77434	0.77434	0.43493	0.43494
1.2	0.97946	0.97946	0.61464	0.61464	0.27932	0.27934
1.6	0.96416	0.96417	0.48705	0.48706	0.19933	0.19935
2.0	0.94531	0.94531	0.39335	0.39332	0.15338	0.15340
2.4	0.92341	0.92342	0.32543	0.32542	0.12415	0.12419
5.0	0.74443	0.74448	0.14607	0.14609	0.05460	0.05479
8.0	0.55316	0.55334	0.08821	0.08840	0.03249	0.03315

To show the accuracy obtainable with the single-equation solution, the steady-state efficiency for different r_2/r_1 ratios, for a range of values of m are shown in Table 10.6. The data compare well with the exact solution; usually up to five significant figures. All entries in Table 10.6 are for $N = 9$. Table 10.7 contains the efficiency calculated for different values of N. Only one value of r_2/r_1 is used in this presentation. Table 10.7 shows that, when $N = 5$, sufficient accuracy is achieved for nearly all practical applications.

TABLE 10.7
Fin Efficiency Using GBI Method with Different N Values for Straight Cylindrical Fins, $r_2 / r_1 = 2$

	GBI Solution			
m^*	$N = 2$	$N = 5$	$N = 7$	Exact Solution
0.0	1.00000	1.00000	1.00000	1.00000
0.4	0.90400	0.93008	0.93021	0.93024
0.8	0.74667	0.77418	0.77432	0.77434
1.2	0.58523	0.61449	0.61463	0.61464
1.6	0.45599	0.48688	0.48706	0.48706
2.0	0.36067	0.39312	0.39332	0.39332
2.4	0.29133	0.32516	0.32541	0.32542
5.0	0.10673	0.14515	0.14603	0.14609
8.0	0.04999	0.08590	0.08807	0.08840

10.6 CONCLUSIONS

The Galerkin-based GF solution method discussed in this chapter has many advantages. Generally, for one-dimensional problems, a two-term solution provides results accurate enough for most applications. The methodology, especially for one-dimensional bodies, is universal, and a single computer program can be used for different-shaped bodies with different boundary conditions. This is a unique feature that is not shared by the exact solution.

Two solution methods were covered in this chapter: a GF solution and an alternative GF solution. The general solution method requires time partitioning of the GF to achieve a high degree of accuracy. The alternative GF solution, when available, is simple and accurate if time is not extremely small. If time is extremely small, the small-time solution yields the temperature and time partitioning of the GF is not a prerequisite. The time-partitioned Galerkin-based GF solution has flexibility, accuracy, and computational speed, which are the features of an efficient computational method. The unique feature is that a single solution is used for nearly all one-dimensional conduction problems. This feature was successfully incorporated in a computer program.

Many GFs for two- or three-dimensional solutions of regular bodies are the products of appropriate one-dimensional GFs; others must be computed in multidimensional space, see Chapter 11.

The derivation of the GF solution presented in this chapter applies to heterogeneous as well as homogeneous bodies. However, all examples given in this chapter are for homogeneous and regular bodies. Further discussion of this solution method and additional examples are presented in Chapter 11.

PROBLEMS

10.1 Verify Equation 10.33.

10.2 Repeat Example 10.2, except assume the surface at $x = 0$ is insulated.

10.3 Find the eigenvalues and write a two-term expression for temperature distribution in a solid cylinder with radius r_0. The initial temperature is T_0 and the surface temperature is suddenly reduced to T_s. (Answer: $\gamma_1 = 5.784\alpha / r_0^2, \gamma_2 = 36.88\alpha / r_0^2$.)

10.4 Consider the GF in Problem 3 to be the large-time GF. Find a small-time GF to solve for temperature distribution using the time partitioning of the GF.

10.5 A solid cylinder with radius r_0 is initially at temperature T_0. If there is a prescribed heat flux at the rate $q(t)$, find the basis functions, Retaining only two eigenvalues, $N = 2$, calculate the eigenvalues, and derive an expression for the temperature distribution. (Answer: $\gamma_1 = 0, \gamma_2 = 15\alpha / r_0^2$.)

10.6 Use the GF partitioning to derive an equation for the temperature at the surface of the cylinder described in Problem 5.

10.7 Show that the functions $\psi_1, \psi_2, \psi_3, \ldots,$ are orthogonal, that is,

$$\int_V \rho c_p \psi_m \psi_n dV = 0$$

when m and n are not equal. (Hint: substitute ψ_m and ψ_n in Equation 10.5, then use Identities 1 through 3 in Note 1.)

10.8 Verify Equations 10.39 and 10.40.

10.9 Verify Equation 10.77.

10.10 Find the eigenvalues and write a two-term expression for temperature distribution in a solid sphere with radius r_0. The initial temperature is T_0 and the surface temperature is suddenly reduced to T_s. (Answer: $\gamma_1 = 9.875\alpha / r_0^2, \gamma_2 = 50.12\alpha / r_0^2$.)

10.11 Find the eigenvalues and write a two-term expression for temperature distribution in a hollow cylinder whose inner radius is r_1 and outer radius is r_2. The initial temperature is T_0, the surface temperature is suddenly reduced to T_s at $r = r_1$, and the surface at $r = r_2$ is insulated. (Answer: $\gamma_1 = 7.407\alpha / r_2^2, \gamma_2 = 88.61\alpha / r_2^2$.)

10.12 Find the eigenvalues and write a two-term expression for temperature distribution in a hollow sphere whose inner radius is r_1 and outer radius is r_2. The initial temperature is T_0 and the surface at $r = r_2$ is insulated. There is convection to a zero temperature fluid at $r = r_1$ surface and $hr_2 / k = 1$. (Answer: $\gamma_1 = 7.480\alpha / r_2^2, \gamma_2 = 47.02\alpha / r_2^2$.)

10.13 A straight fin with constant cross-sectional area A and length L is insulated at the tip while the base temperature is T_b. Find T^* and use the alternative GFSE to derive a solution. Compare the fin efficiency for $N = 2$ with its exact value when $L(hP/kA)^{0.5}$ I; P is the perimeter. (Answer: Fin efficiency when $N = 2$ is 0.75312.)

10.14 For a circular pin fin write the temperature solution. The radius r varies as x^2, the tip at $x = 0$ may be considered insulated, and the base is at a constant temperature.

NOTE 1: MATHEMATICAL IDENTITIES

Consider v to be a scalar and $\mathbf{W}$ to be a vector.

Identity 1

(Hay, 1953, p. 117)

$$\nabla \cdot (\mathrm{v}\mathbf{W}) = \mathrm{v}(\nabla \cdot \mathbf{W}) + (\nabla \mathrm{v}) \cdot \mathbf{W}$$

Identity 2

$$\nabla \mathrm{v} \cdot \mathbf{n} = \frac{\partial \mathrm{v}}{\partial n}$$

Identity 3

The generalization of the Green's theorem for line integrals is called the Green's theorem in space (Hay, 1953, p. 143), the Green's theorem, the divergence theorem, or Gauss's theorem (Kaplan, 1956, p. 269).

$$\int_V \nabla \cdot \mathbf{W}\, dv = \int_S \mathbf{W} \cdot \mathbf{n}\, dS$$

Identity 4

$$\delta(z - b) = 0 \qquad \text{when } z \neq b \text{ and } \int_{-\infty}^{+\infty} \delta(z)dz = 1$$

Identity 5

$$\mathrm{v}(z)\delta(z - b) = \mathrm{v}(b)\delta(z - b)$$

Identity 6

$$\int_{-\infty}^{+\infty} \mathrm{v}(z)\delta(z - b)dz = \mathrm{v}(b)$$

NOTE 2: A MATHEMATICA PROGRAM FOR DETERMINATION OF TEMPERATURE IN EXAMPLE 10.2

This Mathematica program (Wolfram, 2005) below determines eigenvalues, eigenvectors, and temperature as described in Example 2.

```
(*INPUT DATA, Part 1*)
n=15;l1=0;lb=1;cap=1;k=1;
(*Basis functions*)
```

```
fj=x*(1-x)*x^(j-1); fi=x*(1-x)*x^(i-1);
(*Determination of the Matrices A and B, Part 2*)
Amat=Table[Integrate[fi*(D[k*D[fj,x],x]),x,l1,lb],i,1,n,j,1,n];
Bmat=Table[Integrate[cap*fi*fj,x,l1,lb],i,1,n,j,1,n];
(*Calculation of Eigenvalues and Matrices D and P, Part 3*)
Amat=N[Amat,48]; Bmat=N[Bmat,48];
Lmat=CholeskyDecomposition[Bmat];
LmatT=Transpose[Lmat];
linv=Inverse[bbt]; linvt=Inverse[bb1];
Abar=-(Inverse[LmatT]).Amat.(Inverse[Lmat]);
Eigv=N[Eigenvalues[Abar],20];
Dmat=Transpose[(Inverse[Lmat]).Transpose[Eigenvectors[Abar]]];
Pmat=Inverse[Transpose[Dmat.Bmat]];
(*Calculation of Temperature, Part 4*)
temp=0;
Do[psi[ne]=Sum[Dmat[[ne,j]]*fj,j,1,n],ne,1,n];
temp=Sum[psi[ne]*Exp[-Eigv[[ne]]*t]*Sum[Pmat[[ne,i]]*Integrate[cap*fi,x,l1,lb],{i,1,n}],{ne,1,n}];
```

Below is a modification for Part 3 when Equation 10.11 is written as for determination of the eigenvalues and eigenvectors. To get proper matrix inversion in this simplification, listed below, every term within an assigned 15´15 matrix **B** must be computed with more that 640 significant figures, instead of 48 in the program above that uses the Cholesky decomposition technique.

```
Amat=N[Amat,641];Bmat=N[Bmat,641];
Abar=-Inverse[Bmat].Amat; Eigv=N[Eigenvalues[Abar],20]
Dmat=Eigenvectors[Abar]; Pmat=Inverse[Transpose[Dmat.Bmat]];
```

REFERENCES

Carnahan, B., Luther, H. A., and Wilkes, J. O., 1969, *Applied Numerical Methods*, John Wiley, New York, p. 250.

Forsythe, G. and Moler, C. B., 1967, *Computer Solution of Linear Algebraic Systems*, Prentice-Hall, Englewood Cliffs, N.J.

Haji-Sheikh, A., 1988, Heat diffusion in heterogeneous media using heat-flux-conserving basis functions, *ASME J. Heat Transfer*, vol. 110, pp. 276–282.

Haji-Sheikh, A. and Beck, J.V., 1988, Green's function partitioning in Galerkin-based solution of the diffusion equation, in *Proceedings of the 1988 National Heat Transfer Conference*, ed. H. R. Jacobs, ASME HTD-96, pp. 257–266 in vol. 3.

Haji-Sheikh, A. and Lakshminarayanan, R., 1987, Integral solution of diffusion equation: Part 2 – Boundary conditions of second and third kinds, *ASME J. Heat Transfer*, vol. 109, pp. 557–562.

Haji-Sheikh, A. and Mashena, M., 1987, Integral solution of diffusion equation: Part 1 – General solution, *ASME J. Heat Transfer*, vol. 109, pp. 551–556.

Haji-Sheikh, A., Yan, L., and Kinsey, S. P., 1991, A single-equation solution for conduction in fins, *Int. J Heat Mass Transfer*, vol. 34, no. 1, pp. 159–165.

Hay, G. E., 1953, *Vector and Tensor Analysis*, Dover, New York.

Kantorovich, L. V. and Krylov, V. I., 1960, *Approximate Methods of Higher Analysis*, John Wiley, New York.

Kaplan, W., 1956, *Advanced Calculus*, Addison-Wesley, Reading, Mass.

LeCroy, R. C. and Eraslan, A. H., 1969, The solution of temperature development in the entrance region of an MHD channel by the G. B. Galerkin method, *ASME J. Heat Transfer*, vol. 91, pp. 212–220.

Ozisik, M. N., 1993, *Heat Conduction*, John Wiley, New York.

Wolfram, S., 2005, *The Mathematica Book*, 5th edition, Cambridge University Press, Cambridge, UK.

11 Applications of the Galerkin-Based Green's Functions

11.1 INTRODUCTION

The Galerkin-based Green's function (GF) solution of the diffusion equation is presented in Chapter 10, which also contains simple one-dimensional examples that demonstrate the method of solution and that discuss the accuracy of the results. In this chapter, the Galerkin-based GF solution method is extended to more advanced problems. Thermal conduction in multidimensional bodies is presented in Section 11.2. In Section 11.3 the basis functions are modified so that conduction in heterogeneous materials can be accommodated. Then, in Section 11.4, the GF solution is developed and applied to steady-state conduction problems. Finally, a study of heat transfer in the thermal entrance region of ducts is included in Section 11.5.

Sections 11.2 through 11.6 each include one or more examples that demonstrate the procedure. Except for selection of basis functions, the same mathematical procedure applies to simple and complex problems. The major difficulty in dealing with complex problems is the selection of a complete and linearly independent set of basis functions that satisfy the boundary conditions. Unlike the one-dimensional problems studied in Chapter 10, there is no generalized form for the basis functions; hence, each multidimensional body must be treated differently.

The GF for a few orthogonal multidimensional bodies are products of one-dimensional GFs. However, it is not difficult to define basis functions for most orthogonal multidimensional bodies. The basis functions are usually the products of one-dimensional basis functions. It is also possible to find basis functions for irregular bodies when boundary conditions are of the first kind. However, the basis functions for nonorthogonal bodies with boundary conditions of second and third kinds are sometimes difficult to obtain. Once the basis functions are defined, the computation of matrices **A** and **B** may require numerical integration. After the matrices **A** and **B** are determined, the calculation of parameters in the GF is exactly the same as that for one-dimensional bodies. Indeed, the matrix algebra, the GFs, and the GF solution method are the same for one-dimensional or multidimensional, and orthogonal or nonorthogonal bodies. In this chapter, emphasis is placed on finding the basis functions.

11.2 BASIS FUNCTIONS IN SOME COMPLEX GEOMETRIES

As discussed earlier, the derivation of the GF solution method is in Chapter 10. The algebraic steps leading to the computation of the GF and the GF solution method are the same for bodies of different shapes. The method of finding the basis functions and the analytical (or numerical) efforts needed to compute matrices **A** and **B** elucidate the complexity of the problem. The procedure for finding the basis functions is not unique and any properly defined basis functions, as discussed in Section 10.2, are acceptable; a few methods of selecting basis functions for nonorthogonal bodies are discussed. The procedure includes the basis functions that satisfy boundary conditions of the first kind (prescribed temperature, $f_j = 0$), the second kind (prescribed heat flux, $\partial f_j / \partial n = 0$), or the third kind (convective, $-k\partial f_j / \partial n = hf_j$).

11.2.1 Boundary Conditions of the First Kind

A universal relation to give a_{ij} and b_{ij} for many regular geometries is derived in Chapter 10. For more complex geometries, the necessary integrations using Equations 10.12 and 10.13 may require a symbolic software or numerical quadrature. By using time partitioning, it may be possible to reduce the computations in these cases. The remaining matrix operation is independent of the dimensions of the body and the boundary conditions. When a multidimensional body has a regular shape, the GF is a product of one-dimensional GFs. To obtain a reasonably accurate solution for irregular multidimensional bodies, the number of basis functions is usually larger than 2. Numerical matrix operation becomes necessary when dealing with complex multidimensional problems.

The method of selecting the basis functions for boundary conditions of the first kind is available in the literature (Kantorovich and Krylov, 1960; Ozisik, 1993; Haji-Sheikh and Mashena, 1987). If a region is bounded by M surfaces $\phi_1, \phi_2, \ldots, \phi_M$ (Figure 11.1), the first member of the set of basis functions is

$$f_1(\mathbf{r}) = \phi_1\phi_2\phi_3 \cdots \phi_M \tag{11.1}$$

Each subsequent member of the set of basis functions is obtained by multiplying $f_1(\mathbf{r})$ by an element of a complete set, for example, in a Cartesian coordinate system

$$f_2(\mathbf{r}) = f_1(\mathbf{r})x \tag{11.2a}$$

$$f_3(\mathbf{r}) = f_1(\mathbf{r})y \tag{11.2b}$$

$$f_4(\mathbf{r}) = f_1(\mathbf{r})z \tag{11.2c}$$

$$f_5(\mathbf{r}) = f_1(\mathbf{r})x^2 \tag{11.2d}$$

$$f_6(\mathbf{r}) = f_1(\mathbf{r})xy \tag{11.2e}$$

$$f_7(\mathbf{r}) = f_1(\mathbf{r})xz \tag{11.2f}$$

$$f_8(\mathbf{r}) = f_1(\mathbf{r})y^2 \tag{11.2g}$$

$$f_9(\mathbf{r}) = f_1(\mathbf{r})yz \tag{11.2h}$$

$$f_{10}(\mathbf{r}) = f_1(\mathbf{r})z^2 \tag{11.2i}$$

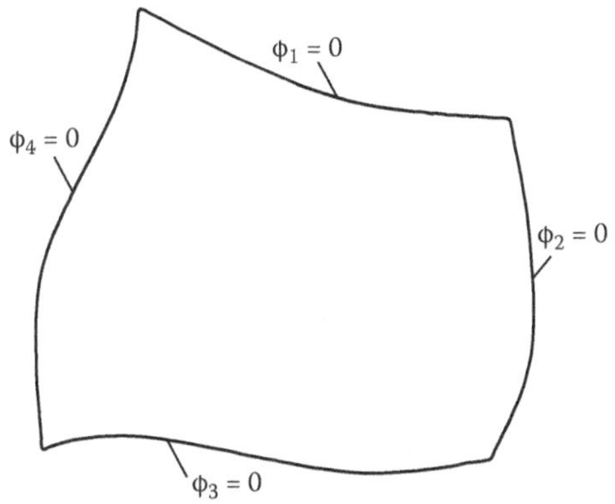

FIGURE 11.1 Generalized configuration for use with Equation 11.1.

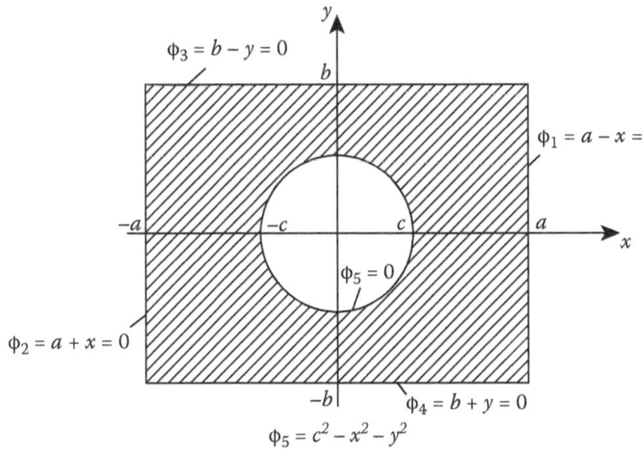

FIGURE 11.2 Two-dimensional body for Example 11.1 with boundary conditions of the first kind.

Each basis function is required to vanish only over the exterior boundaries. Some, but not all, basis functions may become zero at any interior point. This can be ensured if $f_1(\mathbf{r})$ is not zero within the region. Whenever all basis functions vanish at an interior point, the region can be subdivided into different subregions. The basis functions are constructed for each subregion and then are matched at the common boundary of the subregions (Kantorovich and Krylov, 1960).

Example 11.1:

Consider a two-dimensional solid bounded by the surfaces $a - x = 0, a + x = 0, b - y = 0, b + y = 0$, and a circular surface $c^2 - x^2 - y^2 = 0$ (Figure 11.2) and find the basis functions.

Solution

The first basis function, for this example, is a product of all the functions representing the surfaces of this body,

$$f_1 = (a^2 - x^2)(b^2 - y^2)(c^2 - x^2 - y^2) \tag{11.3a}$$

The other basis functions are obtained when f_1 is multiplied by polynomial terms in the ascending order (1 has already been used),

$$f_2 = (a^2 - x^2)(b^2 - y^2)(c^2 - x^2 - y^2)x \tag{11.3b}$$

$$f_3 = (a^2 - x^2)(b^2 - y^2)(c^2 - x^2 - y^2)y \tag{11.3c}$$

$$f_4 = (a^2 - x^2)(b^2 - y^2)(c^2 - x^2 - y^2)x^2 \tag{11.3d}$$

$$f_5 = (a^2 - x^2)(b^2 - y^2)(c^2 - x^2 - y^2)xy \tag{11.3e}$$

$$f_6 = (a^2 - x^2)(b^2 - y^2)(c^2 - x^2 - y^2)y^2 \tag{11.3f}$$

In the absence of the circular hole within this rectangular body, clearly the basis functions are the products of one-dimensional basis functions; see Problem 11.1 in this chapter.

Example 11.2:

Calculate temperature distribution in a spheroidal body (see the inset of Figure 11.3) whose contour is defined by

$$1 - r^2 - \frac{z^2}{b^2} = 0 \tag{11.4a}$$

The initial temperature is zero and, at $t \geq 0$, the surface temperature is equal to one.

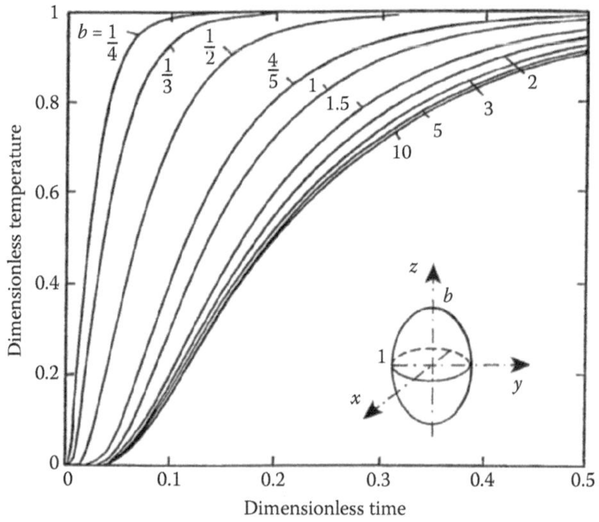

FIGURE 11.3 Example 11.2 results for temperature at $r = 0$ versus $\alpha t/a^2$.

Solution

First, the function f_1 is chosen as

$$f_1 = 1 - r^2 - \frac{z^2}{b^2}$$

and then the remaining basis functions are

$$f_2 = \left(1 - r^2 - \frac{z^2}{b^2}\right) r^2 \tag{11.4b}$$

$$f_3 = \left(1 - r^2 - \frac{z^2}{b^2}\right) z^2 \tag{11.4c}$$

$$f_4 = \left(1 - r^2 - \frac{z^2}{b^2}\right) r^4 \tag{11.4d}$$

$$f_5 = \left(1 - r^2 - \frac{z^2}{b^2}\right) r^2 z^2 \tag{11.4e}$$

$$f_6 = \left(1 - r^2 - \frac{z^2}{b^2}\right) z^4 \tag{11.4f}$$

The spheroidal solid is homogeneous with constant thermophysical properties. The coordinates and time are viewed as dimensionless. Using Equations 10.12 and 10.13, the elements of the matrices **A** and **B** for a three-term solution are (Haji-Sheikh, 1986)

$$a_{11} = \frac{4(2b^2+1)}{5b^2} \quad a_{12} = \frac{8(2b^2+1)}{35b^2} \quad a_{13} = \frac{4(2b^2+1)}{35}$$

$$a_{22} = \frac{32(4b^2+1)}{(315b^2)} \quad a_{23} = \frac{8(2b^2+1)}{315} \quad a_{33} = \frac{4b^2(4b^2+11)}{315}$$

and

$$b_{11} = \frac{8}{35} \quad b_{12} = \frac{16}{315} \quad b_{13} = \frac{8b^2}{315} \quad b_{22} = \frac{64}{3465}$$

$$b_{23} = \frac{16b^2}{3465} \quad b_{33} = \frac{8b^4}{1155}$$

Symbolic computer programming was used to calculate the integrals. Each of these values is divided by the volume of the spheroid. Additionally, the integration of the function f_i over the volume is needed in the GF solution method, Equation 10.77. The corresponding values, after they are divided by the volume of the spheroid, are 2/5, 4/35, and $2b^2/35$.

The steps for a one-eigenvalue solution are discussed mainly to show that the procedure is independent of the shape of the domain and complexity of the problem. Equation 10.11 yields the eigenvalue $\gamma_1 = a_{11}/b_{11} = 7(2b^2+1)/(2b^2)$ and the eigenvector $d_{11} = 1$. Equations 10.41 and 10.45 yield $p_{11} = 35/8$ and the GF using Equation 10.60 is

$$G(r', z', t|r, z, \tau) = \frac{35}{8} f_1(r, z) f_1(r', z') \exp[-\gamma_1(t - \tau)]$$

Finally, the solution using Equation 10.77 is

$$T(r, z, \tau) = 1 - \frac{7}{4}\left(1 - r^2 - \frac{z^2}{b^2}\right) \exp\left[-\frac{7(2b^2 + 1)t}{2b^2}\right]$$

A one-eigenvalue solution is a crude approximation to the exact solution. A 10-term solution yields four accurate digits except at small time. For example when $b = 2$ and $t = 0.3$ the exact solution is 0.7758, while a one-term solution gives 0.835 and a three-term solution is 0.780.

A desktop computer is adequate to perform similar calculations using more eigenvalues. For instance, the calculated values of temperature at the point (0, 0) are computed with speed and efficiency using a small personal computer, and the results are plotted in Figure 11.3. As many as 21 eigenvalues are used to compute the data. The large-time data agree with the exact solution within five significant digits; this accuracy diminishes as t becomes small (Haji-Sheikh, 1986).

11.2.2 Boundary Conditions of the Second Kind

We now focus on the insulated boundaries. The selection of the basis functions becomes simple if a flat section of boundary is insulated. As an illustration, Figure 11.4b shows a flat section of the boundary described by $\phi_1 = 0$, which is insulated. For this planar surface, a condition of symmetry about that surface is implied. Then, the original region can be replaced by a new region that includes itself and its mirror image (Figure 11.4a). If, for instance, x is selected perpendicular to the $\phi_1 = 0$ surface, then $\phi_3 = 0$ is $\phi_2 = 0$ except the variable x is replaced by $-x$. Therefore, the basis function f_1 is obtained using the boundary conditions of the first kind, by utilizing Equation 11.1, as $f_1 = \phi_2\phi_3$. Then, the remaining basis functions are defined by

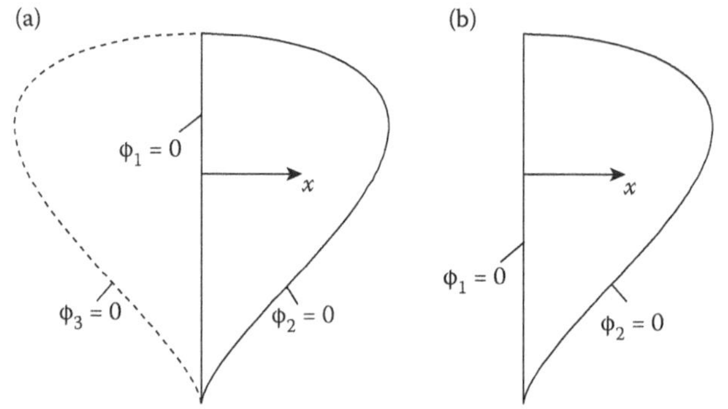

FIGURE 11.4 Body with flat surface insulated at $\phi_1 = 0$.

using Equation 11.2 and retaining the terms with x to the power of even numbers. This will automatically result in $\partial f_j / \partial n = 0$ along the x-axis on the $\phi_1 = 0$ line for all the basis functions.

There are other methods of finding the basis functions that satisfy the boundary conditions of the second kind (Lee and Haji-Sheikh, 1991). One way of finding the basis functions for boundary conditions of the second kind can be illustrated through a simple example. Consider that the geometry depicted in Figure 11.4b has the following boundary conditions: $f_j = 0$ on the $\phi_1 = 0$ line and $\partial f_j / \partial n = 0$ on the $\phi_2 = 0$ line. The basis function $f_j^{(2)}$ that satisfies the boundary conditions of the second kind is considered to be of the form

$$f_j^{(2)} = f_j^{(1)}(\phi_2 H - 1) \tag{11.5}$$

The term -1 in the parentheses is for the convenience of analysis and has no effect on the final solution because $f_j^{(2)}$ can be multiplied by a constant without loss of generality, and H is yet to be determined. The function $f_j^{(1)}$ satisfies the boundary condition of the first kind everywhere except on the $\phi_2 = 0$ surface, which is insulated

$$f_j^{(1)} = \phi_1 x^{m_j} y^{n_j} z^{l_j} \tag{11.6}$$

Since the surface $\phi_2 = 0$ is insulated, then the relation $\partial f_j^{(2)} / \partial n = 0$ on the $\phi_2 = 0$ surface requires that

$$\frac{-\partial f_j^{(1)}}{\partial n} + f_j^{(1)} \left(\frac{\partial \phi_2}{\partial n} \right) H = 0 \qquad \text{on} \quad \phi_2 = 0 \text{ surface} \tag{11.7}$$

which yields a relation for function H to be used in Equation 11.5 as

$$H = \left(\frac{\partial f_j^{(1)} / \partial n}{f_j^{(1)} \partial \phi_2 / \partial n} \right) \Bigg|_{\phi_2=0} = \left(\frac{\nabla f_j^{(1)} \cdot \nabla \phi_2}{f_j^{(1)} \nabla \phi_2 \cdot \nabla \phi_2} \right) \Bigg|_{\phi_2=0} \tag{11.8}$$

For some geometric configurations, it is possible to define a set of basis functions in the polynomial form with free constants. The free constants can be evaluated so that $\nabla f_j \cdot \nabla \phi = 0$ on the $\phi = 0$ surface. This procedure is described in a forthcoming example. However, no established method is presently available to determine the basis functions for all different-shaped bodies with some (or all) walls insulated.

Example 11.3:

Consider a homogeneous spheroidal solid, Figure 11.5, whose boundary in the cylindrical coordinate system is given by the equation $\phi_1 = 1 - r^2 / a^2 - z^2 / b^2$ and find the basis functions when the external surface is insulated.

Solution

Although this is a regular or orthogonal body in spheroidal coordinates, it is a non-orthogonal body in the cylindrical coordinates. The temperature is independent of

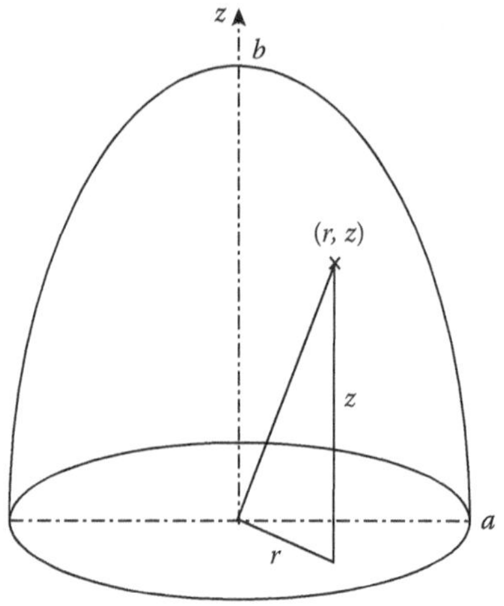

FIGURE 11.5 Spheroidal body for Examples 11.3 and 11.4.

the angular coordinate and is symmetric about the $z = 0$ plane. Basis functions for boundary conditions of the second kind are (Haji-Sheikh and Lakshminarayanan, 1987)

$$f_j^{(2)} = r^{m_j} z^{n_j} (B_1 r^2 + B_2 z^2 + B_3) \qquad j = 1, 2, \ldots, N \tag{11.9}$$

where $m_j = 0, 2, 4, \ldots$, and $n_j = 0, 2, 4, \ldots$. However, if the temperature is not symmetric about the $z = 0$ plane, the odd n_j's must be included. The basis functions, Equation 11.9, must satisfy the boundary condition

$$\frac{\partial f_j^{(2)}}{\partial n} = 0 \qquad \text{when } \phi_1 = 0 \tag{11.10}$$

which can be written as

$$\nabla f_j^{(2)} \cdot \nabla \phi_1 = 0 \qquad \text{when } \phi_1 = 0 \tag{11.11}$$

Introducing Equation 11.9 into Equation 11.11 and deleting z using the relation $\phi_1 = 0$, the following second-degree polynomial equation is obtained:

$$\left[B_1\left(\frac{m_j+2}{a^2} + \frac{n_j}{b^2}\right) - B_2\left(\frac{m_j}{a^2} + \frac{n_j+2}{b^2}\right)\frac{b^2}{a^2}\right] r^2 + \left[B_2\left(\frac{m_j}{a^2} + \frac{n_j+2}{b^2}\right) b^2 + B_3\left(\frac{m_j}{a^2} + \frac{n_j}{b^2}\right)\right] r^0 = 0 \tag{11.12}$$

Since Equation 11.12 is satisfied at all r's, then the coefficients that multiply by r to any power must be zero, or

$$B_2 = -\frac{B_3(m_j/a^2 + n_j/b^2)}{b^2(m_j/a^2 + (n_j+2)/b^2)} = B_3\left[\frac{2/b^2}{m_j b^2/a^2 + n_j + 2} - \frac{1}{b^2}\right] \tag{11.13}$$

and

$$B_1 = \frac{B_2 b^2/a^2[m_j/a^2 + (n_j+2)/b^2]}{(m_j+2)/a^2 + n_j/b^2}$$
$$= B_3 \frac{1}{a^2}\left[\frac{2b^2/a^2}{(m_j+2)b^2/a^2 + n_j} - 1\right] \qquad (11.14)$$

One of the three coefficients can be selected arbitrarily (e.g., $B_3 = 1$).

11.2.3 Boundary Conditions of the Third Kind

We will demonstrate that the basis functions satisfying the boundary conditions of the third kind can be constructed from the basis functions that satisfy the boundary conditions of the second kind. For this presentation, $f_j^{(2)}$ will designate the basis functions that satisfy the boundary conditions of the second kind on the $\phi_1 = 0$ surface. The basis functions satisfying the boundary conditions of the third kind are obtained from the simple relation,

$$f_j^{(3)} = f_j^{(2)}\left(\phi_1 H - \frac{k}{h}\right) \qquad j = 1, 2, \ldots, N \qquad (11.15)$$

The method of calculating H in Equation 11.15 is similar to that for Equation 11.5. The function $f_j^{(3)}$ must satisfy the relation $-k\partial f_j^{(3)}/\partial n = hf_j^{(3)}$ on the surface $\phi_1 = 0$. This leads to

$$-kf_j^{(2)}\frac{H\partial\phi_1}{\partial n} = hf_j^{(2)}\left(\frac{-k}{h}\right) \qquad \text{when } \phi_1 = 0 \qquad (11.16)$$

The function H to be used in Equation 11.15 then becomes

$$H = \left(\frac{1}{\partial\phi_1/\partial n}\right)\Bigg|_{\phi_1=0} \qquad (11.17)$$

in which $\phi_1 = 0$ designates the convective surface. It is also possible to obtain, for some geometries, the basis functions using series expansion as discussed for boundary conditions of the second kind.

Example 11.4:

Spheroidal bodies with a convective surface have many interesting applications in aerospace, food, and agricultural industries. The spheroidal body defined in Example 11.3 is subject to convective boundary conditions. The initial temperature is T_0 and the ambient temperature is T_∞ when $t \geq 0$. Calculate the temperature at the point $r = 0$ and $z = 0$.

Solution

Although, in theory, a spheroid with a convective surface submits to an exact solution, such an exact solution has not been found. This is due to the complexity of the exact mathematical and subsequent numerical procedures. Using Equation 11.15, it is possible to find a set of basis functions that satisfy the convective boundary conditions. After $f_j^{(2)}$ from Equation 11.9 is inserted into Equation 11.15, the basis functions for convective spheroids are

$$f_j^{(3)} = f_j^{(2)} \left\{ \frac{\text{Bi}[(b^2/a^2)(1 - r^2/a^2) - z^2/a^2]}{2(b/a)[(b^2/a^2)(r^2/a^2) - r^2/a^2 + 1]^{1/2}} - 1 \right\} \tag{11.18}$$

Next, the function $f_j^{(3)}$ must replace f_j in Equations 10.12 and 10.13 to compute matrices **A** and **B**. The analytical integrations of the resulting equations, if possible, are complicated and are not cost effective. Numerical quadrature was used by Haji-Sheikh and Lakshminarayanan (1987) to compute a_{ij}'s and b_{ij}'s. The remaining steps are identical to those described in Example 11.2. In fact, the same computer program is used to solve for temperature here and for Example 11.2, except matrices **A** and **B** in this example are computed numerically, while symbolic computer algebra was used to calculate **A** and **B** in Example 10.2. The computed temperature results for a range of Biot numbers, ha/k, and aspect ratios are shown in Figure 11.6. The solid lines are generally in good agreement with

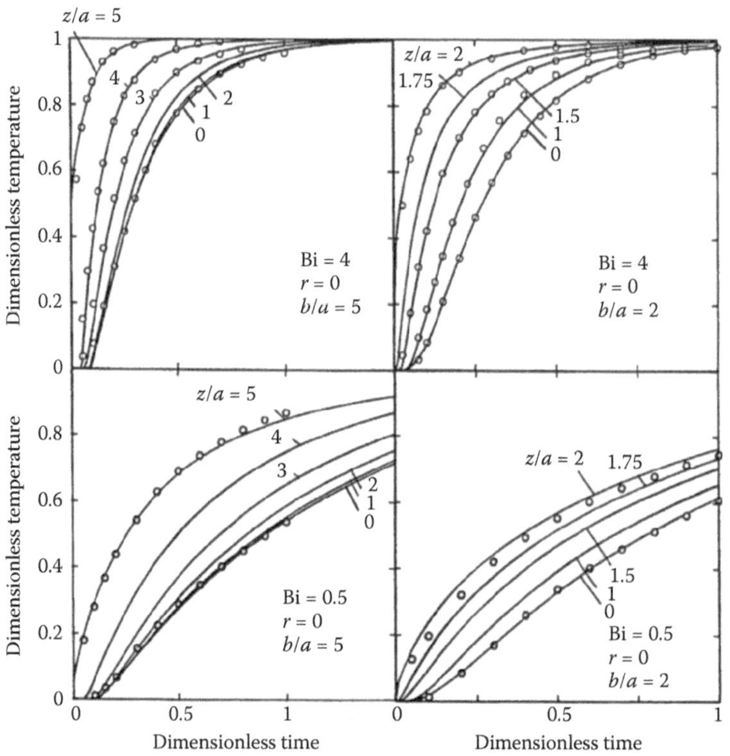

FIGURE 11.6 Example 11.4 results for temperature, $(T - T_0)/(T_\infty - T_0)$, at $r = 0$ versus $\alpha t/a^2$.

the Monte Carlo data (Haji-Sheikh and Sparrow, 1967). Previous comparisons, in Example 10.2, with the exact solution imply that any small discrepancy can be attributed to the sampling error in the Monte Carlo solution.

11.3 HETEROGENEOUS SOLIDS

The derivation of the GF solution, Equation 10.68 or the alternative GF solution, Equation 10.77 permits the computation of thermal conduction in heterogeneous bodies. The only difference between solutions for homogeneous and heterogeneous solids is the selection of a set of basis functions. First, a set of basis functions must be defined that satisfies the boundary conditions on the external surfaces and perfect or imperfect contact relations between adjacent materials. Then, the basis functions are used to solve a numerical example.

Let the subscript e identify an inclusion of different material enclosed in the main body (Figure 11.7), and let m denote the main domain. The basis function $f_{j,m}$, which satisfies the boundary conditions of the main body, is selected ignoring the inclusion; therefore, f_j is $f_{j,m}$ in the main body. However, the basis function should be modified as it crosses the boundary of the inclusion. The formulation of the basis functions in the absence of contact conductance is given by Haji-Sheikh (1988). The formulation is then modified to include the effect of finite contact conductance as (Haji-Sheikh and Beck, 1990)

$$f_j = f_{j,m} \qquad \text{(in the main domain)} \tag{11.19a}$$

and

$$f_j = f_{j,m} + U + \phi_e H \qquad \text{(in the } i\text{th inclusion)} \tag{11.19b}$$

for $j = 1, 2, \ldots, N$. The continuity condition that $k_m(\partial f_j / \partial n)_m = k_e(\partial f_j / \partial n)_e$ and the jump condition $f_{j,e} = f_{j,m} - (k_m / C)(\partial f_{j,m} / \partial n)$ at the boundary of the inclusion ($\phi_e = 0$ surface is different for different inclusions) permit the calculation of U and H as

$$U = -\left(\frac{k_m}{C}\right)\left(\frac{\partial f_{j,m}}{\partial n}\right)\bigg|_{\phi_e=0} \tag{11.20}$$

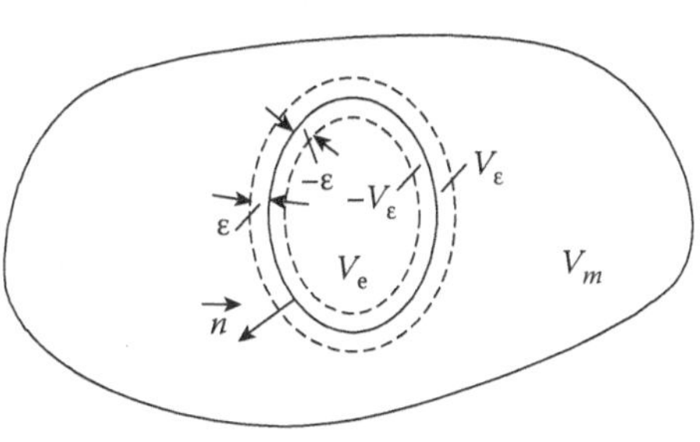

FIGURE 11.7 Composite body with inclusion.

and

$$H = \frac{[(\nabla f_{j,m} \cdot \nabla \phi_e)|_{\phi_e=0}(k_m / k_e - 1) - \nabla U \cdot \nabla \phi_e|_{\phi_e=0}]}{(\nabla \phi_e \cdot \nabla \phi_e)|_{\phi_e=0}} \tag{11.21}$$

where C is the contact conductance. A linear combination of the basis function f_j satisfies the continuity of heat flux, $k_m(\partial T / \partial n)_m = k_e(\partial T / \partial n)_e$, and temperature jump, $T_e = T_m - (k_m / C)(\partial T_{j,m} / \partial n)$, on the boundary of inclusion e. When the inclusion has other boundaries in addition to the $\phi_e = 0$ surface, other modifications to the values of U and H become necessary (e.g., see Example 11.5).

Example 11.5:

To illustrate the method for accommodating the contribution of contact conductance, consider two plates: one has a thickness of a and the other $L - a$ (see Figure 11.8). It is convenient to let subscripts e and m stand for the regions designated using these letters in Figure 11.8. The composite slab is initially at temperature T_0 and has the following boundary conditions

$$-k_m \frac{\partial T}{\partial x} = hT \qquad \text{at } x = L \text{ and when } t > 0 \tag{11.22a}$$

and

$$\frac{\partial T}{\partial x} = 0 \qquad \text{at } x = 0 \text{ and when } t > 0 \tag{11.22b}$$

Write an equation for temperature distribution.

Solution

The basis function, f_j, that satisfies the conditions $-k_m \partial f_{j,m}/\partial x = h f_{j,m}$ at $x = L$ and $\partial f_{j,m}/\partial x = 0$ at $x = 0$, for $j = 1, 2, 3, \ldots,$ is

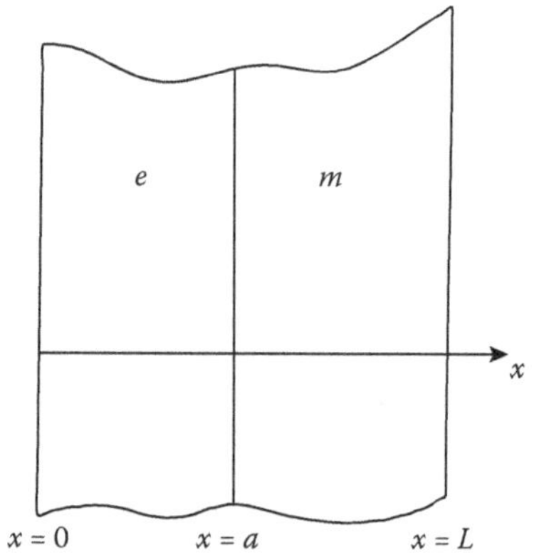

FIGURE 11.8 Two-layer wall insulated at $x = 0$ and convective surface at $x = L$ for Example 11.5.

$$f_j = f_{j,m} = \left(\delta_j - \frac{x^2}{a^2}\right)\left(\frac{x}{a}\right)^{2(j-1)} \qquad a < x < L \tag{11.23a}$$

where $\delta_j = (L/a)^2(L/a + 2j/\text{Bi})/[L/a + 2(j-1)/\text{Bi}]$ and $\text{Bi} = ha/k_m$. A function $\phi(x) = 1 - x^2/a^2$ is selected to satisfy the boundary condition given by Equation 11.22b and to vanish at $x = a$; then

$$f_j = f_{j,e} = \left(\delta_j - \frac{x^2}{a^2}\right)\left(\frac{x}{a}\right)^{2(j-1)} + U + \left(1 - \frac{x^2}{a^2}\right)H \qquad 0 < x < a \tag{11.23b}$$

The value of $U = 2[j - (j-1)\delta_j]/(Ca/k_m)$ is computed so that the equation $f_{j,e} = f_{j,m} - (k_m/C)(\partial f_{j,m}/\partial x)$ at $x = a$ is satisfied, where C is the contact conductance. The continuity of heat flux at $x = a$ yields $H = (k_m/k_e - 1)[j - (j-1)\delta_j]$. The solution when $N = 2, L/a = 2, k_e/k_m = 2, ha/k_m = 1, Ca/k_m = 1$, and $\rho_e c_{pe}/\rho_m c_{pm} = 1$ is

$$\frac{T}{T_0} = 0.15048\psi_1(x)\exp\left(-\frac{1.5407\alpha t}{a^2}\right) + 0.006762\psi_2(x)\exp\left(-\frac{11.449\alpha t}{a^2}\right) \tag{11.24}$$

where

$$\psi_1(x) = f_1 - 0.040305 f_2 \tag{11.25a}$$

$$\psi_2(x) = f_1 + 20.999 f_2 \tag{11.25b}$$

When $x \leq a$ (that is, in region e), the functions $f_{1,e}$ and $f_{2,e}$ replace the f_1 and f_2 functions. However, when $x \geq a$, the functions $f_{1,m}$ and $f_{2,m}$ replace the f_1 and f_2 functions.

Whenever the boundary condition at $x = 0$ is convective, U must also satisfy the convective boundary condition at $x = 0$. In addition, the coefficient $(1 - x^2/a^2)$ that multiplies H should be replaced by a function that becomes 0 when $x = a$ and satisfies the convective condition at $x = 0$.

When calculating a_{ij} from Equation 10.12, the function f_j suffers a step change at and along the contact surface. The derivatives of f_j across the inclusion boundary are singular, and it can be shown that the value of the volume integral over the singularity zone is zero. The integral given in Equation 10.12 for this example is

$$\int_V f_i \nabla\cdot(k\nabla f_j)dV = \int_{V_e - V_\epsilon} f_{i,e}\nabla\cdot(k\nabla f_{j,e})\,dV + \int_{V_e - V_\epsilon}^{V_e + V_\epsilon} f_i\,\nabla\cdot(k\nabla f_j)\,dV + \int_{V - V_e - V_\epsilon} f_{i,m}\nabla\cdot(k\nabla f_{j,m})\,dV \tag{11.26}$$

The contact zone (Figure 11.7) is divided into $+V_\epsilon$ and $-V_\epsilon$. The first integration on the right side is over the inclusion up to the contact zone. The second integration on the right side is over the contact zone, and the last integration is over the entire domain less the inclusion and the contact zone. It can be shown that, as $\epsilon \to 0$, the first and third integrals yield the value of a_{ij} if the second integral (over the contact zone) on the right side of Equation 11.26 vanishes. Assuming the thickness of the contact zone is extremely small, it is possible to ignore the derivatives of f_j in the directions perpendicular to the normal to the contact surface **n** (see Figure 11.7). Then, the second term on the right side is integrated by parts

$$\int_{-\epsilon}^{+\epsilon} f_i \frac{d}{dn}\left(k\frac{df_j}{dn}\right) dn = f_i\left(\frac{kdf_j}{dn}\right)\bigg|_{-\epsilon}^{+\epsilon} - \int_{-\epsilon}^{+\epsilon} k\left(\frac{df_i}{dn}\right)\left(\frac{df_j}{dn}\right) dn \tag{11.27}$$

At the limit as $\epsilon \to 0$, Equation 11.27 reduces to

$$\lim_{\epsilon\to 0} q_j[(f_{i,m}(\epsilon) - f_{i,e}(-\epsilon)] - q_j[(f_{i,m}(\epsilon) - f_{i,e}(-\epsilon)] = 0 \tag{11.28}$$

where $q_j = k_m \partial f_{j,m}/\partial n = k_e \partial f_{j,e}/\partial n$ is a constant in V_ϵ.

When f_j or its normal derivative on the exterior surfaces is zero, it is possible to substitute Equation 10.14 in Equation 10.12 to obtain

$$a_{ij} = -\int_V k\nabla f_i \cdot \nabla f_j \, dV \tag{11.29}$$

If this equation is utilized to compute a_{ij} instead of Equation 10.12, the value of the integral over the contact zone, where the derivatives of f_j are singular, is not zero and should be evaluated.

At this stage, it is appropriate to solve a three-dimensional nonorthogonal problem to illustrate the strength of this Galerkin-based integral (GBI) solution. The following example does not have an exact solution and numerical computation of the temperature is a formidable task. The procedure discussed in Example 11.5 is applied to a more complex problem to demonstrate the possibility of accommodating difficult thermal conduction problems.

Example 11.6:

Consider a spherical inclusion whose radius is equal to a, centrally located in a cubical body with dimensions $2b \times 2b \times 2b$ (Figure 11.9). The initial temperature is 0 and the external surface temperature is maintained at 1 when $t \geq 0$. Find the temperature distribution.

Solution

Although the shape of the body is simple, it contains all the complexities one expects in a conduction problem. The procedure described in this example and the previous example can be applied to other geometries. The basis functions are

$$f_{j,m} = (b^2 - x^2)(b^2 - y^2)(b^2 - z^2)x^{m_j}y^{n_j}z^{l_j} \tag{11.30}$$

in which m_j, n_j, and l_j take values of $0, 1, 2, \ldots$. The function ϕ_e is given by

$$\phi_e = a^2 - x^2 - y^2 - z^2 \tag{11.31}$$

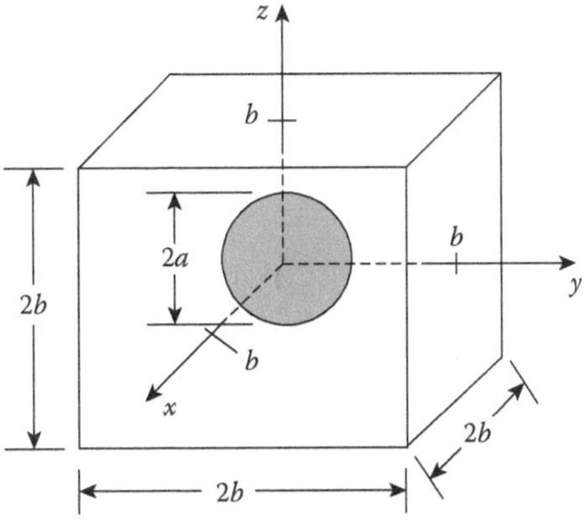

FIGURE 11.9 Cubical body with centrally located spherical inclusion.

which unconditionally vanishes on the surface of the inclusion. The function f_j in the inclusion is $f_{j,e}$ obtained from Equation 11.19b assuming a perfect contact between materials ($U = 0$). Due to symmetry, only even values of m_j, n_j, and l_j need be considered. However, for a one-term solution, $m_j = n_j = l_j = 0$, the functions $f_{1,m}$, $f_{1,e}$ are

$$f_{1,m} = (b^2 - x^2)(b^2 - y^2)(b^2 - z^2) \tag{11.32a}$$

$$\begin{aligned} f_{1,e} = f_{1,m} &+ \left(\frac{k_m}{k_e} - 1\right)(a^2 - x^2 - y^2 - z^2) \\ &\times [x^2(b^2 - y^2)(b^2 - z^2) + y^2(b^2 - x^2)(b^2 - z^2) \\ &+ z^2(b^2 - x^2)(b^2 - y^2)] \end{aligned} \tag{11.32b}$$

A solution with a higher degree polynomial is a four-term solution, and the next higher degree polynomial yields a 10-term solution. Problems with high degree polynomials are ideally suited to symbolic algebra software because the exact integrations leading to the computation of matrices **A** and **B** are repetitive and lengthy. To show the mathematical steps, the elements of matrices **A** and **B**, when $N = 1$, are evaluated using Equations 10.12 and 10.13:

$$a_{11} = I_{a1} + \left(\frac{k_e}{k_m} - 1\right)(I_{a2} - I_{a3}) + \frac{k_e}{k_m}\left(\frac{k_m}{k_e} - 1\right)^2 I_{a4} \tag{11.33}$$

and

$$b_{11} = I_{b1} + \left(\frac{C_e}{C_m} - 1\right) I_{b2} + \frac{C_e}{C_m}\left[\left(\frac{k_m}{k_e} - 1\right) I_{b3} + \left(\frac{k_m}{k_e} - 1\right)^2 I_{b4}\right] \tag{11.34}$$

where C_m and C_e stand for ρc_p of the main region and inclusion, respectively, and the values of I_{a2}, I_{a3}, I_{a4}, I_{b1}, I_{b2}, I_{b3}, and I_{b4} in Equations 11.33 and 11.34 are in Table 11.1. The alternative GF solution, Equation 10.77 is used to compute the

TABLE 11.1
Values of the Integrals in Equations 11.33 through 11.35

Integrals[a]	$b/a = 1.5$	$b/a = 2.0$	$b/a = 2.5$	$b/a = 3$	$b/a = 5$
I_{a1}	-2.214×10^2	-9.321×10^3	-1.695×10^5	-1.814×10^6	-1.389×10^9
I_{a2}	-1.138×10^2	-2.488×10^3	-2.545×10^4	-1.657×10^5	-2.945×10^7
I_{a3}	-1.098×10^2	-2.444×10^3	-2.517×10^4	-1.644×10^5	-2.936×10^7
I_{a4}	-9.063×10^0	-1.014×10^2	-6.374×10^3	-2.820×10^3	-1.751×10^5
I_{b1}	6.643×10^1	4.971×10^3	1.413×10^5	2.177×10^6	4.630×10^9
I_{b2}	3.895×10^1	1.576×10^3	2.566×10^4	2.430×10^5	1.217×10^8
I_{b3}	7.199×10^0	1.505×10^2	1.508×10^3	9.713×10^3	1.699×10^6
I_{b4}	3.828×10^{-1}	4.165×10^0	2.582×10^1	1.134×10^2	6.962×10^3
I_{c1}	1.139×10^1	1.517×10^2	1.130×10^3	5.832×10^3	5.787×10^5
I_{c2}	4.471×10^0	2.864×10^1	1.158×10^2	3.564×10^2	7.981×10^3
I_{c3}	4.104×10^{-1}	1.357×10^0	3.383×10^1	7.092×10^0	5.560×10^1

[a] $I_{a1} = -256a^{13}/255$, $I_{b1} = (8a^5/15)^3$, and $I_{c1} = (2a^3/3)^3$.

temperature distribution

$$\frac{T - T_0}{T_s - T_0} = 1 - p_{11}\left[I_{c1} + \left(\frac{C_e}{C_m} - 1\right) I_{c2} + \frac{C_e}{C_m}\left(\frac{k_m}{k_e} - 1\right) I_{c3}\right] f_1 \exp(-\gamma_1 t) \tag{11.35}$$

The integrals I_{c1}, I_{c2}, and I_{c3} are also presented in Table 11.1. The function f_1 is f_j when $j = 1$ given by Equation 11.30 outside of the inclusion and by Equation 11.32 inside of the inclusion. When the initial temperature $F(\mathbf{r}) = T_0$, the surface temperature T_s is a constant, $b/a = 3$, $k_e/k_m = 10$, and $C_e/C_m = 1$, the following dimensionless parameters are obtained:

$$a_{11} = 1.848 \times 10^6 \tag{11.36a}$$

$$b_{11} = 2.168 \times 10^6 \tag{11.36b}$$

$$\frac{a^2\gamma_1}{\alpha_m} = -\frac{a_{11}}{b_{11}} = 0.8525 \tag{11.36c}$$

$$d_{11} = 1 \tag{11.36d}$$

$$p_{11} = \frac{1}{b_{11}} \tag{11.36e}$$

and the temperature solution using Equations 11.36c through e is

$$\frac{T - T_0}{T_s - T_0} = 1 - 2.687 \times 10^{-3} f_1 \exp\left(-\frac{0.8525\alpha t}{a^2}\right) \tag{11.37}$$

Similar calculations, using many basis functions, were carried out by Nomura and Haji-Sheikh (1988). The computed temperature when $k_e/k_m = 10$ and for

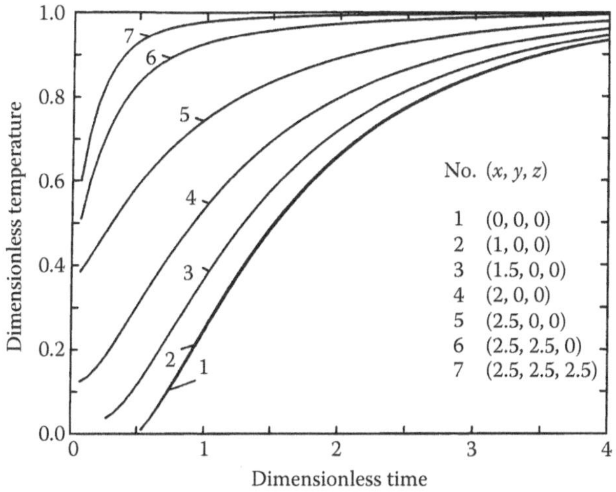

FIGURE 11.10 Dimensionless temperature, $(T - T_0)/(T_s - T_0)$, versus dimensionless time $\alpha t/a^2$ for a sphere in cubical body.

$a = 3$ is shown in Figure 11.10. It is evident that, because of the high thermal conductivity of the inclusion, the temperature change within the inclusion is extremely small. Figure 11.10 shows the temperature at the center of the inclusion, point (0, 0, 0), is nearly the same as the temperature at the contact point (1, 0, 0). The differentiations and integrations required for calculating a_{ij} can be done manually; however, manual integrations are too time consuming. Nomura and Haji-Sheikh (1988) performed the integrations with the aid of the symbolic software, REDUCE-3 (Hearn, 1983). Note that it is mathematically and numerically feasible to add inclusions of various shapes to the main domain.

11.4 STEADY-STATE CONDUCTION

The GFs and GF solutions for steady-state conduction can be deduced by modifying the GF and GF solutions for transient conduction. The steady state is defined as being independent of time. The modification is equally applicable to the GF solution, Equation 10.68, and the alternative GF solution, Equation 10.77. The transient solution approaches the steady-state solution as $t \to \infty$. Accordingly, the contribution of the initial temperature distribution in the GF solution will not influence the steady-state solution. If G_{ss} is defined as the steady-state GF, then, using Equation 10.60,

$$\begin{aligned} G_{ss} &= G(\mathbf{r}'|\mathbf{r}) \\ &= \lim_{t\to\infty} \int_{\tau=0}^{t} G(\mathbf{r}', -\tau|\mathbf{r}, t)\, d\tau \\ &= \sum_{n=1}^{N}\sum_{j=1}^{N}\sum_{i=1}^{N} \frac{d_{nj}\, p_{ni}\, \rho(\mathbf{r}) c_p(\mathbf{r}) f_j(\mathbf{r}') f_i(\mathbf{r})}{\gamma_n} \end{aligned} \tag{11.38}$$

The GF solution then becomes

$$\rho(\mathbf{r})c_p(\mathbf{r})T(\mathbf{r}) = \int_V g(\mathbf{r}')G_{ss}dV' + \int_S k(S')\left(G_{ss}\frac{\partial T}{\partial n} - T\frac{\partial G_{ss}}{\partial n}\right)_{S'} dS' \quad (11.39)$$

Similarly, the alternative GF solution reduces to

$$\rho(\mathbf{r})c_p(\mathbf{r})T(\mathbf{r}) = \rho(\mathbf{r})c_p(\mathbf{r})T^*(\mathbf{r}) + \int_V G_{ss}[g(\mathbf{r}') + f^*]dV' \quad (11.40)$$

One can show analytically that Equation 11.40 reduces to the standard Galerkin solution (Kantorovich and Krylov, 1960)

$$T = T^* - [\{\mathbf{A}^{-1} \cdot \{g^*\}\}^T]\{\mathbf{f}\} \quad (11.41)$$

where $\{\mathbf{f}\}$ is a column vector with elements $f_1, f_2, \ldots, f_N$ and $\{\mathbf{g}^*\}$ is another column vector whose members are

$$g_i^* = \int_V [g(\mathbf{r}') + f^*(\mathbf{r}')]\, f_i(\mathbf{r}')\, dV' \quad (11.42)$$

When the boundary conditions are nonhomogeneous, the standard Galerkin solution of Poisson's equation is possible if an auxiliary function, T^*, exists.

Example 11.7:

Consider a cylindrical pipe with radius $r = a$ centrally placed in a long square box $2b \times 2b$ (Figure 11.11). The boundary conditions are $T = T_0$ at $r = a$ and $T = 0$ at $x = b$ and at $y = b$. Calculate the temperature field and plot the isotherms.

Solution

This example shows the method of calculating the steady-state temperature using the quasisteady temperature T^*. The method is applicable to numerous conduction problems for which an exact solution does not exist. The computation begins by utilizing Equation 10.72b which satisfies the boundary conditions $T = T^* = T_0$ at $r = a$ and $T = T^* = 0$ on the surface of the square box

$$\frac{T^*}{T_0} = 1 - \frac{\ln[(x^2 + y^2)/a^2]}{\ln[(b^2 + x^2y^2/b^2)/a^2]} \quad (11.43)$$

The basis functions are

$$f_j = \left(1 - \frac{x^2}{b^2}\right)\left(1 - \frac{y^2}{b^2}\right)\left(\frac{x^2}{b^2} + \frac{y^2}{b^2} - \frac{a^2}{b^2}\right)\left(\frac{x}{b}\right)^{m_j}\left(\frac{y}{b}\right)^{n_j} \quad (11.44)$$

When $N = 3, a/b = 0.75, g = 0$, and f^* is defined by substituting T^* from Equation 11.43 in Equation 10.73, then Equation 11.42 yields

$$\mathbf{g}^* = \{0.10750, 0.050747, 0.050747\} \quad (11.45)$$

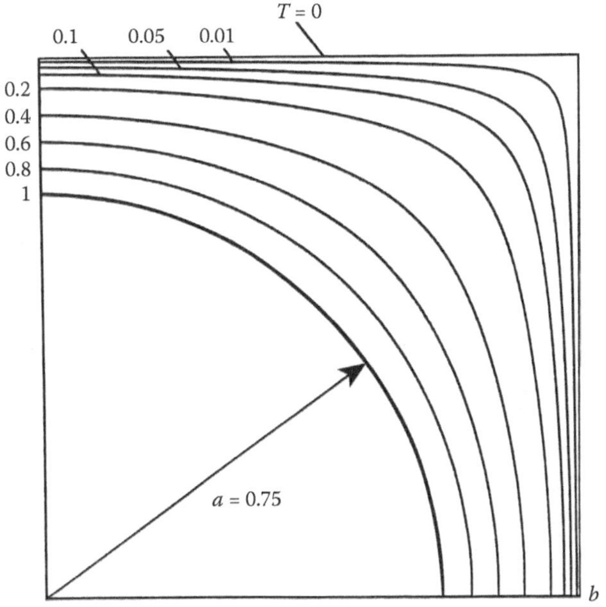

FIGURE 11.11 Long pipe in square box with isotherms in Example 11.7.

Note that when $j = 1, m_j = n_j = 0$; when $j = 2, m_j = 2$ and $n_j = 0$; when $j = 3, m_j = 0$ and $n_j = 2$. After the basis functions defined by Equation 11.44 are inserted in Equation 10.12, matrix **A** becomes (fin effect is neglected)

$$\mathbf{A} = \begin{bmatrix} 0.13268 & 0.065463 & 0.065463 \\ 0.065463 & 0.046750 & 0.024989 \\ 0.065463 & 0.024989 & 0.046750 \end{bmatrix} \tag{11.46}$$

Here, matrices **A** and **B** can be evaluated analytically, but numerical quadrature is needed to evaluate the elements of $\mathbf{g}^*$. Then Equation 11.41 provides the temperature distribution

$$\frac{T}{T_0} = T^* + (b^2 - x^2)(b^2 - y^2)(x^2 + y^2 - a^2)(d_1 + d_2x^2 + d_3y^2) \tag{11.47}$$

where d_2 and d_3 are identical. The isotherms are computed by this method and plotted in Figure 11.11. Table 11.2 also supplies temperature distribution for other values of $b/a = 0.25, 0.5$, and 1.

The availability of an auxiliary function T^* eliminates the need to compute matrix **B** because Equation 11.41 yields the same results as Equation 11.40, yet the number of algebraic and matrix operations are substantially less. However, when T^* is not available, the steady-state formulation of the GF solution, Equation 11.39, should be used.

TABLE 11.2
Coefficients d_1, d_2, and $d_3 = d_2$ in Equation 11.47

d's	$a/b = 0.25$	$a/b = 0.5$	$a/b = 0.75$	$a/b = 0.9$	$a/b = 1$
d_1	0.15837	0.33238	1.1268	4.7624	19.092
d_2	−0.055156	−0.087483	−0.32086	−2.0544	−10.001

11.5 FLUID FLOW IN DUCTS

A knowledge of heat transfer in the entrance region of ducts is essential in the design of compact heat exchangers. The analytical steps described in this section apply to entrance flow in ducts with various cross-sectional shapes; hence, the geometric restriction to obtain a solution is essentially eliminated. The restrictions are that the flow must be hydrodynamically fully developed. The velocity profile is the solution of the momentum equation written for hydrodynamically fully developed, laminar, and Newtonian flow as

$$\frac{\partial^2 W}{\partial X^2} + \frac{\partial^2 W}{\partial Y^2} + 1 = 0 \tag{11.48a}$$

where

$$W = \frac{-w}{(a^2/\mu)(\partial P/\partial z)} \tag{11.48b}$$

P is pressure, w is local velocity in the z-direction, a is the characteristic length, $X = x/a, Y = y/a$, and μ is the viscosity coefficient. After defining the basis functions so that $f_i = 0$ at the wall (note $w = 0$ at the wall), Equation 11.48a yields the value of W using Equation 11.41. The parameters g_i^* are obtained from Equation 11.42 after substituting $g = 1$ and $f^* = 0$ as

$$g_i^* = \frac{1}{A_c}\int_{A_c} f_i \, dA \tag{11.49}$$

where A_c is the cross-sectional area of the duct. Equation 11.41, which is the standard Galerkin method, is used to solve for the velocity distribution. The auxiliary function T^* is zero since the boundary conditions are homogeneous. The $\mathbf{A}^{-1} \cdot \{\mathbf{g}^*\}$ in Equation 11.41 results in coefficients $d_1, d_2, \ldots, d_N$, and the solution for W is

$$W = \sum_{i=1}^{N} d_i f_i \tag{11.50}$$

The standard definition for average velocity is used to calculate the value of W_{av} as

$$W_{av} = \frac{1}{A_c}\int_{A_c} W dA = \sum_{i=1}^{N} d_i g_i^* \tag{11.51}$$

The friction factor C_f is defined as $-D_e(\partial P/\partial z)/(\rho\, w_{\text{av}}^2/2)$ and it can be written as

$$C_f\ \text{Re} = \frac{2D_e^2}{aW_{\text{av}}} = \frac{2\overline{D_e^2}}{W_{\text{av}}} \tag{11.52}$$

where $\text{Re} = \rho D_e w_{\text{av}}/\mu$, and D_e/a is designated as the dimensionless hydraulic diameter. Then, the dimensionless velocity is

$$\frac{W}{W_{\text{av}}} = \frac{w}{w_{\text{av}}} = \frac{C_f \text{Re}}{2D_e^2/a^2} \sum_{j=1}^{N} d_j\, f_j \tag{11.53}$$

After calculating w (or W), attention must be focused on the computation of temperature. The value of $w(\mathbf{r})$ replaces $u(\mathbf{r})$ and z replaces t in Equation 10.11 to yield the energy equation for incompressible fluid flowing at a constant rate in a duct [$g(\mathbf{r}, t) = 0$ and $m(\mathbf{r}) = 0$] as

$$\rho(\mathbf{r})c_p(\mathbf{r})w(\mathbf{r})\frac{\partial T}{\partial z} = \frac{\partial}{\partial x}\left[k(\mathbf{r})\frac{\partial T}{\partial x}\right] + \frac{\partial}{\partial y}\left[k(\mathbf{r})\frac{\partial T}{\partial y}\right] \tag{11.54}$$

Here, the effect of axial conduction is neglected. A solution method that includes the effect of axial conduction is reported in Lakshminarayanan (1988) and Lakshminarayanan and Haji-Sheikh (1988).

The volume integrals in Equations 10.12 and 10.13 become surface integrals once V is replaced by the cross-sectional area A_c, and the variable t is replaced by the axial coordinate z. Equations 10.12 and 10.13 yield the elements of matrices **A** and **B**

$$a_{ij} = \frac{1}{A_c}\int_{A_c} f_i \nabla \cdot (\nabla f_j)\, dA \tag{11.55a}$$

which, for boundary conditions of the first or second kind, can be written as

$$a_{ij} = -\frac{a^2}{A_c}\int_{A_c} \nabla f_i \cdot \nabla f_j\, dA \tag{11.55b}$$

and

$$b_{ij} = \frac{1}{A_c}\int_{A_c} \frac{w}{w_{\text{av}}} f_i\, f_j\, dA \tag{11.56}$$

The thermophysical properties in the definition of a_{ij} and b_{ij} are omitted to make a_{ij} and b_{ij} dimensionless. The quantity

$$\overline{\gamma}_n = \frac{a^2 w_{\text{av}} \gamma_n}{\alpha} \tag{11.57}$$

is now the dimensionless eigenvalue, since dimensionless variables in the mathematical formulations of a_{ij} and b_{ij} are being used. The conservation of energy, $dQ_s = hdA_s(T_s - T_b) = \rho w_{\text{av}} A_c c_p dT_b$, at any z, dictates that

$$\frac{4h}{\rho c_p w_{\text{av}} D_e} = -\frac{d(T_b - T_s)/dz}{T_b - T_s} \tag{11.58}$$

where T_b is the bulk temperature defined by

$$T_b = \frac{1}{A_c} \int_{A_c} (w/w_{av}) T \, dA \tag{11.59}$$

Also, T_s, Q_s, and A_s are the surface temperature, surface heat flux, and surface area, respectively. Equation 10.77 when $g = 0$ and $T^* = T_s$ provides the temperature distribution which can be substituted in Equation 11.59 to obtain the bulk temperature T_b. The substitution of the bulk temperature in Equation 11.58 results in the value of the circumferentially averaged heat transfer coefficient, $h = h(z)$. As z approaches infinity, the contribution of all eigenvalues will diminish except the first eigenvalue and the left side of Equation 11.58 becomes γ_n. Then, using Equation 11.57 to evaluate γ_n, Equation 11.58 reduces to

$$\text{Nu} = \frac{hD_e}{k} = \frac{\overline{D_e^2 \gamma_1}}{4} \tag{11.60}$$

Therefore, the first eigenvalue is proportional to the thermally fully developed Nusselt number. Table 11.3 gives the analytical expressions of the components of matrices **A** and **B** and vector $\mathbf{g}^*$ needed to solve the velocity and temperature fields in selected ducts. The entries in Table 11.3 are for the prescribed surface temperature. Equation 11.55b is used to calculate the elements of matrix **A**.

Example 11.8:

For a laminar and fully developed flow of an incompressible and Newtonian fluid in a circular pipe, calculate temperature distribution and the heat transfer coefficient. Fluid at temperature of 0 enters a heated pipe and the surface temperature of the pipe is maintained at temperature of 1.

Solution

The well-known Graetz problem is selected to illustrate how to use the integral method; it leads to the solution of the heat transfer coefficient for flow in circular pipes. The basis functions are

$$f_j = \left(1 - \frac{r^2}{r_0^2}\right)\left(\frac{r}{r_0}\right)^{2(j-1)} \quad \text{for } j = 1, 2, \ldots, N \tag{11.61}$$

The velocity profile for fully developed flow is $w / w_{av} = 2(1 - r^2 / r_0^2)$. This velocity profile can be obtained from the exact solution or the Galerkin method. The elements of matrix **B** that use this parabolic velocity profile and the elements of matrix **A** are in Table 11.3. Again, Equation 10.11 yields the eigenvalues. As discussed earlier, the first eigenvalue in the duct flow problems is of special importance. It provides the fully developed heat transfer coefficient. When $N = 1$, $\text{Nu} = 3$, whereas a two-term solution (i.e., $N = 2$) yields $\text{Nu} = 20[1 - (2/3)^{1/2}] = 3.6701$; this is very close to the value of 3.6568 obtained from the exact solution. When N is increased to 3, a very accurate value of the $\text{Nu} = 3.6570$ is obtained.

TABLE 11.3
Matrices A and B and Vector g* for Selected Ducts

Circular duct

$w/w_{av} = 2(1-r^2), \phi = 1 - r^2$

$$a_{ij} = -(8i-4)\left(\frac{1}{i+j-1} - \frac{1}{i+j}\right) + 4(i-1)^2\left(\frac{1}{i+j-2} - \frac{2}{i+j-1} + \frac{1}{i+j}\right)$$

and

$$b_{ij} = 2\left(\frac{1}{i+j-1} - \frac{3}{i+j} + \frac{3}{i+j+1} - \frac{1}{i+j+2}\right)$$

Right triangular ducts

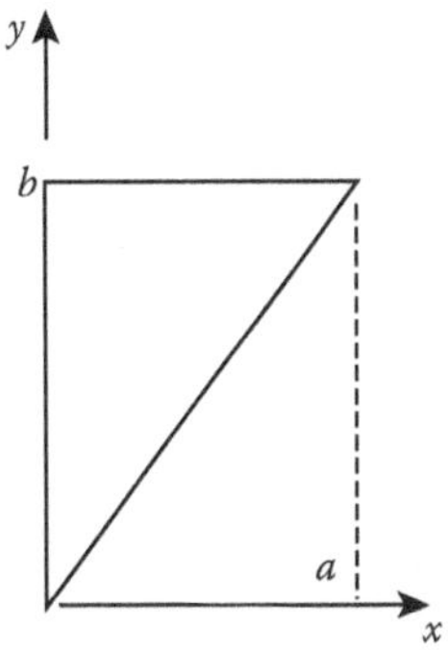

$f_j = (x/a)(y/a - b/a)[y/a - (b/a)(x/a)](x/a)^{m_j}(y/b)^{n_j}$

$j = 1;\quad m_1 = 0 \quad \text{and} \quad n_1 = 0$
$j = 2;\quad m_2 = 1 \quad \text{and} \quad n_2 = 0$
$j = 3;\quad m_3 = 0 \quad \text{and} \quad n_3 = 1$
$j = 4;\quad m_4 = 2 \quad \text{and} \quad n_4 = 0$
$j = 5;\quad m_4 = 1 \quad \text{and} \quad n_4 = 1$

$$a_{ij} = -2\left(\frac{b}{a}\right)^{n_i+n_j+4}\left[\frac{(m_i+1)(m_j+1)}{m_i+m_j+1} - \frac{2m_im_j + 3(m_i+m_j)+4}{m_i+m_j+2}\right.$$
$$\left. + \frac{(m_i+2)(m_j+2)}{m_i+m_j+3}\right]\left(\frac{1}{l+6} - \frac{2}{l+5} + \frac{1}{l+4}\right)$$
$$-2\left(\frac{b}{a}\right)^{n_i+n_j+2}\left(\frac{G_1}{m_i+m_j+3} - \frac{G_2}{m_i+m_j+4} + \frac{G_3}{m_i+m_j+5}\right)$$

where

$$G_1 = \frac{n_in_j + 2(n_i+n_j)+4}{l+6} - \frac{2n_in_j + 3(n_i+n_j)+4}{l+5} + \frac{n_in_j + (n_i+n_j)+1}{l+4}$$

$$G_2 = \frac{2n_in_j + 3(n_i+n_j)+4}{l+6} - \frac{4n_in_j + 4(n_i+n_j)+2}{l+5} + \frac{2n_in_j + n_i + n_j}{l+4}$$

$$G_3 = \frac{n_in_j + n_i + n_j + 1}{l+6} - \frac{2n_in_j + n_i + n_j}{l+5} + \frac{n_in_j}{l+4}$$

$$b_{ij} = 2\frac{C_f Re}{2D_e^2/a^2}\sum_{k=1}^{M} d_k\left(\frac{b}{a}\right)^{\nu_1+6}\left(\frac{1}{\mu_1+4} - \frac{3}{\mu_1+5} + \frac{3}{\mu_1+6} - \frac{1}{\mu_1+7}\right)$$
$$\times\left(\frac{1}{\nu+11} - \frac{3}{\nu+10} + \frac{3}{\nu+9} - \frac{1}{\nu+8}\right)$$

$$\psi_j = \frac{2(b/a)^{n_j+2}}{(m_j+2)(m_j+3)(m_j+n_j+4)(m_j+n_j+5)}$$

(Continued)

TABLE 11.3
Matrices A and B and Vector g* for Selected Ducts (Continued)

Isosceles triangular ducts

$$f_j = (y/a - b/a)[(y/a)^2 - (b/a)^2(x/a)^2](x/a)^{m_j}(y/b)^{n_j}$$

$j = 1;\quad m_1 = 0$ and $n_1 = 0$
$j = 2;\quad m_2 = 0$ and $n_2 = 1$
$j = 3;\quad m_3 = 2$ and $n_3 = 0$
$j = 4;\quad m_4 = 0$ and $n_4 = 2$

$$a_{ij} = -2\left(\frac{b}{a}\right)^{n_i+n_j+6}\left[\frac{m_i m_j}{m_i + m_j - 1} - \frac{2(m_i m_j + m_i + m_j)}{m_i + m_j + 1} + \frac{(m_i+2)(m_j+2)}{m_i + m_j + 3}\right]\left(\frac{1}{l+6} - \frac{2}{l+5} + \frac{1}{l+4}\right) - 2\left(\frac{b}{a}\right)^{n_i+n_j+4} \times \left(\frac{G_1}{m_i + m_j + 1} - \frac{G_2}{m_i + m_j + 3} + \frac{G_3}{m_i + m_j + 5}\right)$$

where

$$G_1 = \frac{(n_i+3)(n_j+3)}{l+6} - \frac{(n_i+2)(n_j+3) + (n_i+3)(n_j+2)}{l+5} + \frac{(n_i+2)(n_j+2)}{l+4}$$

$$G_2 = 2\left[\frac{n_i n_j + 2(n_i+n_j) + 3}{l+6} - \frac{2n_i n_j + 3(n_i+n_j) + 2}{l+5} + \frac{n_i n_j + n_i + n_j}{l+4}\right]$$

$$G_3 = \frac{n_i n_j + n_i + n_j + 1}{l+6} - \frac{2n_i n_j + n_i + n_j}{l+5} + \frac{n_i n_j}{l+4}$$

$$b_{ij} = 2\frac{C_f Re}{2D_e^2/a^2}\sum_{k=1}^{M} d_k \left(\frac{b}{a}\right)^{\nu_1+9}\left(\frac{1}{\mu_1+1} - \frac{3}{\mu_1+3} + \frac{3}{\mu_1+5} - \frac{1}{\mu_1+7}\right) \times \left(\frac{1}{\nu+11} - \frac{3}{\nu+10} + \frac{3}{\nu+9} - \frac{1}{\nu+8}\right)$$

$$g_j^* = \frac{4(b/a)^{n_j+3}}{(m_j+1)(m_j+3)(m_j+n_j+4)(m_j+n_j+5)}$$

Nomenclature of indices

i, j, k, m, n	indices
l	$m_i + n_i + m_j + n_j$
μ_1	$m_i + m_j + m_k$
ν_1	$n_i + n_j + n_k$
ν	$\mu_1 + \nu_1$

(b)

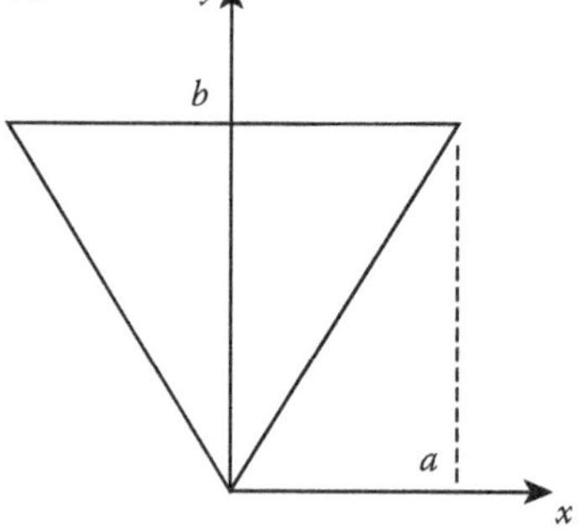

TABLE 11.4
Local Nusselt Number in Circular Ducts

		Results	
$\frac{z/D_e}{Pe}$	Integral method, $N = 12$	Kays and Perkins (1973)	Shah and London (1978)
0.00001	59.621	–	61.877
0.0001	28.148	–	28.254
0.001	12.824	12.86	12.824
0.004	8.036	7.91	8.036
0.01	6.002	5.99	6.002
0.04	4.172	4.18	4.172
0.08	3.769	3.79	3.769
0.1	3.710	3.71	3.710
0.2	3.658	3.66	3.658
0.5	3.657	3.66	3.657

After computation of the eigenvalues and matrices **D** and **P**, the alternative GF solution, Equation 10.77, yields the temperature distribution. The values of the Nusselt number within the entrance region of the pipe are computed by Lakshminarayanan (1988) and Lakshminarayanan and Haji-Sheikh (1986), and compared with the exact solution in Table 11.4. The agreement between the two solutions is generally excellent. The GF solution method permits the inclusion of position dependent wall temperature and locally variable volumetric heat generation in the solution. The boundary condition of second and third kinds can be accommodated using the one-dimensional basis functions already defined in Chapter 10. The calculation can be extended to ducts with more complex cross-sections. For instance, the heat transfer coefficients for various isosceles and right triangular ducts are calculated and reported by Lakshminarayanan (1988) and Lakshminarayanan and Haji-Sheikh (1986). The matrices **A** and **B** for the above-mentioned ducts are in Table 11.3.

11.6 CONCLUSION

The multidimensional applications discussed in this chapter show that many complex geometries can be accommodated using the Galerkin-based GF. The success of this method depends on the availability of the basis functions for a given application or one's ability to find a set of basis functions. Because the number of basis functions needed to provide an accurate solution is usually small, numerical computation can be used to compute the elements of matrices **A** and **B**. Various symbolic software programs are widely available and are valuable tools to assist in the mathematical differentiation of the basis functions. Also, the symbolic integration, whenever possible, results in high-speed computer operation by providing virtually error-free mathematical equations.

We showed that the application of the Galerkin-based GF solution to heterogeneous bodies is possible and the generalized formulation of the GF can be used once the basis functions are available. In addition, we showed that the generalized GF solution can be modified for steady-state conduction problems. However, the steady-state solution, using the alternative formulation of the GF, reduces to the standard Galerkin method.

The Galerkin-based solution can also be used to solve for the heat transfer coefficient in the entrance region of ducts. The usefulness of the GF solution method given in Chapter 10 and utilized in this chapter is limited to the case when the thermal conduction in the flow direction is negligible (large Péclet number). However, it is possible to modify the Galerkin-based integral method so that the effect of axial conduction can be included in the analysis.

PROBLEMS

11.1 A square bar has dimensions 1×1. When the boundary conditions are of the first kind, use the product method to compute the basis functions. Repeat the steps using the method used in Example 11.1.

11.2 A finite cylinder with radius r_0 is subject to convective heat transfer at $r = r_0$ while the temperature is prescribed on other surfaces. Find the GF using the product method.

11.3 A hemisphere of radius r_0 has prescribed convection on $r = r_0$ surface while the temperature is prescribed at the other surface. Use the product method to define the basis functions. Comment on the case when convection is prescribed for all surfaces.

11.4 Consider a spheroidal solid whose surface is given by equation $r^2/a^2 + z^2/b^2 = 1$. For boundary conditions of the first kind, show that $a_{11} = -96(19b^2 + 13)V/945b^2$ and $b_{11} = 384V/2079$, where V is the volume of the spheroid. Find matrices **D** and **P** and the GF. Propose a small-time GF for the purpose of partitioning.

11.5 Equations 11.5 and 11.9 give the basis functions for a spheroidal solid with insulated external surface. A spheroid, $a = 1$ and $b = 6$, receives heat from a heat source at the rate of $q(t)$. Is it possible to have a one-term solution using Equation 10.68? What is the smallest number of terms for a reasonable solution? Show that the first eigenvalue is $\gamma_1 = 0$.

11.6 Repeat Example 11.5, except let $T = 0$ at $x = 0$. Redefine $f_{j,m}, U$, and H so that the boundary conditions are satisfied.

11.7 An isosceles right triangular solid bar is externally insulated. The central portion of this long bar, in a circular zone, has thermophysical properties different from the rest of the bar. Find the parametric relations for the basis functions.

11.8 Use a one-term solution to show that the alternative GF solution becomes identical to the Galerkin solutions as $t \to \infty$. (Hint: when $j = 1$, Equations 11.40 and 11.41 are identical.)

11.9 Show that Equations 11.40 and 11.41 produce the same results for any number of terms.

11.10 Calculate the Nusselt number for a fully developed laminar flow in an elliptical duct. The duct's wall temperature is constant. Find a solution that uses the GF for an arbitrarily selected surface temperature.

11.11 Reproduce the data near the entry point of a circular duct using time partitioning of the GF and compare with the entries in Table 11.4.

11.12 The GFs in solid right-triangular rods are needed. Show that the elements of matrix **A** for boundary conditions of the first kind are the same as those given in Table 11.3. Calculate a similar relation for matrix **B**. (Caution: The entries in Table 11.3 are from Equations 11.51b and 11.52.)

11.13 Fluid passes through an annulus whose external surface is elliptical and the internal surface is circular. Consider that the flow is laminar and the boundary conditions are of the first kind. Find the GFs assuming: (a) slug flow, and (b) viscous flow.

11.14 Use Example 11.8 and find a two-term temperature solution when heat generates at the rate of g W/m^2. The inlet and wall temperatures are maintained at zero.

11.15 Repeat Example 11.8, except, now the surface heat flux is prescribed instead of the wall temperature.

11.16 Repeat Problem 11.14, except now the surface heat flux instead of surface temperature is prescribed.

11.17 A 10-cm diameter steel pipe 3 mm thick, $k = 60$ W/mK, $\rho = 7850$ kg/m^3, and $c_p = 434$ J/kgK, is carrying a gas. It has a 5-cm-thick insulation with thermophysical properties $k = 0.04$ W/mk, $\rho = 100$ kg/m^3, and $c_p = 1200$ J/kgK. Inside and outside fluid temperatures are 600 K and 300 K, the corresponding heat transfer coefficients are 100 W/m^{2}K and 50 W/m^{2}K, and the contact conductance between two layers is 10 W/m^{2}K. If the initial temperature is 300 K, using the COND program, display the surface temperatures and calculate the variation of external and internal heat flux with time.

11.18 A straight fin has dimensionless quantities $L = 0, T = 1$ at $x = 0$, $q = 0$ at $x = 1$. When $m = \sqrt{hP/kA} = 0.5$ and fin is initially at zero temperature, use the COND program to calculate the dimensionless heat flux per unit area of the base at $\alpha t/L^2 = 0.2, 0.4, 0.8, 1, 2, \infty$.

11.19 The radius r of a pin fin varies as x^2 when $0.5 < x/x_2 < 1$. Also, the perimeter varies as x^2. The initial and ambient temperatures are 0. The boundary condition at $x = x1 = x_2/2$ is convective so that $h_1 x_2/k = 0.02$. The heat transfer coefficient, h, on the fin surface varies as $r^{-0.25}$ so that $x_2^2(hP/kA) = 2x^{-2.5}$. Find temperature distribution as a function of $\alpha t/L^2$ at $x = x_1$ using the COND program.

11.20 When initial temperature is $1, a = 0, b = 1, q_1/k = -1, q_2/k = -2$, and $\alpha = 1$, for Example 10.7, use the COND program to display temperature distribution at $x = 0, 0.2, 0.4, 0.6, 0.8$, and 1. Explain the nature of steady-state solution. Repeat the calculations but consider that the plate has a uniform volumetric heat source, $g = 3$.

REFERENCES

Haji-Sheikh, A., 1986, On the solution of parabolic partial differential equations using Galerkin functions, in *Integral Methods in Science and Engineering*, ed. F. R. Payne, C. C. Corduneanu, A. Haji-Sheikh, and T. Huang, Hemisphere, Washington, DC.

Haji-Sheikh, A., 1988, Heat diffusion in heterogeneous media using heat-flux-conserving basis functions, *ASME J. Heat Transfer*, vol. 110, pp. 276–282.

Haji-Sheikh, A. and Beck, J. V., 1990, Green's function partitioning in Galerkin-based solution of the diffusion equation, *ASME J. Heat Transfer*, vol. 112, pp. 28–34.

Haji-Sheikh, A. and Lakshminarayanan, R., 1987, Integral solution of diffusion equation: Part 2 – Boundary conditions of second and third kinds, *ASME J. Heat Transfer*, vol. 109, pp. 557–562.

Haji-Sheikh, A. and Sparrow, E. M., 1967, The solution of heat conduction problems by probability methods, *ASME J. Heat Transfer*, vol. 89, pp. 121–131.

Haji-Sheikh, A., Mashena, M., and Haji-Sheikh, M. J., 1983, Heat transfer coefficient in ducts with constant wall temperature, *ASME J. Heat Transfer*, vol. 105, pp. 878–883.

Hearn, A., 1983, *REDUCE-3 User's Manual*, Rand Corporation, Santa Monica, CA.

Kantorovich, L. V. and Krylov, V. I., 1960, *Approximate Methods of Higher Analysis*, John Wiley, New York.

Kays, W. M. and Perkins, H. C., 1973, Forced convection, internal flow in ducts, in *Handbook of Heat Transfer*, ed. W. M. Rohsenow and J. P. Hartnett, McGraw-Hill, New York, p. 22.

Lakshminarayanan, R., 1988, Integral solutions to thermal and hydrodynamic entrance problems in ducts, Ph.D. thesis, University of Texas at Arlington, Arlington, TX.

Lakshminarayanan, R. and Haji-Sheikh, A., 1986, A generalized closed-form solution to laminar thermal entrance problems, in *Heat Transfer 86*, ed. C. L. Tien, V. P. Carey, and L. K. Ferrel, Hemisphere, Washington, D.C., pp. 861–876.

Lakshminarayanan, R. and Haji-Sheikh, A., 1988, Extended Graetz problems in irregular ducts, in *Proceedings of the 1988 National Heat Transfer Conference*, ed. H. R. Jacobs, ASME HTD-96, pp. 475–482 in vol. 1.

Lee, Y.-M. and Haji-Sheikh, 1991, Temperature field in heterogeneous bodies: A non-orthogonal solution, in *Proceedings of the National Heat Transfer Conference*, eds. M. Imber and M.M. Yovanovich, ASME HTD-vol. 173, pp. 1–9.

Nomura, S. and Haji-Sheikh, A., 1988, Analysis of transient heat conduction in complex shaped composite materials, *ASME J. Eng. Materials Tehnol.*, vol. 110, pp. 110–112.

Ozisik, M. N., 1993, *Heat Conduction*, John Wiley, New York.

Shah, R. K. and London, A. L., 1978, Laminar flow forced convection in ducts, in *Advances in Heat Transfer*, ed. T. F. Irvine and J. P. Hartnett, Academic Press, New York.

12 Unsteady Surface Element Method

12.1 INTRODUCTION

The unsteady surface element (USE) method is a boundary discretization method for solution of linear transient two- and three-dimensional heat transfer problems. Its development originated with the need to calculate interface temperatures and heat fluxes for similar and dissimilar geometries connected over a relatively small portion of their surface boundaries. Examples of bodies connected over a small area occur in contact conductance problems such as the case of two semi-infinite cylinders in contact over only a central circular region, as shown in Figure 12.1a. An example involving dissimilar geometries is the intrinsic thermocouple problem which involves a semi-infinite cylinder attached to a semi-infinite body (Figure 12.1b) or to an infinite plate (Figure 12.1c). Other related examples are those associated with the electrical contacts, cooling of electronic systems, fins, and conjugated problems.

The above-mentioned problems may involve transient heat transfer and differing thermophysical properties. The solution is difficult because the separate regions are coupled by simultaneous interfacial boundary conditions that may vary with time in some unspecified manner. Numerical methods are the primary means to solve such problems, even though for certain problems it is sometimes possible to obtain approximate solutions by relaxing the conditions that the coupled regions must satisfy.

Closely related to the USE method is the boundary element (BE) method, which has been used in a variety of engineering problems such as solid mechanics, fluid flow, soil mechanics, water waves, heat conduction, electrical problems and a broad range of other applications (Banerjee and Butterfield, 1979; Banerjee and Shaw, 1982; Banerjee and Mukherjee, 1984; and Banerjee and Watson, 1985). The BE method involves Green's theorem to formulate the problem described by a partial differential equation in a given region with some specific boundary conditions as an integral equation which applies only to the boundary of the region. Basic building blocks used in the BE method are source solutions (Green's functions, GFs) for infinite homogeneous bodies.

The BE method is well suited for solving steady-state problems with infinite domain and irregular-shaped boundaries. A number of papers have been written for steady-state heat conduction problems (Schneider, 1979; Schneider and LeDain, 1979; Khader, 1980; Khader and Hanna, 1981).

Application of the BE method to transient problems has received less attention compared to the steady-state problems. This is due to the complexity of having the independent variable of time. There are two basic ways of handling the effects of time. One is to temporarily eliminate time as an independent variable by utilizing the Laplace transform and then solving the problem in the transform space by using

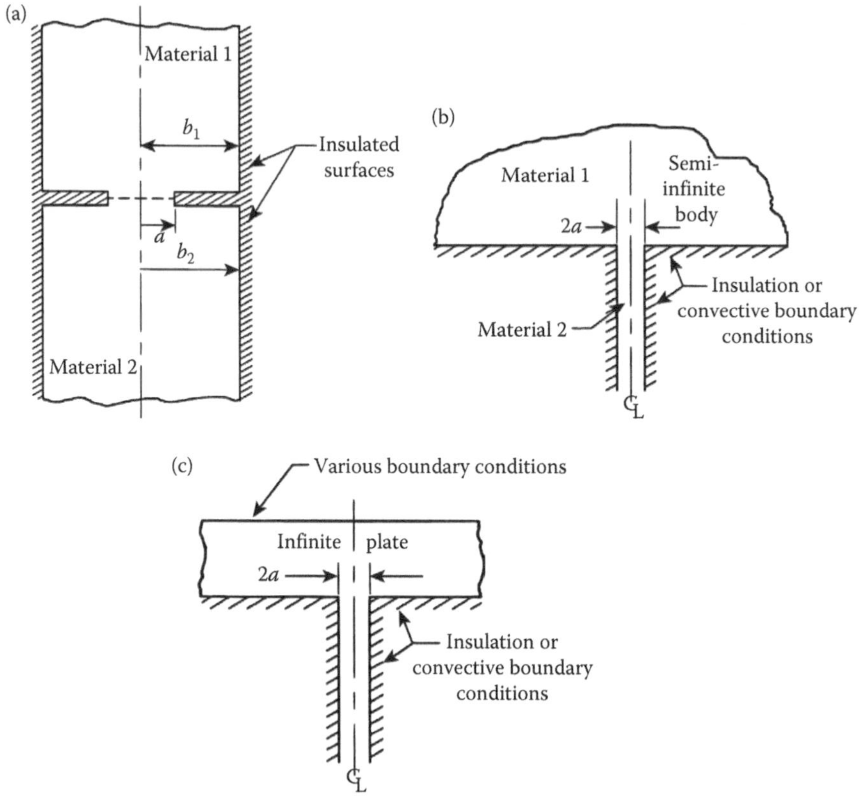

FIGURE 12.1 (a) Two connected semi-infinite cylinders simulating contact conductance problem. (b, c) Some geometries for intrinsic thermocouple problem.

the BE method. The time solution is then obtained by numerical transform inversion. This is the approach taken by Rizzo and Shippy (1970) to solve the problem of heat conduction in an infinite cylinder of an isotropic medium. The other approach is to treat the time directly in the same manner as the spatial coordinates are treated, integrating numerically over the time as well as over the boundary of the body. Shaw (1974) utilized the direct approach to investigate heat conduction in a circular sector of an isotropic medium. A similar approach was taken by Chang et al. (1973) to treat anisotropic heat conduction in the transient case with heat generation. Wrobel and Brebbia (1981) employed this approach to solve three-dimensional axisymmetric transient heat conduction problems of a solid cylinder, a prolate spheroid, and a solid sphere, all with time-dependent boundary conditions.

In the USE method, only the interface between the contacted bodies (or the active part of the boundary) requires discretization as compared to the discretization of the whole domain required in the finite-difference and finite-element methods or discretization of the whole boundary in the BE method. This, in turn, reduces the size of numerical computations, especially for three-dimensional problems. Another aspect of the USE method is that, unlike the above-mentioned alternative methods, it does not require any modifications or special handling of points near the domain boundaries. The USE method uses Duhamel's theorem and involves the inversion of a set of Volterra integral equations, one for each surface element. Though the method is limited to linear regions it can be used for nonlinear boundary conditions.

Two types of kernels ("building blocks" or influence functions) can be employed in the USE method: temperature based and heat flux based. The method requires that these kernels or influence functions be known for the basic geometries under consideration. For many geometries, the influence functions are known or can be obtained by analytical methods or through the use of GFs.

Yovanovich and Martin (1980) suggested the name "surface element method" and did early work on a steady-state form of this method. Keltner and Beck (1981) were the first to employ the surface element method for transient problems. They considered only one element along the interface and utilized the Laplace transform technique to obtain "early" and "late" time approximate analytical solutions for two arbitrary bodies suddenly brought into thermal contact over a small area. The multinode form of USE (numerical approach) was originally developed by Litkouhi and Beck (1985, 1986) and applied to contact between large bodies over a small circular area and the intrinsic thermocouple problem. Cole and Beck (1987, 1988) have extended the USE method to a conjugated heat transfer problem.

The objective of this chapter is to introduce the basic mathematical concepts and formulations of the USE method and to demonstrate its applications by presenting some example problems. Duhamel's theorem and its relation to the GF function method is presented in Section 12.2. Section 12.3 is devoted to formulation and development of the USE equations and the related numerical solutions. The approximate analytical solutions of the USE equations are discussed in Section 12.4, and finally, to illustrate the application of the USE method, some example problems are given and discussed in Section 12.5.

12.2 DUHAMEL'S THEOREM AND GREEN'S FUNCTION METHOD

When the boundary condition is a function of time, solution of a linear heat conduction problem may be deduced from the well-known Duhamel's theorem. Duhamel's theorem employs a fundamental (or a "building block") solution which is used with the superposition principle to obtain temperature at any point $\mathbf{r}$, and time t. Briefly, it states that if $\psi(\mathbf{r}, t)$ is the solution to a linear system initially at zero temperature, due to a unit stepwise input, then the solution to the same system initially at zero temperature due to a time-varying input $F(t)$ (instead of unit step) is given by

$$T(\mathbf{r}, t) = \frac{\partial}{\partial t} \int_0^t F(\tau)\psi(\mathbf{r}, t - \tau)d\tau \tag{12.1a}$$

where $\mathbf{r}$ is the position vector, t is time, and τ is a dummy variable for integration. The input function $F(t)$ can be any type of time-dependent boundary condition (such as prescribed surface temperature, ambient temperature, or prescribed surface heat flux) or heat generation. An alternative form of Equation 12.1a can be obtained with Leibniz's rule for differentiation of an integral,

$$T(\mathbf{r},t) = \int_0^t F(\tau)\frac{\partial\psi(\mathbf{r},t-\tau)}{\partial t}d\tau \tag{12.1b}$$

Equation 12.1a and b represent a form of Duhamel's theorem where the input function $F(t)$ varies only with time. The derivation of this form of Duhamel's theorem is given by several authors using different approaches. The approach presented by Ozisik (1993, p. 195) and Luikov (1968, p. 344) uses Laplace transformations. Myers (1987, p. 153) uses the concept of superposition to derive Duhamel's theorem for prescribed surface temperature boundary condition; while Beck et al. (1985a, p. 81) employ the same principle to derive Duhamel's theorem for heat flux boundary conditions. It is also conventional to treat problems with spatially varying boundary conditions by using Duhamel's theorem with integration over space (Eckert and Drake, 1972, p. 322) and (Kays et al., 2005, p. 111). The following derivation of Duhamel's theorem involves simultaneous variation of both time and space conditions for an arbitrary two-dimensional geometry.

12.2.1 Derivation of Duhamel's Theorem for Time- and Space-Variable Boundary Conditions

Consider the boundary value problem of heat conduction for an arbitrary two-dimensional region R initially at zero temperature, with a time- and space-variable heat flux over boundary S as shown in Figure 12.2. For simplicity, it is assumed that $q(s,t)$ is nonzero only over the portion of boundary S from $s = 0$ to $s = L$, and the other portion of the boundary is insulated $[q(s,t) = 0$, for $s > L]$. The objective is to find an expression for the solution of above problem using Duhamel's theorem.

In the first step the solution to the fundamental problem is found. The fundamental problem is identical to the above problem with the exception that the variable flux boundary condition, $q(s,t)$, is replaced by a special unit step function. It is described by the following equations:

$$\nabla^2\psi_q = \frac{1}{\alpha}\frac{\partial\psi_q}{\partial t} \tag{12.2}$$

$$\psi_q(x,y,0) = 0 \tag{12.3}$$

$$\begin{aligned} k\frac{\partial\psi_q}{\partial n_s} &= 0 \qquad \text{for } t < 0 \qquad \text{or} \qquad s < \eta \\ &= 1 \qquad \text{for } t > 0 \qquad \text{and} \qquad \eta < s < L \end{aligned} \tag{12.4}$$

where η is a dummy length variable along the boundary S between $s = 0$ to $s = L$, and $\psi_q(x, y, \eta, t)$ is the temperature rise at position (x, y) and time t caused by a unit step change of heat flux at time $t = 0$, from $s = \eta$ to $s = L$ as shown in Figure 12.2 by the cross-hatched portion. It is called the flux-based fundamental solution (FBFS). Notice that, for fixed (x, y) and t, $\psi_q(x, y, \eta, t)$ decreases as η increases, that is,

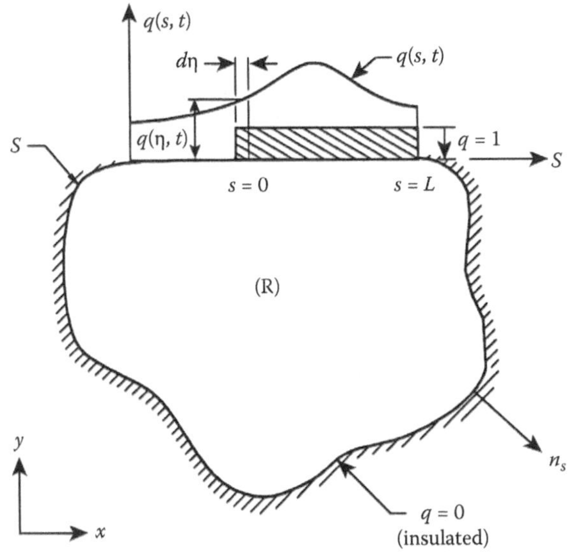

FIGURE 12.2 Geometry showing two-dimensional region heated by arbitrary heat flux.

$$\psi_q(x, y, \eta, t) > \psi_q(x, y, \eta + d\eta, t) \tag{12.5}$$

Temporarily, let η be fixed and consider the variation of the heat flux with time only. From the fundamental solution, the temperature rise at position (x, y) and time t due to a unit step change of heat flux at time τ is

$$\psi_q(x, y, \eta, t - \tau) \tag{12.6a}$$

where $t - \tau$ is the time that has elapsed since the step at τ. Also the temperature rise at time t due to a unit step change of heat flux at time $\tau + d\tau$ is

$$\psi_q[x, y, \eta, t - (\tau + d\tau)] \tag{12.6b}$$

Then from Equation 12.6a and b, the temperature rise at position (x, y) and time t due to a unit step change in q for $\tau < t < \tau + d\tau$ is

$$-d_\tau \psi_q(x, y, \eta, t - \tau) = \psi_q(x, y, \eta, t - \tau) - \psi_q[x, y, \eta, t - (\tau + d\tau)] \tag{12.7}$$

where d_τ is a differentiation operator for τ. Notice that $\psi_q(x, y, \eta, t - \tau)$ is greater than $\psi_q[x, y, \eta, t - (\tau + d\tau)]$. Using Equation 12.7, the temperature rise at position (x, y) and time t due to the value $q(\eta, t)$ for $\tau < t < \tau + d\tau$ and η being fixed is

$$-q(\eta, \tau) d_\tau \psi_q(x, y, \eta, t - \tau) = -q(\eta, \tau) \frac{\partial \psi_q(x, y, \eta, t - \tau)}{\partial \tau} d\tau \tag{12.8}$$

for small $d\tau$. Since the problem is linear, superposition can be employed and the total effect of all step changes of heat flux over small $d\tau$'s from time zero to time t is simply found by integrating Equation 12.8 from 0 to t. Denoting the result $\psi'_q(x, y, \eta, t)$, one can write

$$\psi'_q(x, y, \eta, t) = -\int_0^t q(\eta, \tau)\frac{\partial \psi_q(x, y, \eta, t-\tau)}{\partial \tau}d\tau \tag{12.9a}$$

From the relation

$$\frac{\partial \psi_q(x, y, \eta, t-\tau)}{\partial \tau} = -\frac{\partial \psi_q(x, y, \eta, t-\tau)}{\partial t}$$

Equation 12.9a can be written as

$$\psi'_q(x, y, \eta, t) = \int_0^t q(\eta, \tau)\frac{\partial \psi_q(x, y, \eta, t-\tau)}{\partial t}d\tau \tag{12.9b}$$

Note that ψ'_q is the temperature rise for the case that the time-variable q is zero for $s < \eta$, and is uniformly distributed over space for $\eta < s < L$.

In a similar way, one can show that the temperature rise for the case that the flux q is zero for $s < \eta + d\eta$, and is uniformly distributed for $\eta + d\eta < s < L$, is

$$\psi'_q(x, y, \eta + d\eta, t) = \int_0^t q(\eta, \tau)\frac{\partial \psi_q(x, y, \eta + d\eta, t-\tau)}{\partial t}d\tau \tag{12.10}$$

Using Equations 12.9b and 12.10, the temperature rise due to a uniform heat flux q, between $s = \eta$ and $s = \eta + d\eta$ and for $t > 0$ is

$$\begin{aligned} -d_\eta \psi'_q(x, y, \eta, t) &= \psi'_q(x, y, \eta, t) - \psi'_q(x, y, \eta + d\eta, t) \\ &= -\frac{\partial \psi'_q(x, y, \eta, t)}{\partial \eta}d\eta \end{aligned} \tag{12.11}$$

where d_η is a differentiation operator for η. Notice that $\psi'_q(x, y, \eta, t)$ is greater than $\psi'_q(x, y, \eta + d\eta, t)$. Introducing Equations 12.9b and 12.10 into Equation 12.11 yields

$$-d_\eta \psi'_q(x, y, \eta, t) = -\int_0^t q(\eta, \tau)\frac{\partial^2 \psi_q(x, y, \eta, t-\tau)}{\partial t\,\partial \eta}d\tau\, d\eta \tag{12.12}$$

Again superposition can be employed and the total effect of the variation of heat flux from $s = 0$ to $s = L$ can be found by integrating Equation 12.12 over space from 0 to L, to give

$$T(x, y, t) = -\int_0^L \int_0^t q(\eta, \tau)\frac{\partial^2 \psi_q(x, y, \eta, t-\tau)}{\partial t\,\partial \eta}d\tau\, d\eta \tag{12.13}$$

In this problem, it was assumed that only a portion of the surface boundary is exposed to heat flux with the remainder being insulated. However, if none of the boundary S is insulated, the first integral in Equation 12.13 extends over the entire boundary S. Furthermore, if the initial temperature of the system is T_0 instead of being zero, the solution becomes

$$T(x, y, t) - T_0 = -\int_S \int_0^t q(\eta, \tau)\frac{\partial^2 \psi_q(x, y, \eta, t-\tau)}{\partial t\,\partial \eta}d\tau\, d\eta \tag{12.14}$$

In Equation 12.14, the input function $q(\eta, \tau)$ is the heat flux along the boundary (surface heat flux) which varies with both space and time, and the solution is in terms of the FBFS, ψ_q. If, however, the surface temperature is known along the boundary as the input function (instead of heat flux), then in a similar manner to that described above, the solution in terms of the temperature-based fundamental solution (TBFS), ψ_T, can be obtained as

$$T(x,y,t) - T_0 = -\int_S \int_0^t [T_s(\eta,\tau) - T_0] \frac{\partial^2 \psi_T(x,y,\eta,t-\tau)}{\partial\eta\, \partial t}\, d\tau\, d\eta \qquad (12.15)$$

Equations 12.14 and 12.15 are rather general expressions for the case that the input function varies with both space and time in a two-dimensional region. Both equations can be employed to obtain the temperature history at any position (x, y) of the region. However, depending on the type of boundary condition, one might be more appropriate than the other. To compare the two approaches and discuss their utility for each particular type of boundary condition, both forms of solutions are examined below.

For problems with boundary conditions of the first kind, where the temperature is specified everywhere along the boundary, the right-hand side of Equation 12.15 is known, and one can solve for the temperature history of any interior point of R, by direct integration. If, however, Equation 12.14 is employed instead of Equation 12.15, the direct evaluation of $T(x, y, t)$ is not possible because of the unknown heat flux q in the right-hand side of this equation. In this case, an inverse integration must first be performed to solve for the unknown surface heat flux which is the information needed by Equation 12.14 to find $T(x, y, t)$ at any interior point. Therefore, in problems with the first kind boundary conditions, Equation 12.15 is more appropriate than Equation 12.14.

On the other hand, if the boundary condition is of the second kind where q is specified along the boundary S, then the right-hand side of Equation 12.14 is known which leads to evaluation of a direct integral. In this case Equation 12.14 is more appropriate than Equation 12.15.

For boundary conditions of the third kind where neither the surface temperature nor its normal derivative are completely known over the entire boundary S (mixed boundary conditions), none of the above equations can be used directly to obtain temperature history for any interior point. An example is given to illustrate this case better.

Consider the homogeneous convective boundary condition given by

$$k\frac{\partial T(s,t)}{\partial \eta_s} + h_s T_s(s,t) = 0 \qquad \text{on } S \qquad (12.16)$$

where $\partial/\partial\eta_s$ denotes differentiation with respect to the outward pointing normal to the surface boundary S as shown in Figure 12.2. Substituting for q in Equation 12.14 from Equation 12.16, one can write

$$T(x,y,t) - T_0 = -\int_S \int_0^t h_s T_s(\eta,\tau) \frac{\partial^2 \psi_q(x,y,\eta,t-\tau)}{\partial t\, \partial\eta} d\tau\, d\eta \qquad (12.17)$$

Equation 12.17 cannot directly be integrated for $T(x, y, t)$, since $T_s(s, t)$ inside the integral is unknown. In other words, the number of unknown functions in Equation 12.17 is more than one, $T_s(s, t)$ and $T(x, y, t)$. However, for a point along the boundary S, Equation 12.17 reduces to

$$T_s(s,t) - T_0 = -\int_S \int_0^t h_s T_s(\eta, \tau) \frac{\partial^2 \psi_q(\eta, t-\tau)}{\partial t \partial \eta} d\tau \, d\eta \tag{12.18}$$

which is a Volterra integral equation of the second kind with the only unknown function, $T_s(s, t)$, both inside and outside the integral. In an inverse manner, Equation 12.18 can be solved numerically for $T_s(s, t)$. Once the surface temperature, $T_s(s, t)$, has been determined, the solution to the interior temperature history, $T(x, y, t)$, can be obtained by substituting $T_s(s, t)$ into Equation 12.17.

Hence, for the problems with mixed or convective boundary conditions, the temperature history at any interior point can be determined in two steps:

1. Find the boundary information by solving an inverse integral equation.
2. Using the boundary data obtained in step 1, find the interior temperature history by using a direct integration.

Equations 12.14 and 12.15 are the flux-based and temperature-based forms of Duhamel's theorem. They are used as the basic building blocks in the development of the USE formulation in the following sections.

12.2.2 Relation to the Green's Function Method

The Duhamel's theorem (sometimes called Duhamel's integral) approach given herein is related to the GF method. One advantage of the Duhamel's theorem approach is that it follows from the well-known concepts of superposition in a more direct manner than the GF method. Another advantage is that there are no singularities in the fundamental solutions; that is, the $\psi(\cdot)$ functions are finite for $t - \tau \to 0$, while the GFs go to infinity. The GF method, however, has the advantage that the GFs are more accessible and easier to obtain than the $\psi(\cdot)$ functions.

To show the relationship between the Duhamel's theorem approach and the GF method, two examples are given below. One example demonstrates the application of Equation 12.1b, where the boundary condition is only a function of time (no spatial variation), while the other example shows the applications of Equations 12.14 and 12.15, where the boundary conditions vary with both time and space.

Example 12.1: *X11B-0T0* Case

Consider a one-dimensional flat plate geometry initially at zero temperature with an arbitrary time-dependent prescribed surface temperature at $x = 0$ as shown in Figure 12.3. For $t > 0$, the surface temperature at $x = L$ is kept at its initial value $T_0 = 0$. The describing equations are

$$\frac{\partial^2 T(x,t)}{\partial x^2} = \frac{1}{\alpha} \frac{\partial T(x,t)}{\partial t} \tag{12.19}$$

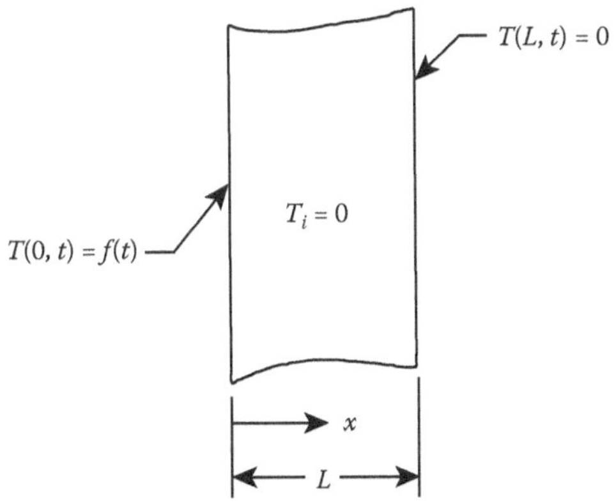

FIGURE 12.3 A flat plate with prescribed surface temperatures.

$$T(0,t) = f(t) \qquad \text{for } t > 0 \tag{12.20a}$$

$$T(L,t) = 0 \qquad \text{for } t > 0 \tag{12.20b}$$

$$T(x,0) = 0 \tag{12.20c}$$

The fundamental solution for this problem is $\psi_T(x, t)$, which represents the temperature at point x and at time t in the flat plate geometry, with zero initial temperature and with a unit-step temperature at the boundary $x = 0$. It is a solution of the problem

$$\frac{\partial^2 \psi_T(x,t)}{\partial x^2} = \frac{1}{\alpha}\frac{\partial \psi_T(x,t)}{\partial t} \tag{12.21}$$

$$\psi_T(0,t) = 0 \qquad \text{for } t < 0$$

$$= 1 \qquad \text{for } t > 0 \tag{12.22a}$$

$$\psi_T(L,t) = 0 \qquad \text{for } t > 0 \tag{12.22b}$$

$$\psi_T(x,0) = 0 \tag{12.22c}$$

Note that the subscript T in $\psi_T(\cdot)$ function indicates that it is a TBFS.

From the Duhamel's integral Equation 12.1b, the transient temperature distribution in the plate is given by

$$T(x,t) = \int_{\tau=0}^{t} f(\tau)\frac{\partial \psi_T(x,\ t-\tau)}{\partial t} d\tau \tag{12.23}$$

The GF solution to this problem is given by Equation 3.46 with zero initial condition and no energy generation within the body,

$$T(x,t) = \alpha \int_{\tau=0}^{t} f(\tau)\ \frac{\partial G(x,t|x',\tau)}{\partial x'}\bigg|_{x'=0} d\tau \tag{12.24}$$

Note that since n_i' in Equation 3.46 represents the *outward normal* from the body, the $(\partial G/\partial n_i')|_{x'=x_i}$ term in this equation is replaced by $-(\partial G/\partial x')|_{x'=0}$ in Equation 12.24. Also, since $T(L, t)$ is equal to zero for $t > 0$, there is no contribution due to the boundary condition at $x = L$.

A comparison of the GF solution, Equation 12.24, and the Duhamel's theorem solution, Equation 12.23, reveals that

$$\frac{\partial \psi_T(x, t-\tau)}{\partial t} = -\frac{\partial \psi_T(x, t-\tau)}{\partial \tau} = \alpha \left. \frac{\partial G(x, t|x', \tau)}{\partial x'} \right|_{x'=0} \tag{12.25}$$

and, then, integrating Equation 12.25 over times gives

$$\psi_T(x, t-\tau) = -\int_{t'=\tau}^{t} \alpha \left. \frac{\partial G(x, t|x', t')}{\partial x'} \right|_{x'=0} dt' \tag{12.26}$$

Thus, Duhamel's theorem for this case is the same as the GF equation for a specified boundary temperature, where the fundamental solution is related to the GF function by Equations 12.25 and 12.26.

Example 12.2: *X20B*(*x*-*t*-)*T*0 Case

Consider a semi-infinite body initially at zero temperature exposed to a time- and space-variable heat flux boundary condition over a portion of its surface boundary from $x = 0$ to $x = L$, with the rest of the surface boundary being insulated (see Figure 12.4). From the flux-based Duhamel's integral, Equation 12.14, the temperature at any point (x, z) of the semi-infinite body and at any time t is given by

$$T(x, z, t) = -\int_0^L \int_0^t q(\eta, \tau) \frac{\partial^2 \psi_q(x, z, \eta, t-\tau)}{\partial t \, \partial \eta} d\tau \, d\eta \tag{12.27}$$

Here η is a dummy length variable along the surface boundary between $x = 0$ to $x = L$, and ψ_q is the flux-based fundamental solution for this problem, described by the following equations:

$$\nabla^2 \psi_q = \frac{1}{\alpha} \frac{\partial \psi_q}{\partial t} \tag{12.28}$$

$$k\frac{\partial \psi_q}{\partial z} = 0 \quad \text{for } t<0 \quad \text{or} \quad x<\eta \quad z=0$$

$$= 1 \quad \text{for } t>0 \quad \text{and} \quad \eta<x<L \quad z=0 \tag{12.29a}$$

$$\psi_q(x, \infty, \eta, t) = 0 \tag{12.29b}$$

$$\psi_q(x, z, \eta, 0) = 0 \tag{12.29c}$$

Similar to that of the previous example, the solution to this example problem can also be obtained from Equation 3.46 for the boundary condition of the second kind, with zero initial temperature and no energy generation as

$$T(\mathbf{r}, t) = \frac{\alpha}{k} \int_{\tau=0}^{t} \int_{S_i} q(\mathbf{r}'_i, \tau) G(\mathbf{r}, t|\mathbf{r}'_i, \tau) \, ds_i \, d\tau \tag{12.30}$$

where S_i is the heated surface. For the coordinates $\mathbf{r} = (x, z)$, the heat flux at $z = 0$, and heating from $x' = \eta = 0$ to L, Equation 12.30 can be written as

$$T(x, z, t) = \int_{\eta=0}^{L} \int_{\tau=0}^{t} \frac{\alpha}{k} q(\eta, \tau) G(x, z, t|\eta, 0, \tau) \, d\tau \, d\eta \tag{12.31}$$

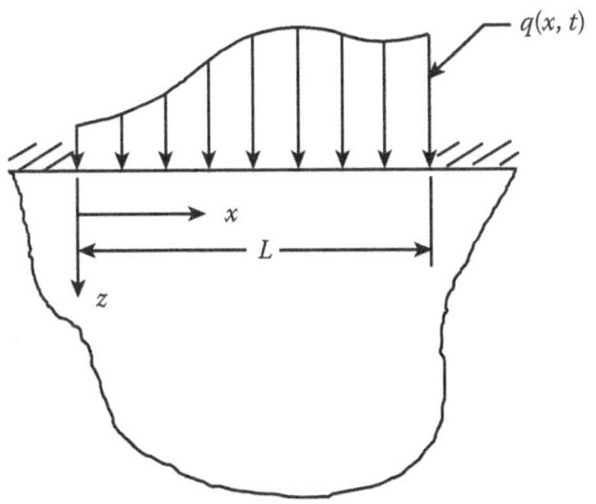

FIGURE 12.4 Semi-infinite body exposed to time- and space-variable boundary condition.

Comparing Equations 12.27 and 12.31 yields

$$-\frac{\partial^2 \psi_q(x, z, \eta, t-\tau)}{\partial t\, \partial \eta} = \frac{\alpha}{k} G(x, z, t|\eta, 0, \tau) \tag{12.32}$$

and then, integrating Equation 12.32 twice (over time and space), gives

$$\psi_q(x, z, \eta, t-\tau) = \int_{t'=\tau}^{t} \int_{x'=\eta}^{L} \frac{\alpha}{k} G(x, z, t|x', 0, t')\, dx'\, dt' \tag{12.33}$$

which demonstrates the relationship between the flux-based fundamental solution and the GF for the semi-infinite body problem given above.

Both Duhamel's theorem and the GF equation are convolution integrals because they involve a product of two functions, one a function of τ and the other a function of $t - \tau$. Duhamel's theorem can be thought of as a boundary condition term of the GF equation, a special case of the general method of GF.

12.3 UNSTEADY SURFACE ELEMENT FORMULATIONS

There are two different formulations of the USE method. One is the single-node formulation, which uses the Laplace transform technique to obtain an approximate analytical solution. The other one is the multinode formulation (numerical solution) which is more general and can be applied to a variety of problems. The multinode formulation allows for spatial variation of surface heat flux and temperature by dividing the surface boundary into several surface elements, while in the single-node formulation, the surface heat flux and temperature are considered spatially constant.

Both formulations may be used with either heat flux-based or TBFS. The multinode formulation is given in this section. In Section 12.4, the single-node analytical solution is given as a special case where there is only one element along the surface boundary.

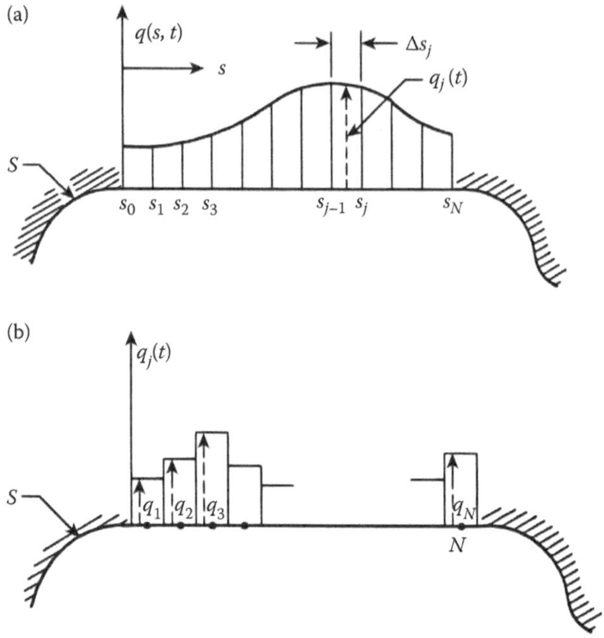

FIGURE 12.5 (a) Geometry showing discretization over heated portion of surface boundary. (b) Uniform heat flux assumption over each surface element.

12.3.1 Surface Element Discretization

To numerically solve the Duhamel's integral Equations 12.14 and 12.15, the surface boundary is divided into N finite surface elements, Δs_j, as shown in Figure 12.5a and b. Notice that only the parts of the boundary with nonzero values of heat flux (for Equation 12.14) and with a temperature different from the initial temperature (for Equation 12.15) need to be discretized. Then, Equations 12.14 and 12.15 can be written as

$$T(x, y, t) - T_0 = -\int_0^t \left[\sum_{j=1}^{N} \int_{\Delta s_j} q(\eta, \tau) \frac{\partial^2 \psi_q(x, y, \eta, t-\tau)}{\partial t \, \partial \eta} d\eta \right] d\tau \quad (12.34a)$$

and

$$T(x, y, t) - T_0 = -\int_0^t \left\{ \sum_{j=1}^{n} \int_{\Delta s_j} [T_s(\eta, \tau) - T_0] \frac{\partial^2 \psi_T(x, y, \eta, t-\tau)}{\partial t \, \partial \eta} d\eta \right\} d\tau \quad (12.34b)$$

By assuming uniform heat flux and temperature over each surface element in Equations 12.34a and 12.34b, respectively, one can write, for flux-based equations,

$$T(x, y, t) - T_0 = -\int_0^t \left\{ \sum_{j=1}^{N} q_j(\tau) \frac{\partial}{\partial t} \left[\psi_q(x, y, \eta, t-\tau)\Big|_{s_{j-1}}^{s_j} \right] \right\} d\tau$$

$$= -\int_0^t \left\{ \sum_{j=1}^{N} q_j(\tau) \frac{\partial}{\partial t} [\Delta\psi_{qj}(x, y, t-\tau)] \right\} d\tau \quad (12.35a)$$

and, for temperature-based equations,

$$T(x, y, t) - T_0 = -\int_0^t \left\{ \sum_{j=1}^{N} [T_{sj}(\tau) - T_0] \frac{\partial}{\partial t} \left[\psi_{Tj}(x, y, t-\tau)\Big|_{s_{j-1}}^{s_j} \right] \right\} d\tau$$

$$= -\int_0^t \left\{ \sum_{j=1}^{N} \left[T_{sj}(\tau) - T_0 \right] \frac{\partial}{\partial t} [\Delta\psi_{Tj}(x, y, t-\tau)] \right\} d\tau \quad (12.35b)$$

where

$$\Delta\psi_{qj} = \psi_q(x, y, s_j, t) - \psi_q(x, y, s_{j-1}, t) \quad (12.36a)$$

$$\Delta\psi_{Tj} = \psi_T(x, y, s_j, t) - \psi_T(x, y, s_{j-1}, t) \quad (12.36b)$$

Further, if the temperature rise at position (x, y) due to a unit step increase in heat flux and temperature at the element j are denoted $\phi_j(x, y, t)$ and $\theta_j(x, y, t)$, respectively, then it can be shown that

$$-\Delta\psi_{qj}(x, y, t) \equiv \phi_j(x, y, t) \quad (12.37a)$$

$$-\Delta\psi_{Tj}(x, y, t) \equiv \theta_j(x, y, t) \quad (12.37b)$$

Using Equation 12.37a in Equation 12.35a, and Equation 12.37b in Equation 12.35b gives,

$$T(x, y, t) - T_0 = \sum_{j=1}^{N} \int_0^t q_j(\tau) \frac{\partial \phi_j(x, y, t-\tau)}{\partial t} d\tau \quad (12.38a)$$

and

$$T(x, y, t) - T_0 = \sum_{j=1}^{N} \int_0^t \left[T_{sj}(\tau) - T_0 \right] \frac{\partial \theta_j(x, y, t-\tau)}{\partial t} d\tau \quad (12.38b)$$

Equations 12.38a and b are the Duhamel's integral forms of the flux-based and the temperature-based USE equations for a single two-dimensional body.

Equation 12.38a gives the temperature rise at location (x, y) and time t due to the effect of N surface heat flux histories $q_1(t), q_2(t), \ldots, q_N(t)$; while Equation 12.38b gives the temperature rise at the same location and time due to the effect of N time-varying surface temperatures $T_{s1}(t), T_{s2}(t), \ldots, T_{sN}(t)$. The functions ϕ_j and θ_j are the basic building-block solutions needed in the above expressions and are termed as the flux-based and the temperature-based *influence functions*. They are called influence functions because they give the influence of the jth surface element on the body.

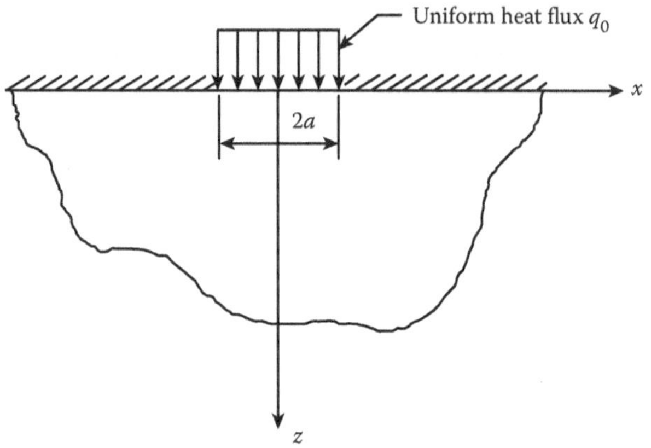

FIGURE 12.6 Geometry of semi-infinite body heated by a uniform heat flux over an infinite strip.

The USE method requires that the influence functions be known for the geometries under consideration. For instance, for the geometry of Figure 12.4, the flux-based influence function ϕ is the solution to the problem of a semi-infinite body heated by a constant heat flux over an infinite strip as shown in Figure 12.6.

The USE Equations 12.38a and b can be written in their general forms by replacing (x, y) in these equations with a position vector $\mathbf{r}$; that is,

$$\text{Flux-based USE: } T(\mathbf{r}, t) - T_0 = \sum_{j=1}^{N} \int_0^t q_j(\tau) \frac{\partial \phi_j(\mathbf{r}, t-\tau)}{\partial t} d\tau \quad (12.39a)$$

$$\text{Temperature-based USE: } T(\mathbf{r}, t) - T_0 = \sum_{j=1}^{N} \int_0^t [T_{sj}(\tau) - T_0] \frac{\partial \theta_j(\mathbf{r}, t-\tau)}{\partial t} d\tau \quad (12.39b)$$

12.3.2 Green's Function Form of the USE Equations

The flux-based and temperature-based Equations 12.39a and b represent the Duhamel's integral forms of the USE equations. The GF forms of the USE equations can be obtained by discretizing Equation 3.46 over N surface elements. For convenience, it is assumed that there is only one nonhomogeneous boundary (with a nonzero value of heat flux or a temperature different from the initial temperature). Accordingly, the summation terms in Equation 3.46 which are for more than one nonhomogeneous boundary condition are dropped and one can write, for boundary condition of the second kind (flux-based equations),

$$T(\mathbf{r}, t) - T_0 = \int_0^t \left[\sum_{j=1}^{N} \frac{\alpha}{k} \int_{\Delta s_j} q(\mathbf{r}'_j, \tau) G(\mathbf{r}, t|\mathbf{r}'_j, \tau) ds_j \right] d\tau \quad (12.40a)$$

and for boundary condition of the first kind (temperature-based equations):

$$T(\mathbf{r},t) - T_0 = \int_0^t \left\{ \sum_{j=1}^{N} \alpha \int_{\Delta s_j} [T_s(\mathbf{r}',\tau) - T_0] \frac{\partial G(\mathbf{r},t|r'_j,\tau)}{\partial n'_j} ds_j \right\} d\tau \quad (12.40b)$$

By assuming uniform heat flux and temperature over each surface element in Equation 12.40a and b, respectively, one can get for the flux-based equation:

$$T(\mathbf{r},t) - T_0 = \sum_{j=1}^{N} \int_0^t q_j(\tau)\overline{G}_j(\mathbf{r},t|\mathbf{r}'_j,\tau)\, d\tau \quad (12.41a)$$

for the temperature-based equation:

$$T(\mathbf{r},t) - T_0 = \sum_{j=1}^{N} \int_0^t [T_{sj}(\tau) - T_0]\overline{G'_j}(\mathbf{r},t|\mathbf{r}'_j,\tau)\, d\tau \quad (12.41b)$$

The notations $\overline{G}_j$ and $\overline{G'_j}$ are defined as

$$\overline{G}_j(\mathbf{r},t|\mathbf{r}'_j,\tau) = \frac{\alpha}{k} \int_{\Delta s_j} G(\mathbf{r},t|\mathbf{r}'_j,\tau)\, ds_j \quad (12.42a)$$

$$\overline{G'_j}(\mathbf{r},t|\mathbf{r}'_j,\tau) = -\alpha \int_{\Delta s_j} \frac{\partial G(\mathbf{r},t|\mathbf{r}',\tau)}{\partial n'_j}\, ds_j \quad (12.42b)$$

Equation 12.41a and b are the GF forms of the flux-based and temperature-based USE equations. Note that $\overline{G}_j$ and $\overline{G'_j}$ appearing in these equations correspond to the time derivatives of the flux-based and temperature-based influence functions, ϕ_j and θ_j, given in Equation 12.39a and b, respectively.

In the derivation of Equation 12.41a and b, it was assumed that there is only one nonhomogeneous boundary in the problem under consideration. For the problems with more than one nonhomogeneous boundary condition, a summation term should be added to each of these equations. Furthermore, it should be noted that Equation 12.41a and b are, respectively, for the problems with the second kind (prescribed surface heat flux) and first kind (prescribed surface temperature) boundary conditions. For the problems with mixed boundary conditions, these two equations can be superimposed.

The USE Equations 12.39a, b and 12.41a, b can be applied to two- or three-dimensional geometries in the rectangular, cylindrical, or spherical coordinate systems. In the two-dimensional USE equations, the surface elements are infinite strips that may be treated as one-dimensional elements (or line elements). In the three-dimensional USE equations, the surface elements are two-dimensional elements and can be chosen in different shapes such as triangular, rectangular, or circular, depending on the nature of the problem under consideration.

12.3.3 Time Integration of the USE Equations

The time integration of the USE Equations 12.39a, b and 12.41a, b, can be performed directly by dividing the entire time domain into M equal small time intervals, Δt, that is, for Duhamel's integral fluxed-based equation:

$$T(\mathbf{r},t) - T_0 = \sum_{j=1}^{N}\sum_{i=1}^{M}\int_{t_{i-1}}^{t_i} q_j(\tau)\frac{\partial\phi_j(\mathbf{r},t-\tau)}{\partial t}d\tau \tag{12.43a}$$

for Duhamel's integral temperature-based equation:

$$T(\mathbf{r},t) - T_0 = \sum_{j=1}^{N}\sum_{i=1}^{M}\int_{t_{i-1}}^{t_i} [T_{sj}(\tau) - T_0]\frac{\partial\theta_j(\mathbf{r},t-\tau)}{\partial t}\,d\tau \tag{12.43b}$$

for the GF flux-based equation:

$$T(\mathbf{r},t) - T_0 = \sum_{j=1}^{N}\sum_{i=1}^{M}\int_{t_{i-1}}^{t_i} q_j(\tau)\overline{G}_j(\mathbf{r},t|\mathbf{r}'_j,\tau)\,d\tau \tag{12.44a}$$

for the GF temperature-based equation:

$$T(\mathbf{r},t) - T_0 = \sum_{j=1}^{N}\sum_{i=1}^{M}\int_{t_{i-1}}^{t_i} [T_{sj}(\tau) - T_0]\overline{G'_j}(\mathbf{r},t|\mathbf{r}'_j,\tau)\,d\tau \tag{12.44b}$$

By assuming the elemental heat flux and temperature histories being uniform over each time interval, for temperature at time $t_M = M\Delta t$, one can write, for the Duhamel's integral flux-based USE equation:

$$T(\mathbf{r},t_M) - T_0 = \sum_{j=1}^{N}\sum_{i=1}^{M} q_{ji}\Delta\phi_{j,M-i}(\mathbf{r}) \tag{12.45a}$$

for the Duhamel's integral temperature-based equation:

$$T(\mathbf{r},t_M) - T_0 = \sum_{j=1}^{N}\sum_{i=1}^{M}[T_{sji} - T_0]\Delta\theta_{j,M-i}(\mathbf{r}) \tag{12.45b}$$

for the GF flux-based equation:

$$T(\mathbf{r},t_M) - T_0 = \sum_{j=1}^{N}\sum_{i=1}^{M} q_{ji}\Delta\overline{G}_{j,M-i}(\mathbf{r}) \tag{12.46a}$$

for the GF temperature-based equation:

$$T(\mathbf{r},t_M) - T_0 = \sum_{j=1}^{N}\sum_{i=1}^{M}(T_{sji} - T_0)\Delta\overline{G}'_{j,M-i}(\mathbf{r}) \tag{12.46b}$$

where q_{ji} and T_{sji} represent the jth surface heat flux and temperature evaluated at time t_i, and

$$\Delta\phi_{j,M-i}(\mathbf{r}) = \phi_j(\mathbf{r}, t_{M+1-i}) - \phi_j(\mathbf{r}, t_{M-i}) \tag{12.47a}$$

$$\Delta\theta_{j,M-i}(\mathbf{r}) = \theta_j(\mathbf{r}, t_{M+1-i}) - \theta_j(\mathbf{r}, t_{M-i}) \tag{12.47b}$$

$$\Delta\overline{G}_{j,M-i}(\mathbf{r}) = \int_{t_{i-1}}^{t_i} \overline{G}(\mathbf{r}, t_M|\mathbf{r}'_j, \tau)d\tau \tag{12.47c}$$

$$\Delta\overline{G}'_{j,M-i}(\mathbf{r}) = \int_{t_{i-1}}^{t_i} \overline{G}'(\mathbf{r}, t_M|\mathbf{r}'_j, \tau)d\tau \tag{12.47d}$$

Notice that $\Delta\overline{G}$ and $\Delta\overline{G}'$ each involve two integrations, one over the element and the other over the time step. For certain geometries, these integrations can be performed analytically (see Problem 12.3). With problems for which the analytical evaluations of these integrals are not possible, the GF USE equations could still be used by replacing the integrals with suitable quadrature formulas.

12.3.4 Flux-Based USE Equations for Bodies in Contact

For convenience, in the further development of the USE formulation, given in the rest of this section, only the flux-based Equation 12.38a and 12.39a are considered. The temperature-based USE formulation can be developed in a similar manner as the flux-based case.

Consider two arbitrary bodies, initially at uniform but different temperatures (T_{01} and T_{02}), brought into perfect contact over a portion of their boundaries, as shown in Figure 12.7. The bodies may have different conductivities, k, and different density-specific heats ρc. For simplicity, a two-dimensional geometry is assumed; there is no variation of temperature or heat flux in the z-direction. To apply the USE method, the interface is divided into N surface elements, each being an infinite strip, Δs_j. It is assumed that the flux and temperature are uniform over each surface element. The temperature associated with element j will be taken as the temperature at the center of the element, located at point $s'_j = s_j - \Delta s_j/2$. The average temperature over the element may also be used as the temperature associated with the element but it complicates the analysis slightly and will not be discussed here. The heat flux $q_j(t)$, associated with element j, which leaves body 2 in Figure 12.7, is the same heat flux that enters body 1 over the region $s = s_{j-1}$ to $s = s_j$, that is,

$$-k_1\frac{\partial T_1}{\partial n_j} = -k_2\frac{\partial T_2}{\partial n_j} \qquad \text{for } t > 0, s_{j-1} \le s \le s_j \qquad \text{on } S \tag{12.48}$$

where n_j is the outward normal to element j. Using Equation 12.38a, the temperature at element k in body 1 and time t can be given by

$$T_{k1}(t) = T_{01} + \sum_{j=1}^{N}\int_0^t q_j(\tau)\frac{\partial\phi_{kj}^{(1)}(t-\tau)}{\partial t}d\tau \tag{12.49}$$

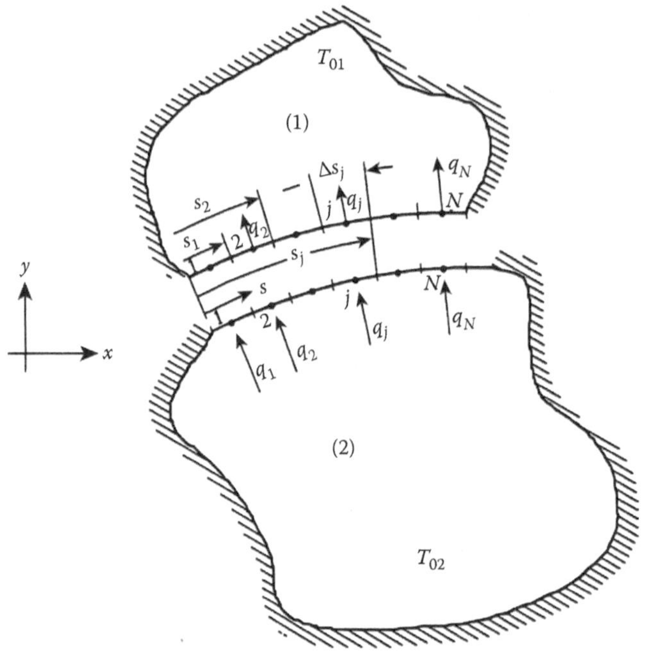

FIGURE 12.7 Possible distribution of surface elements for connected geometries.

where $\phi_{kj}^{(1)}(t)$ is the temperature rise at element k and time t due to unit step heat flux over element j of surface 1. Similar to Equation 12.49, an integral equation can be given for the kth surface element of body 2.

$$T_{k2}(t) = T_{02} - \sum_{j=1}^{N} \int_0^t q_j(\tau) \frac{\partial \phi_{kj}^{(2)}(t-\tau)}{\partial t} d\tau \tag{12.50}$$

where $\phi_{kj}^{(2)}(t)$ is the influence function for body 2. Note that the minus sign before the summation in Equation 12.50 is used because the heat flux is pointing outward from body 2. For the case where the bodies are in perfect contact, one can write

$$T_{k1}(t) = T_{k2}(t) \qquad \text{for} \qquad k = 1, 2, \ldots, N \tag{12.51}$$

Then, by introducing Equations 12.49 and 12.50 into Equation 12.51, a set of integral equations for $k = 1, 2, \ldots, N$ is obtained as

$$T_{02} - T_{01} = \sum_{j=1}^{N} \int_0^t q_j(\tau) \frac{\partial \phi_{kj}(t-\tau)}{\partial t} d\tau \qquad \text{for } k = 1, 2, 3, \ldots, N \tag{12.52}$$

where

$$\phi_{kj} = \phi_{kj}^{(1)}(t) + \phi_{kj}^{(2)}(t) \tag{12.53}$$

Equation 12.52 is the flux-based USE equation for two bodies in perfect contact. It represents a set of Volterra equations of the first kind with the unknown heat fluxes, $q_k(t)$'s, appearing inside the integrals.

Even though the perfect contact is a common interface assumption, it will only be valid for very intimate contact, such as a soldered joint. For a more general case of imperfect contact, Equation 12.51 is replaced by

$$q_k(t) = h_k(t)[T_{k2}(t) - T_{k1}(t)] \qquad \text{for } k = 1, 2, \ldots, N \tag{12.54}$$

where $h_k(t)$ is the time-variable contact conductance for surface element k. The above relation tends to the case of perfect contact as $h_k \to \infty$. It also includes the cases of convection, prescribed heat flux, and prescribed temperature boundary conditions. By introducing Equations 12.49 and 12.50 into Equation 12.54, a set of integral equations for $k = 1, 2, \ldots, N$, is obtained:

$$T_{02} - T_{01} = \frac{q_k(t)}{h_k(t)} + \sum_{j=1}^{N} \int_0^t q_j(\tau) \frac{\partial \phi_{kj}(t-\tau)}{\partial t} d\tau \qquad \text{for } k = 1, 2, \ldots, N \tag{12.55}$$

Equation 12.55 is the flux-based USE equation for two bodies with imperfect contacts. It represents a set of Volterra equations of the second kind with the unknown heat fluxes, $q_k(t)$'s, appearing both inside and outside the integrals.

The sets of integral equations presented by the USE equations (12.52) and (12.55) can be solved simultaneously for N unknown heat flux histories $q_1(t), q_2(t), \ldots, q_N(t)$. The method of solution is described for the case of imperfect contact which includes the other cases as well.

12.3.5 Numerical Solution of the USE Equations for Bodies in Contact

In a similar manner to that discussed in Section 12.3.3, the flux-based USE equation (12.55) can be approximated by a system of linear algebraic equations by replacing the integrals with summations. As the first step, the time region 0 to t is divided into M equal small time intervals, Δt, so that t_M represents the value of t at the end point of the Mth interval ($t_M = M\Delta t$). Then, Equation 12.55 can be written as

$$T_{02} - T_{01} = \frac{q_k(t_M)}{h_k(t_M)} + \sum_{j=1}^{N} \sum_{i=1}^{M} \int_{t_{i-1}}^{t_i} q_j(\tau) \frac{\partial \phi_{kj}(t_M-\tau)}{\partial t} d\tau$$

$$\text{for } k = 1, 2, \ldots, N \tag{12.56a}$$

where

$$t_0 \equiv 0 \tag{12.56b}$$

In the simplest form of approximation the heat flux histories $q_j(t)$ are assumed to have constant values in each time interval so that

$$T_0 = \frac{q_{kM}}{h_{kM}} + \sum_{j=1}^{N} \sum_{i=1}^{M} q_{ji} \Delta \phi_{kj,M-i} \qquad \text{for } k = 1, 2, \ldots, N \tag{12.57a}$$

where

$$T_0 = T_{02} - T_{01} \qquad \Delta\phi_{kj,M-i} = \phi_{kj,M+1-i} - \phi_{kj,M-i} \tag{12.57b, c}$$

and

$$q_{ji} \equiv q_j(t_i) \qquad \phi_{kj,i} \equiv \phi_{kj}(t_i) \tag{12.57d, e}$$

In the form given by Equation 12.57a, the heat fluxes q_{jM}'s (for $j = 1, 2, \ldots, N$) can be determined at different time intervals one after another, by marching forward in time for $M = 1, 2, 3, \ldots$. While calculating each new time component, the fluxes at previous times, $q_{j1}, q_{j2}, q_{j3}, \ldots, q_{j,M-2}, q_{j,M-1}$ are known for $j = 1, 2, \ldots, N$. Thus for each time step, Equation 12.57a represents a system of N equations with N unknowns $q_{1M}, q_{2M}, q_{3M}, \ldots, q_{NM}$. The objective is to solve this system for the unknowns q_{jM}, for $j = 1, 2, \ldots, N$. Rearranging Equation 12.57a in standard form with unknowns, q_{jM}'s, on the left, and knowns on the right, and noting that $\phi_{kj0} = 0$, one can write

$$\frac{q_{kM}}{h_{kM}} + \sum_{j=1}^{N} q_{jM}\phi_{kj1} = T_0 - \sum_{j=1}^{N}\sum_{i=1}^{M-1} q_{ji}\Delta\phi_{kj,M-i} \tag{12.58}$$

Expressing Equation 12.58 in matrix form gives

$$(\overline{\overline{H}}_M + \overline{\overline{\Phi}}_1)\overline{q}_M = \overline{T}_0 - \sum_{i=1}^{M-1} \Delta\overline{\overline{\Phi}}_{M-i}\,\overline{q}_i \tag{12.59}$$

where $\overline{T}_0$ is the initial temperature vector, $\overline{\overline{H}}_M$ is the conductance matrix, $\overline{\overline{\Phi}}_i$ and $\overline{q}_i$ are the influence matrix and the heat flux vector at time t_i, respectively.

$$\overline{\overline{\Phi}}_i \equiv \begin{bmatrix} \phi_{11i} & \phi_{12i} & \cdots & \phi_{1Ni} \\ \phi_{21i} & \phi_{22i} & & \phi_{2Ni} \\ \vdots & & & \\ \phi_{N1i} & \phi_{N2i} & & \phi_{NNi} \end{bmatrix} \tag{12.60a}$$

$$\overline{\overline{H}}_M \equiv \text{diag}\left[\frac{1}{h_{1M}}\,\frac{1}{h_{2M}}\cdots\frac{1}{h_{NM}}\right] \tag{12.60b}$$

$$\overline{q}_i \equiv \begin{bmatrix} q_{1i} \\ q_{2i} \\ \vdots \\ q_{Ni} \end{bmatrix} \qquad \overline{T}_0 \equiv \begin{bmatrix} T_0 \\ T_0 \\ \vdots \\ T_0 \end{bmatrix} \tag{12.60c, d}$$

If further $\overline{\overline{C}}_M$ and $\overline{D}_M$ are defined to be the matrices

$$\overline{\overline{C}}_M = \overline{\overline{H}}_M + \overline{\overline{\Phi}}_1 \qquad \overline{D}_M = \overline{T}_0 + \overline{E}_M - \overline{F}_M \tag{12.61a, b}$$

where

$$\overline{E}_M = \sum_{i=1}^{M-1} \overline{\overline{\Phi}}_{M-i}\, \overline{q}_i \tag{12.62a}$$

and

$$\overline{F}_M = \sum_{i=1}^{M-1} \overline{\overline{\Phi}}_{M+1-i}\, \overline{q}_i \tag{12.62b}$$

Then Equation 12.59 can be written as

$$\overline{\overline{C}}_M \overline{q}_M = \overline{D}_M \tag{12.63}$$

Solving Equation 12.63 for $\overline{q}_M$, gives

$$\overline{q}_M = \overline{\overline{C}}_M^{-1}\, \overline{D}_M \tag{12.64}$$

The $\overline{\overline{C}}_M$ matrix, multiplier of $\overline{q}_M$, has to be calculated at each time step if the diagonal matrix $\overline{\overline{H}}_M$ is a function of time. However, if contact conductances do not change with time, the $\overline{\overline{C}}_M$ matrix needs to be calculated only once during the entire solution and an alternative form of solution can be given as (see Note 1 at end of the chapter).

$$q_1 = \overline{\overline{C}}^{-1}\overline{T}_0 \tag{12.65a}$$

$$\overline{q}_M = M\overline{q}_1 + \overline{\overline{B}}\left[\sum_{i=1}^{M-1} \overline{q}_i\right] - \overline{\overline{C}}^{-1}\overline{F}_M \qquad \text{for } M = 2, 3, \ldots \tag{12.65b}$$

where

$$\overline{\overline{B}} = \overline{\overline{H}}^{-1}\overline{\overline{\Phi}}_1 \tag{12.66}$$

Notice that, since $\overline{\overline{C}}$ and $\overline{\overline{H}}$ are not functions of time in Equations 12.65 and 12.66, the subscript M is dropped. For the case of perfect contact where $h_{kM} \to \infty$, the diagonal conductance matrix, $\overline{\overline{H}}_M$, becomes zero, which implies that

$$\overline{\overline{C}} = \overline{\overline{\Phi}}_1 \tag{12.67}$$

Introducing Equations 12.61b, 12.62a, b and 12.67 into Equation 12.64 results in a simpler form of solution as

$$\overline{q}_1 = \overline{\overline{C}}^{-1}\overline{T}_0 \tag{12.68a}$$

$$\overline{q}_M = M\overline{q}_1 - \overline{\overline{C}}^{-1}\overline{F}_M \qquad \text{for } M = 2, 3, \ldots \tag{12.68b}$$

The elements of the $[N \times N]$ influence matrix $\overline{\overline{\Phi}}_i$ are

$$\phi_{kji} = \phi_{kji}^{(1)} + \phi_{kji}^{(2)} \tag{12.69}$$

If the two bodies in contact have the same geometry and thermal properties, then

$$\phi_{kji} = 2\phi_{kji}^{(1)} = 2\phi_{kji}^{(2)} \tag{12.70}$$

It is helpful to display the expression for $\overline{q}_M$ more explicitly. To illustrate, the case of perfect contact at the interface with only two elements is considered ($N = 2$). In other words there are two heat flux histories, $q_1(t)$ and $q_2(t)$, to be determined. For simplicity, only three time steps are considered ($M = 3$). At the first time step, Equation 12.68a becomes

$$\begin{bmatrix} q_{11} \\ q_{12} \end{bmatrix} = \begin{bmatrix} C_{11} & C_{12} \\ C_{21} & C_{22} \end{bmatrix}^{-1} \begin{bmatrix} T_0 \\ T_0 \end{bmatrix} \tag{12.71}$$

where

$$C_{kj} = \phi_{kj1} = \phi_{kj1}^{(1)} + \phi_{kj1}^{(2)} \tag{12.72}$$

Solving the above system, Equation 12.71 for q_{11} and q_{21} yields

$$q_{11} = \frac{T_0(C_{22} - C_{12})}{\Delta} \tag{12.73a}$$

$$q_{21} = \frac{T_0(C_{11} - C_{21})}{\Delta} \tag{12.73b}$$

where

$$\Delta = C_{11}C_{22} - C_{12}C_{21} \tag{12.73c}$$

For the second time step, $M = 2$, Equation 12.68b becomes

$$\begin{bmatrix} q_{12} \\ q_{22} \end{bmatrix} = 2\begin{bmatrix} q_{11} \\ q_{21} \end{bmatrix} - \begin{bmatrix} C_{11} & C_{12} \\ C_{21} & C_{22} \end{bmatrix}^{-1} \begin{bmatrix} F_{12} \\ F_{22} \end{bmatrix} \tag{12.74}$$

Solving Equation 12.74 for q_{12} and q_{22} yields

$$\begin{aligned} q_{12} &= \frac{(2T_0 - F_{12})C_{22} - (2T_0 - F_{22})C_{12}}{\Delta} \\ &= 2q_{11} - \frac{F_{12}C_{22} - F_{22}C_{12}}{\Delta} \end{aligned} \tag{12.75a}$$

$$\begin{aligned} q_{22} &= \frac{(2T_0 - F_{22})C_{11} - (2T_0 - F_{12})C_{21}}{\Delta} \\ &= 2q_{21} - \frac{F_{22}C_{11} - F_{12}C_{21}}{\Delta} \end{aligned} \tag{12.75b}$$

where

$$F_{12} = \phi_{112}q_{11} + \phi_{122}q_{21} \tag{12.76a}$$

$$F_{22} = \phi_{212}q_{11} + \phi_{222}q_{21} \tag{12.76b}$$

In a similar manner, for the third time step, $M = 3$, one can write

$$q_{13} = 3q_{11} - \frac{F_{13}C_{22} - F_{23}C_{12}}{\Delta} \tag{12.77a}$$

$$q_{23} = 3q_{21} - \frac{F_{23}C_{11} - F_{13}C_{21}}{\Delta} \tag{12.77b}$$

where

$$F_{13} = \sum_{i=1}^{2}(\phi_{11,4-i}\, q_{1i} + \phi_{12,4-i}\, q_{2i}) \tag{12.78a}$$

$$F_{23} = \sum_{i=1}^{2}(\phi_{21,4-i}\, q_{1i} + \phi_{22,4-i}\, q_{2i}) \tag{12.78b}$$

Notice that F_{1M} and F_{2M} are the only terms that should be evaluated at each time step.

Because of convolution behavior of the summations given in Equations 12.62a and b, the influence matrices, $\overline{\overline{\Phi}}_i$'s, need to calculated at each time step. Consequently, most of the computation effort is in the evaluation of column matrix $\overline{D}$, particularly as the value of M becomes larger.

12.3.6 Influence Functions

An influence function, $\phi_{kj}(t)$, is the temperature rise at time t and element k due to a unit step heat flux at $t = 0$ at element j. When providing the influence functions, there are two cases to consider: when $k = j$ (temperature rise at location of heating) and $k \neq j$ (temperature rise at other than the heating location). The more important and more difficult to obtain is for $k = j$, particularly for small times. A number of influence functions for $\phi_{kk}(t)$ are described and referenced in this section. For the case of $k \neq j$, the $\Delta\phi_{kj.M}$ values, given by Equation 12.57c, are efficiently obtained through the use of GFs.

The simplest influence functions are for one-dimensional cases, such as shown in Figure 12.8a, b, c. The first is for constant heat flux q_0 equal to unity over the surface of a semi-infinite body. The $\phi_{kk}(t)$ expression at the heated surface of a semi-infinite body shown in Figure 12.8a is simply

$$\phi_{kk}(t) = 2\left(\frac{t}{\pi k\rho c}\right)^{1/2} \tag{12.79}$$

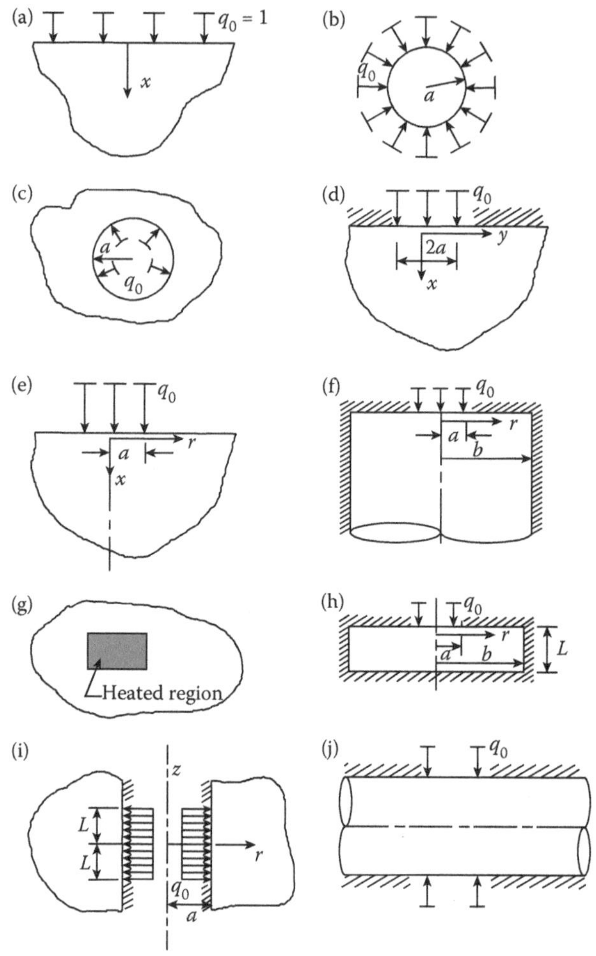

FIGURE 12.8 Geometries and boundary conditions for various influence functions.

Figure 12.8b is for a solid cylinder or sphere. Another basic case is for the region outside a radius of a and with the heat flux of $q_0 = 1$ at $r = a$; this can be for both cylindrical and spherical geometries and is illustrated by Figure 12.8c. For early times, the geometries shown in Figure 12.8b and c have $\phi_{kj}(t)$ values that contain additive curvature corrections (Beck et al., 1985b) to Equation 12.79.

Two cases having two-dimensional heat transfer for the semi-infinite geometries are shown in Figure 12.8d and e. Figure 12.8d is for a heated strip $2a$ wide (Litkouhi and Beck, 1982), and Figure 12.8e has a circular source of radius a (Beck, 1980). These two cases have "edge" corrections for small times that are additive to Equation 12.79. Large-time behavior of these two cases are also known; Figure 12.8d approaches an $\ln(t)$ variation that is typical of a line source. The circular heat source case of

Figure 12.8e goes to steady state for large times. There is also a principle of additivity for large times, which is discussed by Beck et al. (1985b) in detail. Another case of circular source is when it is centered in the surface of a semi-infinite cylinder (Beck, 1981b); see Figure 12.8f. A case of rectangular heat source on the surface of a semi-infinite body (Keltner et al., 1988) is depicted by Figure 12.8g. A finite geometry is shown by Figure 12.8h; a circular source is applied at the end of finite cylinder (Beck and Keltner, 1987). Figure 12.8i and j show two solutions that can be constructed by the principles of additivity that are discussed by Beck et al. (1984). Many other solutions can be constructed in a similar manner.

12.4 APPROXIMATE ANALYTICAL SOLUTION (SINGLE ELEMENT)

It is sometimes possible to obtain approximate analytical solutions by considering only one surface element along the interface between the connected bodies. This is known as the single-node USE approach. In this approach the coupling interfacial boundary conditions is relaxed so that neither temperature nor heat flux need simultaneously match for all points along the interface and at all times. Instead, a less stringent requirement equates average heat fluxes between the coupled regions while still requiring simultaneous matching of area-average interfacial temperatures.

For one surface element, the sets of integral equations represented by Equation 12.39a and b reduce to two single integral equations given, for flux-based equations, by

$$T(\mathbf{r},t) = \int_0^t q(\tau)\frac{\partial\phi(\mathbf{r},t-\tau)}{\partial t}d\tau \tag{12.80a}$$

and, for temperature-based equations:

$$T(\mathbf{r},t) - T_0 = \int_0^t [T_s(\tau) - T_0]\,\frac{\partial\theta(\mathbf{r},t-\tau)}{\partial t}d\tau \tag{12.80b}$$

These Duhamel's integral equations can be written in their alternative forms (see Equation 12.1a and b) as

$$\text{flux-based equation:} \quad T(\mathbf{r},t) = \frac{\partial}{\partial t}\int_0^t q(\tau)\phi(\mathbf{r},t-\tau)d\tau \tag{12.81a}$$

$$\text{temperature-based equation:} \quad T(\mathbf{r},t) - T_0 = \frac{\partial}{\partial t}\int_0^t [T_S(\tau) - T_0] \times \theta(\mathbf{r},t-\tau)d\tau \tag{12.81b}$$

Taking the Laplace transform of Equation 12.81a and b, analytical solutions may be obtained that yield relatively accurate results for a certain class of problems. The ease or difficulty of obtaining such solutions depends entirely on the particular expressions for the influence functions ϕ or θ. The procedure is best illustrated with the following example.

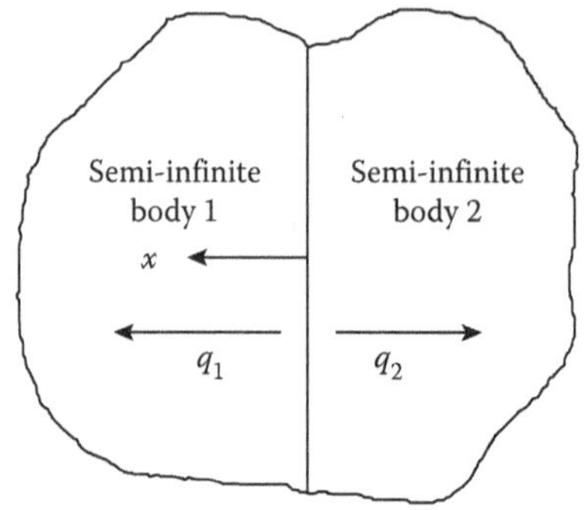

FIGURE 12.9 Two homogeneous semi-infinite bodies at different initial temperatures brought into thermal contact.

Example 12.3:

Consider the specific classic case of two homogeneous semi-infinite bodies initially at different temperatures T_{01} and T_{02} brought together as shown in Figure 12.9. The objective is to find approximate analytical solutions for the interface temperature and/or heat flux by utilizing the Laplace transformations. Both temperature-based and heat flux-based solutions are considered here.

Temperature-Based Solution

From the temperature-based Equation 12.81b, the temperature at position x in body 1 and at time t is given by

$$T_1(x,t) = T_{01} + \frac{\partial}{\partial t}\int_0^t [T_1(0,\tau) - T_{01}]\theta^{(1)}(x, t-\tau)\, d\tau \qquad (12.82)$$

where $\theta^{(1)}(x, t)$ is the temperature-based influence function for body 1. It represents the temperature rise at position x in body 1 due to a unit step increase in temperature at the surface of the body. The heat flow through the surface region of body 1 is given by

$$q_1(t)A = \frac{\partial}{\partial t}\left\{A\int_0^t [T_1(0,\tau) - T_{01}]\theta_q^{(1)}(0, t-\tau) d\tau\right\} \qquad (12.83)$$

where A is the surface area and $q_1(t)$ is the surface (or interface) heat flux. The function $\theta_q^{(1)}$ is the area average heat flux for a unit increase in surface temperature. It is given by

$$\theta_q^{(1)}(0,t) = \frac{1}{A}\int_A -k\left.\frac{\partial\theta^{(1)}(x,t)}{\partial x}\right|_{x=0} dA \qquad (12.84)$$

If there are no interface heat sources, the same average heat flux that enters body 1 then leaves body 2 so that

$$Aq_1(t) = -Aq_2(t) \qquad (12.85)$$

For perfect or imperfect contact the influence function $\theta^{(i)}(x, t)$ is (Carslaw and Jaeger, 1959)

$$\theta^{(i)}(x, t) = \text{erfc}\left(\frac{|x|}{2\sqrt{\alpha_i t}}\right) \tag{12.86}$$

The related heat flux is

$$\theta_q^{(i)}(0, t) = \pm \frac{k_i}{(\pi\alpha_i t)^{1/2}} \exp\left(\frac{-x^2}{4\alpha_i t}\right)\Bigg|_{x=0} \tag{12.87}$$

where the plus sign is for $i = 1$ (body 1) and the minus sign is for $i = 2$ (body 2).

For perfect contact, the interface temperature (at $x = 0$) is identical for both bodies so that $T_1(0, t) = T_2(0, t) = T(0, t)$. Substituting Equation 12.83 in Equation 12.85 gives

$$\begin{aligned}&\frac{\partial}{\partial t}\int_0^t [T(0, \tau) - T_{01}]\theta_q^{(1)}(0,\ t - \tau)d\tau \\ &\quad = \frac{\partial}{\partial t}\int_0^t [T(0, \tau) - T_{02}]\theta_q^{(2)}(0, t - \tau)d\tau\end{aligned} \tag{12.88}$$

and then using Equation 12.87 gives

$$\begin{aligned}&\frac{\partial}{\partial t}\int_0^t [T(0, \tau) - T_{01}]\frac{k_1}{\sqrt{\pi\alpha_1(t - \tau)}}\, d\tau \\ &\quad = -\frac{\partial}{\partial t}\int_0^t [T(0, \tau) - T_{02}]\frac{k_2}{\sqrt{\pi\alpha_2(t - \tau)}}\, d\tau\end{aligned} \tag{12.89}$$

This is a Volterra equation of the first kind where the unknown function is $T(0, t)$. Taking the Laplace transform of Equation 12.89 gives

$$s\mathcal{L}[T(0, t) - T_{01}] \cdot \frac{k_1}{\sqrt{\alpha_1 s}} = -s\mathcal{L}[T(0, t) - T_{02}] \cdot \frac{k_2}{\sqrt{\alpha_2 s}} \tag{12.90}$$

where $\mathcal{L}[T(0, t)]$ is the Laplace transform of $T(0,\ t)$ and s is the Laplace transform parameter. For convenience, let $\hat{T}(0, s) = \mathcal{L}[T(0, t)]$. Without loss of generality, let $T_{01} = T_0$ and $T_{02} = 0$, then

$$\left[\hat{T}(0, s) - \frac{T_0}{s}\right] \cdot \frac{k_1}{\sqrt{\alpha_1 s}} = -\hat{T}(0, s) \cdot \frac{k_2}{\sqrt{\alpha_2 s}} \tag{12.91}$$

Solving for $\hat{T}(0, s)$ gives

$$\hat{T}(0, s) = \frac{T_0}{s}\frac{k_1}{(\alpha_1 s)^{1/2}}\left[\frac{k_1}{(\alpha_1 s)^{1/2}} + \frac{k_2}{(\alpha_2 s)^{1/2}}\right]^{-1} \tag{12.92}$$

which has the inverse Laplace transform of

$$T(0, t) = T_0(1 + \beta)^{-1} \tag{12.93a}$$

where

$$\beta = \left(\frac{k_2 \rho_2 c_2}{k_1 \rho_1 c_1}\right)^{1/2} \tag{12.93b}$$

This is the desired exact solution for the surface temperature for both bodies for the case of perfect contact (Carslaw and Jaeger, 1959).

Next, consider the more complex case of imperfect contact of two semi-infinite bodies. Let there be a contact conductance h between the bodies. The heat fluxes are related by

$$q_1(t) = -q_2(t) = h[T_1(0,t) - T_2(0,t)] \tag{12.94}$$

where now both $T_1(0,t)$ and $T_2(0,t)$ are unknown functions.

Solving Equation 12.94 for $T_1(0,t)$, in terms of $q_2(t)$, substituting in Equation 12.83 and again equating the heat flows gives

$$\begin{aligned} &\frac{\partial}{\partial t}\left(\int_0^t [T_2(0,\tau) - T_{01}]\theta_q^{(1)}(0, t-\tau)d\tau + \int_0^t \left\{\frac{1}{h}\frac{\partial}{\partial t}\int_0^\tau [T_2(0,\gamma)\right.\right. \\ &\left.\left. - T_{02}]\theta_q^{(2)}(0,\tau-\gamma)d\gamma\right\}\theta_q^{(1)}(0,t-\tau)d\tau\right) \\ &= -\frac{\partial}{\partial t}\int_0^t [T_2(0,\tau) - T_{02}]\theta_q^{(2)}(0,t-\tau)d\tau \end{aligned} \tag{12.95}$$

This is an integral equation of the Volterra type, except now a double integration is present; the unknown function is $T_2(0,t)$.

The integral equation given by Equation 12.95 would in most cases be solved numerically, but fortunately in this case the exact solution can be found using the Laplace transform.

$$s\left[\left(\hat{T}_2 - \frac{T_{01}}{s}\right)\hat{\theta}_q^{(1)} + \frac{s}{h}\left(\hat{T}_2 - \frac{T_{02}}{s}\right)\hat{\theta}_q^{(2)}\hat{\theta}_q^{(1)}\right] = -s\left(\hat{T}_2 - \frac{T_{02}}{s}\right)\hat{\theta}_q^{(2)} \tag{12.96}$$

where $\hat{\theta}_q^{(i)} = k_i/(\alpha_i s)^{1/2}$.

Factoring out common terms and letting $T_{01} = T_0$ and $T_{02} = 0$ gives

$$\begin{aligned} \hat{T}_2 &= \frac{T_0}{s}\left(\frac{k_1/\sqrt{\alpha_1 s}}{k_1/\sqrt{\alpha_1 s} + k_1 k_2/h\sqrt{\alpha_1\alpha_2} + k_2/\sqrt{\alpha_2 s}}\right) \\ &= \frac{T_0 h}{\sqrt{k_2\rho_2 c_2}} \cdot \frac{1}{s[\sqrt{s} + h(1/\sqrt{k_1\rho_1 c_1} + 1/\sqrt{k_2\rho_2 c_2})]} \end{aligned} \tag{12.97}$$

Taking the inverse Laplace transform yields the desired exact interface temperature of

$$T_2(0,t) = \frac{T_0[1 - e^{h^2b^2t}\operatorname{erfc}(hb\sqrt{t})]}{1+\beta} \tag{12.98a}$$

where β is defined in Equation 12.93b and

$$b = (k_1\rho_1 c_1)^{-1/2} + (k_2\rho_2 c_2)^{-1/2} \tag{12.98b}$$

Interior temperatures can now be found by introducing the expression given by Equation 12.98a into a Duhamel's integral similar to Equation 12.82 with $\theta(x, t)$ given by Equation 12.86. The heat flow across the interface can be found by using Equation 12.98a in Duhamel's integral similar to Equation 12.83.

Heat Flux-Based Solution

The heat flux-based solution can be obtained in a similar manner as the temperature-based solution. From the heat flux-based integral Equation 12.81a, the temperature at position x in body 1 for a time variable surface heat flux $q(t)$ is given by

$$T_1(x, t) = T_{01} + \frac{\partial}{\partial t}\int_0^t q(\tau)\phi^{(1)}(x, t-\tau)d\tau \tag{12.99}$$

where $\phi^{(1)}(x, t)$ is the flux-based influence function for body 1. It represents the temperature rise at position x in body 1 and at time t due to unit step increase in the surface heat flux at time zero. Similarly, the temperature at any position x in body 2 and at time t is given by

$$T_2(x, t) = T_{02} - \frac{\partial}{\partial t}\int_0^t q(\tau)\phi^{(2)}(x, t-\tau)d\tau \tag{12.100}$$

where $\phi^{(2)}(x, t)$ is the influence function for body 2. The assumption of a spatially uniform heat flux is not always compatible with a spatially uniform temperature. The statement in terms of an average heat flux given by Equation 12.85 is always true.

For the special geometry of two-infinite bodies coming into uniform contact over the complete interface, the interface temperatures are not functions of position. The influence function $\phi^{(i)}(x, t)$ is a function of a single space dimension and time,

$$\phi^{(i)}(x, t) = \frac{2t^{1/2}}{(k_i\rho_i c_i)^{1/2}}\,\text{ierfc}\left[\frac{x}{2(\alpha_i t)^{1/2}}\right] \tag{12.101}$$

where x is directed inward in each body and the i is 1 or 2. At the surface of the body, $x = 0$ and ierfc $(0) = \pi^{-1/2}$.

Consider the first case of perfect contact for which the interface temperature must be the same for both bodies. Equating Equations 12.99 and 12.100 with $T_{01} = T_0$ and $T_{02} = 0$ at $x = 0$ yields

$$T_0 + \frac{\partial}{\partial t}\int_0^t q(\tau)\frac{2(t-\tau)^{1/2}d\tau}{(\pi k_1\rho_1 c_1)^{1/2}} = -\frac{\partial}{\partial t}\int_0^t q(\tau)\frac{2(t-\tau)^{1/2}d\tau}{(\pi k_2\rho_2 c_2)^{1/2}} \tag{12.102}$$

which can be rearranged to the form

$$\frac{\partial}{\partial t}\int_0^t q(\tau)2b\pi^{-1/2}(t-\tau)^{1/2}d\tau = -T_0 \tag{12.103}$$

This is again a Volterra integral equation of the first kind and can be solved for $q(t)$ using the numerical methods in (Beck, 1968). For this simple case, the solution can be obtained as above by utilizing the Laplace transform to get

$$q(t) = -T_0 b^{-1}(\pi t)^{-1/2} \tag{12.104}$$

Utilizing Equation 12.101 (with $x = 0$) and Equation 12.104 in Equation 12.99 yields

$$T_1(0, t) = T_0(1 + \beta)^{-1} \tag{12.105}$$

which is the same as Equation 12.93 which was derived using the temperature form of Duhamel's theorem.

A comparison of the above procedures for the T- and q-based solutions for the perfect contact example considered shows both approaches yield a Volterra integral equation of the first kind. In the temperature case, the solution is for the interface temperature while in the heat flux case the solution is for $q(t)$. The solutions are similar, although different quantities are found. For the T case, the $q(t)$ function is found by solving Equation 12.83 given $T_1(0, t)$ and, for the heat flux case, the $T_i(0, t)$ function is found by solving Equations 12.99 or 12.100 given $q(t)$. If only the interface temperature is desired, then the T-based method is more direct.

Next, for the q-based approach, consider the imperfect contact case. Equation 12.94 still applies but utilizing Equations 12.99 through 12.101 yields

$$-T_0 + \frac{q(t)}{h} = \frac{2b}{\sqrt{\pi}} \frac{\partial}{\partial t} \int_0^t q(\tau)(t - \tau)^{1/2} d\tau \tag{12.106}$$

which is again a Volterra integral equation of the second kind since $q(t)$ appears both inside and outside the integral. Equation 12.106 is simpler than the comparable T-based Equation 12.95 which has a double integral. A solution of Equation 12.106 for $q(t)$ utilizing the Laplace transform is

$$q(t) = -hT_0 e^{h^2 b^2 t} \text{erfc}(hbt^{1/2}) \tag{12.107}$$

where b is defined by Equation 12.98b. If h goes to infinity, Equation 12.107 reduces to Equation 12.104. The next step is to use Equations 12.99 and 12.100 with Equation 12.107 to determine the surface temperature histories. Though the integrals are not easy to evaluate, the same results are found as by the temperature-based approach.

From a comparison of the T- and q-based USE integral equations (12.95) and (12.106), the q-based equation has a simpler form and poses less difficulty in numerical solution (which might be required for more complex geometries). Furthermore, the q-based Equation 12.106, is derived in a much more straightforward manner. Hence, based on the above example, the q-based approach is to be recommended over the T-based approach.

12.5 EXAMPLES

To demonstrate the utility of the USE method, two well-known problems are solved in this section. The first problem is a semi-infinite body with the mixed boundary conditions of a step change of the surface temperature over a disk of radius **a** and insulated elsewhere, as shown in Figure 12.10a. This problem is similar to the problem of two semi-infinite bodies initially at different uniform temperatures suddenly brought together over a circular area, as shown in Figure 12.10b (a contact conductance problem). The second problem involves a semi-infinite cylinder attached perpendicularly

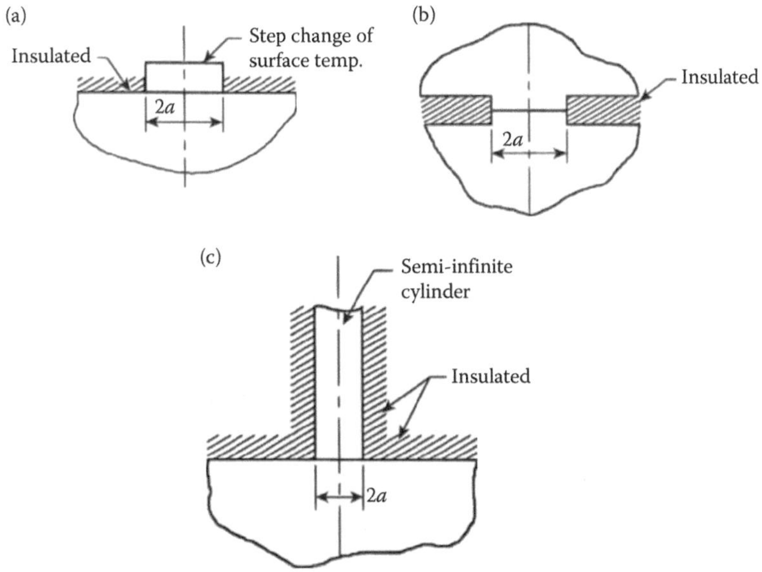

FIGURE 12.10 (a) A semi-infinite body with step change of surface temperature over a circular area and insulated elsewhere. (b) Two homogeneous semi-infinite bodies at different initial temperatures brought into thermal contact over a circular area (a contact conductance problem). (c) Semi-infinite cylinder attached to semi-infinite body simulating intrinsic thermocouple problem.

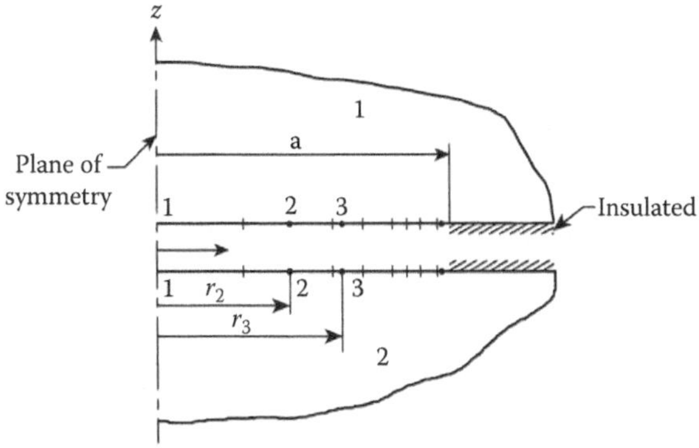

FIGURE 12.11 Distribution of surface elements for connected semi-infinite bodies.

to a semi-infinite body (an intrinsic thermocouple problem); see Figure 12.10c. Both single-node and multinode USE solutions are given and the results are compared with other existing analytical and numerical solutions.

Due to the axisymmetric nature of these problems, in each case the interface area is divided into 10 annular variable-spaced surface elements with smaller elements being closer to the edge of the contact area, as shown in Figures 12.11 and 12.12. The inner

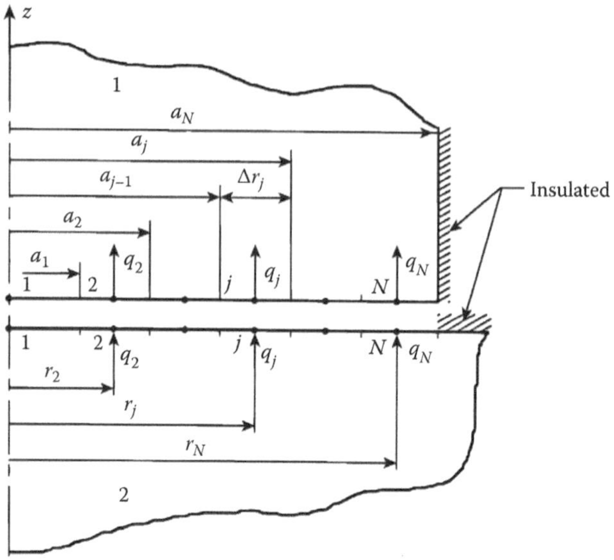

FIGURE 12.12 Distribution of surface elements for connected semi-infinite cylinder and semi-infinite body.

and outer radii of each element are denoted by a_{j-1} and a_j, respectively ($j = 1, 10$ and $a_0 = 0$). The heat flux and temperature are approximated to be constant over each surface element and are specified at the points.

$$r_1 = 0 \tag{12.108a}$$

$$r_j = \frac{a_j - a_{j-1}}{2} \quad \text{for } j = 2, 10 \tag{12.108b}$$

Since, in each problem, the connected bodies are assumed to be in perfect contact, the simplified form of solution given by Equation 12.68a and b are used. Substituting Equations 12.62b and 12.67 into Equation 12.68a, b yields

$$\overline{q}_1 = \overline{\overline{\Phi}}_1^{-1}\overline{T}_0 \tag{12.109a}$$

$$\overline{q}_M = M\overline{q}_1 - \overline{\overline{\Phi}}_1^{-1}\sum_{i=1}^{M-1}\overline{q}_i \qquad \text{for } M = 2, 3, \ldots \tag{12.109b}$$

At each time step, Equation 12.109a, b are solved for unknown elemental heat fluxes $q_{1M}, q_{2M}, \ldots, q_{10M}$.

The required influence functions for the above problems are shown in Figure 12.13a and b. They are evaluated from the available exact closed-form solutions of a semi-infinite body heated by a constant disk heat source (Beck, 1981a, Figure 12.14a) and a semi-infinite insulated cylinder heated by a constant heat flux over a disk area centered at the end (Beck, 1981b, Figure 12.14b), by simple superposition. That is,

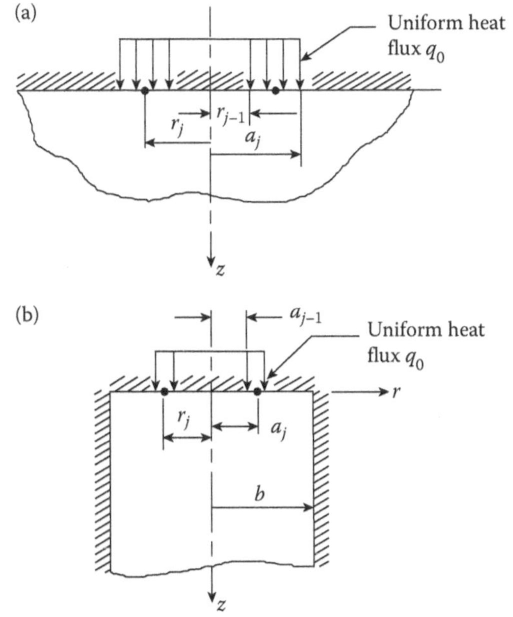

FIGURE 12.13 (a) Semi-infinite body heated at surface over annular-shaped region. (b) Semi-infinite insulated cylinder heated with constant annular-shaped heat source.

$$\phi_{kji} = \gamma_{kji} - \sum_{n=1}^{j-1} \phi_{kni} \quad \text{for } j,k = 1,2,\ldots,10 \tag{12.110}$$

where γ_{kji} represents the temperature rise at element $k(r = r_k)$ due to a unit heat flux at the disk with radius a_j and at time t_i. See Figure 12.15.

For the first problem, the USE solutions are compared with other available solutions (Schneider et al., 1977; Sadhel, 1980; Keltner, 1973) on the basis of the dimensionless thermal constriction resistance across the interface area. The transient thermal constriction resistance is defined as "the difference between the average temperature of the contact area and the temperature far from the contact area divided by the total instantaneous heat flow through the contact area" (Schnider, 1979), and is given by

$$R_{c1}(t) = \frac{T_c(t) - T_{01}}{Q_c(t)} \qquad R_{c2}(t) = \frac{T_{02} - T_c(t)}{Q_c(t)} \tag{12.111a, b}$$

where $T_c(t)$ is the average temperature of the contact area, $Q_c(t)$ is the total heat flow through the contact area, and the $R_{c1}(t)$ and $R_{c2}(t)$ are the thermal constriction resistances for bodies 1 and 2, respectively. The total thermal constriction resistance for the two semi-infinite bodies is then determined by

$$R_c(t) = R_{c1}(t) + R_{c2}(t) = \frac{T_{02} - T_{01}}{Q_c(t)} \tag{12.112}$$

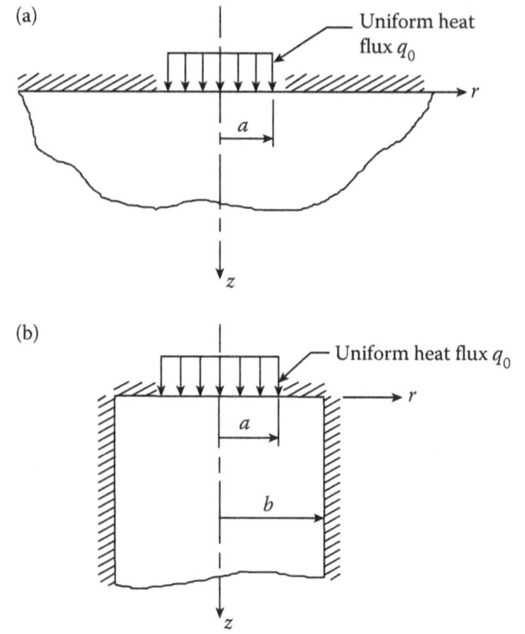

FIGURE 12.14 (a) Semi-infinite body heated over circular area. (b) Semi-infinite insulated cylinder with constant disk heat source.

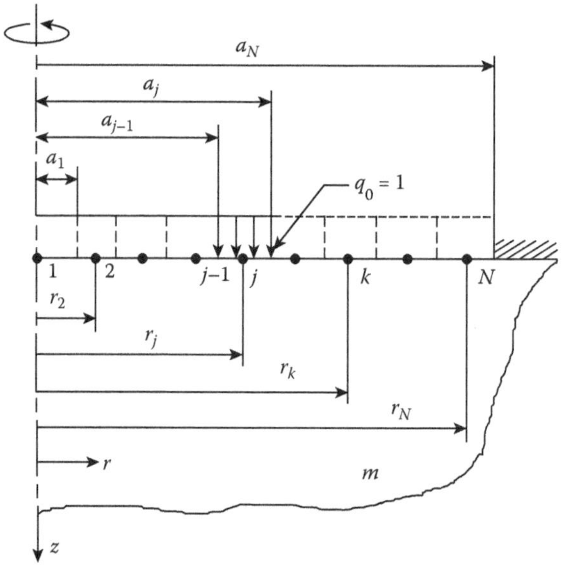

FIGURE 12.15 Geometry describing the influence functions for semi-infinite body.

The average contact area temperature T_c is obtained by summing the products of the elemental temperature and the fraction of the total contact area occupied by the element.

$$T_c(t_M) = \sum_{j=1}^{N} T_{jM} \left(\frac{A_j}{A_c} \right) \tag{12.113}$$

where T_{jM} is the temperature at the center of element j at time t_M, and

$$A_c = \pi a^2 \qquad A_j = \pi(a_j^2 - a_{j-1}^2) \tag{12.114a, b}$$

The total heat flow through the contact area, Q_c, is determined by summing up the heat flows over all elements,

$$Q_c(t_M) = \sum_{j=1}^{N} q_{jM} A_j \tag{12.115}$$

Substituting Equation 12.115 into Equation 12.112 yields

$$R_c(t_M) = \frac{T_{02} - T_{01}}{\sum_{j=1}^{N} q_{jM} A_j} \tag{12.116}$$

With the values of k's = 1, α', s = 1, a = 1, T_{02} = 2, and T_{01} = 0, the first problem was solved for elemental heat fluxes using Equations 12.109a, b. The fluxes were then introduced into Equation 12.116 to evaluate the thermal constriction resistance across the contact area. The results are shown in Table 12.1. The first column in this table is the dimensionless time ($t^+ = \alpha t / a^2$) which extends over many decades. The results from the finite-difference solution of Schneider et al. (1977) are provided in the second column which are most accurate at the late times and least accurate at the early times. The third column comes from an exact solution given by Sadhel (1980) which is claimed to be valid for only large times. The next two columns are for the T-based and the q-based single-node USE solutions. The T-based solution is appropriate for late times and the q-based solution is accurate at early times. The results from the multinode USE solution are displayed in the sixth column. The last column is from a one-dimensional approximate solution given by Keltner (1973) which closely matches the multinode USE solution at short times and retains good accuracy for the mid to late times.

For the second problem, the intrinsic thermocouple problem, the USE solutions are compared with other existing solutions (Henning and Parker, 1967; Keltner, 1973; Shewen, 1976), based on the normalized area averaged interface temperature, defined as

$$T_c^+ = \frac{T_c - T_{01}}{T_{02} - T_{01}} \tag{12.117}$$

where T_c is the area averaged interface temperature given by Equation 12.112, and T_{01} and T_{02} are the initial temperatures of the wire (semi-infinite cylinder) and the substrate (semi-infinite body), respectively. Note that at the initial moment

TABLE 12.1
Results for Dimensionless Constriction Resistance, for an Isothermal Disk Region on the Surface of a Semi-Infinite Body. $R_c^+ = R_c$. a.k.

t^+	SSY (1977)	Sadhal (1980)	Beck and Keltner, 1982 Eq. 56	Eq. 22	MUSE	Keltner (1973)
0.001	0.0386		0.0202	0.0172	0.0166	0.0162
0.002	0.0409		0.0277	0.0240	0.0230	0.0223
0.005	0.0463		0.0411	0.0368	0.0349	0.0340
0.01	0.0532		0.0544	0.0503	0.0471	0.0473
0.02	0.0637		0.0706	0.0678	0.0625	0.0641
0.05	0.0851	4.8988	0.0959	0.0972	0.0879	0.0914
0.1	0.1074	0.2029	0.1171	0.1226	0.1102	0.1142
0.2	0.1336	0.1685	0.1386	0.1468	0.1336	0.1382
0.5	0.1695	0.1752	0.1658	0.1634	0.1631	0.1685
1	0.1933	0.1879	0.1839	0.1500	0.1824	0.1895
2	0.2120	0.2010	0.1994	0.1071	0.1984	0.2074
10	0.2368	0.2247	0.2245		0.2242	0.2347
100	0.2475	0.2413	0.2413		0.2414	0.2477
1000	0.2495	0.2472	0.2472		0.2472	
10000	0.2499	0.2491	0.2491		0.2491	
∞	0.2500	0.2500	0.2500		0.2500	

when the substrate undergoes a step change of temperature, there is no spatial variation in the interface temperature. The normalized value of this instantaneous initial interface temperature is given by Keltner and Beck (1981) as

$$T_{0c}^+ = \frac{T_{0c} - T_{01}}{T_{02} - T_{01}} = (1 + \beta')^{-1} \tag{12.118}$$

where T_{0c} is the instantaneous initial interface temperature and β' is the reciprocal of β defined by Equation 12.93b; that is,

$$\beta' = \left(\frac{k_1\rho_1 c_1}{k_2\rho_2 c_2}\right)^{1/2} \tag{12.119}$$

Using the thermal properties of a chromel substrate ($k = 19.21$ W/m K, $\alpha = 0.492 \times 10^{-5}$ m^2/s) and an alumel wire ($k = 29.76$ W/m K, $\alpha = 0.663 \times 10^{-5}$ m^2/s), the USE results for normalized area averaged interface temperatures compared with other existing values are as shown in Table 12.2. For a chromel and an alumel combination, the normalized instantaneous initial interface temperature, given by Equation 12.118, is equal to 0.4285. The results presented are actually valid for any combination of the materials with the ratios of $k_2/k_1 = 0.645$ and $\alpha_2/\alpha_1 = 0.742$. The first column in this table is the dimensionless time which extends from $t^+ = 0.001$ to $t^+ = 500$. The dimensionless time is based on the thermal diffusivity of the substrate (body 2),

TABLE 12.2
Normalized Area Averaged Interface Temperature Histories for Chromel Semi-Infinite Body and Alumel Semi-Infinite Cylinder

	FD Solutions			T-based (1981)		q-based		
t^+	1973	1976	Henning and Parker, 1967	Early Time	Late Time	Early Time	Late Time	MUSE
.001			.6084	.4489		.4342		.4335
.002		.4421	.6118	.4500		.4366		.4364
.005		.4480	.6185	.4521		.4413		.4422
.01	.4402	.4510	.6257	.4545		.4467		.4488
.02	.4505	.4599	.6356			.4546		.4581
.05	.4700	.4782	.6540			.4709		.4765
.1	.4916	.4991	.6731		.4907	.4904		.4973
.2	.5215	.5283	.6972		.5280			.5263
.5	.5770	.5826	.7373		.5910			.5805
1	.6302	.6338	.7729		.6452			.6328
2	.6896	.6921	.8109		.7042			.6915
5	.7688	.7714	.8602		.7810			.7700
10	.8202	.8246	.8933		.8327		.8091	.8236
20		.8694	.9207		.8757		.8614	.8687
50		.9139	.9482		.9186		.9108	.9137
100		.9382	.9689		.9417		.9365	.9381
200			.9786		.9585		.9550	.9559
500			.9832		.9737		.9715	.9719

Note: $k_{ch} = 19.21$ and $k_{al} = 29.76$ w / m-K, $\alpha_{ch} = .492 \times 10^{-5}$ and $\alpha_{al} = .663 \times 10^{-5}$ m^2 / s.

$$t^+ = \frac{\alpha_2 t}{a^2} \tag{12.120}$$

The second and the third columns give the results of the finite-difference solutions given by Keltner (1973) and Shewen (1967), respectively. The fourth column is evaluated from the analytical solution given by Henning and Parker (1967) which is only good for late times ($t^+ > 20$). The early and late times results of the T-based and the q-based single-node USE solutions are displayed in the next four columns. The last column represents the multinode USE solution.

As can be seen from Table 12.2, there is a very good agreement between the finite-difference solutions and the USE solutions for the time range covered. However, note that both finite-difference solutions have difficulty regarding the computational effort and cost, particularly for the early times, $t^+ < 0.01$, and the late times, $t^+ > 10$. The T-based and the q-based single-node USE solutions are convenient in that the mathematics is not difficult and the expressions are simple to evaluate. Each solution provides two expressions; one for early times and the other for late times. The q-based solution is more appropriate for the early times. It approaches the exact solution (0.4285) as t^+ goes to zero, and closely matches the multinode USE solution up to dimensionless time $t^+ = 0.1$. It also provides relatively good results for the late times, $t^+ > 10$. The T-based solution does not approach the exact solution as t^+ goes to zero, and consequently, is less accurate than the q-based solution for early times. Because of the uniform interface temperature assumption, however, it yields very good results for the times of $t^+ > 0.1$. Even though neither the T-based solution nor the q-based solution is solely suitable for the complete time domain, a combination of the early-time q-based solution and the late-time T-based solution provides very good results over the entire time domain. These two solutions match very closely at the dimensionless time $t^+ = 0.1$.

A study of the Tables 12.1 and 12.2 shows that for each of the above problems, the USE method performed very well. The single-node solutions represent relatively accurate results for certain ranges of time. The advantage of this approach is in its simplicity. The multinode solution is superior to other analytical and numerical solutions in terms of accuracy and ability to treat the complete time range. Also, there is no restriction regarding the choice of the time step in the multinode approach. For instance, in the above problems, the elemental surface heat fluxes are determined for various values of t^+, in 20 time steps. This means that for larger times, larger time steps are considered. [To evaluate, $q_j(t^+)$'s, at times of $t^+ = 0.01, 1$, and 1000, the time steps of $\Delta t^+ = 0.0005, 0.05$, and 50, were used, respectively.] This substantially reduces the computational work compared with the case where a small constant time step in used for the entire time range.

PROBLEMS

12.1 A semi-infinite body is initially at zero temperature. For times $t > 0$, the surface at $x = 0$ is subjected to a temperature which varies linearly with time as $T(0, x) = at$. Using the Duhamel's theorem, find the transient temperature distribution in the body.

12.2 Consider a semi-infinite body initially at zero temperature subjected to a uniform surface temperature over an infinite strip of width $2a$ with the rest of the surface being insulated. See Figure 12.10a.

(a) By considering only two elements along the active part of the surface, give the appropriate USE equations for three time steps ($M = 3$).

(b) What is the required influence function for this problem?

12.3 Show that for Problem 12.2 the integration of the corresponding GF over the surface element and the time step results in the corresponding flux-based influence function.

12.4 Solve Problem 12.2 by utilizing the GF USE formulation.

12.5 Starting with Equation 12.46a, derive the appropriate GF form of the flux-based USE Equation 12.64 for two bodies in perfect contact over a portion of their boundaries with the rest of the boundaries being insulated.

12.6 Consider two semi-infinite bodies initially at different temperatures, T_{i1} and T_{i2}, brought together in perfect contact over an infinite strip of width $2a$ with the rest of the boundaries being insulated. Using the single-node USE method, obtain an approximate analytical solution for the interface heat flux.

12.7 A thermal property probe can be made by placing a small, thin rectangular electrical resistance heater on a large flat body, named body 1, for which k_1, and α_1, are known. If body 1 and the heater are put into good thermal contact with some body 2 with unknown thermal properties, a transient experiment may be carried out to find k_2 and α_2. The matching conditions between body 1 and body 2 are $\overline{T}_1(t) = \overline{T}_2(t)$ and

$$\overline{q}_1(t) + \overline{q}_2(t) = \begin{cases} q_0 & \text{for } t > 0 \text{ on heater} \\ 0 & \text{otherwise} \end{cases}$$

where the overbar ($^-$) denotes spatial average over the rectangular heater and where q_1 and q_2 are the surface heat flux into bodies 1 and 2, respectively. Initially body 1 and body 2 are at zero temperature.

(a) Formulate the flux-based one-node USE method for two bodies with a heater between them. The heater has negligible mass and negligible temperature gradients perpendicular to the interface.

(b) Using names $\overline{\phi_1}(t)$ and $\overline{\phi_2}(t)$ for the dimensionless spatial average influence functions on the heater, solve the one-node USE equation in the Laplace transform domain.

(c) If $\overline{\phi_1} = 2(t^+/\pi)^{1/2} - (t^+/\pi)[1 + 1/(b/a) - 2/3(t^+/\pi)^{1/2}/b]$ where $t^+ = \alpha_1 t/a^2$ and b/a is the length/width of the rectangular heater, find $\overline{T}(t)$, the spatial average temperature on the heater at early time (an approximate inverse transform is required). If $\overline{T}(t)$ and q_0 are measured, is it possible to deduce k_2 and α_2?

NOTE 1: DERIVATION OF EQUATIONS 12.65A AND 12.65B

Equation 12.65a can readily be obtained by considering that for the first time step ($M = 1$), the vectors $\overline{E}_M$ and $\overline{F}_M$ (given by Equation 12.62a and b, respectively) are zero.

$$\overline{E}_M = \overline{F}_M = 0 \tag{1}$$

Substituting Equation 1 into Equation 12.61b and then its results into Equation 12.64 yields

$$\overline{q}_1 = \overline{\overline{C}}^{-1}\overline{T}_0 \tag{2}$$

which is the same as Equation 12.65a.

To show how Equation 12.65b is derived, Equation 12.63 is expanded for different values of M. By introducing Equations 12.61b and 12.62a and b into Equation 12.63, for $M = 1, 2, 3, \ldots$, one can write

$$\text{for } M = 1, \quad \overline{\overline{C}}\overline{q}_1 = \overline{T}_0 \tag{3.1}$$

$$\text{for } M = 2, \quad \overline{\overline{C}}\overline{q}_2 = \overline{T}_0 + \overline{E}_2 - \overline{F}_2 \tag{3.2}$$

$$\text{for } M = 3, \quad \overline{\overline{C}}\overline{q}_3 = \overline{T}_0 + \overline{E}_3 - \overline{F}_3 \tag{3.3}$$

$$\text{for } M - 1, \quad \overline{\overline{C}}\overline{q}_{M-1} = \overline{T}_0 + \overline{E}_{M-1} - \overline{F}_{M-1} \tag{3.M-1}$$

$$\text{for } M \quad \overline{\overline{C}}\overline{q}_M = \overline{T}_0 + \overline{E}_M - \overline{F}_M \tag{3.M}$$

By adding all M Equations together, 3.1 through 3.M, and noticing that

$$\overline{E}_M = \overline{F}_{M-1} + \overline{\overline{\Phi}}_1\overline{q}_{M-1} \tag{4}$$

it can be shown that

$$\overline{\overline{C}}\left\{\sum_{i=1}^{M}\overline{q}_i\right\} = M\overline{T}_0 + \overline{\overline{\Phi}}_1\sum_{i=1}^{M-1}\overline{q}_i - \overline{F}_M \tag{5}$$

or

$$\overline{q}_M + \sum_{i=1}^{M-1}\overline{q}_i = M\overline{\overline{C}}^{-1}\overline{T}_0 + \overline{\overline{C}}^{-1}\overline{\overline{\Phi}}_1\sum_{i=1}^{M-1}\overline{q}_i - \overline{\overline{C}}^{-1}\overline{F}_M \tag{6}$$

Substituting for $\overline{\overline{C}}, \overline{F}_M$ and $\overline{\overline{C}}^{-1}\overline{T}_0$ from Equations 12.61a, 12.62b, and 2, respectively, yields

$$\overline{q}_M = M\overline{q}_1 + \overline{\overline{B}}\left\{\sum_{i=1}^{M-1}\overline{q}_i\right\} - \overline{\overline{C}}^{-1}\overline{F}_M \tag{7}$$

where the matrix $\overline{\overline{B}}$ is defined as

$$\overline{\overline{B}} = \overline{\overline{H}}^{-1}\overline{\overline{\Phi}}_1 \tag{8}$$

Equation 7 is the same as Equation 12.65b and is valid for $M \geq 2$. Notice that the vector $\overline{F}_M$ is a function of time and should be calculated at each time step.

REFERENCES

Bannerjee, P. K. and Butterfield, R. (Eds.), 1979, *Developments in Boundary Element Methods – 1*, Applied Science Publishers, London.

Bannerjee, P. K. and Mukherjee, S. (Eds.), 1984, *Developments in Boundary Element Methods – 3*, Applied Science Publishers, London.

Bannerjee, P. K. and Shaw, R. P. (Eds.), 1982, *Developments in Boundary Element Methods – 2*, Applied Science Publishers, London.

Bannerjee, P. K. and Watson, J. O. (Eds.), 1985, *Developments in Boundary Element Methods – 4*, Applied Science Publishers, London.

Beck, J. V., 1968, Surface heat flux determination using an integral method, *Nuclear Eng. Design*, vol. 7, pp. 170–178.

Beck, J. V., 1980, Average transient temperature within a body heated by a disk heat source, in *Heat Transfer Thermal Control and Heat Pipes*, ed. W. B. Olstad, vol. 70, Progress in Astronautics and Aeronautics, American Institute of Aeronautics and Astronautics (AIAA), pp. 3–24.

Beck, J. V., 1981a, Large time solutions for temperatures in a semi-infinite body with a disk heat source, *Int. J. Heat Mass Transfer*, vol. 24, pp. 155–164.

Beck, J. V., 1981b, Transient temperatures in a semi-infinite cylinder heated by a disk heat source, *Int. J. Heat Mass Transfer*, vol. 24, pp. 1631–1640.

Beck, J. V. and Keltner, N. R., 1987, Green's function partitioning method applied to foil heat flux gage, *J. Heat Transfer*, vol. 109, pp. 274–280.

Beck, J. V., Blackwell, B., and St. Clair, Jr., C. R., 1985, *Inverse Heat Conduction*, John Wiley, New York.

Beck, J. V., Keltner, N. R., and Schisler, I. P., 1985, Influence functions for the unsteady surface element method, *AIAA J.*, vol. 23, pp. 1978–1982.

Carslaw, H. S. and Jaeger, J. C., 1959, *Conduction of Heat in Solids*, 2nd edn, Oxford University Press, New York.

Chang, Y. P., Kang, C. S., and Chen, D. J., 1973, The use of fundamental Green's functions for solution of problems of heat conduction in anisotropic media, *J. Heat Mass Transfer*, vol. 16, pp. 1905–1918.

Cole, K. D. and Beck, J. V., 1987, Conjugated heat transfer from a strip heater with the unsteady surface element method, *J. Thermophys. Heat Transfer*, vol. 1, pp. 348–354.

Cole, K. D. and Beck, J. V., 1988, Conjugated heat transfer from a hot-film probe for transient air flow, *J. Heat Transfer*, vol. 110, pp. 290–296.

Eckert, E. R. and Drake, Jr., R. M., 1972, *Analysis of Heat and Mass Transfer*, McGraw-Hill, New York.

Henning, C. D. and Parker, R., 1967, Transient response of an intrinsic thermocouple, *J. Heat Transfer, Trans. ASME, Ser. C*, vol. 89, pp. 146–154.

Kays, W. M. Crawford, M. E., and Wiegand, B., 2005, *Convection Heat and Mass Transfer*, 4nd edn, McGraw-Hill, New York.

Keltner, N. R., 1973, Heat transfer in intrinsic thermocouples applications to transient measurement errors, Report SC-RR-72-0719, Sandia National Laboratory, Albuquerque, NM.

Keltner, N. R. and Beck, J. V., 1981, Unsteady surface element method, *J. Heat Transfer*, vol. 103, pp. 759–764.

Keltner, N. R., Bainbridge, B. L., and Beck, J. V., 1988, Rectangular heat source on a semi-infinite solid – An analysis for thin-film heat flux gage calibration, *J. Heat Transfer*, vol. 110, pp. 42–48.

Khader, M. S., 1980, Heat conduction with temperature dependent thermal conductivity, presented at the 19th National Heat Transfer Conference, ASME paper No. 80-Ht-4.

Khader, M. S. and Hanna, M. C., 1981, An iterative boundary integral numerical solution for general steady state heat conduction problems, *J. Heat Transfer*, vol. 103, pp. 26–31.

Litkouhi, B. L. and Beck, J. V., 1982, Temperatures in semi-infinite body heat by constant heat flux over half space, in *Heat Transfer 1982*, ed. U. Grigull, E. Hahne, K. Stephan, and J. Straub, vol. 2, pp. 21–36, Hemisphere, Washington, DC.

Litkouhi, B. L. and Beck, J. V., 1985, Intrinsic thermocouple analysis using multinode unsteady surface element method, 1985, *AIAA J.*, vol. 23, no. 10, pp. 1609–1614.

Litkouhi, B. L. and Beck, J. V., 1986, Multinode unsteady surface element method with application contact conductance problem, *J. Heat Transfer*, vol. 108, pp. 257–263.

Luikov, A. V., 1968, *Analytical Heat Diffusion Theory*, Academic Press, New York.

Myers, G. E., 1987, *Analytical Methods in Conduction Heat Transfer*, Genium Publishing, Schenectady, NY.

Ozisik, M. N., 1993, *Heat Conduction*, John Wiley, New York.

Rizzo, F. J. and Shippy, D. J., 1970, A method of solution for certain problems of transient heat conduction, *AIAA J.*, vol. 8, no. 11, pp. 2004–2009.

Sadhal, S. S., 1980, Transient thermal response of two solids in contact over a circular disk, *Int. J. Heat Mass Transfer*, vol. 23, pp. 731–733.

Schneider, G. E., 1979, Thermal constriction resistance due to arbitrary contacts on a half-space-numerical solution of the Dirichlet problem, in *Progress in Astronautics and Aeronautics: Thermophysics and Thermal Control*, American Institute of Aerounatics and Astronautics, vol. 65, ed. R. Viskanta, New York, pp. 103–119.

Schneider, G. E. and LeDain, B. L., 1979, The boundary integral equation method applied to steady heat conduction with special attention given to the corner problem, AIAA paper no. 79-0176, presented at the 17th Aerospace Sciences Meeting, New Orleans, LA.

Schneider, G. E., Strong, A. B., and Yovanivich, M. M., 1977, Transient thermal response of two bodies communicating through a small circular contact area, *Int. J. Heat Mass Transfer*, vol. 23, pp. 731–733.

Shaw, R. P., 1974, An integral equation approach to diffusion, *Int. J. Heat Mass Transfer*, vol. 17, pp. 693–699.

Shewen, E. C., 1967, A transient numerical analysis of conduction between contacting circular cylinders and halfspaces applied to a biosensor, M.S. thesis, University of Waterloo, Canada.

Wrobel, L. C. and Brebbia, C. A., 1981, A formulation of the boundary element method for axisymmetric transient heat conduction, *Int. J. Heat Mass Transfer*, vol. 24, no. 5, pp. 843–850.

Yovanovich, M. M. and Martin, K. A., 1980, Some basic three-dimensional influence coefficients for the surface element method, AIAA paper no. 80–1491, AIAA 15th Thermophysics Conference, Snowmass, CO.

B Bessel Functions

The differential equation

$$\frac{d^2R}{dz^2} + \frac{1}{z}\frac{dR}{dz} + \left(1 - \frac{\nu^2}{z^2}\right)R = 0 \tag{B.1}$$

is called the Bessel equation of order ν. Two linearly independent solutions of this equation for all values of ν are $J_\nu(z)$, the Bessel function of the first kind of order ν and $Y_\nu(z)$, the Bessel function of the second kind of order ν. Thus, the general solution of Equation B.1 is written as (Hildebrand, 1949; McLachlan, 1961; Watson, 1966)

$$R(z) = c_1 J_\nu(z) + c_2 Y_\nu(z) \tag{B.2}$$

The Bessel function $J_\nu(z)$ in series form is defined as

$$J_\nu(z) = \left(\frac{1}{2}z\right)^\nu \sum_{k=0}^{\infty} (-1)^k \frac{[(1/2)\,z]^{2k}}{k!\Gamma(\nu + k + 1)} \tag{B.3}$$

and

$$Y_\nu(z) = \frac{J_\nu(z)\cos(\nu\pi) - J_{-\nu}(z)}{\sin(\nu\pi)}$$

where $\Gamma(x)$ is the gamma function. The differential equation

$$\frac{d^2R}{dz^2} + \frac{1}{z}\frac{dR}{dz} - \left(1 + \frac{\nu^2}{z^2}\right)R = 0 \tag{B.4}$$

is called the modified Bessel equation of order ν. Two linearly independent solutions of this equation for all values of ν are $I_\nu(z)$ (the modified Bessel function of the first kind of order ν) and $K_\nu(z)$ (the modified Bessel function of the second kind of order ν). Thus, the general solution of Equation B.4 is written as

$$R(z) = c_1 I_\nu(z) + c_2 K_\nu(z) \tag{B.5}$$

$I_\nu(z)$ and $K_\nu(z)$ are real and positive when $\nu > -1$ and $z > 0$. The Bessel function $I_\nu(z)$ in series form is given by

$$I_\nu(z) = \left(\frac{1}{2}z\right)^\nu \sum_{k=0}^{\infty} \frac{[(1/2)z]^{2k}}{k!\Gamma(\nu + k + 1)} \tag{B.6}$$

When ν is neither zero nor a positive integer, the general solutions B.2 and B.5 can be taken, respectively, in the form

$$R(z) = c_1 J_\nu(z) + c_2 J_{-\nu}(z) \tag{B.7a}$$

$$R(z) = c_1 I_\nu(z) + c_2 I_{-\nu}(z) \tag{B.7b}$$

TABLE B.1
First Ten Roots of $J_n(z) = 0$; $n = 0, 1, 2, 3, 4, 5$

n	J_0 (*R*01 Case)	J_1 (*R*02 Case)	J_2	J_3	J_4	J_5
1	2.4048	3.8317	5.1356	6.3802	7.5883	8.7715
2	5.5201	7.0156	8.4172	9.7610	11.0647	12.3386
3	8.6537	10.1735	11.6198	13.0152	14.3725	15.7002
4	11.7915	13.3237	14.7960	16.2235	17.6160	18.9801
5	14.9309	16.4706	17.9598	19.4094	20.8269	22.2178
6	18.0711	19.6159	21.1170	22.5827	24.0190	25.4303
7	21.2116	22.7601	24.2701	25.7482	27.1991	28.6266
8	24.3525	25.9037	27.4206	28.9084	30.3710	31.8117
9	27.4935	29.0468	30.5692	32.0649	33.5371	34.9888
10	30.6346	32.1897	33.7165	35.2187	36.6990	38.1599

When $\nu = n$ is a positive integer, the solutions $J_n(z)$ and $J_{-n}(z)$ are not independent (see Tables B.1 through B.5); they are related by

$$J_n(z) = (-1)^n J_{-n}(z) \qquad \text{and} \qquad J_{-n}(z) = J_n(-z) \tag{B.8}$$

(n = integer). Similarly, when $\nu = n$ is a positive integer, the solutions $I_n(z)$ and $I_{-n}(z)$ are not independent.

We summarize various forms of solutions of Equation B.1 as

$$R(z) = c_1 J_\nu(z) + c_2 Y_\nu(z) \qquad \text{always} \tag{B.9a}$$

$$R(z) = c_1 J_\nu(z) + c_2 J_{-\nu}(z) \qquad \nu \text{ is not zero or a positive integer} \tag{B.9b}$$

and the solutions of Equation B.4 as

$$R(z) = c_1 I_\nu(z) + c_2 K_\nu(z) \qquad \text{always} \tag{B.10a}$$

$$R(z) = c_1 I_\nu(z) + c_2 I_{-\nu}(z) \qquad \nu \text{ is not zero or a positive integer} \tag{B.10b}$$

B.1 GENERALIZED BESSEL EQUATION

Sometimes a given differential equation, after suitable transformation of the independent variable, yields a solution that is a linear combination of Bessel functions. A convenient way of finding out whether a given differential equation possesses a solution in terms of Bessel functions is to compare it with the generalized Bessel equation (Sherwood and Reed, 1939, p. 65)

$$\frac{d^2R}{dx^2} + \left(\frac{1-2m}{x} - 2\alpha\right)\frac{dR}{dx} + \left[p^2a^2x^{2p-2} + \alpha^2 + \frac{\alpha(2m-1)}{x} + \frac{m^2 - p^2\nu^2}{x^2}\right]R = 0 \tag{B.11a}$$

TABLE B.2
First Six Roots of $\beta J_1(\beta) - cJ_0(\beta) = 0$ (Case $R03$, where $c = h\,b/k$)[a]

c	β_1	β_2	β_3	β_4	β_5	β_6
0.00	0.0000	3.8317	7.0156	10.1735	13.3237	16.4706
0.01	0.1412	3.8343	7.0170	10.1745	13.3244	16.4712
0.02	0.1995	3.8369	7.0184	10.1754	13.3252	16.4718
0.04	0.2814	3.8421	7.0213	10.1774	13.3267	16.4731
0.06	0.3438	3.8473	7.0241	10.1794	13.3282	16.4743
0.08	0.3960	3.8525	7.0270	10.1813	13.3297	16.4755
0.10	0.4417	3.8577	7.0298	10.1833	13.3312	16.4767
0.15	0.5376	3.8706	7.0369	10.1882	13.3349	16.4797
0.20	0.6170	3.8835	7.0440	10.1931	13.3387	16.4828
0.30	0.7465	3.9091	7.0582	10.2029	13.3462	16.4888
0.40	0.8516	3.9344	7.0723	10.2127	13.3537	16.4949
0.50	0.9408	3.9594	7.0864	10.2225	13.3611	16.5010
0.60	1.0184	3.9841	7.1004	10.2322	13.3686	16.5070
0.70	1.0873	4.0085	7.1143	10.2419	13.3761	16.5131
0.80	1.1490	4.0325	7.1282	10.2516	13.3835	16.5191
0.90	1.2048	4.0562	7.1421	10.2613	13.3910	16.5251
1.00	1.2558	4.0795	7.1558	10.2710	13.3984	16.5312
1.50	1.4569	4.1902	7.2233	10.3188	13.4353	16.5612
2.00	1.5994	4.2910	7.2884	10.3658	13.4719	16.5910
3.00	1.7887	4.4634	7.4103	10.4566	13.5434	16.6499
4.00	1.9081	4.6018	7.5201	10.5423	13.6125	16.7073
5.00	1.9898	4.7131	7.6177	10.6223	13.6786	16.7630
6.00	2.0490	4.8033	7.7039	10.6964	13.7414	16.8168
7.00	2.0937	4.8772	7.7797	10.7646	13.8008	16.8684
8.00	2.1286	4.9384	7.8464	10.8271	13.8566	16.9179
9.00	2.1566	4.9897	7.9051	10.8842	13.9090	16.9650
10.00	2.1795	5.0332	7.9569	10.9363	13.9580	17.0099
15.00	2.2509	5.1773	8.1422	11.1367	14.1576	17.2008
20.00	2.2880	5.2568	8.2534	11.2677	14.2983	17.3442
30.00	2.3261	5.3410	8.3771	11.4221	14.4748	17.5348
40.00	2.3455	5.3846	8.4432	11.5081	14.5774	17.6508
50.00	2.3572	5.4112	8.4840	11.5621	14.6433	17.7272
60.00	2.3651	5.4291	8.5116	11.5990	14.6889	17.7807
80.00	2.3750	5.4516	8.5466	11.6461	14.7475	17.8502
100.00	2.3809	5.4652	8.5678	11.6747	14.7834	17.8931
∞	2.4048	5.5201	8.6537	11.7915	14.9309	18.0711

[a]From Carslaw and Jaeger (1959).

TABLE B.3
First Five Roots of $J_0(\beta)Y_0(\lambda\beta) - Y_0(\beta)J_0(\lambda\beta)$ (Case *R*11, where $\lambda = b/a$, $\lambda > 1$)

λ^{-1}	1	2	3	4	5
0.80	12.55847 031	25.12877	37.69646	50.26349	62.83026
0.60	4.69706 410	9.41690	14.13189	18.84558	23.55876
0.40	2.07322 886	4.17730	6.27537	8.37167	10.46723
0.20	0.76319 127	1.55710	2.34641	3.13403	3.92084
0.10	0.33139 387	0.68576	1.03774	1.38864	1.73896
0.08	0.25732 649	0.53485	0.81055	1.08536	1.35969
0.06	0.18699 458	0.39079	0.59334	0.79522	0.99673
0.04	0.12038 637	0.25340	0.38570	0.51759	0.64923
0.02	0.05768 450	0.12272	0.18751	0.25214	0.31666
0.00	0.00000 000	0.00000	0.00000	0.00000	0.00000

TABLE B.4
First Five Roots of $J_1(\beta)Y_0(\lambda\beta) - Y_1(\beta)J_0(\lambda\beta)$ (Cases *R*12 or *R*21, where $\lambda = b/a$, $\lambda > 1$)

λ^{-1}	1	2	3	4	5
0.80	6.56973 310	18.94971	31.47626	44.02544	56.58224
0.60	2.60328 138	7.16213	11.83783	16.53413	21.23751
0.40	1.24266 626	3.22655	5.28885	7.36856	9.45462
0.20	0.51472 663	1.24657	2.00959	2.78326	3.56157
0.10	0.24481 004	0.57258	0.90956	1.25099	1.59489
0.08	0.19461 772	0.45251	0.71635	0.98327	1.25203
0.06	0.14523 798	0.33597	0.53005	0.72594	0.92301
0.04	0.09647 602	0.22226	0.34957	0.47768	0.60634
0.02	0.04813 209	0.11059	0.17353	0.23666	0.29991
0.00	0.00000 000	0.00000	0.00000	0.00000	0.00000

and the corresponding solution of which is

$$R = x^m e^{\alpha x}[c_1 J_\nu(ax^p) + c_2 Y_\nu(ax^p)] \tag{B.11b}$$

where c_1 and c_2 are arbitrary constants. For example, by comparing the differential equation

$$\frac{d^2R}{dx^2} + \frac{1}{x}\frac{dR}{dx} - \frac{\beta}{x}R = 0 \tag{B.12}$$

with the above generalized Bessel equation, we find

$$\alpha = 0 \qquad m = 0 \qquad p = \frac{1}{2} \qquad a = 2i\sqrt{\beta} \qquad \nu = 0$$

TABLE B.5
First Five Roots of $J_1(\beta)Y_1(\lambda\beta) - Y_1(\beta)J_1(\lambda\beta)$ (Case $R22$, where $\lambda = b/a$, $\lambda > 1$)

λ^{-1}	1	2	3	4	5
0.80	12.59004 151	25.14465	37.70706	50.27145	62.83662
0.60	4.75805 426	9.44837	14.15300	18.86146	23.57148
0.40	2.15647 249	4.22309	6.30658	8.39528	10.48619
0.20	0.84714 961	1.61108	2.38532	3.16421	3.94541
0.10	0.39409 416	0.73306	1.07483	1.41886	1.76433
0.08	0.31223 576	0.57816	0.84552	1.11441	1.38440
0.06	0.23235 256	0.42843	0.62483	0.82207	1.02001
0.04	0.15400 729	0.28296	0.41157	0.54044	0.66961
0.02	0.07672 788	0.14062	0.20409	0.26752	0.33097
0.00	0.00000 000	0.00000	0.00000	0.00000	0.00000

Hence, the solution of differential equation B.12 is in the form

$$R = c_1 J_0(2i\sqrt{\beta x}) + c_2 Y_0(2i\sqrt{\beta x}) \tag{B.13a}$$

or

$$R = c_1 I_0(2\sqrt{\beta x}) + c_2 K_0(2\sqrt{\beta x}) \tag{B.13b}$$

which involves Bessel functions.

B.2 LIMITING FORM FOR SMALL z

For small values of $z (z \to 0)$, the retention of the leading terms in the series results in the following approximations for the values of Bessel functions (Abramowitz and Stegun, 1964, p. 360)

$$J_\nu(z) \approx \left(\frac{1}{2}z\right)^\nu \frac{1}{\Gamma(\nu+1)} \qquad \nu \neq -1, -2, -3, \ldots \tag{B.14a}$$

$$Y_\nu(z) \approx -\frac{1}{\pi}\left(\frac{2}{z}\right)^\nu \Gamma(\nu) \qquad \nu \neq 0 \qquad \text{and} \qquad Y_0(z) = \frac{2}{\pi}\ln z \tag{B.14b}$$

$$I_\nu(z) \approx \left(\frac{1}{2}z\right)^\nu \frac{1}{\Gamma(\nu+1)} \qquad \nu \neq -1, -2, -3, \ldots \tag{B.15a}$$

$$K_\nu(z) \approx \frac{1}{2}\left(\frac{2}{z}\right)^\nu \Gamma(\nu) \qquad \nu \neq 0 \qquad K_0(z) \approx -\ln z \tag{B.15b}$$

B.3 LIMITING FORM FOR LARGE z

For large values of z ($z \to \infty$), the values of Bessel functions can be approximated as (Abramowitz and Stegun, 1964, pp. 364 and 377)

$$J_\nu(z) \approx \left(\frac{2}{\pi z}\right)^{1/2} \cos\left(z - \frac{\pi}{4} - \frac{\nu\pi}{2}\right) \tag{B.16a}$$

$$Y_\nu(z) \approx \left(\frac{2}{\pi z}\right)^{1/2} \sin\left(z - \frac{\pi}{4} - \frac{\nu\pi}{4}\right) \tag{B.16b}$$

$$I_\nu(z) \approx \frac{e^z}{\sqrt{2\pi z}} \quad \text{and} \quad K_\nu(z) \approx \left(\frac{2}{\pi z}\right)^{1/2} e^{-z} \tag{B.16c}$$

B.4 DERIVATIVES OF BESSEL FUNCTIONS (HILDEBRAND, 1949, PP. 161–163)

$$\frac{d}{dz}[z^\nu W_\nu(\beta z)] = \begin{cases} \beta z^\nu W_{\nu-1}(\beta z) & \text{for } W = J, Y, I \quad \text{(B.17a)} \\ -\beta z^\nu W_{\nu-1}(\beta z) & \text{for } W = K \quad \text{(B.17b)} \end{cases}$$

$$\frac{d}{dz}[z^{-\nu} W_\nu(\beta z)] = \begin{cases} -\beta z^{-\nu} W_{\nu+1}(\beta z) & \text{for } W = J, Y, K \quad \text{(B.18a)} \\ \beta z^{-\nu} W_{\nu+1}(\beta z) & \text{for } W = I \quad \text{(B.18b)} \end{cases}$$

For example, by setting $\nu = 0$, we obtain

$$\frac{d}{dz}[W_0(\beta z)] = \begin{cases} -\beta W_1(\beta z) & \text{for } W = J, Y, K \quad \text{(B.19a)} \\ \beta W_1(\beta z) & \text{for } W = I \quad \text{(B.19b)} \end{cases}$$

B.5 RECURRENCE RELATIONS

The recurrence formulas for the Bessel functions are given as (Abramowitz and Stegun, 1964, p. 361; Watson, 1966, pp. 45 and 66)

$$W_{\nu-1}(z) + W_{\nu+1}(z) = \frac{2\nu}{z} W_\nu(z) \tag{B.20a}$$

$$W_{\nu-1}(z) - W_{\nu+1}(z) = 2W_\nu'(z) \tag{B.20b}$$

$$W_{\nu-1}(z) - \frac{\nu}{z} W_\nu(z) = W_\nu'(z) \tag{B.20c}$$

$$-W_{\nu+1}(z) + \frac{\nu}{z} W_\nu(z) = W_\nu'(z) \tag{B.20d}$$

where $W = J$ or Y or any linear combination of these functions the coefficients in which are independent of z and ν.

B.6 INTEGRALS OF BESSEL FUNCTIONS

$$\int x J_0(x)dx = x J_1(x) \tag{B.21}$$

$$\int J_1(x)dx = -J_0(x) \tag{B.22}$$

$$\int_0^z J_0(x)dx = z \sum_{k=0}^{\infty} J_{2k+1}(z) \tag{B.23}$$

Note: Numerical values for $\int J_0(x)$ and $\int Y_0(x)$ are tabulated in Abramowitz and Stegun (1964, pp. 491–493).

$$\int x Y_0(x)dx = x Y_1(x) \tag{B.24}$$

$$\int Y_1(x)dx = -Y_0(x) \tag{B.25}$$

$$(\beta^2 - \alpha^2) \int_0^1 x J_k(\alpha x) J_k(\beta x)dx = \frac{\alpha\beta}{2k}[J_{k-1}(\alpha)J_{k+1}(\beta) - J_{k+1}(\alpha)J_{k-1}(\beta)] \tag{B.26}$$

$$\int x J_k^2(\alpha x)dx = \frac{x^2}{2}[J_k^2(\alpha x) - J_{k-1}(\alpha x)J_{k+1}(\alpha x)] \tag{B.27}$$

In the following formulas, $C_k(x)$ and $\overline{C}_k(x)$ denote two general Bessel functions, (i.e., linear combinations):

$$C_k(x) = aJ_k(x) + bY_k(x) \qquad \overline{C}_k(x) = \overline{a}J_k(x) + \overline{b}Y_k(x)$$

with arbitrary constant, $a, b, \overline{a}, \overline{b}$.

$$\int x^{k+1} C_k(x)dx = x^{k+1} C_{k+1}(x) \tag{B.28}$$

$$\int x^{1-k} C_k(x)dx = -x^{1-k} C_{k-1}(x) \tag{B.29}$$

$$\int x C_k(hx)\overline{C}_k(gx)dx = (h^2 - g^2)^{-1} x \left[hC_{k+1}(hx)\overline{C}_k(gx) - gC_k(hx)\overline{C}_{k+1}(gx)\right] \tag{B.30}$$

$$\int x C_k(hx)\overline{C}_k(hx)dx = -\frac{1}{4}x^2 \left[C_{k-1}(hx)\overline{C}_{k+1}(hx) - 2C_k(hx)\overline{C}_k(hx) + C_{k+1}(hx)\overline{C}_{k-1}(hx)\right] \tag{B.31}$$

$$\int x^{-1} C_m(hx)\overline{C}_k(hx)dx = (m^2 - k^2)^{-1} \left[(m-k)C_m(hx)\overline{C}_{k+1}(hx) - hxC_{m+1}(hx)\overline{C}_k(hx) + hxC_m(hx)\overline{C}_{k+1}(hx)\right] \tag{B.32}$$

REFERENCES

Abramowitz, M. and Stegun, I. A. (Eds), 1964, *Handbook of Mathematical Functions with Formulas, Graphs, and Mathematical Tables*, Applied Mathematics Series 55, National Bureau of Standards, Washington, DC.

Carslaw, H. S. and Jaeger, J. C., 1959, *Conduction of Heat in Solids*, 2nd edn, Oxford University Press, New York.

Hildebrand, F. B., 1949, *Advanced Calculus for Engineers*, Prentice-Hall, Englewood Cliffs, NJ.

McLachlan, N. W., 1961, *Bessel Functions for Engineers*, 2nd edn, Clarendon Press, Oxford.

Sherwood, T. K. and Reed, C. E., 1939, *Applied Mathematics in Chemical Engineering*, McGraw-Hill, New York.

Watson, G. N., 1966, *A Treatise on the Theory of Bessel Functions*, 2nd edn, Cambridge University Press, London.

D Dirac Delta Function

The Dirac delta function (sometimes called the *unit impulse function*) plays a central role in the method of Green's functions. The Dirac delta function $\delta(x)$ is defined to be zero when $x \neq 0$, and infinite at $x = 0$ in such a way that the area under the function is unity. A concise definition is the following: given nonzero numbers $\eta_1 > 0$ and $\eta_2 > 0$,

$$\delta(x) = 0 \text{ if } x \neq 0; \quad \int_{-\eta_1}^{\eta_2} \delta(x)\, dx = 1 \tag{D.1}$$

This is a "weak" definition of $\delta(x)$, since the limits of integration are never allowed to be precisely zero. This definition is sufficient for work with Green's functions. See Barton (1989, p. 11) for a discussion of "weak" and "strong" definitions.

D.1 PROPERTIES OF THE DIRAC DELTA FUNCTION

1. Sifting property. Given function $f(x)$ continuous at $x = x'$,

$$\int_a^b f(x')\, \delta(x - x')\, dx' = \begin{cases} f(x) & \text{if } a < x < b \\ 0 & \text{if } (a, b) \text{ does not contain } x \end{cases} \tag{D.2}$$

When integrated, the product of any (well-behaved) function and the Dirac delta yields the function evaluated where the Dirac delta is singular. The sifting property also applies if the arguments of functions f and δ are exchanged: $f(x')\, \delta(x - x')dx' = f(x)\, \delta(x' - x)dx$.
Next the sifting property will be proved. Let $\gamma = x - x'$. Then

$$\int_a^b f(x')\, \delta(x - x')\, dx' = \int_{\gamma = x-b}^{x-a} f(x - \gamma)\, \delta(\gamma)\, d\gamma \tag{D.3}$$

From the definition, $\delta(x) = 0$ for any $x \neq 0$ so the limits on γ may be replaced by $(-\epsilon, \epsilon)$ for some small $\epsilon > 0$.

$$\int_a^b f(x)\, \delta(x - x')\, dx = \int_{-\epsilon}^{\epsilon} f(x - \gamma)\, \delta(\gamma)\, d\gamma$$

Now take ϵ to be very small, so that over the interval $(-\epsilon, \epsilon)$ function f is essentially constant. That is, $f(x - \gamma) \approx f(x)$ so that $f(x)$ may be removed from the integral over γ

$$\begin{aligned} \int_a^b f(x)\, \delta(x - x')\, dx &= f(x) \int_{-\epsilon}^{\epsilon} \delta(\gamma)\, d\gamma \\ &= f(x) \cdot 1 \end{aligned} \tag{D.4}$$

The integral of the delta function completes the proof.

2. Relationship with the step function.

$$\int_{-\infty}^{t} \delta(\tau)\,d\tau = H(t); \qquad \frac{dH(t-\tau)}{dt} = \delta(t-\tau) \tag{D.5}$$

where $H(t)$ is the Heaviside unit step function defined as

$$H(t) = \begin{cases} 0 & \text{if } t < 0 \\ 1 & \text{if } t > 0 \end{cases} \tag{D.6}$$

3. Units. Since the definition of the Dirac delta requires that the product $\delta(x)dx$ is dimensionless, the units of the Dirac delta are the inverse of those of the argument x. That is, $\delta(x)$ has units meters^{-1}, and $\delta(t)$ has units sec^{-1}.
4. Definition for radial, 2D, and 3D geometries. For two- and three-dimensional problems with vector coordinate $\vec{\mathbf{r}}$, the Dirac delta function is defined:

$$\delta(\vec{\mathbf{r}}) = 0 \text{ if } \vec{\mathbf{r}} \neq 0$$

$$\int_{\Omega} \delta(\vec{\mathbf{r}})\,dv = \begin{cases} 1 & \text{if } \Omega \text{ contains } \vec{\mathbf{r}} \\ 0 & \text{if } \Omega \text{ does not contain } \vec{\mathbf{r}} \end{cases} \tag{D.7}$$

where dv is differential volume. The units of $\delta(\vec{\mathbf{r}})$ are given by $[dv]^{-1}$, and three important cases are the listed below.
   a. 1D radial cylindrical coordinates: $dv = 2\pi r dr$, and units of $\delta(\vec{\mathbf{r}})$ are [meters]$^{-2}$.
   b. 1D radial spherical coordinates: $dv = 4\pi r^2 dr$,and units of $\delta(\vec{\mathbf{r}})$ are [meters]$^{-3}$.
   c. 2D Cartesian coordinates: $dv = dxdy$, and units of $\delta(\vec{\mathbf{r}})$ are [meters]$^{-2}$.

D.2 REPRESENTATIONS OF δ

In use, the Dirac delta function is never evaluated without multiplying by a test function and integrating over some domain. Equations involving Dirac delta functions without such integrations are a convenient half-way stage that nevertheless have enormous utility. Properly speaking, the Dirac delta function is not a function at all (it is a generalized function), however it can be represented as the limit of a sequence of ordinary functions.

Representations of the Dirac delta with ordinary functions provide a way to visualize the Dirac delta. Let $F(x,\epsilon)$ be a function that has a peak near $x = 0$, and the shape of the peak is controlled by parameter ϵ. If the integral of $F(x,\epsilon)$ is unity, that is,

$$\int_{-\infty}^{\infty} F(x,\epsilon)\,dx = 1$$

for any value of parameter $\epsilon > 0$, then the Dirac delta function may be represented in the limit:

$$\delta(x) = \lim_{\epsilon \to 0} F(x,\epsilon)$$

The following example representations are illustrated in Figure D.1.

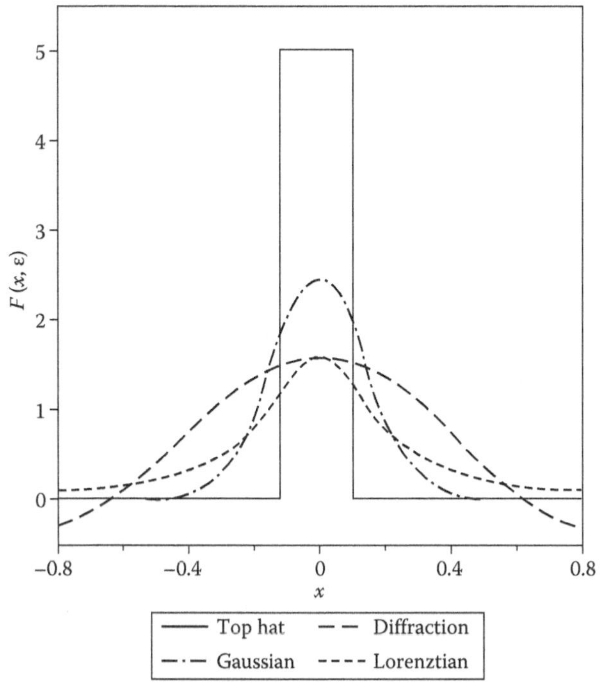

FIGURE D.1 Representations of the Dirac delta function, with $\epsilon = 0.2$.

1. Top-hat function (square step).

$$F(x,\epsilon) = \begin{cases} 1/\epsilon & -\epsilon/2 < x < \epsilon/2 \\ 0 & \text{otherwise} \end{cases} \tag{D.8}$$

2. Diffraction peak.

$$F(x,\epsilon) = \frac{\sin(x/\epsilon)}{\pi x} \tag{D.9}$$

3. Lorentzian.

$$F(x,\epsilon) = \frac{\epsilon/\pi}{(x^2+\epsilon^2)} \tag{D.10}$$

4. Gaussian.

$$F(x,\epsilon) = \frac{1}{2\epsilon}\exp(-x^2/\epsilon^2) \tag{D.11}$$

Although all of the above functions $F(x,\epsilon)$ are symmetric, symmetry is not essential. Nonsymmetric functions produce perfectly good representations of the Dirac delta function.

D.3 SERIES FORM OF δ

In this section the Dirac delta function will be stated in the form of an infinite series. This discussion will begin with the definition of a complete orthogonal set.

A set of functions $\{\phi_n(x)\}$, defined over a finite interval R, and subject to the usual kinds of homogeneous boundary conditions, is *complete* if any (appropriately restricted) function $f(x)$ may be expressed as a linear combination of the set. That is,

$$f(x) = \sum_n c_n \phi_n(x) \tag{D.12}$$

The set $\{\phi_n(x)\}$ is *orthogonal* if, integrating over interval R,

$$\int_R \phi_m^*(x)\,\phi_n(x)\,dx = \begin{cases} N_n & n = m \\ 0 & n \neq m \end{cases} \tag{D.13}$$

where $(\cdot)^*$ is the complex conjugate. Quantity N_m is the (square of the) norm, given by

$$N_n = \int_R [\phi_n(x)]^2\,dx$$

The coefficients c_n are found by multiplying both sides of Equation D.12 by $\phi_m^*(x)$ and integrating over the interval R

$$\int_R f(x)\,\phi_m^*(x)\,dx = \int_R \left(\sum_n c_n \phi_n(x)\right)\,\phi_m^*(x)\,dx \tag{D.14}$$

$$= 0 + 0 + \cdots + 0 + c_n N_n + 0 + \cdots \tag{D.15}$$

Then the coefficient is given by

$$c_n = \frac{1}{N_n}\int_R f(x)\,\phi_n^*(x)\,dx \tag{D.16}$$

There are many complete orthogonal sets, including the familiar Fourier series. Many are composed of real-valued functions for which the complex conjugate plays no role.

The importance of complete orthogonal sets is that each one provides a different series representation of the Dirac delta function. Given a complete orthogonal set $\{\phi_n(x)\}$, then the Dirac delta may be expressed as the following series:

$$\delta(x - x') = \sum_n \frac{\phi_n^*(x')\,\phi_n(x)}{N_n} \tag{D.17}$$

The proof follows from the definition of a complete orthogonal set. Function $f(x)$ may be written

$$f(x) = \sum_n c_n \phi_n(x) \tag{D.18}$$

$$= \sum_n \left[\frac{\int_R f(x')\,\phi_n^*(x')\,dx'}{N_n}\right]\phi_n(x) \tag{D.19}$$

Now swap the order of the integral and the sum

$$f(x) = \int_R f(x') \left[\sum_n \frac{\phi_n^*(x')\phi_n(x)}{N_n} \right] dx. \quad \text{(D.20)}$$

Compare this to the sifting property, and this equation can only be true if the quantity in brackets is the Dirac delta function $\delta(x - x')$.

The series form of the Dirac delta function, customized as it is to specific boundary conditions on interval R, can be used to construct series forms of Green's functions on finite intervals.

D.4 INTEGRAL FORM OF δ AND THE FOURIER TRANSFORM

In this section two integral expressions will be developed for the Dirac delta function. Consider the following integral, which can be found in standard integral tables (for example, Gradshteyn and Rhzhik, 2007, number 2.663.3):

$$\frac{1}{\pi} \int_0^\infty e^{-\epsilon k} \cos(kx) dk = \frac{\epsilon / \pi}{x^2 + \epsilon^2}$$

Because the cosine is an even function, we can write

$$\frac{1}{2\pi} \int_{-\infty}^\infty e^{-\epsilon|k|} \cos(kx) dk = \frac{\epsilon / \pi}{x^2 + \epsilon^2}$$

Notice that the right-hand side of the above equation is the Lorentz representation of δ.

$$F(x, \epsilon) = \frac{1}{2\pi} \int_{-\infty}^\infty e^{-\epsilon|k|} \cos(kx) dk = \frac{\epsilon / \pi}{x^2 + \epsilon^2}$$

The limit as ($\epsilon \to 0$) can be explicitly evaluated to obtain

$$\lim_{\epsilon \to 0} F(x, \epsilon) = \delta(x) = \frac{1}{2\pi} \int_{-\infty}^\infty \cos(kx)\, dk \quad \text{(D.21)}$$

This is an integral form of the Dirac delta function constructed from the Lorentz representation. A similar result may be obtained from the diffraction peak representation. Consider the following integral, for $a > 0$ (recall $e^{i\theta} = \cos\theta + i \sin\theta$):

$$\int_{-a}^a e^{ikx} dk = \frac{2 \sin(ax)}{x}$$

Now let $a = 1/\epsilon$ and divide by 2π to get

$$\frac{\sin(x/\epsilon)}{\pi x} = \frac{1}{2\pi} \int_{-1/\epsilon}^{1/\epsilon} e^{ikx} dk \quad \text{(D.22)}$$

That is, the diffraction peak representation of the Dirac delta function also has an integral form. In the limit as $\epsilon \to 0$, we find

$$\lim_{\epsilon \to 0} F(x, \epsilon) = \delta(x) = \frac{1}{2\pi} \int_{-\infty}^{\infty} e^{ikx}\, dk \tag{D.23}$$

These representations are important for the Fourier transform, defined by the following transform pair:

$$\overline{F}(k) = \int_{-\infty}^{\infty} F(x)\, e^{-ikx}\, dx$$
$$F(x) = \frac{1}{2\pi} \int_{-\infty}^{\infty} \overline{F}(k) e^{ikx}\, dk \tag{D.24}$$

Note $\delta(x)$ and 1 are Fourier transforms of each other; likewise $\delta(x - x')$ and $\exp(-ikx')$ are Fourier transforms of each other.

REFERENCES

Barton, G., 1989, *Elements of Green's Functions and Propagation*, Oxford University Press, New York.

Gradshteyn, I. S. and Ryzhik, I. M., 2007, *Tables of Integrals, Series and Products*, 7th edn, Elsevier, London.

E Error Function and Related Functions

E.1 DEFINITION

The error function is denoted erf(x) and is defined by

$$\operatorname{erf}(x) = \frac{2}{\pi^{1/2}} \int_0^x e^{-t^2} dt \tag{E.1}$$

and the complementary error function, erfc(x), is defined by

$$\operatorname{erfc}(x) = \frac{2}{\pi^{1/2}} \int_x^\infty e^{-t^2} dt \tag{E.2}$$

It can be shown that

$$\operatorname{erf}(\infty) = 1 \tag{E.3a}$$

$$\operatorname{erf}(-x) = -\operatorname{erf}(x) \tag{E.3b}$$

$$\operatorname{erf}(x) + \operatorname{erfc}(x) = 1 \tag{E.3c}$$

$$\operatorname{erfc}(-x) = 1 + \operatorname{erf}(x) = 2 - \operatorname{erfc}(x) \tag{E.3d}$$

Alternative definitions of erf (x) are

$$\operatorname{erf}(x) = \frac{\operatorname{sign}(x)}{\pi^{1/2}} \int_0^{x^2} t^{-1/2} e^{-t} dt \tag{E.4a}$$

$$\operatorname{erf}(x) = \frac{2}{\pi} \int_0^\infty t^{-1} e^{-t^2} \sin(2xt)\, dt \tag{E.4b}$$

$$\operatorname{erf}(x) = \frac{2x}{\pi^{1/2}} \int_0^1 e^{-x^2 t^2} dt \tag{E.4c}$$

E.2 SERIES EXPRESSIONS

Two ways to expand erf(x) are

$$\operatorname{erf}(x) = \frac{2}{\pi^{1/2}} \left[x - \frac{x^3}{3} + \frac{x^5}{10} - \cdots \right] = \frac{2x}{\pi^{1/2}} \sum_{j=0}^\infty \frac{(-x^2)^j}{j!(2j+1)} \tag{E.5a}$$

$$\operatorname{erf}(x) = \frac{2}{\pi^{1/2}} e^{-x^2} \left(x + \frac{2x^3}{3} + \frac{4x^5}{15} + \cdots \right) = e^{-x^2} \sum_{j=0}^\infty \frac{x^{2j+1}}{\Gamma[(2j+3)/2]} \tag{E.5b}$$

Here $\Gamma(\cdot)$ is the gamma function. The above expressions are most useful for small values of x. For large values of x, erfc(x) is expandable asymptotically as

$$\operatorname{erfc}(x) \sim \frac{\exp(-x^2)}{x\pi^{1/2}}\left(1 - \frac{1}{2x^2} + \frac{1\cdot 3}{2^2x^4} - \frac{1\cdot 3\cdot 5}{2^3x^6} + \cdots\right) \tag{E.6}$$

Care must be exercised in using Equation E.6 numerically because the error is only less than the absolute value of the last term retained. Also for large x, the continued fraction expression given below may be used:

$$\operatorname{erfc}(x) = \cfrac{\pi^{-1/2}\exp(-x^2)}{x + \cfrac{1}{2x + \cfrac{2}{x + \cfrac{3}{2x + \cfrac{4}{x + \cfrac{5}{2x + \cfrac{6}{x + \cdots}}}}}}} \tag{E.7}$$

See Press et al. (1992, p. 163) for an efficient way to evaluate continued fractions.

E.3 RELATED FUNCTIONS

A set of functions is defined by the integral

$$\mathrm{i}^n\operatorname{erfc}(x) = \frac{2}{\pi^{1/2}}\int_x^\infty \frac{(t-x)^n}{n!}e^{-t^2}dt \qquad n = 2, 3, 4, \ldots \tag{E.8}$$

The notation is usually extended to embrace

$$\mathrm{i}^1\operatorname{erfc}(x) = \operatorname{ierfc}(x) = \int_x^\infty \operatorname{erfc}(t)dt \tag{E.9a}$$

$$\mathrm{i}^0\operatorname{erfc}(x) = \operatorname{erfc}(x) \tag{E.9b}$$

$$\mathrm{i}^{-1}\operatorname{erfc}(x) = \frac{2}{\pi^{1/2}}e^{-x^2} \tag{E.9c}$$

The ierfc and i^n erfc functions are known as the complementary error function integral, and the repeated integrals of the error function complement, respectively. Plots of some of these functions are given in Figure E.1. Some numerical values of erfc and related functions are given in Table E.1.

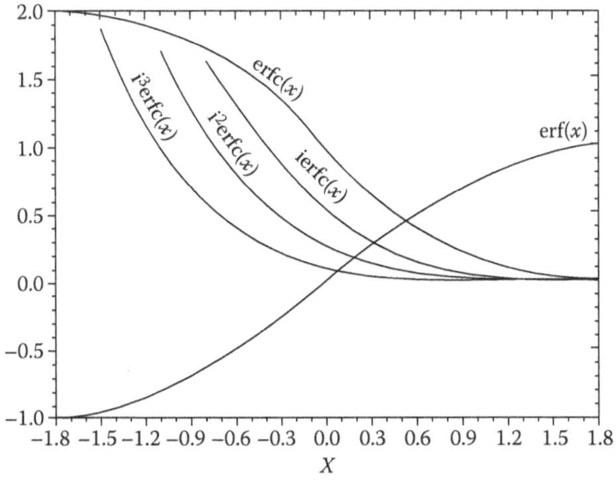

FIGURE E.1 Error function erf(x) and related functions.

The power series expansion is

$$\mathrm{i}^n\mathrm{erfc}(x) = \frac{1}{2^n}\sum_{j=0}^{\infty}\frac{(-2x)^j}{j!\Gamma[(2+n-j)/2]} \tag{E.10}$$

which shows that

$$\mathrm{i}^n\mathrm{erfc}(0) = \left[2^n\Gamma\left(1+\frac{n}{2}\right)\right]^{-1} \tag{E.11}$$

of which the first few values are shown in Table E.2. The behavior of i^n erfc(x) for large values of x is described by

$$\begin{aligned}\mathrm{i}^n\mathrm{erfc}(x) \sim \frac{2\exp(-x^2)}{\pi^{1/2}(2x)^{n+1}}\Bigg[1 &- \frac{(n+1)(n+2)}{4x^2} \\ &+ \frac{(n+1)(n+2)(n+3)(n+4)}{32x^4} \\ &- \cdots + \frac{(n+2j)!}{n!\,j!(-4x^2)^j} + \cdots\Bigg]\end{aligned} \tag{E.12}$$

E.4 RECURSION RELATION

The $\mathrm{i}^n\mathrm{erfc}(x)$ functions obey the relation

$$\mathrm{i}^n\mathrm{erfc}(x) = -\frac{x}{n}\mathrm{i}^{n-1}\,\mathrm{erfc}(x) + \frac{1}{2n}\mathrm{i}^{n-2}\,\mathrm{erfc}(x) \qquad n = 1, 2, 3, \ldots, \tag{E.13}$$

TABLE E.1
Error Functions for Argument Common in Heat Conduction

z	erf $[(4z)^{-1/2}]$	erfc $[(4z)^{-1/2}]$	$(4z)^{1/2}$ ierfc $[(4z)^{-1/2}]$
0.01	1.000000	0.000000	0.000000
0.02	0.999999	0.000001	0.000000
0.03	0.999955	0.000045	0.000002
0.04	0.999593	0.000407	0.000029
0.05	0.998435	0.001565	0.000135
0.06	0.996108	0.003892	0.000393
0.07	0.992474	0.007526	0.000867
0.08	0.987581	0.012419	0.001603
0.09	0.981578	0.018422	0.002625
0.10	0.974653	0.025347	0.003943
0.20	0.886154	0.113846	0.030732
0.30	0.803294	0.196706	0.071893
0.40	0.736448	0.263552	0.118437
0.50	0.682689	0.317311	0.166631
0.60	0.638690	0.361310	0.214891
0.70	0.601975	0.398025	0.262515
0.80	0.570805	0.429195	0.309190
0.90	0.543943	0.456057	0.354791
1.00	0.520500	0.479500	0.399282
2.00	0.382925	0.617075	0.791186
3.00	0.316909	0.683091	1.115053
4.00	0.276326	0.723674	1.396355
5.00	0.248170	0.751830	1.648248
6.00	0.227170	0.772830	1.878325
7.00	0.210732	0.789268	2.091402
8.00	0.197413	0.802587	2.290758
9.00	0.186336	0.813664	2.478736
10.00	0.176937	0.823063	2.657085
20.00	0.125633	0.874367	4.109212
40.00	0.089021	0.910979	6.181053
60.00	0.072736	0.927264	7.776780
80.00	0.063013	0.936987	9.124053
100.00	0.056372	0.943628	10.311989
200.00	0.039878	0.960122	14.977634
400.00	0.028204	0.971796	21.581687
600.00	0.023030	0.976970	26.651048
800.00	0.019945	0.980055	30.925355
1000.00	0.017840	0.982160	34.691403
2000.00	0.012615	0.987385	49.468958
3000.00	0.010300	0.989700	60.809023
4000.00	0.008920	0.991080	70.369425
5000.00	0.007979	0.992021	78.792445
6000.00	0.007284	0.992716	86.407516
7000.00	0.006743	0.993257	93.410346
8000.00	0.006308	0.993692	99.928455
9000.00	0.005947	0.994053	106.050420
10000.00	0.005642	0.994358	111.840738

TABLE E.2
Values of i^nerfc(0)

n	i^n erfc (0)
−1	$2\pi^{-1/2}$
0	1
1	$\pi^{-1/2}$
2	$\frac{1}{4}$
3	$\frac{\pi^{-1/2}}{6}$
4	$\frac{1}{32}$

Sufficient applications of this formula permits any of the $i^n\text{erfc}(x)$ functions to be expressed in terms of $\text{erfc}(x)$ and $\exp(-x^2)$ and hence evaluated. Some examples are

$$\text{ierfc}(x) = \pi^{-1/2}e^{-x^2} - x\,\text{erfc}(x) \tag{E.14a}$$

$$i^2\,\text{erfc}(x) = \frac{1+2x^2}{4}\,\text{erfc}(x) - \frac{x}{2\pi^{1/2}}e^{-x^2} \tag{E.14b}$$

$$i^3\,\text{erfc}(x) = \frac{1+x^2}{6\pi^{1/2}}e^{-x^2} - \frac{3x+2x^3}{12}\,\text{erfc}(x) \tag{E.14c}$$

E.5 INTEGRALS AND DERIVATIVES

Differentiation gives

$$\frac{d}{dx}\,\text{erf}\,(bx+c) = -\frac{d}{dx}\,\text{erfc}(bx+c) = \frac{2b}{\pi^{1/2}}e^{-(bx+c)^2} \tag{E.15a}$$

Integration gives

$$\int_0^x \text{erf}\,(bt)dt = x\,\text{erf}\,(bx) - \frac{1-\exp(-b^2x^2)}{b\pi^{1/2}} \tag{E.15b}$$

$$\int_x^\infty \text{erfc}(bt)dt = \frac{1}{b}\,\text{ierfc}(bx) \tag{E.15c}$$

Also for the $i^n\text{erfc}(x)$ function the following relations are valid:

$$\frac{d}{dx}i^n\text{erfc}(x) = -i^{n-1}\text{erfc}(x) \tag{E.16a}$$

$$\int_x^\infty i^n\text{erfc}(t)dt = i^{n+1}\text{erfc}(x) \tag{E.16b}$$

Notice that

$$\frac{d}{dx}\,\mathrm{erfc}(x) = -\frac{2}{\pi^{1/2}}e^{-x^2} \tag{E.16c}$$

E.6 COMPLEX ARGUMENT

For x replaced by $x + iy$, the complex function is given by

$$\begin{aligned}\mathrm{erf}(x+iy) &= \frac{2}{\pi^{1/2}}\left[e^{y^2}\int_0^x e^{-t^2}\cos(2yt)dt\right.\\ &\quad\left.+e^{-x^2}\int_0^y e^{t^2}\sin(2xt)dt\right]\\ &\quad-\frac{2i}{\pi^{1/2}}\left[e^{y^2}\int_0^x e^{-t^2}\sin(2yt)dt\right.\\ &\quad\left.-e^{-x^2}\int_0^y e^{t^2}\cos(2xt)dt\right]\end{aligned} \tag{E.17}$$

These four integrals cannot be evaluated in simpler terms. Confusingly, the function generally known as the "error function of the complex argument" is denoted

$$W(x+iy) = W(z) = e^{-z^2}\left[1 + \frac{2i}{\pi^{1/2}}\int_0^z e^{t^2}dt\right] = e^{-z^2}\,\mathrm{erfc}(-iz) \tag{E.18}$$

REFERENCE

Press, W. H., Flannery, B. P., Teukolsky, S. A., and Vetterling, W. T., 1992, *Numerical Recipes: The Art of Scientific Computing*, Cambridge University Press, New York.

F Functions and Series

No.	Equation	Reference
1	$\sum_{n=1}^{\infty} \frac{\cos(n\pi x)}{n} = -\ln\left(2 \sin \frac{\pi x}{2}\right) \quad (0 < x < 1)$	(A + S, p. 1005)
	$= \frac{1}{2} \ln \frac{1}{2[1 - \cos(\pi x)]} \quad (0 < x < 1)$	(G + R, p. 38)
2	$\sum_{n=1}^{\infty} \frac{\cos(n\pi x)}{n^2} = \frac{\pi^2}{2} \left[\frac{1}{2}x^2 - x + \frac{1}{3}\right] \quad (0 \le x \le 1)$	(A + S, p. 1005)
3	$\sum_{n=1}^{\infty} (-1)^{n-1} \frac{\cos(n\pi x)}{n^2} = \frac{\pi^2}{4} \left(\frac{1}{3} - x^2\right) \quad -1 \le x \le 1$	(G + R, p. 38)
4	$\frac{2}{\pi^3} \sum_{n=1}^{\infty} \frac{1}{n^3} \cos(n\pi x) \sin(n\pi\delta) = \begin{cases} \left[-\frac{1}{2}x^2 + \frac{\delta}{3} - \frac{\delta^2}{6}\right](1-\delta) & 0 \le x \le \delta \\ \frac{1}{2}\delta x(x-2) + \frac{\delta^3}{6} + \frac{1}{3}\delta & \delta \le x \le 1 \end{cases}$	
5.1	$\frac{2}{\pi} \sum_{n=1}^{\infty} \frac{\cos(n\pi x) \sin(n\pi\delta)}{n} = \begin{cases} 1-\delta & 0 \le x < \delta \\ -\delta & \delta < x \le 1 \end{cases}$	
5.2	$\frac{2}{\pi} \sum_{n=1}^{\infty} \frac{\sin(n\pi x) \cos(n\pi\delta)}{n} = \begin{cases} -x & 0 \le x < \delta \\ 1-x & \delta < x \le 1 \end{cases}$	
6	$\frac{4}{\pi^3} \sum_{m=1}^{\infty} \sum_{n=1}^{\infty} \frac{1}{m^3} \left[1 + \left(\frac{n}{m}\frac{L_x}{L_y}\right)^2\right]^{-1} \cos\left(m\pi \frac{x}{L_x}\right) \cos\left(n\pi \frac{y}{L_y}\right) \sin\left(m\pi \frac{\delta}{L_x}\right) (-1)^n$ $= -\frac{2}{\pi^3} \sum_{m=1}^{\infty} \frac{1}{m^3} \cos\left(m\pi \frac{x}{L_x}\right) \sin\left(m\pi \frac{\delta}{L_x}\right) + \frac{L_y}{L_x} \frac{2}{\pi^3} \sum_{n=1}^{\infty} \frac{\sin(n\pi\delta / L_x) \cos(n\pi x / L_x) \cosh(n\pi y / L_x)}{n^2 \sinh(n\pi L_y / L_x)}$ $0 \le x \le L_x$, $0 \le y \le L_y$, $0 \le \delta \le L_x$	

(*Continued*)

No.	Equation	Reference
7	$\frac{2}{\pi}\sum_{m=1}^{\infty}\frac{\sin(m\pi y/L_y)}{m\{1+[(n/m)(L_y/L_x)]^2\}}=\frac{\sinh[n\pi(L_y-y)/L_x]}{\sinh(n\pi L_y/L_x)}$	
8	$\sum_{n=-\infty}^{\infty} e^{-(x-x'+2nL)^2/4\alpha t}=\frac{(\pi\alpha t)^{1/2}}{L}\times\left[1+2\sum_{n=1}^{\infty}\cos\frac{n\pi(x-x')}{L}e^{-n^2\pi^2\alpha t/L^2}\right]$ (Poisson's summation formula, Carslaw and Jaeger, 1959, p. 275) x' can be positive or negative	
9	$\sum_{n=1}^{\infty}(-1)^n n\sin\left(n\pi\frac{x}{L}\right)e^{-n^2\pi^2\alpha t/L^2}=-\frac{L}{2\pi^{3/2}(\alpha t/L^2)^{1/2}}\frac{d}{dx}\sum_{n=-\infty}^{\infty}\exp\left\{-\frac{[x-(2n+1)L]^2}{4\alpha t}\right\}$	(M + F, p. 1587)
10	$\sum_{n=1}^{\infty}\frac{\sin(n\pi x)}{n}=\frac{\pi}{2}(1-x)\quad 0<x<2$	(A + S, p. 1005)
11	$\sum_{n=1}^{\infty}(-1)^n\frac{\sin(n\pi x)}{n}=-\frac{\pi}{2}x$	(Ozisik, p. 203)
12	$\sum_{n=1}^{\infty}\frac{(-1)^n\cos(n\pi y/L_y)}{[(m/L_x)^2+(n/L_y)^2]}=-\frac{1}{2}(L_x/m)^2+L_y\frac{\pi}{2}(L_x/m)\frac{\cosh(m\pi y/L_x)}{\sinh(m\pi L_y/L_x)}$	
13	$\sum_{n=1}^{\infty}(-1)^n\{1+[(n/m)(L_x/L_y)]^2\}^{-1}\cos\left(n\pi\frac{y}{L_y}\right)=\frac{\pi}{2}\frac{mL_y}{L_x}\frac{\cosh(m\pi y/L_x)}{\sinh(m\pi L_y/L_x)}$	

REFERENCES

Abramowitz, M. and Stegun, I. A. (Eds), 1964, *Handbook of Mathematical Functions with Formulas, Graphs, and Mathematical Tables*, Applied Mathematics Series 55, National Bureau of Standards, Washington, D.C.

Carslaw, H. S. and Jaeger, J. C., 1959, *Conduction of Heat in Solids*, 2nd edn, Oxford University Press, New York.

Gradshteyn, I. S. and Ryzhik, I. M., 2007, *Tables of Integrals, Series, and Products*, 7th edn, Elsevier, New York.

Morse, P. M. and Feshbach, H., 1953, *Methods of Theoretical Physics*, McGraw-Hill, New York.

Ozisik, M. N., 1993, *Heat Conduction*, John Wiley, New York.

I Integrals

TABLE I.1
Expressions Involving Exponential Functions and Integrals

No.	Equation	Reference
0	$E_0(z) = z^{-1}e^{-z}$	(A + S)
1	$E_1(z) = \int_z^{\infty} \frac{e^{-t}}{t}\,dt = \int_{z/c^2}^{\infty} \frac{e^{-c^2 t}}{t}\,dt \ (\lvert\arg z\rvert < \pi)$	
2	$E_1(x) = -Ei(-x)$	(A + S)
3	$E_n(z) = \int_1^{\infty} \frac{e^{-zt}}{t^n}\,dt \ (Rz > 0)$	
3.1	$E_n(x) = x^{n-1} \int_x^{\infty} \frac{e^{-w}}{w^n}\,dw$	
3.2	$\int_0^t u^j e^{-a/u}\,du = t^{j+1} E_{j+2}(at^{-1})$	
3.3	$\int_x^{\infty} \frac{1}{t^{i+1}} e^{-c^2 t} dt = x^{-i} E_{i+1}(c^2 x)$	
4	$E_{n+1}(z) = \frac{1}{n}[e^{-z} - zE_n(z)] \qquad n = 1, 2, 3, \ldots$	(A + S)
4.1	$E_2(z) = e^{-z} - zE_1(z)$	
4.2	$E_3(z) = \frac{1}{2}[z^2 E_1(z) + e^{-z}(1 - z)]$	
5	$\int \frac{1}{r^{2m-1}} e^{-r^2/(4t)}\,dr = -\frac{1}{2} r^{2(1-m)} E_m\left(\frac{r^2}{4t}\right) \quad$ for $m = 1, 2, 3, \ldots$	
5.1	$\int \frac{1}{r} e^{-r^2/(4t)}\,dr = -\frac{1}{2} E_1\left(\frac{r^2}{4t}\right)$	
5.2	$\int \frac{1}{r^3} e^{-r^2/(4t)}\,dr = -\frac{1}{2} r^{-2} E_2\left(\frac{r^2}{4t}\right)$	
6	$\int r^{2n+1} e^{-r^2/(4t)}\,dr = \frac{1}{2}(4t)^{n+1} \int^{r^2/4t} V^n e^{-V}\,dV$ $= -\frac{1}{2}(4t)^{n+1} e^{-r^2/(4t)} \sum_{i=0}^{n} \frac{n!}{(n-i)!} \left(\frac{r^2}{4t}\right)^{n-i}$ for $n = 0, 1, 2, \ldots$	
6.1	$\int r e^{-r^2/(4t)}\,dr = -2te^{-r^2/(4t)}$	
6.2	$\int r^3 e^{-r^2/(4t)}\,dr = -8t^2\left(1 + \frac{r^2}{4t}\right) e^{-r^2/(4t)}$	

(Continued)

TABLE I.1
Expressions Involving Exponential Functions and Integrals (Continued)

No.	Equation	Reference
6.3	$\int r^5 e^{-r^2/(4t)}\, dr = -32t^3 e^{-r^2/(4t)}\left[2 + \frac{2r^2}{4t} + \left(\frac{r^2}{4t}\right)^2\right]$	
7	$\int E_n(u)du = -E_{n+1}(u)$	
8	$\int_0^b xE_1(x^2)dx = \frac{1}{2}\int_0^{b^2} E_1(u)du = \frac{1}{2}[1 - E_2(b^2)]$	
9	$\int_a^\infty x^2 E_1(x^2)dx = \frac{1}{6}[\sqrt{\pi}\,\text{erfc}(a) + 2aE_2(a^2)]$	
10a	$\int_z^\infty E_1(u^2)du = \sqrt{\pi}\,\text{erfc}(z) - zE_1(z^2)$	
10b	$\int_0^\infty E_1(u^2)du = \sqrt{\pi}$	
11	$E_{3/2}(x^2) = 2\sqrt{\pi}\,\text{ierfc}(x),\ x > 0$	
12	$\int_a^\infty \frac{1}{w(1+bw)} e^{-w}\, dw = E_1(a) - e^{b^{-1}} E_1(a + b^{-1})$	

TABLE I.2
Integrals Involving erf(*x*) and erfc(*x*)

No.	Equation	Reference
1	$\int \text{erf}(ax)dx = x\,\text{erf}(ax) + \frac{1}{a\sqrt{\pi}}e^{-a^2x^2} = x + \frac{1}{a}\,\text{ierfc}(ax)$	(N + G 4.1.1)
2	$\int \text{erfc}(ax)dx = x\,\text{erfc}(ax) - \frac{1}{a\sqrt{\pi}}e^{-a^2x^2} = -\frac{1}{a}\text{ierfc}(ax)$	(N + G 4.1.2)
3	$\int x\,\text{erf}(ax)dx = \left(\frac{x^2}{2} - \frac{1}{4a^2}\right)\text{erf}(ax) + \frac{x}{2a\sqrt{\pi}}e^{-a^2x^2}$	
	$= \frac{x^2}{2} - \frac{1}{4a^2}\,\text{erf}(ax) + \frac{x}{2a}\,\text{ierfc}(ax)$	(N + G 4.1.4)
4	$\int x\,\text{erfc}(ax)dx = \frac{1}{4a^2}\,\text{erf}(ax) + \frac{x^2}{2}\,\text{erfc}(ax) - \frac{x}{2a\sqrt{\pi}}e^{-x^2a^2}$	
	$= \frac{1}{4a^2}\,\text{erf}(ax) - \frac{x}{2a}\,\text{ierfc}(ax)$	(N + G 4.1.5)
5	$\int x^{-1}\,\text{erf}(ax)dx = ln(x)\,\text{erf}(ax) - \frac{2a}{\sqrt{\pi}}\int ln(x)e^{-a^2x^2}dx$	(N + G 4.1.12)

TABLE I.2
Integrals Involving erf(x) and erfc(x) (Continued)

No.	Equation	Reference
6	$\int x^{-1}\,\mathrm{erfc}(ax)dx = ln(x)\,\mathrm{erfc}(ax) + \frac{2a}{\sqrt{\pi}}\int ln(x)e^{-a^2x^2}dx$	(N + G 4.1.13)
7	$\int x^{-n}\,\mathrm{erf}(ax)dx = \frac{-\mathrm{erf}(ax)}{(n-1)x^{n-1}} + \frac{2a}{(n-1)\sqrt{\pi}} \times \int \frac{e^{-a^2x^2}}{x^{n-1}}dx \quad n \geq 2$	(N + G 4.1.14)
8	$\int x^{-n}\,\mathrm{erfc}(ax)dx = \frac{-\mathrm{erfc}(ax)}{(n-1)x^{n-1}} - \frac{2a}{(n-1)\sqrt{\pi}} \times \int \frac{e^{-a^2x^2}}{x^{n-1}}dx \quad n \geq 2$	(N + G 4.1.15)
9	$\int e^{bx}\,\mathrm{erf}(ax+c)dx = \frac{e^{bx}}{b}\,\mathrm{erf}(ax+c) - \frac{e^{(b^2/4a^2-bc/a)}}{b}\,\mathrm{erf}\left(ax+c-\frac{b}{2a}\right)$	
10	$\int e^{bx}\,\mathrm{erfc}(ax)dx = \frac{e^{bx}}{b}\,\mathrm{erfc}(ax) + \frac{e^{b^2/4a^2}}{b}\,\mathrm{erf}\left(ax-\frac{b}{2a}\right)$	(N + G 4.2.2)
11	$\int x\,\mathrm{erf}^3(x)dx = \left(\frac{x^2}{2}-\frac{1}{4}\right)\mathrm{erf}^3(x) + \frac{3x}{2\sqrt{\pi}}e^{-x^2}\,\mathrm{erf}^2(x) + \frac{3}{2\pi}e^{-2x^2}\,\mathrm{erf}(x) - \frac{\sqrt{3}}{2\pi}\,\mathrm{erf}(x\sqrt{3})$	(Cho 2.3.6)
12	$\int_z^{\infty}\frac{1}{x^4}\,\mathrm{ierfc}(ax)dx = \frac{2}{z^3}\mathrm{i}^3\,\mathrm{erfc}(az)$	
13	$\int_z^{\infty} e^{-a^2x^2}\,\mathrm{erf}(x)dx = \frac{\pi^{1/2}}{2a}H(z,a)$ (See Section 6.8)	

TABLE I.3
Integrals Involving erf($a\sqrt{x}$) or erfc($a\sqrt{x}$)

No.	Equation	Reference
1	$\int \mathrm{erf}(a\sqrt{x})dx = \left(x-\frac{1}{2a^2}\right)\mathrm{erf}(a\sqrt{x}) + \frac{1}{a}\sqrt{\frac{x}{\pi}}e^{-a^2x} \quad a \neq 0$ $= x - \frac{1}{2a^2}(\mathrm{erf}(a\sqrt{x})-1) + \sqrt{\frac{x}{a^2}}\,\mathrm{ierfc}(a\sqrt{x})$	(Cho 2.4.1)
2	$\int x\,\mathrm{erf}(a\sqrt{x})dx = \frac{x^2}{2}\,\mathrm{erf}(a\sqrt{x}) - \frac{3}{8a^4}\,\mathrm{erf}(a\sqrt{x}) + \left(\frac{x^{3/2}}{2a\sqrt{\pi}} + \frac{3x^{1/2}}{4a^3\sqrt{\pi}}\right)e^{-a^2x} \quad a \neq 0$	(Cho 2.4.2)

(Continued)

TABLE I.3
Integrals Involving erf($a\sqrt{x}$) or erfc($a\sqrt{x}$) (Continued)

No.	Equation	Reference
3	$\int \frac{\mathrm{erf}(a\sqrt{x})}{\sqrt{x}}dx = 2\sqrt{x}\,\mathrm{erf}(a\sqrt{x}) + \frac{2}{a\sqrt{\pi}}e^{-a^2x}$	
	$= 2\sqrt{x} + \frac{2}{a}\,\mathrm{ierfc}(a\sqrt{x}) \qquad a \neq 0$	(Cho 2.4.3)
4	$\int \frac{x-b}{\sqrt{x}(x+b)^2}\,\mathrm{erf}(a\sqrt{x})dx = \frac{2\sqrt{x}}{x+b}\,\mathrm{erf}(a\sqrt{x})$	
	$- 2e^{a^2b}\left(\frac{a}{\sqrt{\pi}}\right)E_1(a^2b + a^2x)$	
5	$\int e^x\,\mathrm{erf}(\sqrt{x})dx = e^x\mathrm{erf}\sqrt{x} - 2(x/\pi)^{1/2}$	(Cho 2.4.4)
5.1	$\int e^x\,\mathrm{erfc}(x^{1/2})dx = e^x\mathrm{erfc}(x^{1/2}) + 2(x/\pi)^{1/2}$	
6	$\int e^{ax}\,\mathrm{erf}(\sqrt{bx})dx = \frac{1}{a}e^{ax}\mathrm{erf}(\sqrt{bx}) - \frac{1}{a}\sqrt{\frac{b}{b-a}}$	
	$\times\,\mathrm{erf}\{[(b-a)x]^{1/2}\} \qquad a \neq 0 \qquad b \neq 0 \qquad a \neq b$	(Cho 2.4.5)
7	$\int_0^t u^{a/2}e^u\,\mathrm{erf}(u^{1/2})du$	
	$= \frac{2}{\pi^{1/2}}\sum_{i=0}^{\infty}2^i\frac{t^{i+(a+3)/2}}{1\cdot 3\cdot 5\cdots(2i+1)[i+(a+3)/2]} \qquad a > -3$	
	$\approx \frac{2}{\pi^{1/2}}\left(\frac{t^2}{2} + \frac{2t^3}{9} + \frac{1}{15}t^4\right)$ for $a = 1$ and small t values	
	$\approx \frac{2}{\pi^{1/2}}\left(\frac{t^3}{3} + \frac{t^4}{6} + \frac{4t^5}{75}\right)$ for $a = 3$ and small t values	
8	$\int_0^t u^{a/2}e^u\,\mathrm{erfc}(u^{1/2})du \approx t^{(a+2)/2}\left[\frac{2}{a+2} - \frac{4}{a+3}\left(\frac{t}{\pi}\right)^{1/2} + \cdots\right]$	
	for small t values and $a > -2$	
9	$\int_0^t u^{-1/2}e^u\,\mathrm{erfc}(u^{1/2})du = 2\int_0^{t^{1/2}} e^{V^2}\,\mathrm{erfc}(V)dV \approx 2t^{1/2}\left[1 - \left(\frac{t}{\pi}\right)^{1/2} + \frac{1}{3}t\right]$	
	for small t values	
10	$\int u^{1/2}e^u\,\mathrm{erfc}(u^{1/2})du = 2\int V^2e^{V^2}\,\mathrm{erfc}(V)dV$	
	$= \frac{u}{\pi^{1/2}} + u^{1/2}e^u\,\mathrm{erfc}(u^{1/2})$	
	$- \frac{1}{2}\int u^{-1/2}e^u\,\mathrm{erfc}(u^{1/2})du$	

TABLE I.3
Integrals Involving erf($a\sqrt{x}$) or erfc($a\sqrt{x}$) (Continued)

No.	Equation	Reference
11	$\int_0^t (1 - e^{-1/4u}) e^{B^2 u} \operatorname{erfc}(B u^{1/2}) du$ $\approx t\left[1 - E_2\left(\frac{1}{4t}\right)\right]$ $- \frac{B}{\pi^{1/2}}\left[\frac{4}{3}t^{3/2} - \frac{1}{3}\pi^{1/2}\operatorname{erfc}\left(\frac{1}{2t^{1/2}}\right) + \frac{2}{3}t^{1/2}e^{-1/4t}(1-2t)\right]$ $+ \frac{1}{2}B^2 t^2\left[1 - 2E_3\left(\frac{1}{4t}\right)\right]$ for $Bt^{1/2} << 1$	
12	$\int_0^t \frac{1}{(t-u)^{1/2}}\left\{\frac{1}{u^{1/2}} - \frac{(\pi\alpha)^{1/2}}{a}e^{\alpha u/a^2}\operatorname{erfc}\left[\frac{(\alpha u)^{1/2}}{a}\right]\right\} du$ $= \pi e^{\alpha t/a^2}\operatorname{erfc}\left[\frac{(\alpha t)^{1/2}}{a}\right]$	(Levine)

TABLE I.4
Integrals Involving erf($a/\sqrt{x}$) or erfc($a/\sqrt{x}$)

No.	Equation	Reference
1	$\int \operatorname{erfc}\left(\frac{a}{\sqrt{x}}\right) dx = (x + 2a^2)\operatorname{erfc}\left(\frac{a}{\sqrt{x}}\right) - 2a\sqrt{\frac{x}{\pi}}e^{-a^2/x}$ $= x\operatorname{erfc}\left(\frac{a}{\sqrt{x}}\right) - 2a\sqrt{x}\operatorname{ierfc}\left(\frac{a}{\sqrt{x}}\right)$	(Cho 2.5.1)
2	$\int x \operatorname{erfc}(a/\sqrt{x}) dx = \left(\frac{x^2}{2} - \frac{2a^4}{3}\right)\operatorname{erfc}(a\sqrt{x})$ $-(x^{3/2} - 2a^2 x^{1/2})\frac{a}{3\sqrt{\pi}}e^{-a^2/x}$ $= \frac{x^2}{2}\operatorname{erfc}\left(\frac{a}{\sqrt{x}}\right) - \frac{a}{3}\sqrt{\frac{x^3}{\pi}}e^{-a^2/x}$ $+ \frac{2a^3\sqrt{x}}{3}\operatorname{ierfc}\left(\frac{a}{\sqrt{x}}\right)$	(Cho 2.5.2)
3a	$\int_{u=0}^t \frac{1}{u^2}\operatorname{erf}\left(\frac{x}{(4\alpha u)^{1/2}}\right)e^{-y^2/4\alpha u} du = \frac{4\alpha}{y^2}\left\{e^{-y^2/4\alpha t}\operatorname{erf}\left[\frac{x}{(4\alpha t)^{1/2}}\right]\right.$ $+ \frac{x}{(x^2+y^2)^{1/2}}$ $\left.\times \operatorname{erfc}\left[\frac{(x^2+y^2)^{1/2}}{(4\alpha t)^{1/2}}\right]\right\}$ $y \neq 0$	

(*Continued*)

TABLE I.4
Integrals Involving erf($a/\sqrt{x}$) or erfc($a/\sqrt{x}$) (Continued)

No.	Equation	Reference
3b	$\int_{u=0}^{t} \frac{1}{u^2} \operatorname{erfc}\left(\frac{x}{(4\alpha u)^{1/2}}\right) e^{-y^2/4\alpha u}\, du = \frac{4\alpha}{y^2}\left\{e^{-y^2/4\alpha t} \operatorname{erfc}\left[\frac{x}{(4\alpha t)^{1/2}}\right] - \frac{x}{(x^2+y^2)^{1/2}} \times \operatorname{erfc}\left[\frac{(x^2+y^2)^{1/2}}{(4\alpha t)^{1/2}}\right]\right\}$ $\quad y \neq 0$	
4	$\int_{u=0}^{t} \frac{1}{u^{3/2}} e^{-y^2/4\alpha u} \operatorname{erf}\left[\frac{x}{(4\alpha u)^{1/2}}\right] du = \frac{2\sqrt{\alpha\pi}}{y} H\left(\frac{x}{\sqrt{4\alpha t}}, \frac{y}{x}\right)$ (see Section 6.8)	
5	$\int_{u=0}^{t} \frac{1}{(\pi\alpha u)^{1/2}} \operatorname{erfc}\left[\frac{C_1}{(4\alpha u)^{1/2}}\right] \operatorname{erfc}\left[\frac{C_2}{(4\alpha u)^{1/2}}\right] du$ $\approx \frac{1}{\pi^{3/2}} \frac{(C_1^2+C_2^2)^{3/2}}{2\alpha C_1 C_2} \left\{\Gamma\left(-\frac{3}{2}, \frac{C_1^2+C_2^2}{4\alpha t}\right) - \frac{1}{2}\left[\frac{C_1^2+C_2^2}{C_1 C_2}\right]^2 \Gamma\left(-\frac{5}{2}, \frac{C_1^2+C_2^2}{4\alpha t}\right)\right\}$ $\approx \frac{1}{\pi^{3/2}} \frac{(C_1^2+C_2^2)^{1/2}}{C_1 C_2} \exp\left[-\frac{C_1^2+C_2^2}{4\alpha t}\right] \times \left(\frac{4\alpha t}{C_1^2+C_2^2}\right)^{5/2} \left\{1 - \frac{5}{2}\frac{4\alpha t}{C_1^2+C_2^2} - \frac{1}{2}\left[\frac{C_1^2+C_2^2}{C_1 C_2}\right]^2 \left[\frac{4\alpha t}{C_1^2+C_2^2}\right]\right\}$ for small values of $\frac{4\alpha t}{C_1^2}$ and $\frac{4\alpha t}{C_2^2}$	
6	$\int_{u=0}^{t} [4\pi\alpha u]^{-1/2} \exp\left(-\frac{z^2}{4\alpha u}\right) \operatorname{erf}\left[\frac{a}{(4\alpha u)^{1/2}}\right] \operatorname{erf}\left[\frac{b}{(4\alpha u)^{1/2}}\right] du$ $\approx \left(\frac{t}{\alpha}\right)^{1/2} \operatorname{ierfc}\left[\frac{\|z\|}{(4\alpha t)^{1/2}}\right] - \frac{t}{\pi}\left[\frac{1}{a} E_2\left(\frac{z^2+a^2}{4\alpha t}\right) + \frac{1}{b} E_2\left(\frac{z^2+b^2}{4\alpha t}\right)\right]$ for small $\frac{\alpha t}{a^2}$ and $\frac{\alpha t}{b^2}$ values	
7a	$\int_{0}^{t} \frac{1}{(\pi\alpha u)^{1/2}} \frac{(4\alpha u)^{(m+n+2)/2}}{C_1^{m+1} C_2^{n+1}} e^{-(C_1^2+C_2^2)/(4\alpha u)}\, du$ $= \frac{1}{2\alpha\pi^{3/2}} \frac{(C_1^2+C_2^2)^{(m+n+3)/2}}{C_1^{m+1} C_2^{n+1}} \Gamma\left(-\frac{m+n+3}{2}, \frac{C_1^2+C_2^2}{4\alpha t}\right)$	

TABLE I.4
Integrals Involving erf($a/\sqrt{x}$) or erfc($a/\sqrt{x}$) (Continued)

No.	Equation	Reference
7b	$\int_0^t \frac{1}{(\pi\alpha u)^{1/2}}(4\alpha u)^{(n+2)/2} e^{-C^2/(4\alpha u)}\, du = \frac{1}{2\alpha\pi^{3/2}} C^{n+3}\Gamma\left(-\frac{n+3}{2}, \frac{C^2}{4\alpha t}\right)$	
8	$\frac{\alpha}{a^2}\int_{u=0}^{t}\left(\frac{\alpha u}{a^2}\right)^{n/2} \text{ierfc}\left[\frac{a}{(\alpha u)^{1/2}}\right] du$ $= 2\int_{a/(\alpha t)^{1/2}}^{\infty} \frac{1}{w^{n+3}} \text{ierfc}(w) dw$ $= \frac{2}{n+2}\left\{\left(\frac{\alpha t}{a^2}\right)^{n/2+1} \text{ierfc}\left[\frac{a}{(\alpha t)^{1/2}}\right] - \frac{1}{n+1}\left(\frac{\alpha t}{a^2}\right)^{n/2+1/2}\right.$ $\left.\times \text{erfc}\left[\frac{a}{(\alpha t)^{1/2}}\right] + \frac{1}{n+1}\frac{1}{\pi^{1/2}}\Gamma\left(-\frac{n}{2}, \frac{a^2}{\alpha t}\right)\right\} \quad n = 0, 1, 2, \ldots$	
9	$\int_0^t \left(\frac{\tau}{t_0}\right)^{m/2} \text{erfc}\left\{\frac{z}{[4\alpha(t-\tau)]^{1/2}}\right\} d\tau$ $= t_0\Gamma\left(\frac{m}{2}+1\right)\left(\frac{4t}{t_0}\right)^{(m+2)/2} \text{i}^{m+2}\,\text{erfc}\left[\frac{z}{(4\alpha t)^{1/2}}\right], \; m = -1, 0, 1, 2, \ldots$	
10	$\int_0^t \left(\frac{\tau}{t_0}\right)^{m/2} \frac{\alpha(t-\tau)}{L^2} \text{erfc}\left\{\frac{z}{[4\alpha(t-\tau)]^{1/2}}\right\} d\tau = t_0\frac{\alpha t_0}{L^2}\left(\frac{4t}{t_0}\right)^{(m+4)/2}$ $\times\left\{\frac{1}{4}\Gamma\left(\frac{m}{2}+1\right) \text{i}^{m+2}\,\text{erfc}\left[\frac{z}{(4\alpha t)^{1/2}}\right]\right.$ $\left.-\Gamma\left(\frac{m}{2}+1\right) \text{i}^{m+4}\,\text{erfc}\left[\frac{z}{(4\alpha t)^{1/2}}\right]\right\}, \; m = -1, 0, 1, 2, \ldots$	

Below are some $\Gamma(-m, x^2)$ relations:

$$\Gamma(-m, x^2) = x^{-2m}E_{m+1}(x^2) \qquad m = 0, 1, 2, 3, \ldots$$

$$\Gamma\left(-\frac{1}{2}, x^2\right) = \frac{2\pi^{1/2}}{x}\text{ierfc}(x)$$

$$\Gamma\left(-\frac{3}{2}, x^2\right) = \frac{4}{3}\left[\pi^{1/2}\text{erfc}(x) - \frac{1}{x}e^{-x^2}\left(1 - \frac{1}{2x^2}\right)\right]$$

(See also Appendix E)

TABLE I.5
Integrals Involving $\mathrm{erfc}(a\sqrt{x}+b/\sqrt{x})$

No.	Equation	Reference
1	$\int \mathrm{erfc}\left(a\sqrt{x}+\frac{b}{\sqrt{x}}\right)dx = -\left[\frac{1}{4a^2}\left(\mathrm{erfc}(a\sqrt{x}+b/\sqrt{x}) + e^{-4ab}\,\mathrm{erfc}(a\sqrt{x}-b/\sqrt{x})\right) + \frac{\sqrt{x}}{a}\,\mathrm{ierfc}\left(a\sqrt{x}+\frac{b}{\sqrt{x}}\right)\right] \quad a \neq 0$	(Cho 2.6.1)
2	$\int e^{x}\,\mathrm{erfc}\left(a\sqrt{x}+\frac{b}{\sqrt{x}}\right)dx = e^{x}\,\mathrm{erfc}\left(a\sqrt{x}+\frac{b}{\sqrt{x}}\right) - \frac{1}{2}\left(1+\frac{a}{(a^2-1)^{1/2}}\right)e^{-2b[a-(a^2-1)^{1/2}]} \times \mathrm{erfc}\left[(a^2-1)^{1/2}x^{1/2}+\frac{b}{\sqrt{x}}\right] + \frac{1}{2}\left(1-\frac{a}{(a^2-1)^{1/2}}\right)e^{-2b[a+(a^2-1)^{1/2}]} \times \mathrm{erfc}\left[(a^2-1)^{1/2}x^{1/2}-\frac{b}{\sqrt{x}}\right] \quad a>1$	(Cho 2.6.3)
3	$\int e^{(a^2-b^2)x}\,\mathrm{erfc}\left(a\sqrt{x}+\frac{c}{\sqrt{x}}\right)dx = \frac{1}{a^2-b^2}e^{(a^2-b^2)x}\,\mathrm{erfc}\left(a\sqrt{x}+\frac{c}{\sqrt{x}}\right) - \frac{1}{2(a^2-b^2)}\left(1+\frac{a}{b}\right)e^{-2(a-b)c} \times \mathrm{erfc}\left(b\sqrt{x}+\frac{c}{\sqrt{x}}\right) + \frac{1}{2(a^2-b^2)} \times \left(1-\frac{a}{b}\right)e^{-2(a+b)c} \times \mathrm{erfc}\left(b\sqrt{x}-\frac{c}{\sqrt{x}}\right) \quad a^2 \neq b^2$	(Cho 2.6.4)
4	$\int_0^t \tau^{m/2}\exp\left[\frac{hx}{k}+\frac{h^2}{k^2}\alpha(t-\tau)\right]\mathrm{erfc}\left\{\frac{x}{[4\alpha(t-\tau)]^{1/2}}+\frac{h}{k}[\alpha(t-\tau)]^{1/2}\right\}d\tau = \frac{\Gamma\left(\frac{m}{2}+1\right)}{\alpha^{m/2+1}}\left(-\frac{k}{h}\right)^{m+2}\left\{\exp\left[\frac{hx}{k}+\frac{h^2}{k^2}\alpha t\right]\mathrm{erfc}\left[\frac{x}{(4\alpha t)^{1/2}}+\frac{h}{k}(\alpha t)^{1/2}\right] - \sum_{j=0}^{m+1}\left[-\frac{h}{k}(4\alpha t)^{1/2}\right]^j \mathrm{i}^j\,\mathrm{erfc}\left[\frac{x}{(4\alpha t)^{1/2}}\right]\right\},\ m=-1,0,1,2,3,\ldots$	

Note: $\Gamma(z)$ is the gamma function (Abramowitz and Stegun, 1964, p. 255).

TABLE I.6
Integrals Involving the Exponential Function

No.	Equation	Reference
1	$\int e^{-(ax^2+2bx+c)}\,dx = \frac{1}{2}\sqrt{\frac{\pi}{a}}e^{(b^2-ac)/a}\operatorname{erf}\left(\sqrt{a}x + \frac{b}{\sqrt{a}}\right) \qquad a > 0$	(A + S 7.4.32)
2	$\int e^{-a^2x^2+bx}\,dx = \frac{\sqrt{\pi}}{2a}e^{b^2/4a^2}\operatorname{erf}\left(ax - \frac{b}{2a}\right)$	
3	$\int x^{3/2}e^{-a^2x}\,dx = \frac{3\sqrt{\pi}}{4a^5}\operatorname{erf}(a\sqrt{x}) - \frac{3\sqrt{x}}{2a^4}e^{-a^2x} - \frac{x^{3/2}}{a^2}e^{-a^2x}$ $= \frac{3\sqrt{\pi}}{4a^5}\operatorname{erf}(a\sqrt{x}) - \left(\frac{3\sqrt{x}}{2a^4} + \frac{x^{3/2}}{a^2}\right)e^{-a^2x} \qquad a \neq 0$	(Cho 2.7.2)
4	$\int \sqrt{x}e^{-a^2x}\,dx = \frac{\sqrt{\pi}}{2a^3}\operatorname{erf}(a\sqrt{x}) - \frac{\sqrt{x}}{a^2}e^{-a^2x} \qquad a \neq 0$	(Cho 2.7.3)
5	$\int x^{-1/2}e^{-a^2x}\,dx = \frac{\sqrt{\pi}}{a}\operatorname{erf}(a\sqrt{x}) \qquad a \neq 0$	(Cho 2.7.4)
6	$\int x^{-3/2}e^{-a^2x}\,dx = -2a\sqrt{\pi}\operatorname{erf}(a\sqrt{x}) - \frac{2}{\sqrt{x}}e^{-a^2x}$ $= -2a\sqrt{\pi}\left(1 + \frac{1}{a\sqrt{x}}\operatorname{ierfc}(a\sqrt{x})\right) \qquad a \neq 0$	(Cho 2.7.5)
7	$\int \sqrt{x}e^{-a^2/x}\,dx = \left(\frac{2x^{3/2}}{3} - \frac{4a^2\sqrt{x}}{3}\right)e^{-a^2/x} - \frac{4a^3\sqrt{\pi}}{3}\operatorname{erf}\left(\frac{a}{\sqrt{x}}\right)$ $= \frac{2}{3}x^{3/2}e^{-a^2/x} - \frac{4}{3}a^2\sqrt{\pi}\left[a + \sqrt{x}\operatorname{ierfc}\left(\frac{a}{\sqrt{x}}\right)\right]$	(Cho 2.8.2)
8	$\int x^{3/2}e^{-a^2/x}\,dx = \left(\frac{2x^{5/2}}{5} - \frac{4a^2x^{3/2}}{15} + \frac{8a^4x^{1/2}}{15}\right)e^{-a^2/x}$ $+\frac{8a^5\sqrt{\pi}}{15}\operatorname{erf}\left(\frac{a}{\sqrt{x}}\right)$ $= \frac{x^{3/2}}{15}(6x - 4a^2)e^{-a^2/x} + \frac{8a^4\sqrt{\pi}}{15}$ $\times\left[a + \sqrt{x}\operatorname{ierfc}\left(\frac{a}{\sqrt{x}}\right)\right]$	(Cho 2.8.1)
9	$\int x^{-1/2}e^{-a^2/x}\,dx = 2a\sqrt{\pi}\operatorname{erf}\left(\frac{a}{\sqrt{x}}\right) + 2\sqrt{x}e^{-a^2/x}$ $= 2\pi^{1/2}\left[a + \sqrt{x}\operatorname{ierfc}\left(\frac{a}{\sqrt{x}}\right)\right]$	(Cho 2.8.3)
10	$\int x^{-3/2}e^{-a^2/x}\,dx = -\frac{\sqrt{\pi}}{a}\operatorname{erf}\left(\frac{a}{\sqrt{x}}\right) \qquad a \neq 0$	(Cho 2.8.4)

(Continued)

TABLE I.6
Integrals Involving the Exponential Function (Continued)

No.	Equation	Reference
11	$\int e^{-a^2x^2-b^2/x^2}dx = \frac{\sqrt{\pi}}{4a}\left[e^{2ab}\operatorname{erf}\left(ax+\frac{b}{x}\right) + e^{-2ab}\operatorname{erf}\left(ax-\frac{b}{x}\right)\right] \quad a \neq 0$	(Cho 2.9.1)
12	$\int x^{-1/2}e^{-a^2x-b^2/x}\,dx = \frac{\sqrt{\pi}}{2a}\left[e^{2ab}\operatorname{erf}\left(a\sqrt{x}+\frac{b}{\sqrt{x}}\right) + e^{-2ab}\operatorname{erf}\left(a\sqrt{x}-\frac{b}{\sqrt{x}}\right)\right] \quad a \neq 0$	(Cho 2.9.4)
13	$\int x^{-3/2}e^{-a^2x-b^2/x}\,dx = \frac{\sqrt{\pi}}{2b}\left[e^{-2ab}\operatorname{erf}\left(a\sqrt{x}-\frac{b}{\sqrt{x}}\right) - e^{2ab}\operatorname{erf}\left(a\sqrt{x}+\frac{b}{\sqrt{x}}\right)\right] \quad b \neq 0$	(Cho 2.9.5)
14	$\int x^{1/2}e^{-a^2x-b^2/x}\,dx = \frac{\sqrt{\pi}}{2a^2}\left(\frac{1}{2a}-b\right)e^{2ab}\operatorname{erf}\left(a\sqrt{x}+\frac{b}{\sqrt{x}}\right) - \frac{\sqrt{x}}{a^2}e^{-a^2x-b^2/x} + \frac{\sqrt{\pi}}{2a^2}\left(\frac{1}{2a}+b\right)e^{-2ab} \times \operatorname{erf}\left(a\sqrt{x}-\frac{b}{\sqrt{x}}\right) \quad a \neq 0$	(Cho 2.9.3)
15	$\int x^{3/2}e^{-a^2x-b^2/x}\,dx = \left(b^2+\frac{3}{4a^2}-\frac{3b}{2a}\right)\frac{\sqrt{\pi}}{2a^3}e^{2ab}\operatorname{erf}\left(a\sqrt{x}+\frac{b}{\sqrt{x}}\right) - \frac{3\sqrt{x}}{2a^4}e^{-a^2x-b^2/x} + \left(b^2+\frac{3}{4a^2}+\frac{3b}{2a}\right)\frac{\sqrt{\pi}}{2a^3}e^{-2ab} \times \operatorname{erf}\left(a\sqrt{x}-\frac{b}{\sqrt{x}}\right) - \frac{x^{3/2}}{a^2}e^{-a^2x-b^2/x} \quad a \neq 0$	(Cho 2.9.2)
16	$\int_0^t e^{-a^2(t-\tau)}\,d\tau = \frac{1}{a^2}\left[1-e^{-a^2t}\right]$	
17	$\int_0^t e^{-C\tau}e^{-a^2(t-\tau)}\,d\tau = \frac{1}{a^2-C}\left\{e^{-Ct}-e^{-a^2t}\right\}; \quad a^2 \neq C$	
18	$\int_0^t \frac{\tau}{t_0}e^{-a^2(t-\tau)}\,d\tau = t_0\left(\frac{1}{a^2t_0}\right)^2\left\{(a^2t-1)+e^{-a^2t}\right\}$	
19	$\int_0^t \left(\frac{\tau}{t_0}\right)^{-1/2}e^{-a^2(t-\tau)}\,d\tau = t_0\left(\frac{1}{a^2t_0}\right)^{1/2}F\left((a^2t)^{1/2}\right)$	

Note that $F(z) \equiv e^{-z^2}\int_0^z e^{x^2}\,dx$, the Dawson integral.

TABLE I.7
Spatial Integrals with $K(z - x', u)$, the Fundamental Heat Conduction Solution

No.	Equation
1	$\int_a^b K(z-x',u)\,dx' = \frac{1}{2}\left(\text{erfc}\left\{\frac{z-b}{[4\alpha u]^{1/2}}\right\} - \text{erfc}\left\{\frac{z-a}{[4\alpha u]^{1/2}}\right\}\right)$
2	$\int_a^b \frac{x'}{L} K(z-x',u)dx' = \frac{z}{2L}\left(\text{erfc}\left\{\frac{z-b}{[4\alpha u]^{1/2}}\right\} - \text{erfc}\left\{\frac{z-a}{[4\alpha u]^{1/2}}\right\}\right) + \frac{2\alpha u}{L}[-K(z-b,u)+K(z-a,u)]$
3	$\int_a^b \left(\frac{x'}{L}\right)^2 K(z-x',u)\,dx' = \left[\frac{1}{2}\left(\frac{z}{L}\right)^2 + \frac{\alpha u}{L^2}\right]\left(\text{erfc}\left\{\frac{z-b}{[4\alpha u]^{1/2}}\right\} - \text{erfc}\left\{\frac{z-a}{[4\alpha u]^{1/2}}\right\}\right) + \frac{2\alpha u}{L^2}[-(z+b)K(z-b,u)+(z+a)K(z-a,u)]$
4	$\int_a^b \exp\left(\frac{-Bx'}{2\alpha}\right) K(z-x',u)\,dx' = \frac{1}{2}\exp\left[\frac{B^2u}{4\alpha} - \frac{Bz}{2\alpha}\right] \text{erf}\left\{\frac{x'-z}{[4\alpha u]^{1/2}} + \frac{Bu^{1/2}}{2\alpha^{1/2}}\right\}\Bigg\vert_{x'=a}^{b}$

TABLE I.8
Time Integrals Related to the Short-Cotime GFs

No.	Equation
1	$\int_{t-\Delta t}^{t} K(z,t-\tau)d\tau = \left(\frac{\Delta t}{\alpha}\right)^{1/2} \text{ierfc}\left[\frac{\vert z\vert}{(4\alpha\Delta t)^{1/2}}\right]$
2	$\int_{t-\Delta t}^{t} \frac{\alpha(t-\tau)}{L^2} K(z,t-\tau)d\tau = \frac{\vert z\vert}{L^2}\Delta t \times\left[\left(\frac{4\alpha\Delta t}{z^2}\right)^{1/2} \text{i}^3\,\text{erfc}\,\frac{\vert z\vert}{(4\alpha\Delta t)^{1/2}} + \text{i}^2\,\text{erfc}\,\frac{\vert z\vert}{(4\alpha\Delta t)^{1/2}}\right]$
3	$\int_{t-\Delta t}^{t} \frac{1}{t-\tau} K(z,t-\tau)d\tau = \frac{1}{\vert z\vert}\,\text{erfc}\,\frac{\vert z\vert}{(4\alpha\Delta t)^{1/2}}$
4	$\int_{t-\Delta t}^{t} \text{erfc}\left\{\frac{z}{[4\alpha(t-\tau)]^{1/2}}\right\}d\tau = 4\Delta t\,\text{i}^2\,\text{erfc}\left[\frac{\vert z\vert}{(4\alpha\Delta t)^{1/2}}\right]$
5	$\int_{t-\Delta t}^{t} \frac{\alpha(t-\tau)}{L^2}\,\text{erfc}\left\{\frac{z}{[4\alpha(t-\tau)]^{1/2}}\right\}d\tau = \frac{\alpha}{L^2}(4\Delta t)^2\left\{\frac{1}{4}\,\text{i}^2\,\text{erfc}\left[\frac{\vert z\vert}{(4\alpha\Delta t)^{1/2}}\right] - \text{i}^4\,\text{erfc}\left[\frac{\vert z\vert}{(4\alpha\Delta t)^{1/2}}\right]\right\}$

(Continued)

TABLE I.8
Time Integrals Related to the Short-Cotime GFs (Continued)

No.	Equation
6	$\int_0^t \left(\frac{\tau}{t_0}\right)^{m/2} K(z,t-\tau)d\tau = \frac{1}{2\alpha^{1/2}}\Gamma\left(\frac{m}{2}+1\right)\left(\frac{4t}{t_0}\right)^{(m+1)/2} t_0^{1/2}\,\mathrm{i}^{m+1}\operatorname{erfc}\left[\frac{\lvert z\rvert}{(4\alpha t)^{1/2}}\right], \quad m=-1,0,1,2,\ldots$
7	$\int_0^t \left(\frac{\tau}{t_0}\right)^{m/2} \frac{\alpha(t-\tau)}{L^2} K(z,t-\tau)d\tau = t_0\frac{\lvert z\rvert}{L^2}\Gamma\left(\frac{m}{2}+1\right)2^m\left(\frac{t}{t_0}\right)^{(m+2)/2} \times\left[\left(\frac{4\alpha t}{z^2}\right)^{1/2}\mathrm{i}^{m+3}\operatorname{erfc}\frac{\lvert z\rvert}{(4\alpha t)^{1/2}} + \mathrm{i}^{m+2}\operatorname{erfc}\frac{\lvert z\rvert}{(4\alpha t)^{1/2}}\right], \quad m=-2,-1,0,1,2,\ldots$
8	$\int_0^t \left(\frac{\tau}{t_0}\right)^{m/2} \frac{1}{t-\tau} K(z,t-\tau)d\tau = \frac{1}{\lvert z\rvert}\left(\frac{4t}{t_0}\right)^{m/2}\mathrm{i}^m\operatorname{erfc}\left[\frac{\lvert z\rvert}{(4\alpha t)^{1/2}}\right]\Gamma\left(\frac{m}{2}+1\right), \quad m=0,1,2,\ldots$
9	$\int_0^t \left(\frac{\tau}{t_0}\right)^{m/2} \operatorname{erfc}\left\{\frac{z}{[4\alpha(t-\tau)]^{1/2}}\right\}d\tau = t_0\Gamma\left(\frac{m}{2}+1\right)\left(\frac{4t}{t_0}\right)^{(m+2)/2}\mathrm{i}^{m+2}\operatorname{erfc}\left[\frac{z}{(4\alpha t)^{1/2}}\right], \quad m=-1,0,1,2,\ldots$
10	$\int_0^t \left(\frac{\tau}{t_0}\right)^{m/2} \frac{\alpha(t-\tau)}{L^2}\operatorname{erfc}\left\{\frac{z}{[4\alpha(t-\tau)]^{1/2}}\right\}d\tau = t_0\frac{\alpha t_0}{L^2}\left(\frac{4t}{t_0}\right)^{(m+4)/2} \times\left\{\frac{1}{4}\Gamma\left(\frac{m}{2}+1\right)\mathrm{i}^{m+2}\operatorname{erfc}\left[\frac{z}{(4\alpha t)^{1/2}}\right] - \Gamma\left(\frac{m}{2}+1\right) \times\mathrm{i}^{m+4}\operatorname{erfc}\left[\frac{z}{(4\alpha t)^{1/2}}\right]\right\}, \quad m=-1,0,1,2,\ldots$
11	$\int_0^t \tau^{m/2}\exp\left[\frac{hx}{k}+\frac{h^2}{k^2}\alpha(t-\tau)\right]\operatorname{erfc}\left\{\frac{x}{[4\alpha(t-\tau)]^{1/2}}+\frac{h}{k}[\alpha(t-\tau)]^{1/2}\right\}d\tau = \frac{\Gamma\left(\frac{m}{2}+1\right)}{\alpha^{m/2+1}}\left(-\frac{k}{h}\right)^{m+2}\left\{\exp\left[\frac{hx}{k}+\frac{h^2}{k^2}\alpha t\right]\operatorname{erfc}\left[\frac{x}{(4\alpha t)^{1/2}}+\frac{h}{k}(\alpha t)^{1/2}\right] - \sum_{j=0}^{m+1}\left[-\frac{h}{k}(4\alpha t)^{1/2}\right]^j\mathrm{i}^j\operatorname{erfc}\left[\frac{x}{(4\alpha t)^{1/2}}\right]\right\}, \quad m=-1,0,1,2,3,\ldots$

Note: $\Gamma(z)$ is the gamma function; see Abramowitz and Stegun (1964, p. 255).

REFERENCES

Abramowitz, M. and Stegun, I. A. (Eds), 1964, *Handbook of Mathematical Functions with Formulas, Graphs, and Mathematical Tables*, Applied Mathematics Series 55, National Bureau of Standards, Washington, DC.

Cho, S. H., 1971, *A Short Table of Integrals Involving the Error Function*, Department of Mechanics, Korean Military Academy, Seoul, Korea.

Levine, H., 1987, On the solution of an unsteady diffusion problem and its ultimate behavior, *Q. J. Mech. Applied Math.*, vol. 40 (Pt. 2), pp. 213–223.

Ng, E. W. and Geller, M., 1969, A table of integrals of the error functions, *J. Res. National Bureau of Standards – B. Mathematical Sciences*, vol. 73B, pp. 1–20.

Ozisik, M. N., 1993, *Heat Conduction*, John Wiley, New York.

L Laplace Transform Method

In this appendix, the highlights of the Laplace transform method are given without presenting detailed mathematical theorems. A table of Laplace transform pairs is also given.

L.1 DEFINITION

For a piecewise continuous function of t, the Laplace transform is defined as (Churchill, 1958)

$$\int_0^\infty e^{-st} f(t)dt = \overline{f}(s)$$
$$= \mathscr{L}[f(t)] \tag{L.1}$$

The notation $\mathscr{L}[f(t)]$ refers to the Laplace transform of function $f(t)$. The Laplace transform exists if the integral in Equation L.1 converges uniformly; that is,

$$\int_0^\infty e^{-st}|f(t)|dt < \infty \tag{L.2}$$

Alternatively, the following notation is used to define the inverse Laplace transform

$$f(t) = \mathscr{L}^{-1}[\overline{f}(s)] \tag{L.3}$$

where $\mathscr{L}^{-1}[\cdot]$ is the inverse Laplace transform operator. The Cauchy's integral formula in a complex plane, Figure L.1, (Kreyszig, 1979) for complex variable z and complex constant s,

$$\overline{f}(s) = \frac{1}{2\pi i}\int_c \frac{\overline{f}(z)}{z-s}\,dz \tag{L.4}$$

can be written for the region $R \geq \gamma$ where γ is a real number and c is the contour of region R. One assumes that $\overline{f}(s)$ is an analytic function of the order $O(s^{-m})$ in the complex half plane $x \geq \gamma$ as $z \to \infty$, where $m > 0$ is a real constant. The contour integral reduces to

$$\overline{f}(s) = \frac{1}{2\pi i}\lim_{\beta\to\infty}\int_{\gamma-i\beta}^{\gamma+i\beta} \frac{\overline{f}(z)}{s-z}\,dz \tag{L.5}$$

The inverse Laplace transformation for the transformed variable s on either side of this equation is obtained by applying Equation L.3 to both sides of Equation L.5,

$$\mathscr{L}^{-1}[\overline{f}(s)] = \frac{1}{2\pi i}\lim_{\beta\to\infty}\int_{\gamma-i\beta}^{\gamma+i\beta} \mathscr{L}^{-1}\left[\frac{\overline{f}(z)}{s-z}\right]dz \tag{L.6a}$$

TABLE L.1
Table of Laplace Transforms

No.	$\bar{f}(s)$	$f(t)$
1	$\frac{1}{s}$	1
2	$\frac{1}{s^2}$	$\frac{1}{t}$
3	$\frac{n!}{s^{n+1}}\ (n = 0, 1, \dots)$	t^n
4	$\frac{a}{s^2 + a^2}$	$\sin at$
5	$\frac{s}{s^2 + a^2}$	$\cos at$
6	$\frac{a}{s^2 - a^2}$	$\sinh at$
7	$\frac{s}{s^2 - a^2}$	$\cosh at$
8	$\frac{1}{\sqrt{s}}$	$\frac{1}{\sqrt{\pi t}}$
9	$s^{-3/2}$	$2\sqrt{t/\pi}$
10	$s^{-(n+1/2)}\ (n = 1, 2, \dots)$	$\frac{2^n}{[1 \cdot 3 \cdot 5 \cdot \dots \cdot (2n-1)]} \frac{t^{n-1/2}}{\sqrt{\pi}}$
11	$\frac{1}{s^n}\ (n > 0)$	$\frac{1}{\Gamma(n)} t^{n-1}$
12	$\frac{1}{s + a}$	e^{-at}
13	$\frac{1}{(s + a)^n}\ (n = 1, 2, 3, \dots)$	$\frac{t^{n-1}e^{-at}}{(n-1)!}$
14	$\frac{\Gamma(k)}{(s + a)^k}\ (k > 0)$	$t^{k-1}e^{-at}$
15	$\frac{1}{(s + a)(s + b)}\ (a \neq b)$	$\frac{e^{-at} - e^{-bt}}{(b - a)}$
16	$\frac{s}{(s + a)(s + b)}\ (a \neq b)$	$\frac{ae^{-at} - be^{-bt}}{(b - a)}$
17	$\frac{1}{s(s^2 + a^2)}$	$\frac{1}{a^2}(1 - \cos at)$
18	$\frac{1}{s^2(s^2 + a^2)}$	$\frac{1}{a^3}(at - \sin at)$
19	$\frac{1}{(s^2 + a^2)^2}$	$\frac{1}{2a^3}(\sin at - at \cos at)$

TABLE L.1
Table of Laplace Transforms (Continued)

No.	$\bar{f}(s)$	$f(t)$
20	$\dfrac{s}{(s^2+a^2)}$	$\dfrac{t}{2a}\sin at$
21	$\dfrac{s^2}{(s^2+a^2)}$	$\dfrac{1}{2a}(\sin at + at\cos at)$
22	$\dfrac{s^2-a^2}{(s^2+a^2)}$	$t\cos at$
23	$\dfrac{1}{\sqrt{s}+a}$	$\dfrac{1}{\sqrt{\pi t}} - ae^{a^2t}\operatorname{erfc} a\sqrt{t}$
24	$\dfrac{\sqrt{s}}{s-a^2}$	$\dfrac{1}{\sqrt{\pi t}} + ae^{a^2t}\operatorname{erf} a\sqrt{t}$
25	$\dfrac{\sqrt{s}}{s+a^2}$	$\dfrac{1}{\sqrt{\pi t}} - \dfrac{2a}{\sqrt{\pi}}e^{-a^2t}\int_0^{a\sqrt{t}} e^{\lambda^2}\,d\lambda$
26	$\dfrac{1}{\sqrt{s}\,(s-a^2)}$	$\dfrac{1}{a}e^{a^2t}\operatorname{erf} a\sqrt{t}$
27	$\dfrac{1}{\sqrt{s}\,(s+a^2)}$	$\dfrac{2}{a\sqrt{\pi}}e^{-a^2t}\int_0^{a\sqrt{t}} e^{\lambda^2}\,d\lambda$
28	$\dfrac{b^2-a^2}{(s-a^2)(b+\sqrt{s})}$	$e^{a^2t}[b - a\operatorname{erf} a\sqrt{t}] - b\,e^{b^2t}\operatorname{erfc} b\sqrt{t}$
29	$\dfrac{1}{\sqrt{s}\,(\sqrt{s}+a)}$	$e^{a^2t}\operatorname{erfc} a\sqrt{t}$
30	$\dfrac{1}{(s+a)\sqrt{s+b}}$	$\dfrac{1}{\sqrt{b-a}}e^{-at}\operatorname{erf}(\sqrt{b-a}\,\sqrt{t})$
31	$\dfrac{\sqrt{s+2a}}{\sqrt{s}} - 1$	$a\,e^{-at}[I_1(at) + I_0(at)]$
32	$\dfrac{1}{\sqrt{s+a}\,\sqrt{s+b}}$	$e^{-(a+b)t/2}\,I_0\left(\dfrac{a-b}{2}\,t\right)$
33	$\dfrac{1}{\sqrt{s^2+a^2}}$	$J_0(at)$
34	$\dfrac{(\sqrt{s^2+a^2}-s)^{\nu}}{\sqrt{s^2+a^2}}\ (\nu > -1)$	$a^{\nu}J_{\nu}(at)$

(Continued)

TABLE L.1
Table of Laplace Transforms (Continued)

No.	$\bar{f}(s)$	$f(t)$
35	$\dfrac{(s-\sqrt{s^2+a^2})^{\nu}}{\sqrt{s^2-a^2}}\ (\nu > -1)$	$a^{\nu} I_{\nu}(at)$
36	$\dfrac{1}{s}\, e^{-ks}$	$H(t-k)^a$
37	$\dfrac{1}{s^2}\, e^{-ks}$	$(t-k)\, H(t-k)$
38	$\dfrac{1}{s}\, e^{-k/s}$	$J_0(2\sqrt{kt})$
39	$\dfrac{1}{s^{\mu}}\, e^{-k/s}\ (\mu > 0)$	$\left(\dfrac{t}{k}\right)^{(\mu-1)/2} J_{\mu-1}(2\sqrt{kt})$
40	$\dfrac{1}{s^{\mu}}\, e^{+k/s}\ (\mu > 0)$	$\left(\dfrac{t}{k}\right)^{(\mu-1)/2} I_{\mu-1}(2\sqrt{kt})$
41	$e^{-k\sqrt{s}}\ (k > 0)$	$\dfrac{k}{2\sqrt{\pi t^3}} \exp\left(-\dfrac{k^2}{4t}\right)$
42	$\dfrac{1}{s}\, e^{-k\sqrt{s}}\ (k \geq 0)$	$\operatorname{erfc}\left(\dfrac{k}{2\sqrt{t}}\right)$
43	$\dfrac{1}{\sqrt{s}}\, e^{-k\sqrt{s}}\ (k \geq 0)$	$\dfrac{1}{\sqrt{\pi t}} \exp\left(-\dfrac{k^2}{4t}\right)$
44	$\dfrac{1}{s^{3/2}}\, e^{-k\sqrt{s}}\ (k \geq 0)$	$2\sqrt{\dfrac{t}{\pi}} \exp\left(-\dfrac{k^2}{4t}\right) - k \operatorname{erfc} \dfrac{k}{2\sqrt{t}}$ $= 2\sqrt{t}\ \operatorname{i erfc} \dfrac{k}{2\sqrt{t}}$
45	$\dfrac{e^{-k\sqrt{s}}}{s^{1+n/2}}\ (n = 0, 1, 2, \ldots, k \geq 0)$	$(4t)^{n/2}\ \mathrm{i}^n \operatorname{erfc} \dfrac{k}{2\sqrt{t}}$
46	$\dfrac{e^{-k\sqrt{s}}}{a+\sqrt{s}}\ (k \geq 0)$	$\dfrac{1}{\sqrt{\pi t}} \exp\left(-\dfrac{k^2}{4t}\right) - a e^{ak} e^{a^2 t}$ $\times \operatorname{erfc}\left(a\sqrt{t} + \dfrac{k}{2\sqrt{t}}\right)$
47	$\dfrac{e^{-k\sqrt{s}}}{\sqrt{s}\,(a+\sqrt{s})}\ (k \geq 0)$	$e^{ak} e^{a^2 t} \operatorname{erfc}\left(a\sqrt{t} + \dfrac{k}{2\sqrt{t}}\right)$
48	$\dfrac{e^{-k\sqrt{s^2+as}}}{\sqrt{s(s+a)}}\ (k \geq 0)$	$e^{-at/2}\, I_0\left(\dfrac{a}{2}\sqrt{t^2-k^2}\right) H(t-k)$

TABLE L.1
Table of Laplace Transforms (Continued)

No.	$\bar{f}(s)$	$f(t)$
49	$\dfrac{e^{-k\sqrt{s^2+a^2}}}{\sqrt{s^2+a^2}}\ (k \geq 0)$	$J_0(a\sqrt{t^2-k^2})\,H(t-k)$
50	$\dfrac{e^{-k\sqrt{s^2+a^2}}}{\sqrt{s^2-a^2}}\ (k \geq 0)$	$I_0(a\sqrt{t^2-k^2})\,H(t-k)$
51	$\dfrac{e^{-k\sqrt{s}}}{s(a+\sqrt{s})}\ (k \geq 0)$	$-e^{ak}e^{a^2t}\operatorname{erfc}\left(a\sqrt{t}+\dfrac{k}{2\sqrt{t}}\right)+\operatorname{erfc}\dfrac{k}{2\sqrt{t}}$
52	$\dfrac{1}{s^2}e^{-k\sqrt{s}}$	$4t\,\mathrm{i}^2\operatorname{erfc}\left(\dfrac{k}{2\sqrt{t}}\right)=\left(t+\dfrac{k^2}{2}\right)$
		$\times\operatorname{erfc}\left(\dfrac{k}{2\sqrt{t}}\right)-k\left(\dfrac{t}{\pi}\right)^{1/2}\exp\left(-\dfrac{k^2}{4t}\right)$
53	$\dfrac{1}{s}\ln s$	$-\gamma-\ln t\ (\gamma=0.57721\ 56649\ldots,$ Euler's constant)
54	$\ln\dfrac{s+a}{s+b}$	$\dfrac{1}{t}(e^{-bt}-e^{-at})$
55	$\ln\dfrac{s^2+a^2}{s^2}$	$\dfrac{2}{t}(1-\cos at)$
56	$\ln\dfrac{s^2-a^2}{s^2}$	$\dfrac{2}{t}(1-\cosh at)$
57	$K_0(ks)\ (k>0)$	$\dfrac{1}{\sqrt{t^2-k^2}}H(t-k)$
58	$K_0(k\sqrt{s})\ (k>0)$	$\dfrac{1}{2t}\exp\left(-\dfrac{k^2}{4t}\right)$
59	$\dfrac{1}{\sqrt{s}}K_1(k\sqrt{s})\ (k>0)$	$\dfrac{1}{k}\exp\left(-\dfrac{k^2}{4t}\right)$

[a] $H(t)$ is the Heaviside unit-step function

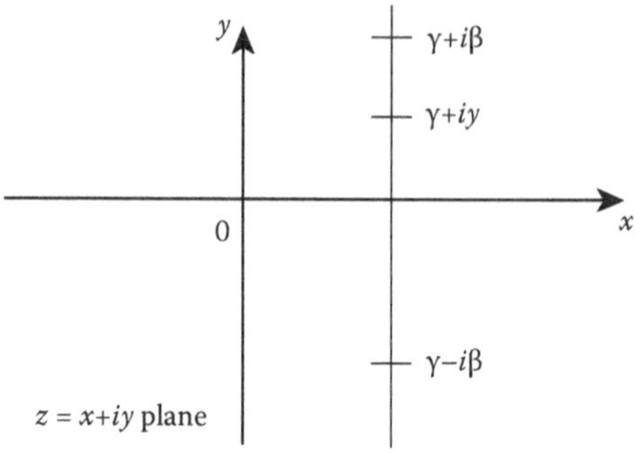

FIGURE L.1 Complex z-plane.

or

$$f(t) = \frac{1}{2\pi i} \lim_{\beta\to\infty} \int_{\gamma-i\beta}^{\gamma+i\beta} \overline{f}(z)e^{zt}dz \tag{L.6b}$$

The specific conditions to be satisfied are: $\overline{f}(z)$ is analytic in the half plane $x \geq \gamma$ and there exists two constants m and M such that

$$\overline{f}(z) < \frac{M}{|z|^{m+1}} \quad \text{when } x \geq c \tag{L.7}$$

It is possible to prove that the inversion integral, Equation L.6b, converges to the function $f(t)$, and $f(t)$ is continuous when $t \geq 0$ when the real part of z or s is larger than γ and $F(0) = 0$ (Churchill, 1958). The conditions set forth in the inverse transform formula are often more severe than necessary to insure the convergence of the inversion integral to the inverse transformation, e.g., the condition $f(0) = 0$.

The second derivation of the inversion formula begins by using the Fourier integral,

$$G(t) = \frac{1}{2\pi} \lim_{\beta\to\infty} \int_{-\beta}^{+\beta} e^{iyt} \int_{-\infty}^{+\infty} G(\tau)e^{-iy\tau}d\tau dy \tag{L.8}$$

The function $G(t)$ and $G'(t)$ are sectionally continuous on each finite integral along the t-axis between $-\infty$ and $+\infty$, and $G(t)$ takes its average value across each jump at $t = t_o$,

$$G(t_o) = \frac{1}{2}[G(t_o - 0) + G(t_o + 0)] \tag{L.9}$$

Moreover, the integral

$$\int_{-\infty}^{+\infty} |G(t)|dt \tag{L.10}$$

converges. Now, we define a function $f(t)$ so that $f(t)$ and $f'(t)$ are continuous for $t \geq 0$ and of exponential order for large t, that is,

$$|f(t)| < Me^{at} \tag{L.11}$$

Then, the Laplace transformation of $f(t)$, or $\overline{f}(s)$ exists when $Re(s) > a$. Define the function G in Equation L.8 by the relation

$$\begin{aligned} G(t) &= 0 \text{ when } t < 0 \\ &= e^{-\gamma t} f(t) \text{ when } t > 0 \\ &= \tfrac{1}{2} f(0) \text{ when } t = 0 \end{aligned} \tag{L.12}$$

where $\gamma > a$. Equation L.8 then becomes (Churchill, 1958),

$$\begin{aligned} \frac{1}{2\pi} \lim_{\beta\to\infty} \int_{-\beta}^{+\beta} e^{iyt} \int_{-\infty}^{+\infty} e^{-(\gamma+iy)\tau} f(\tau) d\tau dy &= e^{-\gamma t} f(t) \text{ when } t > 0 \\ &= 0 \text{ when } t > 0 \\ &= \frac{1}{2} f(0) \text{ when } t = 0 \end{aligned} \tag{L.13}$$

The inside integral is $\overline{f}(z)$ where $z = \gamma + iy$, and if $z = \gamma + iy$, then for all t values,

$$\begin{aligned} e^{\gamma t}\, G(t) &= \lim_{\beta\to\infty} \frac{1}{2\pi i} \int_{\gamma-i\beta}^{\gamma+i\beta} e^{zt} \overline{f}(z) dz; \quad t > 0 \\ &= f(t) \end{aligned} \tag{L.14}$$

This equation is known as the Cauchy principal value of the inversion integral and, as $\beta \to \infty$, it becomes

$$f(t) = \frac{1}{2\pi i} \int_{\gamma-i\beta}^{\gamma+i\beta} e^{zt} \overline{f}(z) dz; \quad t > 0 \tag{L.15}$$

Therefore, if $\overline{f}(s)$ is the transform of $f(t)$, and $f(t)$ and $f'(t)$ are sectionally continuous and of exponential order, then the inverse Laplace transformation converges to

$$\begin{aligned} \mathscr{L}^{-1}[\overline{f}(s)] &= f(t) \text{ when } t > 0 \\ &= \frac{1}{2} \text{ when } t = 0 \\ &= 0 \text{ when } t < 0 \end{aligned} \tag{L.16}$$

L.2 PROPERTIES OF LAPLACE TRANSFORMATION

This section contains a few properties of the Laplace transformation that are useful when solving heat conduction problems. First a brief list of properties is given, followed by a discussion of transforms of simple functions, derivatives, and integrals.

(a) *Linear property.* For c_1 and c_2 arbitrary constants,

$$\mathscr{L}[c_1 f(t) + c_2 g(t)] = c_1 \mathscr{L}[f(t)] + c_2 \mathscr{L}[g(t)] = c_1 \overline{f}(s) + c_2 \overline{g}(s) \tag{L.17a}$$

(b) *Multiplication by* t^n. For n any positive integer,

$$\mathscr{L}[t^n f(t)] = (-1)^n \frac{d^n \overline{f}(s)}{ds^n} = (-1)^n \overline{f}^{(n)}(s) \tag{L.17b}$$

(c) *Division by* t.

$$\mathscr{L}\left[\frac{f(t)}{t}\right] = \int_s^\infty \overline{f}(s')\, ds' \tag{L.17c}$$

(d) *Transform of derivatives.* If $n > 0$ is an integer and $\lim f(t)e^{-st} = 0$ as $t \to \infty$, then for $t > 0$,

$$\mathscr{L}[f^{(n)}(t)] = s^n \overline{f}(s) - s^{n-1} f(0) - s^{n-2} f'(0) - \cdots\cdots - f^{(n-1)}(0) \tag{L.17d}$$

(e) *Transform of integrals.* If $\lim e^{-st} \int_0^t f(u)\, du = 0$ as $t \to \infty$, then

$$\mathscr{L}\left[\int_0^t f(u)\, du\right] = \frac{1}{s}\, \overline{f}(s) \tag{L.17e}$$

(f) *Change of scale.* If a is any positive constant, then

$$\mathscr{L}[f(at)] = \frac{1}{a}\, \overline{f}\left(\frac{s}{a}\right) \tag{L.17f}$$

(g) *Transform of convolution.* If $\overline{f}(s)$ is $\mathscr{L}[f(t)]$ and $\overline{g}(s)$ is $\mathscr{L}[g(t)]$, then

$$\mathscr{L}\left[\int_0^t f(u)\, g(t-u)\, du\right] = \mathscr{L}\left[\int_0^t f(t-u)\, g(u)\, du\right] = \overline{f}(s) \cdot \overline{g}(s) \tag{L.17g}$$

Transformation of polynomials. The transform of powers of t are given by:

$$\mathscr{L}[t] = \int_0^\infty (t)e^{-st} dt = -\left(\frac{t}{s} + \frac{1}{s^2}\right) e^{-st}\Big|_0^\infty = \frac{1}{s^2} \tag{L.18a}$$

$$\mathscr{L}[t^2] = \int_0^\infty (t^2)e^{-st} dt = -\left(\frac{t^2}{s} + \frac{2t}{s^2} + \frac{2}{s^3}\right) e^{-st}\Big|_0^\infty = \frac{2}{s^3} \tag{L.18b}$$

and

$$\mathscr{L}[t^n] = \int_0^\infty (t^n)e^{-st} dt = \frac{n!}{s^{n+1}} \tag{L.18c}$$

Transformation of derivatives. Using Equation L.1, the transformation of a derivative is

$$\begin{aligned}\mathscr{L}\left[\frac{d}{dt} f(t)\right] &= \int_0^\infty e^{-st} \left[\frac{d}{dt} f(t)\right] dt \\ &= f(t)e^{-st}\big|_0^\infty + s \int_0^\infty e^{-st} f(t) dt\end{aligned}$$

Therefore,

$$\mathscr{L}\left[\frac{d}{dt}f(t)\right] = s\,\overline{f}(s) - f(0) \tag{L.19a}$$

$$\mathscr{L}\left[\frac{d^2}{dt^2}f(t)\right] = s^2\,\overline{f}(s) - s\,f(0) - f'(0) \tag{L.19b}$$

$$\mathscr{L}\left[\frac{d^3}{dt^3}f(t)\right] = s^3\,\overline{f}(s) - s^2\,f(0) - s\,f'(0) - f''(0) \tag{L.19c}$$

and

$$\mathscr{L}\left[\frac{d^n}{dt^n}f(t)\right] = s^n\,\overline{f}(s) - s^{n-1}\,f(0) - s^{n-2}\,f'(0) - s^{n-3}\,f''(0) - \cdots - f^{(n-1)}(0) \tag{L.19d}$$

where $f(0)$ is the initial condition and $f'(0)$, $f''(0)$, $\cdots$, $f^{(n-1)}(0)$ are first, second, $\cdots$, $(n-1)$th derivatives evaluated at $t = 0$.

Transform of Integrals. The transform of definite integrals is obtained using the definition of the Laplace transformation, Equation L.1,

$$\mathscr{L}\left[\int_0^t f(t)dt\right] = \int_0^\infty \left[\int_0^t f(t)dt\right] e^{-st}dt \tag{L.20a}$$

After integrating by parts and applying the limits from 0 to ∞, the following relation is obtained,

$$\mathscr{L}\left[\int_0^t f(t)dt\right] = \frac{1}{s}\overline{f}(s) \tag{L.20b}$$

Any indefinite integral can be replaced by an equivalent definite integral of the form,

$$\int f(t)dt = \int_0^t f(t)dt + C \tag{L.21}$$

where C represents the value of the indefinite integral at $t = 0$. The transform of the first term on the right side of Equation L.21 is given by Equation L.20b while the transform of C is C/s; accordingly,

$$\mathscr{L}\left[\int_0^t f(t)dt + C\right] = \frac{1}{s}[\overline{f}(s) + C] \tag{L.22}$$

L.3 LAPLACE TRANSFORM THEOREMS

There are theorems of general interest and a few are especially useful for calculating the Green's function.

First Shift Theorem. This theorem describes the situation when s in the Laplace transformation is replaced by $s - a$,

$$\begin{aligned}\overline{f}(s-a) &= \int_0^\infty e^{-(s-a)t} f(t)dt \\ &= \int_0^\infty e^{-st} e^{at} f(t)dt \\ &= \int_0^\infty e^{-st} \left[e^{at} f(t)\right] dt \\ &= \mathscr{L}[e^{at} f(t)] \end{aligned} \tag{L.23}$$

Second-Shift Theorem. The second-shift theorem concerns the replacement of t by $t - a$. It seeks the Laplace transformation of $f(t-a)$ subject to the condition that $f(t-a) = 0$ when $t < a$. Then, using the definition of the Laplace transformation, one obtains,

$$\begin{aligned}\mathscr{L}[f(t-a)] &= \int_0^\infty e^{-st} f(t-a)dt \\ &= \int_a^\infty e^{-st} f(t-a)dt \end{aligned} \tag{L.24a}$$

Now, replacing t by $\tau + a$ yields

$$\begin{aligned}\mathscr{L}[f(t-a)] &= \int_0^\infty e^{-s(\tau+a)} f(\tau)d\tau \\ &= e^{-as} \int_0^\infty e^{-s\tau} f(\tau)d\tau \\ &= e^{-as}\overline{f}(s) \end{aligned} \tag{L.24b}$$

This theorem is useful when mathematically describing functions with built-in delays.

Initial-Value Theorem. The initial-value theorem enables one to calculate the value of $f(t)$ at time $0+$ from the transform function $\overline{f}(s)$. The function $f(t)$ has a zero value when $t \leq 0$ and its value suffers a jump at $0+$. The transform of the function $f'(t) = df(t)/dt$ can be expressed as

$$\begin{aligned}\mathscr{L}[f'(t)] &= \int_0^\infty e^{-st} f'(t)dt \\ &= \int_0^{0+} f'(t)dt + \int_{0+}^\infty e^{-st} f'(t)dt \end{aligned} \tag{L.25}$$

Note that the quantity e^{-st} in the first term o n the right side of Equation L.25 vanishes since s is finite and $0 \leq t \leq 0+$. As $s \to \infty$, the second term on the right side of Equation L.25 will vanish since $e^{-\infty} = 0$. Moreover, Equation L.19a yields the limit for the term $\mathscr{L}[f'(t)]$, in Equation L.25, when $s \to \infty$, as

$$\lim_{s\to\infty} \mathscr{L}[f'(t)] = \lim_{s\to\infty} [s\,\overline{f}(s) - f(0)] \tag{L.26}$$

Therefore, the limit of Equation L.25 as $s \to \infty$ is

$$\lim_{s\to\infty} [s\,\overline{f}(s) - f(0)] = f(0+) - f(0)$$

or

$$\lim_{s\to\infty} [s\,\overline{f}(s))] = f(0+) \tag{L.27}$$

The initial-value theorem described by Equation L.27 yields the initial condition at $t = 0+$ following an initial jump condition for function $f(t)$. In the absence of a jump at $t = 0$, the function $f(0+)$ becomes $f(0)$.

Final-Value Theorem. This theorem yields the value of $\overline{f}(s)$, the Laplace transformation of $f(t)$ as $t \to \infty$. The theorem applies when function $f(t)$ approaches a finite limit as $t \to \infty$; otherwise, the final-value theorem does not apply. The derivation of this theorem follows examination of $\mathscr{L}[f'(t)]$,

$$\begin{aligned}\mathscr{L}[f'(t)] &= \int_0^\infty e^{-st} f'(t)dt \\ &= s\,\overline{f}(s) - f(0)\end{aligned} \tag{L.28}$$

As $s \to 0$, the integral in Equation L.28 reduces to

$$\int_0^\infty e^{-st} f'(t)dt = \int_0^\infty f'(t)dt = f(\infty) - f(0)$$

that can be inserted in Equation L.28 to give

$$f(\infty) - f(0) = \lim_{s\to 0}[s\,\overline{f}(s) - f(0)]$$

or

$$\lim_{s\to 0}[s\,\overline{f}(s)] = f(\infty) \tag{L.29}$$

Equation L.29 mathematically describes the final-value theorem.

L.4 TABLE OF LAPLACE TRANSFORMS

Table L.1 contains a list of Laplace – transform pairs, several of which are useful for heat conduction and for Green's functions.

REFERENCES

Churchill, R. V., 1958, *Operational Mathematics*, McGraw-Hill, New York.
Kreyszig, E., 1979, *Advanced Engineering Mathematics*, John Wiley, New York.

P Properties of Selected Materials

This table is intended to indicate the order of magnitude of property values likely to occur in practice. For additional information, and for the variation of property values with temperature, consult standard works. The values for nonmetals should be regarded as rough averages, as there is considerable variation between different samples of the same substance.

TABLE P.1
Table of Properties of Selected Materials at 300 K

Substance	ρ kg/m^3	c J/kg/K	k W/m/K	$\alpha \times 10^6$ m^2/s
Metals				
silver	10,500	235	429	174
copper	8933	385	401	117
gold	19,300	129	317	127
aluminum	2702	903	237	97.1
magnesium	1740	1024	156	87.6
brass (0.3 Zn)	8530	380	111	34.2
nickel	8900	444	91	23
mild steel (0.1 % C)	7830	434	64	18.8
stainless steel (AISI 316)	8238	468	13	3.4
lead	11,340	129	35	24.1
titanium	4500	522	21.9	9.32
bismuth	9780	122	7.9	6.59
Nonmetals				
silicon	2330	712	148	89.2
alumina ($Al_2 0_3$)	3970	765	36	11.9
carbon (graphite)	1810	1300	98	42
teflon	2200	1050	0.35	0.15
polyethylene (high-dens.)	960	2090	0.33	0.16
polyamide (nylon)	1140	1670	0.24	0.13
ice (273K)	910	1930	2.22	1.26
snow (fresh, 273K)	110	1930	0.049	0.23
water	996	4178	0.611	0.147
air (1 atm)	1.177	1005	0.0267	22.5

(Continued)

TABLE P.1
Table of Properties of Selected Materials at 300 K (Continued)

Substance	ρ kg/m^3	c J/kg/K	k W/m/K	$\alpha \times 10^6$ m^2/s
Building materials				
brick, common	1920	835	0.72	0.45
concrete (1:2:4)	2100	880	1.4	0.75
glass, silica	2220	745	1.38	0.83
fiberglass batting	16	835	0.046	3.4
polystyrene, rigid foam	30–60	1210	0.028	0.4–0.8
soil, dry	1500	1900	1.0	0.35
soil, wet	1900	2200	2.0	0.50
white pine (with grain)	500	2800	0.24	0.17
white pine (across grain)	500	2800	0.10	0.071

R Green's Functions for Radial-Cylindrical Coordinates (r)

$$dv' = 2\pi r' dr'$$
$$ds' = 2\pi a \text{ or } 2\pi b$$

The partial differential equation for transient, cylindrical radial heat conduction is

$$\frac{1}{r}\frac{\partial}{\partial r}\left(r\frac{\partial T}{\partial r}\right) = \frac{1}{\alpha}\frac{\partial T}{\partial t}$$

*R*00 INFINITE BODY

For steady Green's functions (GFs) in cylindrical coordinates see Table R.1. The GF is

$$G_{R00}(r,\ t|r',\ \tau) = \frac{1}{4\pi\alpha(t-\tau)}\exp\left[-\frac{r^2+r'^2}{4\alpha(t-\tau)}\right] I_0\left[\frac{rr'}{2\alpha(t-\tau)}\right] \quad \text{(R00.1a)}$$

$$G_{R00}(r,\ t|r',\ \tau) = \frac{1}{4\pi\alpha(t-\tau)}\exp\left[-\frac{(r-r')^2}{4\alpha(t-\tau)}\right] \times \exp\left[-\frac{rr'}{2\alpha(t-\tau)}\right] I_0\left[\frac{rr'}{2\alpha(t-\tau)}\right] \quad \text{(R00.1b)}$$

(units of $1/\text{m}^2$) (Carslaw and Jaeger, 1959, pp. 259 and 368). See Figure R00.1 which shows $r'^2 G_{R00}(\cdot)$ versus r/r' for fixed values of $\alpha(t-\tau)/r'^2$. A similar plot is given by Figure R00.2 for $r'^2 G_{R00}(\cdot)$ versus r/r' for fixed values of $\alpha(t-\tau)/rr'$. The integral of $G_{R00} 2\pi r\, dr'$ for $r' = 0$ to ∞ is unity,

$$\int_0^\infty G_{R00}(r,\ t|r',\ \tau) 2\pi r'\, dr' = 1 \quad \text{(R00.2)}$$

A special case is for the source at $r' = 0$ (or equivalently for a source at $r' = r$ and the observation at $r = 0$):

$$G_{R00}(r,\ t|0,\ \tau) = \frac{1}{4\pi\alpha(t-\tau)} e^{-r^2/[4\alpha(t-\tau)]} \quad \text{(R00.3a)}$$

$$G_{R00}(r,\ t|0,\ \tau) = G_{X00}(x,\ t|0,\ \tau) G_{Y00}(y,\ t|0,\ t) \quad \text{(R00.3b)}$$

since $r^2 = x^2 + y^2$.

TABLE R.1 Steady Green's Functions, Radial Cylindrical Coordinates G Satisfies: $\frac{1}{r}\frac{d}{dr}\left(r\frac{dG}{dr}\right) = -\frac{\delta(r-r')}{2\pi r}$

Case	Boundary Conditions (Note $B_1 = h_1 a/k$, $B_2 = h_2 b/k$)	$2\pi G(r\|r')$
$R01$	$\partial G(0\|r')/\partial r = 0$; $G(b\|r') = 0$	$\ln(b/r'); r < r'$ $\ln(b/r); r > r'$
$R03$	$\partial G(0\|r')/\partial r = 0$ $k\partial G(b\|r')/\partial r + h_2 G(b\|r') = 0$	$\ln(b/r') + 1/B_2; r < r'$ $\ln(b/r) + 1/B_2; r > r'$
$R11$	$G(a\|r') = 0$; $G(b\|r') = 0$	$\ln(b/r')\ln(r/a)/\ln(b/a); r < r'$ $\ln(b/r)\ln(r'/a)/\ln(b/a); r > r'$
$R12$	$G(a\|r') = 0$; $\partial G(b\|r')/\partial r = 0$	$\ln(r/a); r < r'$ $\ln(r'/a); r > r'$
$R13$	$G(a\|r') = 0$ $k\partial G(b\|r')/\partial r + h_2 G(b\|r') = 0$	$(\ln r/a)(1 + B_2 \ln b/r')/(1 + B_2 \ln b/a); r < r'$ $(\ln r'/a)(1 + B_2 \ln b/r)/(1 + B_2 \ln b/a); r > r'$
$R21$	$\partial G(a\|r')/\partial r = 0$ $G(b\|r') = 0$	$\ln(b/r'); r < r'$ $\ln(b/r); r > r'$
$R23$	$\partial G(a\|r')/\partial r = 0$ $k\partial G(b\|r')/\partial r + h_2 G(b\|r') = 0$	$(1/B_2 + \ln b/r'); r < r'$ $(1/B_2 + \ln b/r); r > r'$
$R31$[a]	$k\partial G(a\|r')/\partial n + h_1 G(a\|r') = 0$ $G(b\|r') = 0$	$(\ln b/r')(1/B_1 + \ln r/a)/(1/B_1 + \ln b/a); r < r'$ $(\ln b/r)(1/B_1 + \ln r'/a)/(1/B_1 + \ln b/a); r > r'$
$R32$[a]	$k\partial G(a\|r')/\partial n + h_1 G(a\|r') = 0$ $\partial G(b\|r')/\partial r = 0$	$(1/B_1 + \ln r/a); r < r'$ $(1/B_1 + \ln r'/a); r > r'$
$R33$[a]	$k\partial G(a\|r')/\partial n + h_1 G(a\|r') = 0$ $k\partial G(b\|r')/\partial r + h_2 G(b\|r') = 0$	$[(B_1 B_2(\ln b/r')(\ln r/a) + B_1 \ln r/a + B_2 \ln b/r' + 1)]/D;\ r < r'$ $[(B_1 B_2(\ln b/r)(\ln r'/a) + B_1 \ln r'/a + B_2 \ln b/r + 1)]/D;\ r > r'$ where $D = B_1 + B_2 + B_1 B_2 \ln b/a$

[a] $\partial/\partial n\big|_{r=a} = -\,\partial/\partial r\big|_{r=a}$.

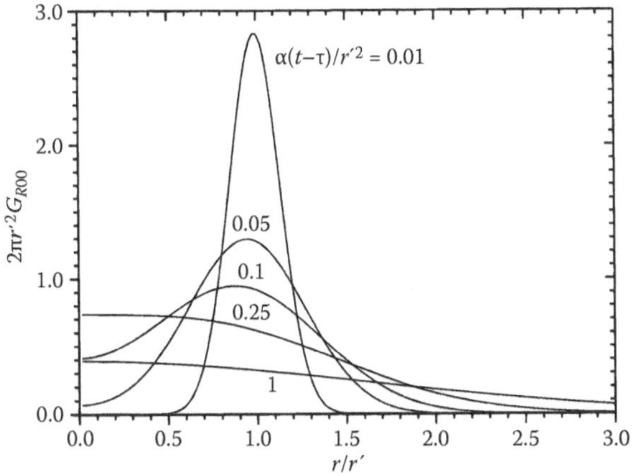

FIGURE R00.1 Green's function G_{R00} (multiplied by $2\pi r'^2$) versus r/r' for several values of $\alpha(t-\tau)/r'^2$.

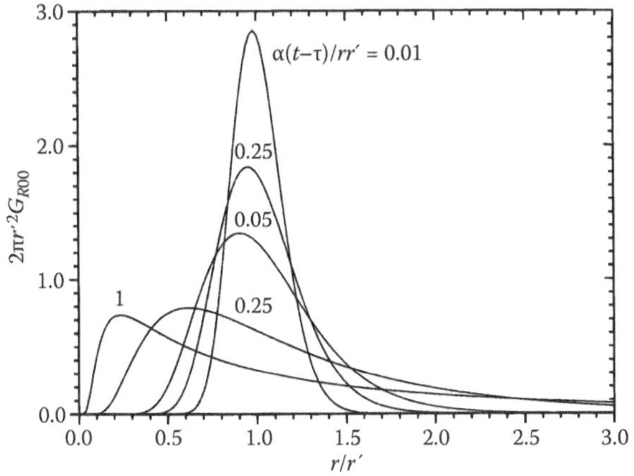

FIGURE R00.2 Green's function G_{R00} (multiplied by $2\pi r'^2$) versus r/r' for several values of $\alpha(t-\tau)/(rr')$.

R.1 SMALL- AND LARGE-TIME APPROXIMATIONS FOR $G_{R00}(\cdot)$

For small values of $\alpha(t-\tau)/(rr')$, $G_{R00}(r, t|r', t)$ can be approximated by

$$G_{R00}(r, t|r', \tau) \approx \frac{1}{2\pi[4\pi rr'\alpha(t-\tau)]^{1/2}} \exp\left[-\frac{(r-r')^2}{4\alpha(t-\tau)}\right]$$
$$\times \left\{1 + \frac{\alpha(t-\tau)}{4rr'} + \frac{9[\alpha(t-\tau)]^2}{32(rr')^2} + 1.5\frac{75[\alpha(t-\tau)]^3}{128(rr')^3}\right\} \qquad \text{(R00.4)}$$

The coefficient 1.5 in the last term is used to improve the accuracy. (The series expansion has a unity coefficient instead of 1.5.) The percent errors for Equation R00.4 are given by:

$\frac{\alpha(t-\tau)}{rr'}$	0	0.05	0.1	0.125	0.167	0.25
% error	0	0.0022	−0.005	−0.038	−0.14	0.016

For large values of $\alpha(t-\tau)/rr'$, $G_{R00}(\bullet)$ can be approximated by:

$$G_{R00}(r, t|r', \tau) \approx \frac{1}{4\alpha(t-\tau)} e^{-r'^2/[4\alpha(t-\tau)]} \left\{ 1 + \frac{(rr')^2}{[4\alpha(t-\tau)]^2} + \frac{1}{4}\frac{(rr')^4}{[4\alpha(t-\tau)]^4} + \frac{1}{36}\frac{(rr')^6}{[4\alpha(t-\tau)]^6} \right\} \tag{R00.5}$$

where the errors are given by

$\frac{\alpha(t-\tau)}{rr'}$	0.25	0.33	0.5	0.625	∞
% error	−0.08	−0.012	−0.0005	−0.00009	0

For large values of both $\alpha(t-\tau)/r^2$ and $\alpha(t-\tau)/r'^2$, $G(r, t|r', \tau)$ can be approximated by

$$G_{R00}(r, t|r', \tau) \approx \frac{1}{4\pi\alpha(t-\tau)} \left[1 - \frac{r^2 + r'^2}{4\alpha(t-\tau)} + \frac{r^4 + 3r^2r'^2 + r'^4}{[4\alpha\,(t-\tau)]^2} \right] \tag{R00.6}$$

R.2 INTEGRAL FROM $r' = 0$ TO $r' = a$

For small times, an approximate result for the integral over r' from $r' = 0$ to a for $r^+ \geq 1$ is

$$\int_0^a G_{R00}(r, t|r', \tau) 2\pi r' dr'$$
$$\approx \frac{1}{2}(r^+)^{-1/2} \left\{ \text{erfc}\left[\frac{r^+ - 1}{(4u)^{1/2}}\right] - \frac{1}{4}\left(\frac{1}{r^+} + 3\right) u^{1/2}\, \text{ierfc}\left[\frac{r^+ - 1}{(4u)^{1/2}}\right] + \frac{3}{32}\left[\frac{3}{(r^+)^2} + \frac{2}{r^+} - 5\right] u\, \text{i}^2\text{erfc}\frac{r^+ - 1}{(4u)^{1/2}} \right\} \tag{R00.7}$$

where

$$u \equiv \frac{\alpha(t-\tau)}{a^2} \qquad r^+ \equiv \frac{r}{a} \tag{R00.8a, b}$$

This expression is accurate for $u < 0.1$. For $u < (r^+ - 1)^2/36$, the integral is nearly zero.

For $r^+ = 1$, Equation R00.7 gives

$$\int_0^a G_{R00}(a,\ t|r',\ \tau)\ 2\pi r'\ dr \approx \frac{1}{2}\left[1 - \left(\frac{u}{\pi}\right)^{1/2} - \frac{1}{4\pi^{1/2}}u^{3/2}\right] \tag{R00.9}$$

At $u = 0.25$, this gives the value of 0.35014 which is 1.3% high and is more accurate for smaller u values.

For $a/2 < r < a$ and for small times u, the integral over r' is

$$\begin{aligned}&\int_0^a G_{R00}(r,\ t|r',\ \tau)2\pi r' dr'\\ &\quad\approx 1 - \frac{1}{2}(r^+)^{-1/2}\left\{\operatorname{erfc}\left[\frac{1-r^+}{(4u)^{1/2}}\right] + \frac{1}{4}\left(\frac{1}{r^+}+3\right)u^{1/2}\operatorname{ierfc}\frac{1-r^+}{(4u)^{1/2}}\right.\\ &\quad\left.+\frac{3}{32}\left(\frac{3}{r^{+2}}+\frac{2}{r^+}-5\right)u\ \mathrm{i}^2\operatorname{erfc}\frac{1-r^+}{(4u)^{1/2}}\right\}\end{aligned} \tag{R00.10}$$

Note the similarity with Equation R00.7. Equation R00.10 is accurate for $u < 0.01$. For $0 < r < a/2$, the value of the integral is nearly unity. For $r^+ = 0.5$ and $u = 0.01$, Equation R00.10 gives 0.9997. For large times u, an approximate result for the integral over r' from $r' = 0$ to a is

$$\begin{aligned}&\int_0^a G_{R00}(r,\ t|r',\ \tau)2\pi r'\ dr'\\ &\quad\approx e^{-r^{+2}/4u}\left[P\left(1,\frac{1}{4u}\right)+\frac{r^{+2}}{4u}P\left(2,\frac{1}{4u}\right)\right.\\ &\quad\left.+\frac{1}{2!}\frac{r^{+4}}{(4u)^2}P\left(3,\frac{1}{4u}\right)+\frac{1}{3!}\frac{r^{+6}}{(4u)^3}P\left(4,\frac{1}{4u}\right)\right]\end{aligned} \tag{R00.11}$$

where $P(n, x)$ is given by

$$P(n,\ x) = 1 - e_{n-1}(x)e^{-x} \qquad n = 1, 2, \ldots \tag{R00.12a}$$

$$e_{n-1}(x) = \sum_{j=0}^{n-1}\frac{x^j}{j!} \tag{R00.12b}$$

Equation R00.11 was derived using the four-term expression for $I_0(x)$ for large x for $x = \alpha(t-\tau)/rr'$.

For the center point, $r^+ = 0$, the integral is

$$\int_0^a G_{R00}(0,\ t|r',\ \tau)2\pi r'\ dr' = 1 - e^{-1/4u} = P\left(1,\frac{1}{4u}\right) \tag{R00.13}$$

which is exact. For $r^+ = 1$ and $u = 0.25$, the value of given by Equation R00.11 is 0.34569 which is −0.016% in error. Some values of the integral versus u for $r^+ = 0, 0.5^{1/2}, 1, 2$, and 4 are given in Table R00.1.

TABLE R00.1
Integral of $G_{R00}(r, t|r', \tau)$ From $r' = 0$ to a for Various Radii

	Values for Radii				
u[a]	$r^+ = 0$	$r^+ = 2^{-1/2}$	$r^+ = 1$	$r^+ = 2$	$r^+ = 4$
0.020	0.999996	0.910694	0.459902	0.000000	0.000000
0.100	0.917915	0.646447	0.408230	0.008333	0.000000
0.200	0.713495	0.515406	0.364977	0.035206	0.000000
0.500	0.393469	0.324351	0.267120	0.081892	0.000590
1.000	0.221199	0.198146	0.177482	0.091529	0.006337
10.000	0.024690	0.024387	0.024088	0.022368	0.016633
100.000	0.002497	0.002494	0.002491	0.002472	0.002399

[a] $u \equiv \alpha(t-\tau)/a^2$.

Another expression for large times is (Beck, 1981)

$$\int_0^a G_{R00}(r,\ t|r',\ \tau) 2\pi r'\, dr' \approx \frac{1}{4u}\left\{1 - \frac{1+2r^{+2}}{2!}\frac{1}{4u} + \frac{1+6r^{+2}+3r^{+4}}{3!}\frac{1}{(4u)^2}\right\} \qquad \text{(R00.14)}$$

which requires larger u values as r^+ increases.

R.3 AVERAGE GREEN'S FUNCTION FOR CIRCULAR REGION

The exact expression for the average GF for a circular region is (Amos, 1979)

$$\overline{G}_{R00}(t,\tau) \equiv \frac{4}{a^4}\int_{r=0}^{a}\int_{r'=0}^{a} G_{R00}(r,\ t|r',\ \tau) r'\, r\, dr'\, dr = \frac{1}{\pi a^2}\left\{1 - e^{-1/2u}\left[I_0\left(\frac{1}{2u}\right) + I_1\left(\frac{1}{2u}\right)\right]\right\} \qquad \text{(R00.15)}$$

where $u \equiv \alpha(t-\tau)/a^2$. Because this expression is not an easy one for subsequent integration, some approximate relations are given. One expression is

$$\pi a^2 \overline{G}_{R00}(t,\ \tau) = 1 - \left(\frac{u}{\pi}\right)^{1/2}\left(2 - \frac{u}{2} - \frac{u^2}{4}\right) \quad \text{for } 0 < u < 0.5 \qquad \text{(R00.16a)}$$

$$\pi a^2 \overline{G}_{R00}(t,\ \tau) = \frac{1}{4u}\left(1 - \frac{1}{4u} + \frac{5}{96u^2} - \frac{1}{128u^3}\right) \quad \text{for } u > 0.5 \qquad \text{(R00.16b)}$$

This is about 0.5% in error for $u = 0.4$ and more accurate elsewhere. Much more accurate relations are, for $0 < u < 0.15$:

$$\overline{G}_{R00}(t,\ \tau) \approx \frac{1}{\pi a^2}\left[1 - \left(\frac{u}{\pi}\right)^{1/2}\left(2 - \frac{u}{2} - \frac{3u^2}{16} - \frac{15u^3}{64} - \frac{525u^4}{1024} - \frac{6615u^5}{2048}\right)\right] \quad \text{(R00.17a)}$$

for $0.15 < u < 0.55$:

$$\overline{G}_{R00}(t,\ \tau) \approx \frac{1}{\pi a^2}\sum_{n=1}^{6} A_n e^{-\beta_n^2 u/9} \quad \text{(R00.17b)}$$

where the A_n and β_n values are:

n	A_n	β_n
1	0.350307417	2.4048256
2	0.383866200	5.5200781
3	0.10146524	8.6537279
4	0.0008193642	11.7915344
5	0.044399184	14.9309177
6	0.0257058527	18.0710640

for $u > 0.55$:

$$\overline{G}_{R00}(t,\ \tau) \approx \frac{V}{\pi a^2}\left(1 - V + \frac{5V^2}{6} - \frac{7V^3}{12} + \frac{7V^4}{20} - \frac{11V^5}{60} + \frac{143V^6}{1680} - \frac{143V^7}{8064}\right) \quad \text{(R00.17c)}$$

where $V \equiv 1/4u = a^2/[4\alpha(t-\tau)]$.

The Equation R00.17a expression is only -0.001% in error at $u = 0.15$ and is better for smaller u values. The Equation R00.17b expression is accurate to six significant figures at $u = 0.15$ and is -0.009% in error at $u = 0.55$. At $u = 0.55$, Equation R00.17c is 0.007% in error and the accuracy improves with increasing u.

Values of $\pi a^2 \overline{G}_{R00}$ are given in Table R00.2 along with values for $\pi a^2 G_{R0I}$, $I = 1, 2, 3$ for $b^+ = b/a = 2$.

R.4 DERIVATIVE OF G_{R00} WITH RESPECT TO r

Using Equation R00.4, for small values of $\alpha(t-\tau)/rr'$, the derivative of G_{R00} with respect to r is

$$\frac{\partial G_{R00}}{\partial r} \approx -\frac{1}{2\pi[4\pi r r'\alpha(t-\tau)]^{1/2}}\frac{1}{2\alpha(t-\tau)}\exp\left[-\frac{(r-r')^2}{4\alpha(t-\tau)}\right] \times \left[(r-r') + \frac{\alpha(t-\tau)}{4rr'}(r+3r') + \frac{[\alpha(t-\tau)]^2}{32(rr')^2}(9r+63r')\right] \quad \text{(R00.18)}$$

TABLE R00.2
Comparison of $\pi a^2 \overline{G}_{R0l}$ Values for $b^+ = 2$ with Values of $\pi a^2 \overline{G}_{R00}$

				$\pi a^2 \overline{G}_{R03}$	
u	$\pi a^2 \overline{G}_{R00}$	$\pi a^2 \overline{G}_{R01}$	$\pi a^2 \overline{G}_{R02}$	$B = 0.0002$	$B = 10^4$
0.001	0.964326				
0.01	0.887445				
0.1	0.652487	0.652480	0.652487	0.652488	0.652279
0.2	0.523369			0.523620	
1	0.198544	0.150571	0.264423	0.264415	0.150534
10	0.024388	0.00000034	0.25	0.249769	3.3E–7
100	0.002494		0.25	0.247531	
1000	0.000250		0.25	0.226227	

If $r = r'$, the first term inside the braces disappears and makes no contribution for $t - \tau$ greater than zero. However, as $t - \tau$ goes to zero, there can be a Dirac delta function at $r = r'$. See the $X00$ case.

*R*01 SOLID CYLINDER, $G = 0$ AT $r = b$

$$G_{R01}(r, t|r', \tau) = \frac{1}{\pi b^2} \sum_{m=1}^{\infty} e^{-\beta_m^2 \alpha(t-\tau)/b^2} \frac{J_0(\beta_m r/b) J_0(\beta_m r'/b)}{J_1^2(\beta_m)} \tag{R01.1}$$

Eigenvalues are found from

$$J_0(\beta_m) = 0 \tag{R01.2}$$

The derivative of G with respect to n' and evaluated at $r' = b$ is

$$-\left.\frac{\partial G_{R01}}{\partial n'}\right|_{r'=b} = \frac{1}{\pi b^3} \sum_{m=1}^{\infty} e^{-\beta_m^2 \alpha(t-\tau)/b^2} \frac{\beta_m J_0(\beta_m r/b)}{J_1(\beta_m)} \tag{R01.3}$$

Also the cross derivative at $r = r' = b$ is

$$-\left.\frac{\partial^2 G_{R01}}{\partial r \, \partial r'}\right|_{r=r'=b} = \frac{1}{2\pi b} \frac{1}{\{4\pi[\alpha(t - \tau)]^3\}^{1/2}} \left[1 + \frac{\alpha(t - \tau)}{8b^2}\right] \tag{R01.4}$$

For small values of $\alpha(t - \tau)/b^2$ and r and r' not near b,

$$G_{R01}(r, t|r', \tau) \approx G_{R00}(r, t|r', \tau)$$

For small $\alpha(t-\tau)/b^2$ values and r and r' not near zero, $G(\cdot)$ is

$$G_{R01}(r,\ t|r',\ \tau)$$
$$\approx \frac{1}{4\pi[\pi\alpha rr'(t-\tau)]^{1/2}}\left\{\exp\left[-\frac{(r-r')^2}{4\alpha(t-\tau)}\right]-\exp\left[-\frac{(2b-r-r')^2}{4\alpha(t-\tau)}\right]\right\}$$
$$+\frac{1}{32\pi(rr')^{1/2}}\left(\left(\frac{1}{r'}-\frac{1}{r}\right)\operatorname{erfc}\left\{\frac{r-r'}{[4\alpha(t-\tau)]^{1/2}}\right\}\right.$$
$$\left.-\left(\frac{1}{r'}+\frac{1}{r}-\frac{2}{b}\right)\operatorname{erfc}\left\{\frac{2b-r-r'}{[4\alpha(t-\tau)]^{1/2}}\right\}\right) \tag{R01.5}$$

for $0<r'<r$. For $0<r<r'$, exchange r and r'. For small $\alpha(t-\tau)/b^2$ values and r not near zero, the derivative is

$$-\left.\frac{\partial G_{R01}}{\partial n'}\right|_{r'=b}\approx\frac{1}{4\pi b^2\sqrt{rb}}\frac{r-b}{[\pi\alpha(t-\tau)]^{1/2}}e^{-(b-r)^2/[4\alpha(t-\tau)]}\left[\frac{b^2}{\alpha(t-\tau)}+\frac{1}{8}\frac{b}{r}\right] \tag{R01.6}$$

The average GF for a circular region of radius a is given by Equation R00.19 for $u\equiv\alpha(t-\tau)/a^2$ less than $(b^+-1)^2/12$ (with $b^+\equiv b/a$) and for larger u values by

$$\overline{G}_{R01}(t,\ \tau)\equiv\frac{4}{a^2}\int_{r=0}^{a}\int_{r'=0}^{a}G_{R01}(r,\ t|r',\ \tau)rr'\,dr\,dr'$$
$$=\frac{4}{\pi a^2}\sum_{m=1}^{\infty}e^{-\beta_m^2\alpha(t-\tau)/b^2}\left[\frac{J_1(\beta_m a/b)}{\beta_m J_1(\beta_m)}\right]^2 \tag{R01.7}$$

R02 SOLID CYLINDER, $\partial G/\partial r = 0$ AT $r = b$

$$G_{R02}(r,t|r',\tau)=\frac{1}{\pi b^2}\left[1+\sum_{m=1}^{\infty}e^{-\beta_m^2\alpha(t-\tau)/b^2}\frac{J_0(\beta_m r/b)J_0(\beta_m r'/b)}{J_0^2(\beta_m)}\right] \tag{R02.1}$$

Eigenvalues are found from

$$J_1(\beta_m)=0 \tag{R02.2}$$

Some values are given in Table R02.

TABLE R02

m	β_m	$J_0(\beta_m)$
1	3.83170597	−0.40275940
2	7.01558667	0.30011575
3	10.17346814	−0.24970488
4	13.32369194	0.21835941
5	16.47063005	−0.19646537
6	19.61585851	0.18006338

For small times ($\alpha t/b^2 < .01$) and r/b and r'/b not near zero, $G_{R02}(\bullet)$ is

$$G_{R02}(r,t|r',\tau) \approx G_{R00}(r,t|r',\tau) + \frac{1}{4\pi b(rr')^{1/2}}\left(\frac{b}{[\pi\alpha(t-\tau)]^{1/2}}\right.$$

$$\times\exp\left[-\frac{(2b-r-r')^2}{4\alpha(t-\tau)}\right]$$

$$+\frac{1}{8}\left(6+\frac{b}{r}+\frac{b}{r'}\right)\operatorname{erfc}\left\{\frac{2b-r-r'}{[4\alpha(t-\tau)]^{1/2}}\right\}$$

$$+\frac{1}{64}\left[36+12\frac{b}{r}+12\frac{b}{r'}+9\left(\frac{b}{r}\right)^2+9\left(\frac{b}{r'}\right)^2+2\frac{b}{r}\frac{b}{r'}\right]$$

$$\left.\times\left[\frac{\alpha(t-\tau)}{b^2}\right]^{1/2}\operatorname{ierfc}\left\{\frac{2b-r-r'}{[4\alpha(t-\tau)]^{1/2}}\right\}\right) \tag{R02.3}$$

For $r=r'=b$, this expression with $\tau=0$ reduces to

$$G_{R02}(b,\ t|b,\ 0) \approx \frac{1}{2\pi b^2}\left[\frac{b}{(\pi\alpha t)^{1/2}}+\frac{1}{2}+\frac{3}{4(\pi)^{1/2}}\left(\frac{\alpha t}{b^2}\right)^{1/2}\right] \tag{R02.4}$$

A more accurate expression is given by Equation R02.5. For $\alpha t/b^2 < 0.1$ and at $r=b'=b(-0.1\%$ at $\alpha t/b^2=0.04$ and -0.8% at $0.1)$,

$$G(b,\ t|b,\ 0) \approx \frac{1}{2\pi b^2}\left[\frac{b}{(\pi\alpha t)^{1/2}}+\frac{1}{2}+\frac{3}{4\sqrt{\pi}}\left(\frac{\alpha t}{b^2}\right)^{1/2}+\frac{3}{8}\left(\frac{\alpha t}{b^2}\right)\right] \tag{R02.5}$$

For $\alpha t/b^2 > 0.04$ at $r=r'=b$ (0.002% error at $\alpha t/b^2=0.04$),

$$G_{R02}(b,\ t|b,\ 0) \approx \frac{1}{\pi b^2}\left(1+\sum_{m=1}^{4}e^{-\beta_m^2\alpha t/b^2}\right) \tag{R02.6}$$

The β_m values are given in Table R02.

For $\alpha t/b^2 < 0.02$ for $r=0$ and $r'=b$ (about 0.0015% error at $\alpha t/b^2=0.02$),

$$b^2G(0,\ t|b,\ 0) \approx b^2[4\pi\alpha(t-\tau)]^{-1}\exp\left[-\frac{b^2}{4\alpha(t-\tau)}\right] \tag{R02.7}$$

and for $\alpha t/b^2 > 0.02$ for $r=0$ and $r'=b$ (about 0.005% error at $\alpha t/b^2=0.02$)

$$b^2G(0,\ t|b,\ 0) = \frac{1}{\pi}\left[1+\sum_{m=1}^{6}\frac{1}{J_0(\beta_m)}\exp\left(-\beta_m^2\frac{\alpha t}{b^2}\right)\right] \tag{R02.8}$$

where the β_m and $J_0(\beta_m)$ values are given in Table R02.

The integral of $G_{R02}(r,\ t|r',\ \tau)2\pi r'dr'$ from $r'=0$ to a is

$$\int_0^a G_{R02}(r,\ t|r',\ \tau)2\pi r'\,dr'$$

$$=\left(\frac{a}{b}\right)^2+\frac{2a}{b}\sum_{m=1}^{\infty}e^{-\beta_m^2\alpha(t-\tau)/b^2}\frac{J_0(\beta_m r/b)J_1(\beta_m a/b)}{\beta_m J_0^2(\beta_m)} \tag{R02.9}$$

Though this expression applies for all $\alpha(t-\tau)/b^2$ values equal to or greater than zero, for small dimensionless times the expression for the integral over G_{R00} can be used. The average GF for a circular region of radius a is given by Equation R00.15 for $u \equiv \alpha(t-\tau)/a^2$ less than $(b^+ - 1)^2/12$ (with $b^+ \equiv b/a$) and for larger values of u by

$$\overline{G}_{R02}(t,\ \tau) \equiv \frac{4}{a^4}\int_{r=0}^{a}\int_{r'=0}^{a} G_{R02}(r,\ t|r',\ \tau) r r'\,dr'\,dr$$
$$= \frac{1}{\pi b^2}\left\{1 + 4\left(\frac{b}{a}\right)^2 \sum_{m=1}^{\infty} e^{-\beta_m^2\alpha(t-\tau)/b^2}\left[\frac{J_1(\beta_m a/b)}{\beta_m J_0(\beta_m)}\right]^2\right\} \tag{R02.10}$$

For $a/b = 1, \overline{G}_{R02}(t,\ \tau)$ is given by

$$\overline{G}_{R02}(t,\ \tau) = \frac{1}{\pi b^2} \tag{R02.11}$$

The integral of $G_{R02}\ 2\pi r'\,dr'$ for $r' = 0$ to b is unity:

$$\int_0^b G_{R02}(r,\ t|r',\ \tau) 2\pi r'\,dr' = 1 \tag{R02.12}$$

*R*03 SOLID CYLINDER, $k\partial G/\partial r + hG = 0$ AT $r = b$

$$G_{R03}(r, t|r', \tau) = \frac{1}{\pi b^2}\sum_{m=1}^{\infty} e^{-\beta_m^2\alpha(t-\tau)/b^2}\frac{\beta_m^2 J_0(\beta_m r/b) J_0(\beta_m r'/b)}{J_0^2(\beta_m)(B^2+\beta_m^2)} \tag{R03.1}$$

The eigencondition is

$$-\beta_m J_1(\beta_m) + B J_0(\beta_m) = 0 \tag{R03.2}$$
$$B = \frac{hb}{k} \tag{R03.3}$$

The average GF for a circular region of radius a is given by Equation R00.15 for $u \equiv \alpha(t-\tau)/a^2$ less than $(b^+ - 1)^2/12$ (with $b^+ \equiv b/a$) and for larger values of u by

$$\overline{G}_{R03}(t,\ \tau) = \frac{4}{\pi a^2}\sum_{m=1}^{\infty} e^{-\beta_m^2\alpha(t-\tau)/b^2}\frac{J_1^2(\beta_m a/b)}{J_0^2(\beta_m)(B^2+\beta_m^2)} \tag{R03.4}$$

*R*10 OUTSIDE THE CYLINDRICAL REGION $r = a$, $G = 0$ AT $r = a$

$$G_{R10}(r,\ t|r',\ \tau) = \frac{1}{2\pi a^2}\int_{\beta=0}^{\infty} e^{-\beta^2\alpha(t-\tau)/a^2}$$
$$\times\ \beta[J_0(\beta r/a)Y_0(\beta) - Y_0(\beta r/a)J_0(\beta)]$$
$$\times\ \frac{[J_0(\beta r'/a)Y_0(\beta) - Y_0(\beta r'/a)J_0(\beta)]}{J_0^2(\beta) + Y_0^2(\beta)}d\beta \tag{R10.1}$$

$$-\left.\frac{\partial G_{R10}}{\partial n'}\right|_{r'=a} = -\frac{1}{\pi^2 a^3}\int_{\beta=0}^{\infty} e^{-\beta^2\alpha(t-\tau)/a^2} \times \frac{\beta[J_0(\beta r/a)Y_0(\beta) - Y_0(\beta r/a)J_0(\beta)]}{J_0^2(\beta)+Y_0^2(\beta)}d\beta \tag{R10.2}$$

$$-\left.\frac{\partial^2 G_{R10}}{\partial r\,\partial n'}\right|_{r'=r=a} = \frac{2}{\pi^3 a^4}\int_{\beta=0}^{\infty} e^{-\beta^2\alpha(t-\tau)/a^2}\frac{\beta}{J_0^2(\beta)+Y_0^2(\beta)}d\beta \tag{R10.3}$$

*R*11 HOLLOW CYLINDER, $G = 0$ AT $r = a$ AND b

$$G_{R11}(r,\, t|r',\, \tau) = \frac{\pi}{4a^2}\sum_{m=1}^{\infty} e^{-\beta_m^2\alpha(t-\tau)/a^2} \times \frac{\beta_m^2 J_0^2(\beta_m)[J_0(\beta_m r/a)Y_0(\beta_m b/a) - J_0(\beta_m b/a)Y_0(\beta_m r/a)]}{J_0^2(\beta_m) - J_0^2(\beta_m b/a)} \times \left[J_0\left(\beta_m\frac{r'}{a}\right)Y_0\left(\beta_m\frac{b}{a}\right) - J_0\left(\beta_m\frac{b}{a}\right)Y_0\left(\beta_m\frac{r'}{a}\right)\right] \tag{R11.1}$$

The eigenvalues are found from

$$J_0(\beta_m)Y_0\left(\beta_m\frac{b}{a}\right) - J_0\left(\beta_m\frac{b}{a}\right)Y_0(\beta_m) = 0 \tag{R11.2}$$

The normal derivatives at $r' = a$ and b are

$$-\left.\frac{\partial G_{R11}}{\partial n'}\right|_{r'=a} = \frac{\pi}{4a^3}\sum_{m=1}^{\infty} e^{-\beta_m^2\alpha(t-\tau)/a^2} \times \beta_m^3 J_0^2(\beta_m)[J_0(\beta_m r/a)Y_0(\beta_m b/a) - J_0(\beta_m b/a)Y_0(\beta_m r/a)] \times \frac{[J_1(\beta_m)Y_0(\beta_m b/a) - J_0(\beta_m b/a)Y_1(\beta_m)]}{J_0^2(\beta_m) - J_0^2(\beta_m b/a)} \tag{R11.3}$$

$$-\left.\frac{\partial G_{R11}}{\partial n'}\right|_{r'=b} = \frac{1}{2a^3}\frac{a}{b}\sum_{m=1}^{\infty} e^{-\beta_m^2\alpha(t-\tau)/a^2} \times \frac{\beta_m^2 J_0^2(\beta_m)[J_0(\beta_m r/a)Y_0(\beta_m b/a) - J_0(\beta_m b/a)Y_0(\beta_m r/a)]}{J_0^2(\beta_m) - J_0^2(\beta_m b/a)} \tag{R11.4}$$

*R*12 HOLLOW CYLINDER, $G = 0$ AT $r = a$, $\partial G/\partial r = 0$ AT $r = b$

$$G_{R12}(r,\, t|r',\, \tau) = \frac{\pi}{4a^2}\sum_{m=1}^{\infty} e^{-\beta_m^2\alpha(t-\tau)/a^2}\frac{\beta_m^2 J_0^2(\beta_m)}{J_0^2(\beta_m) - J_1^2(\beta_m b/a)} \times [J_0(\beta_m r/a)Y_1(\beta_m b/a) - J_1(\beta_m b/a)Y_0(\beta_m r/a)] \times [J_0(\beta_m r'/a)Y_1(\beta_m b/a) - J_1(\beta_m b/a)Y_0(\beta_m r'/a)] \tag{R12.1}$$

where the eigenvalues are found from

$$J_0(\beta_m)Y_1(\beta_m b/a) - J_1(\beta_m b/a)Y_0(\beta_m) = 0 \tag{R12.2}$$

The normal derivative at $r' = a$ is

$$\begin{aligned} -\frac{\partial G_{R12}}{\partial n'}\bigg|_{r'=a} &= \frac{\pi}{4a^3}\sum_{m=1}^{\infty} e^{-\beta_m^2\alpha(t-\tau)/a^2}\frac{\beta_m^3 J_0^2(\beta_m)}{J_0^2(\beta_m) - J_1^2(\beta_m b/a)} \\ &\times \left[J_0\left(\beta_m\frac{r}{a}\right)Y_1\left(\beta_m\frac{b}{a}\right) - J_1\left(\beta_m\frac{b}{a}\right)Y_0\left(\beta_m\frac{r}{a}\right)\right] \\ &\times \left[J_1(\beta_m)Y_1\left(\beta_m\frac{b}{a}\right) - J_1\left(\beta_m\frac{b}{a}\right)Y_1(\beta_m)\right] \end{aligned} \tag{R12.3}$$

*R*13 HOLLOW CYLINDER, $G = 0$ AT $r = a$, $k\partial G/\partial r + hG = 0$ AT $r = b$

$$\begin{aligned} G_{R13}(r,\ t|r',\ \tau) &= \frac{\pi}{4a^2}\sum_{m=1}^{\infty} e^{-\beta_m^2\alpha(t-\tau)/a^2}\frac{\beta_m^2 J_0^2(\beta_m)}{(B^2+\beta_m^2)J_0^2(\beta_m) - V_0^2} \\ &\times \left[S_0 J_0\left(\beta_m\frac{r}{a}\right) - V_0 Y_0\left(\beta_m\frac{r}{a}\right)\right] \\ &\times \left[S_0 J_0\left(\beta_m\frac{r'}{a}\right) - V_0 Y_0\left(\beta_m\frac{r'}{a}\right)\right] \end{aligned} \tag{R13.1}$$

where

$$V_0 = -\beta_m J_1\left(\beta_m\frac{b}{a}\right) + BJ_0\left(\beta_m\frac{b}{a}\right) \qquad B = \frac{ha}{k} \tag{R13.2a, b}$$

$$S_0 = -\beta_m Y_1\left(\beta_m\frac{b}{a}\right) + BY_0\left(\beta_m\frac{b}{a}\right) \tag{R13.3}$$

and the eigencondition is

$$S_0 J_0(\beta_m) - V_0 Y_0(\beta_m) = 0 \tag{R13.4}$$

The normal derivative at $r' = a$ is

$$\begin{aligned} -\frac{\partial G_{R13}}{\partial n'}\bigg|_{r'=a} &= \frac{\pi}{4a^3}\sum_{m=1}^{\infty} e^{-\beta_m^2\alpha(t-\tau)/a^2}\frac{\beta_m^3 J_0^2(\beta_m)}{(B^2+\beta_m^2)J_0^2(\beta_m) - V_0^2} \\ &\times \left[S_0 J_0\left(\beta_m\frac{r}{a}\right) - V_0 Y_0\left(\beta_m\frac{r}{a}\right)\right] \\ &\times [S_0 J_1(\beta_m) - V_0 Y_1(\beta_m)] \end{aligned} \tag{R13.5}$$

*R*20 OUTSIDE THE CYLINDRICAL REGION $r = a$, $\partial G/\partial r = 0$ AT $r = a$

$$G_{R20}(r,\, t|r',\, \tau) = \frac{1}{2\pi a^2}\int_{\beta=0}^{\infty} e^{-\beta^2\alpha(t-\tau)/a^2}\, d\beta \times \beta[J_0(\beta r/a)Y_1(\beta) - Y_0(\beta r/a)J_1(\beta)] \times \frac{[J_0(\beta r'/a)Y_1(\beta) - Y_0(\beta r'/a)J_1(\beta)]}{J_1^2(\beta) + Y_1^2(\beta)} \tag{R20.1}$$

Note that $J_0(z)Y_1(z) - Y_0(z)J_1(z) = -2/(\pi z)$ (see Carslaw and Jaeger, 1959, p. 489)

For $r = r' = a$

$$G_{R20}(a,\, t|a,\, \tau) = \frac{1}{2\pi a^2}\frac{4}{\pi^2}\int_0^{\infty} \frac{\exp[-\beta^2\alpha(t-\tau)/a^2]d\beta}{\beta[J_1^2(\beta) + Y_1^2(\beta)]} \tag{R20.2}$$

Approximate values for $G_{R20}(a,\, t|a,\, \tau)$ are given below.
For small $t^+ = \alpha(t-\tau)/a^2$ values,

$$G_{R20}(a,\, t|a,\, \tau) \approx \frac{1}{2\pi a^2}\left[(\pi t^+)^{-1/2} - \frac{1}{2}C_1 + \frac{3}{4}\left(\frac{t^+}{\pi}\right)^{1/2} C_2 - \frac{3}{8}t^+ C_3 + \frac{21}{32}\frac{1}{\pi^{1/2}}(t^+)^{3/2}C_4\right] \tag{R20.3}$$

The series expansion has $C_1 = C_2 = C_3 = C_4 = 1$ which is accurate only for $\alpha(t-\tau)/a^2 << 1.0$. If Euler's transformation is used (Abramowitz and Stegun, 1964, p. 16), $C_1 = 15/16$, $C_2 = 11/16$, $C_3 = 5/16$, and $C_4 = 1/16$ and the accuracy is much improved, better than 0.15% for $t^+ < 0.4$. See Table R20 where (R20.3) denotes Equation R20.3 for $C_1 = 15/16$ and so on. More accurate values are obtained using the polynomial fit of

$$G_{R20}(a,\, t|a,\, \tau) \approx \frac{1}{2\pi a^2}\left[(\pi t^+)^{-1/2} - 0.5 + 0.413434(t^+)^{1/2} - 0.299877t^+ + 0.154483(t^+)^{3/2} - 0.045263(t^+)^2 + 0.005484(t^+)^{5/2}\right] \tag{R20.4}$$

For large t^+ values,

$$G_{R20}(a,\, t|a,\, \tau) \approx \frac{1}{4\pi a^2}\frac{1}{t^+}\left\{1 - \frac{1}{2t^+}L\left[1 + \frac{3}{4t^+}(1-L)\right] - (\pi^2+4)\frac{C}{16t^{+2}}\right\} \quad L = \ln 4t^+ - \gamma \tag{R20.5}$$

TABLE R20

Exact $2\pi a^2 G_{R20}(a,\ t|a,\ \tau)$ and Approximate Equations R20.4, R20.3, and R20.5. $u \equiv \alpha(t-\tau)/a^2$.

u	(R20.4)	(R20.3)	(R20.5)	Exact
0.5	0.4846	0.4844		0.484220
1.0	0.2924	0.2923		0.292633
1.5	0.2143	0.2149		0.214567
2.0	0.1708	0.1727		0.170938
2.5	0.1428	0.1465		0.142723
3.0	0.1230			0.122844
3.5	0.1082		0.1053	0.108019
4.0	0.0966		0.0950	0.096506
4.5	0.0872		0.0864	0.087288
5.0	0.0796		0.0792	0.079730
5.5	0.0733		0.0731	0.073414
6.0	0.0683		0.0679	0.068058
6.5	0.0646		0.0633	0.063440
7.0	0.0621		0.0594	0.059429
7.5	0.0609		0.0559	0.055907
		Percent Errors		
0.5	0.07	0.04		
1.0	−0.06	−0.11		
1.5	−0.13	0.16		
2.0	−0.08	1.01		
2.5	0.03	2.67		
3.0	0.15			
3.5	0.18		−2.51	
4.0	0.09		−1.56	
4.5	−0.07		−0.98	
5.0	−0.21		−0.63	
5.5	−0.15		−0.40	
6.0	0.38		−0.25	
6.5	1.77		−0.14	
7.0	4.46		−0.07	
7.5	9.00		−0.02	

where γ = Euler's constant = 0.57722. This equation with $C = 0.5$ is accurate to +0.1% for $t^+ > 10$, to −0.6% at $t^+ = 5$. A comparison of results is given in Table R20. Equation R20.4 is recommended for $t^+ < 6$ and Equation R20.5 for $t^+ > 6$.

The integral of $G_{R20}(r,\ t|r',\ \tau)2\pi r'dr'$ from $r' = a$ to ∞ is unity,

$$\int_a^\infty G_{R20}(r,\ t|r',\ \tau)2\pi r'dr' = 1 \qquad \text{(R20.6)}$$

*R*21 HOLLOW CYLINDER, $\partial G/\partial r = 0$ AT $r = a$, $G = 0$ AT $r = b$

$$G_{R21}(r, t|r', \tau) = \frac{\pi}{4a^2}\sum_{m=1}^{\infty} e^{-\beta_m^2\alpha(t-\tau)/a^2}\frac{\beta_m^2 J_1^2(\beta_m)}{J_1^2(\beta_m) - J_0^2(\beta_m b/a)}$$
$$\times\left[J_0\left(\beta_m\frac{r}{a}\right)Y_0\left(\beta_m\frac{b}{a}\right) - J_0\left(\beta_m\frac{b}{a}\right)Y_0\left(\beta_m\frac{r}{a}\right)\right]$$
$$\times\left[J_0\left(\beta_m\frac{r'}{a}\right)Y_0\left(\beta_m\frac{b}{a}\right) - J_0\left(\beta_m\frac{b}{a}\right)Y_0\left(\beta_m\frac{r'}{a}\right)\right] \quad \text{(R21.1)}$$

Eigencondition

$$J_1(\beta_m)Y_0\left(\beta_m\frac{b}{a}\right) - J_0\left(\beta_m\frac{b}{a}\right)Y_1(\beta_m) = 0 \quad \text{(R21.2)}$$

$$-\left.\frac{\partial G_{R21}}{\partial n'}\right|_{r'=b} = -\frac{1}{2a^2}\frac{1}{b}\sum_{m=1}^{\infty} e^{-\beta_m^2\alpha(t-\tau)/a^2}\frac{\beta_m^2 J_0^2(\beta_m)}{J_1^2(\beta_m) - J_0^2(\beta_m b/a)}$$
$$\times\left[J_0\left(\beta_m\frac{r}{a}\right)Y_0\left(\beta_m\frac{b}{a}\right) - J_0\left(\beta_m\frac{b}{a}\right)Y_0\left(\beta_m\frac{r}{a}\right)\right] \quad \text{(R21.3)}$$

*R*22 HOLLOW CYLINDER, $\partial G/\partial r = 0$ AT $r = a$ AND b

$$G_{R22}(r, t|r', \tau) = \frac{1}{\pi(b^2 - a^2)} + \frac{\pi}{4a^2}\sum_{m=1}^{\infty} e^{-\beta_m^2\alpha(t-\tau)/a^2}$$
$$\times\frac{\beta_m^2 J_1^2(\beta_m)}{J_1^2(\beta_m) - J_1^2(\beta_m b/a)}$$
$$\times\left[J_0\left(\beta_m\frac{r}{a}\right)Y_1\left(\beta_m\frac{b}{a}\right) - J_1\left(\beta_m\frac{b}{a}\right)Y_0\left(\beta_m\frac{r}{a}\right)\right]$$
$$\times\left[J_0\left(\beta_m\frac{r'}{a}\right)Y_1\left(\beta_m\frac{b}{a}\right) - J_1\left(\beta_m\frac{b}{a}\right)Y_0\left(\beta_m\frac{r'}{a}\right)\right] \quad \text{(R22.1)}$$

Eigencondition

$$J_1(\beta_m)Y_1\left(\beta_m\frac{b}{a}\right) - J_1\left(\beta_m\frac{b}{a}\right)Y_1(\beta_m) = 0 \quad \text{(R22.2)}$$

For $r = r' = b$ and $\alpha(t-\tau)/(b-a)^2 < 1/12$,

$$G_{R22}(b, t|b, \tau) \approx \frac{1}{2\pi b^2}\left\{\frac{b}{[\pi\alpha(t-\tau)]^{1/2}} + \frac{1}{2}\right.$$
$$\left. + \frac{3}{4\sqrt{\pi}}\left[\frac{\alpha(t-\tau)}{b^2}\right]^{1/2} + \frac{3}{8}\left[\frac{\alpha(t-\tau)}{b^2}\right]\right\} \quad \text{(R22.3)}$$

For $r = r' = a$ and $\alpha(t - \tau)/(b - a)^2 < 1/12$ and $\alpha(t - \tau)/a^2 < 0.4$ (or for $b/a > 3$, only $\alpha(t - \tau)/a^2 < 0.4$), use the small t^+ expression of the $R20$ case. The integral of $G_{R22}(r, t|r', \tau)2\pi r'\,dr'$ from $r' = a$ to b is unity,

$$\int_a^b G_{R22}(r, t|r', \tau)2\pi r'\,dr' = 1 \tag{R22.4}$$

*R*23 HOLLOW CYLINDER, $\partial G/\partial r = 0$ AT $r = a$, $k\partial G/\partial r + hG = 0$ AT $r = b$

$$G_{R23}(r,t|r',\tau) = \frac{\pi}{4a^2}\sum_{m=1}^{\infty} e^{-\beta_m^2\alpha(t-\tau)/a^2}\frac{\beta_m^2 J_1^2(\beta_m)}{(B^2+\beta_m^2)J_1^2(\beta_m) - V_0^2}$$
$$\times\left[S_0 J_0\left(\beta_m\frac{r}{a}\right) - V_0 Y_0\left(\beta_m\frac{r}{a}\right)\right]$$
$$\times\left[S_0 J_0\left(\beta_m\frac{r'}{a}\right) - V_0 Y_0\left(\beta_m\frac{r'}{a}\right)\right] \tag{R23.1}$$

where S_0 and V_0 are defined for $R13$.

Eigencondition

$$S_0 J_1(\beta_m) - V_0 Y_1(\beta_m) = 0 \tag{R23.2}$$

For $r = r' = a$ and $\alpha(t - \tau)/(b - a)^2 < 1/12$ and $\alpha(t - \tau)/a^2 < 0.4$ [or for $b/a > 3$, only $\alpha(t - \tau)/a^2 < 0.4$], use the small t^+ expression of the $R20$ case.

*R*30 REGION OUTSIDE $r = a$, $-k\partial G/\partial r + hG = 0$ AT $r = a$

$$G_{R30}(r, t|r', \tau) = \frac{1}{2\pi a^2}\int_{\beta=0}^{\infty} e^{-\beta^2\alpha(t-\tau)/a^2}$$
$$\times\,\beta[W_0 J_0(\beta r/a) - U_0 Y_0(\beta r/a)]$$
$$\times\,\frac{[W_0 J_0(\beta r'/a) - U_0 Y_0(\beta r'/a)]}{U_0^2 + W_0^2}d\beta \tag{R30.1}$$

where

$$W_0 = -\beta Y_1(\beta) - BY_0(\beta) \tag{R30.2}$$

$$U_0 = -\beta J_1(\beta) - BJ_0(\beta) \qquad B = \frac{ha}{k} \tag{R30.3a, b}$$

$$G_{R30}(a,t|a,\tau) \approx \frac{kB}{\alpha}\left\{\left[\frac{\alpha}{\pi a^2(t-\tau)}\right]^{1/2} + \left(B + \frac{1}{2}\right)\frac{\alpha}{a^2}\right.$$
$$\left. - 2\left(B^2 + B + \frac{3}{8}\right)\left[\frac{\alpha^3(t-\tau)}{a^6}\right]^{1/2}\right\} \tag{R30.4}$$

for small $\alpha(t - \tau)/a^2$ values and $B \le 1$.

*R*31 HOLLOW CYLINDER, $-k\partial G/\partial r + hG = 0$ AT $r = a$, $G = 0$ AT $r = b$

$$G_{R31}(r,\ t|r',\ \tau) = \frac{\pi}{4a^2}\sum_{m=1}^{\infty} e^{-\beta_m^2\alpha(t-\tau)/a^2}$$
$$\times \frac{\beta_m^2 U_0^2}{U_0^2 - (B^2+\beta_m^2)J_0^2(\beta_m b/a)}$$
$$\times \left[J_0\left(\beta_m\frac{r}{a}\right)Y_0\left(\beta_m\frac{b}{a}\right) - J_0\left(\beta_m\frac{b}{a}\right)Y_0\left(\beta_m\frac{r}{a}\right)\right]$$
$$\times \left[J_0\left(\beta_m\frac{r'}{a}\right)Y_0\left(\beta_m\frac{b}{a}\right) - J_0\left(\beta_m\frac{b}{a}\right)Y_0\left(\beta_m\frac{r'}{a}\right)\right] \quad \text{(R31.1)}$$

where

$$U_0 = -\beta_m J_1(\beta_m) - BJ_0(\beta_m)$$
$$W_0 = -\beta_m Y_1(\beta_m) - BY_0(\beta_m) \qquad B = \frac{ha}{k} \quad \text{(R31.2a, b, c)}$$

Eigencondition:

$$U_0 Y_0\left(\beta_m\frac{b}{a}\right) - W_0 J_0\left(\beta_m\frac{b}{a}\right) \quad \text{(R31.3)}$$

$$-\frac{\partial G_{R31}}{\partial n'}\bigg|_{r'=b} = \frac{1}{2a^2}\frac{1}{b}\sum_{m=1}^{\infty} e^{-\beta_m^2\alpha(t-\tau)/a^2}$$
$$\times \frac{\beta_m^2 U_0^2[J_0(\beta_m r/a)Y_0(\beta_m b/a) - J_0(\beta_m b/a)Y_0(\beta_m r/a)]}{U_0^2 - (B^2+\beta_m^2)J_0^2(\beta_m b/a)} \quad \text{(R31.4)}$$

*R*32 HOLLOW CYLINDER, $-k\partial G/\partial r + hG = 0$ AT $r = a$, $\partial G/\partial r = 0$ AT $r = b$

$$G_{R32}(r,t|r',\tau) = \frac{\pi}{4a^2}\sum_{m=1}^{\infty} e^{-\beta_m^2\alpha(t-\tau)/a^2}$$
$$\times \frac{\beta_m^2 U_0^2}{U_0^2 - (B^2+\beta_m^2)J_1^2(\beta_m b/a)}$$
$$\times \left[J_0\left(\beta_m\frac{r}{a}\right)Y_1\left(\beta_m\frac{b}{a}\right) - J_1\left(\beta_m\frac{b}{a}\right)Y_0\left(\beta_m\frac{r}{a}\right)\right]$$
$$\times \left[J_0\left(\beta_m\frac{r'}{a}\right)Y_1\left(\beta_m\frac{b}{a}\right) - J_1\left(\beta_m\frac{b}{a}\right)Y_0\left(\beta_m\frac{r'}{a}\right)\right] \quad \text{(R32.1)}$$

For U_0 and W_0 see *R*31.

Eigencondition

$$U_0 Y_1\left(\beta_m\frac{b}{a}\right) - W_0 J_1\left(\beta_m\frac{b}{a}\right) = 0 \quad \text{(R32.2)}$$

$R33$ HOLLOW CYLINDER, $-k\partial G/\partial r + h_1 G = 0$ AT $r = a$, $k\partial G/\partial r + h_2 G = 0$ AT $r = b$

$$G_{R33}(r,t|r',\tau) = \frac{\pi}{4a^2}\sum_{m=1}^{\infty} e^{-\beta_m^2\alpha(t-\tau)/a^2} \times \frac{\beta_m^2 U_0^2}{(B_2^2+\beta_m^2)U_0^2 - (B_1^2+\beta_m^2)V_0^2} \times \left[S_0 J_0\left(\beta_m \frac{r}{a}\right) - V_0 Y_0\left(\beta_m \frac{r}{a}\right)\right] \times \left[S_0 J_0\left(\beta_m \frac{r'}{a}\right) - V_0 Y_0\left(\beta_m \frac{r'}{a}\right)\right] \quad \text{(R33.1)}$$

where

$$S_0 \equiv -\beta_m Y_1\left(\beta_m \frac{b}{a}\right) + B_2 Y_0\left(\beta_m \frac{b}{a}\right) \quad \text{(R33.2)}$$

$$U_0 \equiv -\beta_m J_1(\beta_m) - B_1 J_0(\beta_m) \quad \text{(R33.3)}$$

$$V_0 \equiv -\beta_m J_1\left(\beta_m \frac{b}{a}\right) + B_2 J_0\left(\beta_m \frac{b}{a}\right) \quad \text{(R33.4)}$$

$$W_0 \equiv -\beta_m Y_1(\beta_m) - B_1 Y_0(\beta_m) \quad \text{(R33.5)}$$

$$B_1 \equiv \frac{h_1 a}{k} \qquad B_2 \equiv \frac{h_2 a}{k} \quad \text{(R33.6a, b)}$$

Eigencondition

$$S_0 U_0 - V_0 W_0 = 0 \quad \text{(R33.7)}$$

REFERENCES

Abramowitz, M. and Stegun, I. A. (Eds), 1964, *Handbook of Mathematical Functions with Formulas, Graphs, and Mathematical Tables*, Applied Mathematics Series 55, National Bureau of Standards, Washington, DC.

Amos, D., 1979, Personal communication, Sandia Laboratories, Albuquerque, NM.

Beck, J. V., 1981, Large time solutions for temperature in a semi-infinite body with a disk heat source, *Int. J. Heat Mass Transfer*, vol. 24, pp. 155–164.

Carslaw, H. S. and Jaeger, J. C., 1959, *Conduction of Heat in Solids*, 2nd edn, Oxford University Press, New York.

RΦ Green's Functions for Cylindrical Coordinates (r, ϕ)

$$dV' = r'\,dr'\,d\phi'$$

$$ds' = a\,d\phi' \text{ at } r = a \text{ and } ds' = b\,d\phi' \text{ at } r = b$$

$$ds' = dr' \text{ at } \phi = 0 \text{ or } \phi_0$$

The partial differential equation for transient conduction with cylindrical coordinates (r, ϕ) is

$$\frac{1}{r}\frac{\partial}{\partial r}\left(r\frac{\partial T}{\partial r}\right) + \frac{1}{r^2}\frac{\partial^2 T}{\partial \phi^2} = \frac{1}{\alpha}\frac{\partial T}{\partial t}$$

*R*00Φ11 WEDGE FOR $0 < \phi < \phi_0 < 2\pi$ AND WITH $G = 0$ AT $\phi = 0$ AND $\phi = \phi_0$

$$G(r, \phi, t|r', \phi', \tau) = \frac{1}{\phi_0 \alpha(t-\tau)} e^{-(r^2+r'^2)/[4\alpha(t-\tau)]} \times \sum_{n=1}^{\infty} n I_s\left(\frac{rr'}{2\alpha(t-\tau)}\right) \sin\left(n\pi\frac{\phi}{\phi_0}\right) \times \sin\left(n\pi\frac{\phi'}{\phi_0}\right) \quad \text{(R00Φ11.1)}$$

where $s = n\pi/\phi_0$ (Carslaw and Jaeger, 1959, p. 379).

$$-\left.\frac{\partial G}{\partial n'}\right|_{\phi'=0} = \frac{\pi}{r'\phi_0^2 \alpha(t-\tau)} e^{-(r^2+r'^2)/[4\alpha(t-\tau)]} \times \sum_{n=1}^{\infty} n I_s\left[\frac{rr'}{2\alpha(t-\tau)}\right] \sin\left(n\pi\frac{\phi}{\phi_0}\right) \quad \text{(R00Φ11.2)}$$

$$-\left.\frac{\partial G}{\partial n'}\right|_{\phi'=\phi_0} = -\frac{\pi}{r'\phi_0^2 \alpha(t-\tau)} e^{-(r^2+r'^2)/[4\alpha(t-\tau)]} \times \sum_{n=1}^{\infty} n I_s\left[\frac{rr'}{2\alpha(t-\tau)}\right] \sin\left(n\pi\frac{\phi}{\phi_0}\right)(-1)^n \quad \text{(R00Φ11.3)}$$

*R*00Φ12 WEDGE FOR $0 < \phi < \phi_0 < 2\pi$ AND WITH $G = 0$ AT $\phi = 0$ AND $\partial G/\partial\phi = 0$ AT $\phi = \phi_0$

$$G(r,\phi,t|r',\phi',\tau) = \frac{1}{\phi_0\alpha(t-\tau)} e^{-(r^2+r'^2)/[4\alpha(t-\tau)]} \times \sum_{m=1}^{\infty} I_{\beta_m}\left(\frac{rr'}{2\alpha(t-\tau)}\right) \sin(\beta_m\phi)\sin(\beta_m\phi') \qquad (\text{R00}\Phi12.1)$$

where $\beta_m = (2m-1)\dfrac{\pi}{2\phi_0}, m = 1, 2, \ldots$.

*R*00Φ22 WEDGE FOR $0 < \phi < \phi_0 < 2\pi$ AND WITH $\partial G/\partial\phi = 0$ AT $\phi = 0$ AND AT $\phi = \phi_0$

$$G(r,\phi,t|r',\phi',\tau) = \frac{1}{2\phi_0\alpha(t-\tau)} e^{-(r^2+r'^2)/[4\alpha(t-\tau)]} \left\{ I_0\left[\frac{rr'}{2\alpha(t-\tau)}\right] + 2\sum_{n=1}^{\infty} \cos\left(n\pi\frac{\phi}{\phi_0}\right)\cos\left(n\pi\frac{\phi'}{\phi_0}\right) \times I_s\left[\frac{rr'}{2\alpha(t-\tau)}\right]\right\} \qquad (\text{R00}\Phi22.1)$$

where $s = n\pi/\phi_0$ (Carslaw and Jaeger, 1959, p. 379).

*R*01Φ00 SOLID CYLINDER WITH RADIAL AND ANGULAR DEPENDENCE; $G = 0$ AT $r = a$

$$G(r,\phi,t|r',\phi',\tau) = \frac{2}{a^2}\sum_{n=0}^{\infty}\frac{1}{\pi}\cos[n(\phi-\phi')]\sum_{m=1}^{\infty} e^{-\beta_{mn}^2\alpha(t-\tau)/a^2} \times \frac{J_n(\beta_{mn}r/a)\,J_n(\beta_{mn}r'/a)}{[J_n'(\beta_{mn})]^2} \qquad (\text{R01}\Phi00.1)$$

where β_{mn} for $m, n = 1, 2, \ldots$ are the positive roots of $J_n(\beta_{mn}) = 0$. Replace π for 2π for $n = 0$ (Carslaw and Jaeger, 1959, p. 377, Equation 6; Ozisik, 1993, p. 134).

*R*01Φ11 SECTOR OF RADIUS *b*; $0 \le \phi \le \phi_0 < 2\pi$; $G = 0$ AT $r = b$, $\phi = 0$ AND $\phi = \phi_0$

$$G(r,\phi,t|r',\phi',\tau) = \frac{4}{b^2\phi_0}\sum_{m=1}^{\infty}\sum_{\nu}^{\infty} e^{-\beta_{m\nu}^2\alpha(t-\tau)/b^2} \times \frac{J_\nu(\beta_{m\nu}r/b)\,J_\nu(\beta_{m\nu}r'/b)\sin(\nu\phi)\sin(\nu\phi')}{J_\nu'^2(\beta_{m\nu})} \qquad (\text{R01}\Phi11.1)$$

where

$$\nu = \frac{n\pi}{\phi_0} \quad n = 1, 2, 3, \ldots \tag{R01Φ11.2}$$

and the β_{mv} eigenvalues are given by the positive roots of

$$J_v(\beta_{mv}) = 0 \quad \text{for the above } v \text{ values} \tag{R01Φ11.3}$$

R01Φ12 SECTOR OF RADIUS b; $0 \le \phi \le \phi_0 < 2\pi$; $G = 0$ AT $r = b$ AND $\phi = 0$, $\partial G/\partial\phi = 0$ AT $\phi = \phi_0$

$$G(r, \phi, t|r', \phi', \tau) = \frac{4}{b^2\phi_0} \sum_{m=1}^{\infty} \sum_{\nu}^{\infty} e^{-\beta_{m\nu}^2 \alpha(t-\tau)/b^2} \times \frac{J_\nu(\beta_{m\nu} r/b)\, J_\nu(\beta_{m\nu} r'/b) \sin v\phi \sin \nu\phi'}{J_\nu'^2(\beta_{m\nu})} \tag{R01Φ12.1}$$

where

$$v = (2n-1)\frac{\pi}{2\phi_0} \quad n = 1, 2, 3, \ldots \tag{R01Φ12.2}$$

and the β_{mv} eigenvalues are given by the positive roots of

$$J_v(\beta_{mv}) = 0 \quad \text{for the above } v \text{ values} \tag{R01Φ12.3}$$

R01Φ22 SECTOR OF RADIUS b; $0 \le \phi \le \phi_0 < 2\pi$; $G = 0$ AT $r = b$, $\partial G/\partial\phi = 0$ AT $\phi = 0$ AND $\phi = \phi_0$

$$G(r, \phi, t|r', \phi', \tau) = \frac{1}{b^2\phi_0} \sum_{m=1}^{\infty} \sum_{v}^{\infty} 4e^{-\beta_{m\nu}^2 \alpha(t-\tau)/b^2} \times \frac{J_v(\beta_{mv} r/b)\, J_v(\beta_{mv} r'/b) \cos(v\phi) \cos(v\phi')}{J_v'^2(\beta_{mv})} \tag{R01Φ22.1}$$

where

$$\nu = \frac{n\pi}{\phi_0} \quad n = 0, 1, 2, 3, \ldots \tag{R01Φ22.2}$$

and the β_{mv} eigenvalues are given by the positive roots of

$$J_v(\beta_{mv}) = 0 \quad \text{for the above } v \text{ values.} \tag{R01Φ22.3}$$

Also replace the 4 coefficient for $v = 0$ by the value of 2.

*R*02Φ00 SOLID CYLINDER WITH RADIAL AND ANGULAR DEPENDENCE; $\partial G/\partial r = 0$ AT $r = a$

$$G(r,\phi,t|r',\phi',\tau) = \frac{2}{a^2}\left\{\frac{1}{2\pi} + \sum_{n=0}^{\infty}\frac{1}{\pi}\cos\left[n(\phi-\phi')\right]\sum_{m=1}^{\infty} e^{-\beta_{mn}^2\alpha(t-\tau)/a^2} \times \frac{\beta_{mn}^2 J_n(\beta_{mn}r/a)\, J_n(\beta_{mn}r'/a)}{(\beta_{mn}^2 - n^2)\, J_n^2(\beta_{mn})}\right\} \quad \text{(R02}\Phi\text{00.1)}$$

where β_{mn} are the positive roots of $J_n'(\beta_{mn}) = 0$. (For $\beta = 0$, the r-direction equation from the separation of variables is $r^2R'' + rR' - n^2R = 0$ which has the solution $R = C_1r^{-n} + C_2r^n, n \neq 0$. The solution is $R = 0$. For $n = 0, R = C$.) Replace π inside the summation by 2π for $n = 0$ (Carslaw and Jaeger, 1959, p. 378, Equation 7).

*R*02Φ11 SECTOR OF RADIUS b; $\partial G/\partial r = 0$ AT $r = b$, AT $G = 0$ AT $\phi = 0$ AND ϕ_0

$$G(r,\phi,t|r',\phi',\tau) = \frac{4}{b^2\phi_0}\left\{\frac{1}{2} + \sum_{m=1}^{\infty}\sum_{v}^{\infty} e^{-\beta_{mv}^2\alpha(t-\tau)/b^2} \times \frac{\beta_{mv}^2 J_v(\beta_{mv}r/b)\, J_v(\beta_{mv}r'/b)\sin(v\phi)\sin(v\phi')}{(\beta_{mv}^2 - v^2)\, J_v^2(\beta_{mv})}\right\} \quad \text{(R02}\Phi\text{11.1)}$$

where

$$v = \frac{n\pi}{\phi_0} \qquad n = 1, 2, 3, \ldots \quad \text{(R02}\Phi\text{11.2)}$$

and the β_{mv} eigenvalues are given by the positive roots of

$$J_v'(\beta_{mv}) = 0 \quad \text{(R02}\Phi\text{11.3)}$$

*R*02Φ12 SECTOR OF RADIUS b; $\partial G/\partial r = 0$ AT $r = b$, $G = 0$ AT $\phi = 0$ AND $\partial G/\partial r = 0$ AT $\phi = \phi_0$

$$G(r,\phi,t|r',\phi',\tau) = \frac{4}{b^2\phi_0}\left[\frac{1}{2} + \sum_{m=1}^{\infty}\sum_{v}^{\infty} e^{-\beta_{mv}^2\alpha(t-\tau)/b^2} \times \frac{\beta_{mv}^2 J_v(\beta_{mv}r/b)\, J_v(\beta_{mv}\, r'/b)\sin(v\phi)\sin(v\phi')}{(\beta_{mv}^2 - v^2)\, J_v^2(\beta_{mv})}\right] \quad \text{(R02}\Phi\text{12.1)}$$

where

$$v = (2n-1)\frac{\pi}{2\phi_0} \qquad n = 1, 2, 3, \ldots \quad \text{(R02}\Phi\text{12.2)}$$

and the β_{mv} eigenvalues are given by the positive roots of

$$J_v'(\beta_{mv}) = 0 \qquad \text{(R02}\Phi\text{12.3)}$$

*R*02Φ22 SECTOR OF RADIUS *b*; ∂*G*/∂*r* = 0 AT *r* = *b*, ∂*G*/∂*r* = 0 AT φ = 0 AND ϕ_0

$$G(r, \phi, t|r', \phi', \tau) = \frac{1}{b^2\phi_0}\left[2 + \sum_{m=1}^{\infty}\sum_{v}^{\infty} 4e^{-\beta_{mv}^2\alpha(t-\tau)/b^2} \times \frac{\beta_{mv}^2 J_v(\beta_{mv}r/b)\, J_v(\beta_{mv}r'/b)\cos(v\phi)\cos(v\phi')}{(\beta_{mv}^2 - v^2)\, J_v^2(\beta_{mv})}\right] \qquad \text{(R02}\Phi\text{22.1)}$$

where

$$v = \frac{n\pi}{\phi_0} \qquad n = 0, 1, 2, \ldots \qquad \text{(R02}\Phi\text{22.2)}$$

and the β_{mv} eigenvalues are given by the positive roots of

$$J_v'(\beta_{mv}) = 0 \quad \text{for the above } v \text{ values} \qquad \text{(R02}\Phi\text{22.3)}$$

Also replace the 4 coefficient for $v = 0$ by the value of 2.

*R*11Φ00 ANNULUS WITH RADIAL AND ANGULAR DEPENDENCE; *G* = 0 AT *r* = *a* AND *b*

$$G(r, \phi, t|r', \phi', \tau) = \frac{1}{b^2}\sum_{m=1}^{\infty}\sum_{n=0}^{\infty}\frac{e^{-\beta_{mn}^2\alpha(t-\tau)/b^2}}{\pi N(\beta_{mn})} R_n(\beta_{mn}, r)\, R_n(\beta_{mn}, r') \times \cos[n(\phi - \phi')]$$

Replace π by 2π for $n = 0$. Also the following relations are given:

$$R_n(\beta_{mn}, r) = J_n(\beta_{mn}r/b)\, Y_n(\beta_{mn}) - J_n(\beta_{mn})\, Y_n(\beta_{mn}r/b)$$

$$\frac{1}{N(\beta_{mn})} = \frac{\pi^2}{2}\frac{\beta_{mn}^2 J_n^2(\beta_{mn}a/b)}{J_n^2(\beta_{mn}a/b) - J_n^2(\beta_{mn})}$$

and β_{mn}'s are the positive roots of

$$J_n(\beta_{mn}a/b)\, Y_n(\beta_{mn}) - J_n(\beta_{mn})\, Y_n(\beta_{mn}a/b) = 0$$

*R*12Φ00 ANNULUS WITH RADIAL AND ANGULAR DEPENDENCE; *G* = 0 AT *r* = *a* AND ∂*G*/∂*x* = 0 AT *r* = *b*

$$G(r,\phi,t|r',\phi',\tau) = \frac{1}{b^2}\sum_{m=1}^{\infty}\sum_{n=0}^{\infty}\frac{e^{-\beta_{mn}^2 a(t-\tau)/b^2}}{\pi N(\beta_{mn})} R_n(\beta_{mn},r)\,R_n(\beta_{mn},r') \times \cos\left[n(\phi-\phi')\right]$$

Replace π by 2π for $n = 0$. Also the following relations are given:

$$R_{mn}(\beta_{mn},r) = J_n(\beta_{mn}r/b)\,Y_n'(\beta_{mn}) - J_n'(\beta_{mn})\,Y_n(\beta_{mn}r/b)$$

$$\frac{1}{N(\beta_{mn})} = \frac{\pi^2}{2}\frac{\beta_{mn}^2 J_n^2(\beta_{mn}a/b)}{[1-(n/\beta_{mn})^2]\,J_n^2(\beta_{mn}a/b) - J_n'^2(\beta_{mn})}$$

and the β_{mn}'s are the positive roots of

$$J_n(\beta_{mn}a/b)\,Y_n'(\beta_{mn}) - J_n'(\beta_{mn})\,Y_n(\beta_{mn}a/b) = 0$$

for $m = 1, 2, \ldots,$ and $n = 0, 1, 2, \ldots$

*R*13Φ00 ANNULUS WITH RADIAL AND ANGULAR DEPENDENCE; *G* = 0 AT *r* = *a* AND *k*∂*G*/∂*r* + *hG* = 0 AT *r* = *b*

$$G(r,\phi,t|r',\phi',\tau) = \frac{1}{b^2}\sum_{m=1}^{\infty}\sum_{n=0}^{\infty}\frac{e^{-\beta_{mn}^2 \alpha(t-\tau)/b^2}}{\pi N(\beta_{mn})} R_n(\beta_{mn},r)\,R_n(\beta_{mn},r') \times \cos\left[n(\phi-\phi')\right]$$

Replace π by 2π for $n = 0$. Also the following relations are given:

$$R_{mn}(\beta_{mn},r) = S_{mn}J_n(\beta_{mn}r/b) - V_{mn}Y_n(\beta_{mn}r/b)$$

$$S_{mn} \equiv \beta_{mn}Y_n'(\beta_{mn}) + BY_n(\beta_{mn}) \quad B \equiv hb/k$$

$$V_{mn} \equiv \beta_{mn}J_n'(\beta_{mn}) + BJ_n(\beta_{mn})$$

$$\frac{1}{N(\beta_{mn})} = \frac{\pi^2}{2}\frac{\beta_{mn}^2 J_n^2(\beta_{mn}a/b)}{C_{mn}J_n^2(\beta_{mn}a/b) - V_{mn}^2}$$

$$C_{mn} \equiv B^2 + \beta_{mn}^2\left[1 - \left(\frac{n}{\beta_{mn}}\right)^2\right]$$

and the β_{mn}'s are the positive roots of

$$S_{mn}J_n(\beta_{mn}a/b) - V_{mn}Y_n(\beta_{mn}a/b) = 0$$

for $m = 1, 2, \ldots,$ and $n = 0, 1, 2, \ldots$.

$R21\Phi00$ ANNULUS WITH RADIAL AND ANGULAR DEPENDENCE; $\partial G/\partial r = 0$ AT $r = a$ AND $G = 0$ AT $r = b$. SAME AS $R12\Phi00$ WITH $a \to b$ AND $b \to a$

$R22\Phi00$ ANNULUS WITH RADIAL AND ANGULAR DEPENDENCE; $\partial G/\partial r = 0$ AT $r = a$ AND $r = b$

$$G(r, \phi, t|r', \phi', \tau) = \frac{1}{b^2}\left\{\frac{1}{\pi[1-(a/b)^2]} + \sum_{m=1}^{\infty}\sum_{n=0}^{\infty}\frac{e^{-\beta_{mn}^2\alpha(t-\tau)/b^2}}{\pi N(\beta_{mn})}\right.$$
$$\left.\times R_n(\beta_{mn}, r)\, R_n(\beta_{mn}, r')\cos[n(\phi-\phi')]\right\}$$

For $n = 0$, replace π inside the summation by 2π. Also the following relations are given:

$$R_n(\beta_{mn}, r) = J_n(\beta_{mn} r/b)\, Y_n'(\beta_{mn}) - J_n'(\beta_{mn})\, Y_n(\beta_{mn} r/b)$$
$$\frac{1}{N(\beta_{mn})} = \frac{\pi^2}{2}\frac{\beta_{mn}^2 J_n'^2(\beta_{mn} a/b)}{[1-(n/\beta_{mn})^2]\, J_n'^2(\beta_{mn} a/b) - \{1-[nb/(\beta_{mn} a)]^2\}\, J_n'^2(\beta_{mn})}$$

and the β_{mn}'s are the positive roots of

$$J_n'(\beta_{mn} a/b)\, Y_n'(\beta_{mn}) - J_n'(\beta_{mn})\, Y_n'(\beta_{mn} a/b) = 0$$

for $m = 1, 2, \ldots,$ and $n = 0, 1, 2, \ldots$.

$R23\Phi00$ ANNULUS WITH RADIAL AND ANGULAR DEPENDENCE; $\partial G/\partial r = 0$ AT $r = a$ AND $+k\partial G/\partial r + hG = 0$ AT $r = b$

$$G(r, \phi, t|r', \phi', \tau) = \frac{1}{b^2}\sum_{m=1}^{\infty}\sum_{n=0}^{\infty}\frac{e^{-\beta_{mn}^2\alpha(t-\tau)/b^2}}{\pi N(\beta_{mn})} R_n(\beta_{mn}, r)\, R_n(\beta_{mn}, r')$$
$$\times \cos[n(\phi-\phi')]$$

Replace π by 2π for $n = 0$. Also the following relations are given:

$$R_{mn}(\beta_{mn}, r) = S_{mn} J_m(\beta_{mn} r/b) - V_{mn} Y(\beta_{mn} r/b)$$

See $R13\Phi00$ for S_{mn} and V_{mn}.

$$\frac{1}{N(\beta_{mn})} = \frac{\pi^2}{2}\frac{\beta_{mn}^2 J_n'^2(\beta_{mn} a/b)}{B J_n'^2(\beta_{mn} a/b) - \{1-[nb/(\beta_{mn} a)]^2\} V_{mn}^2} \qquad B \equiv \frac{hb}{k}$$

and the β_{mn}'s are the positive roots of

$$S_{mn} J_n'(\beta_{mn} a/b) - V_{mn} Y_n'(\beta_{mn} a/b) = 0$$

for $m = 1, 2, \ldots,$ and $n = 0, 1, 2, \ldots$.

TABLE RΦ.1

IJ	$R_0(\beta_{00}, r/b)$	$R_v(\beta_{mv}, r/b)$, $m \neq 0$
01	0	$J_v(\beta_{mv} r/b)$
02	1	$J_v(\beta_{mv} r/b)$
03	0	$J_v(\beta_{mv} r/b)$
11	0	$J_v\left(\beta_{mv}\frac{r}{b}\right) Y_v(\beta_{mv}) - J_v(\beta_{mv}) Y_v\left(\beta_{mv}\frac{r}{b}\right)$
12	0	$J_v\left(\beta_{mv}\frac{r}{b}\right) Y_v'(\beta_{mv}) - J_v'(\beta_{mv}) Y_v\left(\beta_{mv}\frac{r}{b}\right)$
13	0	$S_{mv} J_v\left(\beta_{mv}\frac{r}{b}\right) - V_{mv} Y_v\left(\beta_{mv}\frac{r}{b}\right)$
21	0	Same as $R11$
22	1	Same as $R12$
23	0	Same as $R13$
31	0	Same as $R11$
32	0	Same as $R12$
33	0	Same as $R13$

*R*31Φ00 ANNULUS WITH RADIAL AND ANGULAR DEPENDENCE: $-k\partial G/\partial r + hG = 0$ AT $r = a$ AND $G = 0$ AT $r = b$. SAME AS *R*13Φ00 WITH $a \to b$, $b \to a$ AND $h \to -h$

*R*32Φ00 ANNULUS WITH RADIAL AND ANGULAR DEPENDENCE: $-k\partial G/\partial r + hG = 0$ AT $r = a$ AND $\partial G/\partial r = 0$ AT $r = b$. SAME AS *R*23Φ00 WITH $a \to b, b \to a$, AND $h \to -h$

*R*33Φ00 ANNULUS WITH RADIAL AND ANGULAR DEPENDENCE: $-k\partial G/\partial r + h_1 G = 0$ AT $r = a$ AND $k\partial G/\partial r + h_2 G = 0$ AT $r = b$ (SEE TABLES RΦ.1 THROUGH RΦ.4 FOR A SUMMARY OF THE RΦ CASES)

$$G(r, \phi, t|r', \phi', \tau) = \frac{1}{b^2}\sum_{m=1}^{\infty}\sum_{n=0}^{\infty}\frac{e^{-\beta_{mn}^2\alpha(t-\tau)/b^2}}{\pi N(\beta_{mn})} R_n(\beta_{mn}, r)\, R_n(\beta_{mn}, r') \times \cos[n(\phi - \phi')]$$

Replace π by 2π for $n = 0$. Also the following relations are given:

$$R_{mn}(\beta_{mn}, r) = S_{mn} J_n(\beta_{mn} r/b) - V_{mn} Y_n(\beta_{mn} r/b)$$

$$S_{mn} = \beta_{mn} Y_n'(\beta_{mn}) + B_2 Y_n(\beta_{mn}) \qquad B_2 = \frac{h_2 b}{k}$$

$$V_{mn} = \beta_{mn} J_n'(\beta_{mn}) + B_2 J_n(\beta_{mn})$$

$$\frac{1}{N(\beta_{mn})} = \frac{\pi^2}{2}\frac{\beta_{mn}^2 U_{mn}^2}{C_2 U_{mn}^2 - C_1 V_{mn}^2}$$

$$C_1 = B_1^2 + \beta_{mn}^2\left[1 - \left(\frac{n}{\beta_{mn} a/b}\right)^2\right] \qquad B_1 = h_1 b/k$$

TABLE RΦ.2

IJ	$1/N(\beta_{00})$	$1/N(\beta_{m\nu})$, $m \neq 0$
01	–	$2/[b^2 J_\nu'^2(\beta_{m\nu})]$
02	$2/b^2$	$2\beta_{m\nu}^2[b^2 J_\nu^2(\beta_{m\nu})(\beta_{m\nu}^2 - \nu^2)]^{-1}$
03	–	$2\beta_{m\nu}^2[b^2 J_\nu^2(\beta_{m\nu})(B_2^2 + \beta_{m\nu}^2 - \nu^2)]^{-1}$
11	–	$[\pi^2/(2b^2)][\beta_{m\nu}^2 J_\nu^2(\beta_{m\nu}a/b)][J_\nu^2(\beta_{m\nu}ab) - J_\nu^2(\beta_{m\nu})]^{-1}$
12	–	$[\pi^2/(2b^2)][\beta_{m\nu}^2 J_\nu^2(\beta_{m\nu}a/b)][A_{2\nu} J_\nu^2(\beta_{m\nu}a/b) - J_\nu'^2(\beta_{m\nu})]^{-1}$
13	–	$[\pi^2/(2b^2)][\beta_{m\nu}^2 J_\nu^2(\beta_{m\nu}a/b)][(B_2^2 + \beta_{m\nu}^2 A_{2\nu})J_\nu^2(\beta_{m\nu}a/b) - V_{m\nu}^2]^{-1}$
21	–	$[\pi^2/(2b^2)][\beta_{m\nu}^2 J_m'^2(\beta_{m\nu}a/b)][J_\nu'^2(\beta_{m\nu}a/b) - A_{1\nu} J_\nu^2(\beta_{m\nu})]^{-1}$
22	$2/(b^2 - a^2)$	$[\pi^2/(2b^2)][\beta_{m\nu}^2 J_\nu'^2(\beta_{m\nu}a/b)][A_{2\nu} J_\nu'^2(\beta_{m\nu}a/b) - A_{1\nu} J_\nu'^2(\beta_{m\nu})]^{-1}$
23	–	$[\pi^2/(2b^2)][\beta_{m\nu}^2 J_\nu'^2(\beta_{m\nu}a/b)][(B_2^2 + A_{2\nu}\beta_{m\nu}^2)J_\nu'^2(\beta_{m\nu}a/b) - A_{1\nu} V_{m\nu}^2]^{-1}$
31	–	$[\pi^2/(2b^2)][\beta_{m\nu}^2 U_{m\nu}^2][U_{m\nu}^2 - (B_1^2 + A_{1\nu}\beta_{m\nu}^2)J_\nu^2(\beta_{m\nu})]^{-1}$
32	–	$[\pi^2/(2b^2)][\beta_{m\nu}^2 U_{m\nu}^2][A_{2\nu} U_{m\nu}^2 - (B_1^2 + A_{1\nu}\beta_{m\nu}^2)J_\nu'^2(\beta_{m\nu})]^{-1}$
33	–	$[\pi^2/(2b^2)][\beta_{m\nu}^2 U_{m\nu}^2][(B_2^2 + A_{2\nu}\beta_{m\nu}^2)U_{m\nu}^2 - (B_1^2 + A_{1\nu}\beta_{m\nu}^2)V_{m\nu}^2]^{-1}$

TABLE RΦ.3

IJ	β_{00}	Eigencondition (Positive Roots of), $m \neq 0$
01	–	$J_\nu(\beta_{m\nu}) = 0$
02	0	$J_\nu'(\beta_{m\nu}) = 0$
03	–	$\beta_{m\nu} J_\nu'(\beta_{m\nu}) + B_2 J_\nu(\beta_{m\nu}) = 0$
11	–	$J_\nu(\beta_{m\nu}a/b) Y_\nu(\beta_{m\nu}) - J_\nu(\beta_{m\nu}) Y_\nu(\beta_{m\nu}a/b) = 0$
12	–	$J_\nu(\beta_{m\nu}a/b) Y_\nu'(\beta_{m\nu}) - J_\nu'(\beta_{m\nu}) Y_\nu(\beta_{m\nu}a/b) = 0$
13	–	$S_{m\nu} J_\nu(\beta_{m\nu}a/b) - V_{m\nu} Y_\nu(\beta_{m\nu}a/b) = 0$
21	–	$J_\nu'(\beta_{m\nu}a/b) Y_\nu(\beta_{m\nu}) - J_\nu(\beta_{m\nu}) Y_\nu'(\beta_{m\nu}a/b) = 0$
22	0	$J_\nu'(\beta_{m\nu}a/b) Y_\nu'(\beta_{m\nu}) - J_\nu'(\beta_{m\nu}) Y_\nu'(\beta_{m\nu}a/b) = 0$
23	–	$S_{m\nu} J_\nu'(\beta_{m\nu}a/b) - V_{m\nu} Y_\nu'(\beta_{m\nu}a/b) = 0$
31	–	$U_{m\nu} Y_\nu(\beta_{m\nu}) - W_{m\nu} J_\nu(\beta_{m\nu}) = 0$
32	–	$U_{m\nu} Y_\nu'(\beta_{m\nu}) - W_{m\nu} J_\nu'(\beta_{m\nu}) = 0$
33	–	$S_{m\nu} U_{m\nu} - V_{m\nu} W_{m\nu} = 0$

$$C_2 = B_2^2 + \beta_{mn}^2\left[1 - \left(\frac{n}{\beta_{mn}}\right)^2\right]$$

$$U_{mn} = \beta_{mn} J_n'(\beta_{mn}a/b) - B_1 J_n(\beta_{mn}a/b)$$

and the β_{mn}'s are the positive roots of

$$S_{mn}U_{mn} - V_{mn}W_{mn} = 0$$

TABLE RΦ.4

KL	$\Phi(\nu,\phi)$	$1/N(\nu)$	Eigencondition
11	$\sin v\phi$	$2/\phi_0$	$\sin v\phi_0 = 0$
12	$\sin v\phi$	$2/\phi_0$	$\cos v\phi_0 = 0$
13	$\cos v\phi$	$2/\phi_0$	$\cos v\phi_0 = 0$
22	$v = 0 : 1$	$v = 0 : 1/\phi_0$	$v = 0$
	$v \neq 0 : \cos(v\phi)$	$v \neq 0 : 2/\phi_0$	$v \neq 0 :\ \sin(v\phi_0) = 0$

for $m = 1, 2, \ldots,$ and $n = 0, 1, 2, \ldots.$

$$W_{mn} = \beta_{mn} Y_n'\left(\beta_{mn}\frac{a}{b}\right) - B_1 Y_n\left(\beta_{mn}\frac{a}{b}\right)$$

SUMMARY FOR CASES *RIJΦKL J* ≠ 0, *K*, *L* ≠ 0

$$G(r,\phi,t|r',\phi',\tau) = \sum_{m=0}^{\infty}\sum_{v}^{\infty} e^{-\beta_{mv}^2\alpha(t-\tau)/b^2} \times \frac{R_v(\beta_{mv}, r/b)\, R_v(\beta_{mv}, r'/b)}{N(\beta_{mv})} \frac{\Phi(v,\phi)\,\Phi(v,\phi')}{N(v)}$$

$$S_{mv} = \beta_{mv} Y_v'(\beta_{mv}) + B_2 Y_v(\beta_{mv})$$
$$V_{mv} = \beta_{mv} J_v'(\beta_{mv}) + B_2 J_v(\beta_{mv})$$

$$B_1 = h_1 b/k;\ B_2 = h_2 b/k;\ A_{1v} = 1 - \left(\frac{v}{\beta_{mn} a/b}\right)^2;\ A_{2v} = 1 - \left(\frac{v}{\beta_{mn}}\right)^2$$

$$V_{mv} = \beta_{mv} J_v'(\beta_{mv} a/b) - B_1 J_v(\beta_{mv} a/b);\ W_{mv} = \beta_{mv} Y_v'(\beta_{mv} a/b) - B_1 Y_v(\beta_m a/b)$$

where R_v, $N(\beta_{mn})$, β_{mn}, $N(v)$ and Φ are given in Tables RΦ.1 through RΦ.4.

REFERENCES

Carslaw, H. S. and Jaeger, J. C., 1959, *Conduction of Heat in Solids,* 2nd edn, Oxford University Press, New York.

Ozisik, M. N., 1993, *Heat Conduction*, John Wiley, New York.

Φ Cylindrical Polar Coordinate, ϕ Thin Shell Case

$$dv' = \delta a d\phi' \cdot ds' = \delta$$

Partial differential equation:

$$\frac{1}{a^2}\frac{\partial^2 T}{\partial \phi^2} = \frac{1}{\alpha}\frac{\partial T}{\partial t}$$

Φ00 COMPLETE CYLINDRICAL SHELL OF RADIUS a

$$G(\phi, t|\phi', \tau) = \frac{1}{\delta a}\sum_{m=0}^{\infty}\frac{1}{\pi}e^{-m^2\alpha(t-\tau)/a^2}\cos[m(\phi - \phi')]$$

where π is replaced by 2π for $m = 0$.

Φ11 PARTIAL CYLINDRICAL SHELL OF RADIUS a

$$G(\phi, t|\phi', \tau) = \frac{2}{a\phi_0\delta}\sum_{m=1}^{\infty} e^{-\frac{m^2\pi^2\alpha(t-\tau)}{\phi_0^2 a^2}}\sin\left(m\pi\frac{\phi}{\phi_0}\right)\sin\left(m\pi\frac{\phi'}{\phi_0}\right)$$

The Φ12, Φ13, ..., Φ33 cases are the same as the standard X12, X13, ..., X33 cases with L replaced by $a\phi_0$, x by $a\phi$ and x' by $a\phi'$. Also the G_{X--} expression is divided by δ.

RS Green's Functions for Radial Spherical Geometries

$$dv' = 4\pi r'^2\, dr',$$
$$ds' = 4\pi a^2 \text{ or } 4\pi b^2$$

STEADY GREEN'S FUNCTIONS

The spherical-radial steady heat conduction equation is

$$\frac{1}{r^2}\frac{d}{dr}\left(r^2\frac{dT}{dr}\right) = 0$$

The steady GF associated with this heat conduction equation, for boundary conditions of the first, second, and third kinds, are given in Table RS.1 for the infinite body and solid spheres, and in Table RS.2 for hollow spheres.

TRANSIENT GREEN'S FUNCTIONS

The partial differential equation for transient, radial-spherical heat conduction can be written as

$$\frac{1}{r}\frac{\partial^2(rT)}{\partial r^2} = \frac{1}{\alpha}\frac{\partial T}{\partial t} \quad \text{or} \quad \frac{1}{r^2}\frac{\partial}{\partial r}\left(r^2\frac{\partial T}{\partial r}\right) = \frac{1}{\alpha}\frac{\partial T}{\partial t}$$

*RS*00 INFINITE REGION WITH RADIAL SPHERICAL SYMMETRY

$$G(r, t|r', t) = \frac{1}{8\pi r r'[\pi\alpha(t-\tau)]^{1/2}}\left\{\exp\left[-\frac{(r-r')^2}{4\alpha(t-\tau)}\right] - \exp\left[-\frac{(r+r')^2}{4\alpha(t-\tau)}\right]\right\} \quad \text{(RS00.1)}$$

(Carslaw and Jaeger, 1959, p. 259, Equation 6).

*RS*01 SOLID SPHERE, $G = 0$ AT $r = b$

There are two expressions, one better for small times and one better for large times. For small times, the better expression to use is (see Carslaw and Jaeger, 1959, pp. 275 and 367)

TABLE RS.1

Steady Green's Functions, Radial Spherical Coordinates.

G Satisfies: $\dfrac{1}{r^2}\dfrac{d}{dr}\left(r^2\dfrac{dG}{dr}\right) = -\dfrac{\delta(r-r')}{4\pi r^2}$

Case	Boundary Conditions	$4\pi G(r\|r')$ (m^{-1})
Infinite Body		
*RS*00	$\partial G(0\|r')/\partial r = 0$	$1/r';\ r < r'$
	$G(r \to \infty, r') = 0$	$1/r;\ r > r'$
Infinite Body with Spherical Void		
*RS*10	$G(a\|r') = 0$	$(1 - a/r)/r';\ r < r'$
	$G(r \to \infty\|r') = 0$	$(1 - a/r')/r;\ r > r'$
*RS*20	$\partial G(a\|r')/\partial r = 0$	$1/r';\ r < r'$
	$G(r \to \infty\|r') = 0$	$1/r;\ r > r'$
*RS*30	$k\partial G(a\|r')/\partial r - h_1 G(a\|r') = 0$	$1/r' - B_1 a/((1+B_1)rr');\ r < r'$
	$G(r \to \infty\|r') = 0$	$1/r - B_1 a/((1+B_1)rr');\ r > r'$
		where $B_1 = h_1 a/k$
Solid Sphere of Radius b		
*RS*01	$\partial G(0\|r')/\partial r = 0$	$1/r' - 1/b;\ r < r'$
	$G(b, r') = 0$	$1/r - 1/b;\ r > r'$
*RS*02[a]	$\partial G(0\|r')/\partial r = 0$	$1/r' + [r^2 + (r')^2]/(2b^3);\ r < r'$
	$\partial G(b\|r')/\partial r = 0$	$1/r + [r^2 + (r')^2]/(2b^3);\ r > r'$
*RS*03	$\partial G(0\|r')/\partial r = 0$	$1/r' + (1/B_2 - 1)/b;\ r < r'$
	$k\partial G(b\|r')/\partial r + h_2 G(b\|r') = 0$	$1/r + (1/B_2 - 1)/b;\ r > r'$
		where $B_2 = h_2 b/k$

[a] Special temperature solution needed with this pseudo-GF.

$$G(r,\ t|r',\ \tau) = (4\pi rr')^{-1}\,[4\pi\alpha(t-\tau)]^{-1/2}\sum_{n=-\infty}^{\infty}\times\left\{\exp\left[-\frac{(2nb + r - r')^2}{4\alpha(t-\tau)}\right] - \exp\left[-\frac{(2nb + r + r')^2}{4\alpha(t-\tau)}\right]\right\} \tag{RS01.1}$$

$$-\left.\frac{\partial G}{\partial n'}\right|_{r'=b} = \frac{1}{rb}[4\pi\alpha(t-\tau)]^{-3/2}\sum_{n=-\infty}^{\infty}|(2n-1)b + r|\times\exp\left[-\frac{(2nb + r - b)^2}{4\alpha(t-\tau)}\right] \tag{RS01.2}$$

TABLE RS.2
Steady Green's Functions for the Hollow Sphere, Where $B_1 = h_1 a/k$, $B_2 = h_2 b/k$. G Satisfies: $\frac{1}{r^2}\frac{d}{dr}\left(r^2\frac{dG}{dr}\right) = -\frac{\delta(r-r')}{4\pi r^2}$; $a < r < b$

Case	Boundary Conditions	$4\pi G(r\|r')$ (m^{-1})
$RS11$	$G(a\|r') = 0$	$(b-r')(1-a/r)/[r'(b-a)]$; $r < r'$
	$G(b\|r') = 0$	$(b-r)(1-a/r')/[r(b-a)]$; $r > r'$
$RS12$	$G(a\|r') = 0$	$1/a - 1/r$; $r < r'$
	$\partial G(b\|r')/\partial r = 0$	$1/a - 1/r'$; $r > r'$
$RS13$	$G(a\|r') = 0$	$\left[B_2\left(\frac{a}{r} - \frac{ba}{rr'} - 1 + \frac{b}{r'}\right) - \frac{a}{r} + 1\right]$ $\div(B_2 b - B_2 a + a)$; $r < r'$
	$k\partial G(b\|r')/\partial r + h_2 G(b\|r') = 0$	$\left[B_2\left(\frac{a}{r'} - \frac{ba}{rr'} - 1 + \frac{b}{r}\right) - \frac{a}{r'} + 1\right]$ $\div(B_2 b - B_2 a + a)$; $r > r'$
$RS21$	$\partial G(a\|r')/\partial r = 0$	$1/r' - 1/b$; $r < r'$
	$G(b\|r') = 0$	$1/r - 1/b$; $r > r'$
$RS22$[a]	$\partial G(a\|r')/\partial r = 0$	$\left[\frac{r^2 + (r')^2}{2} + \frac{a^3}{r} + \frac{b^3}{r'}\right]/(b^3 - a^3)$; $r < r'$
	$\partial G(b\|r')/\partial r = 0$	$\left[\frac{r^2 + (r')^2}{2} + \frac{a^3}{r'} + \frac{b^3}{r}\right]/(b^3 - a^3)$; $r > r'$
$RS23$	$\partial G(a\|r')/\partial r = 0$	$1/r' + (1/B_2 - 1)/b$; $r < r'$
	$k\partial G(b\|r')/\partial r + h_2 G(b\|r') = 0$	$1/r + (1/B_2 - 1)/b$; $r > r'$
$RS31$	$k\partial G(a\|r')/\partial r - h_1 G(a\|r') = 0$	$\left[B_1\left(\frac{b}{r'} - \frac{ba}{rr'} - 1 + \frac{a}{r}\right) + \frac{b}{r'} - 1\right]$ $\div(B_1 b - B_1 a + b)$; $r < r'$
	$G(b\|r') = 0$	$\left[B_1\left(\frac{b}{r} - \frac{ba}{rr'} - 1 + \frac{a}{r'}\right) + \frac{b}{r'} - 1\right]$ $\div(B_1 b - B_1 a + b)$; $r > r'$
$RS32$	$k\partial G(a\|r')/\partial r - h_1 G(a\|r') = 0$	$-1/r + (1/B_1 + 1)/a$; $r < r'$
	$\partial G(b\|r')/\partial r = 0$	$-1/r' + (1/B_1 + 1)/a$; $r > r'$
$RS33$	$k\partial G(a\|r')/\partial r - h_1 G(a\|r') = 0$	$\left[B_1\left(1 - \frac{a}{r}\right) + B_1 B_2\left(\frac{a}{r} - \frac{ab}{rr'} + \frac{b}{r'} - 1\right) - B_2\left(1 - \frac{b}{r'}\right) + 1\right] \div D$; $r < r'$
	$k\partial G(b\|r')/\partial r + h_2 G(b\|r') = 0$	$\left[B_1\left(1 - \frac{a}{r'}\right) + B_1 B_2\left(\frac{a}{r'} - \frac{ab}{rr'} + \frac{b}{r} - 1\right) - B_2(1 - \frac{b}{r}) + 1\right] \div D$; $r > r'$
		where $D = [B_2 b + B_1 a + B_1 B_2(b - a)]$.

[a] Special temperature solution needed with this pseudo-GF.

For large times, the better expression is (Carslaw and Jaeger, 1959, pp. 233 and 366)

$$G(r,t|r',\tau) = \frac{1}{2\pi b r r'} \sum_{m=1}^{\infty} e^{-m^2\pi^2\alpha(t-\tau)/b^2} \times \sin\left(m\pi\frac{r}{b}\right) \sin\left(m\pi\frac{r'}{b}\right) \tag{RS01.3}$$

$$-\frac{\partial G}{\partial n'}\bigg|_{r'=b} = -\frac{1}{2b^3 r} \sum_{m=1}^{\infty} e^{-m^2\pi^2\alpha(t-\tau)/b^2} \times (-1)^m m \sin\left(m\pi\frac{r}{b}\right) \tag{RS01.4}$$

Note the similarity with the $X11$ case with $L_x \to b,\ x \to r,\ x' \to r'$ and the G_{X11} expression divided by $4\pi r r'$.

*RS*02 SOLID SPHERE, $\partial G/\partial r = 0$ AT $r = b$

$$G(r,\ t|r',\ \tau) = \frac{3}{4\pi b^3} + \frac{1}{2\pi b r r'} \times \sum_{m=1}^{\infty} e^{-\beta_m^2\alpha(t-\tau)/b^2} \frac{\beta_m^2 + 1}{\beta_m^2} \times \sin\left(\beta_m\frac{r}{b}\right) \sin\left(\beta_m\frac{r'}{b}\right) \tag{RS02.1}$$

The eigenvalues are found from the positive roots of

$$\beta_m \cot \beta_m = 1 \tag{RS02.2}$$

For the small times of $\alpha(t-\tau)/b^2 \le 0.022$

$$\begin{aligned} G(r,t|r',\tau) \approx{}& \frac{1}{4\pi r r'} \frac{1}{[4\pi\alpha(t-\tau)]^{1/2}} \left\{ \exp\left[-\frac{(r-r')^2}{4\alpha(t-\tau)}\right] \right. \\ & \left. - \exp\left[-\frac{(r+r')^2}{4\alpha(t-\tau)}\right] + \exp\left[-\frac{(2b-r-r')^2}{4\alpha(t-\tau)}\right] \right\} \\ & - \frac{B_2}{4\pi b r r'} \exp\left[B_2\frac{(2b-r-r')}{b} + B_2^2\frac{\alpha(t-\tau)}{b^2}\right] \\ & \times \operatorname{erfc}\left\{\frac{2b-r-r'}{[4\alpha(t-\tau)]^{1/2}} + \frac{B_2}{b}[\alpha(t-\tau)]^{1/2}\right\} \end{aligned} \tag{RS02.3}$$

$$B_2 = -1 \tag{RS02.4}$$

*RS*03 SOLID SPHERE, $k\partial G/\partial r + h_2 G = 0$ AT $r = b$

$$G(r, t|r', \tau) = \frac{1}{2\pi b r r'} \sum_{m=1}^{\infty} e^{-\beta_m^2 \alpha(t-\tau)/b^2} \frac{\beta_m^2 + B_2^2}{\beta_m^2 + B_2^2 + B_2} \times \sin\left(\beta_m \frac{r}{b}\right) \sin\left(\beta_m \frac{r'}{b}\right) \quad \text{(RS03.1)}$$

where

$$B_2 = \frac{h_2 b}{k} - 1 \quad \text{(RS03.2)}$$

and β_m, $m = 1, 2, \ldots$, are the positive roots of

$$\beta_m \cot \beta_m = -B_2 \quad \text{(RS03.3)}$$

Carslaw and Jaeger, 1959, p. 367). For small times $\alpha(t-\tau)/b^2 \leq 0.022$, $G(r, t|r', \tau)$ is approximated by Equation RS02.3 with B_2 defined by Equation RS03.2.

*RS*10 INFINITE REGION OUTSIDE THE SPHERICAL CAVITY, $r > a$; $G = 0$ AT $r = a$

$$G(r, t|r', \tau) = \frac{1}{4\pi r r' [4\pi\alpha(t-\tau)]^{1/2}} \times \left(e^{-(r-r')^2/[4\alpha(t-\tau)]} - e^{-(r+r'-2a)^2/[4\alpha(t-\tau)]}\right) \quad \text{(RS10.1)}$$

(Carslaw and Jaeger, 1959, p. 247)

$$-\left.\frac{\partial G}{\partial n'}\right|_{r'=a} = \frac{r-a}{ra[4\pi\alpha(t-\tau)]^{3/2}} e^{-(r-a)^2/[4\alpha(t-\tau)]} \quad \text{(RS10.2)}$$

*RS*11 HOLLOW SPHERE WITH $G = 0$ AT $r = a$ AND b

The better expression for small times is:

$$G(r, t|r', \tau) = \frac{1}{4\pi r r' [4\pi\alpha(t-\tau)]^{1/2}} \times \sum_{n=-\infty}^{\infty} \left\{\exp\left[-\frac{(2n(b-a)+r-r')^2}{4\alpha(t-\tau)}\right] - \exp\left[-\frac{(2n(b-a)+r+r'-2a)^2}{4\alpha(t-\tau)}\right]\right\} \quad \text{(RS11.1)}$$

$$-\left.\frac{\partial G}{\partial n'}\right|_{r'=a} = \frac{1}{ra[4\pi\alpha(t-\tau)]^{3/2}} \sum_{n=-\infty}^{\infty} |2n(b-a)+r-a| \times \exp\left[-\frac{(2n(b-a)+r-a)^2}{4\alpha(t-\tau)}\right] \tag{RS11.2}$$

$$-\left.\frac{\partial G}{\partial n'}\right|_{r'=b} = \frac{1}{rb[4\pi\alpha(t-\tau)]^{3/2}} \sum_{n=-\infty}^{\infty} |2n(b-a)+r-b| \times \exp\left[-\frac{(2n(b-a)+r-b)^2}{4\alpha(t-\tau)}\right] \tag{RS11.3}$$

The better expression for large times is

$$G(r,\, t|r',\, \tau) = \frac{1}{2\pi(b-a)rr'} \sum_{m=1}^{\infty} e^{-m^2\pi^2\alpha(t-\tau)/(b-a)^2} \times \sin\left(m\pi\frac{r-a}{b-a}\right) \sin\left(m\pi\frac{r'-a}{b-a}\right) \tag{RS11.4}$$

$$-\left.\frac{\partial G}{\partial n'}\right|_{r'=a} = \frac{1}{2(a-b)^2 ra} \sum_{m=1}^{\infty} e^{-m^2\pi^2\alpha(t-\tau)/(b-a)^2} \times m \sin\left(m\pi\frac{r-a}{b-a}\right) \tag{RS11.5}$$

$$-\left.\frac{\partial G}{\partial n'}\right|_{r'=b} = \frac{1}{2(b-a)^2 rb} \sum_{m=1}^{\infty} e^{-m^2\pi^2\alpha(t-\tau)/(b-a)^2} \times m \sin\left(m\pi\frac{r-a}{b-a}\right)(-1)^m \tag{RS11.6}$$

*RS*12 HOLLOW SPHERE WITH $G = 0$ AT $r = a$ AND $\partial G/\partial r = 0$ AT $r = b$

For large times, a convenient expression is (Shakir, 1982)

$$G(r,\, t|r',\, \tau) = \frac{1}{2\pi(b-a)rr'} \sum_{m=1}^{\infty} e^{-\beta_m^2\alpha(t-\tau)/(b-a)^2} \times \frac{\beta_m^2 + H_2^2}{\beta_m^2 + H_2^2 + H_2} \sin\left(\beta_m\frac{r-a}{b-a}\right) \sin\left(\beta_m\frac{r'-a}{b-a}\right) \tag{RS12.1}$$

where β_m are the positive roots of

$$\beta_m \cot\beta_m = -H_2 \qquad H_2 = B_2 R_2 \tag{RS12.2a, b}$$

$$B_2 = -1 \qquad R_2 = 1 - \frac{a}{b} \tag{RS12.3a, b}$$

$$-\left.\frac{\partial G}{\partial n'}\right|_{r'=a} = \frac{1}{2\pi(b-a)^2 ra}\sum_{m=1}^{\infty} e^{-\beta_m^2\alpha(t-\tau)/(b-a)^2}$$
$$\times \frac{\beta_m\left(\beta_m^2+H_2^2\right)}{\beta_m^2+H_2^2+H_2}\sin\left(\beta_m\frac{r-a}{b-a}\right) \qquad \text{(RS12.4)}$$

For small times, $\alpha(t-\tau)/(b-a)^2 \le 0.022$, $G(r,t|r',\tau)$ is efficiently given by

$$G(r,\ t|r',\ \tau) \approx \frac{1}{4\pi rr'}\frac{1}{[4\pi\alpha(t-\tau)]^{1/2}}\left\{\exp\left[-\frac{(r-r')^2}{4\alpha(t-\tau)}\right]\right.$$
$$\left. -\exp\left[-\frac{(r+r'-2a)^2}{4\alpha(t-\tau)}\right] + \exp\left[-\frac{(2b-r-r')^2}{4\alpha(t-\tau)}\right]\right\}$$
$$-\frac{B_2}{4\pi rr'b}\exp\left[B_2\frac{(2b-r'-r)}{b} + B_2^2\frac{\alpha(t-\tau)}{b^2}\right]$$
$$\times\ \text{erfc}\left\{\frac{2b-r-r'}{[4\alpha(t-\tau)]^{1/2}} + \frac{B_2}{b}[\alpha(t-\tau)]^{1/2}\right\} \qquad \text{(RS12.5)}$$

*RS*13 HOLLOW SPHERE WITH $G = 0$ AT $r = a$ AND $k\partial G/\partial r + h_2 G = 0$ AT $r = b$

For large times, the $G(r,t|r',\tau)$ relations are found using Equations RS12.1 through RS12.3 with

$$B_2 = \frac{h_2 b}{k} - 1 \qquad \text{(RS13.1)}$$

For small times, $G(r,\ t|r',\ \tau)$ is approximated by Equation RS12.5 with B_2 given by Equation RS13.1.

*RS*20 INFINITE REGION OUTSIDE A SPHERICAL CAVITY AT $r = a$ WITH $\partial G/\partial r = 0$ AT $r = a$

$$G(r,\ t|r',\ \tau) = \frac{1}{4\pi rr'[4\pi\alpha(t-\tau)]^{1/2}}$$
$$\times\left\{\exp\left[-\frac{(r-r')^2}{4\alpha(t-\tau)}\right] + \exp\left[-\frac{(r+r'-2a)^2}{4\alpha(t-\tau)}\right]\right\}$$
$$-\frac{B_1}{4\pi rr'a}\exp\left[B_1\frac{r+r'-2a}{a} + B_1^2\frac{\alpha(t-\tau)}{a^2}\right]$$
$$\times\ \text{erfc}\left\{\frac{r+r'-2a}{[4\alpha(t-\tau)]^{1/2}} + \frac{B_1}{a}[\alpha(t-\tau)]^{1/2}\right\} \qquad \text{(RS20.1)}$$

where B_1 is equal to 1. See $X30$ case for approximate values.

*RS*21 HOLLOW SPHERE WITH $\partial G/\partial r = 0$ AT $r = a$ AND $G = 0$ AT $r = b$

$$G(r,\ t|r',\ \tau) = \frac{1}{2\pi(b-a)rr'}\sum_{m=1}^{\infty} e^{-\beta_m^2\alpha(t-\tau)/(b-a)^2}$$
$$\times \frac{(\beta_m^2 + H_1^2)\sin[\beta_m(b-r)/(b-a)]\sin[\beta_m(b-r')/(b-a)]}{\beta_m^2 + H_1^2 + H_1} \quad \text{(RS21.1)}$$

where β_m are the positive roots of

$$\beta_m \cot\beta_m = -H_1 \quad H_1 = B_1 R_1 \quad \text{(RS21.2a, b)}$$
$$B_1 = 1 \quad R_1 = \frac{b}{a} - 1 \quad \text{(RS21.3a, b)}$$

A useful derivative is

$$-\left.\frac{\partial G}{\partial n'}\right|_{r'=b} = \frac{1}{2\pi(b-a)^2 br}\sum_{m=1}^{\infty} e^{-\beta_m^2\alpha(t-\tau)/(b-a)^2}$$
$$\times \frac{\beta_m(\beta_m^2 + H_1^2)\sin[\beta_m(b-r)/(b-a)]}{\beta_m^2 + H_1^2 + H_1} \quad \text{(RS21.4)}$$

For small times, $\alpha(t-\tau)/(b-a)^2 \le 0.022$, $G(\cdot)$ is approximated by

$$G(r,\ t|r',\ \tau) \approx \frac{1}{4\pi rr'[4\pi\alpha(t-\tau)]^{1/2}}\left\{\exp\left[-\frac{(r-r')^2}{4\alpha(t-\tau)}\right]\right.$$
$$\left. + \exp\left[-\frac{(r+r'-2a)^2}{4\alpha(t-\tau)}\right] - \exp\left[-\frac{(2b-r-r')^2}{4\alpha(t-\tau)}\right]\right\}$$
$$- \frac{B_1}{4\pi rr'a}\exp\left[B_1\frac{r+r'-2a}{a} + B_1^2\frac{\alpha(t-\tau)}{a^2}\right]$$
$$\times \operatorname{erfc}\left\{\frac{r+r'-2a}{[4\alpha(t-\tau)]^{1/2}} + \frac{B_1}{a}[\alpha(t-\tau)]^{1/2}\right\} \quad \text{(RS21.5)}$$

*RS*22 HOLLOW SPHERE WITH $\partial G/\partial r = 0$ AT $r = a$ AND b

For large times (Shakir, 1982)

$$G(r,\ t|r',\ \tau) = \frac{3B_0}{4\pi(b^3-a^3)} + \frac{1}{2\pi(b-a)rr'}\sum_{m=1}^{\infty} e^{-\beta_m^2\alpha(t-\tau)/(b-a)^2}$$
$$\times \{\beta_m\cos[\beta_m(r-a)/(b-a)] + H_1\sin[\beta_m(r-a)/(b-a)]\}$$
$$\times \frac{\{\beta_m\cos[\beta_m(r'-a)/(b-a)] + H_1\sin[\beta_m(r'-a)/(b-a)]\}}{(\beta_m^2 + H_1^2)[1 + H_2/(\beta_m^2 + H_2^2)] + H_1} \quad \text{(RS22.1)}$$

$$B_0 = 1 \qquad H_1 = B_1 R_1 \qquad H_2 = B_2 R_2 \qquad \text{(RS22.2a, b)}$$
$$B_1 = 1 \qquad B_2 = -1$$

$$R_1 = \frac{b}{a} - 1 \qquad R_2 = \frac{a}{b} - 1 \qquad \text{(RS22.3a, b, c, d)}$$

The eigenvalues β_m are the positive roots of

$$\tan \beta_m = \frac{\beta_m (H_1 + H_2)}{\beta_m^2 - H_1 H_2} \qquad \text{(RS22.4)}$$

For small times such that, $\alpha(t - \tau)/(b - a)^2 \leq 0.022$, $G(\cdot)$ is approximated by

$$\begin{aligned} G(r, t|r', \tau) \approx & \frac{1}{4\pi r r' [4\pi\alpha(t - \tau)]^{1/2}} \left\{ \exp\left[-\frac{(r - r')^2}{4\alpha(t - \tau)} \right] \right. \\ & \left. + \exp\left[-\frac{(r + r' - 2a)^2}{4\alpha(t - \tau)} \right] + \exp\left[-\frac{(2b - r - r')^2}{4\alpha(t - \tau)} \right] \right\} \\ & - \frac{B_1}{4\pi r r' a} \exp\left[B_1 \frac{r + r' - 2a}{a} + B_1^2 \frac{\alpha(t - \tau)}{a^2} \right] \\ & \times \operatorname{erfc}\left\{ \frac{r + r' - 2a}{[4\alpha(t - \tau)]^{1/2}} + \frac{B_1}{a} [\alpha(t - \tau)]^{1/2} \right\} \\ & - \frac{B_2}{4\pi r r' b} \exp\left[B_2 \frac{2b - r - r'}{b} + B_2^2 \frac{\alpha(t - \tau)}{b^2} \right] \\ & \times \operatorname{erfc}\left\{ \frac{(2b - r - r')}{[4\alpha(t - \tau)]^{1/2}} + \frac{B_2}{b} [\alpha(t - \tau)]^{1/2} \right\} \end{aligned} \qquad \text{(RS22.5)}$$

*RS*23 HOLLOW SPHERE WITH $\partial G/\partial r = 0$ AT $r = a$ AND $k\partial G/\partial r + h_2 G = 0$ AT $r = b$

For large times, $G(r, t|r', \tau)$ is found using Equations RS22.1 through RS22.4 with

$$B_0 = 0 \qquad B_1 = 1 \qquad B_2 = \frac{h_2 b}{k} - 1 \qquad \text{(RS23.1)}$$

For small times, $G(r, t|r', \tau)$ is given by Equation RS22.5.

*RS*30 INFINITE REGION OUTSIDE A SPHERICAL CAVITY ($r \geq a$) WITH $k\partial G/\partial r - h_1 G = 0$ AT $r = a$

$G(r, t|r', \tau)$ is given by Equation RS20.1 with B_1 given by

$$B_1 = \frac{h_1 a}{k} + 1 \qquad \text{(RS30.1)}$$

See *X*30 case for approximate values.

*RS*31 HOLLOW SPHERE WITH $-k\partial G/\partial r + h_1 G = 0$ AT $r = a$ AND $G = 0$ AT $r = b$

For large times, the $G(r, t|r', \tau)$ relations are found from using Equations RS21.1 through RS21.4 with

$$B_1 = \frac{ha}{k} + 1 \qquad \text{(RS31.1)}$$

For small times, $G(r, t|r', \tau)$ is found from Equation RS21.5.

*RS*32 HOLLOW SPHERE WITH $-k\partial G/\partial r + h_1 G = 0$ AT $r = a$ AND $\partial G/\partial r = 0$ AT $r = b$

For large times, the $G(r, t|r', \tau)$ relations are given by Equation RS22.1 through RS22.4 with

$$B_0 = 0 \qquad B_1 = \frac{h_1 a}{k} + 1 \qquad B_2 = -1 \qquad \text{(RS32.1a, b, c)}$$

For small times, $\alpha(t - \tau)/(b - a)^2 \leq 0.022$, $G(r, t|r', \tau)$ is approximated by Equation RS22.5.

*RS*33 HOLLOW SPHERE WITH $-k\partial G/\partial r + h_1 G = 0$ AT $r = a$ AND $k\partial G/\partial r + h_2 G = 0$ AT $r = b$

For large times, the $G(r, t|r', \tau)$ relations are given by Equations RS22.1 through RS22.4 with

$$B_0 = 1 \qquad B_1 = \left(\frac{h_1 a}{k} + 1\right)\frac{b}{a} \qquad B_2 = \frac{h_2 b}{k} - 1 \qquad \text{(RS33.1a, b, c)}$$

For small times, $\alpha(t - \tau)/(a - b)^2 \leq 0.022$, $G(r, t|r', \tau)$ is approximated by Equation RS22.5.

*RS*01Θ00 SOLID SPHERE WITH RADIAL AND AZIMUTHAL DEPENDENCE; $G = 0$ AT $r = b$

$$dv' = 2\pi r^2 dr' d\mu' \quad \mu' = \cos\theta' \quad -1 < \mu < 1$$
$$ds' = 2\pi b^2 d\mu'$$

$$G_{RS01\Theta00}(r, \theta, t|r', \theta', \tau)$$
$$= \frac{1}{2\pi(rr')^{1/2} b^2} \sum_{n=0}^{\infty}\sum_{p=1}^{\infty} e^{-\beta_{np}^2 \alpha(t-\tau)/b^2}$$
$$\times \frac{(2n+1)\, J_{n+1/2}(\beta_{np} r/b)\, J_{n+1/2}(\beta_{np} r'/b)\, P_n(\mu)\, P_n(\mu')}{[J'_{n+1/2}(\beta_{np})]^2} \qquad \text{(RS01Θ00.1)}$$

where the β_{np}'s are the positive roots of

$$J_{n+1/2}(\beta_{np}) = 0 \qquad \text{(RS01θ00.2)}$$

and $P_n(\mu)$ is the nth Legendre polynomial.

Note that

$$J_{1/2}(x) = \left(\frac{2}{\pi x}\right)^{1/2} \sin x; \quad J_{-1/2}(x) = \left(\frac{2}{\pi x}\right)^{1/2} \cos x \qquad \text{(RS01θ00.3a, b)}$$

$$J_{n+1/2}(x) = \frac{2n-1}{x} J_{n-1/2}(x) - J_{n-3/2}(x) \qquad \text{(RS01θ00.4)}$$

*RS*01θ01 HEMISPHERE WITH RADIAL AND AZIMUTHAL DEPENDENCE; $G = 0$ AT $r = b$ AND $G = 0$ AT $\mu = 0$ (OR $\theta = \pi/2$) ($0 < \mu < 1$)

$$G_{RS01\theta01}(r, \theta, t|r', \theta', \tau) = \frac{1}{\pi(rr')^{1/2}b^2} \sum_{n=1,3,\ldots}^{\infty} \sum_{p=1}^{\infty} \times e^{-\beta_{np}^2 \alpha(t-\tau)/b^2} \frac{2n+1}{-J_{n-1/2}(\beta_{np}) J_{n+3/2}(\beta_{np})} \times J_{n+1/2}\left(\beta_{np}\frac{r}{b}\right) J_{n+1/2}\left(\beta_{np}\frac{r'}{b}\right) \times P_n(\mu) P_n(\mu') \qquad \text{(RS01θ01.1)}$$

where the eigenvalues β_{np} are the positive roots of

$$J_{n+1/2}(\beta_{np}) = 0 \qquad \text{(RS01θ01.2)}$$

*RS*00Φ00θ00 INFINITE REGION WITH (r, ϕ, θ) DEPENDENCE (BUTKOVSKIY, P. 171)

$$G(r, \phi, \theta, t|r', 0, 0, \tau) = \frac{1}{[4\pi\alpha(t-\tau)]^{3/2}} \exp\left[-\frac{r^2 + r'^2 - 2rr' \cos\theta}{4\alpha(t-\tau)}\right]$$

*RS*00Φ00 INFINITE REGION WITH (r, ϕ) DEPENDENCE (BUTKOVSKIY, P. 140)

$$G(r, \phi, t|r', \phi', \tau) = \frac{1}{4\pi\alpha(t-\tau)} \exp\left[-\frac{r^2 - r'^2 + 2rr' \cos(\phi - \phi')}{4\alpha(t-\tau)}\right]$$

*RSIJ*Φ00, $J \neq 0$ FINITE REGION WITH (r, ϕ) DEPENDENCE

$$G(r, \phi, t|r', \phi', \tau) = \sum_{m=0}^{\infty} \sum_{n=0}^{\infty} e^{-\beta_{mn}^2 \alpha(t-\tau)/b^2} \times \frac{R_n(\beta_{mn}, r/b)\, R_n(\beta_{mn}, r'/b)}{\pi N(\beta_{mn})} \cos[n(\phi - \phi')]$$

Replace π by 2π for $n = 0$. In Appendix $R\Phi$, see Table RΦ.1 for $R_n(\beta_{mn}, r/b)$, Table RΦ.2 for $N(\beta_{mn})$ and Table RΦ.3 for eigenconditions.

REFERENCES

Butkovskiy, A. G., 1982, *Green's Functions and Transfer Functions Handbook*, Ellis Horwood, Chichester, U.K.

Carslaw, H. S. and Jaeger, J. C., 1959, *Conduction of Heat in Solids*, 2nd edn, Oxford University Press, New York.

Shakir, S., 1982, Personal communication, Department of Mechanical Engineering, Michigan State University, East Lansing, Michigan.

X Green's Functions: Rectangular Coordinates

STEADY GREEN'S FUNCTIONS

Two types of steady one-dimensional GF are given here, ordinary conduction and conduction with fin losses. The one-dimensional heat conduction equation is

$$\frac{d^2T}{dx^2} = 0$$

The GF associated with this heat conduction equation, for boundary conditions of the first, second, and third kinds, are given in Table X.1 for infinite and semi-infinite bodies, and in Table X.3 for slab bodies.

The one-dimensional fin equation is given by

$$\frac{d^2T}{dx^2} - m^2T = 0$$

where m^2 is constant. The GF associated with this equation are listed in Table X.2 for infinite and semi-infinite bodies and in Table X.4 for slab bodies. These GF also find use as kernel functions in two- and three-dimensional bodies in rectangular coordinates. For this reason these GF are given in the form of exponentials with negative exponents, which are computationally better behaved, for large arguments, than the hyperbolic trigonometric functions (i.e., cosh and sinh) that are given elsewhere.

TRANSIENT GREEN'S FUNCTIONS

The transient Green's functions (GFs) are for the equation

$$\frac{\partial^2 T}{\partial x^2} = \frac{1}{\alpha}\frac{\partial T}{\partial t}$$

where α = constant.

The solutions are arranged using a numbering system for the x-coordinate with an X being the first letter. The X is followed by two numbers, the first is for the $x = 0$ boundary and the second is for the $x = L$ boundary. If the boundary goes to infinity, the digit zero is used. See Chapter 2 for more information on the heat conduction numbering system.

Products of two or three one-dimensional GFs for boundary conditions of the zeroth, first, second, and third kinds can be used to get two- and three-dimensional GFs in rectangular coordinates. Hence, it is necessary to give the GFs for only one rectangular coordinate for these cases.

TABLE X.1

Steady Green's Function in the Infinite and Semi-Infinite Body, Satisfies

$$\frac{d^2G}{dx^2} = -\delta(x - x'); \quad 0 < x < \infty$$

Case	Boundary Conditions	$G(x\|x')$ (in meters)
X00[a]	$\frac{dG(x \to \pm\infty\|x')}{dx}$ is bounded	$-\frac{1}{2}\|x - x'\|$
X10	$G(0\|x') = 0$; $\frac{dG(\infty\|x')}{dx}$ is bounded	$-\frac{1}{2}\|x - x'\| + \frac{1}{2}\|x + x'\|$
X20[a]	$\frac{dG(0\|x')}{dx} = 0$; $\frac{dG(\infty\|x')}{dx}$ is bounded	$-\frac{1}{2}\|x - x'\| - \frac{1}{2}\|x + x'\|$
X30	$-\frac{dG(0\|x')}{dx} + hG(0\|x') = 0$ $\frac{dG(\infty\|x')}{dx}$ is bounded	$-\frac{1}{2}\|x - x'\| + \frac{1}{2}\|x + x'\| + \frac{k}{h}$

[a]Pseudo-GF; special temperature solution needed.

TABLE X.2

Steady Green's Function with Fin Term, (m^2 = Constant), for Infinite and Semi-Infinite Body, Satisfies

$$\frac{d^2G}{dx^2} - m^2G = -\delta(x - x'); \quad 0 < x < \infty$$

Case	Boundary Conditions	$G(x\|x')$ (in meters)
X00	$G(-\infty\|x') = 0$ $G(+\infty\|x') = 0$	$\frac{e^{-m\|x-x'\|}}{2m}$
X10	$G(0\|x') = 0$ $G(+\infty\|x') = 0$	$\frac{e^{-m\|x-x'\|} - e^{-m(x+x')}}{2m}$
X20	$dG(0\|x')/dx = 0$ $G(+\infty\|x') = 0$	$\frac{e^{-m\|x-x'\|} + e^{-m(x+x')}}{2m}$
X30	$-k\frac{dG}{dx}\Big\|_0 + h_1 G\|_0 = 0$ $G(+\infty\|x') = 0$	$\frac{(km + h_1)\, e^{-m\|x-x'\|} + (km - h_1)\, e^{-m(x+x')}}{2m\,(km + h_1)}$

TABLE X.3
Steady Green's Functions, One-Dimensional Rectangular Coordinates, Where $B_1 = h_1L/k$, $B_2 = h_2L/k$

$$G \text{ Satisfies: } \frac{d^2G}{dx^2} = -\delta(x - x'); \quad 0 < x < L$$

Case	Boundary Conditions	$G(x\|x')$ (in meters)
X11	$G(0\|x') = 0$ $G(L\|x') = 0$	$x(1 - x'/L); x < x'$ $x'(1 - x/L); x > x'$
X12	$G(0\|x') = 0$ $\frac{dG}{dx}(L\|x') = 0$	$x; x < x'$ $x'; x > x'$
X13	$G(0\|x') = 0$ $k\frac{dG}{dx}(L\|x') + h_2G(L\|x') = 0$	$x[1 - B_2(x'/L)/(1 + B_2)]; x < x'$ $x'[1 - B_2(x/L)/(1 + B_2)]; x > x'$
X21	$\frac{dG(0\|x')}{dx} = 0$ $G(L\|x') = 0$	$L - x'; x < x'$ $L - x; x > x'$
X22[a]	$\frac{dG(0\|x')}{dx} = 0; \frac{dG(L\|x')}{dx} = 0$	$[(x')^2 + x^2]/(2L) - x' + L/3; \quad x < x'$ $[x^2 + (x')^2]/(2L) - x + L/3; \quad x > x'$
X23	$\frac{dG(0\|x')}{dx} = 0$ $k\frac{dG(L\|x')}{dx} + h_2G(L\|x') = 0$	$L(1 + 1/B_2 - x'/L); x < x'$ $L(1 + 1/B_2 - x/L); x > x'$
X31[b]	$k\frac{dG(0\|x')}{dn} + h_1G(0\|x') = 0$ $G(L\|x') = 0$	$\frac{B_1x - B_1x'x/L + L - x'}{1 + B_1}; x < x'$ $\frac{B_1x' - B_1x'x/L + L - x}{1 + B_1}; x > x'$
X32[b]	$k\frac{dG(0\|x')}{dn} + h_1G(0\|x') = 0$ $\frac{dG(L\|x')}{dx} = 0$	$L(1/B_1 + x/L); x < x'$ $L(1/B_1 + x'/L); x > x'$
X33[b]	$k\frac{dG(0\|x')}{dn} + h_1G(0\|x') = 0$ $k\frac{dG(L\|x')}{dx} + h_2G(L\|x') = 0$	$(B_1B_2x + B_1x - B_1B_2xx'/L - B_2x' + B_2L + L)/(B_1B_2 + B_1 + B_2); x < x'$ $(B_1B_2x' + B_1x' - B_1B_2xx'/L - B_2x + B_2L + L)/(B_1B_2 + B_1 + B_2); x > x'$

[a] Special temperature solution needed with this pseudo-GF.
[b] Note $d/dn|_{x=0} = -d/dx|_{x=0}$; $d/dn|_{x=L} = +d/dx|_{x=L}$.

TABLE X.4

Steady Green's Function with Fin Term (m^2 = Constant), Satisfies[a]

$$\frac{d^2G}{dx^2} - m^2G = -\delta(x - x'); \quad 0 < x < L$$

Case	$G(x\|x')$ (in meters)
X11	$\dfrac{e^{-m(2L-\|x-x'\|)} - e^{-m(2L-x-x')}}{2m\left(1 - e^{-2mL}\right)} + \dfrac{e^{-m\|x-x'\|} - e^{-m(x+x')}}{2m\left(1 - e^{-2mL}\right)}$
X12	$\dfrac{-e^{-m(2L-\|x-x'\|)} + e^{-m(2L-x-x')}}{2m\left(1 + e^{-2mL}\right)} + \dfrac{e^{-m\|x-x'\|} - e^{-m(x+x')}}{2m\left(1 + e^{-2mL}\right)}$
X13	$\left\{\dfrac{(km - h_2)\left(-e^{-m(2L-\|x-x'\|)} + e^{-m(2L-x-x')}\right)}{2m\left(km + h_2 + (km - h_2)e^{-2mL}\right)} + \dfrac{(km + h_2)\left(e^{-m\|x-x'\|} - e^{-m(x+x')}\right)}{2m(km + h_2 + (km - h_2)e^{-2mL})}\right\}$
X21	$\dfrac{-e^{-m(2L-\|x-x'\|)} - e^{-m(2L-x-x')}}{2m\left(1 + e^{-2mL}\right)} + \dfrac{e^{-m\|x-x'\|} + e^{-m(x+x')}}{2m\left(1 + e^{-2mL}\right)}$
X22	$\dfrac{e^{-m(2L-\|x-x'\|)} + e^{-m(2L-x-x')}}{2m\left(1 - e^{-2mL}\right)} + \dfrac{e^{-m\|x-x'\|} + e^{-m(x+x')}}{2m\left(1 - e^{-2mL}\right)}$
X23	$\left\{\dfrac{(km - h_2)\left(e^{-m(2L-\|x-x'\|)} + e^{-m(2L-x-x')}\right)}{2m(km + h_2 - (km - h_2)e^{-2mL})} + \dfrac{(km + h_2)\left(e^{-m\|x-x'\|} + e^{-m(x+x')}\right)}{2m(km + h_2 - (km - h_2)e^{-2mL})}\right\}$
X31[b]	Let $x = L - \xi$ and $x' = L - \xi'$ in case X13.
X32[b]	Let $x = L - \xi$ and $x' = L - \xi'$ in case X23.
X33	$\left\{\dfrac{(km - h_2)\left((km - h_1)e^{-m(2L-\|x-x'\|)} + (km + h_1)e^{-m(2L-x-x')}\right)}{2m[(km + h_1)(km + h_2) - (km - h_1)(km - h_2)e^{-2mL}]} + \dfrac{(km + h_2)\left((km + h_1)e^{-m\|x-x'\|} + (km - h_1)e^{-m(x+x')}\right)}{2m[(km + h_1)(km + h_2) - (km - h_1)(km - h_2)e^{-2mL}]}\right\}$

[a] Boundary conditions are given in Table X.3.
[b] See also the GF Library internet site (http://www.greensfunction.unl.edu).

For a one-dimensional case, $dv' = dx'$.
For a two-dimensional case, $dv' = dx'dy'$.
For a three-dimensional case, $dv' = dx'dy'dz'$.

In most *finite-body* cases (XIJ, I, and J not equal to zero) two forms of the GFs are given, one best for small values of $\alpha(t-\tau)/L^2$, sometimes referred to as "small cotime," and one for large values of $\alpha(t-\tau)/L^2$ (large cotime). If an *infinite* summation is used in an expression, it is actually valid for all cotimes, both small and large.

*X*00 INFINITE REGION

$$\begin{aligned} G_{X00}(x,t|x',\tau) &= G_{X00}(x-x',t-\tau) \\ &= [4\pi\alpha(t-\tau)]^{-1/2}\exp\left[-\frac{(x-x')^2}{4\alpha(t-\tau)}\right] \end{aligned} \tag{X00.1}$$

Notice that

$$\int_{-\infty}^{\infty} G_{X00}(x,t|x',\tau)dx' = 1 \tag{X00.2a}$$

$$\frac{\partial G_{X00}(x-x',t-\tau)}{\partial x} = -\frac{\partial G_{X00}(x-x',t-\tau)}{\partial x'} \tag{X00.2b}$$

Notice that the integral over x' from a to b is

$$\begin{aligned} &[4\pi\alpha(t-\tau)]^{-1/2}\int_a^b \exp\left[-\frac{(x-x')^2}{4\alpha(t-\tau)}\right]dx' \\ &= \frac{1}{2}\left(\operatorname{erfc}\left\{\frac{x-b}{[4\alpha(t-\tau)]^{1/2}}\right\} - \operatorname{erfc}\left\{\frac{x-a}{[4\alpha(t-\tau)]^{1/2}}\right\}\right) \end{aligned} \tag{X00.3}$$

and thus

$$\int_0^{\infty} G_{X00}(x,t|x',\tau)dx' = 1 - \frac{1}{2}\operatorname{erfc}\frac{x}{[4\alpha(t-\tau)]^{1/2}} \tag{X00.4}$$

A relation involving differentiation and integration is

$$\frac{\partial}{\partial x}\int_a^b G_{X00}(x-x',t-\tau)dx' = G_{X00}(x-a,t-\tau) - G_{X00}(x-b,t-\tau) \tag{X00.5}$$

The average from $x = c$ to d for integration over x' from a to b is

$$\begin{aligned} &\frac{1}{d-c}\int_{x=c}^{d}\int_{x'=a}^{b} G_{X00}(x,t|x',\tau)dx'dx \\ &= \frac{[\alpha(t-\tau)]^{1/2}}{d-c}\left\{\operatorname{ierfc}\frac{c-b}{[4\alpha(t-\tau)]^{1/2}} - \operatorname{ierfc}\frac{d-b}{[4\alpha(t-\tau)]^{1/2}}\right. \\ &\quad \left. - \operatorname{ierfc}\frac{c-a}{[4\alpha(t-\tau)]^{1/2}} + \operatorname{ierfc}\frac{d-a}{[4\alpha(t-\tau)]^{1/2}}\right\} \end{aligned} \tag{X00.6}$$

The average over $a < x < b$ is

$$\overline{G}_{X00} = \frac{1}{b-a}\int_{x=a}^{b}\int_{x'=a}^{b} G_{X00}(x-x', t-\tau)dx'dx$$
$$= 1 - \frac{[4\alpha(t-\tau)]^{1/2}}{b-a}\left(\pi^{-1/2} - \text{ierfc}\left\{\frac{b-a}{[4\alpha(t-\tau)]^{1/2}}\right\}\right) \quad \text{(X00.7)}$$

For accurate, approximate expressions, see the $X20$ case.
Let $4\alpha(t-\tau)/(b-a)^2$ for the $X00$ case be equal to u in the $\overline{G}_{X20}$ approximations.
The integral of G_{X00} over τ is

$$\int_0^t G_{X00}(x-x', t-\tau)d\tau = \frac{(\alpha t)^{1/2}}{\alpha}\,\text{ierfc}\left[\frac{|x-x'|}{(4\alpha t)^{1/2}}\right] \quad \text{(X00.8a)}$$

An integral from t_1 to t_2 over τ and at $x = x'$ is

$$\int_{t_1}^{t_2} G_{X00}(x, t|x, \tau)d\tau$$
$$= (\pi\alpha)^{-1/2}[(t-t_1)^{1/2} - (t-t_2)^{1/2}] \qquad t_1 < t_2 \leq t \quad \text{(X00.8b)}$$

A general integral is

$$\int_0^t \tau^{n/2} G_{X00}(x-x', t-\tau)d\tau$$
$$= \Gamma\left(\frac{n}{2}+1\right)\frac{1}{2\alpha^{1/2}}(4t)^{(n+1)/2}\,\text{i}^{n+1}\,\text{erfc}\left[\frac{|x-x'|}{(4\alpha t)^{1/2}}\right] \quad \text{(X00.9)}$$

The integral over τ from 0 to t of $\partial G/\partial x$ is

$$\int_0^t \frac{\partial G_{X00}}{\partial x}(x, t|x', \tau)d\tau = -\int_0^t \frac{2(x-x')}{\pi^{1/2}[4\alpha(t-\tau)]^{3/2}} e^{-(x-x')^2/[4\alpha(t-\tau)]}d\tau$$
$$= -\,\text{sgn}\,(x-x')\frac{1}{2\alpha}\,\text{erfc}\left[\frac{|x-x'|}{(4\alpha t)^{1/2}}\right] \quad \text{(X00.10)}$$

where sgn $(x-x')$ means the sign of $(x-x')$. Note that

$$-k\int_0^t \frac{\partial G_{X00}(x, t|x', \tau)}{\partial x}d\tau\Bigg|_{x\to x'^-}^{x\to x'^+} = -k\left(-\frac{1}{2\alpha}\right)[1-(-1)] = \rho c \quad \text{(X00.11)}$$

*X*10 SEMI-INFINITE REGION WITH *G* = 0 AT *x* = 0

$$G_{X10}(x, t|x', \tau) = \frac{1}{[4\pi\alpha(t-\tau)]^{1/2}}$$
$$\times\left\{\exp\left[-\frac{(x-x')^2}{4\alpha(t-\tau)}\right] - \exp\left[-\frac{(x+x')^2}{4\alpha(t-\tau)}\right]\right\} \quad \text{(X10.1)}$$

$$-\frac{\partial G_{X10}}{\partial n'}\bigg|_{x'=0} = \frac{x}{\{4\pi[\alpha(t-\tau)]^3\}^{1/2}} \exp\left[-\frac{x^2}{4\alpha(t-\tau)}\right] \tag{X10.2}$$

$$\alpha \int_0^t \left(-\frac{\partial^2 G_{X10}}{\partial x\, \partial n'}\bigg|_{x'=0}\right) d\tau \bigg|_{x=0} = -(\pi\alpha t)^{-1/2} \tag{X10.3}$$

A relation between the $X00$ and $X10$ GFs is

$$\begin{aligned} G_{X10}(x,t|x',\tau) &= G_{X10}(x,x',t-\tau) \\ &= G_{X00}(x-x',t-\tau) - G_{X00}(x+x',t-\tau) \end{aligned} \tag{X10.4}$$

A relation between the $X00$, $X10$, and $X20$ GFs is

$$2G_{X00}(x-x',t-\tau) = G_{X10}(x,x',t-\tau) + G_{X20}(x,x',t-\tau) \tag{X10.5}$$

An integral from $x'=0$ to b gives

$$\begin{aligned} \int_0^b G_{X10}(x,t|x',\tau)dx' = \frac{1}{2}\Bigg(&\text{erfc}\left\{\frac{x-b}{[4\alpha(t-\tau)]^{1/2}}\right\} \\ &- 2\,\text{erfc}\left\{\frac{x}{[4\alpha(t-\tau)]^{1/2}}\right\} + \text{erfc}\left\{\frac{x+b}{[4\alpha(t-\tau)]^{1/2}}\right\}\Bigg) \end{aligned} \tag{X10.6}$$

and for $b \to \infty$,

$$\begin{aligned} \int_0^\infty G_{X10}(x,t|x',\tau)dx' &= 1 - \text{erfc}\left\{\frac{x}{[4\alpha(t-\tau)]^{1/2}}\right\} \\ &= \text{erf}\left\{\frac{x}{[4\alpha(t-\tau)]^{1/2}}\right\} \end{aligned} \tag{X10.7}$$

The average of the integral over x of the integral over x' is

$$\begin{aligned} \overline{G}_{X10}(t|\tau) &= \frac{1}{b}\int_0^b \int_0^b G_{X10}(x,t|x',\tau)dx'\,dx \\ &= 1 - \left[\frac{\alpha(t-\tau)}{b^2}\right]^{1/2}\Bigg(\frac{3}{\sqrt{\pi}} + \text{ierfc}\left\{\frac{b}{[\alpha(t-\tau)]^{1/2}}\right\} \\ &\quad - 4\,\text{ierfc}\left\{\frac{b}{[4\alpha(t-\tau)]^{1/2}}\right\}\Bigg) \end{aligned} \tag{X10.8}$$

For $\alpha(t-\tau)/b^2$ less than 0.0625, the error in $\overline{G}(t|\tau)$ is less than 0.05% using

$$\overline{G}_{X10}(t|\tau) \approx 1 - 3\left[\frac{\alpha(t-\tau)}{\pi b^2}\right]^{1/2} + \frac{8}{\pi^{1/2}}\left[\frac{\alpha(t-\tau)}{b^2}\right]^{3/2} e^{-b^2/[4\alpha(t-\tau)]} \tag{X10.9}$$

and for $\alpha(t-\tau)/b^2 > 1$, $\overline{G}(t|\tau)$ is within 0.1% using

$$\overline{G}_{X10}(t|\tau) \approx \frac{1}{\pi^{1/2}}\frac{1}{8}\left[\frac{b^2}{\alpha(t-\tau)}\right]^{3/2}\left\{1-\frac{1}{4}\frac{b^2}{\alpha(t-\tau)} + \frac{3}{64}\left[\frac{b^2}{\alpha(t-\tau)}\right]^2 - \frac{17}{2304}\left[\frac{b^2}{\alpha(t-\tau)}\right]^3\right\} \tag{X10.10}$$

X11 PLATE WITH $G = 0$ AT $x = 0$ AND L

Two expressions are available: one is computationally better for "small" $\alpha(t-\tau)/L^2$ values and the other for "large" values. See Figure X11.1 for plots of G_{X11}. The expression best for small cotimes (see Equation X11.10 for long-cotime expression) is

$$G_{X11}(x,t|x',\tau) = [4\pi\alpha(t-\tau)]^{-1/2}\sum_{n=-\infty}^{\infty}\times\left\{\exp\left[-\frac{(2nL+x-x')^2}{4\alpha(t-\tau)}\right]-\exp\left[-\frac{(2nL+x+x')^2}{4\alpha(t-\tau)}\right]\right\} \tag{X11.1}$$

(see Carslaw and Jaeger, 1959, p. 274). For $\alpha(t-\tau)/L^2 < 0.022$ use

$$G_{X11}(x,t|x',\tau) \approx [4\pi\alpha(t-\tau)]^{-1/2}\left\{\exp\left[-\frac{(x-x')^2}{4\alpha(t-\tau)}\right] - \exp\left[-\frac{(x+x')^2}{4\alpha(t-\tau)}\right] - \exp\left[-\frac{(2L-x-x')^2}{4\alpha(t-\tau)}\right]\right\} \tag{X11.2}$$

Some important derivatives are

$$-\left.\frac{\partial G_{X11}}{\partial n'}\right|_{x'=0} = \{4\pi[\alpha(t-\tau)]^3\}^{-1/2}\sum_{n=-\infty}^{\infty}(2nL+x)\times\exp\left[-\frac{(2nL+x)^2}{4\alpha(t-\tau)}\right] \tag{X11.3}$$

$$-\left.\frac{\partial G_{X11}}{\partial n'}\right|_{x'=L} = \{4\pi[\alpha(t-\tau)]^3\}^{-1/2}\sum_{n=-\infty}^{\infty}[(2n-1)L+x]\times\exp\left[-\frac{[(2n-1)L+x]^2}{4\alpha(t-\tau)}\right] \tag{X11.4}$$

$$-\left.\frac{\partial^2 G_{X11}}{\partial x\,\partial n'}\right|_{x'=x=0} = \{4\pi[\alpha(t-\tau)]^3\}^{-1/2}\sum_{n=-\infty}^{\infty}\left(1-\frac{2n^2L^2}{\alpha(t-\tau)}\right)\times e^{-n^2L^2/[\alpha(t-\tau)]}, \qquad t-\tau>0 \tag{X11.5}$$

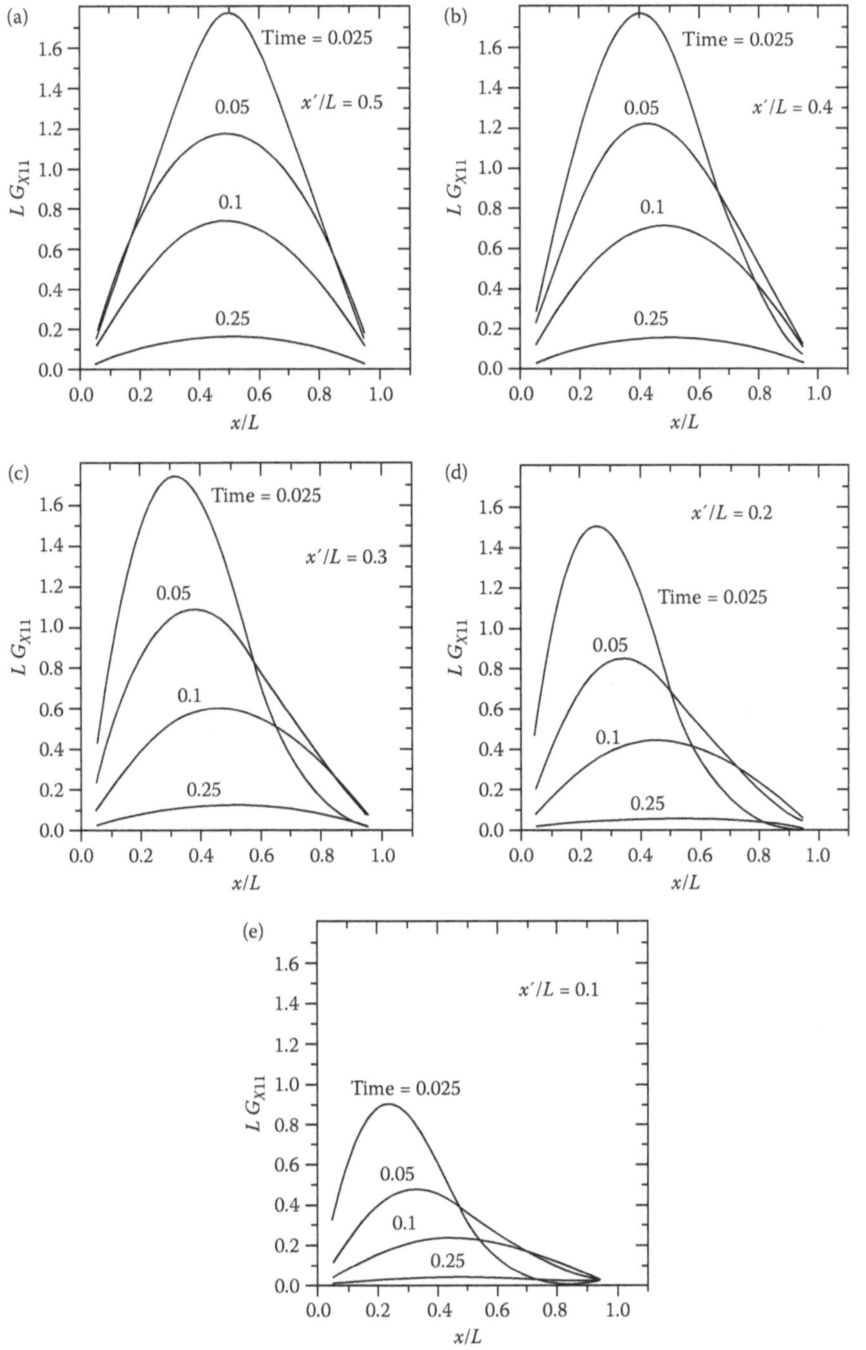

FIGURE X11.1 $LG_{X11}(x, t|x', T)$ versus x/L for $\alpha(t - r)/L^2 = 0.025$, 0.05, 0.10, and 0.25 and for five different heat-source locations x'/L.

For $\alpha(t-\tau)/L^2 < 0.05$, the approximations

$$-\left.\frac{\partial G_{X11}}{\partial n'}\right|_{x'=0} \approx x\{4\pi[\alpha(t-\tau)]^3\}^{-1/2}e^{-x^2/[4\alpha(t-\tau)]} \tag{X11.6}$$

$$-\left.\frac{\partial G_{X11}}{\partial n'}\right|_{x'=L} \approx (L-x)\{4\pi[\alpha(t-\tau)]^3\}^{-1/2}\exp\left[-\frac{(L-x)^2}{4\alpha(t-\tau)}\right] \tag{X11.7}$$

are quite accurate. For $\alpha(t-\tau)/L^2 < 0.2$, with errors less than $(1E\text{–}6)/L^3$, the mixed-derivative at $x = x' = 0$ is

$$-\left.\frac{\partial^2 G_{X11}}{\partial x\,\partial n'}\right|_{x'=x=0} \approx \{4\pi[\alpha(t-\tau)]^3\}^{-1/2} \times \left\{1+\left[2-\frac{4L^2}{\alpha(t-\tau)}\right]e^{-L^2/[\alpha(t-\tau)]}\right\},\quad t-\tau>0 \tag{X11.8}$$

and for $\alpha(t-\tau)/L^2 < 0.067$, the error is less than 0.002% using

$$-\left.\frac{\partial^2 G_{X11}}{\partial x\,\partial n'}\right|_{x'=x=0} \approx \left\{4\pi[\alpha(t-\tau)]^3\right\}^{-1/2},\qquad t-\tau>0 \tag{X11.9}$$

The expression best for large cotimes is

$$G_{X11}(x,t|x',\tau) = \frac{2}{L}\sum_{m=1}^{\infty} e^{-m^2\pi^2\alpha(t-\tau)/L^2}\sin\left(m\pi\frac{x}{L}\right)\sin\left(m\pi\frac{x'}{L}\right) \tag{X11.10}$$

For $\alpha(t-\tau)/L^2 > 0.1$, the errors are less than about $0.0003/L$ for the maximum m value of 2; for the maximum $m = 3$, the error is less than $(3E\text{–}7)/L$. For $\alpha(t-\tau)/L^2 > 0.05$ and the maximum $m = 5$, the error is less than $(4E\text{–}8)/L$. Some important derivatives are

$$-\left.\frac{\partial G_{X11}}{\partial n'}\right|_{x'=0} = \frac{2\pi}{L^2}\sum_{m=1}^{\infty} e^{-m^2\pi^2\alpha(t-\tau)/L^2}\, m\sin\left(m\pi\frac{x}{L}\right) \tag{X11.11}$$

$$-\left.\frac{\partial G_{X11}}{\partial n'}\right|_{x'=L} = -\frac{2\pi}{L^2}\sum_{m=1}^{\infty} e^{-m^2\pi^2\alpha(t-\tau)/L^2}\, m(-1)^m\sin\left(m\pi\frac{x}{L}\right) \tag{X11.12}$$

$$-\left.\frac{\partial^2 G_{X11}}{\partial x\,\partial n'}\right|_{x'=x=0} = -\frac{2\pi^2}{L^3}\sum_{m=1}^{\infty} m^2 e^{-m^2\pi^2\alpha(t-\tau)/L^2},\qquad t-\tau>0 \tag{X11.13}$$

For $\alpha(t-\tau)/L^2 > 0.2$, with errors less than 0.0002%, the mixed-derivative is

$$-\left.\frac{\partial^2 G_{X11}}{\partial x\,\partial n'}\right|_{x'=x=0} \approx \frac{2\pi^2}{L^3}\left[e^{-\pi^2\alpha(t-\tau)/L^2}+4e^{-4\pi^2\alpha(t-\tau)/L^2}\right],\qquad t-\tau>0 \tag{X11.14}$$

and for $\alpha(t-\tau)/L^2 > 0.067$, with errors less than 0.0004%, the maximum m needed is 4. An integral of $G_{X11}(\cdot)$ from $x' = 0$ to b for small cotimes is best given by

$$\int_0^b G_{X11}(x,t|x',\tau)\,dx' = \frac{1}{2}\sum_{n=-\infty}^{\infty}\left(\operatorname{erfc}\left\{\frac{2nL+x-b}{[4\alpha(t-\tau)]^{1/2}}\right\} - 2\operatorname{erfc}\left\{\frac{2nL+x}{[4\alpha(t-\tau)]^{1/2}}\right\} + \operatorname{erfc}\left\{\frac{2nL+x+b}{[4\alpha(t-\tau)]^{1/2}}\right\}\right) \tag{X11.15}$$

and for large cotimes by

$$\int_0^b G_{X11}(\cdot)dx' = \frac{2}{\pi}\sum_{m=1}^{\infty} e^{-m^2\pi^2\alpha(t-\tau)/L^2}\frac{1}{m}\sin\left(m\pi\frac{x}{L}\right) \times\left[1-\cos\left(m\pi\frac{b}{L}\right)\right] \tag{X11.16}$$

For $b = L$, the integral for small cotimes is

$$\int_0^L G_{X11}(x,t|x',\tau)dx'$$

$$= \frac{1}{2}\sum_{n=-\infty}^{\infty}\left(\operatorname{erfc}\left\{\frac{(2n-1)L+x}{[4\alpha(t-\tau)]^{1/2}}\right\} - 2\operatorname{erfc}\left\{\frac{2nL+x}{[4\alpha(t-\tau)]^{1/2}}\right\} + \operatorname{erfc}\left\{\frac{(2n+1)L+x}{[4\alpha(t-\tau)]^{1/2}}\right\}\right) \tag{X11.17a}$$

$$= \operatorname{erf}\left\{\frac{x}{[4\alpha(t-\tau)]^{1/2}}\right\} - \sum_{n=1}^{\infty}\left(\operatorname{erfc}\left\{\frac{2nL+x}{[4\alpha(t-\tau)]^{1/2}}\right\} + \operatorname{erfc}\left\{\frac{(2n-1)L-x}{[4\alpha(t-\tau)]^{1/2}}\right\} - \operatorname{erfc}\left\{\frac{2nL-x}{[4\alpha(t-\tau)]^{1/2}}\right\} - \operatorname{erfc}\left\{\frac{(2n-1)L+x}{[4\alpha(t-\tau)]^{1/2}}\right\}\right) \tag{X11.17b}$$

For large cotimes, use

$$\int_0^L G_{X11}(\cdot)dx' = \frac{4}{\pi}\sum_{m=1}^{\infty}\frac{\sin[(2m-1)\pi x/L]}{2m-1}e^{-(2m-1)^2\pi^2\alpha(t-\tau)/L^2} \tag{X11.18}$$

The average of the integral from $x' = 0$ to L is

$$\overline{G}_{X11}(t|\tau) \equiv \frac{1}{L}\int_{x=0}^{L}\int_{x'=0}^{L} G_{X11}(x,t|x',\tau)dx'dx$$
$$= 1 - 4\left[\frac{\alpha(t-\tau)}{L^2}\right]^{1/2}\left[\frac{1}{\pi^{1/2}} - 2\sum_{n=1}^{\infty}\left(\text{ierfc}\left\{\frac{(2n-1)L}{[4\alpha(t-\tau)]^{1/2}}\right\}\right.\right.$$
$$\left.\left. - \text{ierfc}\left\{\frac{2nL}{[4\alpha(t-\tau)]^{1/2}}\right\}\right)\right] \qquad \text{(X11.19a)}$$
$$= \frac{8}{\pi^2}\sum_{m=1}^{\infty}\frac{1}{(2m-1)^2}e^{-(2m-1)^2\pi^2\alpha(t-\tau)/L^2} \qquad \text{(X11.19b)}$$

where Equation X11.19a is best for small cotimes and Equation X11.19b is best for large cotimes. For $\alpha(t-\tau)/L^2 < 0.03$, an accurate expression is simply

$$\overline{G}_{X11}(t|\tau) \approx 1 - 4\left[\frac{\alpha(t-\tau)}{\pi L^2}\right]^{1/2} \qquad \text{(X11.19c)}$$

The error is less than 0.0016%. For $\alpha(t-\tau)/L^2 > 0.03$, only two terms in the large-cotime expression, Equation X11.19b, are needed for an error of less than 0.003%.

X12 PLATE WITH $G = 0$ AT $x = 0$ AND $\partial G/\partial x = 0$ AT L

Two expressions are available: one is more computationally efficient for small $\alpha(t-\tau)/L^2$ values and the other for large values.

Expression best for small cotimes

$$G_{X12}(x,t|x',\tau) = [4\pi\alpha(t-\tau)]^{-1/2}\sum_{n=-\infty}^{\infty}(-1)^n\left\{\exp\left[-\frac{(2nL+x-x')^2}{4\alpha(t-\tau)}\right]\right.$$
$$\left. - \exp\left[-\frac{(2nL+x+x')^2}{4\alpha(t-\tau)}\right]\right\} \qquad \text{(X12.1)}$$

For $\alpha(t-\tau)/L^2 < 0.2$ and for a maximum $n = 2$, the errors are less than $(2\text{E}{-}14)/L$. For $\alpha(t-\tau)/L^2 < 0.022$, use

$$G_{X12}(x,t|x',\tau) \approx [4\pi\alpha(t-\tau)]^{-1/2}\left\{\exp\left[-\frac{(x-x')^2}{4\alpha(t-\tau)}\right]\right.$$
$$\left. - \exp\left[-\frac{(x+x')^2}{4\alpha(t-\tau)}\right] + \exp\left[-\frac{(2L-x-x')^2}{4\alpha(t-\tau)}\right]\right\} \qquad \text{(X12.2)}$$

For $\alpha(t-\tau)/L^2 < 0.2$, with errors less than (3E–9) / L:

$$G_{X12}(L,t|L,\tau) \approx [\pi\alpha(t-\tau)]^{-1/2}\left(1 - 2e^{-L^2/[\alpha(t-\tau)]}\right) \quad \text{(X12.3)}$$

An expression for $-\partial G/\partial n'|_{x'=0}$ is

$$-\left.\frac{\partial G_{X12}}{\partial n'}\right|_{x'=0} = \left\{4\pi[\alpha(t-\tau)]^3\right\}^{-1/2} \sum_{n=-\infty}^{\infty} (-1)^n (2nL+x) \times \exp\left[-\frac{(2nL+x)^2}{4\alpha(t-\tau)}\right] \quad \text{(X12.4)}$$

For $\alpha(t-\tau)/L^2 < 0.022$, use

$$-\left.\frac{\partial G_{X12}}{\partial n'}\right|_{x'=0} \approx \frac{x}{\{4\pi[\alpha(t-\tau)]^3\}^{1/2}} e^{-x^2/[4\alpha(t-\tau)]} \quad \text{(X12.5)}$$

The mixed-derivative evaluated at $x = x' = 0$ is

$$-\left.\frac{\partial^2 G_{X12}}{\partial x\,\partial n'}\right|_{x=x'=0} = \left\{4\pi[\alpha(t-\tau)]^3\right\}^{-1/2} \sum_{n=-\infty}^{\infty} (-1)^n \times \left[1 - \frac{2n^2L^2}{\alpha(t-\tau)}\right] e^{-n^2L^2/[\alpha(t-\tau)]} \quad \text{(X12.6)}$$

For $\alpha(t-\tau)/L^2 < 0.2$, with errors less than (5E–7)/L^3,

$$-\left.\frac{\partial^2 G_{X12}}{\partial x\,\partial n'}\right|_{x=x'=0} \approx \{4\pi[\alpha(t-\tau)]^3\}^{-1/2} \times \left\{1 - 2\left[1 - \frac{2L^2}{\alpha(t-\tau)}\right] e^{-L^2/[\alpha(t-\tau)]}\right\} \quad \text{(X12.7)}$$

Expression best for large cotimes:

$$G_{X12}(x,t|x',\tau) = \frac{2}{L}\sum_{m=1}^{\infty} e^{-\beta_m^2\alpha(t-\tau)/L^2} \sin\left(\beta_m\frac{x}{L}\right)\sin\left(\beta_m\frac{x'}{L}\right) \quad \text{(X12.8)}$$

where $\beta_m = (2m-1)(\pi/2)$, $m = 1, 2, \ldots$.

For $\alpha(t-\tau)/L^2 > 0.2$ and for a maximum $m = 2$, the errors are less than (9E–6)/L. For $\alpha(t-\tau)/L^2 > 0.2$, with errors less than (5E–6) / L, at $x = x' = L$, G is

$$G_{X12}(L,t|L,\tau) \approx \frac{2}{L}\left(e^{-\pi^2\alpha(t-\tau)/(4L^2)} + e^{-9\pi^2\alpha(t-\tau)/(4L^2)}\right) \quad \text{(X12.9)}$$

An expression for $-\partial G/\partial n'|_{x'=0}$ is

$$-\left.\frac{\partial G_{X12}}{\partial n'}\right|_{x'=0} = \frac{2}{L^2}\sum_{m=1}^{\infty} e^{-\beta_m^2\alpha(t-\tau)/L^2}\beta_m \sin\left(\beta_m\frac{x}{L}\right) \quad \text{(X12.10)}$$

$$-\left.\frac{\partial^2 G_{X12}}{\partial x\,\partial n'}\right|_{x'=x=0} = \frac{2}{L^3}\sum_{m=1}^{\infty} e^{-\beta_m^2\alpha(t-\tau)/L^2}\beta_m^2 \quad \text{(X12.11)}$$

For $\alpha(t-\tau)/L^2 > 0.2$, with errors less than $0.0006/L^3$ (i.e., 0.02%) a mixed-derivative is

$$-\left.\frac{\partial^2 G_{X12}}{\partial x\, \partial n'}\right|_{x'=x=0} \approx \frac{\pi^2}{2L^3}\left(e^{-\pi^2\alpha(t-\tau)/(4L^2)} + 9e^{-9\pi^2\alpha(t-\tau)/(4L^2)}\right) \qquad \text{(X12.12)}$$

X13 PLATE WITH $G = 0$ AT $x = 0$ AND $k\partial G/\partial x + hG = 0$ AT $x = L$

For small values of $\alpha(t-\tau)/L^2$ (≤ 0.022) use

$$\begin{aligned} G_{X13}(x,t|x',\tau) \approx{}& [4\pi\alpha(t-\tau)]^{-1/2}\left\{\exp\left[-\frac{(x-x')^2}{4\alpha(t-\tau)}\right]\right. \\ &\left. - \exp\left[-\frac{(x+x')^2}{4\alpha(t-\tau)}\right] + \exp\left[-\frac{(2L-x-x')^2}{4\alpha(t-\tau)}\right]\right\} \\ &- \frac{h}{k}\exp\left[\frac{h(2L-x-x')}{k} + \frac{h^2\alpha(t-\tau)}{k^2}\right] \\ &\times \operatorname{erfc}\left\{\frac{2L-x-x'}{[4\alpha(t-\tau)]^{1/2}} + \frac{h}{k}[\alpha(t-\tau)]^{1/2}\right\} \end{aligned} \qquad \text{(X13.1)}$$

Also, for small $\alpha(t-\tau)/L^2$ values, use

$$-\left.\frac{\partial G_{X13}}{\partial n'}\right|_{x'=0} \approx \frac{x}{\{4\pi[\alpha(t-\tau)]^3\}^{1/2}}\exp\left[-\frac{x^2}{4\alpha(t-\tau)}\right] \qquad \text{(X13.2)}$$

$$-\left.\frac{\partial^2 G_{X13}}{\partial x\, \partial n'}\right|_{x=x'=0} \approx \{4\pi[\alpha(t-\tau)]^3\}^{-1/2} \qquad \text{(X13.3)}$$

For any time, but best for large $\alpha(t-\tau)/L^2$ values, use

$$\begin{aligned} G_{X13}(x,t|x',\tau) ={}& \frac{2}{L}\sum_{m=1}^{\infty} e^{-\beta_m^2\alpha(t-\tau)/L^2} \\ &\times \frac{(\beta_m^2 + B^2)\sin(\beta_m x/L)\sin(\beta_m x'/L)}{\beta_m^2 + B^2 + B} \end{aligned} \qquad \text{(X13.4)}$$

Eigencondition:

$$\beta_m \cot \beta_m = -B \qquad B \equiv \frac{hL}{k} \tag{X13.5a, b}$$

$$G_{X13}(L, t|L, \tau) = \frac{2}{L} \sum_{m=1}^{\infty} e^{-\beta_m^2 \alpha(t-\tau)/L^2} \frac{\beta_m^2}{\beta_m^2 + B^2 + B} \tag{X13.6}$$

$$-\left.\frac{\partial G_{X13}}{\partial n'}\right|_{x'=0} = \frac{2}{L^2} \sum_{m=1}^{\infty} e^{-\beta_m^2 \alpha(t-\tau)/L^2} \frac{\beta_m(\beta_m^2 + B^2)\sin(\beta_m x/L)}{\beta_m^2 + B^2 + B} \tag{X13.7}$$

*X*14 PLATE WITH $G = 0$ AT $x = 0$ AND $k\partial G/\partial x + (\rho c b)_2\, \partial G/\partial t = 0$ AT $x = L$

For small values of $\alpha(t-\tau)/L^2(\leq 0.022)$, use

$$\begin{aligned} G_{X14}(x, t|x', \tau) \approx [4\pi\alpha(t-\tau)]^{-1/2} \Bigg\{ &\exp\left[-\frac{(x-x')^2}{4\alpha(t-\tau)}\right] \\ &- \exp\left[-\frac{(x+x')^2}{4\alpha(t-\tau)}\right] - \exp\left[-\frac{(2L-x-x')^2}{4\alpha(t-\tau)}\right]\Bigg\} \\ &+ \frac{1}{LC_2} \exp\left[\frac{1}{C_2}\frac{2L-x-x'}{L} + \frac{1}{C_2^2}\frac{\alpha(t-\tau)}{L^2}\right] \\ &\times \operatorname{erfc}\left\{\frac{2L-x-x'}{[4\alpha(t-\tau)]^{1/2}} + \frac{1}{C_2}\frac{[\alpha(t-\tau)]^{1/2}}{L}\right\} \end{aligned} \tag{X14.1}$$

Also, for small $\alpha(t-\tau)/L^2$ values,

$$-\left.\frac{\partial G_{X14}}{\partial n'}\right|_{x'=0} \approx \frac{x}{\{4\pi[\alpha(t-\tau)]^3\}^{1/2}} \exp\left[-\frac{x^2}{4\alpha(t-\tau)}\right] \tag{X14.2}$$

$$-\left.\frac{\partial^2 G_{X14}}{\partial x\, \partial n'}\right|_{x=x'=0} \approx \left\{4\pi[\alpha(t-\tau)]^3\right\}^{-1/2} \tag{X14.3}$$

For any time, but best for large $\alpha(t-\tau)/L^2$ values, use

$$\begin{aligned} G_{X14}(x, t|x', \tau) = \frac{2}{L} \sum_{m=1}^{\infty} &e^{-\beta_m^2 \alpha(t-\tau)/L^2} \\ &\times \frac{(C_2^2\beta_m^2 + 1)\sin(\beta_m x/L)\sin(\beta_m x'/L)}{C_2^2\beta_m^2 + C_2 + 1} \end{aligned} \tag{X14.4}$$

$$\text{Eigencondition:} \quad \beta_m \tan \beta_m = \frac{1}{C_2} \qquad \beta_m > 0 \qquad m = 1, 2, \ldots \tag{X14.5}$$

$$C_2 \equiv \frac{(\rho c b)_2}{\rho c L} \tag{X14.6}$$

X15 PLATE WITH $G = 0$ AT $x = 0$ AND $k\partial G/\partial x + h_2 G + (\rho cb)_2\, \partial G/\partial t = 0$ AT $x = L$

For small values of $\alpha(t-\tau)/L^2 (\le 0.022)$, use

$$G_{X15}(x,t|x',\tau) \approx [4\pi\alpha(t-\tau)]^{-1/2}\left\{\exp\left[-\frac{(x-x')^2}{4\alpha(t-\tau)}\right] - \exp\left[-\frac{(x+x')^2}{4\alpha(t-\tau)}\right] - \exp\left[-\frac{(2L-x-x')^2}{4\alpha(t-\tau)}\right]\right\}$$
$$+ \frac{1}{L}\frac{1}{C_2(S_4-S_3)}\left\{\exp\left[\frac{1}{S_4}\frac{2L-x-x'}{L} + \frac{1}{S_4^2}\frac{\alpha(t-\tau)}{L^2}\right] \times \operatorname{erfc}\left[\frac{2L-x-x'}{[4\alpha(t-\tau)]^{1/2}} + \frac{1}{S_4}\frac{[\alpha(t-\tau)]^{1/2}}{L}\right]\right.$$
$$\left. - \exp\left[\frac{1}{S_3}\frac{2L-x-x'}{L} + \frac{1}{S_3^2}\frac{\alpha(t-\tau)}{L^2}\right] \times \operatorname{erfc}\left[\frac{2L-x-x'}{[4\alpha(t-\tau)]^{1/2}} + \frac{1}{S_3}\frac{[\alpha(t-\tau)]^{1/2}}{L}\right]\right\} \quad \text{(X15.1)}$$

for $C_2 < 1/4\, B_2$ and

$$S_3 = \frac{1}{2C_2}\left[1 - (1 - 4\,B_2 C_2)^{1/2}\right] \quad \text{(X15.2a)}$$

$$S_4 = \frac{1}{2C_2}\left[1 + (1 - 4\,B_2 C_2)^{1/2}\right] \quad \text{(X15.2b)}$$

Also, for small $\alpha(t-\tau)/L^2$ values,

$$-\left.\frac{\partial G_{X15}}{\partial n'}\right|_{x'=0} \approx \frac{x}{\{4\pi[\alpha(t-\tau)]^3\}^{1/2}}\exp\left[-\frac{x^2}{4\alpha(t-\tau)}\right] \quad \text{(X15.3)}$$

$$-\left.\frac{\partial^2 G_{X15}}{\partial x\,\partial n'}\right|_{x=x'=0} \approx \left\{4\pi[\alpha(t-\tau)]^3\right\}^{-1/2} \quad \text{(X15.4)}$$

For any time, but best for large $\alpha(t-\tau)/L^2$ values, use

$$G_{X15}(x,t|x',\tau) = \sum_{m=1}^{\infty} e^{-\beta_m^2\alpha(t-\tau)/L^2}\frac{\sin(\beta_m x/L)\sin(\beta_m x'/L)}{N_m}$$

where

$$N_m = \frac{L}{2}\frac{\left(B_2 - C_2\beta_m^2\right)^2 + \beta_m^2 + B_2 + C_2\beta_m^2}{\left(B_2 - C_2\beta_m^2\right)^2 + \beta_m^2} \quad \text{(X15.5)}$$

Eigencondition: $\left(B_2 - C_2\beta_m^2\right)\tan\beta_m = -\beta_m$ (X15.6)

$$\beta_m > 0 \qquad m = 1, 2, \ldots$$

$$C_2 \equiv \frac{(\rho c b)_2}{\rho c L} \qquad B_2 = \frac{h_2 L}{k} \tag{X15.7a, b}$$

X20 SEMI-INFINITE BODY WITH $\partial G/\partial x = 0$ AT $x = 0$

$$G_{X20}(x,t|x',\tau) = \frac{1}{[4\pi\alpha(t-\tau)]^{1/2}}\left\{\exp\left[-\frac{(x-x')^2}{4\alpha(t-\tau)}\right] + \exp\left[-\frac{(x+x')^2}{4\alpha(t-\tau)}\right]\right\} \tag{X20.1}$$

$$G_{X20}(0,t|0,\tau) = [\pi\alpha(t-\tau)]^{-1/2} \tag{X20.2}$$

See Figures X20.1 and X20.2. A relation between the X00 and X20 GFs is

$$G_{X20}(x,t|x',\tau) = G_{X20}(x,x',t-\tau) = G_{X00}(x-x',t-\tau) + G_{X00}(x+x',t-\tau) \tag{X20.3}$$

An integral from $x' = 0$ to ∞ is

$$\int_{x'=0}^{\infty} G_{X20}(x,t|x',\tau)dx' = 1 \tag{X20.4}$$

The integral of G_{X20} from $x' = 0$ to b is

$$\int_0^b G_{X20}(x,t|x',\tau)dx' = \frac{1}{2}\left(\operatorname{erfc}\left\{\frac{x-b}{[4\alpha(t-\tau)]^{1/2}}\right\} - \operatorname{erfc}\left\{\frac{x+b}{[4\alpha(t-\tau)]^{1/2}}\right\}\right) \tag{X20.5}$$

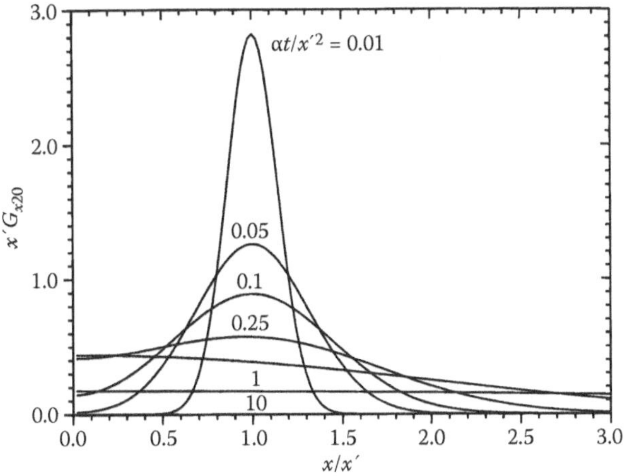

FIGURE X20.1 $x'G_{X20}(x,t|x',0)$ versus x/x' for $\alpha t/(x')^2 = 0.01, 0.05, 0.1, 0.25, 1.0$ and 10.

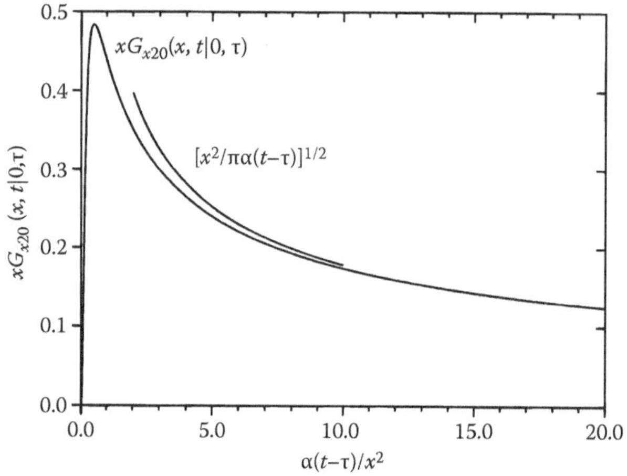

FIGURE X20.2 $xG_{X20}(x, t|0, \tau)$ versus $\alpha(t-\tau)/x^2$.

The average of this integral over $x = 0$ to b is

$$\overline{G}_{X20} \equiv \frac{1}{b}\int_{x=0}^{b}\int_{x'=0}^{b} G_{X20}(x, t|x', \tau)\, dx'\, dx$$
$$= 1 - \left[\frac{\alpha(t-\tau)}{\pi b^2}\right]^{1/2}\left(1 - \pi^{1/2}\, \text{ierfc}\left\{\frac{b}{[\alpha(t-\tau)]^{1/2}}\right\}\right) \qquad \text{(X20.6)}$$

Integrals of $\overline{G}_{X20}$ in this form can be difficult to evaluate analytically due to the ierfc(·) term. Expressions more amenable to analytical integrals are given next. For small values of $u \equiv \alpha(t-\tau)/b^2$, $\overline{G}_{X20}$ is approximated by

$$\overline{G}_{X20} \approx 1 - \left(\frac{u}{\pi}\right)^{1/2} + \frac{u^{3/2}}{2\pi^{1/2}} e^{-1/u}\left[1 - \frac{3}{2}u + \frac{15}{4}u^2 - \frac{105}{8}C_1 u^3\right] \qquad \text{(X20.7)}$$

where the greatest accuracy is found for C_1 near 1/3. Hence, let $C_1 = 1/3$. For large values of u, $\overline{G}_{X20}$ is approximated by

$$\overline{G}_{X20} \approx \frac{1}{(\pi u)^{1/2}}\left(1 - \frac{1}{6u} + \frac{1}{30u^2} - \frac{1}{168u^3} + \frac{1}{1080u^4} - \frac{C_2}{7920u^5}\right) \qquad \text{(X20.8)}$$

where $C_2 = 0.89$ improves accuracy over $C_2 = 1$ which comes from a series approximation. Table X.5 provides a comparison of results. Equation X20.7 is an accurate approximation for $u \leq 0.5$ and Equation X20.8 for $u > 0.5$.

If desired, an even more accurate approximation in the intermediate range can be obtained from the $\overline{G}_{X22}$ equation for $b/L = 0.25$; the result is

$$\overline{G}_{X20} \approx \frac{1}{4} + \frac{8}{\pi^2}\sum_{m=1}^{7}\frac{A_m}{m^2} e^{-m^2\pi^2 u/16} \qquad \text{(X20.9)}$$

TABLE X.5
Comparison of Results for $\overline{G}_{X20}$

u	Exact, Eq. X20.6	Approx., Eq. X20.7	% Error, Eq. X20.7	Approx., Eq. X20.8	% Error, Eq. X20.8
0	1	1			
0.25	0.718394	0.718416	+0.003		
0.4	0.647118	0.647393	+0.042	0.645724	
0.5	0.609548	0.609705	+0.026	0.609265	−0.046
0.6	0.577634	0.575486	−0.37	0.577562	−0.012
0.75	0.537721	0.518095	−3.6	0.537711	−0.002
1	0.486065			0.486065	
2	0.368746			0.368746	
4	0.270903				
10	0.1955				
100	0.0563				

where $A_1 = A_3 = A_5 = A_7 = 0.5$, $A_2 = A_6 = 1$ and $A_4 = 0$. The answers are accurate to six significant figures for $u = 0.25$–0.75. An alternative set of equations can be obtained by restricting Equation X20.7 to the first two terms, namely

$$\overline{G}_{X20} \approx 1 - \left(\frac{u}{\pi}\right)^{1/2} \tag{X20.10}$$

for $u \leq 0.125$ and by using Equation X20.9 for an intermediate range but with the number of terms changed to 9 with $A_8 = 0$ and $A_9 = 0.5$. The errors would be less than 10^{-5}.

X21 PLATE WITH $\partial G/\partial x = 0$ AT $x = 0$ AND $G = 0$ AT $x = L$

General expressions best for small cotimes:

$$G_{X21}(x, t|x', \tau) = [4\pi\alpha(t-\tau)]^{-1/2} \sum_{n=-\infty}^{\infty} (-1)^n \left\{ \exp\left[-\frac{(2nL + x - x')^2}{4\alpha(t-\tau)}\right] + \exp\left[-\frac{(2nL + x + x')^2}{4\alpha(t-\tau)}\right]\right\} \tag{X21.1}$$

$$-\left.\frac{\partial G_{X21}}{\partial n'}\right|_{x'=L} = [4\pi[\alpha(t-\tau)]^3]^{1/2} \sum_{n=-\infty}^{\infty} (-1)^n [(2n+1)L - x] \times \exp\left[-\frac{[(2n+1)L - x]^2}{4\alpha(t-\tau)}\right] \tag{X21.2}$$

For small values of $\alpha(t-\tau)/L^2 (\leq 0.022)$, use

$$G_{X21}(x,t|x',\tau) \approx [4\pi\alpha(t-\tau)]^{-1/2}\left\{\exp\left[-\frac{(x-x')^2}{4\alpha(t-\tau)}\right] + \exp\left[-\frac{(x+x')^2}{4\alpha(t-\tau)}\right] - \exp\left[-\frac{(2L-x-x')^2}{4\alpha(t-\tau)}\right]\right\} \tag{X21.3}$$

$$-\left.\frac{\partial G_{X21}}{\partial n'}\right|_{x'=L} \approx \frac{L-x}{[4\pi[\alpha(t-\tau)]^3]^{1/2}}\exp\left[-\frac{(L-x)^2}{4\alpha(t-\tau)}\right] \tag{X21.4}$$

$$-\left.\frac{\partial^2 G_{X21}}{\partial x\,\partial n'}\right|_{x=x'=L} \approx [4\pi[\alpha(t-\tau)]^3]^{-1/2} \tag{X21.5}$$

General expressions best for large cotimes:

$$G_{X21}(x,t|x',\tau) = \frac{2}{L}\sum_{m=1}^{\infty} e^{-\beta_m^2\alpha(t-\tau)/L^2}\cos\left(\beta_m\frac{x}{L}\right)\cos\left(\beta_m\frac{x'}{L}\right) \tag{X21.6}$$

$$\beta_m = \pi\left(m-\frac{1}{2}\right)$$

$$-\left.\frac{\partial G_{X21}}{\partial n'}\right|_{x'=L} = -\frac{2}{L^2}\sum_{m=1}^{\infty} e^{-\beta_m^2\alpha(t-\tau)/L^2}\beta_m(-1)^m\cos\left(\beta_m\frac{x}{L}\right) \tag{X21.7}$$

*X*22 PLATE WITH $\partial G/\partial x = 0$ AT $x = 0$ AND L

Expression best for small cotimes:

$$G_{X22}(x,t|x',\tau) = [4\pi\alpha(t-\tau)]^{-1/2}\sum_{n=-\infty}^{\infty}\left\{\exp\left[-\frac{(2nL+x-x')^2}{4\alpha(t-\tau)}\right] + \exp\left[-\frac{(2nL+x+x')^2}{4\alpha(t-\tau)}\right]\right\} \tag{X22.1}$$

For $\alpha(t-\tau)/L^2 < 0.25$, the maximum n value needed for four significant figures is 1. For small values of $\alpha(t-\tau)/L^2$ (≤ 0.022), use

$$G_{X22}(x,t|x',\tau) \approx [4\pi\alpha(t-\tau)]^{-1/2}\left\{\exp\left[-\frac{(x-x')^2}{4\alpha(t-\tau)}\right] + \exp\left[-\frac{(x+x')^2}{4\alpha(t-\tau)}\right] + \exp\left[-\frac{(2L-x-x')^2}{4\alpha(t-\tau)}\right]\right\} \tag{X22.2}$$

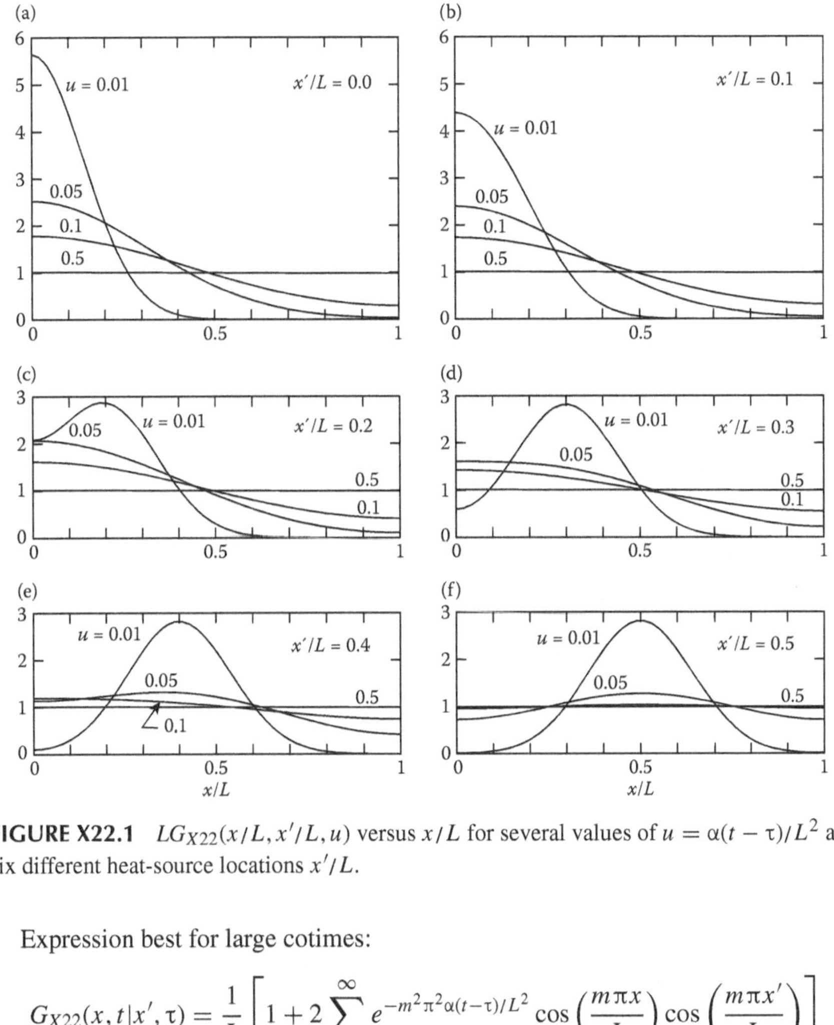

FIGURE X22.1 $LG_{X22}(x/L, x'/L, u)$ versus x/L for several values of $u = \alpha(t-\tau)/L^2$ and six different heat-source locations x'/L.

Expression best for large cotimes:

$$G_{X22}(x,t|x',\tau) = \frac{1}{L}\left[1 + 2\sum_{m=1}^{\infty} e^{-m^2\pi^2\alpha(t-\tau)/L^2}\cos\left(\frac{m\pi x}{L}\right)\cos\left(\frac{m\pi x'}{L}\right)\right] \quad \text{(X22.3)}$$

See Figure X22.1 for $LG_{X22}(\cdot)$ for various values of x'/L and $u(\equiv\alpha(t-\tau)/L^2)$ versus x'/L. Also see Figure X22.2 for $LG_{X2I}(0,t\,/\,0,\tau)$ for $I = 0, 1, 2$, and 3.

For $\alpha(t-\tau)/L^2 > 0.25$, the maximum m value needed for four significant figures is 2. For the locations $x = x' = 0$ or $x = x' = L$,

$$G_{X22}(0,t|0,\tau) = G(L,t|L,\tau)$$

$$\approx [\pi\alpha(t-\tau)]^{-1/2}\left(1 + 2e^{-L^2/[\alpha(t-\tau)]}\right) \quad \text{(X22.4a)}$$

$$\approx \frac{1}{L}\left(1 + 2e^{-\pi^2\alpha(t-\tau)/L^2}\right) \quad \text{(X22.4b)}$$

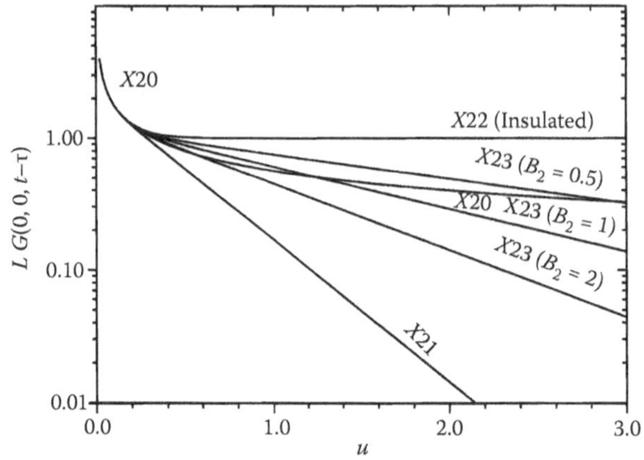

FIGURE X22.2 $LG_{X22}(0, 0, t-\tau)$ versus $u = \alpha(t-\tau)/L^2$ for several geometries.

where Equation X22.4a is used for $\alpha(t-\tau)/L^2 < \pi^{-1}$ and Equation X22.4b for larger values. The error is in the sixth significant figure or less. For example, at $\alpha(t-\tau)/L^2 = \pi^{-1}$, both give $LG_{X22}(0, t|0, \tau) = 1.086428$, while the exact value is 1.086435.

Alternative expressions are

$$G_{X22}(0, t|0, \tau) \approx [\pi\alpha(t-\tau)]^{-1/2} \tag{X22.5a}$$

for $\alpha(t-\tau)/L^2 < 0.08$ and for larger values of $\alpha(t-\tau)/L^2$,

$$G_{X22}(0, t|0, \tau) \approx \frac{1}{L}\left[1 + 2\left(e^{-\pi^2\alpha(t-\tau)/L^2} + e^{-4\pi^2\alpha(t-\tau)/L^2} + e^{-9\pi^2\alpha(t-\tau)/L^2}\right)\right] \tag{X22.5b}$$

which are also accurate to about six decimal places. For $G(L, t|0, \tau)$ $[=G(0, t|L, \tau)]$ approximate expressions accurate to about six decimal places are

$$G(L, t|0, \tau) \approx 0 \tag{X22.6a}$$

for $\alpha(t-\tau)/L^2 < 0.02$ and for greater values of $\alpha(t-\tau)/L^2$, use

$$G(L, t|0, \tau) \approx \frac{1}{L}\left[1 + 2\sum_{m=1}^{7}(-1)^m e^{-m^2\pi^2\alpha(t-\tau)/L^2}\right] \tag{X22.6b}$$

The integral of G_{X22} from $x' = 0$ to b is

$$\int_0^b G_{X22}(x, t|x', \tau)\,dx' = \frac{1}{2}\sum_{n=-\infty}^{\infty}\left(\operatorname{erfc}\left\{\frac{2nL + x - b}{[4\alpha(t-\tau)]^{1/2}}\right\} - \operatorname{erfc}\left\{\frac{2nL + x + b}{[4\alpha(t-\tau)]^{1/2}}\right\}\right) \tag{X22.7a}$$

$$\int_0^b G_{X22}(x,t|x',\tau)dx'$$
$$= \frac{b}{L} + \frac{2}{\pi}\sum_{m=1}^{\infty}\frac{1}{m}e^{-m^2\pi^2\alpha(t-\tau)/L^2}\cos\left(\frac{m\pi x}{L}\right)\sin\left(\frac{m\pi b}{L}\right) \quad \text{(X22.7b)}$$

where Equation X22.7a is better for small cotimes, and Equation X22.7b is better for large cotimes. For $b = L$, the integral is

$$\int_0^L G_{X22}(x,t|x',\tau)\,dx' = 1 \quad \text{(X22.8)}$$

At $x = 0$ for small cotimes,

$$\int_0^b G_{X22}(0,t|x',\tau)\,dx'$$
$$\approx 1 - \operatorname{erfc}\frac{b}{[4\alpha(t-\tau)]^{1/2}} + \operatorname{erfc}\frac{2L-b}{[4\alpha(t-\tau)]^{1/2}} - \operatorname{erfc}\frac{2L+b}{[4\alpha(t-\tau)]^{1/2}} \quad \text{(X22.9)}$$

For $\alpha(t-\tau)/b^2 < 0.02$ and to six significant figures

$$\int_0^b G_{X22}(0,t|x',\tau)\,dx' \approx 1 \quad \text{(X22.10)}$$

For the average of the integral from $x' = 0$ to b, the result for small cotimes is

$$\frac{1}{b}\int_{x=0}^b\int_{x'=0}^b G_{X22}(x,t|x',\tau)\,dx'\,dx$$
$$\approx 1 + \frac{[\alpha(t-\tau)]^{1/2}}{b}\left\{-\frac{1}{\pi^{1/2}} + \operatorname{ierfc}\frac{L-b}{[\alpha(t-\tau)]^{1/2}} + \operatorname{ierfc}\frac{b}{[\alpha(t-\tau)]^{1/2}} - 2\operatorname{ierfc}\frac{L}{[\alpha(t-\tau)]^{1/2}} + \operatorname{ierfc}\frac{L+b}{[\alpha(t-\tau)]^{1/2}}\right\} \quad \text{(X22.11a)}$$

and for large cotimes

$$= \frac{b}{L} + \frac{L}{b}\frac{2}{\pi^2}\sum_{m=1}^{\infty}\frac{1}{m^2}e^{-m^2\pi^2\alpha(t-\tau)/L^2}\left(\sin\frac{m\pi b}{L}\right)^2 \quad \text{(X22.11b)}$$

For $\alpha(t-\tau)/L^2 = 0.25$ and $b/L = 0.25$, the small cotime expression, Equation X22.11a, gives 0.2842 and the large cotime expression, Equation X22.11b, the value of 0.2844 which are in good agreement. For $0.1 < b/L < 0.9$ and $\alpha(t-\tau)/L^2 < 0.09$, only the first two ierfc $(\cdot)$ functions are needed in Equation X22.11a with an error less than in the sixth significant digit.

X23 PLATE WITH $\partial G/\partial x = 0$ AT $x = 0$ AND $k\partial G/\partial x + h_2 G = 0$ AT $x = L$

For small values of $\alpha(t-\tau)/L^2 (\le 0.022)$ use

$$G_{X23}(x,t|x',\tau) \approx [4\pi\alpha(t-\tau)]^{-1/2}\left\{\exp\left[-\frac{(x-x')^2}{4\alpha(t-\tau)}\right]\right.$$

$$\left. + \exp\left[-\frac{(x+x')^2}{4\alpha(t-\tau)}\right] + \exp\left[-\frac{(2L-x-x')^2}{4\alpha(t-\tau)}\right]\right\}$$

$$-\frac{1}{L}B_2 \exp\left[B_2\frac{2L-x-x'}{L} + B_2^2\frac{\alpha(t-\tau)}{L^2}\right]$$

$$\times \operatorname{erfc}\left\{\frac{2L-x-x'}{[4\alpha(t-\tau)]^{1/2}} + B_2\frac{[\alpha(t-\tau)]^{1/2}}{L}\right\} \quad \text{(X23.1)}$$

For any value of $\alpha(t-\tau)/L^2$ but best for $\alpha(t-\tau)/L^2 > 0.022$,

$$G_{X23}(x,t|x',\tau) = \frac{2}{L}\sum_{m=1}^{\infty} e^{-\beta_m^2\alpha(t-\tau)/L^2}\frac{\beta_m^2 + B_2^2}{\beta_m^2 + B_2^2 + B_2}$$

$$\times \cos\left(\beta_m\frac{x}{L}\right)\cos\left(\beta_m\frac{x'}{L}\right) \quad \text{(X23.2)}$$

Eigencondition:

$$\beta_m \tan\beta_m = B_2 \qquad B_2 \equiv \frac{h_2 L}{k} \quad \text{(X23.3)}$$

X24 PLATE WITH $\partial G/\partial x = 0$ AT $x = 0$ AND FILM WITH FINITE HEAT CAPACITY AT $x = L$

The boundary condition at $x = L$ is $k\partial G/\partial x + (\rho cb)_2 \partial G/\partial t = 0$. For small values of $\alpha(t-\tau)/L^2$ (≤ 0.022) use

$$G_{X24}(x,t|x',\tau) \approx [4\pi\alpha(t-\tau)]^{-1/2}\left\{\exp\left[-\frac{(x-x')^2}{4\alpha(t-\tau)}\right]\right.$$

$$\left. + \exp\left[-\frac{(x+x')^2}{4\alpha(t-\tau)}\right] - \exp\left[-\frac{(2L-x-x')^2}{4\alpha(t-\tau)}\right]\right\}$$

$$+\frac{1}{LC_2}\exp\left[\frac{1}{C_2}\frac{2L-x-x'}{L} + \frac{1}{C_2^2}\frac{\alpha(t-\tau)}{L^2}\right]$$

$$\times \operatorname{erfc}\left\{\frac{2L-x-x'}{[4\alpha(t-\tau)]^{1/2}} + \frac{1}{C_2}\frac{[\alpha(t-\tau)]^{1/2}}{L}\right\} \quad \text{(X24.1)}$$

where

$$C_2 \equiv \frac{(\rho cb)_2}{\rho c L} \quad \text{(X24.2)}$$

For any value of $\alpha(t-\tau)/L^2$ but best for $\alpha(t-\tau)/L^2 > 0.022$,

$$G_{X24}(x,t|x',\tau) = \frac{(1/L)}{1+C_2} + \sum_{m=1}^{\infty} e^{-\beta_m^2\alpha(t-\tau)/L^2} \frac{\cos(\beta_m x/L)\cos(\beta_m x'/L)}{N_m} \tag{X24.3}$$

where

$$N_m = \frac{L}{2}\frac{1+C_2^2\beta_m^2+C_2}{1+C_2^2\beta_m^2} \tag{X24.4}$$

Eigencondition: $\tan\beta_m = -C_2\beta_m \quad m = 1, 2, \ldots \quad \beta_m > 0$ (X24.5)

*X*25 PLATE WITH $\partial G/\partial x = 0$ AT $x = 0$ AND FILM WITH FINITE HEAT CAPACITY AND CONVECTION COEFFICIENT AT $x = L$

The boundary condition at $x = L$ is $k\,\partial G/\partial x + h_2 G + (\rho cb)_2 \partial G/\partial t = 0$. For small values of $\alpha(t-\tau)/L^2 (\le 0.022)$, use

$$\begin{aligned} G_{X25}(x,t|x',\tau) \approx {} & [4\pi\alpha(t-\tau)]^{-1/2}\left\{\exp\left[-\frac{(x-x')^2}{4\alpha(t-\tau)}\right]\right. \\ & \left. + \exp\left[-\frac{(x+x')^2}{4\alpha(t-\tau)}\right] - \exp\left[-\frac{(2L-x-x')^2}{4\alpha(t-\tau)}\right]\right\} \\ & + \frac{1}{L(1-4B_2C_2)^{1/2}}\left(\exp\left[S_4\frac{2L-x-x'}{L} + S_4^2\frac{\alpha(t-\tau)}{L^2}\right]\right. \\ & \times \operatorname{erfc}\left\{\frac{2L-x-x'}{[4\alpha(t-\tau)]^{1/2}} + S_4\frac{[\alpha(t-\tau)]^{1/2}}{L}\right\} \\ & - \exp\left[S_3\frac{2L-x-x'}{L} + S_3^2\frac{\alpha(t-\tau)}{L^2}\right] \\ & \left. \times \operatorname{erfc}\left\{\frac{2L-x-x'}{[4\alpha(t-\tau)]^{1/2}} + S_3\frac{[\alpha(t-\tau)]^{1/2}}{L}\right\}\right) \end{aligned} \tag{X25.1}$$

where

$$C_2 \equiv \frac{(\rho cb)_2}{\rho cL} \qquad B_2 \equiv \frac{h_2 L}{k} \tag{X25.2a, b}$$

See Equations X15a, b for S_3 and S_4, For any value of $\alpha(t-\tau)/L^2$ but best for $\alpha(t-\tau)/L^2 > 0.022$,

$$G_{X25}(x,t|x',\tau) = \sum_{m=1}^{\infty} e^{-\beta_m^2\alpha(t-\tau)/L^2} \frac{\cos(\beta_m x/L)\cos(\beta_m x'/L)}{N_m} \tag{X25.3}$$

where

$$N_m = \frac{L}{2}\frac{\left[\beta_m^2 + \left(B_2 - C_2\beta_m^2\right)^2\right](1+2C_2) + \left(B_2 - C_2\beta_m^2\right)\left[1 - 2C_2\left(B_2 - C_2\beta_m^2\right)\right]}{\beta_m^2 + \left(B_2 - C_2\beta_m^2\right)^2} \tag{X25.4}$$

$$\text{Eigencondition}: \qquad \beta_m \tan\beta_m = B_2 - C_2\beta_m^2 \tag{X25.5}$$

$$m = 1, 2, \ldots, \qquad B_2 \neq 0$$

$X30$ SEMI-INFINITE BODY WITH $-k\partial G/\partial x + hG = 0$ AT $x = 0$

$$\begin{aligned} G_{X30}(x,t|x',\tau) = {} & [4\pi\alpha(t-\tau)]^{-1/2}\left\{\exp\left[-\frac{(x-x')^2}{4\alpha(t-\tau)}\right]\right. \\ & \left. + \exp\left[-\frac{(x+x')^2}{4\alpha(t-\tau)}\right]\right\} - \frac{h}{k}\exp[\alpha(t-\tau)h^2k^{-2} \\ & + h(x+x')k^{-1}]\ \text{erfc}\left\{\frac{x+x'}{[4\alpha(t-\tau)]^{1/2}} + \frac{h}{k}[\alpha(t-\tau)]^{1/2}\right\} \end{aligned} \tag{X30.1}$$

Notice that the first two exp(·) terms in Equation X30.1 are equal to $G_{X20}(x,t|x',\tau)$. Then the GF can also be written as

$$\begin{aligned} G_{X30}(x,t|x',\tau) = {} & G_{X20} - \frac{h}{k}\exp\left[-\frac{(x+x')^2}{4\alpha(t-\tau)}\right] \\ & \times\ \text{rerf}\left\{\frac{x+x'}{[4\alpha(t-\tau)]^{1/2}} + \frac{h}{k}[\alpha(t-\tau)]^{1/2}\right\} \end{aligned} \tag{X30.2}$$

where

$$\text{rerf}\,(z) \equiv e^{z^2}\,\text{erfc}\,(z) \tag{X30.3}$$

For $x = x' = 0$,

$$\begin{aligned} G_{X30}(0,t|0,\tau) = {} & [\pi\alpha(t-\tau)]^{-1/2} \\ & - \frac{h}{k}\exp\left[\frac{\alpha(t-\tau)h^2}{k^2}\right]\ \text{erfc}\left\{\frac{h}{k}[\alpha(t-\tau)]^{1/2}\right\} \end{aligned} \tag{X30.4}$$

For small $(h/k)[\alpha(t-\tau)]^{1/2}$ values,

$$G_{X30}(0,t|0,\tau) \approx [\pi\alpha(t-\tau)]^{-1/2} - \frac{h}{k}\left\{1 - 2\frac{h}{k}\left[\frac{\alpha(t-\tau)}{\pi}\right]^{1/2}\right\} \tag{X30.5}$$

For large values of $(h/k)[\alpha(t-\tau)]^{1/2}$:

$$G_{X30}(0,t|0,\tau) \approx \frac{k^2}{2h^2\pi^{1/2}}\frac{1}{[\alpha(t-\tau)]^{3/2}}\left\{1-\frac{3}{2}\frac{k^2}{h^2[\alpha(t-\tau)]}\right\} \quad \text{(X30.6)}$$

For small values of $(h/k)[\alpha(t-\tau)]^{1/2}$ and any x and x' values,

$$G_{X30}(x,t|x',\tau) \approx G_{X20}(x,t|x',\tau) - \frac{h}{k}\left(\operatorname{erfc}\left\{\frac{x+x'}{[4\alpha(t-\tau)]^{1/2}}\right\} - \frac{h}{k}[4\alpha(t-\tau)]^{1/2}\operatorname{ierfc}\left\{\frac{x+x'}{[4\alpha(t-\tau)]^{1/2}}\right\}\right) \quad \text{(X30.7)}$$

For large values of $(h/k)[\alpha(t-\tau)]^{1/2}$ and any x and x' values,

$$G_{X30}(x,t|x',\tau) \approx G_{X10}(x,t|x',\tau) + \frac{x+x'}{2\pi^{1/2}[\alpha(t-\tau)]^{3/2}}\frac{k}{h}\exp\left[-\frac{(x+x')^2}{4\alpha(t-\tau)}\right] \quad \text{(X30.8)}$$

For any time with $h(x+x')/k$ large (about 10 or larger),

$$G_{X30}(x,t|x',\tau) \approx G_{X20}(x,t|x',\tau) - [\pi\alpha(t-\tau)]^{-1/2} \times \left[1+\frac{(x+x')^2}{2\alpha(t-\tau)}\frac{k}{h(x+x')}\right]^{-1}\exp\left[-\frac{(x+x')^2}{4\alpha(t-\tau)}\right] \quad \text{(X30.9)}$$

The integral of G_{X30} from $x'=0$ to b is

$$\begin{aligned}\int_0^b G_{X30}(x,t|x',\tau)\,dx' = \frac{1}{2}\Bigg(&\operatorname{erfc}\left[\frac{x-b}{[4\alpha(t-\tau)]^{1/2}}\right] \\ &+\operatorname{erfc}\left\{\frac{x+b}{[4\alpha(t-\tau)]^{1/2}}\right\} - 2\operatorname{erfc}\left[\frac{x}{[4\alpha(t-\tau)]^{1/2}}\right]\Bigg) \\ &+\exp\left[\frac{hx}{k}+\frac{h^2}{k^2}\alpha(t-\tau)\right] \\ &\times\operatorname{erfc}\left\{\frac{x}{[4\alpha(t-\tau)]^{1/2}}+\frac{h}{k}[\alpha(t-\tau)]^{1/2}\right\} \\ &-\exp\left[\frac{h(x+b)}{k}+\frac{h^2}{k^2}\alpha(t-\tau)\right] \\ &\times\operatorname{erfc}\left\{\frac{x+b}{[4\alpha(t-\tau)]^{1/2}}+\frac{h}{k}[\alpha(t-\tau)]^{1/2}\right\}\end{aligned} \quad \text{(X30.10)}$$

If $b \to \infty$, the integral becomes

$$\int_0^\infty G_{X30}(x,t|x',\tau)dx' = \text{erf}\left[\frac{x}{[4\alpha(t-\tau)]^{1/2}}\right] + \exp\left[\frac{hx}{k} + \frac{h^2}{k^2}\alpha(t-\tau)\right] \times \text{erfc}\left[\frac{x}{[4\alpha(t-\tau)]^{1/2}} + \frac{h}{k}[\alpha(t-\tau)]^{1/2}\right] \quad \text{(X30.11)}$$

*X*31 PLATE WITH $-k\partial G/\partial x + hG = 0$ AT $x = 0$ AND $G = 0$ AT $x = L$

For small values of $\alpha(t-\tau)/L^2(\leq 0.022)$ use

$$G_{X31}(x,t|x',\tau) \approx [4\pi\alpha(t-\tau)]^{-1/2}\left\{\exp\left[-\frac{(x-x')^2}{4\alpha(t-\tau)}\right] + \exp\left[-\frac{(x+x')^2}{4\alpha(t-\tau)}\right] - \exp\left[-\frac{(2L-x-x')^2}{4\alpha(t-\tau)}\right]\right\} - \frac{h}{k}\exp\left[\frac{h(x+x')}{k} + \frac{h^2\alpha(t-\tau)}{k^2}\right] \times \text{erfc}\left\{\frac{x+x'}{[4\alpha(t-\tau)]^{1/2}} + \frac{h}{k}[\alpha(t-\tau)]^{1/2}\right\} \quad \text{(X31.1)}$$

$$-\frac{\partial G_{X31}}{\partial n'}\bigg|_{x'=L} \approx \frac{L-x}{\{4\pi[\alpha(t-\tau)]^3\}^{1/2}}\exp\left[-\frac{(L-x)^2}{4\alpha(t-\tau)}\right] \quad \text{(X31.2)}$$

$$-\frac{\partial G_{X31}}{\partial n'}\bigg|_{x=x'=L} \approx \{4\pi[\alpha(t-\tau)]^3\}^{-1/2} \quad \text{(X31.3)}$$

For larger values of $\alpha(t-\tau)/L^2$ (but valid for all times) use

$$G_{X31}(x,t|x',\tau) = \frac{2}{L}\sum_{m=1}^{\infty} e^{-\beta_m^2\alpha(t-\tau)/L^2} \times \frac{(\beta_m^2 + B^2)\sin[\beta_m(1-x/L)]\sin[\beta_m(1-x'/L)]}{\beta_m^2 + B^2 + B} \quad \text{(X31.4)}$$

Eigencondition:

$$\beta_m \cot\beta_m = -B \qquad B = \frac{hL}{k} \quad \text{(X31.5a, b)}$$

$$-\frac{\partial G_{X31}}{\partial n'}\bigg|_{x'=L} = \frac{2}{L^2}\sum_{m=1}^{\infty} e^{-\beta_m^2\alpha(t-\tau)/L^2}\frac{\beta_m(\beta_m^2 + B^2)\sin[\beta_m(1-x/L)]}{\beta_m^2 + B^2 + B} \quad \text{(X31.6)}$$

X32 PLATE WITH $-k\partial G/\partial x + hG = 0$ AT $x = 0$ AND $\partial G/\partial x = 0$ AT $x = L$

For small values of $\alpha(t-\tau)/L^2 (\leq 0.022)$ use

$$G_{X32}(x,t|x',\tau) \approx [4\pi\alpha(t-\tau)]^{-1/2} \left\{ \exp\left[-\frac{(x-x')^2}{4\alpha(t-\tau)}\right] + \exp\left[-\frac{(x+x')^2}{4\alpha(t-\tau)}\right] + \exp\left[-\frac{(2L-x-x')^2}{4\alpha(t-\tau)}\right]\right\} - \frac{h}{k}\exp\left[\frac{h(x+x')}{k} + \frac{h^2\alpha(t-\tau)}{k^2}\right] \times \operatorname{erfc}\left\{\frac{x+x'}{[4\alpha(t-\tau)]^{1/2}} + \frac{h}{k}[\alpha(t-\tau)]^{1/2}\right\} \quad \text{(X32.1)}$$

For larger values of $\alpha(t-\tau)/L^2$ (but valid for all times), use

$$G_{X32}(x,t|x',\tau) = \frac{2}{L}\sum_{m=1}^{\infty} e^{-\beta_m^2\alpha(t-\tau)/L^2} \frac{\beta_m^2 + B^2}{\beta_m^2 + B^2 + B} \times \cos\left[\beta_m\left(1-\frac{x}{L}\right)\right]\cos\left[\beta_m\left(1-\frac{x'}{L}\right)\right] \quad \text{(X32.2)}$$

$$\text{Eigencondition}: \qquad \beta_m \tan\beta_m = B \qquad B = \frac{hL}{k} \quad \text{(X32.3a, b)}$$

X33 PLATE WITH $-k\partial G/\partial x + h_1 G = 0$ AT $x = 0$ AND $k\partial G/\partial x + h_2 G = 0$ AT $x = L$

For small values of $\alpha(t-\tau)/L^2$ (≤ 0.022) use

$$G_{X33}(x,t|x',\tau) \approx [4\pi\alpha(t-\tau)]^{-1/2} \left\{ \exp\left[-\frac{(x-x')^2}{4\alpha(t-\tau)}\right] + \exp\left[-\frac{(x+x')^2}{4\alpha(t-\tau)}\right] + \exp\left[-\frac{(2L-x-x')^2}{4\alpha(t-\tau)}\right]\right\} - \frac{h_1}{k}\exp\left[\frac{h_1(x+x')}{k} + \frac{h_1^2\alpha(t-\tau)}{k^2}\right] \times \operatorname{erfc}\left\{\frac{x+x'}{[4\alpha(t-\tau)]^{1/2}} + \frac{h_1}{k}[\alpha(t-\tau)]^{1/2}\right\} - \frac{h_2}{k}\exp\left[\frac{h_2(2L-x-x')}{k} + \frac{h_2^2\alpha(t-\tau)}{k^2}\right] \times \operatorname{erfc}\left\{\frac{2L-x-x'}{[4\alpha(t-\tau)]^{1/2}} + \frac{h_2}{k}[\alpha(t-\tau)]^{1/2}\right\} \quad \text{(X33.1)}$$

For larger values of $\alpha(t-\tau)/L^2$ (but valid for all times), use

$$G_{X33}(x,t|x',\tau) = \frac{2}{L}\sum_{m=1}^{\infty} e^{-\beta_m^2\alpha(t-\tau)/L^2}[\beta_m\cos(\beta_m x/L) + B_1\sin(\beta_m x/L)] \times \frac{[\beta_m\cos(\beta_m x'/L) + B_1\sin(\beta_m x'/L)]}{(\beta_m^2+B_1^2)[1+B_2/(\beta_m^2+B_2^2)]+B_1} \quad \text{(X33.2)}$$

where the β_m values are the positive eigenvalues (arranged in increasing order) of

$$\tan\beta_m = \frac{\beta_m(B_1+B_2)}{\beta_m^2 - B_1B_2} \qquad B_1 = \frac{h_1L}{k} \qquad B_2 = \frac{h_2L}{k} \quad \text{(X33.3a, b, c)}$$

X34 PLATE WITH $-k\partial G/\partial x + h_1G = 0$ AT $x = 0$ AND AT $x = L$ THE BOUNDARY CONDITION IS $k\partial G/\partial x + (\rho cb)_2\,\partial G/\partial t = 0$

For small values of $\alpha(t-\tau)/L^2 (\leq 0.022)$ use

$$\begin{aligned} G_{X34}(x,t|x',\tau) &\approx [4\pi\alpha(t-\tau)]^{-1/2}\left\{\exp\left[-\frac{(x-x')^2}{4\alpha(t-\tau)}\right]\right. \\ &+ \exp\left[-\frac{(x+x')^2}{4\alpha(t-\tau)}\right] - \left.\exp\left[-\frac{(2L-x-x')^2}{4\alpha(t-\tau)}\right]\right\} \\ &- \frac{h_1}{k}\exp\left[\frac{h_1(x+x')}{k} + \frac{h_1^2\alpha(t-\tau)}{k^2}\right] \\ &\times \operatorname{erfc}\left\{\frac{x+x'}{[4\alpha(t-\tau)]^{1/2}} + \frac{h_1}{k}[\alpha(t-\tau)]^{1/2}\right\} \\ &+ \frac{\rho c}{(\rho cb)_2}\exp\left[\frac{\rho c(2L-x-x')}{(\rho cb)_2} + \frac{(\rho c)^2\alpha(t-\tau)}{(\rho cb)_2^2}\right] \\ &\times \operatorname{erfc}\left\{\frac{2L-x-x'}{[4\alpha(t-\tau)]^{1/2}} + \frac{\rho c}{(\rho cb)_2}[\alpha(t-\tau)]^{1/2}\right\} \end{aligned} \quad \text{(X34.1)}$$

For larger values of $\alpha(t-\tau)/L^2$ (but valid for all times), use

$$G_{X34}(x,t|x',\tau) = \sum_{m=1}^{\infty} e^{-\beta_m^2\alpha(t-\tau)/L^2}\frac{X_m(x,\beta_m)X_m(x',\beta_m)}{N_m} \quad \text{(X34.2)}$$

where

$$X_m(x,\beta_m) = B_1\sin\left(\beta_m\frac{x}{L}\right) + \beta_m\cos\left(\beta_m\frac{x}{L}\right) \quad \text{(X34.3)}$$

$$\begin{aligned} N_m = L\Bigg(&\frac{1}{2}(B_1^2+\beta_m^2) + \beta_m^2C_2 + \frac{\tan\beta_m}{1+\tan^2\beta_m}\left\{\frac{1}{2\beta_m}(\beta_m^2 - B_1^2)\right. \\ &+ \left. 2C_2B_1\beta_m + \tan(\beta_m)[C_2(B_1^2-\beta_m^2)+B_1]\right\}\Bigg) \end{aligned} \quad \text{(X34.4)}$$

The eigenvalues are the positive roots of

$$\tan\beta_m = \frac{B_1 - C_2\beta_m^2}{\beta_m(1 + B_1C_2)} \tag{X34.5}$$

where

$$B_1 = \frac{h_1L}{k} \qquad C_2 = \frac{(\rho cb)_2}{\rho cL} \tag{X34.6a, b}$$

X35 PLATE WITH $-k\partial G/\partial x + h_1G = 0$ AT $x = 0$ AND AT $x = L$, THE BOUNDARY CONDITION IS $k\partial G/\partial x + h_2G + (\rho cb)_2\,\partial G/\partial t = 0$

For small values of $\alpha(t-\tau)/L^2(\leq 0.022)$ use

$$\begin{aligned}G_{X35}(x,t|x',\tau) \approx{}& [4\pi\alpha(t-\tau)]^{-1/2}\left\{\exp\left[-\frac{(x-x')^2}{4\alpha(t-\tau)}\right]\right.\\&\left.+\exp\left[-\frac{(x+x')^2}{4\alpha(t-\tau)}\right]-\exp\left[-\frac{(2L-x-x')^2}{4\alpha(t-\tau)}\right]\right\}\\&+\frac{1}{L}\left\{-B_1ER(x+x',t-\tau,B_1)\right.\\&+\frac{1}{(1-4B_2C_2)^{1/2}}[S_4ER(2L-x-x',t-\tau,S_4)\\&\left.-S_3ER(2L-x-x',t-\tau,S_3)]\right\}\end{aligned} \tag{X35.1}$$

where for $C_2 < 1/4B_2$

$$S_3 = \frac{1}{2C_2}[1-(1-4B_2C_2)^{1/2}] \tag{X35.2}$$

$$S_4 = \frac{1}{2C_2}[1+(1-4B_2C_2)^{1/2}] \tag{X35.3}$$

$$\begin{aligned}ER(x,t-\tau,B) ={}& \exp\left[\frac{Bx}{L}+\frac{B^2\alpha(t-\tau)}{L^2}\right]\\&\times\operatorname{erfc}\left\{\frac{x}{[4\alpha(t-\tau)]^{1/2}}+B\frac{[\alpha(t-\tau)]^{1/2}}{L}\right\}\end{aligned} \tag{X35.4}$$

For larger times of $\alpha(t-\tau)/L^2$ (but valid for all times), use

$$G_{X35}(x,t|x',\tau) = \sum_{m=1}^{\infty} e^{-\beta_m^2\alpha(t-\tau)/L^2}\frac{X_m(x,\beta_m)X_m(x',\beta_m)}{N_m} \tag{X35.5}$$

where

$$X_m(x, \beta_m) = B_1 \sin\left(\beta_m \frac{x}{L}\right) + \beta_m \cos\left(\beta_m \frac{x}{L}\right) \tag{X35.6}$$

$$N_m = L\left(\frac{1}{2}(B_1^2 + \beta_m^2) + \beta_m^2 C_2 + \frac{\tan\beta_m}{1+\tan^2\beta_m} \times \left\{\frac{1}{2\beta_m}(\beta_m^2 - B_1^2) + 2C_2 B_1 \beta_m + \tan\beta_m [C_2(B_1^2 - \beta_m^2) + B_1]\right\}\right) \tag{X35.7}$$

The eigenvalues are the positive roots of

$$\tan\beta_m = \frac{\beta_m(B_1 + B_2 - C_2\beta_m^2)}{\beta_m^2 - B_1(B_2 - C_2\beta_m^2)} \tag{X35.8}$$

where

$$B_1 = \frac{h_1 L}{k} \qquad B_2 = \frac{h_2 L}{k} \qquad C_2 = \frac{(\rho c b)_2}{\rho c L} \tag{X35.9a, b, c}$$

*X*40 SEMI-INFINITE BODY WITH $-k\partial G/\partial x + (\rho c b)_1 \partial G/\partial t = 0$ AT $x = 0$

$$G_{X40}(x, t|x', \tau) = [4\pi\alpha(t-\tau)]^{-1/2}\left\{\exp\left[-\frac{(x-x')^2}{4\alpha(t-\tau)}\right] - \exp\left[-\frac{(x+x')^2}{4\alpha(t-\tau)}\right]\right\} + \frac{1}{bP}\exp\left[-\frac{(x+x')^2}{4\alpha(t-\tau)}\right] \times \text{rerf}\left\{\frac{x+x'}{2[\alpha(t-\tau)]^{1/2}} + \frac{1}{P}\frac{[\alpha(t-\tau)]^{1/2}}{b}\right\} \tag{X40.1}$$

$$P = \frac{(\rho c)_1}{\rho c} \qquad \text{rerf}\,(z) = e^{z^2}\,\text{erfc}\,(z) \tag{X40.2a, b}$$

*X*41 PLATE WITH $-k\partial G/\partial x + (\rho c b)_1 \partial G/\partial t = 0$ AT $x = 0$ AND $G = 0$ AT $x = L$

For $\alpha(t-\tau)/L^2 < 0.1$, an accurate approximation is

$$\begin{aligned} G_{X41}(x, x'|t, \tau) \approx \frac{1}{L}\{&EX(x-x', t-\tau) - EX(x+x', t-\tau) \\ &- EX(2L - x - x', t-\tau) \\ &+ EX(2L + x - x', t-\tau) \\ &+ EX(2L - x + x', t-\tau) \\ &+ C_1^{-1}[ER(x+x', t-\tau, C_1^{-1}) \\ &- ER(2L + x - x', t-\tau, C_1^{-1}) \\ &- ER(2L - x + x', t-\tau, C_1^{-1})]\} \end{aligned} \tag{X41.1}$$

where $ER(\cdot)$ is defined by Equation X35.4, and

$$EX(z, t-\tau) = [4\pi\alpha(t-\tau)]^{-1/2} \exp\left[-\frac{z^2}{4\alpha(t-\tau)}\right] \tag{X41.2}$$

For all times but best for large times, $G_{X41}(\cdot)$ is

$$G_{X41}(x, t|x', \tau) = \sum_{m=1}^{\infty} \frac{1}{N_m} e^{-\beta_m^2 \alpha(t-\tau)/L^2} X_m(x) X_m(x') \tag{X41.3}$$

where

$$X_m(x) = \cos\left(\beta_m \frac{x}{L}\right) - C_1 \beta_m \sin\left(\beta_m \frac{x}{L}\right) \tag{X41.4}$$

$$C_1 = \frac{(\rho c b)_1}{\rho c L} \tag{X41.5}$$

$$N_m = \frac{L}{2}[(C_1\beta_m)^2 + C_1 + 1] \tag{X41.6}$$

Eigencondition:

$$\beta_m \tan \beta_m = C_1^{-1} \tag{X41.7}$$

X42 PLATE WITH $-k\partial G/\partial x + (\rho c b)_1 \partial G/\partial t = 0$ AT $x = 0$ AND $\partial G/\partial x = 0$ AT $x = L$

$$G(x, t|x', \tau) = \left[\frac{1}{N_0} + \sum_{m=1}^{\infty} \frac{1}{N_m} e^{-\beta_m^2 \alpha(t-\tau)/L^2} X_m(x) X_m(x')\right] \tag{X42.1}$$

where

$$X_m(x) = \cos\left(\beta_m \frac{x}{L}\right) - C_1 \beta_m \sin\left(\beta_m \frac{x}{L}\right) \tag{X42.2}$$

$$C_1 = \frac{(\rho c b)_1}{\rho c L} \tag{X42.3}$$

$$N_0 = L(C_1 + 1) \tag{X42.4}$$

$$N_m = \frac{L}{2}[(C_1\beta_m)^2 + C_1 + 1] \qquad m = 1, 2, \ldots \tag{X42.5}$$

Eigencondition:

$$\beta_m \cot \beta_m = \frac{-1}{C_1} \tag{X42.6}$$

X50 SEMI-INFINITE BODY WITH $-k\partial G/\partial x + hG + (\rho cb)_1 \partial G/\partial t = 0$ AT $x = 0$

$$G_{X50}(x,t|x',\tau) = [4\pi\alpha(t-\tau)]^{-1/2}\left\{\exp\left[-\frac{(x-x')^2}{4\alpha(t-\tau)}\right] - \exp\left[-\frac{(x+x')^2}{4\alpha(t-\tau)}\right]\right\}$$
$$+\frac{1}{2bAP}\exp\left[-\frac{(x+x')^2}{4\alpha(t-\tau)}\right]$$
$$\times\left((1+A)\text{rerf}\left\{\frac{x+x'}{2[\alpha(t-\tau)]^{1/2}} + (1+A)\frac{[\alpha(t-\tau)]^{1/2}}{2bP}\right\}\right.$$
$$\left.-(1-A)\text{rerf}\left\{\frac{x+x'}{2[\alpha(t-\tau)]^{1/2}} + (1-A)\frac{[\alpha(t-\tau)]^{1/2}}{2bP}\right\}\right) \quad \text{(X50.1)}$$

$$P = \frac{(\rho c)_1}{\rho c} \qquad B = \frac{hb}{k} \qquad A = (1-4BP)^{1/2} \qquad \text{for } 4BP < 1 \quad \text{(X50.2a, b, c)}$$

X51 PLATE WITH $-k\partial G/\partial x + hG + (\rho cb)_1\, \partial G/\partial t = 0$ AT $x = 0$ AND $G = 0$ AT $x = L$

For $\alpha(t-\tau)/L^2 < 0.1$, an approximate expression is

$$G_{X51}(x,x'|t,\tau) \approx \frac{1}{L}\{EX(x-x',t-\tau) - EX(x+x',t-\tau)$$
$$- EX(2L-x-x',t-\tau)$$
$$+ EX(2L+x-x',t-\tau)$$
$$+ EX(2L-x+x',t-\tau)$$
$$+ \frac{1}{C_1(S_1-S_2)}[ER(x+x',t-\tau,S_2)$$
$$- ER(x+x',t-\tau,S_1)$$
$$- ER(2L+x-x',t-\tau,S_2)$$
$$+ ER(2L+x-x',t-\tau,S_1)$$
$$- ER(2L-x+x',t-\tau,S_2)$$
$$+ ER(2L-x+x',t-\tau,S_1)]\}$$
$$\text{for } C_1 < (1/4B_1) \quad \text{(X51.1)}$$

where

$$S_1 = \frac{1}{2C_1}\left[-1+(1-4B_1C_1)^{1/2}\right]$$
$$S_2 = \frac{1}{2C_1}\left[-1-(1-4B_1C_1)^{1/2}\right] \quad \text{(X51.2)}$$

For larger $\alpha(t-\tau)/L^2$ (but valid for any time), $G_{X51}(\cdot)$ is given by

$$G_{X51}(x,t|x',\tau) = \sum_{m=1}^{\infty} \frac{1}{N_m} e^{-\beta_m^2 \alpha(t-\tau)/L^2} X_m(x) X_m(x') \tag{X51.3}$$

where

$$X_m(x) = D_m \sin\left(\beta_m \frac{x}{L}\right) + \cos\left(\beta_m \frac{x}{L}\right) \tag{X51.4}$$

$$D_m = \frac{B}{\beta_m} - C\beta_m \qquad B = \frac{hL}{k} \qquad C = \frac{(\rho c)_1 b}{\rho c L} \tag{X51.5}$$

$$N_m = \frac{L}{2}\left(D_m^2 + \frac{D_m}{\beta_m} + 2C + 1\right) \tag{X51.6}$$

Eigencondition:

$$\tan\beta_m = \frac{\beta_m}{C\beta_m^2 - B} \qquad m = 1, 2, \ldots \qquad \beta_m > 0 \tag{X51.7}$$

For eigenvalues, see case X33.

*X*52 PLATE WITH $-k\partial G/\partial x + hG + (\rho cb)_1 \partial G/\partial t = 0$ AT $x = 0$ AND $\partial G/\partial x = 0$ AT $x = L$

$$G(x,t|x',\tau) = \sum_{m=1}^{\infty} \frac{1}{N_m} e^{-\beta_m^2 \alpha(t-\tau)/L^2} X_m(x) X_m(x') \tag{X52.1}$$

where

$$X_m(x) = D_m \sin\left(\beta_m \frac{x}{L}\right) + \cos\left(\beta_m \frac{x}{L}\right) \tag{X52.2}$$

$$D_m = \frac{B}{\beta_m} - C\beta_m \qquad B = \frac{hL}{k} \qquad C = \frac{(\rho c)_1 b}{\rho c L} \tag{X52.3a, b, c}$$

$$N_m = \frac{L}{2}\left(D_m^2 + \frac{1}{\beta_m} D_m + 2C + 1\right) \tag{X52.4}$$

Eigencondition:

$$\tan\beta_m = D_m \qquad m = 1, 2, \ldots \tag{X52.5}$$

$$(\beta_m > 0 \text{ for } B > 0)$$

REFERENCE

Carslaw, H. S. and Jaeger, J. C., 1959, *Conduction of Heat in Solids*, 2nd edn, Oxford University Press, New York.

Index of Solutions by Number System

Number	Equation	Comment
R00	(7.3)	Transient GF, cylindrical case
R00T5	(7.16b)	Temperature at $r = 0$
R01	(7.40)	Transient GF, cylinder
R01B0T1	(7.43)	Solid cylinder
R01B0T5	(7.44)	Piecewise-constant initial condition
R01B1T0	(7.54)	Standard solution
R01B1T0	(7.57)	Alternate solution
R01B0T0G1	(7.66)	Standard form
R01B0T0G1	(7.67)	Improved convergence
R02B1T0	(7.62)	Best for small time
R02B1T0	(7.64)	Spatial average temperature
R02B0T0G(r5)	(7.70)	Piecewise-constant internal heating
R03B0T1	(7.50)	Cylinder, suddenly quenched
R03B0G1	(7.164)	Steady, uniform internal heating
R03B0G5	(7.166)	Steady, piecewise internal heating
R03B0G(x1t6)	(9.49)	Steady periodic, internal heating
R10B1T0	(7.95)	Infinite body with cylindrical hole
R20B–T0	(7.99)	Surface temp. at small time
R20B–T0	(7.100)	Surface temp. at large time
R11B00T1	(7.78)	Hollow cylinder
R21B10	(7.83)	Steady, hollow cylinder
R21B00T-	(7.91)	Initially steady case R21B10
R23B02T0	(10.111)	From Galerkin-based GF
R00 Z20B(r5)T0	(7.131)	Semi-infinite cylinder
R00 Z20B(r5)	(7.133)	Steady, surface temperature
R00 Z20B(r5)T0	(7.136)	Centerline temperature
R00 Z20B(r5t6)	(9.123)	Steady periodic, half space
R01B0 Z11B00T1	(7.116)	Finite cylinder
R01B1 Z11B00	(7.171)	Steady, double-sum form
R01B1 Z11B00	(7.176)	Steady, single-sum form
R02B(z5) Z00T0	(7.122)	Surface temperature
R03B0 Z23B60	(9.107)	Steady periodic, eigenfunctions along z
R03B0 Z23B60	(9.113)	Steady periodic, eigenfunctions along r
R03 ZI0	(9.120)	Steady periodic GF, $I = 0, 1, 2, 3, 4, 5.$
R01B0 Φ00T–	(7.154)	Integral expression
R01B5 Φ00	(7.182)	Steady, piecewise surface temperature
R02B– Φ22T0	(7.160)	Cylindrical sector
R03B0 Z11B00 Φ00G(r7ϕ5t6)	(9.145)	Model of hotfilm sensor
R03B0 Z11B00 Φ00G(r7ϕ5t6)	(9.150)	Alternate form
RS00	(4.157)	Steady point source
RS00	(4.188)	Steady point source
RS00	(8.4)	Transient GF, spherical coordinates

Number	Equation	Comment
RS00	(8.6)	Point source at origin
RS01	(8.61)	Transient GF, sphere
RS01B0T1	(8.65)	Large-time form
RS02	(8.72)	Transient GF, sphere
RS02B1T1	(8.76)	Constant surface heat flux
RS02B2T0	(8.78)	Linear-in-time surface heat flux
RS02B0T0G1	(8.91)	Uniform heating
RS02B0T0G(r2)	(8.94)	Linear-in-radius heating
RS02B0T0G(r4)	(8.103)	Exponential-in-radius heating
RS02B0T0G(t4)	(8.105)	Exponential-in-time heating
RS02B0T0G(r6)	(8.110)	Sinusoidal-in-radius heating
RS03	(8.142)	Steady GF
RS03	(8.40)	Transient GF, sphere
RS03B1T0	(8.82)	Small-time form
RS03B1T0	(8.85)	Large-time form
RS03B0G1	(8.144)	Steady, sphere
RS03B0G4	(8.146)	Steady, spatially-varying heating
RS10	(8.130)	Transient GF, large body with void
RS10B0T0G(r5)	(8.133)	Spatially-varying heating
RS20	(8.125)	Transient GF, large body with void
RS20B1T0	(8.127)	Uniform surface flux
RS30	(4.40)	Transient GF, large body with void
RS11	(8.138)	Steady GF
RS11B10	(8.140)	Steady, hollow sphere
RS11B10T0	(8.122)	Standard solution
RS11B10T0	(8.123)	Alternate solution
RS21	(8.135)	Steady GF, hollow sphere
RS21B10	(8.136)	Steady, hollow sphere
RS21	(8.113)	Transient GF, hollow sphere
RS21B10T0	(8.116)	Standard solution
RS21B10T0	(8.118)	Alternate solution
RS33	(8.54)	Transient GF, hollow sphere
RS40B6	(9.55)	Steady periodic, lumped sphere
X00	(1.72)	Transient GF, infinite body
X00	(1.75)	Fundamental heat conduction solution
X00T5	(1.79)	Example 1.1, one piecewise segment
X00T5	(1.83)	Example 1.2, two piecewise segments
X00T0G(x7t1)	(1.89)	Example 1.3, plane source
X10	(1.101)	Transient GF, semi-infinite body
X10B1T0	(4.13)	By Laplace transform
X10B0T1	(1.109)	Example 1.4
X10B0T5	(6.7)	$I = 1$
X10B1T0	(1.112)	Example 1.5
X10B1T0	(6.16)	Constant boundary temperature
X10B3T0	(6.20)	Boundary varying as $t^{n/2}$
X10B0T0G(t3)	(6.34a)	Generation varies as $t^{n/2}$
X20	(1.107)	Transient GF, semi-infinite body
X20B0T5	(6.7)	$I = 2$
X20B0T2	(6.11)	Linearly-varying initial condition
X20B1T0	(6.24)	Constant boundary heat flux
X20B3T0	(6.25)	Boundary varying as $t^{n/2}$
X20B0T0G(t3)	(6.34b)	Generation varies as $t^{n/2}$
X30	(4.27)	Transient GF, semi-infinite body
X30B1T0	(6.29)	Boundary convection suddenly applied

Number	Equation	Comment
X30B0G(x4t6)	(9.45)	Steady periodic, internal heating
X11	(4.90)	Large-cotime form
X11B00T1	(3.88)	Transient with fin losses
X11B00T1	(6.43)	Small-time form
X11B00T1	(6.59a)	large-time form
X11B00T1	(10.23)	One-term solution
X11B00T1	(10.30)	From polynomial basis functions
X11B00T5	(6.56)	$J = 1$
X11B00G(x7)	(6.66)	Steady, plane source
X11B00T0G(x7t1)	(6.68)	Transient, plane source
X11B06T0	(10.83)	From Galerkin-based GF
X11B06T0	(10.84)	Exact solution
X11B06T0	(3.67)	Standard solution
X11B06T0	(3.72)	Alternate solution
X11B10	(3.99)	Steady fin, T specified at end
X11B10T0	(5.19)	Large-time form
X11B10T0	(5.27)	Steady series removed
X12	(1.52)	Steady GF, plane wall
X12	(4.59)	Small-cotime form
X12	(4.155)	Transient GF, plane wall
X12B00G1	(1.40)	Steady, uniform heat generation
X12B00G4	(1.54)	Steady, heating varies exponentially
X12B00G5	(1.55)	Steady, step in heat generation
X12B00T5	(6.56)	$J = 2$
X12B00G(x7)	(6.67)	Steady, plane source
X12B00T0G(x7t1)	(6.69)	Transient, plane source
X21B1T0	(6.81)	Best for small time
X21B1T0	(6.83)	Best for large time
X21B1T0	(6.85)	Improved convergence
X21B21T1	(5.51)	Standard form
X21B21T1	(5.60)	Improved-convergence form
X21B30T0	(6.52)	boundary varying as $t^{n/2}$
X22	(4.178)	Pseudo GF, steady 1D
X22	(4.109)	Large-cotime form
X22B10T0	(6.87)	Standard form
X22B10T0	(6.94)	Alternate form
X22B30T0	(6.52)	boundary varying as $t^{n/2}$
X23B60	(9.39)	Steady periodic, surface heating
X32B10T0	(6.98)	Standard form
X32B10T0	(6.105)	Improved convergence
X32B00T1	(6.112)	Homogeneous boundary
XIJB00T0G(x7t3)	(6.50)	$I, J = 1, 2$; plane source varying as $t^{n/2}$
XV10B0T1	(3.134)	Moving body, velocity V
X00 Y20B5T0	(6.143)	Heated over half of surface
X00 Y20B5T0	(6.144)	Surface temperature only, half is heated
X00 Y20B5T0	(6.161)	Surface temperature only, strip is heated
X00 Y21	(6.170)	Steady GF, 2D
X00 Y21B(x5)0	(6.172)	Steady, strip heater
X00 Y23B00G(x5y7t6)	(9.77)	Steady periodic, heated strip
X00 Y23B00G(x5y7t6)	(9.80)	Steady periodic, alternate form
X00 YI0	(9.81)	Steady periodic GF, $I = 1, 2, 3, 4, 5$.
X11 Y20	(4.198)	Steady GF, 2D
X11B00 Y21B(x5)0T0	(6.129)	Best for large time
X11B00 Y21B(x5)0T0	(6.133)	For small y and x near a_1

Number	Equation	Comment
X11B00 Y21B(x5)0	(6.166)	Steady, step in wall heating
X12 Y12	(4.145)	Steady GF, 2D
X12 Y12	(5.32)	Steady GF, double-sum form
X12 Y12	(5.40)	Steady GF, single sum form
X12B10 Y12B00	(5.44)	Steady, single-sum form
X12B10 Y12B00	(5.75)	Steady, alternate form
X12B10 Y12B00T0	(5.37)	Large-time form
X12B10 Y12B00T0	(5.47)	Improved-convergence form
X21B10 Y21B01	(6.122)	Best for large time
X22 Y22	(4.180)	Pseudo GF, steady 2D
X22 Y22	(4.184)	Pseudo GF, alternate form
X11 Y11 Z11	(4.192)	Steady 3D GF, triple-sum form
X11 Y11 Z11	(4.193)	Steady 3D GF, double-sum form
X11 Y11 Z11	(6.177)	Steady, triple-sum form
X11 Y11 Z11	(6.180)	Steady, double-sum form
Φ22B10	(7.111)	Steady with fin losses
Φ22B10T0	(7.110)	Transient with fin losses

Author Index

Subject Index

D

E

G

H

M

T